MEDICINAL PROTEIN ENGINEERING

MEDICINAL PROTEIN ENGINEERING

EDITED BY
YURY E. KHUDYAKOV

CRC Press
Taylor & Francis Group
Boca Raton London New York

CRC Press is an imprint of the
Taylor & Francis Group, an **informa** business

CRC Press
Taylor & Francis Group
6000 Broken Sound Parkway NW, Suite 300
Boca Raton, FL 33487-2742

First issued in paperback 2019

ISBN 13: 978-0-367-44609-3 (pbk)
ISBN 13: 978-0-8493-7368-8 (hbk)

Library of Congress Cataloging-in-Publication Data

Medicinal protein engineering / editor, Yury E. Khudyakov.
p. ; cm.
"A CRC title."
Includes bibliographical references and index.
ISBN 978-0-8493-7368-8 (hardback : alk. paper)
1. Protein engineering. 2. Biomedical engineering. I. Khudyakov, Yury E. II. Title.
[DNLM: 1. Protein Engineering. 2. Biomedical Engineering. 3. Quantitative Structure-Activity Relationship. 4. Viral Vaccines. QU 450 M489 2009]

TP248.65.P76M43 2009
610.28--dc22 2008025423

Visit the Taylor & Francis Web site at
http://www.taylorandfrancis.com

and the CRC Press Web site at
http://www.crcpress.com

Contents

Preface

Knowing the small by way of the great, one goes from the shallow to the deep.

Miyamoto Musashi
The Book of Five Rings

The American Engineering Council for Professional Development defines engineering as "the creative application of scientific principles to design or develop structures, machines, apparatus, or manufacturing process..." With dry eloquence this definition clearly recognizes the roots of engineering in science and knowledge. Engineering is one of the most important tools developed by humans for using knowledge to control and adapt the environment to human needs. As a result, we live in the engineered world. Scientific exploration makes us aware of new opportunities to improve our welfare and engineering takes complete advantage of this knowledge to our benefit. Engineering started from inventing basic things for mere survival like making fire, and is now steadfastly penetrating all areas of our lives and life itself. Although engineering of biological objects is gaining a rather careful acceptance and sounds alien or controversial to many people (Chapter 19), it is actually as old as humankind and, since prehistoric times, it has been employed to breed animals and cultivate plants. The development of molecular biology led to the understanding of molecular mechanics of life and opened a new door for engineering of biological objects. Among many areas, this engineering promises unprecedented control over health. Although medicine is a very conservative discipline that cautiously and scrupulously evaluates all innovations before applying them to patients, it eagerly embraced and stimulated the development of genetic and protein engineering.

Proteins derived from pathogenic microorganisms and viruses have a broad array of applications in all three domains of medicine—diagnostics, prophylaxis, and therapy. Human proteins also may serve as diagnostic reagents and prophylactic agents for many complex diseases like cancer. There are three major sources of pathogen and human proteins for medicinal applications: (1) the human body, (2) tissue and cell cultures, and (3) recombinant DNA technology. The first two sources were exploited first. However, significant safety concerns and frequently prohibitive cost limited the use of these sources. On the other hand, the medicinal application of recombinant DNA technology to, for example, viral diseases provided an immediate and safe access to viral proteins that could be used for diagnostics and vaccines. One of the most widely known examples of such use of recombinant DNA is the hepatitis B virus surface antigen obtained from transformed yeast to create an efficient, safe, and affordable recombinant vaccine (Chapter 12). However, it was very rapidly realized that the straightforward approach employed for the development of a recombinant hepatitis B vaccine is not applicable to many other viruses such as the human immunodeficiency virus (Chapter 7) or the hepatitis C virus (Chapter 12). Alan R. Fersht and his colleagues found the solution to this problem around 25 years ago, presenting the concept of protein engineering (Chapter 9) as a new area for the application of molecular techniques and scientific principles to the design and construction of novel proteins with desired properties.

The remarkable developments over the last three decades in the automatic high-throughput chemical syntheses of long peptides, as well as in chemical and enzymatic syntheses of long DNA fragments, have built a solid technological foundation for protein engineering. However, the currently limited understanding of protein structure and of the relationship between structure and function has hindered the application of protein engineering to many complex medical problems.

Rational design of proteins is deep-rooted in the quantitative knowledge of their chemical, physical, and biological properties. In cases when sufficient knowledge of these properties was secured, a significant progress in protein engineering was attained. Parts I and II of this book review some of the success stories in the development of vaccines, diagnostics, and therapeutics as well as problems associated with protein engineering in these fields.

Major strategies for engineering of proteins with predetermined biological properties can be classified into structural, functional, and focused approaches. The structural approach is a major focus of this book. This approach is as close to the real rational design of proteins as the modern state of science allows. However, because comprehensive knowledge of a protein's quantitative structure–activity relationship (QSAR) is often unavailable, this approach can be applied only with caution. Protein engineering requires the availability of simple building blocks with clearly established QSAR. Such blocks became available through research on the antigenic and immunogenic properties of proteins (Chapters 6 through 13). Building blocks for diagnostics and vaccines exist in the form of antigenic epitopes that can be modeled with short protein fragments and in the form of carrier molecules that can efficiently present antigenic epitopes to immunocompetent cells or antibodies.

There are two major approaches to using these building blocks in the engineering of proteins with predetermined immunological properties. One approach, based on construction of artificial multiple epitope proteins (MEP), is used mainly to develop reagents with improved diagnostically relevant properties as described in Chapter 13. Because antibody–antigen binding is a relatively simple protein–protein interaction process, the MEP approach was shown to be somewhat efficient for obtaining artificial diagnostic antigens. The structural design of vaccines is more complicated because the process of eliciting antibodies is significantly more complex. In this case, antigenic epitopes of interest are frequently inserted into carrier proteins that serve to stimulate an immunoresponse against the inserted epitopes. Chapters 6 and 8 through 11 describe this approach using virus-like particles to vaccine development.

The functional approach, also known as "directed evolution," requires no prior knowledge of a protein's QSAR. Rather, it is based on the availability of a representative library of peptides or proteins and a selection procedure for the peptide or protein with the desired activity. It is termed a "functional" approach because proteins are selected based on specific functions without regard to structure. This approach is only briefly reviewed in this book (Chapter 17). It is not protein engineering *per se*, because proteins generated using this approach are discovered rather than engineered. To a significant degree, the functional approach is used as a "magic wand" that, unfortunately, has a very limited efficiency. There are examples of reasonably successful applications of this strategy to antigens and antibodies described in Chapters 9, 15, and 17.

The focused approach applies mathematical modeling to gain new QSAR knowledge and uses this knowledge to design proteins with improved desired properties (Chapter 2). The mathematical QSAR model can be built using a limited set of protein variants displaying variations in their activity. Accurate quantitative models can be used to directly design a desired protein. Modestly accurate models can be used as a guide in a cyclic process of the virtual search of sequence space for proteins with improved properties, experimental construction and testing of novel proteins, and optimization of the model for a more accurate virtual exploration. Part I of this book reviews available Web computational resources (Chapter 1), machine learning (Chapters 2 and 4) and phylogenetic techniques (Chapter 3), as well as techniques for detection of coordinated substitutions (Chapter 5) that can be used to study protein properties and build mathematical models for engineering of novel vaccines, diagnostic reagents, and therapeutics.

Protein engineering is unavoidably connected to genetic engineering. Some protein applications require a careful matching between the designed proteins and genetic constructs. This intricate matching is thoroughly surveyed in Chapter 20. Many chapters in Parts II and III frequently touch on plant systems used for the expression of engineered proteins. The interest of protein engineers in plant expression is quite understandable. Besides many advantages meticulously illustrated in Chapters 18 and 19, plant expression systems allow for circumventing some technological problems related to protein purification, storage, and implementation that are not directly linked to the targeted protein properties. As such, these systems open the door for an intriguing prospect of protein engineering "without proteins."

Medicinal protein engineering is an immense field for molecular and computational exploration. Attempts to describe this field in one book will inevitably be selective. Nevertheless, I hope that many current practitioners of protein engineering as well as many computational and experimental researchers with an aspiration for exceptionally thrilling and extremely challenging tasks will find this volume instructive and stimulating. Protein engineering for medicine is only going to be more exciting. In his book *The Romantic Generation* Charles Rosen wrote, "There are details of music which cannot be heard but only imagined." This is the wonderful way of art. However, it is not what we expect from science. This book was conceived with the hope that all the ideas the readers may "imagine" after going through this volume will be definitely "heard" at the lab bench, at the computer, or, better yet, at the patient's bed.

Since I opened this preface with a quotation from *The Book of Five Rings* by Miyamoto Musashi I would like to close by borrowing another timeless gem from this philosopher and peerless fighter that in my mind appropriately emphasizes the power of knowledge—"When one is in full combat gear, one does not think of small things."

Yury E. Khudyakov
Centers for Disease Control and Prevention
Atlanta, Georgia

Editor

Yury E. Khudyakov, PhD, is chief of the Molecular Epidemiology and Bioinformatics Laboratory, Laboratory Branch (LB), Division of Viral Hepatitis (DVH), Centers for Disease Control and Prevention (CDC), Atlanta, Georgia. He received his MS in genetics from the Novosibirsk State University, Novosibirsk, Russia and his PhD in molecular biology from the D.I. Ivanovsky Institute of Virology, Academy of Medical Sciences, Moscow, Russia. He started his research career in the laboratory of gene chemistry at the M.M. Schemyakin Institute of Bioorganic Chemistry, Academy of Sciences, Moscow, Russia. He was a research fellow in the laboratory of viral biochemistry and chief of the genetic engineering section in the laboratory of chemistry of viral nucleic acids and proteins at the D.I. Ivanovsky Institute of Virology, Moscow, Russia. In 1991, he joined the Hepatitis Branch (HB), Division of Viral and Rickettsial Diseases (DVRD)/CDC, Atlanta, Georgia National Academy of Sciences, United States as a National Research Council research associate. Since 1996, he has served as chief in the Developmental Diagnostic Unit, Molecular and Immunodiagnostic Section/HB/DVRD/NCID/CDC and later as chief of Computational Molecular Biology Activity and deputy chief of Developmental Diagnostic Laboratory, LB/DVH/CDC.

Dr. Khudyakov's main research interests are molecular epidemiology of viral diseases, development of new diagnostics and vaccines, molecular biology and evolution of viruses, and bioinformatics.

Dr. Khudyakov has published over 110 research papers and book chapters. He has also edited a book *Artificial DNA*, CRC Press (2002). He has authored several issued and pending patents. He is a member of the editorial board for the *Journal of Clinical Microbiology*.

Contributors

Michael Aidoo
Laboratory Branch
Division of HIV/AIDS Prevention
National Center for HIV, Hepatitis, STD, and Tuberculosis Prevention
Centers for Disease Control and Prevention
Atlanta, Georgia

Ekaterina Alekseeva
Latvian Biomedical Research and Study Centre
University of Latvia
Riga, Latvia

Martin F. Bachmann
Cytos Biotechnology AG
Schlieren, Switzerland

Rick A. Bright
Novavax, Inc.
Rockville, Maryland

Salvatore Butera
Laboratory Branch
Division of HIV/AIDS Prevention
National Center for HIV, Hepatitis, STD, and Tuberculosis Prevention
Centers for Disease Control and Prevention
Atlanta, Georgia

Stephen A. Cammer
Virginia Bioinformatics Institute
Virginia Polytechnic Institute and State University
Blacksburg, Virginia

David Campo
Molecular Epidemiology and Bioinformatics Laboratory
Division of Viral Hepatitis
Centers for Disease Control and Prevention
Atlanta, Georgia

Michael Czar
Virginia Bioinformatics Institute
Virginia Polytechnic Institute and State University
Blacksburg, Virginia

Zoya Emilova Dimitrova
Molecular Epidemiology and Bioinformatics Laboratory
Division of Viral Hepatitis
Centers for Disease Control and Prevention
Atlanta, Georgia

Dennis Ellenberger
Laboratory Branch
Division of HIV/AIDS Prevention
National Center for HIV, Hepatitis, STD, and Tuberculosis Prevention
Centers for Disease Control and Prevention
Atlanta, Georgia

Lilia Milkova Ganova-Raeva
Molecular Epidemiology and Bioinformatics Laboratory
Division of Viral Hepatitis
Centers for Disease Control and Prevention
Atlanta, Georgia

Alma Gedvilaite
Institute of Biotechnology
Vilnius, Lithuania

María José Gómara
Department of Peptide and Protein Chemistry
Institute for Chemical and Environmental Research (IIQAB-CSIC)
Barcelona, Spain

Elmars Grens
Latvian Biomedical Research and Study Centre
University of Latvia
Riga, Latvia

Isabel Haro
Department of Peptide and Protein Chemistry
Institute for Chemical and Environmental Research (IIQAB-CSIC)
Barcelona, Spain

Reimar Johne
Federal Institute for Risk Assessment
Berlin, Germany

Andris Kazaks
Latvian Biomedical Research and Study Centre
University of Latvia
Riga, Latvia

Tatyana Kozlovska
Latvian Biomedical Research and Study Centre
University of Latvia
Riga, Latvia

James Lara
Molecular Epidemiology and Bioinformatics Laboratory
Division of Viral Hepatitis
Centers for Disease Control and Prevention
Atlanta, Georgia

Alla Likhacheva
NPO Diagnostic Systems
Nizhny Novgorod, Russia

Miguel Angel Gómez Lim
Centro de Investigación y de Estudios Avanzados del I.P.N.
Departamento de Ingeniería Genética de Plantas
Irapuato, Mexico

Baiba Niedre-Otomere
Latvian Biomedical Research and Study Centre
University of Latvia
Riga, Latvia

Anna Obriadina
NPO Diagnostic Systems
Nizhny Novgorod, Russia

Velta Ose
Latvian Biomedical Research and Study Centre
University of Latvia
Riga, Latvia

Néstor O. Perez
Probiomed S.A. de C.V.
Mexico City, Mexico

Paul Pumpens
Latvian Biomedical Research and Study Centre
University of Latvia
Riga, Latvia

Michael Anthony Purdy
Molecular Epidemiology and Bioinformatics Laboratory
Division of Viral Hepatitis
Centers for Disease Control and Prevention
Atlanta, Georgia

Peter M. Pushko
Novavax, Inc.
Rockville, Maryland

Till A. Röhn
Cytos Biotechnology AG
Schlieren, Switzerland

Lívia M.G. Rossi
NL Biotechnologies LLC
Little Rock, Arkansas

Kestutis Sasnauskas
Institute of Biotechnology
Vilnius, Lithuania

Gale E. Smith
Novavax, Inc.
Rockville, Maryland

James M. Smith
Laboratory Branch
Division of HIV/AIDS Prevention
National Center for HIV, Hepatitis, STD, and Tuberculosis Prevention
Centers for Disease Control and Prevention
Atlanta, Georgia

Terrence M. Tumpey
Influenza Division
National Center for Immunization and Respiratory Diseases
Centers for Disease Control and Prevention
Atlanta, Georgia

Tatyana Ulanova
NPO Diagnostic Systems
Nizhny Novgorod, Russia

Rainer G. Ulrich
Friedrich-Loeffler-Institut
Federal Research Institute for Animal Health
Institute for Novel and Emerging Infectious Diseases
Greifswald, Germany

Tatyana Voronkova
Latvian Biomedical Research and Study Centre
University of Latvia
Riga, Latvia

Jonny Yokosawa
NL Biotechnologies LLC
Little Rock, Arkansas

Anna Zajakina
Latvian Biomedical Research and Study Centre
University of Latvia
Riga, Latvia

Andris Zeltins
Latvian Biomedical Research and Study Centre
Riga, Latvia

Part I

Computational Approaches

1 Web Resources for Protein Analysis

Stephen A. Cammer and Michael Czar

CONTENTS

1.1 INTRODUCTION

Biological science has become increasingly dependent on access to and use of distributed resources, including many databases and systems available through the Internet. In the field of protein analysis and engineering, many such web sites are useful for design and analysis within projects of all scales. The myriad of protein resources and the capability for rapid evolution on the Internet defies a complete description in one text. Therefore, we have sought to present a small collection of highly useful sites for the protein researcher.

Ten partially overlapping categories of sites are described here, and Table 1.1 provides a listing of these categories and the number of resources surveyed. There is a plethora of general resources that serve as a good jumping off point for collecting information and sequences. Some of the resources specialize in segments of proteins and the analysis of motifs, domains, and regions of a protein that would be necessary for membrane insertion. While many of these resources aim to provide information on every protein available, others specialize in specific protein types, including enzymes and transcription factors. Other sites specialize in organizing protein mutations and determining how they affect protein function. Still, other sites are concerned with how proteins interact with each other. These resources are described in more detail within the body of this chapter. The types of sites and number reviewed are shown in Table 1.1.

TABLE 1.1
Type of Web Resources Reviewed along with the Number of Resources Reviewed

	Number
General protein resources	2
Protein sequences and database searching	4
Motif and domain resources	11
Structure-based domain identification	2
3D structure classification, comparison, and modeling	6
Membrane protein prediction and analysis	5
Enzyme resources	2
Protein mutation resources	3
Protein–protein interaction resources	4
Transcription factors	1

The collection presented here covers only a small portion of the available Web sites useful in protein science; however, the breadth covered in this chapter should help researchers understand what is available for their use. Hopefully, this set will act as a catalyst, prompting researchers to make better use of the wide array of resources available.

1.2 GENERAL PROTEIN RESOURCES

Databases of annotated proteins are essential in most areas of protein research. In this section, the two most comprehensive resources for protein annotations that have been cross-indexed throughout many of the primary sequence databases are discussed in detail. Both of these resources have played prominent roles in protein research for many years, and both provide a wealth of information covering protein function, classification, and primary structure.

1.2.1 UniProt

UniProt (http://www.pir.uniprot.org/) [1] provides large repositories of protein functional and structural data. These resources are designed to serve as starting points for laboratory and computational experiments. Towards this goal, UniProt supports biological research by providing a comprehensive, annotated, protein sequence database that is cross-referenced with other resources and facilitates queries through a simple interface. This resource is freely accessible by the educational and research communities.

The European Bioinformatics Institute (EBI), the Swiss Institute of Bioinformatics (SIB), and Protein Information Resource (PIR) united as the UniProt Consortium in 2002. The UniProt Consortium maintains the ExPASy (Expert Protein Analysis System) servers, which are a central resource for proteomics tools and databases. PIR, hosted by the National Biomedical Research Foundation (NBRF), maintains the oldest protein sequence database—Margaret Dayhoff's *Atlas of Protein Sequence and Structure* [2].

Consortium members provide protein database maintenance and annotation. Historically, EBI and SIB together produced Swiss-Prot and TrEMBL (Translated EMBL Nucleotide Sequence Data Library), while PIR produced the Protein Sequence database (PIR-PSD). These databases maintained differing protein sequence coverage and annotation efforts. Swiss-Prot is recognized as the preeminent source of protein annotation, with extensive cross-references, literature citations, and computational analyses provided by expert curators. In response to the rate at which sequence data are generated, TrEMBL was created to provide automated annotations for those proteins not in Swiss-Prot. At the same time PIR maintained the PIR-PSD and related databases, including iProClass, a database of curated families of protein sequences. The consortium members pooled their overlapping and complementary resources, efforts, and expertise to create the UniProt databases.

The UniProt database includes three database layers to obtain complete coverage of sequence space at several resolutions. First, the UniProt Archive (UniParc) provides the complete, nonredundant collection of publicly available protein sequence data. An enhanced collection, the UniProt Knowledgebase (UniProtKB), is a centralized database of protein sequences with accurate and comprehensive sequence and functional annotations. The UniProt Reference Clusters (UniRef) databases provide nonredundant reference clusters based on the UniProt knowledgebase and selected UniParc records.

1.2.2 PIR

The Protein Information Resource (PIR) (http://pir.georgetown.edu/pirwww/index.shtml) [3] at the Georgetown University Medical Center (GUMC) is an integrated public bioinformatics resource to support genomic and proteomic research. PIR was established by the National Biomedical Research Foundation (NBRF) to assist researchers in the identification and interpretation of protein sequence information. Before PIR was founded, the NBRF compiled the first comprehensive collection of macromolecular sequences in the *Atlas of Protein Sequence and Structure* [2], edited by Dayhoff. Dayhoff and her research group were instrumental in the development of computer methods for protein sequence comparison, detection of distant homologies, identification of duplications within sequences, and understanding evolutionary trajectories from protein sequence alignments. Over the last four decades, PIR has provided protein sequence databases and analysis tools to aid the research and education communities in the discovery and understanding of proteins.

1.3 PROTEIN SEQUENCES AND DATABASE SEARCHING

At their most basic level, the resources described in this section provide a means for researchers to find protein sequences that are relevant to their work. Additional value is found in the many links to other resources that are available through the NCBI databases and SwissProt. SwissProt also provides in depth manual annotation of proteins, which is based on both bioinformatic analyses and review of published literature. The PIRSF database, which portrays evolutionary relationships between proteins, can serve as a basis for comparative proteomics.

1.3.1 NCBI Protein

The NCBI (http://www.ncbi.nlm.nih.gov/entrez/query.fcgi?db = Protein) system is built around the search tool, Entrez. Protein entries in the Entrez system for search and retrieval have been collected from SwissProt, PIR, PRF, PDB, and from translations of annotated coding regions in GenBank and RefSeq. Each entry in the protein database includes various definitions of the locus and accession number, the database source, keywords, the organism source, taxonomy, authors for any references, publication title, journal, features, including the segment reported, genomic details, various tags, and the actual coding sequence. This information can be displayed in a range of formats, including XMLs. Links for the entry include pages that show the top 200 similar sequences, according to BLAST scores, and links to the other databases listed above.

One of the best features of the protein database is extensive linking to other resources within the Entrez system. Links to Taxonomy Classification, the Nucleotide database, the Conserved Domain database, and literature in the PubMed system are the most numerous. The next most populated links are made with PubMed Central, the Gene database, and the Genome database. Other links are made to three-dimensional (3D) Domains, Structure, PC Substance, PC Compounds, PopSet, HomoloGene, SNP, and OMIM. The free and structured text searches of the database allow the user to quickly research in depth the available knowledge about the protein.

1.3.2 NCBI Protein BLAST

The NCBI BLAST site (http://www.ncbi.nlm.nih.gov/BLAST/) provides an interface to facilitate searching a protein sequence or set of sequences against a large database of potential matches. The databases that can be searched include a complete, nonredundant set of protein sequences and translations of nucleotide sequences from GenBank, the SwissProt database, RefSeq, Patented sequences, the PDB sequences, and the previous month's submissions.

1.3.3 SwissProt

UniProtKB/Swiss-Prot (http://ca.expasy.org/sprot/) [4] is part of the UniProt database that is described above. This database is manually curated, and is designed to provide a high level of protein sequence annotation, including a description of the protein function, structural domains, posttranslational modifications, and variant information. SwissProt aims to provide a minimal level of redundancy and high level of integration with other databases. SwissProt is considered by many to produce the "gold standard" for protein annotation. SwissProt can be searched using the Sequence Retrieval System (SRS), which uses full text queries, or by using more advanced structured queries based on descriptions, gene names, and organism. The database can also be searched using sequences compared via the BLAST program. There is a facility to browse by taxonomy as well. Searches can be used to retrieve a list of UniProtKB entries, and the site provides a user manual, release notes, and indices. A user can also use a service at the site that sends an alert when relevant sequences are added to the database.

UniProtKB/TrEMBL is the computer-annotated supplement of Swiss-Prot that contains all the translations of EMBL nucleotide sequence entries that have not yet been integrated in Swiss-Prot.

1.3.4 PIRSF

PIRSF (http://pir.georgetown.edu/pirwww/dbinfo/pirsf.shtml) aims to provide a comprehensive and nonoverlapping, hierarchical clustering of UniProtKB sequences to reflect evolutionary relationships. PIR has developed this classification strategy based on rules for protein names and functional sites to facilitate the propagation and standardization of protein annotation, as well as the systematic detection of propagated annotation errors. PIRSF aims to improve sensitivity of protein identification and functional inference, and also provides the basis for evolutionary and comparative proteomics. The PIRSF classification system is based on whole proteins and, therefore, allows annotation of generic biochemical and biological functions, as well as classification of proteins without well-known domains.

The primary level of classification is that of the homeomorphic family, whose members demonstrate homology by sharing full-length sequence similarity and a common domain architecture. A lower level of classification contains subfamilies, which are clusters representing functional specificities or domain architecture variations within the family. Distantly related families and orphan proteins are joined into a higher level of classification, superfamilies, based on common domain architectures. Since proteins sometimes belong to more than one domain superfamily, the PIRSF classification is formally a network [5].

1.4 MOTIF AND DOMAIN RESOURCES

Recognition of homology to proteins with known functions and structures is key to identifying proteins, recognizing their structures, and assigning function. Many resources exist that enable researchers to recognize sequence motifs characteristic of specific protein families. Some of the motif searches are based on primary structure, or sequence, while others are designed for recognition of specific patterns derived from comparisons of related 3D structures. Searches are usually done against large databases of known patterns or hidden Markov models (HMMs), and results are often assigned statistical significance values for the identifications. Resources that implement motif-based searches are described in this section. Due to different methodologies used to construct and query these databases, differences of functional prediction occur. When biologists use these tools it is often helpful to run multiple tools to check for agreement in predictions. InterPro facilitates this process by incorporating data from many different tools into its database.

1.4.1 PROSITE

PROSITE (http://ca.expasy.org/prosite/) [6] is a database of protein domains, families, and functional sites that belong to specific classes of structure–function motifs. The collection's purpose is to facilitate annotation by enabling identification of a protein's functional family. The components of an entry include its associated sequence patterns and profiles used for identification.

A complement to PROSITE is ProRule, which is a collection of rules based on profiles and patterns. Additional information about functionally or structurally important amino acids contained in ProRule can help increase the discriminatory utility of the profiles and patterns.

1.4.2 PRINTS

The PRINTS compendium [7] (http://www.bioinf.manchester.ac.uk/dbbrowser/PRINTS/) is a collection of protein motifs, known as fingerprints, which are used together in groups to characterize protein families at the level of amino acid sequence. The fingerprints contained in the PRINTS compendium are iteratively refined by scanning the SWISS-PROT/TrEMBL composite database to improve their effectiveness of identifying the correct protein family. Fingerprint motifs do not usually overlap in sequence space; however, they often represent residues in close proximity in 3D space. Fingerprint motifs are flexible enough to encode protein folds and functionalities in a way that can improve upon use of a single motif for identification of protein family membership. The mutual application of fingerprint motifs yields a diagnostic potential that often surpasses searches for single motifs.

1.4.3 Blocks

Blocks (http://blocks.fhcrc.org/) [8] are multiply aligned, ungapped sequence segments corresponding to highly conserved regions of proteins. The Blocks database is automatically constructed by identifying the most highly conserved sequence regions in groups of related proteins documented in InterPro. The Blocks database was constructed by the PROTOMAT [8] system using the MOTIF algorithm. A tool, Block Maker, is made available for a user to create his/her own blocks. User-created blocks are created from sequences supplied by the user in the same manner as the blocks in the Blocks database. To allow searching, results are reported in the standard Block multiple-sequence alignment format.

Other tools, such as Block Searcher and Get Blocks, serve as aids in the detection and verification of protein sequence homology. These tools compare a protein or DNA sequence to a database of protein blocks, and retrieve the corresponding blocks.

The complete Prints database is also available in Blocks format. The program Blimps can be used to search for Prints data on the Blocks Server, but that data is not available through the Blocks e-mail server. Since different methods are used to construct Prints and Blocks, it is valuable to retrieve the data from both databases; Prints includes more than 300 families not represented in the Blocks database.

The Blocks database is based on InterPro entries with sequences from SWISS_PROT and TrEMBL, and with cross-references to PROSITE, PRINTS, SMART, PFAM, and ProDom entries. To reduce the inclusion of false positive sequences in the InterPro entries by the automated procedure, Blocks were made for each InterPro entry using only sequences in SWISS-PROT, to which TrEMBL sequences were added if they fit the resulting Blocks model.

1.4.4 Pfam

Pfam (http://www.sanger.ac.uk/Software/Pfam/) [9] is a collection of protein families and domains based on multiple protein alignments and derived profile-HMMs. It is a semi-automatically generated protein family database that aims to be comprehensive and accurate. Pfam can be used to view the domain organization of specified proteins. It covers most of the known sequence space −74% of protein sequences have at least one match to Pfam. For a novel protein sequence, one can search against the Pfam database to determine whether it matches any of the domains in Pfam. If there are matches the system will automatically indicate how many domains and where each lies in the sequence. To complement the domain match information, the system provides the associated annotation about the domain.

In addition, one can execute the following tasks for each family in Pfam:

- Look at multiple alignments
- View protein domain architectures
- Examine species distribution of domains
- Follow links to external databases
- View known protein structures

Pfam consists of two major components. The first is the curated part of Pfam, Pfam-A, that contains over 8957 protein families. The second is the automatically generated supplement called Pfam-B. This supplement contains a large number of families taken from the PRODOM database [10] that do not overlap with Pfam-A. The resulting lower quality Pfam-B families can be useful when no Pfam-A families are found for a given sequence.

1.4.5 InterPro

InterPro (http://www.ebi.ac.uk/interpro/) [11] is a database of protein families, domains, and functional sites in which identifiable features found in known proteins can be applied to unknown protein sequences. Several secondary protein databases of functional sites and domains existed prior to InterPro's inception. PROSITE, PRINTS, Pfam, ProDom, etc., are useful for identifying distant relationships in new protein sequences, and thereby facilitate predicting protein function and fold structure; however, these motif signature databases use different formats and terminology and therefore have different utility in protein analysis. To address the need for integration of the motif-recognizing tools, EBI, SIB, University of Manchester, Sanger Institute, GENE-IT, CNRS/INRA, LION Bioscience AG, and University of Bergen combined these databases into a single search tool called InterPro. Subsequently SMART, TIGRFAMs, PIRSF, SUPERFAMILY, PANTHER, and the structure-based Gene3D were integrated. Where applicable, InterPro also cross-references to the Blocks database as well as several other protein family and structure databases. InterPro provides an integrated view of the commonly used signature databases, and has an intuitive interface for text- and sequence-based searches. InterPro is freely available from the EBI Web site as well as from the Web site of each of the member databases.

To help analyze the increasing volume of raw sequence data, InterPro has been applied to improve the functional annotation of UniProtKB. The database can also enhance genome annotation from genome-sequencing projects, where it facilitates whole proteome analysis (http://www.ebi.ac.uk/proteome). The future integration of additional signature recognition methods is expected to increase the coverage of the InterPro database.

1.4.6 TIGRFAMs

TIGRFAMs (http://www.tigr.org/TIGRFAMs/) [12] are protein families based on HMMs. The entries are a collection of protein families, along with curated multiple-sequence alignments, HMMs, and associated information designed to facilitate the automated functional assignment of proteins according to detected sequence homology. The HMMs developed for TIGRFAM represent likelihoods for specific states (residues) from the columns in the multiple alignments. The curated HMMs are created using a seed alignment that can be viewed for each family. In addition, the full alignment of family members and the cutoff scores for the inclusion of each can be viewed. Searches through the TIGRFAMs and HMMs can be based on text in the TIGRFAMs. A user also can search for specific sequences using the TIGRFAMs sequence search.

TIGRFAM relies on classification terms developed specifically for TIGRFAM use. The first important term is equivalog. Equivalogs are subgroups within a set of homologous proteins that share a conserved function since the split from their last common ancestor. Equivalogs are based on as close to full-length sequences as possible. Where subsequences are shared, the term equivalog domain is applied. In cases where homology to a hypothetical protein is shared, the term hypothetical equivalog is applied. Orthologs are designated as proteins related by common ancestry that appear in different species but which may differ in function. Superfamilies occur when homology is shown over the full length of a set of sequences. Subfamilies, or incomplete sets within superfamilies, are more commonly found in TIGRFAM. Subfamily models are the most specific means of functional assignment using TIGRFAM.

1.4.7 ProDom

ProDom (http://prodom.prabi.fr/prodom/current/html/home.php) was first established in 1993 and originally maintained by the Laboratoire de Génétique Cellulaire and the Laboratoire de Interactions Plantes-Microorganismes (INRA/CNRS) in Toulouse. It is now maintained by the PRABI bioinformatics center of Rhone-Alpes. ProDom [10] is intended to be a comprehensive set of protein domain families automatically generated from the SWISS-PROT and TrEMBL sequence databases. The ProDom database is founded on the idea that automated processes are needed to be comprehensive in protein domain analysis. ProDom is generated from the global comparison of available protein sequences that are then automatically clustered according to homologous segments. The ProDom building procedure, MKDOM2, is based on recursive PSI-BLAST searches. The source protein sequences are non-fragment sequences derived from SWISS-PROT and TrEMBL databases. The ProDom database consists of domain family entries that provide a multiple-sequence alignment of homologous domains and a derived consensus sequence. Recently, the use of 3D information from the SCOP database has improved the value of the system. In addition, the system includes ProDom-SG, a ProDom-based server dedicated to the selection and prioritization of candidate proteins for structural proteomics.

The system allows users to browse ProDom and compare their sequences against the database. Users are also able to search by kingdom in the ProDom-CG.

1.4.8 PANTHER

The PANTHER (http://www.pantherdb.org/), Protein ANalysis THrough Evolutionary Relationships, classification system classifies genes according to their functions by using published experimental evidence [13]. In the absence of direct experimental evidence, PANTHER uses evolutionary sequence relationships to predict function. To generate the system, expert biologists classify proteins into families and subfamilies based on shared functions categorized by biochemical or molecular activities and biological process ontologies. Where possible, biochemical interactions forming canonical pathways are made available and can be viewed interactively. PANTHER can be searched using text queries or alternatively, the collection can be browsed by function or through interactive graphs summarizing the distribution of biological functions for whole genomes including human, mouse, rat, and fruit fly. The PANTHER site was designed to facilitate in the analysis of large numbers of gene sequences, proteins, and transcripts.

The PANTHER site allows users to explore protein families, molecular functions, and pathways. Users can find information on protein families and subfamilies, as well as view a multiple-sequence alignment and phylogenetic tree for the family. Users can also browse the PANTHER ontology of molecular functions, biological processes, and pathways, as well as find the genes and proteins that participate in those pathways.

Users are able to generate lists of genes, proteins, or transcripts. These lists can be based on ontology association, protein family or subfamily, genomic region, or simply a list of identifiers. The site also allows users to create and order a list of Applied Biosystems Assays-on-Demand.

In addition to creating lists established on the contents of database queries, PANTHER also lets users analyze gene lists. By supplying a list of genes, proteins, or transcripts based on family, molecular function, biological process, or pathway, gene functions can be graphed in pie charts. These gene lists can identify statistically over- and under-represented functions. Additionally, gene-value pairs can be analyzed on this site.

The system includes a library, PANTHER/LIB, of protein families and subfamilies, and their associated data including phylogenetic trees, multiple-sequence alignments, and HMMs. In addition, the resource provides a database of 130 regulatory and metabolic pathways mapped to protein sequences that are viewable interactively. The system also includes a collection of terms (PANTHER/X) describing protein molecular functions and biological processes.

1.4.9 PROTOMAP

ProtoMap (http://protomap.cornell.edu/) is an automatic hierarchical classification of SWISSPROT and TrEMBL proteins into groups of related proteins [14]. The analysis uses transitivity to identify homologous proteins, whereby within each group, any two members are either directly or transitively related. To prevent unrelated proteins from clustering, the method is applied restrictively. The classification is performed at various levels of confidence, resulting in a hierarchical organization of all the proteins. The resulting classification splits the protein space into well-defined groups of proteins, which are often closely correlated with evolutionarily related families and superfamilies. The hierarchical organization of PROTOMAP can facilitate in the identification of subfamilies that make up the families of protein sequences.

To achieve the clustering of sequences, common measures of similarity between protein sequences (Smith-Waterman, FASTA, BLAST) are combined using two different scoring matrices (blosum 50 and blosum 62) to create an exhaustive list of neighboring sequences for each sequence in the SWISSPROT and TrEMBL databases. This protein space can be represented as a weighted and directed graph with sequences being the vertices and edges representing similarities between near neighbors weighted by their expectation values from the comparisons. Extensively connected components of the graph correspond to clusters of related proteins. The aim is to automatically detect families and thereby obtain a classification of all protein sequences that reflects the topology and connectivity of the protein sequence space.

The analysis begins with a stringent classification based on highly significant similarities represented by expectation values below 10^{-100}, and therefore, yields many initial clusters. Clusters are then merged via a two-step process to accommodate less significant similarities. First, the procedure identifies groups of possibly related clusters based on transitivity and strong connectivity using the local geometry of the network. Second, a global test is applied to identify super-clusters representing strong relationships within groups of clusters, and these clusters are then merged. This process takes place at varying thresholds of E-values, where the algorithm is applied on the clusters of the previous classification at each step, yielding to the more permissive threshold. The analysis starts at the 10^{-100} threshold then subsequently at levels of 10^{-95}, 10^{-90}, $10^{-85}, \ldots, 10^{0}$ (= 1). The procedure results in the hierarchical organization of all proteins used in the analyses.

1.4.10 SBASE—Domain Prediction System

SBASE (http://hydra.icgeb.trieste.it/sbase/) is a collection of protein domain sequences obtained from protein sequence and genomic databases using literature-based searches [15]. Protein domains are defined by their sequence boundaries indicated by the publishing authors or as illustrated in one of the primary sequence databases (Swiss-Prot, PIR, TrEMBL). Domain groups are incorporated when there are well-defined sequence boundaries, and can be distinguished from other sequences using a sequence similarity search. SBASE contains the domain sequences and various statistical parameters of the domain groups. The approach used is similar to memory-based computing. The overall result is a picture of the similarity network of the sequence space.

The SBASE database uses a straightforward method of representing similarities. Sequences are considered similar if they are more similar to members of their group and less similar to other members of the database. Sequences that have a BLAST similarity score above a threshold to at least one member of the group define the neighborhood of the group. Numeric representation of similarity scores for each domain group is designated by two scores. NSD is the number of within-group scores above the default threshold of BLAST. AVS is the average of these values. This method is used to determine the respective quantities of nonmember neighbors in the domain group. A plot of the AVS and NSD values against each other yields a graph that is a local representation of the similarity space around the domain group. The server will analyze each of the local similarities and compare them using a database of self- and nonself-similarities in the known domain groups using various statistical comparisons. If the distribution of local similarities is similar to the similarities within the group, the domain will be assigned to the query. Evaluation of domain content is based on the two quantities, NSD and AVS. SBASE uses threshold values of NSD and AVS for each group, and domain similarities are evaluated when they are above the thresholds. Distributions of NSD and AVS values are stored for known groups as well as for their nonmember neighbors. Using the probability values, a cumulative score is calculated [16] and the query is classified into the group with the highest cumulative score. Alternatively, a neural network can be used.

1.4.11 SMART—Simple Modular Architecture Research Tool

SMART (http://smart.embl.de/) [17] facilitates the identification and annotation of evolutionarily mobile domains as well as the analysis of the domain architectures. SMART allows detection of over 500 domain families from extracellular, signaling, and chromatin-associated proteins, with an emphasis on eukaryotic mobile elements. Domains are annotated with respect to phyletic distribution, functional classes, tertiary structures, and key residues for function. SMART includes predicted proteins from complete genomes and integrates them with predictions of orthology. The database contains information on specific sequence-structure features such as transmembrane regions, coiled-coils, signal peptides, and internal repeats. Domains, search parameters, and taxonomic information are maintained in a relational database system. User interfaces allow searching for proteins with specific domain combinations in defined taxa. Visualization tools are available to allow for the analysis of gene intron–exon structure within the context of protein domain structure.

1.5 STRUCTURE-BASED DOMAIN IDENTIFICATION

While the high homology of a primary sequence can indicate shared protein function, function can be conserved through the maintenance of protein structure. Certain resources make use of protein structure classification databases to enable the assignment of protein structure and putative function. These sites provide searchable models that can be used in protein analysis. Two such resources are discussed in this section.

1.5.1 Superfamily

Superfamily (http://supfam.mrc-lmb.cam.ac.uk/SUPERFAMILY/) [18] provides structure–function assignments to protein sequences at the level of protein superfamily. The system does not distinguish among families within superfamilies; rather, it detects broader and more distant relationships. In this system, a superfamily contains all proteins for which there is evidence of a common ancestral protein through structural relationships. In essence, the system allows sensitive remote homology detection for protein families of known structure. The site allows superfamily assignment for small groups of input sequences, and allows the user to browse genomic superfamily assignments and multiple-sequence alignments for the SCOP-defined superfamilies [19].

The Superfamily database consists of a library of HMMs that covers all proteins with known structure, grouped into 1539 SCOP-defined superfamilies. Query sequences are assigned e-value scores for all of the HMMs and significant hits are returned to the browser. Superfamilies are represented by overlapping HMMs so more than one significant hit to a superfamily can occur for a given query. HMMs are created from a seed sequence that has been aligned to many homologues representing the superfamily. The model is built from the resulting multiple-sequence alignment using the SAM HMM software [20] and release 1.69 of the SCOP [19]. The alignment of query-to-model can be viewed in the browser after a search. To enrich the superfamily model, one can also view the alignment of models to genomic sequences.

1.5.2 GENE3D

Gene3D (http://cathwww.biochem.ucl.ac.uk:8080/Gene3D/) [21] is intended for the structural and functional annotation of protein families. Gene3D is built upon the BioMap sequence database [22], which consists of UniProt (including the genome sets obtained from Integr8 [23]) and extra sequences from various functional resources including KEGG and GO. These sequences are annotated using functional data from GO, COGS, and KEGG. The CATH domain database is scanned against the whole-sequence database and the Pfam domain family data are added for UniProt sequences. Protein–protein interaction data from BIND and MINT are also added, where available. Whole-chain protein families are also formed by clustering the sequences. These families should show good conservation of function and structural features. The Gene3D protein families represent a set of proteins with a consistent domain architecture and similar sequence length. The sequences composing a family are expected to be of common evolutionary origin and possess similar molecular functions.

The site can be searched using several parameters, which include UniProt accession, Gene3D protein family, CATH structural code, Pfam accession, MD5 digest value of amino acid sequence, taxonomy ID, and BLAST. Results can be returned as either summarized or detailed results. Both the detailed and summarized views return a common set of results. The normal mode of operation for this Web site is to return HTML, and it is also possible to get results in XML using Web-services and by downloading files for local processing.

Using a set of sequences obtained through a search, one can construct a multiple-sequence alignment by clicking on the alignment links near the bottom of the results. MUSCLE [24] is used to produce a multiple-sequence alignment.

CATH domains are identified using the CATH HMM library. In addition, the DomainFinder protocol is used to resolve matches to CATH models by creating a "best" architecture. Each domain is assigned a four-part code describing its position in the CATH hierarchy and is taken from the homologous-level superfamily from which the model is generated. A G3DSA: prefix is added to make the identifier unique to CATH/Gene3D.

Pfam domains are taken from the Pfam-to-UniProt mapping (swisspfam) provided by Pfam and assigned by UniProt ID or through an exact sequence match. Each Pfam family consists of three parts: (1) manually curated SEED alignment of sequences, (2) pairs of HMMs for fragments and global sequences, and (3) full alignment against UniProt sequences. There are four types of Pfam families. A domain family is a structured autonomous unit of protein function. A motif family is an unstructured autonomous unit of protein function. A repeat family is a single unit of an aggregate structure (i.e., β-propellors). Finally, the family type for families that do not fall into the above categories.

1.6 3D STRUCTURE CLASSIFICATION: COMPARISON AND MODELING

The number of protein structures available for research has greatly exceeded the number with which most researchers are able to be familiar. To facilitate structure comparisons and provide services for modeling proteins based on homology to a protein with known structure, it has become increasingly necessary to classify and organize the available structures. A collection of sites devoted to cataloging structures, structure comparison, and protein modeling are discussed in this section. These sites enable researchers to make use of the known protein structure data in their own research.

1.6.1 SCOP

The SCOP site (http://scop.mrc-lmb.cam.ac.uk/scop/) [19] maintains a manually created database augmented using automated methods that aim to classify 3D structures of all proteins of known structure. The premise is that nearly all proteins have structural similarities with other proteins and those that are similar often share a common evolutionary origin. SCOP aims to provide a detailed and comprehensive description of these structural and evolutionary relationships. SCOP provides a broad survey of known protein folds, detailed information about the close relatives of each protein, and an approach for protein structure classification and research.

SCOP is divided into the known classes for proteins based on Chothia and Levitt [25]: all-α, all-β, α/β, and $\alpha + \beta$. Other classes include multidomain proteins, small proteins, and membrane-bound proteins. SCOP classes are organized around the concept of a fold, which can be defined as a topologically connected set of secondary structure elements arranged in a specific core pattern in space. SCOP classifies proteins into families and superfamilies. Folds define superfamilies of related proteins that are subdivided into protein families. Families are composed of different proteins with the differential domain content. This hierarchy can be explored interactively at the Web site or computationally using downloadable files.

1.6.2 CATH

The CATH site (http://cathwww.biochem.ucl.ac.uk/latest/index.html) [26] maintains a hierarchical classification of protein domain structures that is clustered at four major levels, class(C), architecture (A), topology(T), and homologous superfamily (H). Class is based on secondary structure content and is assigned for more than 90% of protein structures automatically. Three major classes are defined: mainly α, mainly β, and α–β. The architecture describes the overall orientation of secondary structures, independent of connecting topology, and is assigned manually using simple terms such as three-layer sandwich. At the topology level, groups are composed of structures that form fold clusters according to topological connections and numbers of secondary structures. Topologies are compared and scored for similarity based on the SSAP algorithm [27]. Homologous superfamilies group proteins with highly similar structures and biochemical functions. Assignment to fold groups and homologous superfamilies is made by structure and sequence comparison. The boundaries and assignments for each protein domain are determined using a combination of automated and manual procedures. These include computational techniques, empirical and statistical evidence, literature review, and expert analysis.

The CATH database also includes crystal structures solved to at least 4.0 Å resolution and NMR structures. CATH does not include nonprotein peptides, models, and structures with greater than 30% CA-only. This removal of PDB structures is performed using the SIFT protocol [28].

Structure classification is performed using individual protein domains and multidomain proteins are divided into their constituent domains using automated and manual techniques. Multidomain proteins are subdivided into their constituent domains using a consensus procedure based on three independent algorithms for domain recognition (DETECTIVE [29], PUU [30], and DOMAK [31]). This currently allows approximately 52% of the proteins (i.e., those for which these algorithms agree) to be defined as single or multidomain proteins, automatically. When a protein chain has at least 80% sequence identity to a protein with known boundaries, or the structure similarity score exceeds a predefined threshold, the domain boundary assignment is performed automatically by transferring the boundaries from the other chain using the program ChopClose [32]. Other domain boundaries are assigned manually based on an analysis of results derived from a collection of algorithms (CATHEDRAL [33], SSAP [34], DETECTIVE [29], PUU [30], DOMAK [31]), or by using sequence-based methods including HMMs and relevant literature.

1.6.3 Dali Server

The Dali (http://www.ebi.ac.uk/dali/) server provides a service for comparing protein structures in 3D to reveal biologically interesting similarities that are not detectable by comparing sequences [35]. One submits queries of Protein Data Bank (PDB) format coordinates and Dali then compares them to those of protein structures in the PDB. A structure file containing a multiple alignment of structural neighbors is then e-mailed to the user's address. The Dali database maintains a precompiled list of neighbors for structures already in the PDB. The Dali Domain Dictionary is a numerical taxonomy of known structures in the PDB.

The server provides several other search capabilities. A user can do an SRS search in the Families of Structurally Similar Proteins (FSSP) resource; FSSPs are structural alignments of proteins in the PDB. The DSSP program will perform secondary structure assignment for a user-submitted set of coordinates. HSSP is a searchable database based on the sequence comparison of structures with known alignments; it provides files of alignments for proteins that are similar in structure and sequence. Dalilite is a program that performs pairwise structure comparisons. The program does not update the database when searched, but if the user prefers, Dalilite can be executed locally. Maxsprout uses a fast algorithm to generate a backbone and side chain structure from a coordinate file containing only C-Alpha positions. Backbones are assembled from known structures, and side chains are optimized using a simple potential energy function.

1.6.4 VAST—NCBI

VAST (http://www.ncbi.nlm.nih.gov/Structure/VAST/vast.shtml) is an algorithm that compares 3D protein structures to determine protein structure neighbors in Entrez. There are more than 87,804 domains in the NCBI's Molecular Modeling DataBase (MMDB) [36] and each is compared to every other one in order to map the structural similarity network for proteins of known structure. MMDB structure summary pages retrieved via an Entrez search contain structure neighbors available for protein chains and individual structural domains. Both PDB and MMDB identifiers can be entered to begin an on-the-fly search for 3D neighbors. The results, or neighbor lists, graphically illustrate the segments of protein structure that superimpose between structures.

1.6.5 Swiss-MODEL

The Swiss-MODEL (http://swissmodel.expasy.org/repository/) aim is to enable protein homology-based modeling for molecular biologists, bioinformaticians, and other scientists. SWISS-MODEL is accessible from the ExPASy Web server and provides automated protein structure homology-based modeling based on the DeepView (Swiss PDB-Viewer) molecular modeling program. The SWISS-MODEL Repository is a database of annotated 3D protein structure models generated by the fully automated homology-modeling server SWISS-MODEL. The repository is developed at the Biozentrum Basel within the Swiss Institute of Bioinformatics.

1.6.6 TOPs

The TOPs service (http://www3.ebi.ac.uk/tops/) [37] calculates a 2D topology cartoon based on the submission of a protein structure. The 2D topology cartoon is a simple representation of the arrangement of secondary structure elements in the 3D space. This representation is made to clarify the overall structure of the fold, to illustrate research results, and to produce educational materials. The TOPs representation of structure is particularly useful for comparing two or more 3D structures. Users can search for existing cartoons for PDB structures by entering a four letter PDB code. Cartoons can also be drawn by uploading a new PDB format file, though this takes a few minutes. When a new cartoon has been computed the user will receive an e-mail message containing a unique identifier and magic number with which to access the cartoon. Cartoons are retained on the TOPs server for about 1 month after being generated. Using structural domain definitions will yield better quality cartoons, according to the site literature.

1.7 MEMBRANE PROTEIN PREDICTION AND ANALYSIS

Many interesting proteins are bound to the membranes of cells, often mediating interactions with the cellular environment. Therefore, the identification of the membrane proteins has become an important tool in post-genomic analysis. Most membrane-spanning protein regions are predicted using either membrane region statistics or cell membrane models that predict whether a protein is likely tethered to the cell membrane. Several resources designed for membrane protein prediction are described in this section.

1.7.1 TMHMM Server v. 2.0

The TMHMM server (http://www.cbs.dtu.dk/services/TMHMM/) [38] predicts transmembrane helices in proteins. TMHMM is designed to provide the most accurate predictions of transmembrane segments. TMHMM predicts whether a segment contains a transmembrane helix and whether intervening loops are inside or outside of the cell membrane. Inside or outside location is predicted according to the algorithm called N-best [39] that sums over all paths through the model with the same location and direction of the helices. If the whole sequence is labeled as inside or outside of the membrane, the prediction is that the protein contains no transmembrane helices. The output of this algorithm should not be assumed to be highly accurate in predicting the location of a protein.

The basic features of the TMHMM output include the number of predicted helices, the expected number of amino acid residues in transmembrane helices for the first 60 residues in the sequence, the total probability that the N-terminus is on the cytoplasmic side of the membrane, and the warning of a possible signal peptide sequence in the N-terminus. A plot is shown that displays the posterior probabilities of an inside/outside/TM helix so that one can see possible weak TM helices that were not predicted and get an idea of the certainty of each segment in the prediction. The N-best prediction is shown at the top of the plot.

1.7.2 TMPred

The TMPred program (http://www.ch.embnet.org/software/TMPRED_form.html) [40] predicts membrane-spanning regions and their orientation with respect to the inside/outside of the membrane. The TMPred algorithm is based on a statistical analysis of TMbase [40], a database of natural transmembrane proteins. The prediction is made using a combination of several weight-matrices for scoring.

1.7.3 Additional Transmembrane Prediction Resources

These additional Web resources also identify transmembrane helices.

HMMTOP [41] http://www.enzim.hu/hmmtop/
DAS [42] http://www.sbc.su.se/~miklos/DAS/maindas.html
TopPred 2 [43] http://bioweb.pasteur.fr/seqanal/interfaces/toppred.html

1.8 ENZYME RESOURCES

Enzyme classification has become increasingly necessary to enable researchers to use the vast store of what is known about general enzyme function. The laboratory use of restriction enzymes has also necessitated this classification, so that researchers can find enzymes with the properties needed to accomplish a desired analysis. Two resources that provide searchable databases of enzymes are covered in this section.

1.8.1 BRENDA

BRENDA (www.brenda.uni-koeln.de) is a systematic collection of enzyme functional data made freely available to the research and education community. BRENDA is available free of charge for academic and nonprofit users via the Web. A standalone version of the database is available for commercial users. Data on enzyme function are extracted manually from the literature by expert scientists in either biology or chemistry. Currently, 3500 different types of enzymes are classified according to the Enzyme Commission (EC) [44] list. Formally, data are checked for consistency by computer; however, each entry is checked by at least one biologist and one chemist. Since there are many different literature sources for many enzymes, the literature extraction is not exhaustive. BRENDA aims to develop a resource of metabolic network information with links to expression and regulation data. BRENDA is maintained and developed at the institute of Biochemistry at the University of Cologne.

1.8.2 REBASE—The Restriction Enzyme Database

REBASE (http://rebase.neb.com/rebase/rebase.html) [45] is a collection of information about restriction enzymes and their related proteins. REBASE includes data about recognition sites, cleavage sites, isoschizomers, methylation sensitivity, crystal form, genome sources, sequence data, and commercial availability, along with published and unpublished references. In addition, REBASE includes DNA methyltransferases, homing endonucleases, nicking enzymes, specificity subunits, and control proteins. Using predictions from analyses of genomic sequences, REBASE also includes putative DNA methyltransferases and restriction enzymes. REBASE is updated daily.

1.9 PROTEIN MUTATION RESOURCES

Many natural and engineered protein mutations often result in properties desirable for proteins used in research. For this reason, numerous researchers have conducted experiments making use of mutant proteins and therefore it has become necessary to catalog the growing number of mutants, along with their physical and biological data. Several resources have emerged providing researchers with data on known protein mutants, or with the ability to explore protein mutations virtually. Some of these resources are described in this section.

1.9.1 PMD—Protein Mutant Database

Compilations of protein mutant data are valuable as a basis for protein engineering. They provide information on what kinds of functional and structural influences are brought about by amino acid mutation at a specific position of protein. The Protein Mutant database (PMD) (http://pmd.ddbj.nig.ac.jp/) [46] covers natural and engineered mutants, including random and site-directed ones, for most proteins. However, members of the globin and immunoglobulin families are not included in this database. The PMD is based on literature, rather than proteins; therefore, each entry in the database corresponds to one article which may cover multiple protein mutants. Each database entry is identified by a serial number and is defined as either natural or artificial, depending on the type of mutation. Each entry has an associated set of terms that allow for cross-indexing.

1.9.2 PMR—Protein Mutant Resource Database

The Protein Mutant Resource database [47] is a system for visualizing and analyzing protein conformational change due to point mutations. The PMR interface enables researchers to browse point mutations by different criteria including protein family and sequence-based mapping of available point mutations for specific proteins. Point mutations can be visualized as morphing movies. The interface also allows for some analysis of the structural changes due to mutation.

1.9.3 ProTherm

ProTherm (http://gibk26.bse.kyutech.ac.jp/jouhou/Protherm/protherm.html) [48] is a database of thermodynamic parameters including Gibbs free energy change, enthalpy change, heat capacity change, transition temperature for wild type, and mutant proteins. The data are intended to facilitate an understanding of the structure and stability of proteins. The database also contains information about the secondary structure and solvent accessibility of wild-type residues. In addition, information about experimental conditions including pH, temperature, buffer, ion, and protein concentrations is provided along with methods used for each piece of data and Km and Kcat.

ProTherm is cross-linked with PIR, Swiss-Prot, the Protein Data Bank, the Protein Mutant database, and PubMed. Thermodynamic information is combined with structural and functional information through 3DinSight [49]. The interface enables users to search data based on keywords, and the returned data includes a view of the 3D protein structure with mapped mutation sites and neighboring amino acids.

1.10 PROTEIN–PROTEIN INTERACTION RESOURCES

Nearly all proteins function as multimers or interact with other proteins. Due to the increasing number of experiments conducted to identify these interactions, information about likely or verified protein–protein interactions is growing so rapidly that several resources have been developed to enable researchers to use the existing data in their own projects. Several of the most widely used resources enable researchers to do retrospective and predictive analyses of protein–protein interactions; these sites are described in this section.

1.10.1 DIP—Database of Interacting Proteins

The DIP database (http://dip.doe-mbi.ucla.edu/) [50] catalogs experimentally detected interactions between proteins. DIP integrates information from various sources to generate a single set of protein–protein interactions. Data are in part transferred automatically using methods that are trained on a core of the most reliable protein–protein interactions in the database. In addition, expert curators manually augment data using knowledge obtained from multiple sources. The DIP database includes information on more than 10,000 protein–protein interactions.

Several services are available to facilitate the analysis of data using the DIP database. First, the EPR index can evaluate the quality of large-scale interacting sets by comparing expression profiles of the dataset to that of a high-reliability subset of the DIP database. Next, the paralogous verification method [51] evaluates a putative protein–protein interaction to be probable if the pair has paralogs that are known to interact. Finally, the Domain Pair Verification Method evaluates a protein–protein interaction to be likely if there are likely domain–domain interactions within the pair.

1.10.2 MINT—Molecular INTeraction Database

MINT (http://160.80.34.4/mint/Welcome.do) [52] aims to store data and rich annotation on experimentally observed protein interactions extracted from the literature by expert curators, which number over 95,000. Currently, the resource supports protein–protein interaction data stored in a structured format. MINT supports the Protein Standard Initiative [53] recommendations for data standards, and offers efficient data exploration and analysis using an integrated database structure. The data can be visualized in the context of high throughput data and viewed with the MINT viewer.

In addition, the entire dataset can be accessed online in both interactive and batch modes through the Web interface and from an FTP server. A supplemental database, HomoMINT, catalogs protein–protein interactions between human proteins as inferred from known interactions between orthologous proteins.

1.10.3 BioGRID—General Repository for Interaction Datasets

BioGRID (http://www.thebiogrid.com) [54] is a freely accessible database of protein and genetic interactions that includes >160,000 raw and >10,8000 filtered interactions from *Saccharomyces cerevisiae*, *Caenorhabditis elegans*, *Drosophila melanogaster*, and *Homo sapiens*, as well as other organisms. The site's introduction states that more than 30,000 interactions have recently been added from 5,778 publications through the extensive curation of the *Saccharomyces cerevisiae* literature. The Web interface allows for the quick search and retrieval of interaction data using many identifiers, including GenBank, Entrez Gene, SGD, gene names, SwissProt, RefSeq, and others. Searches can also be limited to a single organism's data. The full dataset or subsets are downloadable as tab-delimited text files and compliant PSI-MI XML. Precomputed graphical images of interaction networks are available in several file formats. A visualization system, Osprey, is linked to BioGRID so that users can generate customized network graphs with embedded protein, gene, and interaction attributes based on GO annotations.

1.10.4 STRING—Search Tool for the Retrieval of Interacting Proteins

STRING (http://string.embl.de/) [55] is a database of known and predicted protein–protein interactions derived from direct physical associations and indirect functional associations. These associations are derived from genomic context, high-throughput experiments, conserved co-expression, and literature. STRING aims to quantitatively integrate interaction data from multiple sources for a large number of organisms, and to transfer information between organisms where appropriate. At the time of writing, the database contained 1,513,782 proteins from 373 species. STRING uses orthology information from the COG database.

STRING produces network models to allow the visualization of protein–protein interactions and it uses a simple mass and spring model to generate network images. Nodes are modeled as masses, and edges as springs, where the final position of the nodes in the model is computed by minimizing the energy score of the system. Higher confidence edges are built using higher spring constants so that they will reach an optimized position before lower confidence edges. A user can set an option to reduce the optimal length of a high confidence edge to force the nodes closer together; this can clarify the network picture for the region around the high confidence connections. High confidence edge lengths are set to 80% of the normal length by default.

Genomic context analysis relies on orthology assignment between genes in different species. Orthology information is mainly derived from the manually curated COG database (Clusters of Orthologous Groups at NCBI). These data are augmented by adding additional species and creating more groups using a similar methodology. Assignments are intended to be replaced with every new update of the COG database.

Predicted associations can be made for a protein when its SWISSPROT identifier or accession is entered, or by entering a sequence of interest at the Web site. If predicted associations are found, a summary view providing the orthologous group and COG-based annotation is displayed. For fused domain proteins, all of the relevant annotations are displayed. Each of the predicted associations is shown below the annotation information and ordered according to score. The individual prediction method scores can be viewed by clicking on the scores in the initial list. Clicking on the text lines reveals the proteins belonging to the orthologous group. By default, only the top 10 associations are shown.

STRING also provides a network view that summarizes the network of predicted associations for a particular group of proteins. In the network view, nodes are groups of orthologous proteins and network edges represent predicted functional associations. Mousing over a node will display its annotation and clicking on it displays the list of proteins. Edges will be drawn with up to three lines of different colors, representing the application of the three types of evidence used in predicting the associations. A red line indicates the presence of domain fusion-based evidence. A green line represents neighborhood evidence. A blue line indicates co-occurrence-based evidence. Each line may be drawn bold to indicate an above-average confidence in the particular evidence. Mousing over an edge will display the different scores, while clicking on it displays the detailed evidence breakdown.

1.11 TRANSCRIPTION FACTORS

Control of most gene expression is influenced by protein factors that increase the efficiency of transcription. Given that many of these transcription factors recognize flexible motifs in gene sequences, it has become possible to predict binding sites and factors for genes of interest. One such site has played a prominent role in enabling the prediction of transcription factor binding sites, and it is described in this section.

1.11.1 TRANSFAC

TRANSFAC (http://www.gene-regulation.com/pub/databases.html#transfac) [56] contains data on transcription factors, experimentally demonstrated binding sites, and regulated genes. TRANSFAC's collection facilitates derivation of positional weight-matrices for prediction of new binding sites. Two programs, Match and Patch, enable searches for new binding sites using TRANSFAC's matrix and pattern-based representations of transcription factor binding sites. The resulting predictions for binding sites are useful for characterizing promoters of a wide array of genes of interest. TRANSFAC also provides a tool for the visualization of gene regulatory networks based on linking transcription factors to gene regulation and expression database entries. In addition to single factor binding sites, TRANSFAC provides a module devoted to composite regulatory elements like those found in eukaryotic promoters and enhancers; the program is called TRANSCompel. The entries in TRANSCompel correspond to composite binding sites made up of sites bound by two different transcription factors. By providing entries for these composite sites, TRANSCompel enables an analysis of signaling integration sites. Searches for composite sites can be performed using the CATCH [57] program, which allows searching a sequence for composite sites represented by their individual signatures. Links to the composite sites corresponding to the types found in a sequence are returned by the program.

A separate TRANSFAC collection, TRANSPro, is composed of entries for promoters from human, mouse, and rat. These signatures are defined using annotated start sites that are clustered to distinguish alternative promoters of the individual genes. In the search results, known sites are reported, and this enables the analysis of promoters of co-regulated genes. TRANSFAC also maintains a collection of pathological, mutant forms of promoter sites, and transcription factors.

ACKNOWLEDGMENTS

M.C. and S.C. were supported by NIAID/NIH Contract HHSN266200400035C, and S.C. was also supported by NIAID/NIH Contract HHSN266200400061C.

REFERENCES

1. Consortium, U., The Universal Protein resource (UniProt), *Nucleic Acids Research*, 35, D193, 2007.
2. Dayhoff, M.O. and National Biomedical Research Foundation, *Atlas of Protein Sequence and Structure*, National Biomedical Research Foundation, Silver Spring, Maryland, 1965.

3. Wu, C.H. et al., The protein information resource, *Nucleic Acids Research*, 31, 345, 2003.
4. Boeckmann, B. et al., The SWISS-PROT protein knowledgebase and its supplement TrEMBL in 2003, *Nucleic Acids Research*, 31, 365, 2003.
5. Wu, C.H. et al., PIRSF: Family classification system at the protein information resource, *Nucleic Acids Research*, 32, D112, 2004.
6. Hulo, N. et al., The PROSITE database, *Nucleic Acids Research*, 34, D227, 2006.
7. Attwood, T. et al., The PRINTS protein fingerprint database: Functional and evolutionary applications., In *Encyclopaedia of Genetics, Genomics, Proteomics and Bioinformatics*, John Wiley & Sons, Hoboken, New Jersy, 2006.
8. Henikoff, J.G. et al., Increased coverage of protein families with the blocks database servers, *Nucleic Acids Research*, 28, 228, 2000.
9. Finn, R.D. et al., Pfam: Clans, web tools and services, *Nucleic Acids Research*, 34, D247, 2006.
10. Bru, C. et al., The ProDom database of protein domain families: More emphasis on 3D, *Nucleic Acids Research*, 33, D212, 2005.
11. Mulder, N.J. et al., New developments in the InterPro database, *Nucleic Acids Research*, 35, D224, 2007.
12. Haft, D.H., Selengut, J.D., and White, O., The TIGRFAMs database of protein families, *Nucleic Acids Research*, 31, 371, 2003.
13. Thomas, P.D. et al., PANTHER: A library of protein families and subfamilies indexed by function, *Genome Research*, 13, 2129, 2003.
14. Yona, G., Linial, N., and Linial, M., ProtoMap: Automatic classification of protein sequences and hierarchy of protein families, *Nucleic Acids Research*, 28, 49, 2000.
15. Vlahovicek, K. et al., The SBASE protein domain library, release 9.0: An online resource for protein domain identification, *Nucleic Acids Research*, 30, 273, 2002.
16. Murvai, J., Vlahovicek, K., and Pongor, S., A simple probabilistic scoring method for protein domain identification, *Bioinformatics* (*Oxford, England*), 16, 1155, 2000.
17. Letunic, I. et al., SMART 4.0: Towards genomic data integration, *Nucleic Acids Research*, 32, D142, 2004.
18. Gough, J. et al., Assignment of homology to genome sequences using a library of hidden Markov models that represent all proteins of known structure, *Journal of Molecular Biology*, 313, 903, 2001.
19. Murzin, A.G. et al., SCOP: A structural classification of proteins database for the investigation of sequences and structures, *Journal of Molecular Biology*, 247, 536, 1995.
20. Hughey, R. and Krogh, A., Hidden Markov models for sequence analysis: Extension and analysis of the basic method, *Computer Application in Bioscience*, 12, 95, 1996.
21. Yeats, C. et al., Gene3D: Modelling protein structure, function and evolution, *Nucleic Acids Research*, 34, D281, 2006.
22. Michael, M. et al., BioMap: Gene family based integration of heterogeneous biological databases using AutoMed Metadata, in *Proceedings of the Database and Expert Systems Applications, 15th International Workshop on* (*DEXA'04*), 2004.
23. Pruess, M., Kersey, P., and Apweiler, R., The Integr8 project—a resource for genomic and proteomic data, *In Silico Biology*, 5, 179, 2005.
24. Edgar, R.C., MUSCLE: A multiple sequence alignment method with reduced time and space complexity, *BMC Bioinformatics*, 5, 113, 2004.
25. Levitt, M. and Chothia, C., Structural patterns in globular proteins, *Nature*, 261, 552, 1976.
26. Orengo, C.A. et al., CATH—a hierarchic classification of protein domain structures, *Structure*, 5, 1093, 1997.
27. Orengo, C.A. and Taylor, W.R., SSAP: Sequential structure alignment program for protein structure comparison, *Methods in Enzymology*, 266, 617, 1996.
28. Michie, A.D., Orengo, C.A., and Thornton, J.M., Analysis of domain structural class using an automated class assignment protocol, *Journal of Molecular Biology*, 262, 168, 1996.
29. Swindells, M.B., A procedure for detecting structural domains in proteins, *Protein Science*, 4, 103, 1995.
30. Holm, L. and Sander, C., The FSSP database of structurally aligned protein fold families, *Nucleic Acids Research*, 22, 3600, 1994.
31. Siddiqui, A.S. and Barton, G.J., Continuous and discontinuous domains: An algorithm for the automatic generation of reliable protein domain definitions, *Protein Science*, 4, 872, 1995.
32. Greene, L.H. et al., The CATH domain structure database: New protocols and classification levels give a more comprehensive resource for exploring evolution, *Nucleic Acids Research*, 35, D291, 2007.

33. Pearl, F.M. et al., The CATH database: An extended protein family resource for structural and functional genomics, *Nucleic Acids Research*, 31, 452, 2003.
34. Taylor, W.R. and Orengo, C.A., Protein structure alignment, *Journal of Molecular Biology*, 208, 1, 1989.
35. Holm, L. and Sander, C., Touring protein fold space with Dali/FSSP, *Nucleic Acids Research*, 26, 316, 1998.
36. Wang, Y. et al., MMDB: Annotating protein sequences with Entrez's 3D-structure database, *Nucleic Acids Research*, 35, D298, 2007.
37. Flores, T.P., Moss, D.S., and Thornton, J.M., An algorithm for automatically generating protein topology cartoons, *Protein Engineering*, 7, 31, 1994.
38. Krogh, A. et al., Predicting transmembrane protein topology with a hidden Markov model: Application to complete genomes, *Journal of Molecular Biology*, 305, 567, 2001.
39. Schwartz, R. and Austin, S., A comparison of several approximate algorithms for finding multiple (N-best) sentence hypotheses, in *International Conference on Acoustics, Speech, and Signal Processing, 1991 (ICASSP-91)*, 701.
40. Hofmann, K. and Stoffel, W., TMbase—A database of membrane spanning protein segments, In *Biological Chemistry Hoppe-Seyler*, 1993, 166.
41. Tusnady, G.E. and Simon, I., Principles governing amino acid composition of integral membrane proteins: Application to topology prediction, *Journal of Molecular Biology*, 283, 489, 1998.
42. Cserzo, M. et al., Prediction of transmembrane alpha-helices in prokaryotic membrane proteins: The dense alignment surface method, *Protein Engineering*, 10, 673, 1997.
43. von Heijne, G., Membrane protein structure prediction. Hydrophobicity analysis and the positive-inside rule, *Journal of Molecular Biology*, 225, 487, 1992.
44. International Union of Biochemistry and Molecular Biology. Nomenclature, C., *Enzyme nomenclature 1992: Recommendations of the Nomenclature Committee of the International Union of Biochemistry and Molecular Biology on the nomenclature and classification of enzymes*, published for the International Union of Biochemistry and Molecular Biology by Academic Press, San Diego, 1992.
45. Roberts, R.J. et al., REBASE—enzymes and genes for DNA restriction and modification, *Nucleic Acids Research*, 35, D269, 2007.
46. Kawabata, T., Ota, M., and Nishikawa, K., The Protein Mutant database, *Nucleic Acids Research*, 27, 355, 1999.
47. Krebs, W.G. and Bourne, P.E., Statistically rigorous automated protein annotation, *Bioinformatics (Oxford, England)*, 20, 1066, 2004.
48. Kumar, M.D. et al., ProTherm and ProNIT: Thermodynamic databases for proteins and protein-nucleic acid interactions, *Nucleic Acids Research*, 34, D204, 2006.
49. An, J. et al., 3DinSight: An integrated relational database and search tool for the structure, function and properties of biomolecules, *Bioinformatics (Oxford, England)*, 14, 188, 1998.
50. Xenarios, I. et al., DIP, the Database of Interacting Proteins: A research tool for studying cellular networks of protein interactions, *Nucleic Acids Research*, 30, 303, 2002.
51. Deane, C.M. et al., Protein interactions: Two methods for assessment of the reliability of high throughput observations, *Molecular Cellular Proteomics*, 1, 349, 2002.
52. Chatr-aryamontri, A. et al., MINT: The Molecular INTeraction database, *Nucleic Acids Research*, 35, D572, 2007.
53. Hermjakob, H. et al., The HUPO PSI's molecular interaction format—a community standard for the representation of protein interaction data, *Nature Biotechnology*, 22, 177, 2004.
54. Stark, C. et al., BioGRID: A general repository for interaction datasets, *Nucleic Acids Research*, 34, D535, 2006.
55. von Mering, C. et al., STRING 7—recent developments in the integration and prediction of protein interactions, *Nucleic Acids Research*, 35, D358, 2007.
56. Matys, V. et al., TRANSFAC and its module TRANSCompel: Transcriptional gene regulation in eukaryotes, *Nucleic Acids Research*, 34, D108, 2006.
57. Kel-Margoulis, O.V. et al., TRANSCompel: A database on composite regulatory elements in eukaryotic genes, *Nucleic Acids Research*, 30, 332, 2002.

2 Artificial Neural Networks for Therapeutic Protein Engineering

James Lara

CONTENTS

2.1 INTRODUCTION

In recent years, molecular medicine and biotechnology have been transformed by discoveries from the studies of the human genome and proteomics. The understanding of disease mechanisms at the cellular level is accelerating the discovery and development of new therapeutic agents. Of these, therapeutic proteins are a major focus of research and pose great challenges in development, delivery, safety, and stability. Over the last few years, there has been a substantial increase in the number of therapeutic peptides and proteins that are reaching the market. This tendency is likely to continue and even escalate in the near future. Approximately \$20 billion is spent on biopharmaceutical research and development annually [1]. Much of this spending has resulted in approximately 2500 biotechnology-based drugs currently in development, 900 in preclinical trials, and over 1600 in clinical trials [2]. It is predicted that over the next decade the major activity in biotechnology will be in human health care, involving both therapeutics and diagnostics [3]. The ability to formulate and manufacture therapeutic proteins is an issue of paramount importance.

Despite the advances in molecular high throughput technologies that have made the construction and analysis of a large number of proteins technically feasible, the extreme difficulty of predicting protein properties and clinical effects from the structure presents a formidable challenge to the development of novel pharmaceutically important proteins. As a result, protein-based drug developers are turning to computational algorithms that associate amino acid sequences with desired properties [4–9]. In this respect, machine-learning techniques are playing an ever greater role. Section 2.2 describes one such approach: artificial neural networks.

2.2 ARTIFICIAL NEURAL NETWORKS: A COMPUTATIONAL APPROACH

"The Electronic Brain" has become a term associated with computers since their invention in the 1940s. However, nothing could be further from the truth. There is very little in common between the way that computers and brains operate. The fact is that for some tasks such as calculating, sorting, filing, and playing games, computers can do better than brains. But in other tasks, such as the recognition and understanding of the world, computers lag a long way. For some years now, enormous research efforts are being devoted to understand the computational principles of the living brain [10–16]. Two reasons compel us to do this. First, we wish to have a better understanding about how the brain achieves its competence, and, second, we want to find ways of transferring this competence to computers beyond that, which can be achieved with current computational methods. The rapid emergence of machine-learning technologies, like artificial neural networks (ANN), is paving the way toward achieving this goal.

2.2.1 Artificial Neural Network Model

Artificial neural networks represent a technology rooted in many disciplines: neurosciences, mathematics, statistics, physics, computer science, and engineering. Neural networks find applications in such diverse fields such as modeling, time series analysis, pattern recognition, signal processing, and control by virtue of an important property: the ability to learn from input data. The artificial neuron model is a form of artificial intelligence (AI) based on brain theory [17–19]. In simple terms, an ANN is a mathematical model of neuron operation that can, in principle, be used to compute any arithmetic or logical function. It consists of two types of components or elements: (1) *processing or computation units* (also called neurons) and (2) *connections* with adjustable "weights" or "strengths." Figure 2.1A shows the model of a single neuron (the basis of an ANN) and the two mathematical terms that describe an artificial neuron k. The fundamental operation of an artificial neuron is carried out in three steps. First, a signal x_1 at the input of synapse j connected to neuron k is multiplied by the synaptic weight w_{kj}. The synaptic weight of an artificial neuron can adopt either a positive or a negative value. Second, the weighted input signals are summed. Third, an activation function $\psi(\bullet)$ limits the amplitude range of the output of the neuron to a finite value. For example, the typical output of a neuron falls within either the closed interval [0,1] or alternatively [−1,1]. Many types of activation functions can be used to train an ANN [20,21]. The model also includes an externally applied bias, denoted by b_k. The bias b_k has the effect of increasing or lowering the net input (u_k) to the activation function, depending on whether it is positive or negative, respectively.

The topology architecture (i.e., layout and connections between the neurons) that can be adopted into the design of an ANN depends on the task and the function for which the system is designed. Figure 2.2 shows the most common architecture, the multiple layer perceptron (MLP), where neurons are classified into three types: (1) input neurons, responsible for receiving and passing the signal vector for mathematical processing; (2) hidden neurons, in which the information received from the input layer undergoes further processing; and (3) output neurons, which produce the system's output or result. The neurons are arranged into layers with corresponding names

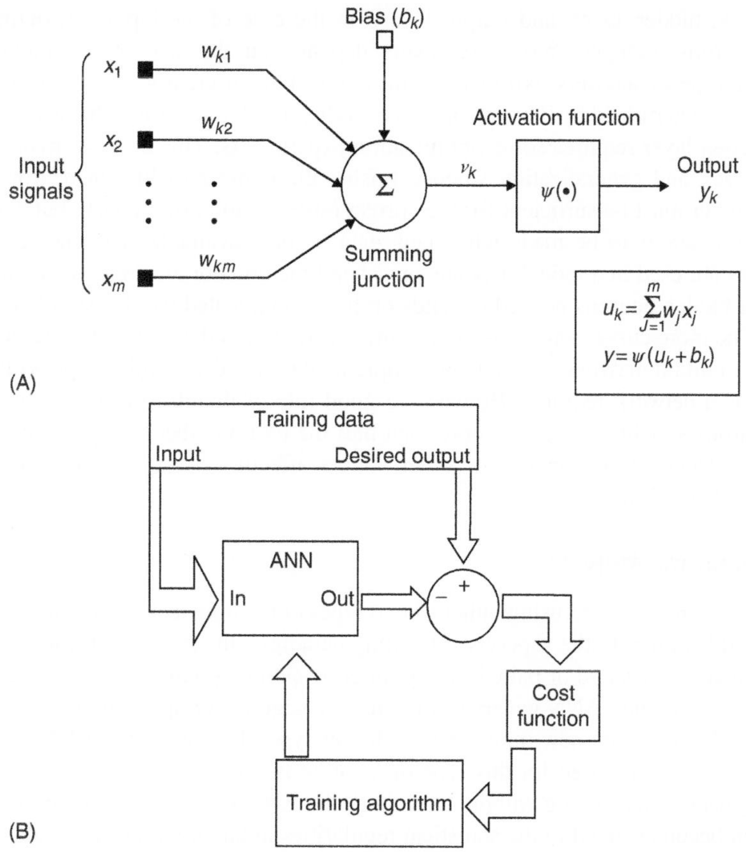

FIGURE 2.1 (A) ANN "single perceptron" model. Inbox shows the two mathematical terms to describe a neuron k, where $x_1, x_2, \ldots, x_m$ are the input signals (e.g., amino acids in the sequence); $w_{k1}, w_{k2}, \ldots, w_{km}$ are the synaptic weights of neuron k; u_k is the linear combiner output due to the input signals; b_k is the bias; $\psi(\bullet)$ is the activation function; and y_k is the output signal of the neuron; the use of bias b_k has the effect of applying an affine transformation to the output u_k of the linear combiner in the model, as shown by $v_k = u_k + b_k$; (B) ANN supervised learning model implementation. The ANN learns to map a set of signal inputs to specified outputs in the training data. The adaptation (value changes) of the weights is achieved through a cost function, for error minimization, and the training algorithm during the training epochs.

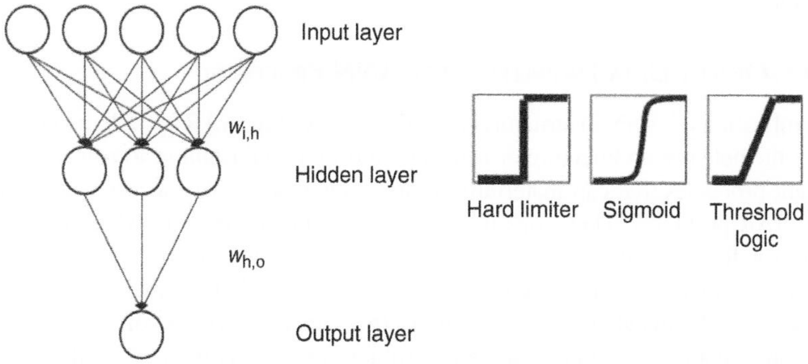

FIGURE 2.2 Schematic representation of a 3-layer feed-forward ANN (also known as a multiple-layer perceptron or MLP). Computational nodes (neurons) are represented as circles with weighted inputs and output shown as arrows. Also shown, are the three common types of nonlinearity that can be used by the neurons to determine output: hard limiter, sigmoid, and threshold.

(i.e., input layer, hidden layer, and output layer). In the case of the input and output layers, the number of neurons arranged into these layers depends on the size of the input vector, and the number or type of answers expected, respectively. Put differently, the number of input and output neurons is determined by the type of data encoding used. In contrast, the determination of the size of the hidden layer requires some optimization experiments. This process involves a trade-off between accuracy and generalization when choosing the number of hidden neurons for a given ANN. The number must be sufficient for the correct representation of the task, but sufficiently low to allow generalizations to be made when new data become available. The choice of how many hidden neurons are used in a model depends on several factors. For example, in classification tasks the number of hidden neurons needed depends on how complicated the decision boundaries in the data set are. The more curves and turns are required to split up a data set, the more hidden neurons are required to build a network. There is no simple method for determining a priori the number of hidden neurons a network requires. However, a good rule of thumb to follow is that the number of hidden neurons should not exceed more than half the total number of input neurons present in the ANN [22]. Detailed descriptions of other factors affecting the size of the hidden layer are discussed elsewhere [22].

2.2.2 Training the ANN

Neural networks can be trained using either of two types of learning techniques: supervised learning or unsupervised learning. In the supervised learning paradigm, the system is presented with a set of input vectors that are matched or paired to a set of corresponding output vectors. Once the learning process is over, the trained ANN generates outputs to categorize or approximate, depending on the nature of the task, any given new input vectors. In this type of learning, the ANN is trained with a teacher. The basic scheme used for this type of learning is shown in Figure 2.1B. In the unsupervised learning paradigm (training without a teacher), the ANN is presented with only input vectors. The ANN then becomes tuned to the statistical regularities in the input data and gains the ability to form internal representations for encoding features of the input, thus creating new classes in an autonomous fashion. The ability to create internal representations or an internal mapping from input data under unsupervised learning is also known as self-organized learning, which is the principal-learning paradigm for the training and operation of another form of ANN known as a self-organizing map (SOM) [23].

Learning tasks for which ANN are typically trained include (1) pattern recognition, whereby the input or received pattern/signal is assigned to one of a predefined number of classes; (2) function approximation or regression, which implies designing an ANN that approximates an unknown function $y = g(x)$; and (3) classification, which typically consists of assigning input vectors to one of two or more classes.

2.2.3 Input–Output Data Engineering for ANN Processing

Successful application of ANN in structure–activity relationship (SAR) problems involves far more than selecting the network architecture or learning algorithm and running it over the data. There are other important processes that can materially improve the chances of success when applying ANN techniques to SAR problems. They constitute a kind of data engineering, which is, engineering the input data into a form suitable for the ANN and the learning scheme chosen and engineering the output model to make it more effective. The task of developing reliable "intelligent" machines depends heavily on knowledge* representation. In a neural network of specified architecture, knowledge representation of the environment (problem or task) is defined by the values taken on

* In the definition of a neural network, knowledge refers to stored information or models used by a person or machine to interpret, predict, and appropriately respond to the outside world [24].

by the parameters (synaptic weights and biases) of the network. The primary characteristics of knowledge representation are twofold: (1) what information is actually made explicit and (2) how the information is physically encoded for subsequent use. For most real-world applications, a good solution often depends on a good representation of knowledge.

Neural networks, like many other machine-learning algorithms, require numerical values for processing. Therefore, biological information, such as protein sequence information, has to be engineered into the appropriate numerical vector representative for its mathematical processing by the neural network. Such engineering is commonly referred to as data preprocessing and postprocessing. They can be looked at as a "bag of tricks" that can be used as a practical way of increasing the chances of success. Sometimes they work, and sometimes they do not. Unfortunately, at the current state of art, it is very difficult to know in advance if they will work or not. So far, trial and error is the only reliable guide, thus, it is important to be resourceful and know what these processes are by having a sound understanding of data mining and data representation techniques.

Data preprocessing consists of data encoding and feature representation. Data encoding addresses the problem of transforming and making discrete data suitable for mathematical processing. Protein sequences are composed of a series of 20 amino acids represented by the characters (for example, A, C, and D for alanine, cysteine, and aspartic acid, respectively). Nucleic acid sequences, on the other hand, are composed of nucleotide bases that are represented with the characters A, T, G, C, and U. However, sequences of alphabetic characters cannot be used for mathematical processing because none of the elements has a numeric value. Thus, numeric representation of alphabetical sequences for mathematical processing requires generation of a numerical code to represent their corresponding elements. Table 2.1 shows some examples of encoding schemes used for protein sequence representation.

Data encoding can be very intricate and greatly impact the performance of the ANN. A common practice is to encode each one of the 20 letters corresponding to the 20 amino acid types of a protein into a numerical binary scheme. For instance, each letter can be represented by a 20-dimensional binary vector, i.e., 20-bin representation. In such case, amino acids are represented by a set of 19 zeros and a single one uniquely positioned to represent each amino acid. For example, A = 00000000000000000001, C = 00000000000000000010, and so forth. Alternatively, a lower dimensional vector can be generated based on the known physicochemical properties of each amino acid type. Both schemes have their advantages and disadvantages. One of the major advantages of binary representation is that very small changes in amino acid composition between sequences can be easily detected and mapped by the ANN. In fact, this type of encoding representation has been successfully used for other tasks, such as clustering aligned protein sequences [25].

Depending on the sequence length, 20-bin encoding representation can greatly impact the size complexity of the ANN model. A critical issue in developing a neural network is generalization, which relates to how well a network makes predictions for cases that are not in the training set. As the size complexity of the network increases (i.e., number input neurons) the number of weights or internal parameters in the network also increases. This fact leads to the problem of overfitting* (i.e., poor generalizations) and data size becomes an important factor. One must have sufficient data for the number of weights in the network. If more data are required to match the required complexity determined by the number of input parameters, then more data must be collected. Otherwise, the chances for the ANN of overfitting and memorizing insignificant patterns unique to the training data increases [18,26]. There are circumstances, however, where collecting more data poses a very difficult proposition. Another way of addressing insufficient data is to manipulate the existing data by artificially expanding it with synthetic data [27]. Further details of this type of manipulation, and

* Overfitting is defined as fitting a statistical model that has too many parameters. An absurd and false model may fit perfectly if the model has enough complexity by comparison to the amount of data available. In cases where training samples are rare, the ANN may adjust to very specific random features of the training data, that may have no causal relation to the target function, which leads to poor prediction performances in data test sets.

TABLE 2.1
Various Schemes Used for Encoding Amino Acids

Amino Acid	Representation (A)			Representation (B)					
	20	6	9	Volume	Bulkiness	Flexibility	Polarity	Aromaticity	Charge
A	10000000000000000000	53500a	100001000	0.1677	0.4433	0.2490	0.3951	0.0	0.5
C	01000000000000000000	45500a	000001001	0.3114	0.5506	0.2048	0.7441	0.0	0.5
D	00100000000000000000	a5001a	001001000	0.3054	0.4532	0.8675	1.0000	0.0	0.5
E	00010000000000000000	a60010	001000100	0.4970	0.5567	0.8112	0.9136	0.0	0.5
F	00001000000000000000	185a0a	100100101	0.7725	0.8976	0.0763	0.0370	0.6667	0.5
G	00000100000000000000	525000	000001000	0.0	0.0	1.0	0.5062	0.0	0.5
H	00000010000000000000	565010	010100111	0.5569	0.5632	0.1124	0.6790	0.5556	0.5
I	00000001000000000000	37500a	100010011	0.6467	0.9852	0.6707	0.0370	0.0	0.5
K	00000000100000000000	a7a02a	010000111	0.6946	0.6738	0.6867	0.7901	0.0	1.0
L	00000000010000000000	375000	100010001	0.6467	0.9852	0.2811	0.0	0.0	0.5
M	00000000001000000000	375005	100000100	0.6108	0.7033	0.0	0.0988	0.0	0.5
N	00000000000100000000	655020	000001000	0.3174	0.5156	0.6747	0.8272	0.0	0.5
P	00000000000010000000	555000	100000010	0.1766	0.7679	0.8594	0.3827	0.0	0.5
Q	00000000000001000000	66502a	000000100	0.4910	0.6048	0.7952	0.6914	0.0	0.5
R	00000000000000100000	a8a046	010000110	0.7246	0.5955	0.9398	0.6914	0.0	1.0
S	00000000000000010000	645010	000001011	0.1737	0.3222	0.8514	0.5309	0.0	0.5
T	00000000000000001000	55501a	000001010	0.3473	0.6771	0.5984	0.4568	0.0	0.5
V	00000000000000000100	365007	100011010	0.4850	0.9945	0.3655	0.1235	0.0	0.5
W	00000000000000000010	0a5a10	100100110	1.0	1.0	0.0402	0.0617	1.0	0.5
Y	00000000000000000001	285a18	000100111	0.7964	0.8008	0.5020	0.1605	0.6667	0.5
X	11111111111111111111	555555	111111111						

Note: (A) binary representations (20-, 6-, and 9-bin), used by Brusic et al. (1995), where rep "6" assigns a 6-place string where each place is a scalar value for a feature (hydrophobicity, volume, charge, aromatic side chain, hydrogen bonds) or a correction bit. "a" stands for 10; "Rep 9" is an intermediate representation using a feature-based grouping of amino acids. (B) used by Milik et al. (1998), where every amino acid is represented by six properties (volume, bulkiness, flexibility, polarity, aromaticity, and charge).

on sampling theory (for dimensionality reduction), can be found elsewhere. Still, in some cases, the data requirements for the ANN might not be satisfied making it necessary to resort to other algorithms, like support vector machines (SVMs)*, capable of handling high dimensionality in sparse data sets.

As mentioned earlier, encoding schemes based on physicochemical properties usually require lower dimensional vectors for amino acid type representations [28–30]. Such schemes allow the encoding of a variety of physicochemical features, such as hydrophobicity, volume, and bulkiness. Furthermore, the number of features that can be encoded does not necessarily impact on the input vector dimensions. Schneider and Wrede (1998) devised an encoding scheme using eigenvalues for each amino acid type derived by principal component analysis (PCA)† of 143 physicochemical scales [31].

* SVMs are a set of related supervised learning methods used for classification and regression. The SVM algorithm is based on the statistical learning theory and the Vapnik–Chervonenkis (VC) dimension.

† PCA is an orthogonal linear transformation technique used to reduce multidimensional data sets to lower dimensions for analysis. It involves the computation of the eigenvalue decomposition of a data set after mean centering the data for each attribute, and is included as part of some ANN software packages.

Feature representation is perhaps the most important aspect of data preprocessing. When analyzing proteins, each letter in the molecular string represents an amino acid. Each amino acid has unique chemical properties associated with specific states or values. A string of characters can be given biological meaning by substituting the characters with their respective chemical values. Many hydrophobic [32–36], chemical [37], physical [38–41], statistical preference [42], biological [43], and mathematical [44] scales of amino acid attributes have been used. Such large number of possible feature representations makes data preprocessing of biological sequences a rather complicated process. The guide at this crucial step in the design and application of an ANN system should be the problem definition, i.e., determination of the desired input and output mapping for the specified task or goal.

The selection of a feature representation and encoding method for a given task limits and specifies the information presented to the ANN. It establishes the parameters of the structures and functions that can improve the accuracy of ANN and allow extraction of statistical consistencies or hidden features in the data. Input vectors that result from the feature representation and encoding method need to have a relative degree of logic to conform to the basic premise that vectors of similar sequences be close together, and vice versa. This is important if input vectors are to carry the biological information of the sequence they represent and maintain both the biological uniqueness and diversity that result from the amino acid composition or sequence length. In contrast, poor feature representation and inadequate encoding methods can result in inadequate vectors, preventing maximal extraction of the statistical features that connect sequences with their structures and functions.

From the above discussion, it becomes obvious that there are two main precepts with which researchers are concerned when choosing a data encoding representation for use with an ANN: the number of variables used (which determines the dimensionality of the input vector) and the extent to which the information required by the network is made explicit. For reasons of computational speed and complexity, one is forced to limit the number of variables (inputs and outputs) with which the ANN needs to contend. This constraint in the number of variables for processing by the ANN hence requires that the used variables contain the information necessary for the problem solution in a manner which is as explicit as possible.

There are many sources that provide encoding schemes for performing the numeric transformation of biological sequences, for example, the amino acid index database (AAindex), available at http://www.genome.ad.jp/aaindex/. In addition to databases and published data, an increasing number of Web-based tools carry out these numerical transformations automatically, which are now available. The Expert Protein Analysis System (ExPASy) (http://us.expasy.org/) has several sequence analysis tools and software packages. For example, the ProtScale program [45], which assigns a numerical value to each type of amino acid as predefined by any of the >50 scales of physicochemical properties and conformational parameters. One may choose to either substitute each individual amino acid character with the corresponding physicochemical values or select a window size to compute the physicochemical profile of a given peptide or protein sequence. Furthermore, software tools for the automatic modeling of a protein's three-dimensional structure are also available on the Internet. Web-based homology modeling servers provide Cartesian coordinates for the three-dimensional representation of proteins (*.pdb files), which may be used for sequence representation and ANN processing. The Geno 3D server (http://pbil.ibcp.fr/htm/index.php) and the Swiss-Model server (http://swissmodel.expasy.org/) are examples of such Web-based homology modeling servers.

Data postprocessing consists in the conversion of the ANN numerical output into a usable form. It comprises mainly of output encoding, which is much more straightforward than preprocessing data encoding. For example, let us say that the task is to classify an input vector as "binding" or "not binding." Then only one output neuron may be used. This output neuron would be configured to map to two possible values, i.e., to either a value of [1] for binding or to a value of [0] for not binding.

2.2.4 Extracting SAR with ANN

Artificial neural networks have been regarded as a "black box" because its operation is neither fully explained nor fully understood. The analysis of a fully trained ANN is in itself a science. Nonetheless, compared to traditional statistical approaches, the ANN has considerable strengths and advantages, including the ability to (1) model complex multidimensional and nonlinear relationships, which make the ANN technique ideal for processing biological information, (2) generalize and find relationships within the data while tolerating erroneous or noisy data, and (3) be retrained with expanded (i.e., new) data sets, which are a plus for model refinement requirements. In fact, an ANN with hidden nodes can extract from input information the higher order features that are ignored by linear models.

During supervised training, the ANN is taught to map a set of input patterns to a corresponding set of output patterns. Hidden neurons form internal representation, which is reflected on the weight values of the connections. Post-training analysis of the network weights provides insight into the sensitivity and relevance of input features or attributes for feature selection. Although, interpretation of the weights of the connections between artificial neurons requires careful consideration, it is generally accepted that the greater the weight value in a connection, the greater the importance of the parameter(s) linked to or associated with that connection. This relationship is analogous to that of the biological counterpart of the artificial neuron, where the strength and number of synapses between biological neurons are believed to be relevant to establishing a particular associative path [17]. Graphical representations of the weights are very helpful and are often the most common form of weight analysis. For example, Holbrook et al. (1990) applied ANN to extract information about the surface accessibility of protein residues from a protein sequence using a database of high-resolution protein structures [46]. The ANN was trained to predict the water accessibility of a central residue in the context of its flanking sequence. The protein sequences were presented to the ANN as sliding windows, of variable sizes up to 13 residues centered on the amino acid of interest. The analysis was performed by scanning the entire sequence with these sliding windows. To predict the surface exposure of the protein residues, the investigators defined two categories for the buried (inaccessible) and exposed (accessible) residues. Examination of the ANN weights allowed identification of the major factors influencing residue exposure. Such analysis made apparent that hydrophobic residues two or three amino acids up or down from the central amino acid residue hinder its surface exposure. In general, they found that high weights for neighboring hydrophobic residues correlated with their trend to favor burial of the central residue, while high weights for neighboring hydrophilic residues correlated with their trend to favor exposure of the central residue.

The capability of the ANN approach to extract relevant factors from potentially any particular biological property is critical to the protein engineering process. The task of engineering therapeutic peptides or proteins with optimal medicinal properties by constructing agents with physicochemical properties that favor such activities requires vital knowledge of the SAR relations involved. To gain understanding and knowledge of the relationships or dependencies between the various physicochemical parameters from a computer model it is necessary first to acquire and then decipher evidence about the nature of the modeling processes that occur within the model itself. But, as in the case of a multilayered neural network solution, the preponderance of local processing operations and weighted relationships, is often too complex, or too demanding, for direct intelligible comprehension. New and powerful analytical techniques for weight analysis, also known as contribution analysis, have advanced the application of ANN towards this task [47–50]. Sections 2.2.4.1 through 2.2.4.3 focus on some of the current methods used.

2.2.4.1 Graphical Weight Analysis

For an ANN without hidden layers, the strength of internal relationships can be determined from direct examination of the connection weights. Many software packages provide tools that are used for analyzing the pattern of weights. A common example is the Hinton diagram. This graphical tool uses squares of different sizes or colors to provide a graphical representation of the weighted connections [51].

2.2.4.2 Sensitivity and Projection Analyses

Examining individual weights in large or complex multilayered networks is time consuming and is not really beneficial. Interaction between the different components is important and with complex structures the exact strength and manner of influence associated with each particular item becomes more difficult to interpret. Traditional ANN model building has a diagnostic procedure, termed sensitivity analysis, which investigates "the rate of change in one factor with respect to change in another." This method of assessment uses positive and negative manipulation of individual inputs to examine the effect of such alterations on the model output, which is quantified using global statistics. Sensitivity analysis can be applied to both traditional and neural models alike. Moreover, when compared to a detailed examination of weights, it has distinct advantages for investigating neural network solutions that have more than one hidden layer [52]. But this technique is subject to potential drawbacks arising from improper combinations or actualizations.

Meanwhile, some software packages also contain tools that can be used to compute a 2.5D visualization of the relationship between two input vectors and either a hidden unit, or an output unit, to provide additional information. This is termed projection analysis [51].

2.2.4.3 Saliency Analysis

Various methods of analysis mentioned above are the standard approaches that can be used to examine a neural network solution. Each existing technique has its merits, but such tools are of limited scope and application, and for complex solutions have insufficient capabilities to enable a purposeful or meaningful evaluation. It is important to recognize that a major difference exists between neurocomputing solutions and standard equation-based tools. Each neural network solution is a distributed information processing structure. This means that the information that is needed to perform data processing operations is stored within the individual processing units (neurons) and the weighted connections that exist between them. Each component within the overall structure is responsible for one small part of the total input–output mapping operation. Therefore, each individual data input can have no more than a marginal influence with respect to the complete solution. The ANN will still function, to generate reasonable mappings, in response to incomplete data or from data that contain noise and fuzziness.

The distributed information processing of the ANN and the power to function with missing data are important with respect to acquiring information on the model. Abrahart, See, and Kneale recently devised a method that takes advantage of these two capabilities of neural networks to determine relevant factors in river flow predictions based on hydrological and meteorological data [53]. They showed that a network (a standard multiple layer perceptron that had 22 inputs, 30 hidden units, and 590 weighted connections) can be disaggregated in terms of its foretelling inputs and will still generate predictions, and the purposeful introduction of missing inputs can be used to provide relevant information on the internal processes. This disaggregation technique, termed saliency analysis, allows assessing the relative importance of different neural network components in terms of the effect that each item has on the error function. The systematic omission or zeroing of input vectors facilitate the examination of the relative magnitude and significance of internal relationships on predicted output. The direct consequence of each input vector within the overall structure is thus established through the effect that each missing item has on the error function, with respect to other input vectors.

2.2.5 ANN Technology

Computational requirements for ANN applications (computation time, hardware requirements, etc.) depend on many factors, which include complexity of the ANN model, size of the data, and the

TABLE 2.2
ANN Software Resources

Program	Architecture	Operations	OS Platform	Distribution	Program Availability
SNNS	S	R, C, A	U, I	F	http://www-ra.informatik.uni-tuebingen.de/SNNS/
JOONE	S	R, C, A	U, W, I	F	http://www.jooneworld.com/
NN Toolbox[a]	S	R, C	U, W, S	$, D	http://www.mathworks.com/products/neuralnet/
James[b]	MLP	A	W, I	F	http://www.bcp.psych.ualberta.ca/~mike/Software/James/
PDP++	MLP, SOM	R, C	O, U, Cy-W	F	http://www.cnbc.cmu.edu/Resources/PDP++/PDP++.html
ESOM	SOM	C	I	F	http://databionic-esom.sourceforge.net/
Tlearn	MLP	R, C	W, Mac, U	F	http://crl.ucsd.edu/innate/tlearn.html
SOM_PAK	SOM	C	U, W	F	http://www.cis.hut.fi/research/som-research/
FANN	MLP	R, C	I	F	http://sourceforge.net/projects/fann/
LNKnet[c]	MLP, RBF	C	O, U, Cy-W	F	http://www.ll.mit.edu/IST/lnknet/index.html
NeuroSolutions	S	R, C	W, Mac	$, D	http://www.neurosolutions.com/
Trajan	MLP, RBF	R, C	W	$, D	http://www.trajan-software.demon.co.uk/index.htm
Tiberius[c]	MLP	R, C	W	$, D	http://www.tiberius.biz/

Note: Links were last accessed on July 25, 2007. Types of supported architectures: single or multiple layer perceptrons, MLP; radial basis function, RBF; self-organized maps, SOM; three or several additional types of architectures, "S" Types of supported operations (i.e., learning tasks): function approximation or regression, "R"; classification or pattern recognition, "C"; pattern association, "A"; Supported Operating System (OS) platforms: Windows, "W"; Unix/Linux, "U"; Cygwin to run under Windows, Cy-W; Solaris and others, "O"; Macintosh, "Mac"; platform independent, "I"; Distribution: free licensing, "F"; commercial licensing, "$"; free demo version available, "D".

[a] Neural network toolbox for use in MATLAB.

[b] This program is used to explore association in a distributed associative memory.

[c] ANN capabilities are available as part of the machine-learning and data-mining software package.

learning task. ANN technology has advanced quickly, generating a market of both proprietary and freeware systems that make ANN fairly accessible for research applications. In most cases, ANN applications can be carried out on a desktop computer. Table 2.2 lists some shareware and freeware ANN simulator software and sources from which they are available.

2.3 PREDICTION OF IMMUNOGENICITY FOR THERAPEUTIC PROTEINS

Therapeutic proteins have made a radical impact on the treatment of many diseases and, because many more therapeutic proteins are likely to become available for an increasingly wide range of indications, scientists and clinicians are becoming more aware of the importance of assessing the immunogenicity of new molecules to avoid the development of treatment resistance and potentially life-threatening immune responses (illustrated, for example, by the responses of some patients to a recombinant human erythropoietin used in the European market) [54]. Although the production of therapeutic proteins has improved with advanced recombinant expression, purification and formulation technologies, the use of recombinant, purely human or humanized proteins has not completely eliminated immunogenicity problems associated with therapeutic proteins.

Immunogenicity of therapeutic proteins can be predicted in advance by in vitro, in vivo, and computational tools, which can identify sequences within the therapeutic protein that, when processed by T cells, elicit an immune response. A limitation in determining the relative immunogenicity of potential therapeutic proteins is the variance in the immunogenicity determined by in vitro or in vivo techniques in animal and human models. Given the sophistication and high-throughput capacity of existing computational tools and the availability of precise in vitro validation assays, accurate prediction of immunogenicity for therapeutic protein products, and more rapid translation of research discoveries into clinical success, may be within reach. Furthermore, *in silico* strategies for assessing immune responses provide drug developers with rapid and cost-effective means for predicting which protein-based drugs will least likely induce adverse immune responses in clinical trials. The following sections describe examples on the application of ANN technology to problem of immunogenicity assessment with focus on the prediction of T-cell and B-cell epitopes.

2.3.1 T-Cell Epitope Prediction

T cells of the immune system continually check for the presence of foreign antigens that may indicate the presence of invading microorganisms (e.g., viruses, bacteria, fungi, parasites), mutated self-cells (e.g., tumors), or non-self-cells and tissues (e.g., transplants). These foreign and self-proteins are processed through specialized mechanisms and degraded to short peptides that are presented on the host cell surface for recognition by the T-cell receptor (TCR) [55,56]. The process involves major histocompatibility complex (MHC) molecules, which bind short peptides and display them to T cells. Peptides presented by MHC molecules originate from intracellular (MHC class I) or extracellular (MHC class II) proteins [57]. MHC class I bound peptides activate cytotoxic T cells, resulting in killing of target cells (e.g., infected, neoplastic, or transplanted tissues), while MHC class II bound peptides serve mainly in regulation (initiation, enhancement, and suppression) of immune responses [57]. The ability of the immune system to respond to a particular antigen varies based on an individual's pattern of MHC genes. In humans, MHC molecules are known as human leukocyte antigen (HLA) class II and I. Each person has only 3–6 HLA class I molecules and at least that many HLA class II molecules [57]. MHC genes show extensive polymorphism; more than 800 variants of HLA class I and more than 500 variants of HLA class II molecules have been characterized [58,59].

MHC molecules have a peptide-binding groove that can bind peptides in a seemingly non-selective and unrestrained manner [60]. For example, the number of peptides that can bind to each HLA class I molecule has been estimated at 1,000–10,000 sequences, or an average of 0.1%–5.0% of all overlapping 9- and 10-mer peptides spanning the entire protein sequence [61]. Binding of a peptide to an MHC molecule is a prerequisite for recognition by T cells, but only certain peptides can bind to a given MHC molecule [62]. However, binding in itself is not enough. Some peptides are good MHC binders in biochemical assays but do not elicit a T-cell immune response [61,63], probably because they are not properly processed and presented by MHC molecules. Thus, T-cell epitopes are a subset of MHC-binding peptides [63]. Determining which peptides bind to a specific MHC molecule is crucial to understanding the mechanisms of immunity and for vaccine and drug design [64]. Fortunately, binding affinities of thousands of MHC ligands have been characterized. Results are available in databases such as MHCPEP [65], JenPep [66], IMGT/HLA [58,59,67], SYFPEITHI [68], and MHCBN [69]. Such vast amount of information serves as basis for empirical computational approaches to elucidate the physicochemical and structural parameters that regulate peptide binding to MHC molecules.

Algorithms for prediction of T-cell epitopes can be classified as direct and indirect methods [70]. Direct methods are based on information on the T-cell epitopes, whereas indirect methods are based on information on the MHC-binding peptides. In the past, direct prediction methods relied on the identification of amphipathic structures [71–74]. The AMPHI algorithm was described by Margalit and coauthors [75]. Another algorithm, SOHHA, was developed based on the assumption that T-cell epitopes consist of 3–5 helical turns with a narrow strip of hydrophobic residues on one side [76]. Algorithms based on secondary structure analysis were superseded by the identification of common motifs among T-cell epitopes in primary structures. Rothbard and Taylor (1988) collected about 57 T-cell epitopes and, based on their patterns, published a list of motifs [77]. They developed an algorithm based on the association between cysteine-containing T-cell epitopes and certain other residues. The system searches for triplets, including CAK, CLV, CKL, and CGS, in the peptide sequence [78]. In 1995, two computational T-cell epitope prediction tools, EpiMer and OptiMer, were developed based on knowledge of MHC-binding motifs [79]. OptiMer predicts amphipathic segments of protein with high motif density, and EpiMer locates segments of protein with high motif density. However, when tested against databases of human and murine T-cell epitopes these direct prediction methods performed with low accuracy [80]. In the last decade, several indirect methods have been developed that predict MHC binders instead of T-cell epitopes [81–84]. Due to more specific interactions of MHC molecules and peptides, these methods are more accurate than direct T-cell epitope prediction methods [70]. These indirect methods are based on (1) structural comparative [85] and ab initio modeling [86], (2) binding motifs [62,81–84,87–90], (3) quantitative matrices (QM) [91–99], and more recently, (4) machine-learning techniques like neural networks [63,70,100–102]. The application of ANN algorithms has been reported to outperform binding motif and QM-based methods [61,70,103]. Predictive performances using ANN methods for the prediction of MHC-binding peptides are in the order of 80% in sensitivity and specificity [61,63,104].

2.3.2 Prediction of MHC Class I and Class II Binders

Brusic and colleagues were the first to report on ANN methods for prediction of MHC-binding peptides [30]. They trained a fully connected 3-layer back-propagation ANN to classify peptides into those predicted to bind and not to bind to either human HLA-A2 or mouse H-2K^b. The number of input nodes was equal to the length of the input signal (9 nodes for 9-mers or 8 nodes for 8-mers representing HLA-A2 and H-2K^b, respectively); the number of hidden neurons was 9, and one output node predicted binding versus nonbinding. Training samples were obtained from the MHCPEP database. The training set consisted of 186 positive examples (binding peptides) and 1071 negative examples (nonbinding peptides) for the HLA-A2 predictions; and 30 binders and 796 nonbinders for the H-2K^b predictions. For the encoding of peptide vectors, three signal input representations (Table 2.1) were used: (1) a 20-bin representation in which each amino acid was encoded by a unique binary string of 0 and 1; (2) Rep6, a 6-number representation based on a scalar value for a physicochemical feature; and (3) Rep9, a 9-number intermediate representation using a feature-based grouping of amino acids. During ANN training, each signal input in the training set was mapped to corresponding binding affinities, represented by IC_{50} values, which were defined as the concentration of peptide required to inhibit the binding of a universal DR4-binding peptide by 50%. The system was validated by testing the ANN performance with selected 9-mers from published data obtained by in vitro assays [95]. The evaluation set consisted of 60 known binders and 68 nonbinders for HLA-A2 and 23 binders and 77 nonbinders for H-2K^b. Of the 128 peptides in the evaluation set for HLA-A2 predictions, 30 were derived from natural proteins and the remainder from synthetic peptides, including poly-glycine and polyalanine sequences. The H-2K^b predictions used two test groups: one with 9 known binders from different proteins and the other with 14 binders and 77 nonbinders from natural proteins. The ANN had a predictive accuracy of 78% for binding to HLA-A2 and 88% for binding to mouse H-2K^b.

Milik and colleagues (1998) [101,102,105] applied the ANN method to the prediction of mouse class I H-2K^b using 8-mer peptides from a phage display library [106–108]. Binding properties were measured by a competitive CTL assay and values expressed as the amount of competitor peptide (OVA 257) necessary to induce 50% inhibition [109]. They obtained sequences of 181 binding and 129 nonbinding phages. For classification procedure, the data were divided into three subsets: training set consisting of 93 binding and 130 nonbinding peptides, validation set 1 (18 binding and 13 nonbinding), and validation set 2 (33 binding and 23 nonbinding). Validation set 1 was used for fine tuning the system during the learning phase and validation set 2 was used for evaluating the performance of the 3-layer feed-forward back-propagation ANN. They tested two ANN models based on two different representation schemes for encoding amino acids. The first scheme (ANN1) consisted of a 10-number representation based on the hierarchical organization of amino acids [110]; the second (ANN2) consisted of a 6-number representation based on normalized physico-chemical properties of amino acids using volume, bulkiness, flexibility, polarity, aromaticity, and charge. ANN1 consisted of one input layer with 180 nodes, one hidden layer with 5 hidden nodes, and one single output neuron, whereas, ANN2 consisted of 48 input neurons, 3 hidden neurons, and 1 output neuron. To evaluate the validity of the analysis they predicted H-2K^b binding peptides from the sequence of chicken albumin, a protein with a well characterized H-2K^b epitope. All peptides that were classified as binding were synthesized and tested for experimentally. ANN1 and ANN2 selected the majority of binding peptides tested, with sensitivities of 70% and 80%, respectively.

Honeyman et al. (1998) [63] and Brusic et al. (1998) [63,70,100–102] reported high accuracy in the prediction of binding peptides (9-mers) to the class II HLA-DR4(B1*0401). Their approach was different from those previously discussed for the class I alleles. In both cases, a hybrid method was developed for predicting binding peptides to MHC class II molecules. Their method, termed PERUN, was implemented by combining available experimental data and knowledge of binding motifs, quantitative matrices (QM) for preprocessing, an evolutionary algorithm (EA) to derive alignment matrices, and an ANN to perform classification. For development of this hybrid system, in the data extraction stage, peptide sequences and their binding affinities were collected from the MHCPEP database, experimental binding data [111], and other private databases (V. Brusic, L.C. Harrison, and M.C. Honeyman, unpublished results). The initial data set consisted of 650 peptides: 338 binders and 312 nonbinders to HLA-DR4(B1*0401). In the preprocessing stage, an EA was used to search for predictive peptide alignment matrices, i.e., quantitative matrices. Through a genetic mechanism (mutation, crossover, and reproduction) and competition, a population of aligned matrices were generated and selected for improved fitness (characterized by good discrimination between binding and nonbinding peptides to HLA-DR4). The fitness function incorporated in the EA algorithm was based on the knowledge of primary anchor positions in reported binding motifs [89], which was used to fix position 1, corresponding to the primary anchor in each matrix, while the rest of the matrix was subjected to the application of above mentioned genetic mechanisms. The end product of the EA search was an alignment matrix, where each residue at each position in the 9-mer was assigned a binding score. The aligned matrix was then used to find and align putative 9-mer cores of known binders for the final stage of classification by the ANN. This set of aligned peptides served as the training set, which comprised of aligned binding 9-mer cores (338 peptides) and nonbinding 9-mers (312 peptides). The aligned peptide sequences were then transformed to numerical signal vectors using a 20-bin amino acid representation. The ANN system was designed as a 3-layer, fully connected, feed-forward back-propagation ANN. Because the binding affinities of the peptides in the training set were not all derived by the same experimental method, affinity measures were grouped into no affinity, low affinity, moderate affinity, and high affinity, and output values represented as 0, 6, 8, and 10, corresponding to these groups, respectively. Honeyman et al. (1998) [63] reported that the ANN-based binding prediction on an experimental binding affinity of synthetic peptides was 77% sensitivity and 80% specificity. Meanwhile, Brusic et al. (1998) [63,70,100–102] compared the PERUN method with the QM-based and binding motif-based methods for prediction of low, moderate, and high affinity binders. PERUN slightly outperformed the QM method, with prediction performance of 73% for low binders, 86% for moderate binders, and 88% for high affinity binders, whereas, the QM performance was 73%, 82%, and 87%, respectively. The binding motif method had the lowest performance (63%, 69%, and 74%, respectively). Such an approach integrates the strengths of the different methods and minimizes their disadvantages. For instance, binding motifs encode the most important rules of peptide–MHC-binding interaction [84,89,96,112]. However, for class II molecules, binding motifs are less well defined [89]. MHC class II molecules show degenerate motifs, which make peptide alignment more difficult. Because of this, except for certain molecules [96,112], binding motif-based methods show poor generalizations. In contrast, QM-based methods, which are in essence refined binding motifs, can predict large subsets of binding peptides reasonably well and can be used when data sets are limited [92,93,95,96,113,114]. However, QM cannot deal with nonlinearity in data and may miss distinct subsets of binding preferences, such as medium binders [100]. Also, QM methods are not adaptive and self-learning; so integration of new data usually requires redesigning of the alignment matrix. On the other hand, ANNs can deal with nonlinearity and are adaptive and self-learning but usually require a large amount of data.

A limitation of ANN predictions based on the indirect methods described above is their inability to discriminate between T-cell epitopes and nonepitope MHC binders, mainly because these methods only predict the MHC binders from antigenic sequences. Bhasin and Raghava (2004) recently developed a direct method for predicting cytotoxic T lymphocyte (CTL) epitopes from antigenic sequences using a hybrid method that combines QM, support vector machine (SVM), and ANN methods. Their system is based on 1137 experimentally proven CTL epitopes and 1134 nonepitopes (786 nonepitopes 9-mers and 348 MHC nonbinders). The system was evaluated with a blind data set consisting of 63 CTL epitopes, 63 nonepitope MHC-binders, and 63 MHC nonbinders. ANN and SVM outperformed the QM method. For training of both the ANN and SVM, peptide sequences were changed to numerical vectors by encoding sequences with 20-bin representation. The ANN performance was 72.2% accuracy, 73.2% sensitivity, and 71.2% specificity. The SVM performance was 75.4%, 73.8%, and 77.0%, respectively, while the QM method performance was 70.0%, 65.2%, and 74.9%, respectively. Also, the ANN and SVM were used to perform consensus and combined prediction of CTL epitopes. In consensus prediction, epitopes predicted by both methods were considered epitopes, otherwise they were considered as nonepitopes. Meanwhile, in combined prediction, epitopes predicted by either method were considered as epitopes. In addition, two models of such implementation can be used, i.e., in one model the ANN can be used as the base method, whereas in the second model, the SVM is used as the base method. The CTLPred server (available at http://www.imtech.res.in/raghava/ctlpred/) is based on such approach. They also developed the ComPred [115] based on a hybrid method using ANN and QM filtering for the combined prediction of MHC-binding peptides or CTL epitopes. Using this prediction approach, they achieved a sensitivity of 79.4% and specificity of 88.4% for discriminating between T-cell epitopes and nonepitope MHC binders. Another program, ANNPred, uses a feed-forward back-propagation ANN for the prediction of MHC binders to 30 MHC alleles. Both ComPred and ANNPred are available at http://bioinformatics.uams.edu/mirror/nhlapred/index.html. Table 2.3 shows a list of T-cell epitope mapping tools presently available.

Representation of peptide sequences using 20-bin numerical representation (0s and 1s), appears to provide an ANN with sufficient discriminative power to predict (or classify) CTL epitopes [70,116] and MHC-binding activities of peptides [30,63,116]. However, this sparse encoding of the amino acids ignores their physicochemical similarities or dissimilarities. Furthermore, in some cases, selection of numerical representation schemes for signal processing is reported to have statistically significant effects on ANN performance [101,102,105]. As previously discussed, Milik et al. (1998) [101,102,105] reported that an ANN derived from the individual representation of amino acids using physicochemical properties (ANN2) outperformed an ANN derived from the clustering of amino acids (ANN1). Also, Nielsen et al. (2003) [101,102,105] report that an ANN trained with a novel sequence encoding scheme using Blosum50 and hidden Markov model sequence encoding, for prediction of T-cell class I epitopes, was superior to the one trained on 20-bin sequence encoding (accuracy of 91.2% and 87.7%, respectively). Another strategy to enhance prediction performance is to combine several ANNs derived from different sequence-encoding schemes instead of using one ANN system derived from a single sequence-encoding scheme [102,105]. Although in some cases using either 20-bin representation or physicochemical features may not affect classification performance, the latter encoding strategy has important advantages over sparse encoding: (1) it allows researchers to weigh in the contribution of specific amino acid parameters and directly extract their correlation to MHC peptide-binding properties and

TABLE 2.3
Overview of T-Cell Epitope Prediction Tools

Tool	Developer	Description	Web Site
EpiMatrix	EpiVax Inc	Matrix-based; pocket profile methods	http://www.epivax.com
Class I/class II binding prediction	Immune Epitope Database (IEDB)	ANN; average relative binding method; stabilized matrix method	http://tools.immuneepitope.org
PREDEP	The Hebrew University of Jerusalem	Motif matrix-based method	http://margalit.huji.ac.il/
ProPred	Institute of Microbial Technology	Quantitative matrices method	http://bic.uams.edu/mirror/propred/index.html
SYFPEITHI	University of Tübingen	Extended anchor method	http://www.syfpeithi.de/
Multipred	Institute for Infocomm Research	ANN; Hidden Markov models	http://discovery.imb.uq.edu.au/AutoantigenDB/MultiPred.shtml
MHC Thread	The Hebrew University of Jerusalem	Protein threading/fold recognition	http://margalit.huji.ac.il
RANKPEP	Molecular Immunology Foundation	Profiles based on Position specific scoring matrix method	http://immunax.dfci.harvard.edu/Tools/rankpep.html
MHCPred	The Edward Jenner Institute for Vaccine Research	Partial least squares-based multivariate statistical method	http://www.jenner.ac.uk/MHCPred
CTLPred	Institute of Microbial Technology	ANN and support vector machine; quantitative matrices method	http://www.imtech.res.in/raghava/ctlpred/index.html
EpiJen	The Edward Jenner Institute for Vaccine Research	Additive approach based on Free-Wilson	http://www.jenner.ac.uk/EpiJen
NetMHC	Institute of Medical Microbiology and Immunology	ANN; weighted matrices	http://www.cbs.dtu.dk/services/NetMHC
NetCTL	Institute of Medical Microbiology and Immunology	ANN; weighted matrices	http://www.cbs.dtu.dk/services/NetCTL/
nHLAPred	Institute of Microbial Technology	ANN; quantitative matrices method	http://www.imtech.res.in/raghava/nhlapred/
SVMHC	University of Tübingen	Support vector machine	http://www-bs.informatik.uni-tuebingen.de/Services/SVMHC
Bimas	National Institutes of Health (NIH)	Matrix-based	http://thr.cit.nih.gov/molbio/hla_bind/

Source: From IEDB, Immune Epitope Database; NIH, National Institutes of Health.
Note: Links were last accessed on August 21, 2008.

CTL epitope SAR, (2) continuous-property scales for signal encoding, unlike sparse encoding, reduces the dimension of the vector input and the danger of generating arbitrary correlations that may lead the ANN to overfit (this is specially important when data is sparse), and (3) in contrast to sparse representation, physicochemical or encoding schemes like Blosum50 provide the neural network with less precise and more general information about the sequence, which allows the system to generalize better, mainly because they encode subtle chemical and evolutionary relationships between the 20 amino acids [102].

The experiments described above provide compelling evidence for two possible applications of ANN technology for developing immunotherapeutic peptide and protein agents: vaccine and drug design. These applications differ, however, in their requirements for specificity and sensitivity. For vaccine design, it is more important to predict all possible binders to a given MHC molecule (specificity), whereas for drug design it is desirable to predict those peptides that will bind with highest affinity (sensitivity). It should be possible to manipulate the performance of the ANN to optimize specificity or sensitivity as required for a given application. Towards this aim, in addition to the programs already mentioned, other systems are available to researchers. The MULTIPRED program [117] (available at http://antigen.i2r.a-star.edu.sg/multipred/) is based on ANN methods to predict HCV T-cell epitopes for MHC class I A3 molecules and human papillomavirus (HPV) type 16 T-cell epitopes for MHC class A2 variants, and is reported to achieve good sensitivity and specificity and prediction capability, with an area under the receiver operating system curve (A_{ROC}) > 0.80. NetChop server [105] (available at http://www.cbs.dtu.dk/services/NetChop/) uses ANN method to predict proteasome cleavage sites responsible for generating CTL epitopes.

ANN-based predictions can also reduce the number of bench experiments needed for T-cell epitope screening. ANN applications allow investigators to reduce and focus the number of candidates for identification of T-cell epitopes while increasing the efficiency of the search space for these candidates. For example, Honeyman and collaborators (1998) [63] synthesized 68 peptides (9-mers) and used a trained ANN to perform binding prediction on this set. After completing experimental binding and T-cell response tests, they show that ANN-predicted binding would have reduced the number of synthesized peptides from 68 to 26, with the potential loss of 4 epitopes. In another case, Schönbach and collaborators (2002) [98] applied ANN methods to the large-scale identification of HIV T-cell epitopes from Pol, Gag, and Env sequences for HLA-A*0201 and found 890 HIV-1 and 232 HIV-2 epitope candidates. The overall sequence coverage of the predicted A*0201 T-cell epitope candidates was 2.7% for HIV-1 and 3.0% for HIV-2. By comparing the ANN-predicted binding with other bioinformatics methods and experimentally confirmed A*0201-restricted epitopes, they were able to extrapolate their results and estimate that approximately 247 predicted HIV-1 are yet to be discovered as active A*0201-restricted epitopes. Finally, with proper encoding, ANNs can also be used for feature extraction. Conclusions can be drawn regarding the relative importance to binding of specific positions in the peptide and chemical properties of amino acids at those positions. Thus, use of these capabilities can facilitate the de novo-driven design of viral vaccine or immunotherapeutic candidates.

2.3.3 B-Cell Epitope Prediction

Unfortunately, computational tools for mapping B-cell epitopes and prediction of antibody–antigen interactions are not as developed as their T-cell epitope counterparts. A B-cell epitope is frequently defined by the three-dimensional structure that can be bound to the variable region of an antibody. The complex nature of antibody–antigen interactions has presented unique challenges for the development of accurate immunoinformatic approaches. At the present state of the art, the best way to predicting the potential risk of a therapeutic protein to induce B-cell mediated immunoresponses is through in vivo experiments (e.g., immunize HLA transgenic mice with the therapeutic protein). Nevertheless, recent efforts toward B-cell epitope prediction started bringing some successful developments. The first attempt to implement ANN technology for this task was made by Saha and Raghava (2006). Their strategy uses a recurrent neural network-based* approach for predicting linear (continuous) B-cell epitopes: the ABCpred Server (http://www.imtech.res.in/raghava/abcpred/) [118]. They trained the ANN on the Bcipep database of B-cell epitopes (http://www.imtech.res.in/raghava/bcipep/), which consists of ~700 B-cell epitopes of 5–30 residues, and ~700 random peptides of maximum length of 20 residues. Their approach is able to predict epitopes with 65.93% accuracy. They also developed the Bcepred algorithm [119], another tool for prediction of continuous B-cell epitopes (available at http://www.imtech.res.in/raghava/bcepred/). The method is based on physiochemical properties and was reported to have slightly lower prediction accuracies (ranging from 52.92% to 58.70%).

* A neural network with feedback connections, in which there are closed-loop signal paths from one unit (neuron) back to itself. A recurrent net is a neurodynamical model used for simulating the behavior of nonlinear dynamical systems. Examples include Hopfield nets, Boltzmann machine, and recurrent backpropagation nets.

As most B-cell epitopes are discontinuous (conformational), these approaches are not very good at mapping (localizing) three-dimensional epitopes nor at predicting antibody–antigen interactions. Currently, there are no reports on the application of neural networks for these types of predictions. However, methods for localizing potential B-cell epitopes on the protein surface are available. Recent examples of such computational tools are the Conformational Epitope Prediction (CEP) server (available at http://202.41.70.74:8080/cgi-bin/cep.pl) [120] and 3D-Epitope-Explorer (3DEX) [121]. There are very few other computational methods available and even a smaller number of these have been used in prospective research studies and validated using in vitro and in vivo studies [122]. Another major drawback is that they are structure-based methods. However, the structural information is not always available for proteins. In general, most B-cell mapping algorithms require further research [123].

2.4 ANN-AIDED DESIGN OF THERAPEUTIC PROTEINS

As therapeutic peptides and proteins, either vaccine or drug molecule candidates, progress down the development pipeline, the ability to predict their immunological, pharmacokinetic, and toxicological properties is becoming increasingly important in reducing the number of expensive late-development failures. As a result, the application of ANN methods in this area of research has gained popularity [124–133]. As described in the previous section, screening therapeutic molecules for their T-cell and B-cell sequence signal content (to design sequences with low antigenic signal content) can avoid potentially serious immunoresponses during the clinical trail phase. In addition to avoiding harmful effects, drug and vaccine developers also search to find therapeutic molecules with optimal efficacy.

Automation of chemical synthesis, tools for designing molecule libraries, and high-throughput technology for biological screening experiments, robustly and quickly enable target candidates to be discovered more effectively. As a result, empirical methods for building predictive models of the relationships between molecular structure and functions of molecular candidates are becoming increasingly important for therapeutic protein discovery and development. SARs are usually complex and nonlinear in nature. The ANN is, therefore, a viable choice because of its flexibility and adaptability for modeling nonlinear relationships and is more noise tolerant compared to other methods [124].

Sections 2.4.1 and 2.4.2 describe examples on how ANN algorithm may be helpful for engineering protein-based molecules with predetermined activity, in particular, protein-binding affinity, protein flexibility, and stability. Although presented as separate problems in ANN applications, in practice any number of properties for a therapeutic protein may be simultaneously searched within a single computational framework. For example, potential therapeutic protein candidates can be screened and selected for their potential immunogenicity and optimal catalytic activity using a computational tool that integrates these two functions.

2.4.1 Focused Design of Protein Binding

Schneider et al. (1998) proposed a systematic focused strategy for peptide design through a combination of rational and evolutionary approaches [134]. They described the use of evolutionary strategies as a cyclic-variation selection of peptide candidates followed by ANN modeling of quantitative structure–activity relationships. They reported a successful application of the approach to the de novo design of a peptide that fully prevents the positive chronotropic effect of anti-b_1-adrenoreceptor autoantibodies from the serum of patients with idiopathic dilated cardiomyopathy (DCM). On the basis of previous reports showing that short synthetic peptides encompassing the natural epitopes were able to neutralize the chronotropic effect of the autoantibodies, their idea was to test whether it was possible to derive novel (artificial) epitope sequences directed against the second extracellular loop of human b_1-adrenoreceptor for potential immunotherapeutic agents. Their design strategy is described in the following four steps.

(1) De novo design and evolutionary approach. The researchers first identified a sequence with a desired activity (b_1-adrenoreceptor loop 2 epitope) that would serve as a "seed" sequence around which an *in silico* library can be developed. A common approach for such searches is through random screening. To obtain a seed peptide, parts of the sequence of human b_1-adrenoreceptor encompassing loop 2 were analyzed by epitope mapping. The ability of the peptide fragments to bind to human anti-b_1-adrenoreceptor antibodies was measured by ELISA. The amino acid sequence ARRCYNDPKC (positions 107–116) was identified as a natural epitope with specific affinity to the antibodies and was therefore selected as the seed peptide.

(2) Focused library strategy. Using the amino acid sequence ARRCYNDPKC as the seed peptide, a set of 90 variants were generated by a simple algorithm describing each residue by the respective property values for hydrophobicity [34] and volume [135]. Thus, a total of 90 20-dimensional vector representations of variants, in terms of these two properties, were generated. These vectors were generated using the Box–Muller formula, which were approximately Gaussian-distributed around the seed peptide vector (standard deviation $(\sigma)=0.1$); where g is a Gaussian-distributed random number and i and j are random numbers in the [0,1] interval:

$$g = \sigma\sqrt{-2\ln(i)}\sin(2\pi j) \tag{2.1}$$

The rationale for obtaining vectors both close and distant from a seed vector, is that peptides with an activity similar to that of the seed peptide are expected to be in close proximity to the seed peptide in sequence space [136], while those that are more distant are expected to have decreased activity or provide new seed vectors of higher activity. The strategy is designed, through a mathematical approach, to focus the search for a desired property in a list of peptide candidates that are more likely to conform to such properties rather than following the conventional random search (epitope mapping) for natural epitopes, which is expensive and time consuming. After generation of the *in silico* library, the property vectors were translated back into amino acid sequences by selecting the most similar residues at each sequence position according to their physicochemical properties. The computer-generated variants were synthesized and tested for their ability to bind to human anti-b_1-adrenoreceptor antibodies. The in vitro assays confirmed the applicability of the mathematical strategy to the generation of a small focusing peptide library. Several non-natural peptides were found to have increased antibody-binding ability when compared with natural epitopes. Thus, the generation and selection of candidates from a focused library provided an efficient starting point from which to extract knowledge. This approach also allowed them to expand the search to a large number of sequences more cost effectively (*in silico*) and in less time (computational process) than would have been possible with conventional techniques such as random mapping or DNA shuffling (REF).

(3) Establishment of SARs with ANN methods. Armed with the information generated from the focused library, a 3-layered, feed-forward network with a sigmoid hidden unit and a single linear output unit was used, under the supervised paradigm, to extract the underlying SAR by mapping the sequence vectors (90 peptides and seed peptide) to their respective semi-quantitative absorbance properties (defined as high, medium, and low binding activities). The trained network was then used as fitness function for selection of subsequently evolved peptide candidates.

(4) Evolutionary/mathematical search for new variants with desired properties. Once the neural network was trained on the information generated from the focused library data, it was used as a heuristic fitness function and focused approach for searching in sequence space. Generally, the expanded sequence space search for the de novo design of peptides

> starts by querying computer-generated sequences. Schneider et al. (1998) [137] opted for an evolutionary strategy for generating peptide libraries. In a cyclic or iterative process, virtual peptide libraries are generated by variation of a seed sequence (parent), the activity of each peptide (offspring) is predicted by the ANN, and the peptide (child) with the highest predicted activity is selected as the parent for the next cycle. This is repeated until no further optimization can be observed. To evaluate the ANN, the selected peptides were tested for their activity and compared to the predicted activities. Despite some conflicts in the degree of binding activity, all peptides predicted as active by the ANN showed activity when tested. The most active peptides shared only one common residue, aspartic acid, at position seven with the parent or seed sequence. Furthermore, a peptide predicted by the ANN as being the most inactive peptide, which contained favorable binding motifs in two positions, in fact tested to be nonbinding. Most striking was the observation of the difference in activities of the de novo peptides when compared to natural-derived peptides at different concentrations. Some of the ANN-designed peptides had similar or significantly higher activities at concentrations 10-fold lower than those of natural peptides.

This example for peptide/protein design shows that a mathematical approach using ANN methods to derive knowledge and search in sequence space can be successful for the identification of mimetopes (antigens mimicking a natural epitope sequence), some with significantly different residue sequences compared to natural-derived peptides. It also shows that the most active peptide represents a good starting point for the expanded search and further optimization. One may describe the virtual strategy presented here as the mathematical "molding" of molecular structural features to a desired function. In fact, computer-based strategies for de novo molecule candidate design has emerged as a complimentary approach to high-throughput techniques and should prove beneficial for development of immunotherapeutic and diagnostic candidates [66,137–139].

2.4.2 Prediction of Protein Flexibility and Stability

Mutations produce changes in the biological activity of proteins [4,140–144]. For protein engineering, making the right changes (mutations) in the root substrate is the key to designing protein products with desired functions. Different protein engineering strategies apply these variations differently. Rational design uses detailed knowledge of the structure and function of the protein to make desired changes. Such changes can be easily introduced by site-directed mutagenesis techniques. However, detailed structural knowledge of a protein is hard to obtain. Even when such knowledge is available, it is very difficult to predict the effects of the various mutations. Only in those cases when sufficient knowledge of these properties was secured a significant progress in protein engineering was attained [145]. Directed evolution strategies, on the other hand, rely on methods that mimic natural evolution through mutations (e.g., random mutagenesis) and recombination processes (e.g., DNA shuffling) and require no prior knowledge of a protein nor is it necessary to predict what effects the given changes have on the function. Through a regime of iterative rounds of mutations and selection, proteins with desired function may be obtained. However, this strategy requires the application of high-throughput technologies. Large amounts of recombinant DNA must be mutated and the products screened for desired qualities. The sheer number of variants often requires expensive robotic equipment to automate the process. Furthermore, not all desired activities can be easily screened for. Computational approaches founded on first principles* and knowledge stored in databases may provide fast and inexpensive ways of predicting changes in the dynamic behavior of proteins that may result from mutations. However, reliable molecular dynamic computations from sequence by approaches based solely on physical and thermodynamic principles is not

* It refers to the calculations that start directly at the level of established laws of physics and thermodynamics and does not make assumptions based on model or fitting parameters.

yet possible. The complexity of the problem by far surpasses today's computational capabilities. Even unlimited computational resources were available; another serious problem is posed by the minute energy differences between the native and mutated protein structures, which are about ~1 kcal/mol [146]. Added to the problem of dealing with minute energy differences, is the uncertainty in estimating constants needed for calculations based on first principles, which makes very hard developing simple and sufficiently accurate approximation approaches. Presently, predictions derived from structure knowledge-based databases and sequence homology is rather more accurate. Such modeling is based on the fact that similar sequences or structures usually have the same properties and display the same dynamical behavior. As detailed knowledge of factors regulating protein motion, better techniques for assessing and relating them to structure and function as well as advancements in high-throughput technology continue to expand, so will the computational protein design algorithms capable of predicting target functions and differentiating between optimal and suboptimal protein-based products.

The inherent instability of many proteins can pose significant challenges for the successful engineering of protein-based therapeutics. The physicochemical stability and biological functions of proteins are believed to be intimately related to their global flexibility, intramolecular fluctuations, and various other dynamic processes. Protein stability problems may lead to product efficacy loss and immune response. Meanwhile, protein flexibility facilitates adaptation and recognition in diverse molecular events. Although conformational dynamics is important for molecular recognition, a balance between rigidity and flexibility must be attained for maintaining both function and stability. Due to the importance that both protein flexibility and stability have on a wide range of biological functions, a great deal of research has been devoted to probing the complex inter-relationships between them [147]. It is now generally accepted that the successful formulation, manufacture, functional optimization, and safety of therapeutic protein dosage forms depend on a thorough understanding of their physicochemical properties including their dynamic behavior.

2.4.2.1 Prediction of Protein Flexibility

The study of the property of protein flexibility is important for developing protein-based therapeutics [148]. Proteins are dynamic molecules that are in constant motion, which is related to structural flexibility. The intrinsic molecular flexibility of proteins is crucial for such basic functions as enzymatic catalysis [149,150], regulation of protein activity [151,152], transport of metabolites [153,154], and protein–ligand affinity [155–157]. Changes in the structural flexibility of a protein can greatly impact its characteristic function. For instance, highly flexible proteins have been either identified or implicated in diseases such as AIDS (HIV gp41) and scrapie [158]. Determination of these regions also plays an important role in structural genomics, since such regions can be a source of problems in protein expression, purification, and crystallization.

The mechanism of molecular motion is difficult to observe directly. Nuclear magnetic resonance spectroscopy (NMR) studies can yield root mean square fluctuations and order parameters [159]. Optical trapping [160] can be used to track the movement of molecular motors. Hydrogen/deuterium exchange can be used to measure changes in the solvent exposure of amide protons [161]. The hinge connecting two independently folded domains in a protein is sometimes a sensitive site for proteolytic cleavage [162]. Many of these experimental techniques, however, require much effort and provide limited information [163]. Given the difficulty of observing motion by experimental means, scientists often rely on computational methods for modeling molecular motion.

Computational simulations have been used for several decades to predict protein dynamic behavior in many types of applications: for docking studies, like protein–ligand docking [164–169], and protein–protein docking [170–175]; for structure-based drug design [168,176–178]; and for combinatorial library design [179]. These types of computational motion simulations are based on information stored in bioinformatics databases of protein motions, such as, the Database of

Macromolecular Motions, MolMovDB.org [180], and DynDom [181,182]. Dynamic behavior simulations help address two major issues: What are the domains of the flexible protein? What is the mechanical nature of the motion of each domain?

In regard to the first issue, the domain* in a flexible protein refers to the rigid portions whose internal coordination does not significantly vary during the simulation (exemplified by the flap domain in HIV protease [183–187]). Much work has been done to solve the related problem of finding domain boundaries, which can be flexible or inflexible. Nagarajan and Yona have shown how to analyze multiple sequence alignments to identify domains [188]. Marsden, McGuffin, and Jones showed that predicted secondary structure could help find domain boundaries [189]. Jones et al. combined PUU [190], DETECTIVE [191], and DOMAC [192] to make a powerful domain boundary predictor [193]. Hayward and Berendsen, showed how domains may be recognized from the difference in the parameters governing their quasi-rigid body motion, and in particular their rotation vectors [194,195]. Flores and Gerstein showed how differences in energetic interactions within and between structural domains may be used to predict flexible hinges [196].

DOMAC [192] is a good example of how neural network techniques have been applied for the task of domain detection. This server combines homology-based and ab initio methods for protein domain predictions. The ab initio part of the DOMAC's hybrid system is based on ANN. Cheng, Sweredoski, and Baldi (2006) developed an ANN-based system capable of parsing protein chains into different domains (known as DOMpro, the precursor of their actual DOMACS system) [197]. Their ANN approach consisted in training a one-dimensional recursive neural network (1D-RNN) using evolutionary information in the form of profiles along with predicted secondary structure, and predicted relative solvent accessibility. The 1D-RNN was trained and tested with a curated data set derived from the CATH domain database, version 2.5.1 and was reported to correctly detect domain boundaries of single-domain proteins with 76% sensitivity and 85% specificity.

After the protein being partitioned into domains, then the issue involving the mechanics of the dynamic behavior of the protein domains can be investigated (how can we describe the motion of each domain? Does it translate in three-dimensional space? Rotate about an axis? Do they shear?). Many computational approaches for modeling (simulating) the dynamic behavior of proteins such as molecular dynamic (MD) and Monte Carlo methods [198], Gaussian network model (GNM) [199], FlexTree [200], and RCI [201] have been developed and some are available. Recently, Schlessinger and Ross (2005) applied neural networks for the prediction of protein flexibility from sequence information [202]. Their method is based on large-scale analysis of experimentally derived B-values.† They used a feed-forward MLP with back-propagation to predict residue flexibility using two different thresholds in the classification of normalized B-values into two classes (flexible/rigid). The ANN was reported as having an accuracy ranging from 77% to 83%, and an accuracy of 90% for the most strongly predicted residues. In addition, they present two case studies to demonstrate possible applications of their ANN-based method: the prediction of the flexible switch II region in Ras, and to the prediction of the rigid region in propeller folds. Switch II regions in the Ras protein are known to be responsible for the GTPase activity of Ras. This 11 residue region in Ras contains a highly conserved residue (Gln61), which is the catalytic residue for the GTP hydrolysis. The hyper-flexibility of this conserved residue is critical for the GTPase catalytic reaction [203]. β-Propeller folds are characterized by 4–8 symmetrical repeats of four-stranded antiparallel and

* Protein domains are structural, functional, and evolutionary units of proteins. The concept of domains in proteins plays a very important role in structural biology, genetics, biochemistry, and evolutionary biology. However, this concept is frequently defined differently in each of these subdisciplines. For example, structural biologists initially defined protein domains as segments of the polypeptide that fold into globular units, whereas, geneticists and biochemists define it as the minimal fragment of a gene capable of performing a given function.

† B-values (also referred to as B-factors), obtained in experimental atomic-resolution structures, provide information about local mobility. They represent the decrease of intensity in diffraction due to both the dynamic disorder caused by the temperature-dependent vibration of the atoms and due to the static disorder, which is related to the orientation of the molecule. In general, the higher the B-factor of a residue the higher is its flexibility.

twisted β-sheets that are arranged around a central tunnel [204,205]. The majority of these proteins use the tunnel or the entrance to it for the coordination of a ligand or as the site of catalytic activity. In this case, the structural rigidity of the propeller domain is believed to be crucial for the function of these proteins [204]. Both, the flexible region in Ras and the rigid region in propeller folds, which are indicative of function, were correctly predicted by the ANN.

2.4.2.2 Prediction of Protein Stability

Another important attribute closely related to the structural flexibility of macromolecules is protein stability. Incorporating changes (by amino acid substitutions or biochemical changes such as glycosylation and methylation) into the naturally occurring sequence of a polypeptide can affect the stability of a protein-based product [206–211]. To be both safe and effective, a therapeutic product must have not only the correct sequence of amino acids (primary structure) but also be stable enough for proper folding of that amino acid chain in three-dimensional space (tertiary structure). One important requirement for fast and accurate therapeutic protein design is the ablility to predict changes of protein stability upon mutation or variation. Understanding the SAR-regulating protein stability is critical for addressing this task and is one of the long-term goals of protein structure analysis [212,213]. Different methods for prediction of protein stability have been described [214–219]. In general, these approaches are based on different energy functions designed to compute the stability free energy changes due to substitution of one residue at a time in the protein sequence. Thermodynamic parameters such as force fields [5,6,220–222], pH, and temperature [223] are used for the derivation of these energy functions.

Over the past decade, development of large databases of thermodynamic data on protein stability changes upon single point mutation, like ProTherm [223], has facilitated the application of machine-learning techniques for predicting free energy stability changes upon mutation starting from the protein sequence. Recently, Capriotti, Fariselli, and Casadio, 2004 [224], developed a sequence-based method using a neural network-based approach for predicting a mutation effect on the protein thermodynamic stability: I-MUTANT (available at http://gpcr.biocomp.unibo.it/). They used a standard feed-forward MLP with back-propagation for predicting protein stability starting from sequence information. Their ANN architecture consisted of one input layer, two hidden nodes, and one output node, which codified protein stability (1 for increased stability, and 0, for decreased stability). They introduced different input encodings that represented the major parameters responsible for changes in protein stability upon mutation: Temperature and pH at which the stability of the mutated protein was measured, relative accessibility of mutated residue, and the three-dimensional residue environment centered at the mutated residue. They trained and tested their ANN-based predictor on a data set consisting of 1615 mutations and reported a classification accuracy of ~81.0%. Their system outperformed other energy-based methods available on the Web like: FOLDX [220], DFIRE [222], and PoPMuSiC [5,225]. When coupled to these energy-based methods, accuracy in the prediction increased up to 90.0%. In addition, they have also developed an SVM-based approach, I-MUTANT 2.0 [226], which was reported to have an accuracy of 77.0%.

2.5 PROSPECTS OF THE ANN ALGORITHM IN THERAPEUTIC PROTEIN DESIGN APPLICATIONS

The application of computational methods for designing optimal and safer therapeutic agents is becoming increasingly more common and important in biotechnology. The studies described herein demonstrate that ANN-based algorithms and machine-learning methods, in general, have the power to learn and predict target functions from sequence. Recent advances in biotechnology and genome sequence projects have generated enormous amounts of information available in databases. With data mining of these databases, knowledge and new insights on SARs have made the prospects of developing novel protein therapeutics much brighter.

As biotechnology's and research scientist's dependence on computational approaches continue to grow, the need to develop sound standards for the evaluation of prediction methods is becoming more important. This is particularly the case with the evaluation of prediction methods that are completely generic, which is to say valid for all prediction methods. As much as many notions and concepts have been introduced to predict SAR and have then been used for other purposes, many mistakes in comparing different methods have also been unraveled. Results can often not be compared between two different publications. This is because prediction methods are often published with estimates of performance that are supported by cross-validation experiments. However, the terms cross-validation or the related term jack-knife are not sufficiently well defined to translate into "estimate is correct." In fact, most publications make some serious mistakes in this respect as is demonstrated by the simple fact that very few estimates for performance have survived. Secondary structure and protein surface accessibility prediction methods may perhaps be the only example of publications with claims to performance accuracy that survived more than 5–10 years, respectively. One of the reasons for this is due to the problem in overlap between the training and test data sets. It is easy to reach very high performance by training on proteins that are very similar to those in the testing set. There are various strategies that have been proposed to deal with the similarity problem [227,228]. Another important matter for consideration is that of using the performance of the test set to choose some parameters by, for example, reporting full cross-validation results for N different parameters and then concluding that the best of those N is the performance of the final method. Instead, performance estimates should always be based on an evaluation data set that was not used in any step of the development. In addition, the more recent the data sets, the more reliable and representative they usually are. Therefore, we also need evaluations based on data sets that are as recent as possible. Furthermore, even when two publications have both used cross-validation correctly, this does not necessarily facilitate direct comparison of the results published by both. First of all, both have to have used the same standard of "truth" (same parameter assignments: for example, same physicochemical scales, secondary structure assignments, etc.). Second, they both must have been based on identical test sets. Frequently, the test sets used by developers are not representative and differ from each other. Proteins vary in their structural complexity and such variation is correlated with prediction difficulty. Appropriate comparisons of prediction methods require large "blind" data sets. One solution to the problem of comparing methods is to use a sufficiently large test set composed of proteins that were neither used nor are similar to any protein that was used for development of any prediction method.

Most of the studies described herein used standard three-layered feed-forward ANN with back-propagation techniques. However, ANN technologies are rapidly evolving and new techniques are now available, such as self-organized maps [23], self-growing neural networks [229,230], associative memory networks [23], and quantum neural networks [231–234]. Application of these innovations to peptidomimetics and de novo protein design should advance the field of protein engineering. In addition, hybrid approaches that combine ANN methods with other well-known techniques such as hidden Markov models (HMM), evolutionary algorithms (EA), SVM, and more recent techniques like Bayesian networks (BN), Particle swarm optimization (PSO) [235], look very promising for developing ANN-based protein prediction algorithms with enhanced prediction performance.

Furthermore, software tools can be used to translate ANN-based solutions into standard computer languages and source code. For example, the Snns2C program, available with the Stuttgart Neural Network Simulator (SNNS), is a tool that compiles a SNNS network into an executable C source. Once the compilation process is done, trained networks may be then implemented as embedded functions within existing modeling functions or assembled into stand-alone computer programs. In addition to their normal use, such embedded functions offer new opportunities for dynamic testing and for internal investigation of the ANN modeling function through techniques such as saliency analysis.

In conclusion, the application of ANN techniques is not just for computer scientists or engineers but comprise a powerful set of tools for wide range of health and biotechnology-related scientists as well. Advancements brought by ANN applications to therapeutic protein design will depend on the ingenuity in formulating problems, neural networks design, reliability of the information stored in databases, and the appropriate testing and evaluation procedures used. Given the fast rate at which computational hardware and software technologies are growing, one may anticipate wide spread emergence of ANN applications for computational design of therapeutic proteins.

REFERENCES

1. Lawrence, S. 2005. Biotech drug market steadily expands. *Nat. Biotechnol.* 23, 1466.
2. Walsh, G. 2006. Biopharmaceutical benchmarks 2006. *Nat. Biotechnol.* 24, 769–776.
3. Consulting Resources Corporation (CRC) 2008. U.S. Biotechnology Product Sales Forecast (http://www.consultingresources.net/biotechnology_pf.html).
4. Cabrita, L.D., Gilis, D., Robertson, A.L., Dehouck, Y., Rooman, M., and Bottomley, S.P. 2007. Enhancing the stability and solubility of TEV protease using in silico design. *Protein Sci.* 16, 2360–2367.
5. Kwasigroch, J.M., Gilis, D., Dehouck, Y., and Rooman, M. 2002. PoPMuSiC, rationally designing point mutations in protein structures. *Bioinformatics* 18, 1701–1702.
6. Pitera, J.W. and Kollman, P.A. 2000. Exhaustive mutagenesis in silico: Multicoordinate free energy calculations on proteins and peptides. *Proteins* 41, 385–397.
7. Chen, Y.Z. and Ung, C.Y. 2002. Computer automated prediction of potential therapeutic and toxicity protein targets of bioactive compounds from Chinese medicinal plants. *Am. J. Chin. Med.* 30, 139–154.
8. Koren, E., De Groot, A.S., Jawa, V., Beck, K.D., Boone, T., Rivera, D., Li, L., Mytych, D., Koscec, M., Weeraratne, D., Swanson, S., and Martin, W. 2007. Clinical validation of the "in silico" prediction of immunogenicity of a human recombinant therapeutic protein. *Clin. Immunol.* 124, 26–32.
9. De Groot, A.S. and Moise, L. 2007. Prediction of immunogenicity for therapeutic proteins: State of the art. *Curr. Opin. Drug Discov. Devel.* 10, 332–340.
10. Edelman, G.M., Reeke, G.N., Jr., Gall, W.E., Tononi, G., Williams, D., and Sporns, O. 1992. Synthetic neural modeling applied to a real-world artifact. *Proc. Natl. Acad. Sci. U S A* 89, 7267–7271.
11. Horwitz, B. and Braun, A.R. 2004. Brain network interactions in auditory, visual and linguistic processing. *Brain Lang.* 89, 377–384.
12. Ichikawa, M. and Matsumoto, G. 2004. The brain-computer: Origin of the idea and progress in its realization. *J. Integr. Neurosci.* 3, 125–132.
13. Krichmar, J.L. and Edelman, G.M. 2002. Machine psychology: Autonomous behavior, perceptual categorization and conditioning in a brain-based device. *Cereb. Cortex* 12, 818–830.
14. Krichmar, J.L. and Edelman, G.M. 2005. Brain-based devices for the study of nervous systems and the development of intelligent machines. *Artif. Life* 11, 63–77.
15. Krichmar, J.L., Nitz, D.A., Gally, J.A., and Edelman, G.M. 2005. Characterizing functional hippocampal pathways in a brain-based device as it solves a spatial memory task. *Proc. Natl. Acad. Sci. U S A* 102, 2111–2116.
16. Seth, A.K., McKinstry, J.L., Edelman, G.M., and Krichmar, J.L. 2004. Visual binding through reentrant connectivity and dynamic synchronization in a brain-based device. *Cereb. Cortex* 14, 1185–1199.
17. Arbib, M.A. 2003. The elements of brain theory and neural networks. In *The Handbook of Brain Theory and Neural Networks* Arbib, M.A. (Ed.), pp. 3–23, The MIT Press, Cambridge, Massachusetts.
18. Fausett, L. 1994. *Fundamentals of Neural Networks: Architectures, Algorithms, and Applications*, pp. 1–38, Prentice Hall, Upper Saddle River, New Jersey.
19. Kecman, V. 2001. Basic tools of soft computing: Neural networks, fuzzy logic systems, and support vector machines. In *Learning and Soft Computing: Support Vector Machines, Neural Networks and Fuzzy Logic Models*, pp. 13–18, The MIT Press, Cambridge, Massachusetts.
20. Hagan, M.T., Demuth, H.B., and Beale, M. 1996. Introduction. In *Neural Network Design*, pp. 1–12, PWS Publishing Company, Boston, Massachusetts.
21. Haykin, S. 1999. Introduction. In *Neural Networks*, pp. 1–49, Prentice Hall, Upper Saddle River, New Jersey.
22. Swingler, K. 1996. Introduction. In *Applying Neural Networks: A Practical Guide*, pp. 3–20, Mourgan Kaufman Publishers Inc., San Francisco, California.
23. Kohonen, T. 2001. *Self-Organizing Maps*, pp. 1–487. Springer-Verlag, New York.

24. Fischler, M.A. and Firschein, O. 1987. Representation of knowledge. In *Intelligence: The Eye, the Brain, and the Computer*, pp. 63–80, Addison-Wesley, Reading, Massachusetts.
25. Casari, G., Sander, C., and Valencia, A. 1995. A method to predict functional residues in proteins. *Nat. Struct. Biol.* 2, 171–178.
26. Principe, J.C., Euliano, N.R., and Lefebvre, W.C. 2000. *Neural and Adaptive Systems: Fundamentals Through Simulations*, pp. 208–219, John Wiley & Sons, Inc., New York.
27. Swingler, K. 1996. Network use and analysis. In *Applying Neural Networks: A Practical Guide*, pp. 165–182, Mourgan Kaufman Publishers Inc., San Francisco, California.
28. Wu, C.H. and McLarty, J.W. 2000. *Neural Networks and Genome Informatics*, pp. 1–158, Elsevier Science, Kidlington, Oxford.
29. Zavaljevski, N., Stevens, F.J., and Reifman, J. 2002. Support vector machines with selective kernel scaling for protein classification and identification of key amino acid positions. *Bioinformatics* 18, 689–696.
30. Brusic, V., Rudy, G., and Harrison, L. 1995. Prediction of MHC binding peptides using artificial neural networks. *Complexity International* 2, 1–9.
31. Schneider, G. and Wrede, P. 1998. Artificial neural networks for computer-based molecular design. *Prog. Biophys. Mol. Biol.* 70, 175–222.
32. Edelman, J. 1993. Quadratic minimization of predictors for protein secondary structure. Application to transmembrane alpha-helices. *J. Mol. Biol.* 232, 165–191.
33. Eisenberg, D., Schwarz, E., Komaromy, M., and Wall, R. 1984. Analysis of membrane and surface protein sequences with the hydrophobic moment plot. *J. Mol. Biol.* 179, 125–142.
34. Engelman, D.M., Steitz, T.A., and Goldman, A. 1986. Identifying nonpolar transbilayer helices in amino acid sequences of membrane proteins. *Annu. Rev. Biophys. Biophys. Chem.* 15, 321–353.
35. Juretic, D., Lucic, B., and Trinajstic, N. 1993. Predicting membrane protein secondary structure. Preference functions method for finding optimal conformational preferences. *Croat. Chem. Acta* 66, 201–208.
36. Kyte, J. and Doolittle, R.F. 1982. A simple method for displaying the hydropathic character of a protein. *J. Mol. Biol.* 157, 105–132.
37. O'Neil, K.T. and DeGrado, W.F. 1990. A thermodynamic scale for the helix-forming tendencies of the commonly occurring amino acids. *Science* 250, 646–651.
38. Grantham, R. 1974. Amino acid difference formula to help explain protein evolution. *Science* 185, 862–864.
39. Woese, C.R., Dugre, D.H., Dugre, S.A., Kondo, M., and Saxinger, W.C. 1966. On the fundamental nature and evolution of the genetic code. *Cold Spring Harb. Symp. Quant. Biol.* 31, 723–736.
40. Mathusamy, R. and Ponnuswamy, P.K. 1990. Variation of amino properties in protein secondary structures, alpha-helices and beta-strands. *Int. J. Pept. Protein Res.* 35, 378–395.
41. Jones, D.D. 1975. Amino acid properties and side-chain orientation in proteins. *J. Theor. Biol.* 21, 167–183.
42. Chou, P.Y. 1989. Prediction of protein structural classes from amino acid composition. In *Prediction of Protein Structure and the Principles of Protein conformation* Fasman, G.D. (Ed.), pp. 549–586, Plenum, New York.
43. Hopp, T.P. and Woods, K.R. 1981. Prediction of protein antigenic determinants from amino acid sequences. *Proc. Natl. Acad. Sci. U S A* 78, 3824–3828.
44. Fauchere, J.L., Charton, M., Kier, L.B., Verloop, A., and Pliska, V. 1988. Amino acid side chain parameters for correlation studies in biology and pharmacology. *Int. J. Pept. Protein Res.* 32, 269–278.
45. Gasteiger, E., Hoogland, C., Gattiker, A., Duvaud, S., Wilkins, M.R., Appel, R.D., and Bairoch, A. 2005. Protein identification and analysis tools on the ExPASy server. In *The Proteomics Protocols Handbook* Walker, J.M. (Ed.), pp. 571–607, Humana Press, Herts, United Kingdom.
46. Holbrook, S.R., Muskal, S.M., and Kim, S.H. 1990. Predicting surface exposure of amino acids from protein sequence. *Protein Eng.* 3, 659–665.
47. Wang, J. 1997. Feature selection using neural networks with contribution measures. *IEEE Trans. Neural Netw.* 8, 645–662.
48. Leray, P. and Gallinary, P. 1999. Feature selection with neural networks. *Behaviormetrika* 26, 145–166.
49. Heckerling, P.S., Gerber, B.S., Tape, T.G., and Wigton, R.S. 2005. Selection of predictor variables for pneumonia using neural networks and genetic algorithms. *Methods Inf. Med.* 44, 89–97.

50. Heckerling, P.S., Gerber, B.S., Tape, T.G., and Wigton, R.S. 2003. Entering the black box of neural networks. *Methods Inf. Med.* 42, 287–296.
51. Zell, A., Mamier, G., Vogt, M., Mache, N., Hubner, R., Doring, S., Herrmann, R., Soyez, T., Schmalzl, M., Sommer, T., Hatzigeorgiou, A., Posselt, D., Schreiner, T., Kett, B., Clemente, G., Wieland, J., and Gatter, J. 2000. SNNS Stuttgart Neural Network Simulator Manual. SNNS Group.
52. Maier, H. and Dandy, G. 1996. The use of artificial neural networks for the prediction of water quality parameters. *Water Resour. Res.* 32, 1013–1022.
53. Abrahart, R.J., See, L., and Kneale, P. 2001. Applying saliency analysis to neural network rainfall-runoff modelling. *Computers and Geosciences* 27, 921–928.
54. Aterini, S., Fusco, I., and Amato, M. 2004. Pure red-cell aplasia in a peritoneal dialysis patient with HCV-related cryoglobulinemia in the absence of neutralizing antierythropoietin antibodies. *J. Nephrol.* 17, 744–746.
55. Davis, S.J., Ikemizu, S., Evans, E.J., Fugger, L., Bakker, T.R., and van der Merwe, P.A. 2003. The nature of molecular recognition by T cells. *Nat. Immunol.* 4, 217–224.
56. van der Merwe, P.A. and Davis, S.J. 2003. Molecular interactions mediating T cell antigen recognition. *Annu. Rev. Immunol.* 21, 659–684.
57. Goldsby, R.A., Kindt, T.J., and Osborne, B.A. 2000. Major histocompatibility complex. In *Immunology* Kuby, J. (Ed.), pp. 173–197, W.H. Freeman and Company, New York.
58. Robinson, J., Malik, A., Parham, P., Bodmer, J.G., and Marsh, S.G. 2000. IMGT/HLA database—a sequence database for the human major histocompatibility complex. *Tissue Antigens* 55, 280–287.
59. Robinson, J., Waller, M.J., Parham, P., de Groot, N., Bontrop, R., Kennedy, L.J., Stoehr, P., and Marsh, S.G. 2003. IMGT/HLA and IMGT/MHC: Sequence databases for the study of the major histocompatibility complex. *Nucleic Acids Res.* 31, 311–314.
60. Bankovich, A.J., Girvin, A.T., Moesta, A.K., and Garcia, K.C. 2004. Peptide register shifting within the MHC groove: Theory becomes reality. *Mol. Immunol.* 40, 1033–1039.
61. Brusic, V., Bajic, V.B., and Petrovsky, N. 2004. Computational methods for prediction of T-cell epitopes—a framework for modelling, testing, and applications. *Methods* 34, 436–443.
62. Buus, S. 1999. Description and prediction of peptide-MHC binding: The human MHC project. *Curr. Opin. Immunol.* 11, 209–213.
63. Honeyman, M.C., Brusic, V., Stone, N.L., and Harrison, L.C. 1998. Neural network-based prediction of candidate T-cell epitopes. *Nat. Biotechnol.* 16, 966–969.
64. Berzofsky, J.A., Ahlers, J.D., and Belyakov, I.M. 2001. Strategies for designing and optimizing new generation vaccines. *Nat. Rev. Imminol.* 1, 209–219.
65. Brusic, V., Rudy, G., and Harrison, L.C. 1998. MHCPEP, a database of MHC-binding peptides: Update 1997. *Nucleic Acids Res.* 26, 368–371.
66. McSparron, H., Blythe, M.J., Zygouri, C., Doytchinova, I.A., and Flower, D.R. 2003. JenPep: A novel computational information resource for immunobiology and vaccinology. *J. Chem. Inf. Comput. Sci.* 43, 1276–1287.
67. Robinson, J., Waller, M.J., Parham, P., Bodmer, J.G., and Marsh, S.G. 2001. IMGT/HLA—a sequence database for the human major histocompatibility complex. *Nucleic Acids Res.* 29, 210–213.
68. Rammensee, H., Bachmann, J., Emmerich, N.P., Bachor, O.A., and Stevanovic, S. 1999. SYFPEITHI: Database for MHC ligands and peptide motifs. *Immunogenetics* 50, 213–219.
69. Bhasin, M., Singh, H., and Raghava, G.P. 2003. MHCBN: A comprehensive database of MHC binding and non-binding peptides. *Bioinformatics* 19, 665–666.
70. Bhasin, M. and Raghava, G.P. 2004. Prediction of CTL epitopes using QM, SVM and ANN techniques. *Vaccine* 22, 3195–3204.
71. DeLisi, C. and Berzofsky, J.A. 1985. T-cell antigenic sites tend to be amphipathic structures. *Proc. Natl. Acad. Sci. U S A* 82, 7048–7052.
72. Pincus, M.R., Gerewitz, F., Schwartz, R.H., and Scheraga, H.A. 1983. Correlation between the conformation of cytochrome c peptides and their stimulatory activity in a T-lymphocyte proliferation assay. *Proc. Natl. Acad. Sci. U S A* 80, 3297–3300.
73. Spouge, J.L., Guy, H.R., Cornette, J.L., Margalit, H., Cease, K., Berzofsky, J.A., and DeLisi, C. 1987. Strong conformational propensities enhance T cell antigenicity. *J. Immunol.* 138, 204–212.
74. Reyes, V.E., Fowlie, E.J., Lu, S., Chin, L.T., Humphreys, R.E., and Lew, R.A. 1990. Comparison of three related methods to select T cell-presented sequences of protein antigens. *Mol. Immunol.* 27, 1021–1027.

75. Margalit, H., Spouge, J.L., Cornette, J.L., Cease, K.B., DeLisi, C., and Berzofsky, J.A. 1987. Prediction of immunodominant helper T cell antigenic sites from the primary sequence. *J. Immunol.* 138, 2213–2229.
76. Stille, C.J., Thomas, L.J., Reyes, V.E., and Humphreys, R.E. 1987. Hydrophobic strip-of-helix algorithm for selection of T cell-presented peptides. *Mol. Immunol.* 24, 1021–1027.
77. Rothbard, J.B. and Taylor, W.R. 1988. A sequence pattern common to T cell epitopes. *EMBO J.* 7, 93–100.
78. Mouritsen, S., Meldal, M., Ruud-Hansen, J., and Werdelin, O. 1991. T-helper-cell determinants in protein antigens are preferentially located in cysteine-rich antigen segments resistant to proteolytic cleavage by cathepsin B, L, and D. *Scand. J. Immunol.* 34, 421–431.
79. Meister, G.E., Roberts, C.G., Berzofsky, J.A., and De Groot, A.S. 1995. Two novel T cell epitope prediction algorithms based on MHC-binding motifs; comparison of predicted and published epitopes from *Mycobacterium tuberculosis* and HIV protein sequences. *Vaccine* 13, 581–591.
80. Deavin, A.J., Auton, T.R., and Greaney, P.J. 1996. Statistical comparison of established T-cell epitope predictors against a large database of human and murine antigens. *Mol. Immunol.* 33, 145–155.
81. Falk, K., Rotzschke, O., Stevanovic, S., Jung, G., and Rammensee, H.G. 1991. Allele-specific motifs revealed by sequencing of self-peptides eluted from MHC molecules. *Nature* 351, 290–296.
82. Guillet, J.G., Hoebeke, J., Lengagne, R., Tate, K., Borras-Herrera, F., Strosberg, A.D., and Borras-Cuesta, F. 1991. Haplotype specific homology scanning algorithm to predict T-cell epitopes from protein sequences. *J. Mol. Recognit.* 4, 17–25.
83. Nijman, H.W., Houbiers, J.G., Vierboom, M.P., van der Burg, S.H., Drijfhout, J.W., D'Amaro, J., Kenemans, P., Melief, C.J., and Kast, W.M. 1993. Identification of peptide sequences that potentially trigger HLA-A2.1-restricted cytotoxic T lymphocytes. *Eur. J. Immunol.* 23, 1215–1219.
84. Sette, A., Buus, S., Appella, E., Smith, J.A., Chesnut, R., Miles, C., Colon, S.M., and Grey, H.M. 1989. Prediction of major histocompatibility complex binding regions of protein antigens by sequence pattern analysis. *Proc. Natl. Acad. Sci. U S A* 86, 3296–3300.
85. Hattotuwagama, C.K., Doytchinova, I.A., and Flower, D.R. 2005. In silico prediction of peptide binding affinity to class I mouse major histocompatibility complexes: A comparative molecular similarity index analysis (CoMSIA) study. *J. Chem. Inf. Model.* 45, 1415–1423.
86. Bordner, A.J. and Abagyan, R. 2006. Ab initio prediction of peptide-MHC binding geometry for diverse class I MHC allotypes. *Proteins* 63, 512–526.
87. Gulukota, K., Sidney, J., Sette, A., and DeLisi, C. 1997. Two complementary methods for predicting peptides binding major histocompatibility complex molecules. *J. Mol. Biol.* 267, 1258–1267.
88. Kast, W.M., Brandt, R.M., Sidney, J., Drijfhout, J.W., Kubo, R.T., Grey, H.M., Melief, C.J., and Sette, A. 1994. Role of HLA-A motifs in identification of potential CTL epitopes in human papillomavirus type 16 E6 and E7 proteins. *J. Immunol.* 152, 3904–3912.
89. Rammensee, H.G., Friede, T., and Stevanoviic, S. 1995. MHC ligands and peptide motifs: First listing. *Immunogenetics* 41, 178–228.
90. Sidney, J., Oseroff, C., del Guercio, M.F., Southwood, S., Krieger, J.I., Ishioka, G.Y., Sakaguchi, K., Appella, E., and Sette, A. 1994. Definition of a DQ3.1-specific binding motif. *J. Immunol.* 152, 4516–4525.
91. Brusic, V., Schonbach, C., Takiguchi, M., Ciesielski, V., and Harrison, L.C. 1997. Application of genetic search in derivation of matrix models of peptide binding to MHC molecules. *Proc. Int. Conf. Intell. Syst. Mol. Biol.* 5, 75–83.
92. Hammer, J., Bono, E., Gallazzi, F., Belunis, C., Nagy, Z., and Sinigaglia, F. 1994. Precise prediction of major histocompatibility complex class II-peptide interaction based on peptide side chain scanning. *J. Exp. Med.* 180, 2353–2358.
93. Kondo, A., Sidney, J., Southwood, S., del Guercio, M.F., Appella, E., Sakamoto, H., Celis, E., Grey, H.M., Chesnut, R.W., Kubo, R.T., and Sette, A. 1995. Prominent roles of secondary anchor residues in peptide binding to HLA-A24 human class I molecules. *J. Immunol.* 155, 4307–4312.
94. Mallios, R.R. 2001. Predicting class II MHC/peptide multi-level binding with an iterative stepwise discriminant analysis meta-algorithm. *Bioinformatics* 17, 942–948.
95. Parker, K.C., Bednarek, M.A., and Coligan, J.E. 1994. Scheme for ranking potential HLA-A2 binding peptides based on independent binding of individual peptide side-chains. *J. Immunol.* 152, 163–175.
96. Rothbard, J.B., Marshall, K., Wilson, K.J., Fugger, L., and Zaller, D. 1994. Prediction of peptide affinity to HLA DRB1*0401. *Int. Arch. Allergy Immunol.* 105, 1–7.

97. Schafer, J.R., Jesdale, B.M., George, J.A., Kouttab, N.M., and De Groot, A.S. 1998. Prediction of well-conserved HIV-1 ligands using a matrix-based algorithm, EpiMatrix. *Vaccine* 16, 1880–1884.
98. Schonbach, C., Kun, Y., and Brusic, V. 2002. Large-scale computational identification of HIV T-cell epitopes. *Immunol. Cell Biol.* 80, 300–306.
99. Udaka, K., Wiesmuller, K.H., Kienle, S., Jung, G., Tamamura, H., Yamagishi, H., Okumura, K., Walden, P., Suto, T., and Kawasaki, T. 2000. An automated prediction of MHC class I-binding peptides based on positional scanning with peptide libraries. *Immunogenetics* 51, 816–828.
100. Brusic, V., Rudy, G., Honeyman, G., Hammer, J., and Harrison, L. 1998. Prediction of MHC class II-binding peptides using an evolutionary algorithm and artificial neural network. *Bioinformatics* 14, 121–130.
101. Milik, M., Sauer, D., Brunmark, A.P., Yuan, L., Vitiello, A., Jackson, M.R., Peterson, P.A., Skolnick, J., and Glass, C.A. 1998. Application of an artificial neural network to predict specific class I MHC binding peptide sequences. *Nat. Biotechnol.* 16, 753–756.
102. Nielsen, M., Lundegaard, C., Worning, P., Lauemoller, S.L., Lamberth, K., Buus, S., Brunak, S., and Lund, O. 2003. Reliable prediction of T-cell epitopes using neural networks with novel sequence representations. *Protein Sci.* 12, 1007–1017.
103. Borras-Cuesta, F., Golvano, J., Garcia-Granero, M., Sarobe, P., Riezu-Boj, J., Huarte, E., and Lasarte, J. 2000. Specific and general HLA-DR binding motifs: Comparison of algorithms. *Hum. Immunol.* 61, 266–278.
104. Adams, H.P. and Koziol, J.A. 1995. Prediction of binding to MHC class I molecules. *J. Immunol. Methods* 185, 181–190.
105. Nielsen, M., Lundegaard, C., Lund, O., and Kesmir, C. 2005. The role of the proteasome in generating cytotoxic T-cell epitopes: Insights obtained from improved predictions of proteasomal cleavage. *Immunogenetics* 57, 33–41.
106. Jackson, M.R., Song, E.S., Yang, Y., and Peterson, P.A. 1992. Empty and peptide-containing conformers of class I major histocompatibility complex molecules expressed in *Drosophila melanogaster* cells. *Proc. Natl. Acad. Sci. U S A* 89, 12117–12121.
107. Matsumura, M., Saito, Y., Jackson, M.R., Song, E.S., and Peterson, P.A. 1992. In vitro peptide binding to soluble empty class I major histocompatibility complex molecules isolated from transfected *Drosophila melanogaster* cells. *J. Biol. Chem.* 267, 23589–23595.
108. Hammer, J., Takacs, B., and Sinigaglia, F. 1992. Identification of a motif for HLA-DR1 binding peptides using M13 display libraries. *J. Exp. Med.* 176, 1007–1013.
109. Jameson, S.C. and Bevan, J. 1992. Dissection of major histocompatibility complex (MHC) and T cell receptor contact residues in a Kb-restricted ovalbumin peptide and assessment of the predictive power of MHC-binding motifs. *Eur. J. Immunol.* 22, 2663–2667.
110. Taylor, W.R. 1986. The classification of amino acid conservation. *J. Theor. Biol.* 119, 205–218.
111. Hammer, J., Nagy, Z.A., and Sinigaglia, F. 1994. Rules governing peptide-class II MHC molecule interactions. *Behring Inst. Mitt.* 124–132.
112. Hammer, J., Bono, E., Givehchi, A., Negri, D.R., and Sinigaglia, F. 1994. Precise prediction of MHC class II-peptide interaction based on peptide side chain scanning. *J. Exp. Med.* 180, 2353–2358.
113. Schonbach, C., Ibe, M., Shiga, H., Takamiya, Y., Miwa, K., Nokihara, K., and Takiguchi, M. 1995. Fine tuning of peptide binding to HLA-B*3501 molecules by nonanchor residues. *J. Immunol.* 154, 5951–5958.
114. Southwood, S., Sidney, J., Kondo, A., del Guercio, M.F., Appella, E., Hoffman, S., Kubo, R.T., Chesnut, R.W., Grey, H.M., and Sette, A. 1998. Several common HLA-DR types share largely overlapping peptide binding repertoires. *J. Immunol.* 160, 3363–3373.
115. Bhasin, M., Singh, H., and Raghava, G.P. 2003. MHCBN: A comprehensive database of MHC binding and non-binding peptides. *Bioinformatics* 19, 665–666.
116. Honeyman, M.C., Brusic, V., and Harrison, L.C. 1997. Strategies for identifying and predicting islet autoantigen T-cell epitopes in insulin-dependent diabetes mellitus. *Ann. Med.* 29, 401–404.
117. Zhang, G.L., Khan, A.M., Srinivasan, K.N., August, J.T., and Brusic, V. 2005. MULTIPRED: A computational system for prediction of promiscuous HLA binding peptides. *Nucleic Acids Res.* 33, W172–W179.
118. Saha, S. and Raghava, G.P. 2006. Prediction of continuous B-cell epitopes in an antigen using recurrent neural network. *Proteins* 65, 40–48.

119. Saha, S. and Raghava, G.P. 2004. BcePred: Prediction of continuous B-cell epitopes in antigenic sequences using physico-chemical properties. In *LNCS*, pp. 197–204, Springer Berlin, Heidelberg, Germany.
120. Kulkarni-Kale, U., Bhosle, S., and Kolaskar, A.S. 2005. CEP: A conformational epitope prediction server. *Nucleic Acids Res.* 33, W168–W171.
121. Schreiber, A., Humbert, M., Benz, A., and Dietrich, U. 2005. 3D-Epitope-Explorer (3DEX): Localization of conformational epitopes within three-dimensional structures of proteins. *J. Comput. Chem.* 26, 879–887.
122. Batori, V., Friis, E.P., Nielsen, H., and Roggen, E.L. 2006. An in silico method using an epitope motif database for predicting the location of antigenic determinants on proteins in a structural context. *J. Mol. Recognit.* 19, 21–29.
123. Greenbaum, J.A., Andersen, P.H., Blythe, M., Bui, H.H., Cachau, R.E., Crowe, J., Davies, M., Kolaskar, A.S., Lund, O., Morrison, S., Mumey, B., Ofran, Y., Pellequer, J.L., Pinilla, C., Ponomarenko, J.V., Raghava, G.P., van Regenmortel, M.H., Roggen, E.L., Sette, A., Schlessinger, A., Sollner, J., Zand, M., and Peters, B. 2007. Towards a consensus on datasets and evaluation metrics for developing B-cell epitope prediction tools. *J. Mol. Recognit.* 20, 75–82.
124. Devillers, J. 1996. *Neural Networks in QSAR and Drug Design*, pp. 1–284. Academic Press, San Diego, California.
125. Lohmann, R., Schneider, G., and Wrede, P. 1996. Structure optimization of an artificial neural filter detecting membrane-spanning amino acid sequences. *Biopolymers* 38, 13–29.
126. Schneider, G., Schuchhardt, J., and Wrede, P. 1994. Artificial neural networks and simulated molecular evolution are potential tools for sequence-oriented protein design. *Comput. Appl. Biosci.* 10, 635–645.
127. Schneider, G., Schuchhardt, J., and Wrede, P. 1995. Development of simple fitness landscapes for peptides by artificial neural filter systems. *Biol. Cybern.* 73, 245–254.
128. Wrede, P., Landt, O., Klages, S., Fatemi, A., Hahn, U., and Schneider, G. 1998. Peptide design aided by neural networks: Biological activity of artificial signal peptidase I cleavage sites. *Biochemistry* 37, 3588–3593.
129. Xiao, Y. and Segal, M.R. 2005. Prediction of genomewide conserved epitope profiles of HIV-1: Classifier choice and peptide representation. *Stat. Appl. Genet. Mol. Biol.* 4, Article25.
130. Zhao, Y., Pinilla, C., Valmori, D., Martin, R., and Simon, R. 2003. Application of support vector machines for T-cell epitopes prediction. *Bioinformatics* 19, 1978–1984.
131. Corbet, S., Nielsen, H.V., Vinner, L., Lauemoller, S., Therrien, D., Tang, S., Kronborg, G., Mathiesen, L., Chaplin, P., Brunak, S., Buus, S., and Fomsgaard, A. 2003. Optimization and immune recognition of multiple novel conserved HLA-A2, human immunodeficiency virus type 1-specific CTL epitopes. *J. Gen. Virol.* 84, 2409–2421.
132. Lauemoller, S.L., Kesmir, C., Corbet, S.L., Fomsgaard, A., Holm, A., Claesson, M.H., Brunak, S., and Buus, S. 2000. Identifying cytotoxic T cell epitopes from genomic and proteomic information: "The human MHC project". *Rev. Immunogenet.* 2, 477–491.
133. Gombar, V.K., Silver, I.S., and Zhao, Z. 2003. Role of ADME characteristics in drug discovery and their in silico evaluation: In silico screening of chemicals for their metabolic stability. *Curr. Top. Med. Chem.* 3, 1205–1225.
134. Schneider, G., Schrodl, W., Wallukat, G., Muller, J., Nissen, E., Ronspeck, W., Wrede, P., and Kunze, R. 1998. Peptide design by artificial neural networks and computer-based evolutionary search. *Proc. Natl. Acad. Sci. U S A* 95, 12179–12184.
135. Harpaz, Y., Gerstein, M., and Chothia, C. 1994. Volume changes on protein folding. *Structure* 2, 641–649.
136. Eigen, M., McCaskill, J., and Schuster, P. 1988. Molecular quasi-species. *J. Phys. Chem.* 92, 6881–6891.
137. Schneider, G. and Fechner, U. 2005. Computer-based de novo design of drug-like molecules. *Nat. Rev. Drug Discov.* 4, 649–663.
138. Flower, D.R., McSparron, H., Blythe, M.J., Zygouri, C., Taylor, D., Guan, P., Wan, S., Coveney, P.V., Walshe, V., Borrow, P., and Doytchinova, I.A. 2003. Computational vaccinology: Quantitative approaches. *Novartis Found Symp.* 254, 102–120.
139. Doytchinova, I.A. and Flower, D.R. 2002. Quantitative approaches to computational vaccinology. *Immunol. Cell Biol.* 80, 270–279.

140. Xiao, Z., Bergeron, H., Grosse, S., Beauchemin, M., Garron, M.L., Shaya, D., Sulea, T., Cygler, C., and Lau, P.C. 2007. Improving the thermostability and activity of a pectate Lyase by single amino acid substitutions using a strategy based on TM guided sequence alignment. *Appl. Environ. Microbiol.* 74, 1183–1189.
141. Schlotawa, L., Steinfeld, R., von, F.K., Dierks, T., and Gartner, J. 2008. Molecular analysis of SUMF1 mutations: Stability and residual activity of mutant formylglycine-generating enzyme determine disease severity in multiple sulfatase deficiency. *Hum. Mutat.* 29, 205.
142. Ye, L., Wu, Z., Eleftheriou, M., and Zhou, R. 2007. Single-mutation-induced stability loss in protein lysozyme. *Biochem. Soc. Trans.* 35, 1551–1557.
143. Guardiani, C., Cecconi, F., and Livi, R. 2007. Stability and kinetic properties of C5-domain of Myosin binding protein C and its mutants. *Biophys. J.* 94, 1403–1411.
144. Lakshmanan, M. and Dhathathreyan, A. 2007. Towards understanding structure-stability and surface properties of laminin peptide YIGSR and mutants. *Biophys. Chem.* 129, 190–197.
145. Pumpens, P. and Grens, E. 2002. Artificial genes for chimeric virus-like particles. In *Artificial DNA: Methods and Applications* Khudyakov, Y. and Fields, H., (Eds.), pp. 249–327, CRC Press LLC, Boca Raton, Florida.
146. Frisch, C., Schreiber, G., Johnson, C.M., and Fersht, A.R. 1997. Thermodynamics of the interaction of barnase and barstar: Changes in free energy versus changes in enthalpy on mutation. *J. Mol. Biol.* 267, 696–706.
147. Kamerzell, T.J. and Middaugh, C.R. 2008. The complex inter-relationships between protein flexibility and stability. *J. Pharm. Sci.* (Epub ahead of print; PMID: 18186490).
148. Teague, S.J. 2003. Implications of protein flexibility for drug discovery. *Nat. Rev. Drug Discov.* 2, 527–541.
149. Bennett, W.S., Jr. and Steitz, T.A. 1978. Glucose-induced conformational change in yeast hexokinase. *Proc. Natl. Acad. Sci. U S A* 75, 4848–4852.
150. Remington, S., Wiegand, G., and Huber, R. 1982. Crystallographic refinement and atomic models of two different forms of citrate synthase at 2.7 and 1.7 A resolution. *J. Mol. Biol.* 158, 111–152.
151. Perutz, M.F. and Brunori, M. 1982. Stereochemistry of cooperative effects in fish an amphibian haemoglobins. *Nature* 299, 421–426.
152. Perutz, M.F. 1989. Mechanisms of cooperativity and allosteric regulation in proteins. *Q. Rev. Biophys.* 22, 139–237.
153. Anderson, B.F., Baker, H.M., Norris, G.E., Rumball, S.V., and Baker, E.N. 1990. Apolactoferrin structure demonstrates ligand-induced conformational change in transferrins. *Nature* 344, 784–787.
154. Spurlino, J.C., Lu, G.Y., and Quiocho, F.A. 1991. The 2.3-A resolution structure of the maltose- or maltodextrin-binding protein, a primary receptor of bacterial active transport and chemotaxis. *J. Biol. Chem.* 266, 5202–5219.
155. Bowman, A.L., Lerner, M.G., and Carlson, H.A. 2007. Protein flexibility and species specificity in structure-based drug discovery: Dihydrofolate reductase as a test system. *J. Am. Chem. Soc.* 129, 3634–3640.
156. Lerner, M.G., Bowman, A.L., and Carlson, H.A. 2007. Incorporating dynamics in *E. coli* dihydrofolate reductase enhances structure-based drug discovery. *J. Chem. Inf. Model.* 47, 2358–2365.
157. Rauh, D., Klebe, G., and Stubbs, M.T. 2004. Understanding protein-ligand interactions: The price of protein flexibility. *J. Mol. Biol.* 335, 1325–1341.
158. Chan, D.C., Fass, D., Berger, J.M., and Kim, P.S. 1997. Core structure of gp41 from the HIV envelope glycoprotein. *Cell* 89, 263–273.
159. Berjanskii, M. and Wishart, D.S. 2006. NMR: Prediction of protein flexibility. *Nat. Protoc.* 1, 683–688.
160. Abbondanzieri, E.A., Greenleaf, W.J., Shaevitz, J.W., Landick, R., and Block, S.M. 2005. Direct observation of base-pair stepping by RNA polymerase. *Nature* 438, 460–465.
161. Lanman, J. and Prevelige, P.E., Jr. 2004. High-sensitivity mass spectrometry for imaging subunit interactions: Hydrogen/deuterium exchange. *Curr. Opin. Struct. Biol.* 14, 181–188.
162. Ahmed, S.A., Fairwell, T., Dunn, S., Kirschner, K., and Miles, E.W. 1986. Identification of three sites of proteolytic cleavage in the hinge region between the two domains of the beta 2 subunit of tryptophan synthase of *Escherichia coli* or *Salmonella typhimurium. Biochemistry* 25, 3118–3124.
163. Kondrashov, D.A., Cui, Q., and Phillips, G.N., Jr. 2006. Optimization and evaluation of a coarse-grained model of protein motion using x-ray crystal data. *Biophys. J.* 91, 2760–2767.

164. Alves, C.N., Marti, S., Castillo, R., Andres, J., Tunon, I., Moliner, V., and Silla, E. 2007. A quantum mechanics/molecular mechanic study of the wild-type and N155S mutant HIV-1 integrase complexed with a diketo acid. *Biophys. J.* 94, 2443–2451.
165. Alves, C.N., Marti, S., Castillo, R., Andres, J., Moliner, V., Tunon, I., and Silla, E. 2007. A quantum mechanics/molecular mechanics study of the protein-ligand interaction for inhibitors of HIV-1 integrase. *Chemistry* 13, 7715–7724.
166. Huang, N., Kalyanaraman, C., Bernacki, K., and Jacobson, M.P. 2006. Molecular mechanics methods for predicting protein-ligand binding. *Phys. Chem. Chem. Phys.* 8, 5166–5177.
167. Kozisek, M., Bray, J., Rezacova, P., Saskova, K., Brynda, J., Pokorna, J., Mammano, F., Rulisek, L., and Konvalinka, J. 2007. Molecular analysis of the HIV-1 resistance development: Enzymatic activities, crystal structures, and thermodynamics of nelfinavir-resistant HIV protease mutants. *J. Mol. Biol.* 374, 1005–1016.
168. Steinbrecher, T., Case, D.A., and Labahn, A. 2006. A multistep approach to structure-based drug design: Studying ligand binding at the human neutrophil elastase. *J. Med. Chem.* 49, 1837–1844.
169. Welsch, C., Domingues, F.S., Susser, S., Antes, I., Hartmann, C., Mayr, G., Schlicker, A., Sarrazin, C., Albrecht, M., Zeuzem, S., and Lengauer, T. 2008. Molecular basis of telapre vir resistance due to V36 and T54 mutations in the NS3–4 A protease of HCV. *Genome Biol.* 9, R16.
170. Heifetz, A., Pal, S., and Smith, G.R. 2007. Protein-protein docking: Progress in CAPRI rounds 6–12 using a combination of methods: The introduction of steered solvated molecular dynamics. *Proteins* 69, 816–822.
171. Chandrasekaran, V., Ambati, J., Ambati, B.K., and Taylor, E.W. 2007. Molecular docking and analysis of interactions between vascular endothelial growth factor (VEGF) and SPARC protein. *J. Mol. Graph. Model.* 26, 775–782.
172. Eyrisch, S. and Helms, V. 2007. Transient pockets on protein surfaces involved in protein-protein interaction. *J. Med. Chem.* 50, 3457–3464.
173. Efremov, R.G., Chugunov, A.O., Pyrkov, T.V., Priestle, J.P., Arseniev, A.S., and Jacoby, E. 2007. Molecular lipophilicity in protein modeling and drug design. *Curr. Med. Chem.* 14, 393–415.
174. Liu, Y., Scolari, M., Im, W., and Woo, H.J. 2006. Protein-protein interactions in actin-myosin binding and structural effects of R405Q mutation: A molecular dynamics study. *Proteins* 64, 156–166.
175. Grunberg, R., Nilges, M., and Leckner, J. 2006. Flexibility and conformational entropy in protein-protein binding. *Structure* 14, 683–693.
176. Karkola, S., Holtje, H.D., and Wahala, K. 2007. A three-dimensional model of CYP19 aromatase for structure-based drug design. *J. Steroid Biochem. Mol. Biol.* 105, 63–70.
177. Steuber, H., Zentgraf, M., Gerlach, C., Sotriffer, C.A., Heine, A., and Klebe, G. 2006. Expect the unexpected or caveat for drug designers: Multiple structure determinations using aldose reductase crystals treated under varying soaking and co-crystallisation conditions. *J. Mol. Biol.* 363, 174–187.
178. Subramanian, J., Sharma, S., and Rao, C. 2007. Modeling and selection of flexible Proteins for structure-based drug design: Backbone and side chain movements in p38 MAPK. *ChemMedChem* 15, 336–344.
179. Andre, S., Pei, Z., Siebert, H.C., Ramstrom, O., and Gabius, H.J. 2006. Glycosyldisulfides from dynamic combinatorial libraries as O-glycoside mimetics for plant and endogenous lectins: Their reactivities in solid-phase and cell assays and conformational analysis by molecular dynamics simulations. *Bioorg. Med. Chem.* 14, 6314–6326.
180. Gerstein, M. and Krebs, W. 1998. A database of macromolecular motions. *Nucleic Acids Res.* 26, 4280–4290.
181. Lee, R.A., Razaz, M., and Hayward, S. 2003. The DynDom database of protein domain motions. *Bioinformatics* 19, 1290–1291.
182. Qi, G., Lee, R., and Hayward, S. 2005. A comprehensive and non-redundant database of protein domain movements. *Bioinformatics* 21, 2832–2838.
183. Harte, W.E., Jr., Swaminathan, S., and Beveridge, D.L. 1992. Molecular dynamics of HIV-1 protease. *Proteins* 13, 175–194.
184. Chen, Z., Li, Y., Schock, H.B., Hall, D., Chen, E., and Kuo, L.C. 1995. Three-dimensional structure of a mutant HIV-1 protease displaying cross-resistance to all protease inhibitors in clinical trials. *J. Biol. Chem.* 270, 21433–21436.
185. Rose, R.B., Craik, C.S., and Stroud, R.M. 1998. Domain flexibility in retroviral proteases: Structural implications for drug resistant mutations. *Biochemistry* 37, 2607–2621.

186. Aruksakunwong, O., Wolschann, P., Hannongbua, S., and Sompornpisut, P. 2006. Molecular dynamic and free energy studies of primary resistance mutations in HIV-1 protease-ritonavir complexes. *J. Chem. Inf. Model.* 46, 2085–2092.
187. Foulkes, J.E., Prabu-Jeyabalan, M., Cooper, D., Henderson, G.J., Harris, J., Swanstrom, R., and Schiffer, C.A. 2006. Role of invariant Thr80 in human immunodeficiency virus type 1 protease structure, function, and viral infectivity. *J. Virol.* 80, 6906–6916.
188. Nagarajan, N. and Yona, G. 2004. Automatic prediction of protein domains from sequence information using a hybrid learning system. *Bioinformatics* 20, 1335–1360.
189. Marsden, R.L., McGuffin, L.J., and Jones, D.T. 2002. Rapid protein domain assignment from amino acid sequence using predicted secondary structure. *Protein Sci.* 11, 2814–2824.
190. Holm, L. and Sander, C. 1994. Parser for protein folding units. *Proteins* 19, 256–268.
191. Swindells, M.B. 1995. A procedure for detecting structural domains in proteins. *Protein Sci.* 4, 103–112.
192. Cheng, J. 2007. DOMAC: An accurate, hybrid protein domain prediction server. *Nucleic Acids Res.* 35, W354–W356.
193. Jones, S., Stewart, M., Michie, A., Swindells, M.B., Orengo, C., and Thornton, J.M. 1998. Domain assignment for protein structures using a consensus approach: Characterization and analysis. *Protein Sci.* 7, 233–242.
194. Hayward, S., Kitao, A., and Berendsen, H.J. 1997. Model-free methods of analyzing domain motions in proteins from simulation: A comparison of normal mode analysis and molecular dynamics simulation of lysozyme. *Proteins* 27, 425–437.
195. Hayward, S. and Berendsen, H.J. 1998. Systematic analysis of domain motions in proteins from conformational change: New results on citrate synthase and T4 lysozyme. *Proteins* 30, 144–154.
196. Flores, S.C. and Gerstein, M.B. 2007. FlexOracle: Predicting flexible hinges by identification of stable domains. *BMC Bioinformatics* 8, 215.
197. Cheng, J., Sweredoski, M., and Baldi, P. 2006. DOMpro: Protein domain prediction using profiles, secondary structure, relative solvent accessibility, and recursive neural networks, *Data Mining and Knowledge Discovery* 13(1), 1–10.
198. Hansson, T., Oostenbrink, C., and van, G.W. 2002. Molecular dynamics simulations. *Curr. Opin. Struct. Biol.* 12, 190–196.
199. Tirion, M.M. 1996. Large amplitude elastic motions in proteins from a single-parameter, atomic analysis. *Phys. Rev. Lett.* 77, 1905–1908.
200. Zhao, Y., Stoffler, D., and Sanner, M. 2006. Hierarchical and multi-resolution representation of protein flexibility. *Bioinformatics*. 22, 2768–2774.
201. Berjanskii, M.V. and Wishart, D.S. 2007. The RCI server: Rapid and accurate calculation of protein flexibility using chemical shifts. *Nucleic Acids Res.* 35, W531–W537.
202. Schlessinger, A. and Rost, B. 2005. Protein flexibility and rigidity predicted from sequence. *Proteins* 61, 115–126.
203. Kosloff, M. and Selinger, Z. 2003. GTPase catalysis by Ras and other G-proteins: Insights from substrate directed superimposition. *J. Mol. Biol.* 331, 1157–1170.
204. Fulop, V. and Jones, D.T. 1999. Beta propellers: Structural rigidity and functional diversity. *Curr. Opin. Struct. Biol.* 9, 715–721.
205. Springer, T.A. 1997. Folding of the N-terminal, ligand-binding region of integrin alpha-subunits into a beta-propeller domain. *Proc. Natl. Acad. Sci. U S A* 94, 65–72.
206. Chi, E.Y., Krishnan, S., Randolph, T.W., and Carpenter, J.F. 2003. Physical stability of proteins in aqueous solution: Mechanism and driving forces in nonnative protein aggregation. *Pharm. Res.* 20, 1325–1336.
207. Karnoup, A.S., Turkelson, V., and Anderson, W.H. 2005. O-linked glycosylation in maize-expressed human IgA1. *Glycobiology* 15, 965–981.
208. Kozlowski, A., Charles, S.A., and Harris, J.M. 2001. Development of pegylated interferons for the treatment of chronic hepatitis C. *BioDrugs* 15, 419–429.
209. Shanafelt, A.B. 2005. Medicinally useful proteins—enhancing the probability of technical success in the clinic. *Expert. Opin. Biol. Ther.* 5, 149–151.
210. Smales, C.M., Pepper, D.S., and James, D.C. 2001. Protein modifications during antiviral heat bioprocessing and subsequent storage. *Biotechnol. Prog.* 17, 974–978.

211. Werner, R.G., Kopp, K., and Schlueter, M. 2007. Glycosylation of therapeutic proteins in different production systems. *Acta Paediatr. Suppl.* 96, 17–22.
212. Daggett, V. and Fersht, A.R. 2003. Is there a unifying mechanism for protein folding? *Trends Biochem. Sci.* 28, 18–25.
213. Gianni, S., Guydosh, N.R., Khan, F., Caldas, T.D., Mayor, U., White, G.W., DeMarco, M.L., Daggett, V., and Fersht, A.R. 2003. Unifying features in protein-folding mechanisms. *Proc. Natl. Acad. Sci. U S A* 100, 13286–13291.
214. Bajaj, K., Madhusudhan, M.S., Adkar, B.V., Chakrabarti, P., Ramakrishnan, C., Sali, A., and Varadarajan, R. 2007. Stereochemical criteria for prediction of the effects of proline mutations on protein stability. *PLoS. Comput. Biol.* 3, e241.
215. Fernandez, M., Caballero, J., Fernandez, L., Abreu, J.I., and Acosta, G. 2008. Classification of conformational stability of protein mutants from 3D pseudo-folding graph representation of protein sequences using support vector machines. *Proteins* 70, 167–175.
216. Gromiha, M.M. 2007. Prediction of protein stability upon point mutations. *Biochem. Soc. Trans.* 35, 1569–1573.
217. Parthiban, V., Gromiha, M.M., Abhinandan, M., and Schomburg, D. 2007. Computational modeling of protein mutant stability: Analysis and optimization of statistical potentials and structural features reveal insights into prediction model development. *BMC. Struct. Biol.* 7, 54.
218. Pey, A.L., Stricher, F., Serrano, L., and Martinez, A. 2007. Predicted effects of missense mutations on native-state stability account for phenotypic outcome in phenylketonuria, a paradigm of misfolding diseases. *Am. J. Hum. Genet.* 81, 1006–1024.
219. Tan, Y.H. and Luo, R. 2008. Protein Stability Prediction: A Poisson-Boltzmann Approach. *J. Phys. Chem. B* 112, 1875–1883.
220. Guerois, R., Nielsen, J.E., and Serrano, L. 2002. Predicting changes in the stability of proteins and protein complexes: A study of more than 1000 mutations. *J. Mol. Biol.* 320, 369–387.
221. Prevost, M., Wodak, S.J., Tidor, B., and Karplus, M. 1991. Contribution of the hydrophobic effect to protein stability: Analysis based on simulations of the Ile-96—-Ala mutation in barnase. *Proc. Natl. Acad. Sci. U S A* 88, 10880–10884.
222. Zhou, H. and Zhou, Y. 2002. Distance-scaled, finite ideal-gas reference state improves structure-derived potentials of mean force for structure selection and stability prediction. *Protein Sci.* 11, 2714–2726.
223. Gromiha, M.M., Uedaira, H., An, J., Selvaraj, S., Prabakaran, P., and Sarai, A. 2002. ProTherm, Thermodynamic Database for Proteins and Mutants: Developments in version 3.0. *Nucleic Acids Res.* 30, 301–302.
224. Capriotti, E., Fariselli, P., and Casadio, R. 2004. A neural-network-based method for predicting protein stability changes upon single point mutations. *Bioinformatics* 20(Suppl 1), i63–i68.
225. Gilis, D. and Rooman, M. 2000. PoPMuSiC, an algorithm for predicting protein mutant stability changes: Application to prion proteins. *Protein Eng.* 13, 849–856.
226. Capriotti, E., Fariselli, P., Calabrese, R., and Casadio, R. 2005. Predicting protein stability changes from sequences using support vector machines. *Bioinformatics* 21(Suppl 2), ii54–ii58.
227. Golbraikh, A. and Tropsha, A. 2002. Predictive QSAR modeling based on diversity sampling of experimental datasets for the training and test set selection. *J. Comput. Aided Mol. Des.* 16, 357–369.
228. Zemla, A., Venclovas, C., Fidelis, K., and Rost, B. 1999. A modified definition of Sov, a segment-based measure for protein secondary structure prediction assessment. *Proteins* 34, 220–223.
229. Alahakoon, D., Halgamuge, S.K., and Srinivasan, B. 2000. Dynamic self-organizing maps with controlled growth for knowledge discovery. *IEEE Trans. Neural Netw.* 2, 601–614.
230. Fritzke, B. 1994. Growing cell structures—a self-organizing network for unsupervised and supervised learning. *Neural Netw.* 7, 1441–1460.
231. Braunheim, B.B., Miles, R.W., Schramm, V.L., and Schwartz, S.D. 1999. Prediction of inhibitor binding free energies by quantum neural networks. Nucleoside analogues binding to trypanosomal nucleoside hydrolase. *Biochemistry* 38, 16076–16083.
232. Gupta, S. and Zia, R.K.P. 2001. Quantum neural networks. *J. Comput. Syst. Sci.* 63, 355–383.
233. Lewenstein, M. and Olko, M. 1992. Storage capacity of "quantum" neural networks. *Phys. Rev. A* 45, 8938–8943.

234. Sahni, V., Sahni, V., Patvardhan, C., and Patvardhan, C. 2006. Iris data classification using quantum neural networks. *AIP Conf. Proc.* 864, 219–227.
235. Gudise, V.G. and Venayagamoorthy, G.K. 2003. *Comparison of Particle Swarm Optimization and Backpropagation as Training Algorithms for Neural Networks*, pp. 110–117, Indianapolis, Indiana, United States.

3 Phylogenetics and Medicinal Protein Engineering

Michael Anthony Purdy

CONTENTS

3.1 INTRODUCTION

With the continuing exponential growth in the number of sequences made available to the biologist, sequence alignment is an essential tool for sequence manipulation. These alignments have diverse applications in sequence assembly, annotation, and database searching as well as structural and functional predictions for proteins and genes, phylogenetic analysis, ancestral sequence reconstruction, and evolutionary analysis. These activities can be used for structure-based protein redesign to engineer proteins with desired novel function, to recreate ancestral proteins, or to create proteins with cross-reactive antigenic properties, which can span a group of proteins with diverse antigenic properties. This chapter examines some of the basic concepts involved in the manipulation of nucleotide sequences and applications of phylogenetics for protein engineering.

3.2 SEQUENCE SELECTION

The researcher usually has a sequence database in hand. However, it may be necessary to add additional sequences to the database. If additional sequences are needed, how does the researcher acquire these sequences? There are no standard guidelines for which sequences should be selected or as to how many sequences to use to obtain useful information [1]. Public databases would seem to be the perfect answer. Public sequence databases contain over 125 billion base pairs from over 200,000 organisms [2]. The main repository for sequence data, the International Nucleotide

Disclaimer: The findings and conclusions in this chapter are those of the author and do not necessarily represent the views of the Centers for Disease Control and Prevention and the Agency for Toxic Substances and Disease Registry.

TABLE 3.1
List of Useful Web Sites for Data Acquisition

Type	Name	URL
General databases	DDBJ	http://www.ddbj.nig.ac.jp/
(INSD)	EMBL	http://www.ebi.ac.uk/Databases/
	GenBank	http://www.ncbi.nlm.nih.gov/
Genomes	Ensembl	http://www.ensembl.org/index.html
	JGI	http://img.jgi.doe.gov/cgi-bin/pub/main.cgi
	NCBI	http://www.ncbi.nlm.nih.gov/Genomes/index.html
	Sanger	http://www.sanger.ac.uk/Projects/Microbes/
	TIGR	http://www.tigr.org/tdb/mdb/mdbcomplete.html
Search engines	BLAST	http://www.ncbi.nlm.nih.gov/blast/
	Entrez	http://www.ncbi.nlm.nih.gov/Entrez/
	SRS	http://srs.ebi.ac.uk/

Sequence Databases (INSD), is a redundant database stored at DDBJ (Japan), EMBL (EU), and GenBank (United States). Each of these databases is updated nightly and compared against each other to ensure their integrity. Genomic data can be found on the Web sites of a number of genome projects (Table 3.1). It is also worthwhile to conduct Internet searches for curated databases. Many organism-specific databases exist on the Internet (e.g., hepatitis C virus, http://hcv.lanl.gov/ and HIV, http://hiv-web.lanl.gov).

There are two basic types of search strategy, keyword and similarity, for finding related sequences. A keyword search uses keywords to search for sequences by searching through the written description (annotation) attached to the sequence. The two main keyword search engines are Entrez and SRS. Entrez is used to search the information at NCBI, and SRS searches other indexed databases. Similarity searches use the sequences themselves to find similar sequences. The main search engine for the similarity search is BLAST [3] (Table 3.1). It is generally better to work with nucleotide rather than amino acid sequences as they usually contain greater potential phylogenetic signals [4].

The next issue is the integrity of the downloaded sequences. There are several potential problems with sequences from databases. Not every researcher uses the same care in annotating their sequences. While most annotations are quite thorough, some are incomplete or incorrect. Because of the rapid growth of these databases it is not always possible for the curators to check all the annotations for newly submitted sequences. Several researchers have noted errors in annotation [5,6]. There are also problems with errors in the sequences themselves [7]. Many times, especially with unpublished sequences, there is no way for the researcher to verify the annotation by going to a journal article. However, it should be noted that a careful search of publication databases will usually discover a publication for an "unpublished" sequence. The researcher needs to check each annotation and sequence to ensure that the sequence is correctly labeled. In some cases, the archived sequence is not from the gene described in the annotation. In other cases, the sequence is an engineered construct. There are instances where the amino acid translation accompanying a nucleotide sequence is incorrect. Some sequences contain ambiguous bases. It is incumbent on the researcher to determine whether these sequences are useful in his/her situation. The anomaly could be in a region that the researcher does not plan to use. The anomaly could be a quasispecies variant or mutation, and provide critical information to the research project.

It is preferable to have a dataset with longer rather than shorter sequences [8], and many sequences from the taxon under investigation rather than fewer [4]. However, it is possible to create a dataset with an overabundance of very closely related sequences. A large dataset may also be intractable for the computer resources on hand. One possible solution is to set a similarity cutoff, where sequences above the cutoff are assumed to be identically aligned and the group is represented in the database by an exemplar [9]. Sequence alignments and phylogenetic trees are useful tools to ensure the sequence database is focused and does not contain outlier sequences. It should also be remembered that each dataset will create its own unique context, which will affect the final outcome of any set of manipulations applied to those data. Any change in the dataset will likely change the final result.

The INDS has a uniform policy of free unrestricted access to the data records in its databases. It is expected that researchers will give due credit to the original submitters [10] but the struggle for scientific credit can be acrimonious. Scientists at large genome centers claim bioinformaticists download the data and publish papers before the genome centers can analyze the data, and it has been suggested that large-scale analyses be cleared with the submitters before publication [11].

3.3 SEQUENCE ALIGNMENT

Sequence alignment is integral to medicinal protein engineering. It may be an end in itself where the alignment alone contains enough information for the biologist to analyze the sequences in question, or it may be the base upon which more complex analyses are conducted. In this chapter, we are interested in phylogenetic analysis and ancestral reconstruction. Other methodologies are described in additional chapters in this book.

The alignment of nucleic or amino acid sequences is sometimes seen as a trivial process [2]. The selection of the right alignment algorithm is expected to yield an acceptable result. In reality, sequence alignment may be one of the hardest things a biologist can attempt. While each of the sequences to be aligned results from specific evolutionary events leading through time to their most recent common ancestor, the history of those processes for any individual sequence is unknowable. We can only apply stochastic algorithms to the sequences to create an alignment. Thus, multiple-sequence alignment is the use of stochastic processes and biological knowledge in an imperfect attempt to reconstruct unknowable evolutionary events [12].

Alignment homology is divided into two states, primary and secondary homology [13,14]. The primary homology of a group of sequences is based on similarity between sequence characters, e.g., are the characters in an alignment analogous? This is the level at which most alignment programs function, whereby an algorithm examines the similarity between text characters in a set of text strings to create the best alignment of these characters. These algorithms use general rules created by biologists indicating how sequence characters are related to each other. Secondary homology is the detection of character states. This is the evolutionary concept of homology and refers to the relationship of shared features among taxa to their most common ancestor. While primary homology looks for similarities between sequences, secondary homology looks for synapomorphies among taxa [13]. After a similarity alignment has been created, the biologist must bring his knowledge of the specific sequences in question and postulate biologically plausible processes that must have occurred to generate the alignment. The biologist examines not the text but an alignment of nucleotides or amino acids and critically adjusts this alignment in light of the biological processes, which he or she posits have occurred [15]. The biologist's alignment is based on the secondary homology between nucleotides or amino acids not the similarity between text characters. Secondary homology will always be the key as primary homology can result from common ancestry, convergence, parallelism, or reversal [12,16].

The choice of an alignment program can be difficult because the problem of sequence alignment has not been solved. Part of the reason for this are the assumptions upon which sequence alignment is based. It is assumed that each position in a sequence is independent of every other position, and mutation at any position in a sequence is independent of mutation at every other position. These assumptions are incorrect [17]. Sequence alignments have diverse applications and attempts continue to create algorithms that align sequences through homology rather than similarity. An examination of the literature will show that the number of alignment programs continues to grow [3]. This is due in part because of the complexity of sequence alignment, which is a nondeterministic polynomial-time (NP) hard problem. Because alignment is an NP-hard problem, computational demands limit our ability to generate an optimal alignment and force us to rely on heuristics to complete the alignment in a reasonable time [2]. Although CLUSTALW [18], which has recently been rewritten in C++, is still one of the most popular alignment programs; more computationally complex programs like DIALIGN [19,20], MAFFT [21,22], MUSCLE [23], PROBCONS [24], and T-COFFEE [25] are now becoming widely used. Some factors affecting the choice of program include computational time; the number of sequences to be aligned; the length of the sequences; the degree of sequence divergence; whether nucleic or amino acids are being aligned; whether the alignment will be local or global; and whether there is gene duplication, deletion, or rearrangement among the sequences. Batzoglou [3] and Edgar and Batzoglou [26] discuss some of these issues and suggest scenarios under which the programs listed above may be useful (see also Table 4 in Ref. [12]).

Because alignments attempt to recreate the evolutionary path from the most common ancestor to each sequence in the alignment many algorithms use an evolutionary rate model to approximate how sequence change over time. There is a trade-off between bias and variance in a rate model. As the number of rate parameters used in a model increases to improve the fit to the data, there is a decrease in the inherent bias due to model selection; however, there is an increase in the variance associated with those parameters [27]. The sequence alignment must be good but not exact [28]. The best model is one that minimizes both bias and variance. Programs like HyPhy [29], MODELFIND (http://hcv.lanl.gov/content/ hcv-db/findmodel/findmodel.html), and MODELTEST [30–33] can test various rate models to help a researcher determine which model is the best for his/her data (Table 3.2).

It has been suggested that computer alignments be considered as first approximations for a final sequence alignment [15,34,35]. Developers of alignment programs have found DNA, especially noncoding regions, to be difficult to align [36]. At present the best quality control in multiple-sequence alignment relies on biological insight. It is important to examine the alignment in light of one's knowledge of the evolutionary processes that most likely acted on the aligned sequences to ensure the alignment is plausible [12]. To do this, the scientist must be able to describe and justify the criteria used to make alignment decisions [37].

The initial alignment of sequences in a dataset should be done using the nucleotide sequences but the refinement of the sequence based on homology should be done by translating the nucleic acids into amino acids. *In vivo* DNA is translated into proteins, which have specific structures and functions. These structures and functions are more conserved through evolutionary history than are the primary nucleic acids that code for these attributes. While nucleic acid sequences carry more potential phylogenetic information than proteins, the proteins are a better record of how structure/function has changed over time. An analysis of structure/function attributes can be used to assess the homology of the nucleotides in the alignment by constraining the alignment based on these

TABLE 3.2
List of Web Sites with Information about Model Testing

Program	Information
FINDMODEL (nucleotide)	http://hcv.lanl.gov/content/hcv-db/findmodel/findmodel.html http://hcv.lanl.gov/content/hcv-db/findmodel/doc.pdf
HyPhy (nucleotide)	http://www.hyphy.org/ http://www.datamonkey.org
MODELTEST (nucleotide)	http://darwin.uvigo.es/software/modeltest.html http://www.rhizobia.co.nz/phylogenetics/modeltest.html
ProtTest (amino acid)	http://darwin.uvigo.es/software/prottest.html

attributes [38]. Within coding regions, reading frames should be maintained, and start and stop codons can be used to align sequences. Catalytic sites, binding sites, and protein motifs can be used to align codons within the coding region [39]. Structural features such as α-helices, β-sheets, turns, and disulfide bridges can be used to align regions of sequence [40]. After an amino acid alignment is conducted, it is necessary to reexamine the nucleotide alignment to ensure the homologies are plausible and parsimonious, as it is possible they may still represent homoplasies.

So far this discussion has assumed that evolutionary changes are single point mutations in nucleic acid sequences. Sequences which are derived from a common ancestor can be aligned to each other through these homologies and a plausible history of the occurrence of these events as a string of point mutations. There are changes which are not homologous but can lead to confusion during alignment. Recombination transfers a block of bases from one organism to another. If this transfer is not from a common ancestor, the nucleotides are not homologous. If the transfer is from a common ancestor, the nucleotides are homologous because they come from the common ancestor; however, as multiple differences result from a single event, they are not homologous. Insertions and deletions (indels) are another problem [41]. Inversions are the replacement of a subsequence with a reversed sequence; translocations are the excision of a subsequence and its insertion to a new location; transpositions result from copying a subsequence to a new location; and tandem duplications result from copying a subsequence to a location immediately adjacent to itself. Each of these processes moves multiple homologous nucleotides to new locations in a single event. It is this verisimilitude of homology that can cause problems. As these events do not result from homologous events, as strictly defined, these regions should not be included in phylogenetic analysis or ancestral reconstruction. Deletions modeled as gaps in sequence alignments cause other problems. While most homology events are modeled as single character events, gaps, like insertions, usually affect more than one nucleotide position in a sequence. Gap costs are usually calculated as a gap opening cost plus a length-dependent gap extension cost [42], or as a fifth character state in a nucleotide sequence [43,44]. Many researchers ignore potential phylogenetic information in gaps by removing gaps from their alignments; however, research has shown gaps can be a source of phylogenetic information [45,46]. Because gaps in a sequence alignment are rarely of equal length, this can lead to a situation where nonhomologous residues are aligned within a gap when they should remain unaligned [47,48]. Because gaps may be caused by single or multiple events, manual realignment may be one of the better ways to properly align gaps [12].

3.4 PHYLOGENETIC ANALYSIS

Although some biologists look on phylogenetic analysis as a trivial matter of selecting a computer program that creates phylogenetic trees and choosing the default values, others recognize the complexity of this stage [49]. Phylogenetics is such a complex field that the researcher may, like Pontius Pilate, find himself asking, "What is Truth?", and obtain a philosophical response [50–52]. On a less philosophical plane, a range of information can be found on phylogenetics from basic tutorials [53,54] to in-depth analysis [55]. For those looking for more practical information, instructional approaches to phylogenetic software and how to create phylogenetic trees are also available [56,57].

There are two general categories of methods for creating phylogenetic trees. The first is the distance methods and the second is the character-based methods. Distance methods convert a sequence alignment into a distance matrix of pairwise distances between the sequences. Distances are calculated based on the differences found between pairs of sequences. Bifurcating trees are created from these distance data, which are used to determine the branching order and branch lengths for each tree. Distance methods tend to be faster computationally than character-based methods. Distance methods include neighbor joining, least squares methods, minimum evolution, and UPGMA (unweighted pair-group method with arithmetic mean) [58]. There are three primary character-based methods: parsimony, maximum likelihood, and Bayesian analysis. These methods compare the characters within each column (position) in the multiple-sequence alignment. Parsimony seeks for the tree or trees created by the minimum number of character changes among the sequence data set. Maximum likelihood uses a model of evolution to find the tree that maximizes the likelihood of the observed data. Bayesian analysis looks for the trees with the greatest likelihoods given the data. Character-based methods are much more information rich than distance methods. There is a hypothesis for each column in the alignment so that it may be possible to trace the evolution of specific sites [53].

Problems with phylogenetic trees include the subjective nature of the data and the fact that stochastic methods are used to estimate historical events. Subjectivity enters the process through the researcher's choice of which sequences to include, which genes to examine, the length of the sequences, the presence of gap inclusion or exclusion, how to treat recombination, the method used to determine the degree of homology, and whether the tree building method is being applied [1,59–63]. The stochastic element is brought into the process through the methods used to determine the degree of homology and for tree construction [64]. A modification in any of the factors will usually lead to the creation of a new set of trees. The possible number of trees created is based on the number of sequences (taxa) used. A set of m taxa leads to the creation of $\frac{(2m-5)!}{2^{m-3}(m-3)!}$ unrooted and $\frac{(2m-3)!}{2^{m-2}(m-2)!}$ rooted trees. It should be obvious that once the number of taxa becomes greater than three the "true" tree will be hidden in this set of possible trees. Eight taxa will generate about 10,000 unrooted trees and 15 taxa will generate about 8×10^{12} unrooted trees. As the number of taxa increases, the search for the true tree becomes increasingly more difficult and because of the subjectivity of the process, a given set of trees may not contain a tree which reflects the historical events leading to the formation of the taxa in the analysis. Because of this, a number of methods have been developed to determine the statistical significance of a tree. The three basic methods for determining the accuracy of a tree are bootstrapping, likelihood estimation, and Bayesian probability [65–67]. The use of these methods has been evaluated [50,67–69] and even though the results are inconclusive, the methods remain mainstays for determining the accuracy of trees and the clades within those trees, and they continue to be evaluated for effectiveness [4,70–74].

There is no single correct phylogenetic method. Sources are available which describe phylogenetic theory [55,58] and methodology [56,57]. The ultimate use of a phylogenetic analysis, the number of sequences, the length of the sequences, time constraints, and computational resources, all have bearing on which method or methods should be used. Among the more popular programs are EMBOSS [75], GCG [76,77], HYPHY [29], MEGA [78,79], PAUP* [80], PHYLIP [81], and PHYML [82].

3.5 MEDICINAL PHYLOGENETICS

Phylogenetic analysis has broad applications in the area of medicinal biology and protein engineering including Darwinian selection, molecular clock calculations, polyvalent vaccine construction, ancestral protein reconstruction, palaeoenvironment inference, adaptive replacements, functional module identification, engineering of functional changes, molecular epidemiology, biopharmaceuticals, and nanotechnology.

The examination of synonymous and nonsynonymous mutations can be useful in determining whether codons have undergone positive Darwinian or purifying selection. As synonymous substitutions are apparently free from natural selection, the synonymous substitution rate is assumed to equal the neutral nucleotide substitution rate. Nonsynonymous substitutions, on the other hand, can experience selective pressures. If the nucleotide substitution rate is neutral across synonymous and nonsynonymous codon positions, nucleotide substitution rates across both types of codons should be equal. If the nonsynonymous substitution rate is higher than the neutral substitution rate, the nonsynonymous substitutions may be inferred to be caused by positive Darwinian selection. However, if the nonsynonymous substitution rate is lower than the neutral substitution rate, the nonsynonymous substitutions may be thought to be due to negative or purifying selection [58].

HYPHY [29], MEGA [78,79], and PAML [83,84] can be used to calculate these rates and estimate their statistical reliability. If selection plays no role in evolution then mutations will accumulate according to a stochastic model of genetic drift. If, however, there is strong selective pressure, the accumulation of polymorphisms becomes more predictable [85,86]. These polymorphisms become important as disease phenotypes have been mapped to these variations [87–89]. A longitudinal study with partial HIV envelope sequences from 28 infected children used Bayesian analysis with Markov-chain Monte Carlo methods to show that although genetic drift affects HIV evolution, Darwinian selection has a strong influence on HIV envelope evolution [85]. An examination for selection can be used to identify proteins, which may be useful targets for protein engineering, as with the identification of 28 of 76 genes associated with schizophrenia, which have undergone positive selection [90]. It can also be used to identify functionally important protein regions [91], such as candidate epitopes for vaccine development [92], or to search for polymorphisms that can lead to the identification of a single mutation associated with a phenotype [93]. Such calculations have been used in molecular epidemiological studies to examine viral virulence [94] and chronicity [85,92].

Although no protein or gene evolves at a constant rate over long evolutionary times, the molecular clock hypothesis assumes that the rate of amino acid or nucleotide substitution is approximately constant over time and is usually most suitable for closely related species [58,95]. In spite of the long history of controversy over this hypothesis, the molecular clock has found use in medicinal biology, particularly in the area of molecular epidemiology [96,97]. Molecular clock calculations can be used to estimate the time of appearance of a most common ancestor [96] and this information may be useful in the construction of ancestral proteins and genes. MEGA [78,79] and PAML [83,84] can be used to calculate times and rates based on the molecular clock.

3.6 ANCESTRAL RECONSTRUCTION

Recreating ancestral proteins allows the study of the evolutionary history of protein function. This reconstruction can provide information not easily obtained from the study of extant proteins. Besides providing information about the evolution of extant proteins, such studies may reveal unexpected insights into biochemical functions lost by present-day proteins and the palaeoenvironments of extinct organisms. The reconstruction of an ancestral viral protein can yield insights into lost functionality and epitopes, or the creation of epitopes, which have broader reactivity among extant quasispecies than any single extant viral protein chosen to serve as a vaccine candidate. The proliferation of viral quasispecies is an evolutionary process albeit on a much shorter timescale than normally used for ancestral reconstruction.

Traditional methods for studying protein structure and function rely on mutagenesis to identify protein residues and regions, which may be important. The choice of mutagenesis targets is often directed by knowledge of three-dimensional structures, kinetics, and biochemical studies. The researcher is, nonetheless, left with a large number of targets, and the problem of the proper choice of targets for mutagenesis becomes a problem in combinatorics and biology. Which residues and mutations are important to the process under study? Which mutations will lead to improper protein folding? Because ancestral reconstruction is based on an evolutionary model, the problem of determining which combinations of mutations is historical and functional is treated by the evolutionary model. Such a model should select those mutations with the highest probability of being authentic (within the limits of the model), and the ancestral construct should be devoid of mutations resulting in misfolded proteins or lost functionality.

Ancestral reconstruction uses parsimony or stochastic processes to reconstruct a history of molecular changes from a progenal sequence back to an ancestor sequence, and has been used to test ecological and evolutionary hypotheses [98,99]. In phylogenetic analysis, while the researcher seeks to minimize both the bias and variance of the model he/she uses, the same is not true for the inference of ancestral states. Sequence alignment and minimal bias are more important for correct reconstruction of ancestral sequences than having an accurate phylogenetic tree topology [67]. The better the accuracy of the sequence alignment the more accurate the reconstruction is likely to be [100,101]. The accuracy of inferred ancestral sequences is not changed much by adding or removing a few sequences [67]. As with other phylogenetic procedures, ancestral reconstruction is not a cookbook process. No method presently available yields an error-free inference of ancestral states. The most predominant methods for ancestral reconstruction are parsimony, maximum likelihood, and Bayesian inference [98,100].

As with other stages of phylogenetics, there is much debate and research on the best methods to use [67,99–103]. Each procedure has advantages and disadvantages. Maximum likelihood and maximum parsimony tend to overestimate thermostability and tend to eliminate variation at positions that would be detrimental to structural stability [101]. The process of ancestral reconstruction will sometimes introduce gaps in the inferred sequences. Parsimony methods can be used to correct these errors and improve the accuracy of the reconstructed sequences [41,100]. Because likelihood allows ambiguity in the parameters of a phylogenetic model such ambiguity can lead to uncertainty in inferences of ancestral states. The empiric Bayesian analysis using the Markov-chain Monte Carlo is less affected by ambiguous parameters [104]. Some of the programs for ancestral reconstruction include ANCESCON [105], MacCLADE [106,107], MEGA [78,79], MRBAYES [108,109] PAML [83,84], PAUP* [80], and PHYLIP [81]. Bayesian analysis appears to be the most popular method at the present and, as implemented in MRBAYES, appears to give good results [100]. Finally, when reconstructing an ancestral gene, it is always preferable to optimize codon usage according to the particular species or cell type in which the gene will be expressed or studied [110].

3.7 ANCESTRAL RECONSTRUCTION AND PROTEIN ENGINEERING

The use of ancestral reconstruction and evolutionary analysis can also help to determine the historical function of proteins no longer obtainable so as to gain insights into the biology, kinetics, function, and environment of extinct organisms and proteins [111–115], or analyze much more modern problems such as the dynamics of viral quasispecies [116]. The reconstruction of ancestral glucagon-like peptide-1 sequences has been used to analyze their therapeutic potential for type-2 diabetes [117]. Ancestral reconstruction has also been used to examine the disease process. This has been done by examination of mutant variants [93,118] and molecular epidemiological analysis of human [119,120] and pathogen [121,122] populations.

3.7.1 Examples of Ancestral Protein Engineering

Long interspersed (repetitive) element 1 (L1) in mammals is the prototype for retroposons that encoding reverse transcriptase, lack long terminal repeats, and appear to transpose via a polyadenylated RNA intermediate. In mice, two abundant L1 subfamilies have their promoters within tandem arrays of two alternative nonhomologous monomer sequences named A and F. Phylogenetic analysis has shown that recently transposed L1s are all members of the A-type subfamily, while F-type L1s were active about 6 million years ago. Noting a sequence divergence of 1%–2% among A-type monomers but an average divergence of 24% among F-type elements, Adey et al. [111] proposed that the F-type elements had become inactivated through the accumulation of mutations. Mouse F-type L1 element sequences were obtained from GenBank, aligned, and the sequence for the ancestral F1 monomer was determined using DNAML from the PHYLIP package. Because of two potential SP1-binding sites a mixture of C and T was included at two positions producing four variant sequences in which each of the potential SP1-binding sites was present or absent. The four variant ancestral F1 sequences were resurrected by chemical synthesis. All four variants were found to be active promoters and SP1 binding did not appear to be essential for the resurrected promoter activity, confirming the hypothesis that F-type elements became inactivated through mutation and showing that an extinct function could be resurrected through ancestral reconstruction.

Vertebrate genomes contain six evolutionarily related nuclear receptors for steroid hormones (SRs). These ligand-activated transcription factors mediate hormonal effects directing sexual differentiation, reproduction, behavior, immunity, and stress response. There are no orthologs in the insect, *Drosophila melanogaster*; the nematode, *Caenorhabditis elegans*; or the urochordate, *Ciona intestinalis*. Thronton et al. [112] used degenerate PCR to identify estrogen-related receptors from the mollusk, *Aplysia californica*. The ligand-binding domain of this receptor was insensitive to estrogen activation. To explain the evolution of function in the steroid hormone receptor gene family, they reconstructed the ancestral steroid receptor using an alignment of 74 receptor genes, including the *Aplysia* estrogen receptor gene (ER) sequence and 18 closely related outgroup gene sequences. The sequences were aligned with ClustalX. The maximum parsimony and Bayesian Markov-chain Monte Carlo (BMCMC) techniques were used to analyze the relationships between the receptor sequences. Parsimony analysis was done with PAUP*, BMCMC analysis was conducted using MrBayes, and likelihood ratios were calculated independently with HYPHY. Both techniques strongly indicated that the *Aplysia* ER gene is an ortholog of the vertebrate SRs with a BMCMC posterior probability of 100%. The maximum likelihood for this phylogeny was >100,000-fold greater than the best tree in which the *Aplysia* ER was outside the vertebrate SR clade. The sequence of the putative ancestral ER was inferred using PAML. Support for the ancestral ER was not strong in a parsimony context but had 100% posterior probability. The best tree with this node had a likelihood 4.7 million-fold greater than the best tree without this node. In reporter gene assays, the DNA-binding domain of the resurrected ancestral ER activated transcription almost as well as human SRs or the *Aplysia* ER. The ligand-binding domain of the resurrected ancestral ER bound three estrogens. Originally, the steroid receptors were thought to have evolved in chordates about 400–500 million years ago. These results indicate that steroid receptors are extremely ancient, having diversified from a primordial gene before the origin of bilaterally symmetric animals about 600–1000 million years ago. Phylogenetic analysis further indicated that this gene was lost in the lineage leading to arthropods and nematodes, and became independent of hormone regulation in the *Aplysia* lineage.

An ancient organism's physical environment can be inferred using reconstructed sequences of ancient proteins made by the organism and measuring their properties. Models of the Precambrian suggested that the earth was either cold and snow-covered, or was hotter than at the present. To resolve this uncertainty, Gaucher et al. [113] used a bacterial protein. The chosen protein had to have a temperature dependent behavior with an optimum at physiological temperature. Because of the timescale involved, the rate of sequence divergence had to be slow and the number of available derived and sibling protein sequences had to be large, so that the ancestral sequences could be reconstructed with minimal ambiguity. The database of sequences needed to be large so that any ambiguous site could be sampled to see whether the interpretation, however made, was robust with respect to the ambiguity. They chose elongation factor (EF) Tu. Because saturation at silent sites in the DNA sequences had occurred, amino acid sequences were used in this analysis. Two phylogenetic trees were created. In the first, EF sequences from mesophiles, thermophiles, and hyperthermophiles were aligned with ClustalW, and minor adjustments were made manually. A maximum likelihood tree was constructed from a distance matrix generated using MEGA. The distance matrix was used to build nine subtrees using the maximum evolution criterion in PAUP* and the relationships between the subtrees were inferred using a maximum likelihood algorithm implemented within the MOLPHY package [123]. For the second tree, a literature survey of multiple protein, DNA, and rRNA sequences was first used to identify plausible alternative bacterial phylogenies. Where these trees differed from the maximum likelihood tree, the differences were incorporated into an alternative tree. Ancestral amino acid states were computed using an empirical Bayesian statistical framework in PAML. Several ancestral proteins were constructed including a last common mesophile ancestor. These ancestral proteins were synthesized in small steps by PCR using 50 bp oligonucleotides with 15–20 overlapping bases, and cloned into a prokaryotic expression vector. The temperature profiles for the ancestral EFs exhibited optimal binding of GDP at 65°C and the last common mesophile ancestor exhibited an optimum of 55°C. The results showed that the ancient environment was warmer than the present with the mesophilic ancestor having a higher temperature optimum than its descendants. These results also demonstrated that the behavior of an ancestral protein need not be the average of the behaviors of its descendants.

The green fluorescent protein (GFP) was first isolated from the jellyfish *Aequorea victoria*. These proteins are used for coloration or fluorescence, or both. They come in four colors: green, yellow, red, and nonfluorescent purple-blue. GFP-like proteins are the only natural pigments in which both chromophore and protein are contained in a single gene, as a result of which they are used extensively as tracking tools. To study the function of GFP-like proteins and understand the evolutionary history of that function, Chang et al. [114] examined GFP-like proteins from *Montastrea cavernosa*. These proteins are composed of four paralogous groups corresponding to cyan, short-wave green, long-wave green, and red. They wanted to determine the common ancestor for all four color phenotypes (ALL), the common ancestor of all the red (R) proteins, and two intermediate nodes; the red/green ancestor (RG) and the pre-red (preR) ancestor. They used a dataset composed of most of the cnidarian GFP-like proteins then known. Three maximum likelihood models were used to reconstruct the desired ancestral proteins. These were an amino acid, a codon, and a nucleotide model. The reconstructions of the four ancestral proteins were quite robust regardless of the model used and were in general agreement, although the nucleotide model was the least robust of the three. A small number of sites were poorly predicted by some or all the models. The codons corresponding to these ambiguous sites were designated to be degenerate and the designed genes for the ALL, RG, preR, and R ancestors contained eight, six, four, and six degenerate codons, respectively. The ancestral genes were synthesized from short oligonucleotides using overlap PCR and cloned into *E. coli*. Because the phenotype under examination is color or fluorescence, which can easily be quantified in bacterial colonies growing on solid media, there was no need to purify the ancestral proteins. The ALL ancestor (the progenitor of all four color phenotypes) generated a short-wave green phenotype, while the intermediate red/green (RG) and pre-red (preR) ancestors both had a long-wave green/red phenotype. The red (R) ancestor had an imperfect red phenotype, that is, it had a dominant red emission peak with a minor long-wave green emission. This research showed that ancestral reconstruction can indicate how evolution may have given rise to the diversity of function seen in present-day GFP-like proteins.

The Family C subtype G protein-coupled receptors likely evolved early in the metazoan lineage concurrent with the evolution of the nervous system. This family of proteins encompasses chemosensory receptors, taste receptors, pheromone receptors, and neurotransmitters. The ligand-binding domain of Family C receptors is distantly related to prokaryotic periplasmic-binding proteins involved in amino acid and nutrient transport in bacteria. Free amino acids act on Family C receptors as either orthosteric agonists or allosteric modulators of receptor activity. These proteins have selectivity for one or several amino acids. Kuang et al. [115] examined the pharmacological origins of Family C receptors. They aligned a set of sequences from the receptors including the metabotropic glutamate receptors (mGluR), calcium-sensing receptors, and a diverse group of sensory receptors. Sequences from the Family C $GABA_B$ receptor were used as outgroup. Phylogenetic trees were constructed using maximum parsimony, neighbor joining, and minimum evolution. Only one node was found to differ among the different trees. This node, within the pheromone/olfactory receptor group, was collapsed to form a polytomy. Because the research concerned amino acid binding among these receptors, Kuang et al. performed ancestral reconstructions of the residues in the receptor-binding pocket using the maximum likelihood methods implemented in PAUP*. Rather than reconstruct the ancestral gene, they used site-directed mutagenesis to modify the goldfish 5.24 chemosensory receptor, which is not activated by mGluR-selective agonists and displays very low affinity for glutamate. The 5.24 receptor was chosen for mutagenesis because of its sequence dissimilarity to mGluRs, and its broad amino acid selectivity as opposed to the narrow amino acid selectivity of the mGluRs. Three native active site amino acids in the 5.24 binding pocket critical for amino acid bonding were replaced with the three amino acids determined to be necessary for amino acid binding in the ancestral Family C receptor. This mutated 5.24 receptor was referred to as the "ancestral receptor." It was found that this predicted binding pocket was highly sensitive to glutamate and group I mGluR-specific agonists. The results suggest that glutamate activation of Family C receptors may have arisen early in metazoan evolution and that it was preadapted as a glutamate receptor for later use as excitatory synapses in glutamate-mediated neurotransmission.

Glucgon-like peptide-1 (GLP-1) is an incretin hormone with therapeutic potential against type-2 diabetes. It is a regulator of blood glucose in mammals. It may also decrease food uptake and lead to loss of body weight. GLP-1 is cleaved by proglucagon and encodes a number of peptides involved in blood sugar regulation. While GLP-1 is highly conserved among mammals, more diversity is seen among amphibian species. Some GLP-1 sequences from amphibian species have been found to bind and activate the human GLP-1 receptor. Little difference is observed for the *in vitro* potency for the human GLP-1 receptor; however, larger differences are found in the enzymatic stability in these peptides. The traditional method for analyzing variability among residues in a peptide is to use mutagenesis to change the amino acids present at many or all positions, usually by introducing the fairly neutral alanine into the same background sequence. This method provides insight into essential residues but misses large parts of sequence space not accessible by single mutations. Skovgaard et al. [117] used ancestral sequence reconstruction to analyze the evolution of GLP-1 peptides, to determine whether such an approach could lead to more information about receptor affinity, and to design of novel peptides based on substitution mapping that may thus serve as GLP-1 receptor agonists in humans. Tetrapod GLP-1 sequences were collected from databases and the literature. Two phylogenetic trees were constructed using maximum likelihood. ProML from the PHYLIP package was used to create the trees. One tree contained the GLP-1 paralog, exendin-4, while the other did not. FASTML [124] was used for ancestral reconstruction of the last common GLP-1 ancestor of all frogs. These ancestral reconstructions maintain the ability to activate the human GLP-1 receptor but did not have increased enzymatic stability. To approach the issue of stability, they used a method called evolutionary mapped peptides (EMP). They constructed a tree containing human GLP-1, *Xenopus laevis* GLP-1A, and exendin-4. GLP-1A and exendin-4 were chosen because of their high enzymatic stability. This could either be due to residues shared by GLP-1A and exendin-4, or to residues unique to exendin-4. Two mapped peptides were created by comparing the human GLP-1, GLP-1A, and exendin-4 sequences, and modifying the human GLP-1 sequence by exchanging its residues with either those residues that were unique to exendin-4 or unique to both GLP-1A and exendin-4. Both EMPs activated human GLP-1 receptor and both exhibited higher enzymatic stability than human GLP-1. Their stability was of the same order as that for GLP-1A but lower than that for exendin-4. The increased stability could not be assigned to a single residue as no residue was unique among the EMPs and GLP-1A. Although ancestral reconstruction yielded ancestral peptides and was found to be able to activate human GLP-1 receptor, it was necessary to use an alternative approach using evolutionary data to engineer peptides with increased enzymatic stability.

3.8 VACCINE DEVELOPMENT

While the reconstruction of the evolutionary history of proteins and their function, and the resurrection of extinct proteins may involve timescales of hundreds of millions of years, vaccine research uses much smaller timescales and usually involves sequences from a single microbial species. Accordingly, it is possible to use methods which would be inappropriate to the study of protein evolution. One method is based on the selection of a particular subtype. A microbial isolate is chosen either because the isolate comes from a geographic region where the vaccine is intended for use or because its distribution indicates that it may be useful as basis for the development of a vaccine. One problem with this approach is the large number of circulating variants seen with some microbes, such as may be found in HIV quasispecies. A number of phylogenetic approaches have been developed to create vaccines able to elicit broad reactivity to the epitopes from multiple variants of a microbe circulating within a host population. An attempt to account for sequence diversity begins with a sequence alignment of variants. From such an alignment representative epitope sequences from the major variants can be selected to construct an artificial mosaic protein [125], or the alignment can be used with a genetic algorithm to generate potential candidates for vaccine development [126].

Another approach is to construct an artificial vaccine sequence which minimizes the degree of sequence divergence between the vaccine strain and contemporary circulating variants. Several vaccine design methods have been adopted to minimize the genetic divergence (distance) between a vaccine antigen and all known extant sequences from a pathogenic microbe. Such sequences have the advantage of being more similar and central to currently circulating strains of interest and should elicit broader immune response with the potential for eliciting cross-reactive responses than any individual native strain. These methods either use a consensus sequence [127], within clade rooting [128,129] or construct an ancestral state [130] (Figure 3.1). Most of this research involves the search for a polyvalent HIV vaccine [128,131,132].

The simplest approach is to use a database of circulating variant sequences and create a consensus construct based on the most common amino acid at each position in the alignment [127,133,134]. The consensus will be influenced by sampling bias, and works best when sequence divergence is low. Such a construct is not based on an evolutionary model. Because evolutionary history is not considered, covariable sites required for proper protein folding may not be retained, and the consensus sequence may not reside on a phylogenetic tree created from the sequences used to construct the consensus [132]. If the circulating population is dominated by variants selected through immune escape and the original epitope has been lost, the consensus method may produce a construct with an advantage over methods that reconstruct ancestral states. The consensus method is also likely to be more responsive in producing constructs bearing epitopes more relevant to emerging epidemics.

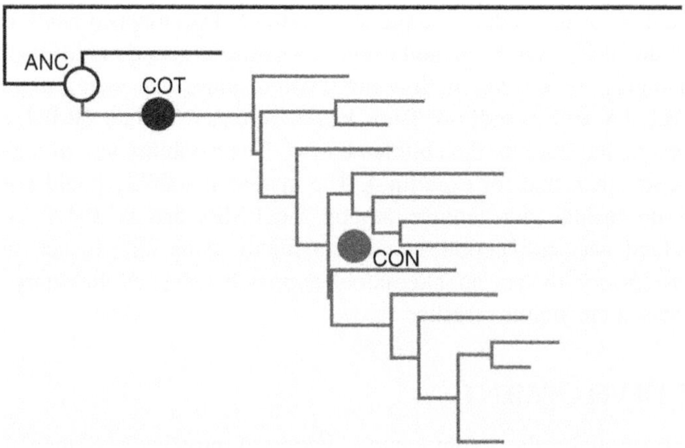

FIGURE 3.1 A schematic showing possible relative positions for artificial sequences constructed from a phylogenetic tree. The constructs are an ancestral sequence (ANC—open circle), a center of tree sequence (COT—solid black circle), and a consensus sequence (CON—solid gray circle). The CON sequence does not have to fall on an evolutionary path, while the ANC and COT sequences will. The COT and CON sequences will have shorter distances to the tips of a clade (the gray section of the tree) than the ANC sequence. (From Nickle D.C. et al., *Science*, 299, 1515, 2003. With permission.)

Ancestral reconstruction is expected to conserve amino acid covariation yielding a protein with increased likelihood of folding like the native protein. Because the ancestral construct is an estimate of an actual sequence that existed in the past, its epitopes should more closely resemble native epitopes than those from consensus sequences. The ancestor is unlikely to change with the addition of new sequences but is more influenced by extra-clade sequences. It is likely to be slightly more distant from within clade sequences than the consensus and may elicit more cross-reactivity with extra-clade sequences than the consensus. While it is assumed that ancestral reconstruction yields a good estimate of an actual sequence, evolutionary models assume neutral evolution and independence of sites. These are simplistic assumptions and do not reflect reality. Evolutionary algorithms generally capture the average evolutionary pattern over the region under consideration rather than at individual sites, correlation between sites, differences in mutational patterns, and different selective pressures in different regions. Thus, the method of rooting a phylogenetic tree and the method of constructing the ancestor contribute to variation in predicted immunological properties. This discordance between biological reality and the evolutionary algorithm used to build maximum likelihood trees can lead to faulty inference of ancestral states [131,135].

A third method has been proposed, which identifies a point on an unrooted phylogeny from which the average evolutionary distance from that point to each tip in the clade is minimized (Figure 3.1). This is done by determining the point on a tree having the least squared distance to all branch tips within the clade. This is called the phylogenetically informed center of tree or COT method [128]. This method is expected to be less sensitive to outliers than ancestral reconstruction, but like consensus sequences should favor heavily sampled sublineages. In contrast to the consensus sequence, the COT sequence resides on an evolutionary path and like the ancestral sequence should capture some of the biological properties of native strains [129]. The COT sequence is expected to occupy the middle ground between consensus and ancestral sequences. It is expected to have some of the advantages of consensus and ancestral sequences while minimizing the biases of both.

3.8.1 HIV Vaccine Development

Optimal selection of a candidate able to elicit a broad cross-reactive response among diverse circulating HIV-1 strains becomes essential due to the time and money that must be invested in human vaccine development [128]. Rather than using a single isolate for vaccine development, the use of an artificially constructed sequence may be a much more attractive alternative.

The infidelity of the HIV reverse transcriptase gives rise to a high viral mutation rate which results in high genetic diversity in HIV. The Los Alamos HIV database contains more than 72,000 sequences. This high level of diversity in HIV sequence poses a major hurdle for AIDS vaccine development. This divergence results in HIV-1 phylogenetic trees that tend to be relatively symmetric and star-like. Because of this, ancestral gene reconstruction is dominated by the effect of the method used to root the tree. Rooting with different outgroups will often cause the ancestral node to be placed differently in each case. Ancestral sequences estimated using outgroups will always be more divergent than those estimated from consensus or using a COT root. In fact, the COT approach minimizes the effect of highly divergent taxa on a tree because the center of such a tree is defined as the point having the least squared distance to the leaves. Taxon sampling may be a source of variation with the variation lying not so much in the number of taxa as in the nature of the input sequences, especially when the tree is rooted with an outgroup. Sampling bias occurs through over- or under-representation of clades. In contrast, constructs created using COT rooting are largely unaffected by taxon sampling. The methods used to reconstruct ancestral sequences and the formats (nucleotides, codons, or amino acids) have only secondary relevance to the variation seen in this setting and the technique used to estimate the phylogenetic tree is irrelevant to the outcome. Although the number of N-glycosylation sites is strongly influenced by the method used to generate ancestral sequences [135].

A small number of HIV consensus constructs have been synthesized. Gao et al. [127] constructed a synthetic group M envelope gene (CON6) using a consensus sequence generated from envelope gene sequences from each HIV-1 subtype from sequences in the 1999 Los Alamos HIV *env* Sequence Database. HIV-1 envelope proteins are the most difficult proteins to use. The hypervariable domains with multiple indels make alignments difficult. The alignments of these regions tend to be subjective as indels do not evolve according to current models for base substitution. Because of this the hypervariable regions (V1, V2, V4, and V5) in the consensus were replaced with the corresponding hypervariable regions from a CRF08_BC recombinant strain (98CN006). The V3 region evolves generally by point mutation rather than insertions and deletions, so it was not replaced. The CON6 envelope protein was biologically functional, preserved many well-studied antigen epitopes, and cross-reacted with patient sera of multiple subtypes. When used as a DNA vaccine followed by a recombinant vaccinia virus boost in BALB/c mice, it elicited γ-interferon-producing T-cell responses that recognized epitopes within overlapping peptide pools from CON6 and subtype B and C envelope proteins, although the neutralization was only effective against a subset of the subtype B peptides. Sera from guinea pigs immunized with CON6 envelope gp120 and gp140CF glycoproteins weakly neutralized selected HIV-1 primary isolates. Expanding this study, Weaver et al. [134] used the CON6 construct in conjunction with specific subtype A, B, and C envelope gene constructs to further analyze the T-cell response of this immunogen. While single wild-type clade A, B, or C immunogens tended to induce subtype specific T-cell immune responses with little cross-reactivity with other subtypes, the CON6 construct induced cross-reactive T-cell responses to all three subtypes. It was noted, however, that some T-cell epitopes were not recognized in mice immunized with CON6 although the same epitopes were recognized after immunization with the subtype specific immunogen. On the basis of the research that showed mutations or deletions of variable loops in the HIV-1 envelope glycoproteins might increase the number or exposure of available neutralization epitopes within the virion-associated viral envelope, Liao et al. [136] reengineered the CON6 construct using the 2001 Los Alamos HIV-1 sequence database, and replaced the hypervariable regions in the CON6 construct with minimal length consensus hypervariable sequences. This new construct was named CON-S. This construct elicited cross-reactive responses as good or better than seen with CON6. As with CON6, it failed to induce neutralizing antibodies only against a subset of subtype B viruses. This may be because CON-S hypervariable regions were purposefully shortened to mimic the shorter variable loops seen in subtypes A and C. Subtype B has longer variable loops with potentially more glycosylation. However, this line of research shows that iterative improvements by rational design can improve the immunogenicity of an artificial gene.

Malm et al. [133] created four consensus constructs each composed of multiple genes. Three constructs covered subtypes A, B, and C, respectively, and the fourth covered subtypes F, G, and H together. These subtypes are all members of the HIV-1 M group. Each construct contained full-length regulatory proteins Rev, Nef, and Tat as well as structural proteins p17 and p24, and the human major histocompatibility complex restricted CTL epitopes from the reverse transcriptase and envelope proteins. Because it was not possible to use a natural HIV-1 challenge in experimental animals, they used substitution methods and inserted the murine *gag* and *env* epitopes into each construct. In the first part of their study, immunization of BALB/c mice with each of the four DNA constructs elicited γ-interferon-producing T-cell responses with high envelope-specific reactivity with no clear dose–response observed. The *gag*-specific immunity was weaker than the envelope response and dose dependent. As with the CON6 study, the humoral response was weak. Cross-clade challenge studies showed that these multigene/multiepitope HIV DNA immunogens are likely to be potent immunogens against the HIV infection of humans.

Doria-Rose et al. [130] used ancestral reconstruction to address the issue of genetic diversity among HIV-1 isolates. They selected 38 subtype B envelope gene sequences from GenBank with three subtype D sequences used as outgroup to root the tree. The sequences were aligned with CLUSTALW. Regions in the alignment that could not be unambiguously aligned were removed. MODELTEST was used to select an appropriate evolutionary model for phylogenetic and ancestral state reconstruction. Evolutionary trees were inferred using maximum likelihood estimation methods as implemented in PAUP*. The ancestral sequence was inferred as the sequence at the basal node of the subtype B clade using inferred phylogenies. The accuracy of this reconstruction was estimated to be in the range of 95%–98%. The resurrected subtype B *env* ancestor, AN1-EnvB, is glycosylated, binds CD4, directs cell–cell fusion, and complements *env*-defective genomes in a virus infectivity assay. Rabbits were inoculated with the ancestral construct and their sera were used for virus neutralization at various times in the course of vaccination. Rabbit sera were shown to neutralize heterologous HIV-1 strains to a modest degree. Purified rabbit IgG was also found to have neutralizing activity. This study did not examine cross-reactive T-cell immune responses as had been done in the CON6 study [127], due to limitations in the rabbit model used. The authors also acknowledged that while the outcome of the two studies was comparable, CON6 sera showed somewhat more breadth than the AN1-EnvB sera.

Rolland et al. [129] examined the use of phylogenetics-informed COT constructs. They used a set of 37–40 of HIV-1 subtype B sequences and three subtype D sequences used as an outgroup. The subtype B sequences were used to create sets of Gag (39 sequences), the first exon of Tat (40 sequences), and Nef (37 sequences) sequences. Consensus sequences were derived from each set without the outgroup. Ancestral reconstructions were estimated from maximum likelihood trees generated with PAUP* using the outgroup. COT constructs were estimated from maximum likelihood trees generated with PAUP* without the outgroup. The average genetic distance between each constructed sequence and all ingroup sequences used to generate the construct was always smallest for the COT constructs and largest for the ancestral sequences. The COT genes were synthesized de novo and expressed in mammalian cells. These COT-derived proteins retained the biological functions of extant HIV-1 proteins. COT Gag was shown to generate virus-like particles, COT Tat transactivated gene expression from the HIV-1 long terminal repeat and COT Nef-mediated downregulation of cell surface major histocompatibility complex I. These proteins were immunogenic as they elicited antigen-specific cytotoxic T-lymphocyte responses in mice and showed that COT-derived proteins can potentially be used as HIV-1 vaccine candidates. However, these results cannot be compared with those from Gao et al. [127] or Doria-Rose et al. [130], as a COT envelope construct was not created in this study, or the results of Malm et al. [133] as different inoculation schedules and targets were used.

All of these methods appear to be useful in generating constructs that are biologically active and immunogenic. Because the assays used and viral isolates tested differ among these studies, it is difficult to compare results directly. At present no vaccine candidate has elicited antibodies with a breadth or potency against primary isolates close to what is found in patients [130].

3.9 CONCLUSION

Phylogenetics has been construed in this chapter to consist of sequence selection, sequence alignment, tree building, and ancestral reconstruction. All or several of these steps may be part of a phylogenetic analysis. The analysis can be an end in itself, or it may be the basis from which other types of analysis, such as those found elsewhere in this book, are conducted.

Phylogenetic analysis is based on a set of assumptions that is overly simplistic. Because of this, phylogenetics is a complex field with multiple methodologies that require intense study. Phylogenetics deals with historical evolutionary processes by using stochastic and heuristic algorithms to approximate historical events, and the potential pitfalls of these methods have been documented. One of the most important things for a biologist to do in this type of analysis is to bring his/her knowledge of biology to the process.

Phylogenetics has proven to be useful in medicinal protein engineering. This is especially true for ancestral reconstruction and the rational design of engineered proteins. Because of the increases in computational power and the development of inexpensive methods for the construction of oligonucleotides and oligopeptides, it has become fairly easy to "resurrect" ancient or extinct proteins. No longer is it necessary to use site-directed mutagenesis and to infer the evolution of protein function from amino acid substitutions thought to be historically accurate. Computer algorithms based on our understanding of evolution and rates of nucleotide substitutions are available to estimate the most probable history for the evolution of a protein. These histories can be analyzed stochastically to determine the probabilities that they do indeed approximate actuality closely. The study of these proteins not only shed light on extant proteins and disease processes but may provide information about how evolution gave rise to present-day diversity, what environmental forces may have shaped that evolution, and how that knowledge can be used in the present for the benefit of mankind. This type of protein engineering has also been useful in the field of vaccinology, helping us understand how antigens from pathogens evolve in small timescales to escape detection, to reconstruct the evolutionary history of these antigens, and to design candidate vaccines that elicit broadly diverse immune responses.

ACKNOWLEDGMENT

The author would like to thank Dr. Chong-Gee Teo for his suggested modifications to the text.

REFERENCES

1. Cummings, M.P. and Meyer, M., Magic bullets and golden rules: Data sampling in molecular phylogenetics, *Zoology*, 108, 329, 2005.
2. Kumar, S. and Filipski, A., Multiple sequence alignment: In pursuit of homologous DNA positions, *Genome Res.*, 17, 127, 2007.
3. Batzoglou, S., The many faces of sequence alignment, *Brief. Bioinform.*, 6, 6, 2005.
4. Simmons, M.P., Pickett, K.M., and Miya, M., How meaningful are Bayesian support values? *Mol. Biol. Evol.*, 21, 188, 2004.
5. Devos, D. and Valencia, A., Intrinsic errors in genome annotation, *Trends Genet.*, 17, 429, 2001.
6. Stevens, J.R. and Schofield, C.J., Phylogenetics and sequence analysis—some problems for the unwary, *Trends Parasitol.*, 19, 582, 2003.
7. Linial, M., How incorrect annotations evolve—the case of the short ORFs, *Trends Biotechnol.*, 21, 298, 2003.
8. Nei, M., Kumar, S., and Takahashi, K., The optimization principle in phylogenetic analysis tends to give incorrect topologies when the number of nucleotides or amino acids used is small, *Proc. Natl. Acad. Sci. U S A*, 95, 12390, 1998.
9. Holm, L. and Sander, C., Removing near-neighbor redundancy from large protein sequence collections, *Bioinformatics*, 14, 423, 1998.
10. Brunak, S., Danchin, A., Hattori, M., Nakamura, H., Shinozaki, K., Matise, T., and Preuss, D., Nucleotide sequence database policies, *Science*, 298, 1333, 2002.
11. Roberts, L., A tussle over the rules for DNA data sharing, *Science*, 298, 1312, 2002.
12. Morrison, D.A., Multiple sequence alignment for phylogenetic purposes, *Aust. Syst. Bot.*, 19, 479, 2006.
13. de Pinna, M.C.C., Concepts and tests of homology in the cladistic paradigm, *Cladistics*, 7, 367, 1991.
14. Phillips, A.J., Homology assessment and molecular sequence alignment, *J. Biomed. Inform.*, 39, 18, 2006.
15. Whiting, A.S., et al., Comparing alignment methods for inferring the history of the new world lizard genus *Mabuya* (Squamata: Scincidea), *Mol. Phylogenet. Evol.*, 38, 719, 2006.
16. Wegnez, M., Letter to the editor, *Cell*, 51, 516, 1987.
17. Dimitrova, Z., et al., Coordinated substitutions within the hepatitis B virus genome. *Mol. Biol. Hep. B Viruses*, September 16–20, 2007, Rome, Italy.
18. Larkin, M.A., et al., Clustal W and Clustal X version 2.0, *Bioinformatics*, 23, 2947, 2007.
19. Morgenstern, B., et al., DIALIGN: Finding local similarities by multiple sequence alignment, *Bioinformatics*, 14, 290, 1998.
20. Morgenstern, B., DIALIGN 2: Improvement of the segment-to segment approach to multiple sequence alignment, *Bioinformatics*, 15, 211, 1999.

21. Katoh, K., et al., MAFFT: A novel method for rapid multiple sequence alignment based on fast Fourier transform, *Nucleic Acids Res.*, 30, 3059, 2002.
22. Katoh, K., et al., Improvement in accuracy of multiple sequence alignment program MAFFT, *Genom. Inform.*, 16, 22, 2005.
23. Edgar, R.C., MUSCLE: Multiple sequence alignment with high accuracy and high throughput, *Nucleic Acids Res.*, 32, 1792, 2004.
24. Do, C.B., et al., ProbCons: Probabilistic consistency-based multiple sequence alignment, *Genome Res.*, 15, 330, 2005.
25. Notredame, C., Higgins, D.G., and Heringa, J., T-Coffee: A novel method for fast and accurate multiple sequence alignment, *J. Mol. Biol.*, 302, 205, 2000.
26. Edgar, R.C. and Batzoglou, S., Multiple sequence alignment, *Curr. Opin. Struct. Biol.*, 16, 368, 2006.
27. Buckley, T.R., Simon, C., and Chambers, G.K., Exploring among-site rate variation models in a maximum likelihood framework using empirical data: Effects of model assumptions on estimates of topology, branch lengths, and bootstrap support, *Syst. Biol.*, 50, 67, 2001.
28. Kelchner, S.A. and Thomas, M.A., Model use in phylogenetics: Nine key questions, *Trends Eco. Evol.*, 22, 87, 2006.
29. Kosakovski Pond, S.L., Frost, S.D.W., and Muse, S.V., HyPhy: Hypothesis testing using phylogenies, *Bioinformatics*, 212, 676, 2005.
30. Posada, D., Using MODELTEST and PAUP* to select a model of nucleotide substitution, In: *Current Protocols in Bioinformatics*, Baxevanis, A.D., Davison, D.B., Page, R.D.M., Petsko, G.A., Stein, L.D., and Stormo, G.D., (Eds.), John Wiley & Sons, Inc., New York, 2003, Unit 6.5.1.
31. Posada, D. and Buckley, T.R., Model selection and model averaging in phylogenetics: Advantages of Akaike information criterion and Bayesian approaches over likelihood ratio tests, *Syst. Biol.*, 53, 793, 2004.
32. Posada, D. and Crandall, K.A., Modeltest: Testing the model of DNA substitution, *Bioinformatics*, 14, 817, 1998.
33. Sullivan, J. and Joyce, P., Model selection in phylogenetics, *Annu. Rev. Ecol. Ecol. Syst.*, 36, 445, 2005.
34. Higgins, D.G., Thompson, J.D., and Gibson, T.J., Using CLUSTAL for multiple sequence alignments, *Methods Enzymol.*, 266, 383, 1996.
35. Poch, O. and Delarue, M., Converting sequence block alignments into structural insights, *Methods Enzymol.*, 266, 662, 1996.
36. Higgins, D.G., Blackshields, G., and Wallace, I.M., Mind the gaps: Progress in progressive alignment, *Proc. Natl. Acad. Sci. U S A*, 102, 10411, 2005.
37. Lebrun, E., et al., The rieske protein: A case study on the pitfalls of multiple sequence alignments and phylogenetic reconstruction, *Mol. Biol. Evol.*, 23, 1180, 2006.
38. Ponting, C.P. and Birney, E., Identification of domains from protein sequences, In: *Protein Structure Prediction: Methods and Protocols*, Webster, D., (Ed.), Humana Press, New York, 2000, 53.
39. Hulo, N., et al., The PROSITE database, *Nucleic Acids Res.*, 34, D227, 2006.
40. Mamitsuka, H., Finding the biological optimal alignment of multiple sequences, *Artif. Intell. Med.*, 35, 9, 2005.
41. Chindelevitch, L., Li, Z., Blais, E., and Blanchette, M., On the inference of parsimonious indel evolutionary scenarios. *J. Bioinform. Comput. Biol.*, 4, 721, 2006.
42. Gotoh, O., An improved algorithm for matching biological sequences, *J. Mol. Biol.*, 162, 705, 1982.
43. McGuire, G., Denham, M.C., and Balding, D.J., Models of sequence evolution for DNA sequences containing gaps, *Mol. Biol. Evol.*, 18, 481, 2001.
44. Simmons, M.P. and Ochcterena, H., Gaps as characters in sequence-based phylogenetic analyses, *Syst. Biol.*, 49, 369, 2000.
45. Kawakita, A., Sota, T., Ascher, J.S., Ito, M., Tanaka, H., and Kato, M., Evolution and phylogenetic utility of alignment gaps within intron sequences of three nuclear genes in bumble bees (Bombus), *Mol. Biol. Evol.*, 20, 87, 2003.
46. Phillips, A., Janies, D., and Wheeler, W., Multiple sequence alignment in phylogenetic analysis, *Mol. Phylogenet. Evol.*, 16, 317, 2000.
47. Barta, J.R., Investigating phylogenetic relationships within the Apicomplexa using sequence data: The search for homology, *Methods*, 13, 81, 1997.

48. Marsden, B. and Abagyan, R., SAD—a normalized structural alignment database: Improving sequence-structure alignments, *Bioinformatics*, 20, 2333, 2004.
49. Bos, D.H. and Posada, D., Using models of nucleotide evolution to build phylogenetic trees, *Dev. Comp. Immunol.*, 29, 211, 2005.
50. Grant, T. and Kluge, A.G., Data exploration in phylogenetic inference: Scientific, heuristic, or neither, *Cladistics*, 19, 379, 2003.
51. Rieppel, O., Gattei on Popper and truth, *Cladistics*, 19, 170, 2003.
52. Rieppel, O., A skeptical look for justification, *Cladistics*, 21, 203, 2005.
53. Baldauf, S.L., Phylogeny for the faint of heart: A tutorial, *Trends Genet.*, 19, 345, 2003.
54. Hall, B.G. and Barlow, M., Phylogenetic analysis as a tool in molecular epidemiology of infectious diseases, *Ann. Epidemiol.*, 16, 157, 2006.
55. Felsenstein, J., *Inferring Phylogenies*, Sinauer Associates, Inc., Sunderland, Massachusetts, 2004.
56. Hall, B.G., *Phylogenetic Trees Made Easy: A How-To Manual*, 3rd ed., Sinauer Associates, Inc., Sunderland, Massachusetts, 2007.
57. Salemi, M. and Vandamme, A.M., *The Phylogenetic Handbook: A Practical Approach to DNA and Protein Phylogeny*, Cambridge University Press, New York, 2003.
58. Nei, M. and Kumar, S., *Molecular Evolution and Phylogenetics*, Oxford University Press, New York, 2000.
59. Amos, W., The hidden value of missing genotypes, *Mol. Biol. Evol.*, 23, 1995, 2006.
60. Bergstein, J., A review of long-branch attraction, *Cladistics*, 21, 163, 2005.
61. Gatsey, J., DeSalle, R., and Wahlberg, N., How many genes should a systematist sample? conflicting insights from a phylogenomic matrix characterized by replicated incongruence, *Syst. Biol.*, 56, 335, 2007.
62. Geuten, K., Massingham, T., Darius, P., Smets, E., and Goldman, N., Experimental design criteria in phylogenetics: Where to add taxa, *Syst. Biol.*, 56, 609, 2007.
63. Rosenberg, M.S. and Kumar, S., Incomplete taxon sampling is not a problem for phylogenetic inference, *Proc. Natl. Acad. Sci. U S A*, 98, 10751, 2001.
64. Rosenberg, M.S., Multiple sequence alignment accuracy and evolutionary distance estimation, *BMC Bioinformatics*, 6, 278, 2005.
65. Czarna, A., Sanjuán, R., Gonzáles-Candelas, F., and Wróbel, B., Topology testing of phylogenies using least squares methods, *BMC Evol. Biol.*, 6, 105, 2006.
66. Soltis, P.S. and Soltis, D.E., Applying the bootstrap in phylogenetic reconstruction, *Stat. Sci.*, 18, 256, 2003.
67. Zhang, J. and Nei, M., Accuracies of ancestral amino acid sequences inferred by the parsimony, likelihood, and distance methods, *J. Mol. Evol.*, 44(Suppl 1), S139, 1997.
68. Hall, B.G. and Salipante, S.J., Measures of clade confidence do not correlate with accuracy of phylogenetic trees, *PLoS Comput. Biol.*, 3, e51, 2007.
69. Hall, B.G. and Salipante, S.J., Retraction: Measures of clade confidence do not correlate with accuracy of phylogenetic trees, *PLoS Comput. Biol.*, 3(7): e158, 2007.
70. Alfaro, M.E., Zoller, S., and Lutzoni, F., Bayes or bootstrap? A simulation study comparing the performance of Bayesian Markov chain Monte Carlo sampling and bootstrapping in assessing phylogenetic confidence, *Mol. Biol. Evol.*, 20, 255, 2003.
71. Brocchieri, L., Phylogenetic inferences from molecular sequences: Review and critique, *Theor. Popul. Biol.*, 59, 27, 2001.
72. Douady, C.J., Delsuc, F., Boucher, Y., Doolittle, W.F., and Douzery, E.J.P., Comparison of Bayesian and maximum likelihood bootstrap measures of phylogenetic reliability, *Mol. Biol. Evol.*, 20, 248, 2003.
73. Freckleton, R.P., Harvey, P.H., and Pagel, M., Phylogenetic analysis and comparative data: A test and review of evidence, *Am. Nat.*, 160, 712, 2002.
74. Philippe, H., Zhou, Y., Brinkmann, H., Rodrigue, N., and Delsuc, F., Heterotachy and long-branch attraction in phylogenetics, *BMC Evol. Biol.*, 5, 50, 2005.
75. Rice, P., Longden, I., and Bleasby, A., EMBOSS: The European molecular biology open software suite, *Trends Genet.*, 16, 276, 2000.
76. Womble, D.D., Web-based interfaces for the GCG sequence analysis programs, *Methods Mol. Biol.*, 132, 23, 2000.
77. GCG version 11.1, Accelrys Inc., San Diego, California.

78. Kumar, S., Tamura, K., and Nei, M., MEGA—molecular evolutionary genetics analysis software for microcomputers, *Comput. Appl. Biosci.*, 10, 189, 1994.
79. Tamura, K., Dudley, J., Nei, M., and Kumar, S., MEGA4: Molecular evolutionary genetics analysis (MEGA) software version 4.0, *Mol. Biol. Evol.*, 24, 1596, 2007.
80. Swofford, D.L., PAUP*, phylogenetic analysis using parsimony (* and other methods), Version 4, Sinauer Associates, Sunderland, Massachusetts, 2002.
81. Felsenstein, J., PHYLIP—phylogeny inference package (version 3.2), *Cladistics*, 5, 164, 1989.
82. Guindon, S. and Gascuel, O., A simple, fast, and accurate algorithm to estimate large phylogenies by maximum likelihood, *Syst. Biol.*, 52, 696, 2003.
83. Yang, Z., PAML: A program package for phylogenetic analysis by maximum likelihood, *Bioinformatics*, 13, 555, 1997.
84. Yang, Z., PAML 4: Phylogenetic analysis by maximum likelihood, *Mol. Biol. Evol.*, 24, 1586, 2007.
85. Edwards, C.T., et al., Evolution of the human immunodeficiency virus envelope gene is dominated by purifying selection, *Genetics*, 174, 1441, 2006.
86. Rouzine, I.M., Rodrigo, A., and Coffin, J.M., Transition between stochastic evolution and deterministic evolution in the presence of selection: General theory and application to virology, *Microbiol. Mol. Biol. Rev.*, 65, 151, 2001.
87. Fujiwara, K., et al., Genetic analysis of hepatitis A virus protein 2C in sera from patients with fulminant and self-limited hepatitis A, *Hepatogastroenterology*, 54, 871, 2007.
88. Naylor, S.L., SNPs associated with prostate cancer risk and prognosis, *Front. Biosci.*, 12, 4111, 2007.
89. Serretti, A., Drago, A., and De Ronchi, D., HTR2A gene variants and psychiatric disorders: A review of current literature and selection of SNPs for future studies, *Curr. Med. Chem.*, 14, 2053, 2007.
90. Crespi, B., Summers, K., and Dorus, S., Adaptive evolution of genes underlying schizophrenia, *Proc. Biol. Sci.*, 274, 2801, 2007.
91. Wagner, A., Rapid detection of positive selection in genes and genomes through variation clusters, *Genetics*, 176, 2451, 2007.
92. Wang, S.Y., et al., Positive selection of hepatitis delta antigen in chronic hepatitis D patients, *J. Virol.*, 81, 4438, 2007.
93. Flanagan, J.M., et al., The fatty acid amide hydrolase 385 A/A (P129T) variant: Haplotype analysis of an ancient missense mutation and validation of risk for drug addiction, *Hum. Genet.*, 120, 581, 2006.
94. Brault, A.C., et al., A single positively selected West Nile viral mutation confers increased virogenesis in American crows, *Nat. Genet.*, 39, 1162, 2007.
95. Kumar, S., Molecular clocks: Four decades of evolution, *Nat. Rev. Genet.*, 6, 654, 2005.
96. Travers, S.A., et al., Timing and reconstruction of the most recent common ancestor of the subtype C clade of human immunodeficiency virus type 1, *J. Virol.*, 78, 10501, 2004.
97. Mor-Cohen, R., et al., Age estimates of ancestral mutations causing factor VII deficiency and Dubin-Johnson syndrome in Iranian and Moroccan Jews are consistent with ancient Jewish migrations, *Blood Coagul. Fibrinolysis*, 18, 139, 2007.
98. Chang, B.S. and Donoghue, M.J., Recreating ancestral proteins, *Trends Ecol. Evol.*, 15, 109, 2000.
99. Cunninghan, C.W., Omland, K.E., and Oakley, T.H., Reconstructing ancestral character states: A critical reappraisal, *Trends Eco. Evol.*, 13, 361, 1998.
100. Hall, B.G., Simple and accurate estimation of ancestral protein sequences, *Proc. Natl. Acad. Sci. U S A*, 103, 5431, 2006.
101. Williams, P.D., Pollock, D.D., Blackburne, B.P., and Goldstein, R.A., Assessing the accuracy of ancestral protein reconstruction methods, *Plos Comput. Biol.*, 2, e69, 2006.
102. Hudek, A.K. and Brown, D.G., Ancestral sequence alignment under optimal conditions, *BMC Bioinformatics*, 6, 273, 2005.
103. Pagel, M., Inferring the historical patterns of biological evolution, *Nature*, 401, 877, 1999.
104. Huelsenbeck, J.P. and Bollback, J.P., Empirical and hierarchal Bayesian estimation of ancestral states, *Syst. Biol.*, 50, 351, 2001.
105. Cai, W., Pei, J., and Grishin, N.V., Reconstruction of ancestral protein sequences and its applications, *BMC Evol. Biol.*, 4, 33, 2004.
106. Maddison, W.P. and Maddison, D.R., Interactive analysis of phylogenetic and character evolution using the computer program MacClade, *Folia Primatol.*, 53, 190, 1989.

107. Maddison, D.R. and Maddison, W.P., MacClade 4: Analysis of Phylogenetic and Character Evolution, Version 4.02, Sinauer Associates, Sunderland, Massachusetts, 2001.
108. Huelsenbeck, J.P. and Ronquist, F., MRBAYES: Bayesian inference of phylogenetic trees, *Bioinformatics*, 17, 754, 2001.
109. Ronquist, F. and Huelsenbeck, J.P., MrBayes 3: Bayesian phylogenetic inference under mixed models, *Bioinformatics*, 19, 1572, 2003.
110. Sharp, P.M., et al., Codon usage patterns in *Escherichia coli, Bacillus subtilis, Saccharomyces cerevisiae, Schizosaccharomyces pombe, Drosophila melanogaster* and *Homo sapiens*—a review of the considerable within-species diversity, *Nucleic Acids Res.*, 16, 8207, 1988.
111. Adey, N.B., et al., Molecular resurrection of an extinct ancestral promoter for mouse L1, *Proc. Natl. Acad. Sci. U S A*, 91, 1569, 1994.
112. Thornton, J.W., Need, E., and Crews, D., Resurrecting the ancestral steroid receptor: Ancient origin of estrogen signaling, *Science*, 301, 1714, 2003.
113. Gaucher, E.A., et al., Inferring the palaeoenvironment of ancient bacteria on the basis of resurrected proteins, *Nature*, 425, 285, 2003.
114. Chang, B.S., Ugalde, J.A., and Matz, M.V., Applications of ancestral protein reconstruction in understanding protein function: GFP-like proteins, *Methods Enzymol.*, 395, 652, 2005.
115. Kuang, D., et al., Ancestral reconstruction of the ligand-binding pocket of Family C G protein-coupled receptors, *Proc. Natl. Acad. Sci. U S A*, 103, 14050, 2006.
116. Herbeck, J.T., et al., Human immunodeficiency virus type 1 env evolves toward ancestral states upon transmission to a new host, *J. Virol.*, 80, 1637, 2006.
117. Skovgaard, M., et al., Using evolutionary information and ancestral sequences to understand the sequence-function relationship in GLP-1 agonists, *J. Mol. Biol.*, 363, 977, 2006.
118. Lang, J., et al., The M53I mutation in CDKN2A is a founder mutation that predominates in melanoma patients with Scottish ancestry, *Genes Chromosomes Cancer*, 46, 277, 2007.
119. Bolnick, D.A., Bolnick, D.I., and Smith, D.G., Asymmetric male and female genetic histories among Native Americans from Eastern North America, *Mol. Biol. Evol.*, 23, 2161, 2006.
120. von Salomé, J., Gyllensten, U., and Bergström, T.F., Full-length sequence analysis of the HLA-DRB1 locus suggests a recent origin of alleles, *Immunogenetics*, 59, 261, 2007.
121. Devi, S.M., et al., Ancestral European roots of Helicobacter pylori in India, *BMC Genomics*, 8, 184, 2007.
122. Mota, A.C., et al., The close relationship between South African and Latin American HTLV type 1 strains corroborated in a molecular epidemiological study of the HTLV type 1 isolates from a blood donor cohort, *AIDS Res. Hum. Retroviruses*, 23, 503, 2007.
123. Adachi, J. and Hasegawa, M., MOLPHY version 2.3: Programs for molecular phylogenetics based on maximum likelihood, *Comput. Sci. Monogr.*, 28, 1, 1996.
124. Pupko, T., et al., A branch-and-bound algorithm for the inference of ancestral amino-acid sequences when the replacement rate varies among sites: Application to the evolution of five gene families, *Bioinformatics*, 18, 1116, 2002.
125. Chang, J.C., et al., Artificial NS4 mosaic antigen of hepatitis C virus, *J. Med. Virol.*, 59, 437, 1999.
126. Fischer, W., et al., Polyvalent vaccines for optimal coverage of potential T-cell epitopes in global HIV-1 variants, *Nature Med.*, 13, 100, 2007.
127. Gao, F., et al., Antigenicity and immunogenicity of a synthetic human immunodeficiency virus type 1 group M consensus envelope glycoprotein, *J. Virol.*, 79, 1154, 2005.
128. Gaschen, B., et al., Diversity considerations in HIV-1 vaccine selection, *Science*, 296, 2354, 2002.
129. Rolland, M., et al., Reconstruction and function of ancestral center-of-tree human immunodeficiency virus type 1 proteins, *J. Virol.*, 81, 8507, 2007.
130. Doria-Rose, N.A., et al., Human immunodeficiency virus type 1 subtype B ancestral envelope protein is functional and elicits neutralizing antibodies in rabbits similar to those elicited by a circulating subtype B envelope, *J. Virol.*, 79, 11214, 2005.
131. Gao, F., et al., Letter to the editor, *Science*, 299, 1517, 2003.
132. Nickle, D.C., et al., Consensus and ancestral state HIV vaccines, *Science*, 299, 1515, 2003.
133. Malm, M., et al., Cross-clade protection induced by human immunodeficiency virus-1 DNA immunogens expressing consensus sequences of multiple genes and epitopes from subtypes A, B, C, and FGH, *Viral. Immunol.*, 18, 678, 2005.

134. Weaver, E.A., et al., Cross-subtype T-cell immune responses induced by a human immunodeficiency virus type 1 group m consensus env immunogen, *J. Virol.*, 80, 6745, 2006.
135. Ross, H.A., et al., Sources of variation in ancestral sequence reconstruction for HIV-1 envelope genes, *Evol. Bioinform. Online*, 2, 18, 2006.
136. Liao, H.X., et al., A group M consensus envelope glycoprotein induces antibodies that neutralize subsets of subtype B and C HIV-1 primary viruses, *Virology*, 353, 268, 2006.

4 Clustering and Classification Approaches to Identifying Protein Properties

Lilia Milkova Ganova-Raeva and Zoya Emilova Dimitrova

CONTENTS

Biology has at least 50 more interesting years.

James D. Watson

Big leaps and big headaches with software, that's one prediction.

Joshua Lederberg

Data! Data! Data!

Sherlock Holmes

4.1 INTRODUCTION

As the ability to acquire information about nucleic acid sequences and proteins has increased due to development of massive parallel sequencing methods, proteomics, and mass spectrometry, the biological diversity we observe is becoming immense. With whole-genome sequencing now becoming more widely adopted, the variety of the biome can be explored like never before. The inability to propagate most of the living organisms in vitro is no longer an obstacle to their discovery [1]. It is estimated that only 10% of the genomic sequences discovered are very closely related or identical to known sequences, an additional 20% may be assigned to a particular family, and more than 70% have no relation to any previously known organisms [1–7]. At the current state of the art, multiple protein sequence alignments from closely related species are mere prerequisites for further analysis based on motif, domain, multidomain, or three-dimensional (3D) relatedness [8–10]. This inevitably leads to the need for development of new algorithms that can reliably uncover true coding sequences, understand the possible structure and function of the newly discovered proteins, evaluate the evolutionary trends [11–14], recognize proteins with beneficial medicinal properties, and ultimately, predict and track emerging pathogens.

In recent years, efforts have been spent on protein sequence analysis by modeling and classification of protein families based on function, structure, and homology. A major assumption of the classification approaches is that the protein sequences for which the families are known a priori can be used to build classification models. These models are then used to solve the challenging problem that arises with the super abundance of new protein sequences without a known family affiliation [15–17]. Clustering is another powerful technique in statistics and computer science, and a subject of extensive investigations. The major goal of clustering is to create a partition of objects such that objects in a given group have similar features. The outcome of the clustering is that a variety of protein sequences may be grouped in such a way as to lead to a better understanding of their nature, function, or both.

The building of a representative database is an important prerequisite to the application of any kind of clustering or classification algorithm. There are multiple valuable databases created and many more tools to query and utilize them. The beta release of Protein Clusters database by National Center for Biotechnology Information NCBI, a collection of reference protein sequence database grouped and annotated based on sequence similarity, made its debut recently (http://www.ncbi.nlm.nih.gov/sites/entrez?db = proteinclusters). It provides easy access to annotations, publications, domains, structures, external links and analytical tools for generating multiple alignments, phylogenetic trees, and genomic neighborhoods. The Clusters of Orthologous Groups (COGs) database (http://www.ncbi.nlm.nih.gov/COG/new/) [18–20] is based on sequenced genomes of prokaryotes and unicellular eukaryotes and the construction of clusters of predicted orthologs which currently consists of nearly 200,000 proteins from 66 whole genomes of organisms from three kingdoms. The integrated documentation resource of protein families, domains and functional sites, (current version InterPro15.1) was created by the European Bioinformatics Institute (http://www.ebi.ac.uk/interpro/) to help identify the functions of unknown protein sequences. InterPro is the collective progeny of the Swiss-Prot database, comprising TrEMBL, PROSITE, PRINTS, Pfam [21], and ProDom. SMART, TIGRFAMs, PIRSF, SUPERFAMILY, and more recently PANTHER [22,23] and Gene3D, have also joined InterPro. The Kyoto Encyclopedia of Genes and Genomes Orthologs (KEGG orthology at http://www.genome.jp/dbget-bin/get_htext?KO) is a "biological systems" database integrating both molecular building block information and higher-level systemic information [24].

4.2 CLUSTERING AND CLASSIFICATION METHODS

4.2.1 Clustering

Cluster analysis is often the first step to data exploration aimed at discovering groups in the data. The result of the analysis is partitioning of the data set in such a way that data points in each group (cluster) are similar to each other, and at the same time are as different as possible from the data points in the other clusters.

The four basic directions in which clustering may be used have been summarized [25] as follows:

- Data reduction: By representing each data point according to the cluster to which it belongs, the size of the data set is reduced, thereby allowing the data set to be more manageable.
- Hypothesis generation: The partitioning of the data revealed by the cluster analysis is used to formulate hypotheses about the nature of the data, which can then be tested and verified with other data sets.
- Hypothesis testing: As a way to verify a hypothesis, the clusters found in the data are compared with the clusters whose existence is hypothesized.
- Prediction based on groups: After partitioning the data set, each cluster is evaluated to find common characteristics of its members. When a data instance with unknown characteristics is assigned to one of the clusters, it can be characterized according to the respective cluster properties.

To be successful in the endeavor of data partitioning, a good measure of similarity between the data points is needed, as well as a way to measure the difference between groups of instances. Many such measures of similarity and dissimilarity are available and it is up to the researcher to choose the most appropriate. The result of clustering is very sensitive to the chosen distance measure between instances and groups.

There is a significant amount of research done in the field of cluster analysis and new methods are continually being developed. The disadvantage inherent to all clustering methods is that they always work, i.e., the algorithms will produce clusters regardless if biologically meaningful grouping really exists in the data set. Although different scoring functions might be helpful in assessing the quality of the clustering [26], it is difficult to quantify the final validation of the clusters in the absence of experimental verification. Rather, it remains dependent on the expert's subjective opinion and the application, and the insight that clustering provides for understanding the nature of the data [26].

Clustering algorithms differ by the restrictions applied to the cluster membership. Hard clustering algorithms assign a data instance to one cluster exclusively. Fuzzy clustering algorithms allow a data instance to belong to several clusters simultaneously.

The most widely used clustering algorithms are hierarchical clustering and partition-based methods like *k*-means.

4.2.1.1 Hierarchical Clustering

Hierarchical clustering obtains not a single clustering of the data, but a hierarchy of nested clusters [25]. There are two main categories of hierarchical clustering algorithms: agglomerative and divisive. Agglomerative algorithms usually start with an initial clustering where each data point constitutes a cluster. At each subsequent step, the two nearest clusters are merged until all data points are assigned to one cluster. The divisive algorithms follow the opposite direction, starting with a single cluster and at each step splitting it into two clusters.

In addition to an appropriate distance function defined for the elements of the data set, hierarchical-clustering methods require a way of measuring the distance between two clusters in order to find the nearest two. Different methods use the distance between different pairs of elements that belong to the two clusters respectively. In the case of single-link clustering, the distance between the closest elements from the different clusters is assigned to be the distance between the clusters. The complete-link clustering uses the farthest possible pair to measure the distance between the clusters. The average-link method calculates the average of the distances of all possible pairs of elements. The centroid clustering uses the distance between the centroids (means) of the two groups.

The agglomerative clustering algorithms are more popular and widely used. Their result is graphically represented in the form of a dendrogram—a tree with data instances represented as leaves and with hierarchical grouping on each level corresponding to the sequence of merging (or splitting) clusters. If a given number of clusters are needed, the tree is intersected at the appropriate level and each data instance is assigned to the cluster starting from the cutoff point of the branch [27].

4.2.1.2 *K*-Means Clustering

K-means is an iterative clustering algorithm that searches for the best splitting of the data into a predetermined number of clusters (k). K random points are chosen to be the centers of the k clusters at the first iteration and each instance is assigned to the closest cluster center. The next step is to recalculate the centers of each cluster to be the centroid (or the mean) of the instances belonging to the cluster. All instances are reassigned again to the closest cluster center. This step is repeated until no more reassigning occurs and the cluster centers are stabilized.

As with all clustering algorithms, a distance function needs to be chosen to measure the closeness of the instances to the cluster centers. One commonly used distance function is the Euclidean distance, but the choice of the function depends primarily on the problem being solved.

For a given value of the parameter k, the k-means algorithm will always find k clusters, regardless of the quality of the clustering. This makes the choice of the parameter k very important. In the cases when there is no a priori knowledge of k, the algorithm might start with a minimum number of two clusters and then the value of k can increase until a certain limit of the distance between the center of a cluster and the instances assigned to it is reached. The value of k may also be assessed by examining the first dimensions of the principal component analysis (PCA) and looking for a grouping of the data instances [26].

The k-means clustering algorithm is very sensitive to the initial choice of the cluster centers. Since we are minimizing the sum of the distances from each instance to its respective cluster center, we may only find a local minimum. To attempt to find the global minimum, we may run the algorithm several times and choose the best clustering.

4.2.2 Classification

The problem of classification involves predicting the membership of data points in the input space in one of K a priori-known classes. The predictive power and the prior knowledge of the classes are what fundamentally distinguish classification from descriptive models such as clustering.

Building a classifier depends on having a training data set in which the class association of each point is already known. By applying this classifier to future data points, we can infer their class membership.

To compare and evaluate different classification methods and specific classifiers, their misclassification rates (the rate of misclassifying future or test data points) need to be considered. The optimal error rate of a classifier would have been achieved if the training data set was representative of the distributions of data points in the input space. In the absence of knowledge of this distribution, the optimal error rate can be estimated by the actual rate that is achieved when using a specific (nonrepresentative of the distributions in the input space) training data set. If a testing data set with known class association is not available, usually a part of the training set is temporarily set aside as a testing set. To minimize the bias, the procedure is repeated by building a classifier using the subset of the training data and testing it with the rest of the set several times. The misclassification rate is averaged after training and testing all sets. In the leave-one-out approach, only one data point is reserved for testing. If the training set is randomly divided into k subsets and each is in turn used as a testing set, the method is called k-fold cross-validation. There are also bootstrap methods for evaluation based on modeling the relationship between the true distributions in the input space and the distributions in the training data.

The misclassification rate on the test set can be reduced by increasing the complexity of the classification model. At the same time the misclassification rate of future data may increase. This is due to overfitting which makes the model generalize poorly to future data. Hence, attention should be paid to the proper balance between model complexity and its ability to generalize.

The most widely used classification methods are artificial neural networks, support vector machines, *k*-nearest neighbor, decision trees, Bayesian networks, and hidden Markov models. Brief descriptions of these methods follow.

4.2.2.1 Artificial Neural Networks

Artificial neural networks (ANNs) are highly parameterized nonlinear statistical models. An ANN transforms a real-valued vector in the input space to a real-valued vector in some output space. They can be used to predict new characteristics (e.g., class membership) of points in the input space.

ANNs first linearly transform the input vector by multiplying it with a weight matrix. Then a nonlinear function is applied to each coordinate of the resulting vector to produce the output. This function is typically a sigmoid function (a differentiable approximation to the step function). This is an example of a single layer network. A multilayer network can be constructed by subsequent applications of another weight matrix and a nonlinear transformation.

Training an ANN involves determining the weight matrix that minimizes the prediction error for a set of training data for which there is knowledge of what the output vectors should be.

4.2.2.2 Support Vector Machines

Support vector machine (SVM) is a type of a linear discrimination method for supervised learning first introduced by Vapnic and his collaborators [28,29]. It takes a training set of data where each point belongs to one of two known classes and is labeled accordingly. It finds a hyperplane in the input space such that all data points from the first class are on one side of the hyperplane and the data points from the other class are on the opposite side. There may be many such hyperplanes. To achieve better generalization, the method looks for the hyperplane that is as far as possible from the points of both classes. The instances from both classes, closest to the hyperplane, are the support vectors of the two classes, and the distance from the support vectors to the hyperplane is called the margin. Hence, the method maximizes the margin to find the optimal separating hyperplane, called SVM. After an SVM classifier is found, the testing data set is assigned to one of the classes based on the relative position of the data points to the separating hyperplane.

In many real-world cases the classes are not linearly separable by a hyperplane in the original input space. One possible way to deal with this problem, called soft-margin optimization, is to look for the separating hyperplane that minimizes the classification error [30]. Another possible approach is to use nonlinear kernel functions to transform the input space into a new space, where the mapped classes are linearly separable and thus the linear model can be used. The most widely applied kernels are polynomials of degree 2, 3, etc., radial-basis functions, and sigmoidal functions [27].

When there are $K > 2$ classes, SVMs can be used to build K hyperplanes each separating one class from the union of all other classes. Classification of a test data point starts with computing its distance to each of the K hyperplanes. The data point is assigned to the class for which this distance is maximal.

The optimization problem in SVM is solved in a dual formulation which makes its size dependent on the number of the support vectors and not on the dimensionality of the input space. This allows the use of kernel functions in the linearly separable case without increasing the algorithmic complexity of the method. These advantages coupled with the sound and well developed theoretical foundation in mathematical optimization make SVM one of the most widely implemented classification methods.

4.2.2.3 *K*-Nearest Neighbor

The *k*-nearest neighbor is one of the most intuitive supervised machine-learning methods. Given a training set with known class association, the class membership of an unknown test instance is predicted to be the same as the majority of the *k* of its closest neighbors. The appropriate choice of a distance function for the feature space plays a significant role for the successful implementation of the method. To avoid tie in the neighbors voting when the problem is a two-class classification, *k* is usually chosen to be an odd number.

4.2.2.4 Decision Trees

Decision trees are representative of the nonparametric group of classification methods. A decision tree is a hierarchical data structure that uses the "divide and conquer" strategy to split the feature space into smaller regions until the local region is identified. The foundation of the decision tree theory for classification and regression is described by Breiman et al. [31]. ID3 [32] and its extension C4.5 [33] are the algorithms on which most of the decision tree implementations are based.

Decision trees are composed of internal decision nodes and terminal leaves. Each decision node implements a test function on one (univariate trees) or many (linear or nonlinear multivariate trees) dimensions of the input data. The branches coming out of each node are labeled with the discrete outcomes of the test function. The terminal leaves are labeled with one of the possible classes. Given a data instance, the algorithm starts with applying the test function at the root node and one of the branches is taken depending on the outcome. This process continues recursively until a leaf node is reached and the data instance is assigned to the class associated with the leaf. This is equivalent to splitting the feature space into regions corresponding to the classes.

Decision trees can handle mixed variables (discrete, continuous, and symbolic) and can be used for classification even when the training data contain errors or missing values. Another advantage is the fast localization of a region covering the input. However, the major appeal of the decision trees lies in their interpretability, i.e., they can be converted into a collection of simple "if–then" rules that are easy to understand and explain. The decision variables closest to the root of the tree have the strongest influence on defining the class participation; thus, they correspond to the significant features of the input data.

4.2.2.5 Sequence Similarity

To apply any of the clustering and classification methods available to researchers to predict the protein properties from protein sequences, it is critical to have a way to express the sequence similarity or dissimilarity. If two objects can be represented by numerical vectors, there are many ways to measure their similarity as a function of the distance between them. The distance can be Euclidian, Hamming, correlation, cosine, etc. However, a protein sequence is a string of alphabetic characters, representing the 20 amino acids, rather than an array of numerical values. The lack of a natural underlying metric for comparing such alphabetic data poses the so-called sequence metric problem and significantly inhibits the use of sophisticated statistical analyses of sequences, modeling, structural and functional aspects of proteins and related problems [34].

The most basic approach to deal with the sequence metric problem is to use sequence alignments and make a conclusion about the sequence similarity from the alignment scores. There are many methods available for pairwise and multiple sequence alignment, including global and local alignment methods based on the Needleman and Wunsch algorithm [35] and multiple sequence alignment methods (FASTA, ClustalW, Pearson). The score for two aligned sequences is calculated as a sum of the log probability of each amino acid of the first sequence to be substituted by its respective aligned amino acid of the second sequence, plus a gap penalty term. These substitution probabilities are based on the propensity for evolutionary change from one amino acid to another. Some of the most widely used substitution matrices, like protein alignment matrix (PAM) [36,37], are derived from an alignment of closely related proteins, while others, like Blocks substitution matrix (BLOSUM) [38,39], are based on more distant proteins. However, the exchangeability of amino acids in a protein context is problematic because phylogenetic relations, mutational bias, and physicochemical effects are difficult to separate [40]. Each amino acid has different physicochemical properties (such as volume, charge, hydrophobicity, etc.) and it is clearly desirable to build distances between the sequences that take them into account. This leads to another approach to the sequence metric problem, which is to replace each amino acid in sequences with its respective value in one or more physicochemical property scales (called amino acid indices) and then express the similarity or dissimilarity of the sequences as the distances between the numerical vectors. Hundreds of amino acid indices exist, some of them representing the same property. An online database, AAIndex, was created for storing different indices and substitution matrices and it is continuously maintained and updated [41–43]. Considering that both amino acid indices and substitution matrices are used to provide measures of similarity of protein sequences, it is an interesting question how they are related to each other. Hierarchical cluster analysis with a correlation distance performed on all available mutation matrices collected in AAindex identified several clusters corresponding to the nature of the data set and the method used for constructing the mutation matrix [42]. This analysis also showed a relationship between the PAM units of Dayhoff's mutation matrix and the volume and hydrophobicity of amino acids. A minimum spanning tree build for the amino acid indices divided them into six groups associated with α and turn propensities, β propensity, amino acid composition, hydrophobicity, and other properties like bulkiness, accessible surface area, and polarity of the amino acid residues.

With so many amino acid indices available, it is difficult to choose the one most relevant for protein sequence representation. Recently Atchley et al. [34] used multivariate statistical analysis on 494 amino acid properties [44] to produce a small set of highly interpretable numerical descriptors that summarize the variation of amino acid physiochemical properties. The resultant factors are linear functions of the original amino acid indices, fewer in number than the original, and reflect clusters of covarying traits that describe the underlying structure of the variables [34].

4.2.2.6 BLAST and PSI-BLAST Profiles

With the current expansion of the available protein databases, it becomes possible to consider not just a single sequence but a whole family of related sequences. When approaching the sequence metric problem, evolutionary information available for a given protein by constructing a multiple sequence alignment can be incorporated, using the additional information that is contained in the observed patterns of variability and location of indels. Such approach has found wide application.

The most commonly used tool for sequence database searching is the Basic Logical Alignment Search Tool (BLAST) [8]. It breaks down the query sequence into words (of four or more characters) and all possible 4-mers of a given query are considered by a sliding window approach. Words that have most common composition are disregarded and the rest are used to search the database. Every time a match is found, it gets extended in both directions until the score (using PAM or BLOSUM matrix) falls below a certain threshold. PSI-BLAST is a more sensitive method for database searching of distance homologies [45]. The algorithm first performs a standard BLAST search and uses the multiple alignment to build a position-specific scoring matrix (PSSM). The next steps are iterations of the search where PSSM is used to evaluate the alignment in every following search. Any new match is then included in construction of a new PSSM. A search like this ends when no more matches to new database sequences are found in the subsequent iterations. Every PSI-BLAST search creates a unique query-specific PSSM that can be saved and used on a different database with the same query. PSI-BLAST is an important tool for protein structure prediction as it enables the collection and generation of large sets of sequences and the corresponding patterns.

4.3 PROTEIN STRUCTURE ANALYSIS

4.3.1 Primary Structure

To better understand the possible and most suitable applications of the various clustering and classification methods, it is important to have a clear perspective of what the protein structure represents and what features and similarity measures it contains. At every level of complexity the protein structure presents a different set of information that can be acquired and used to answer different questions. For example, the primary structure is essential in understanding phylogeny but cannot always be used to infer function. Tertiary and quaternary structures, on the other hand, can be responsible for executing the same function (e.g., glycosylation, phosphorylation) in two distinct organisms but may not have much in common at the primary level. Examples of structural similarity without significant sequence homology are presented by divergent evolution, e.g., TIM barrels [46,47]. Carbonic anhydrases are an example of convergent evolution where identical function is performed by different structures [47].

Clustering has been applied for the analysis of amino acid sequences to predict protein structure and function for well over two decades. Clustering techniques have been used to find primary structure [41,48,49] by discriminating true coding from noncoding sequences [17,48,50], to define protein secondary structure [41,50–54], to identify distant homologs [50,55–57], to define protein families [9,15,58–63] and super families [57,64,65], and more recently, to investigate protein–protein interactions [66–69], residues essential for a particular function [53,70,71] and tertiary structure and function [50,53,72–76].

The protein primary structure by definition is the exact specification of its atomic composition and the chemical bonds connecting those atoms. As for any biopolymer, the primary structure of a protein is determined by the sequence of its monomeric subunits, e.g., amino acid sequence. As the vast majority of available protein sequences are derived from nucleic acid sequences, the ability to identify all true possible coding regions is essential to further study the properties of the putative protein [48,77,78]. The ability to correctly identify the coding regions depends on a number of trivial considerations like the existence of appropriate start codons, length of the open reading frame (ORF), and similarity of the translated sequence to other proteins. Beyond that it is also important to identify the gene's location in the context of the genome, especially in relation to other known genes and is there an effective property of the amino acids that is best preserved by the genetic code. The current belief is that the code has not been "optimized" to favor any particular amino acid characteristic (hydrophobicity, tendency to form residue–residue interactions through hydrophobic link, hydrogen bond, salt bridge, or disulfide bridge, tendency to prefer water–residue hydrogen bonds, etc.) but rather has evolved in such a way as to resist random nucleotide mutations [49].

4.3.2 Secondary Structure

The secondary structure is determined by patterns of hydrogen bonds between backbone amide groups. There are three general groups of amino acid configurations or folding units: α-helix (H), β-sheet (E), and random coil (C). Some groups of amino acids are prone to adopt a helical structure, while others, e.g., with branched side chains or with large aromatic residues, are more often found in β-sheets.

From the perspective of machine learning theory, the secondary structure prediction can be viewed as a classification problem, where the class membership (H, E, or C) of each amino acid residue is determined from sequence features. The performance of different methods can be measured by the three-state overall per-residue accuracy, Q_3, which is the percentage of all residues with correctly predicted secondary structures. Another measure of accuracy used is the segment overlap (SOV) that considers the overlap of predicted and observed segments of secondary structures [79]. The first attempts for predicting secondary structure, like the Chou–Fasman [80] and the Garnier–Osguthorpe–Robson [81] methods, used statistical analysis of the secondary structure formations and the propensity of the amino acids to form them. Brendel et al. [82] included other protein sequence statistics based on the evaluation of the physicochemical properties of amino acids like local clustering of different residue types (e.g., charged residues, hydrophobic residues), long runs of charged or uncharged residues, periodic patterns, counts and distribution of homo oligopeptides, and unusual spacing between particular residue types. These early attempts reached prediction accuracies close to 60%. Later research based on classification methods like neural networks [83–92], SVM [93–100], *k*-nearest neighbor [51,101–104], linear discriminant [105], and hidden Markov models [106–109] achieved accuracies close to 80%. Some authors have proposed hybrid methods, choosing different classifiers for the different stages of the prediction (sequence-to-structure and structure-to-structure) [92,110]. To take advantage of the prediction of several strong performing classifiers, different schemes for combining their predictions were proposed [111–116]. Incorporating prediction of the solvent accessibility into the protein coding scheme also contributed to the increase in the prediction accuracy [117]. When reviewing the progress in the protein secondary structure prediction, it is important to consider the various decisions that the researchers in the field make. These include how to solve the sequence metric problem with an appropriate protein coding scheme, what specific classification method to choose for each step of the prediction algorithm and how to combine the results when hybrid method or several predictors are used together.

Qian and Sejnowski [83] applied neural networks for secondary structure prediction. They used a local coding scheme, for which the input for the neural network is a direct encoding of the amino acids in a sliding window across the sequence and the output is the secondary structure formation corresponding to the middle amino acid in the window. The basic method uses a single sequence-to-structure neural network. A second layer structure-to-structure network is added to account for the fact that the secondary structure at a particular position depends not only on its amino acid residue but also on the context of the structural formation of the sequence, thus bringing the reported Q_3 to 64.3%. Rost and Sander [85] developed the PHD method which involved a two-layered feed-forward neural network and achieved a major improvement of Q_3 to greater than 70%. The novelty of the PHD method was the inclusion of the multiple sequence alignment profile as an input to the neural network, thus incorporating evolutionary information relating the sequences.

Two directions to further improve the accuracy of predictions have been explored. The first is related to protein sequence representation. Jones' PSIPRED method improved the accuracy to 78.3% [86] by utilizing larger sequence databases [118] and more sophisticated sequence profiles generated by PSI-BLAST homology searches and position-specific scoring matrices. The second direction involves the employment of more sophisticated network designs. Peterson et al. [85] used output expansion in which the first layer sequence-to-structure network predicts the secondary structure formation for three neighboring amino acids at once. The additional improvement of Q_3 to over 80% was achieved by training up to 800 networks independently and combining the results by a statistical balloting procedure.

Another innovative approach is the application of recurrent neural networks (NN in which the connections between perceptrons form a directed cycle) by Pollastri et al. [90,91]. A variation of their SSpro method, called SSpro8, was also used to make predictions for all eight possible secondary structure formations (α helix, residue in isolated β bridge, extended strand participating in β ladder, 3-helix, 5-helix, hydrogen-bonded turn, bend and loop or irregular) with a Q_8 accuracy of about 63%.

With the development and increasing popularity of SVM for classification, the method has also been applied for protein secondary structure prediction, first by Hua and Sun [93]. Instead of building a classifier that predicts directly the class membership in one of the three classes H, E, and C, three one-versus-rest binary classifiers (H/~H, E/~E, C/~C) and three one-versus-one (H/E, E/C, and C/H) SVMs with radial basis function kernels were built and then combined into tertiary (three-class) classifiers. The input vectors for the classifiers were multiple sequence alignment profiles, obtained from the HSSP database. The method's three-state Q_3 of 73.5% was comparable with that of the PHD method [85] and its segment overlap accuracy of 76.2% outperformed other current methods. These results were further improved by Ward et al. [94]

who used the PSI-BLAST profiles to train three one-versus-one SVM classifiers with polynomial kernel of degree two and the addition of a second, structure-to-structure SVM with an isotropic Gaussian kernel. An interesting result from this study was that the accuracy of the consensus of the three compared methods—PSIPRED, PROFsec [119] (both based on ANN), and SVM—surpassed the accuracy of the individual methods. Guo et al. [95] used dual-layer SVM, but with a different kernel (radial basis functions) and achieved similar accuracy.

Nguyen and Rajapakse [98] compared the performance of different one-versus-rest and one-versus-one classifiers and their combination into tertiary classifiers, with that of the multiclass SVM (MSVM) formulated by Vapnik and Weston [120] and Crammer and Singer [121]. MSVM had better performance which indicated that it has advantages over the binary classifiers by being able to solve the optimization problem in one step; it was used in both layers of the classification with reported maximum accuracy of 79.5%.

To improve the classification accuracy in the cases for which the test protein sequence does not have sufficiently large set of homologs or good PSI-BLAST generated alignment, Karypis [99] coupled the PSSM profile with the nonposition-specific information from the BLOSUM62 matrix in the coding scheme for his SVM-based method YASSPP. The kernel function combined normalized second-order kernel with an exponential function and was specifically designed to capture the sequence conservation signals around the central residue in each window, decreasing the contribution of the residues situated further. The second-layer SVM was also modified to include the sequence information, thus creating a sequence and structure-to-structure model. Experiments on three commonly used benchmark data sets showed that YASSPP achieved the highest Q_3 of 79.34% in comparison with the previously created neural network and SVM-based classifiers.

Birzele and Kramer [100] proposed a new representation for protein sequences as an input to a two-layer SVM. First, a library of patterns of consecutive amino acids found frequently in the protein database is built. The test sequence is then coded by the weighted frequencies of the patterns found in it and its homologs from a PSI-BLAST generated alignment. Since the library of amino acid patterns may contain hundreds of patterns, the number of features may become intractable and thus requires statistical evaluation and filtering. Despite the relatively small training set, this SVM classifier achieved a Q_3 of 77%, favorably comparable with the other methods for secondary structure prediction. More importantly, it contributed 1.51% to the increased accuracy of consensus prediction when combined with two other classifiers based on neural networks.

Some researchers have applied the *k*-nearest neighbor clustering method to predict the secondary structure of a protein based on the secondary structure of "close" proteins with known structure. Yi and Lander [101] calculated a score for all possible pairs of an amino acid and an environmental class, defined by local structural features like secondary structure formation, solvent accessibility, and polarity to reflect the probability of the amino acid being a class member. Every segment of the proteins with a known structure is coded by the sequence of the environmental classes of its amino acids. The distance between a test protein segment and a segment from a database-coded protein is calculated as the average value of the scores of the respective pairs of amino acid in the test segment and the class of the protein. After finding the nearest segments from the database, the secondary structure for the central residue in the test protein is predicted to be the same as the majority of the central residues in nearest neighbor proteins. This approach was further developed by Salamov and Solovyev [102] in their NNSSP method. To incorporate evolutionary information, the mean scores of the aligned test sequence and its homologs are used, rather than the scores of a single test sequence. The outputs from a number of predictors with different segment size and number of nearest neighbors are combined with a neural network [101] or a majority rule [101,102] and achieved an overall three-state accuracy of 67.1% and 72.2% respectively.

Jiang [51] applied the *k*-nearest neighbor method with a different sequence metric—the pairwise sequence distance with BLOSUM62 similarity matrix. The goodness of any putative cluster was estimated with reliability *z*-score that compares the clustering of the test sequence with real and random proteins. The Q_3 of the method was 68.62% and the reliability *z*-score correlated well with the reliability of prediction. This algorithm was intended to perform well for those secondary structures in a protein whose formation is dominated by the neighboring parts of the sequence and the short-range interactions, and it works with a relatively smaller segment of 21 positions.

Unlike other studies that rely on the secondary structure propensities defined by the local residue interactions within a sliding window, Frishman and Argos [103,104] took into account the nonlocal interactions. The input of their method PREDATOR is a single test sequence and seven types of propensities, calculated for each amino acid. The nonlocal propensities, parallel and antiparallel beta-strand and helix, are based on hydrogen-bonded beta-bridges. The three local propensities of helix, sheet, and coil were calculated based on the known formations of 25 nearest-neighbor sequences in the protein database. Similarly, a beta-turn propensity was calculated. A system of rules was used to predict the secondary structure based on these seven propensities, achieving a Q_3 of 68% [104]. This accuracy has been improved to 75% [103] by using additional information derived from pairwise alignments of the test sequence and its closest homologs from a database, found with a FASTA search. The propensities of the amino acids in the test sequence are computed as weighted average of the propensities of the corresponding amino acids in the aligned sequences. This method is well suited for large-scale sequence analysis projects such as genome characterization, for which precise and significant multiple sequence alignments are not available or achievable.

Some hybrid approaches that combine two or more different classifications methods have also been developed. For example, Ceroni et al. [110] used an SVM as a local predictor for each amino acid and applied a bidirectional recurrent neural network (BRNN) as a filtering stage to refine the results. The authors determined that the SOV accuracy of SVM alone cannot be very high since SVM is trained from individual positions. The BRNN has proved to be effective in smoothing out the final prediction, reaching an SOV of 74%. Lin et al. [92] used a hidden neural network as a first stage predictor with seven output states (helix beginning, helix, helix end, strand beginning, strand, strand end, and coil). The result is then passed through a hidden Markov model to optimally segment the seven-state predictions into three-state secondary structure predictions.

King et al. [112] suggested that combining the strongest suits of several different secondary structure classifiers will improve the prediction accuracy. The authors considered the outputs of PHD [85], DSC [105], NNSSP [102], and PREDATOR [104] methods and combined their predictions using voting, biased voting, linear discriminant, neural network, and decision tree. Voting was able to improve the accuracy of the prediction by an additional 3% compared to the individual classifiers. Decision tree was used by Selbig, Mevissen and Lengauer [111] to form a consensus prediction from six different classifiers. Albreht et al. [113] confirmed that consensus approach based on majority voting of the predictions of several state-of-the-art classifiers is superior to the original predictors. Robles et al. [114] trained three different Bayesian networks (naïve, internal estimation naïve, and Pazzani-EDA) as the consensus voting system to generalize the result of seven predictors and showed an average improvement of 1.21%. Another way of combining two predictors, PSIPRED [86] and a knowledge-based algorithm PROSP [122], into the hybrid method HYPROSP II was successfully tested by Lin et al. [115]. Montgomerie et al. [116] used structure-based sequence alignment in addition to a combination of neural network-based predictors. Their PROTEUS system achieved a Q_3 of 81.3% with sequence-unique test set, even higher accuracies on nonsequence-unique test sets and over 90% accuracy with soluble proteins.

Adamczak, Porollo, and Meller [117] investigated if further improvement can be achieved by incorporating the relative solvent accessibility (RSA) in a neural network predictor of the secondary structure. They extended the traditional protein coding based on PSI-BLAST profile with the entropy, average hydrophobicity, and volume of amino acids observed at each position. The authors first predicted the secondary structure with a two-stage neural network, and then trained networks whose feature space included RSA (predicted by regression-based neural network). The improvement in prediction accuracy of 1.4% when RSA was utilized is probably due to the fact that it captures the effect of the overall packing of the protein at the local structure level. Another proposed measure of structural similarity is based on the profiles of representative local features of C (α) distance matrices of protein structures [123]. A number of representative local feature (LF) patterns are extracts from the distance matrices of all protein fold families by medoid analysis, the medoid being the element closest to the center of a cluster. Each C (α) distance matrix is encoded by labeling all its submatrices with the index of the nearest representative LF patterns. The structure is then represented by the frequency distribution of these indices. The LF frequency profile allows the quick calculation of structural similarity scores among a large number of protein structures. This profile method is also efficient in mapping complex protein structures into a common Euclidean space without prior assignment of secondary structure information or structural alignment.

4.3.3 Tertiary Structure

The tertiary structure of a protein is its 3D structure, as defined by its atomic coordinates. It represents the sum of all its secondary structure elements, is closely related to the primary and secondary structures, and carries out the actual protein function. It has been demonstrated that the rms (root mean square) deviation in atom positions after fitting aligned atoms (a traditional metric of structural similarity) does not perform as well as the structural alignment score. Structural comparison of distantly related proteins detects twice as many distant relationships as the sequence comparison at the same error rate. The number of pairs with significant similarity in terms of sequence, but not structure, are very few, whereas many pairs have significant similarity in terms of structure, but not sequence [15,124]. All tertiary structures can be divided very broadly into two groups: globular and fibrous. The tertiary structure of globular proteins tends to have hydrophilic surface and hydrophobic core. The fibrous proteins tend to be elongated and may consist of multimers. An important feature of the tertiary structure, that is not immediately apparent at the primary and secondary structure levels, is the possibility of modifications. Modifications are any additional biochemical alterations that the protein may acquire prior to or in the process of folding. There are multiple known posttranslational modifications and they involve addition of sugars, alkyl-acetyl- or methyl-groups, oxidation, pegilation, phosphorylation, sulfation, recemization, etc. All these are site-specific and essential to the correct folding and function of the protein. Often, the tertiary structure is supported by additional hydrogen bonds or stable covalent disulfide bonds (bridges), with insulin being the classical example. The tertiary structures are not equally stable and a protein may undergo conformational change with any alteration of its environment. The most common and stable functional state of the protein is its native state. The modifications of the primary structure, prior to the formation of the final configuration, add to the difficulty and uncertainty in developing algorithms for modeling of protein folding. Very often, part of the protein assumes a stable fold that can form independently of the rest of the protein; such structure is recognized as a domain and may have several motifs present in it. The domains are so stable that they can be used as separate components in the construction of a chimera protein, where they can nevertheless retain their individual properties. Large-domain database and search engines are available at: http://www.ncbi.nlm.nih.gov/Structure/cdd/cdd.shtml. The Conserved Domain Database (CDD) is a protein annotation resource, which consists of a collection of well-annotated multiple sequence alignment models for ancient domains and full-length proteins available as PSSMs, for the rapid identification of conserved domains in protein sequences. The Molecular Modeling Database (MMDB) contains experimentally determined biopolymer structures obtained from the Protein Data Bank. The simple modular architecture research tool (SMART) that makes use of simple and genomic nonredundant protein databases can be found at http://smart.embl-heidelberg.de/. Pfam, accessible at http://www.sanger.ac.uk/Software/Pfam/, is a large collection of multiple sequence alignments and hidden Markov models covering many common protein domains and families [21]. The Expert Protein Analysis System (ExPASy) proteomics server, http://ca.expasy.org/ of the Swiss Institute of Bioinformatics, contains databases that provide services which describe protein domains, families, functional sites, and the patterns and profiles associated with them.

The structural classification of proteins (SCOP) is a database of protein domains arranged in hierarchic clusters of families, superfamilies, folds, and finally classes, based on their 3D structure. Proteins within one family have clear evolutionary relationship established by a fixed threshold of sequence similarity (e.g., 30% minimum). Members of the superfamily may have lower sequence similarity but possess structural and functional features consistent with a common evolutionary ancestry. Proteins belong to the same fold if they have the same major secondary structure, the same arrangement, and the same topological connections, but may not have a common evolutionary origin. The highest level of generalization in the accepted hierarchy is the protein class. Currently there are seven defined classes, each comprising of α-proteins, β-proteins, and combinations thereof [125,126]. It is essential to note that the clustering and classification of the protein space based on the homology and structure is a natural prerequisite to function prediction.

Similarity of sequences is currently assessed primarily by using alignments; however, the alignment methods seem inadequate for postgenomic studies since they do not scale well with the data set size and seem to be confined to genomic and proteomic sequences only. Alignment-free similarity measures may need to be pursued further [127–129].

4.3.3.1 Sequence Space

The protein sequence space is defined by all the existing finite variety of protein sequences that are possible in nature. This space is not continuous but has varying granularity and all classification methods aim to organize it in order to identify the structure and function of any given protein. The coarse-grained partitioning of the structure space is based on secondary structure, and the finer fold partitioning on tertiary structure or on function. Protein space becomes "smooth" at the family level. The relation of sequence similarity, as obtained by pairwise alignments to structural or functional properties, has been the subject of many computational approaches. An identity over 30% [130,131] is considered the critical required threshold. The Swiss-Prot data set was used to develop a graph-based clustering approach that uses transitivity and Smith–Waterman local alignment algorithm to assess pairwise similarity. The protein sequences are represented as vertices of a directed graph and an edge is drawn between two vertices that have met the similarity threshold. The algorithm allows for the handling of multidomain proteins and cluster comparisons. SCOP was used for the evaluation of the algorithm and showed 24% improvement for the detection of remote homologs [50]. The algorithm was further improved by using an asymmetric distance measure that scales the similarity values to account for different sequence lengths [132]. ProtoNet is a hierarchical-clustering algorithm applied to the known protein space. It is based on standard similarity measures (as used by gapped BLAST) but in addition uses transitivity and E-scores to define possible merges of clusters and in that manner finds weaker similarities [74]. Fragnostic is a recently proposed Web tool for viewing the protein space and is based on mining for common fragments shared by proteins that are otherwise in different SCOP folds [133]. CLAN is another enriched new algorithm in which protein clustering is based both on sequence similarity and mining of the free annotation text by simple, natural language processing [134].

4.3.3.2 Family/Superfamily Prediction

Since the mid 1990s when information about whole bacterial genome sequences started to accumulate rapidly, the ability to perform whole genome comparisons has become indispensable. Protein structures come in families and these can be closely or loosely knit entities. A measure of relatedness among polymer conformations has also been introduced [135]. It is based on weighted distance maps, is computationally faster and can compare any two proteins, regardless of their relative chain lengths or degree of similarity, thus obviating the requirement for relative alignments. Combined with minimum spanning trees and hierarchical clustering methods, this measure can also be used to define structural families and generate rapid protein structures database searches.

The relationships between genes from different genomes are represented as a system of homologous families that include both orthologs, genes in different species that evolved from common ancestor by speciation, and paralogs, generated by duplications within the genome. Usually the orthologs retain the same functions and the paralogs evolve new functions, so correct ortholog and paralog identification is important for phylogeny, function analysis, and genome-organization analysis. The task of identifying orthologs has been defined as the delineation of clusters of orthologous groups (COGs) [18]. The emerging large-genome sequencing projects have prompted the design of analytical tools that can automatically perform alignment and classification of multiple related protein sequence domains. A procedure that can do this has been developed [136] and it efficiently integrates similarity and alignment techniques. Its successive steps are based on compositional and local sequence similarity searches followed by multiple sequence alignments. Global similarities are detected from the pairwise comparison (e.g., by Smith–Waterman local alignment algorithm) and di-peptide compositions. One sequence from each detected cluster of closely related proteins is compiled in a suffix tree to detect local sequence similarities. Sets of proteins which share similar sequence segments are then weighted according to their closeness and aligned using a fast hierarchical dynamic programming algorithm. This sequence classification has been evaluated for 12,462 primary structures from 341 known families and found to be 93.2% accurate. This low error level compares favorably to the manually constructed PROSITE database (3.4% error rate). The COG system has gained great popularity for studying phylogeny and finding functional patterns for both prokaryotes and eukaryotic organisms [19].

Protein sequence alignment is one of the basic tools in bioinformatics, and accurate alignments are required for any task ranging from derivation of phylogenetic trees to protein-structure prediction. The incorporation of predicted secondary structure information into alignment algorithms improves their performance. Secondary structure predictors have to be trained on a set of defined states (e.g., helix, strand, coil), and it is likely that prediction of other structural features can provide additional improvement. The alignment quality of distantly related proteins can be enhanced by combining two prediction schemes: secondary structure prediction and self-organizing map (an unsupervised clustering method). It has been demonstrated that when the outputs of both schemes are used as inputs to a profile–profile scoring function, the overall secondary structure prediction improves [52]. Phylogenetic analysis of protein sequences is widely used in protein function classification of subfamilies. Many attempts have been made to classify proteins into subfamilies without reconstructing phylogenetic trees (by neighbor joining (NJ) or unweighted pair group method with arithmetic mean), but with simpler distance measures (e.g., hierarchical clustering). It has been demonstrated [137] that there is no uniformly superior algorithm, and that simple protein similarity measures combined with hierarchical clustering produce trees with reasonable and often the most accurate tree-based classification. A tree-building algorithm (TIPS), based on agglomerative clustering with a similarity measure derived from profile scoring, was found comparable with phylogenetic algorithms in classification accuracy. Owing to its time scalability, TIPS is being used in the large-scale protein-classification project PANTHER [22,23].

As noted, sequence identity above some threshold is believed to imply structural similarity due to a common ancestry. This is a necessary, but not sufficient condition for structural similarity; hence, other criteria to identify remote homologs are required [138]. The search for such criteria is becoming more difficult, as the noise level increases with the size of a database. Transitivity allows the deduction of a structural similarity between proteins A and C from the existence of a third protein, B, which has been found to be independently homologous to A as well as to C above a certain threshold. The threshold value is chosen to distinguish between true homolog relationships and random similarities. This threshold is increased to ensure fewer spurious hits in larger databases; this, however, leads inevitably to decreased sensitivity that can be recovered by utilizing refined protocols. It is not fully understood how far the transitivity property can be extended. Graph-based clustering approaches, where transitivity plays a crucial role, have been successfully developed [50,132,139]. Such extended graph-based clustering algorithms use asymmetric distance measure, and scale similarity values based on the length of the compared proteins. The length dependency implied by the self-similarity of the scaling of the alignment scores appears to be an effective criterion to avoid clustering errors associated with multidomain proteins. The significance of the alignment scores is also taken into account and used as filtering step. Further postprocessing merge of the clusters is done based on Profile Hidden Markov Model [132], which is a position-specific probabilistic-based scoring system to capture information about the degree of conservation at various positions in the multiple alignment. This enriched graph-based clustering has been tested on structural classification of proteins (SCOP) sequences and their superfamily level classification as well as with a joint data set containing both SCOP and SWISS-PROT data. It was found to compare very favorably with PSI-BLAST. Usage of transitivity with as many as 12 intermediate sequences was shown crucial to achieving this level of performance.

Both sequence space exploration and correct protein family structure are required for comparative genomics and for better protein function prediction. The emergence of structural proteomics projects requires the development of methods for clustering protein sequences and building families of potentially orthologous sequences. The Ncut algorithm is a clustering strategy capable of grouping related sequences by assessing their pairwise relationships without the need for clustering the full sequence space. This technique was applied with comparable results to the data set used in the construction of the COG database [58]. As aforementioned, a superfamily of proteins can result from divergent evolution of homologs with insignificant similarity in amino acid sequences. Such relationship is established usually after 3-D x-ray analysis or NMR data become available. The SUPFAM database relates homologous protein families in a multiple sequence alignment database of either known or unknown structure. SUPFAM has been derived by relating Pfam families with the families from the PALI database (phylogeny and alignment of homologous protein structures), that contains homologous proteins of known structure derived from SCOP. This first step is followed by matching Pfam families which cannot be associated reliably with a known protein superfamily structure. The IMPALA profile matching procedure (an all-against-all comparison) has been used in all steps. This approach resulted in clustering of 67 homologous protein families of Pfam with apparently no structural information into 28 potential new superfamilies. The same algorithm applied to the genome products of *Mycobacterium tuberculosis* assigned 17 potentially new superfamilies. A newer approach to the problem of automatically clustering protein sequences and discovering protein families, subfamilies etc., is based on the theory of infinite Gaussian mixtures models. In this method the data dictate how many mixture components are required for the modeling, and provide a measure of the probability that two proteins belong to the same cluster. The approach has been tested with globin sequences, globin sequences with known 3D structures and G-protein coupled receptor sequences and had very good consistency of the clusters. This outcome indicates that the method is producing biologically meaningful results. When secondary structure and residue solvent accessibility information are included, the obtained classification of sequences reflects and extends the proteins' SCOP classifications [140].

Grouping proteins into sequence-based clusters is clearly a fundamental step in homology-based prediction of structure or function. Standard clustering methods have only limited practical usefulness because unrelated sequences join clusters as the result of matches to promiscuous domains and this step may precede the formation of biologically meaningful families. The Markov cluster algorithm avoids this nonspecificity, but does not preserve topological or threshold information about protein families. Another sequence-based clustering approach is a hybrid approach that combines the advantages of the standard and Markov cluster algorithms. It has been demonstrated that this algorithm is capable of discerning biologically relevant subclusters within the individual Markov clusters [62]. C-cluster is a good stand-alone clustering algorithm that is able to create large-scale clustering for the identification of protein families within whole proteomes by extraction of maximal cliques. The similarity score's significance is evaluated by a Z-value. The algorithm uses pairwise Smith–Waterman comparison (with PAM 250 matrix) and the Z-values are calculated from all raw alignment scores with value greater than 22. The clustering algorithm that follows is divided into several steps that create core clusters comprised of only maximal cliques (i.e., each member is connected with all others), then adds elements with at least one connection to the core cluster, and, finally, any unidentified sequences [141].

Kelil et al. [142] proposed a new similarity measure based on matching amino acid subsequences to adequately assign unknown protein to a family and to assess differences between proteins that belong to one family but not to a distinct subfamily, i.e., that possesses structural similarity but performs different function. The terms clusters and subfamilies are used interchangeably. The substitution matching similarity (SMS), which can match amino acid subsequences [142] was designed to be applied to nonaligned protein sequences and thus allowed the development of an alignment-free algorithm for clustering of protein sequences (CLUSS). It has been especially designed for application to nonalignable sequences but performs well on alignable ones too. CLUSS was intended for use both for phylogenetic clustering and prediction of biological properties. SMS claims to improve on methods like Kimura's [143] and Felsenstein's [144,145], which in addition to the relative frequencies take into account the existing biological relationships between the amino acid. SMS attempts to better the description of some known properties of the polypeptides or proteins, as it has been determined that certain protein motifs tend to adopt a particular secondary structure. SMS uses experimentally established minimal length of matching subsequences "1" of 4. The CLUSS algorithm has three main stages: building a pairwise similarity matrix based on SMS, building a hierarchical phylogenetic tree, and identifying and extracting the subfamily nodes. The clustering quality measure of this new approach is based on the calculation of the percentage of correctly clustered sequences by known function.

An interesting approach to structure representation is the chaos game representation (CGR) of different families of proteins. CGR is an iterative mapping technique that processes sequences of units, (e.g., amino acids in a protein) in order to find the coordinates for their position in a continuous space. Using a 12-sided regular polygon of concatenated amino acid sequences from proteins of a particular family where the vertices represent a group of amino acid residues that lead to conservative substitutions, the method generates a CGR of the family and a visual representation of the pattern that characterizes it [146]. An estimation of the percentages of plotted grid points in different CGR segments measures the nonrandomness of the generated patterns. Different protein families have been shown to exhibit visually distinct patterns because different functional classes of proteins follow specific statistical biases in the distribution of different mono-, di-, tri-, or higher order peptides along their sequences.

4.3.3.3 Fold

Accurate fold prediction is tightly related to accurate tertiary structure prediction. There are two general approaches to extraction of 3D structures from protein sequences: one is *ab initio*, which predicts directly from the sequences and is based on physicochemical properties; the other is empirical trend, which is based on homology modeling, threading, or taxonomic relationships [147]. Homology modeling compares the query to available 3D structures of similar sequences, threading finds possible fold by sequence-structure alignment without scoring for homology, and the taxonomic approach tries to assign the query to a particular class of fold assuming that the number of folds in nature is limited. Coarse-grained structural classifications (i.e., all alpha-helix, all beta-sheet, etc.) have 70% or better prediction accuracy but are not sufficient for achieving high-resolution (fine–grain) 3D structures. SVM has been applied to the problem of fold assignment. By using six coding schemes to extract structural or physicochemical properties, three descriptors are calculated: composition, transition, and distribution [148,149]. Such an approach gives fairly poor (56%) prediction accuracy using SCOP folds. If the coding scheme is changed to an n-peptide descriptor and the SVM is coupled with jury-voting procedures, the accuracy of prediction improves to 69.6% [147].

RAMP is a suite of programs to aid in the modeling of protein 3D structure. RAMP is the software behind the PROTINFO server and is a *de novo* method for fold generation, developed exclusively for use with protein sequences that are less than 20% identical to any known entities. The method samples the conformational space for native-like conformations, selecting these and filtering them in hierarchically according to different scoring functions. The approach appears to be more suitable for shorter sequences (approx. 100 residues) [150,151]. SVM has also been applied for fold recognition in combination with profile–profile alignment, where the alignment between the query and a template of length n is transformed into a feature vector of length $n + 1$, composed of n profile–profile scores and a raw alignment score. The feature vector is then evaluated by SVM and the output is converted to a posterior probability that the query is related to the template [152]. The method has superior performance at higher levels of specificity probably because it uses intermediate sequence search [153]. Protein fold prediction has also been attempted by using chain graph model or segmentation conditional random fields (SCRFs) [154].

Another protein fold recognition method (MANIFOLD) uses the similarity between target and template proteins, by predicted secondary structure, sequence, and enzyme code, to predict the fold of the target protein. The algorithm has a nonlinear ranking scheme that combines the scores of the three different similarity measures [155]. The functional similarity term increases the prediction accuracy by up to 3% compared to using a combination of secondary structure similarity and PSI-BLAST alone. This demonstrates that using functional and secondary structure information can increase the fold recognition beyond sequence similarity. CLANS (CLuster ANalysis of Sequences) is an alternative Java application that uses a version of the Fruchterman–Reingold graph layout algorithm to display pairwise sequence similarities in two- or three-dimensional space and is useful for large sequence data sets [156].

Amino acid residues critical for the structure and function of a protein are presumed to be conserved throughout evolution and can therefore be used as basis for protein structure prediction. From medium-sized data sets, it was found that the most conserved positions are significantly more clustered than randomly selected sets of positions and this was correct for 92% of the proteins [54]. It could be speculated that this result is biased by differences in amino acid composition between the two "subsets" (i.e., of the conserved and the random residues) or by different sequence separation of the residues. To eliminate those possibilities the degree of clustering of the conserved positions was measured for the native structure and for alternative protein conformations generated by the Rosetta *de novo* structure prediction method. The conserved residues were still found significantly more clustered in the native structure than in the alternative conformations, indicating that the differences in the spatial distribution of conserved residues can indeed be utilized in the prediction.

Proteins are considered to have a similar fold if secondary structure elements are positioned similarly in space and are connected in the same order. Such a common structural scaffold may arise due to either divergent or convergent evolution. Structurally aligned protein regions are separated by less conserved loop regions, where sequence and structure locally deviate from each other and do not superimpose well. Even longer protein loops cannot be viewed as "random coils," and for the majority of protein families there exists a linear dependency between the measures of sequence similarity and the loop structural similarity [157]. A measure for structural (dis)similarity in loop regions, based on the Hausdorff metric, has been introduced to gauge protein relatedness. When tested on homologous and analogous protein structures, it could distinguish them with the same or higher accuracy than the conventional measures that are based on comparing proteins in structurally aligned regions. Distance matrices derived from the loop dissimilarity measure may produce in some cases more reliable cluster trees, so by considering "dissimilar" loop regions rather than conserved core regions only, it is possible to improve understanding of protein evolution [157–159]. Sequence–structure correlation analysis shows that evolutionary plasticity (degrees of structural change per unit of sequence) is significantly different between protein families. Similar analysis performed for protein loop regions of the same 81 homologous protein families showed that evolutionary plasticity of loop regions is greater than the one for the protein core [158].

SVM-Fold is one of the latest tools or servers for discriminative fold and superfamily identification. It combines SVM kernel methods with a new multiclass algorithm. This SVM learns relative weights between one-vs-the-rest classifiers and encodes information about the protein structural hierarchy for multiclass prediction. The method outperforms PSI-BLAST-based nearest neighbor methods in terms of prediction accuracy [160].

4.3.3.4 Class

Predicting a protein's structural class from its amino acid sequence is a fundamental task of computational biology. Small-scale predictions of the structural class of a protein have been achieved by the use of amino acid composition and hydrophobic pattern frequency information as inputs to two types of neural networks. A three-layer back-propagation network and a learning vector quantization network (accuracy of 80.2%) has been proposed and shown to outperform a modified Euclidean statistical clustering algorithm. These results suggest that the information encoded in the protein sequences is both easy to obtain and useful for the prediction of protein structural class by neural networks as well as by standard statistical clustering algorithms [161].

String kernels have been developed for representation of protein sequences with SVM classifiers. SVM approaches are excellent at a binary protein classification problem (i.e., separating objects into two classes), but they have also been used to address the problem that exists in protein science in terms of multiclass, superfamily, or fold recognition. Such multiclass SVM-based server (SVM-Fold) (http://svm-fold.c2b2.columbia.edu) has been introduced recently. The system uses a profile kernel where the underlying feature representation is a histogram of inexact matching k-mer frequencies. Standard binary (one-vs-all) SVM classifiers, trained to recognize individual structural classes, actually create prediction scores that cannot be compared. The one-vs-all SVMs make useful predictions for classes at different levels of the protein structural hierarchy, but do not try to combine these multiple predictions. The proposed SVM method learns the relative weights between one-vs-all classifiers and encodes information about the protein structural hierarchy for multiclass prediction. In large-scale benchmark test based on the SCOP database, this code weighting approach significantly improves on the standard one-vs-all method. Improvements were noted both for superfamily and fold prediction in remote homology setting and for fold recognition. The method strongly outperformed nearest neighbor methods based on PSI-BLAST in terms of prediction accuracy of every structure classification problem [160].

Profile analysis which measures the similarity between a target sequence and a group of aligned sequences (the probe) is an important tool for 3D structure predictions. The probes produce a position-specific scoring table (the profile) that can be aligned with any sequence (the target) by standard dynamic algorithms. Building a library of such profiles, each describing a different structural motif, allows any target sequence to be rapidly scanned for the presence of one of the structures. The success of the prediction is measured by predetermined levels of significance [162]. In the early 1990s there was a several log difference between the available protein sequences and the known tertiary structures. It was proposed that the existing database of known structures can be enriched significantly by the use of sequence homology. At the time, the best relevant method of predicting protein structures was model building by homology inferred from the level of sequence similarity. It was observed that the acceptable threshold of sequence similarity sufficient for structural homology depends strongly on the length of the alignment, so a homology threshold curve as a function of alignment length was used as the first step, followed by homology-derived secondary structure prediction. For each known protein structure, the "derived" database contained the aligned sequences, secondary structure, sequence variability, and sequence profile and implied tertiary structures [163]. Prediction of protein tertiary structure by fold recognition or threading, also known as the Critical Assessment of Structure Prediction (CASP) project was initiated in 1994 [164] and the results recently recapitulated [165,166]. The project was put to the test in series of "blind" structure prediction experiments. Multiple teams around the world were tasked to prepare predictions for multiple different "target" proteins whose structures were soon to be determined.

As the experimental structures became available, the models' accuracy to identify and characterize successful predictions was evaluated. The test targets were categorized in three groups: proteins/folds with known homologs/templates; folds with remote homologs; and a no-template group, proteins with no known homologs. Threading produced specific recognition and accurate models, whenever the structural database contained a template spanning a large fraction of target sequence or not. Presence of conserved sequence motifs was helpful, but not required, and it appeared that threading can succeed whenever similarity to a known structure is sufficiently extensive. Only limited success has been achieved with *de novo* predictions on no-template targets and those predictions have performed fairly well on proteins of only 50–150 amino acid length. Better algorithms appear to be Rosetta [167] and new-fold predictors built on secondary structure, followed by threading to isolate turn structures, followed by clustering to find the best scoring turn fragment [168]. When ANN was used with an amino acid sequence as the input, the accuracy of prediction was about 65%. If evolutionary information (PSI-BLAST) profile, observed secondary structure, and surface accessibility were used as the ANN inputs, the accuracy was reported to be 69.1% successful in β-hairpins prediction. When SVM was used for classification instead of the ANN, the accuracy of the method improved further to 79.2% [169]. It appears that defining the structure of a loop that connects two secondary structures is a central problem in protein structure prediction, and occurs frequently in homology modeling, fold recognition, and in several strategies in *ab initio* structure prediction. The extent to which sequence information in the loop database can be used to predict loop structure has been evaluated by an HMM profile and a PSSM profile derived from PSI-BLAST using a comprehensive classification database of structural motifs ArchDB (containing 451 structural classes). Using the SCOP superfamily definition for loops and a jack-knife test for removing homologous loops, the predictions against the recalculated profiles only take into account the sequence information. Two scenarios were considered: prediction of structural class for comparative modeling and prediction of structural subclass for fold recognition and ab initio modeling. When structural class prediction was made directly over loops with x-ray secondary structure assignment, and only the top 20 classes out of 451 possible were considered, the best accuracy of prediction was 78.5%. If the structural subclass prediction was made over loops using PSI-PRED [86] secondary structure prediction to define loop boundaries, and the top 20 of 1492 subclasses were taken into account, the best accuracy was only 46.7%. The accuracy of loop prediction has also been evaluated by means of RMSD calculations [170]. The performance of the CASP methods has been assessed by specially devised neural network-based methods [171]. Recent CASP highlights are improvements in model accuracy relative to that obtainable from knowledge of a single best-template structure; convergence of the accuracy of models produced by automatic servers toward that produced by human modeling teams; the emergence of methods for predicting the quality of models; and rapidly increasing practical applications of the methods [172].

Prediction of a protein's structural class has also been attempted by a novel, so called supervised fuzzy clustering approach which utilizes the class label information during the training process. As a result, a set of "if-then" fuzzy rules for predicting the protein structural classes are extracted from a training data set. The approach has been tested through two different working data sets and it showed better success than previously introduced unsupervised fuzzy *c*-means methods [75].

Important challenge in genomics is the automatic clustering of homologous proteins from sequence information alone. Most protein clustering methods are local, i.e., based on thresholding a measure related to sequence distance. It has been shown, however, that locality can limit the methods' performance. A spectral clustering-based global method has been introduced to overcome such limitations and was found capable of generating clusters from a set of SCOP proteins close to the number of superfamilies in that set. Fewer elements were singled out, and most remote homologs were grouped correctly. The quality of the clusters measured by sensitivity and specificity was better than hierarchical clustering (84% improvement), Connected Component Analysis (34% improvement), and TribeMCL, also a global method (72% improvement) [56].

By definition, clustering is aimed at partitioning the data in order to find the typical profile of each cluster so that it can be used alone in further analysis, thereby providing great data reduction. The space of all available protein sequences has been used in search of clusters of related proteins in order to automatically detect these sets, and thus obtain a classification of all protein sequences. The analysis used standard measures of sequence similarity as applied to an all-vs-all comparison of SWISS-PROT. The classes correspond to protein subfamilies. Subclasses are subsequently merged using the weaker pairs in a two-phase clustering algorithm. The algorithm makes use of restrictively applied transitivity [76] to identify homologous proteins. This process is repeated at varying levels of statistical significance. Consequently, a hierarchical organization of all proteins is obtained. The resulting classification splits the protein space into well-defined groups of proteins, which are closely correlated with natural biological families and superfamilies. This classification agrees with PROSITE and Pfam databases, domain-based classifications, and it applies for 64.8%–88.5% of the proteins. It also finds many new clusters of protein sequences which were not classified by these databases. The hierarchical organization suggested by this analysis reveals finer subfamilies in families of known proteins as well as many novel relations between protein families [139]. Research on clustering is focused on numerical domains and, in many cases, on assumed metric space. A possible approach to define the similarity between objects in a clear and straightforward way is to use the edit distance between each pair of sequences as the distance measure and then apply some of the existing clustering methods [17]. The computation of pairwise edit distance can be long and inefficient, and usually captures the optimal global alignment between a pair of sequences, but ignores many other local alignments that may represent important features shared by a pair of sequences. An alternative approach is to utilize the suffix tree structure to explore significant patterns of the protein sequences and use them to evaluate the similarity. Such patterns can be used to assess the similarity between a pair of protein sequences and also between a protein sequence and a cluster of protein sequences. The approach can be used to identify and organize significant appearances of segments among cluster of sequences, regardless of the relative positions of these segments within different sequences. These significant patterns imply certain features that are common to a large subset of sequences in the cluster. They reveal important statistical properties of the cluster by inducing a conditional probability distribution of the next symbol given the preceding segments. This probability distribution can be used in estimating the similarity between a sequence and a cluster. Thus, by extracting and maintaining significant patterns that characterize sequence clusters, one can determine whether a sequence should belong to a cluster by calculating the likelihood of reproducing the sequence under the probability distribution that characterizes the given cluster. This algorithm has been tested on a set of 8000 protein sequences of 30 families from the SWISS-PROT database and in 8 min it was able to achieve a high level of accuracy; over 93% of sequences could be correctly clustered with their family members. Therefore, this algorithm can be very useful in clustering a large set of unlabeled protein sequences.

4.4 CONCLUSION

Automatic structural and functional annotations of proteins are fundamental problems of the postgenomic era. The availability of protein interaction networks from many model species has promoted the development of various computational methods to elucidate both protein structure and function. Rigorous measures of similarity for sequence and structure are now well established, yet the problem of defining functional relationships has been particularly challenging. Phylogenetic relatedness data appear indispensable in all current analysis approaches. As we are entering the metagenome era, it is essential to improve and make full use of all bioinformatics tools, and clustering and classification techniques have a lot to offer.

REFERENCES

1. Schmeisser, C., Steele, H., and Streit, W.R. Metagenomics, biotechnology with non-culturable microbes. *Appl. Microbiol. Biotechnol.* 2007, 75(5), 955–962.
2. Breitbart, M., Salamon, P., Andresen, B., Mahaffy, J.M., Segall, A.M., Mead, D., Azam, F., and Rohwer, F. Genomic analysis of uncultured marine viral communities. *Proc. Natl. Acad. Sci. U S A* 2002, 99(22), 14250–14255.
3. Breitbart, M., Miyake, J.H., and Rohwer, F. Global distribution of nearly identical phage-encoded DNA sequences. *FEMS Microbiol. Lett.* 2004, 236(2), 249–256.
4. Breitbart, M., Wegley, L., Leeds, S., Schoenfeld, T., and Rohwer, F. Phage community dynamics in hot springs. *Appl. Environ. Microbiol.* 2004, 70(3), 1633–1640.
5. Kowalchuk, G.A., Speksnijder, A.G., Zhang, K., Goodman, R.M., and van Veen, J.A. Finding the needles in the metagenome haystack. *Microb. Ecol.* 2007, 53(3), 475–485.
6. Rohwer, F. Global phage diversity. *Cell* 2003, 113(2), 141.
7. Schmeisser, C., Stockigt, C., Raasch, C., Wingender, J., Timmis, K.N., Wenderoth, D.F., Flemming, H.C., Liesegang, H., Schmitz, R.A., Jaeger, K.E., and Streit, W.R. Metagenome survey of biofilms in drinking-water networks. *Appl. Environ. Microbiol.* 2003, 69(12), 7298–7309.
8. Altschul, S.F., Gish, W., Miller, W., Myers, E.W., and Lipman, D.J. Basic local alignment search tool. *J. Mol. Biol.* 1990, 215(3), 403–410.
9. Henikoff, S., Henikoff, J.G., and Pietrokovski, S. Blocks +: A non-redundant database of protein alignment blocks derived from multiple compilations. *Bioinformatics* 1999, 15(6), 471–479.
10. Lipman, D.J., Altschul, S.F., and Kececioglu, J.D. A tool for multiple sequence alignment. *PNAS* 1989, 86(12), 4412–4415.
11. Amoutzias, G.D., Weiner, J., and Bornberg-Bauer, E. Phylogenetic profiling of protein interaction networks in eukaryotic transcription factors reveals focal proteins being ancestral to hubs. *Gene* 2005, 347(2), 247–253.
12. Chothia, C. and Gerstein, M. Protein evolution. How far can sequences diverge? *Nature* 1997, 385(6617), 579, 581.
13. Kim, P.M., Lu, L.J., Xia, Y., and Gerstein, M.B. Relating three-dimensional structures to protein networks provides evolutionary insights. *Science* 2006, 314(5807), 1938–1941.
14. Winstanley, H.F., Abeln, S., and Deane, C.M. How old is your fold? *Bioinformatics* 2005, 21(Suppl 1), i449–i458.
15. Elofsson, A. and Sonnhammer, E.L. A comparison of sequence and structure protein domain families as a basis for structural genomics. *Bioinformatics* 1999, 15(6), 480–500.
16. Kunin, V. and Ouzounis, C. Clustering the annotation space of proteins. *BMC Bioinformatics* 2005, 6(1), 24.
17. Yang, J. and Wang, W. Towards automatic clustering of protein sequences. *Proc. IEEE Comput. Soc. Bioinform. Conf.* 2002, 1, 175–186.
18. Tatusov, R.L., Koonin, E.V., and Lipman, D.J. A genomic perspective on protein families. *Science* 1997, 278(5338), 631–637.
19. Tatusov, R.L., Fedorova, N.D., Jackson, J.D., Jacobs, A.R., Kiryutin, B., Koonin, E.V., Krylov, D.M., Mazumder, R., Mekhedov, S.L., Nikolskaya, A.N., Rao, B.S., Smirnov, S., Sverdlov, A.V., Vasudevan, S., Wolf, Y.I., Yin, J.J., and Natale, D.A. The COG database: An updated version includes eukaryotes. *BMC Bioinformatics* 2003, 4, 41.
20. Wheeler, D.L., Barrett, T., Benson, D.A., Bryant, S.H., Canese, K., Chetvernin, V., Church, D.M., DiCuccio, M., Edgar, R., Federhen, S., Geer, L.Y., Kapustin, Y., Khovayko, O., Landsman, D., Lipman, D.J., Madden, T.L., Maglott, D.R., Ostell, J., Miller, V., Pruitt, K.D., Schuler, G.D., Sequeira, E., Sherry, S.T., Sirotkin, K., Souvorov, A., Starchenko, G., Tatusov, R.L., Tatusova, T.A., Wagner, L., and Yaschenko, E. Database resources of the National Center for Biotechnology Information. *Nucleic Acids Res.* 2007, 35(suppl_1), D5–D12.
21. Finn, R.D., Mistry, J., Schuster-Bockler, B., Griffiths-Jones, S., Hollich, V., Lassmann, T., Moxon, S., Marshall, M., Khanna, A., Durbin, R., Eddy, S.R., Sonnhammer, E.L., and Bateman, A. Pfam: Clans, web tools and services. *Nucleic Acids Res.* 2006, 34(Database issue), D247–D251.
22. Thomas, P.D., Kejariwal, A., Campbell, M.J., Mi, H., Diemer, K., Guo, N., Ladunga, I., Ulitsky-Lazareva, B., Muruganujan, A., Rabkin, S., Vandergriff, J.A., Doremieux, O. PANTHER: A browsable

database of gene products organized by biological function, using curated protein family and subfamily classification. *Nucleic Acids Res.* 2003, 31(1), 334–341.
23. Thomas, P.D., Campbell, M.J., Kejariwal, A., Mi, H., Karlak, B., Daverman, R., Diemer, K., Muruganujan, A., and Narechania, A. PANTHER: A library of protein families and subfamilies indexed by function. *Genome Res.* 2003, 13(9), 2129–2141.
24. Kanehisa, M., Goto, S., Hattori, M., oki-Kinoshita, K.F., Itoh, M., Kawashima, S., Katayama, T., Araki, M., and Hirakawa, M. From genomics to chemical genomics: New developments in KEGG. *Nucleic Acids Res.* 2006, 34(Database issue), D354–D357.
25. Theodoridis, S. and Koutroumbas, K. *Pattern Recognition*; Academic Press, San Diego, California: 1999.
26. Hand, D.J., Mannila, H., and Smyth, P. *Principles of Data Mining*; MIT Press, Cambridge, Massasusetts: 2001.
27. Alpaydin, E. *Introduction to Machine Learning*; I ed., The MIT Press, Cambridge, Massasusetts: 2004.
28. Boser, B.E., Guyon, I.M., and Vapnik, V.N. *A Training Algorithm for Optimal Margin Classifiers*; ACM Press, New York: 1992; pp. 144–152.
29. Vapnik, V. *Statistical Learning Theory*; Wiley, New York: 1998.
30. Cristianini, N. and Swawe-Taylor, J. *An Introduction to Support Vector Machines and Other Kernel-Based Learning Methods*; Cambridge University Press, Cambridge, United Kingdom: 2002.
31. Breiman, L., Freidman, J.H., Olshen, R.A., and Stone, C.J. *Classification and Regression Trees*; Wadsworth Statistical Press, Belmont, California: 1984.
32. Quinlan, J.R. Induction of decision trees. *Mach. Learn.* 1986, 1(1), 81–106.
33. Quinlan, J.R. *C4.5: Programs for Machine Learning*; Morgan Kaufmann, San Francisco, California: 1993.
34. Atchley, W.R., Zhao, J., Fernandes, A.D., and Druke, T. Solving the protein sequence metric problem. *Proc. Natl. Acad. Sci. U S A* 2005, 102(18), 6395–6400.
35. Needleman, S.B. and Wunsch, C.D. A general method applicable to the search for similarities in the amino acid sequence of two proteins. *J. Mol. Biol.* 1970, 48(3), 443–453.
36. Dayhoff, M.O. Atlas of protein sequence and structure. *Natl. Biomed. Res. Found.* 1978, 5, 345–358. Ref Type: Serial (Book, Monograph)
37. Dayhoff, M.O., Barker, W.C., and Hunt, L.T. Establishing homologies in protein sequences. *Methods Enzymol.* 1983, 91, 524–545.
38. Henikoff, S. and Henikoff, J.G. Amino acid substitution matrices from protein blocks. *Proc. Natl. Acad. Sci. U S A* 1992, 89(22), 10915–10919.
39. Henikoff, S. and Henikoff, J.G. Performance evaluation of amino acid substitution matrices. *Proteins* 1993, 17(1), 49–61.
40. Yampolsky, L.Y. and Stoltzfus, A. The exchangeability of amino acids in proteins. *Genetics* 2005, 170 (4), 1459–1472.
41. Nakai, K., Kidera, A., and Kanehisa, M. Cluster analysis of amino acid indices for prediction of protein structure and function. *Protein Eng.* 1988, 2(2), 93–100.
42. Tomii, K. and Kanehisa, M. Analysis of amino acid indices and mutation matrices for sequence comparison and structure prediction of proteins. *Protein Eng.* 1996, 9(1), 27–36.
43. Kawashima, S., Pokarowski, P., Pokarowska, M., Kolinski, A., Katayama, T., and Kanehisa, M. AAindex: Amino acid index database, progress report 2008. *Nucleic Acids Res.* 2007, gkm998.
44. Kawashima, S. and Kanehisa, M. AAindex: Amino acid index database. *Nucleic Acids Res.* 2000, 28(1), 374.
45. Altschul, S.F., Madden, T.L., Schaffer, A.A., Zhang, J., Zhang, Z., Miller, W., and Lipman, D.J. Gapped BLAST and PSI-BLAST: A new generation of protein database search programs. *Nucleic Acids Res.* 1997, 25(17), 3389–3402.
46. Das, R. and Gerstein, M. A method using active-site sequence conservation to find functional shifts in protein families: Application to the enzymes of central metabolism, leading to the identification of an anomalous isocitrate dehydrogenase in pathogens. *Proteins* 2004, 55(2), 455–463.
47. Hegyi, H. and Gerstein, M. The relationship between protein structure and function: A comprehensive survey with application to the yeast genome. *J. Mol. Biol.* 1999, 288(1), 147–164.
48. Borodovsky, M., McIninch, J.D., Koonin, E.V., Rudd, K.E., Medigue, C., and Danchin, A. Detection of new genes in a bacterial genome using Markov models for three gene classes. *Nucleic Acids Res.* 1995, 23(17), 3554–3562.

49. Trinquier, G. and Sanejouand, Y.H. Which effective property of amino acids is best preserved by the genetic code? *Protein Eng.* 1998, 11(3), 153–169.
50. Bolten, E., Schliep, A., Schneckener, S., Schomburg, D., and Schrader, R. Clustering protein sequences–structure prediction by transitive homology. *Bioinformatics* 2001, 17(10), 935–941.
51. Jiang, F. Prediction of protein secondary structure with a reliability score estimated by local sequence clustering. *Protein Eng.* 2003, 16(9), 651–657.
52. Ohlson, T., Aggarwal, V., Elofsson, A., and MacCallum, R.M. Improved alignment quality by combining evolutionary information, predicted secondary structure and self-organizing maps. *BMC Bioinformatics* 2006, 7, 357.
53. Ruuskanen, J.O., Laurila, J., Xhaard, H., Rantanen, V.V., Vuoriluoto, K., Wurster, S., Marjamaki, A., Vainio, M., Johnson, M.S., and Scheinin, M. Conserved structural, pharmacological and functional properties among the three human and five zebrafish alpha 2-adrenoceptors. *Br. J. Pharmacol.* 2005, 144(2), 165–177.
54. Schueler-Furman, O. and Baker, D. Conserved residue clustering and protein structure prediction. *Proteins* 2003, 52(2), 225–235.
55. Johnston, C.R. and Shields, D.C. A sequence sub-sampling algorithm increases the power to detect distant homologues. *Nucleic Acids Res.* 2005, 33(12), 3772–3778.
56. Paccanaro, A., Casbon, J.A., and Saqi, M.A. Spectral clustering of protein sequences. *Nucleic Acids Res.* 2006, 34(5), 1571–1580.
57. Pandit, S.B., Gosar, D., Abhiman, S., Sujatha, S., Dixit, S.S., Mhatre, N.S., Sowdhamini, R., and Srinivasan, N. SUPFAM–a database of potential protein superfamily relationships derived by comparing sequence-based and structure-based families: Implications for structural genomics and function annotation in genomes. *Nucleic Acids Res.* 2002, 30(1), 289–293.
58. Abascal, F. and Valencia, A. Clustering of proximal sequence space for the identification of protein families. *Bioinformatics* 2002, 18(7), 908–921.
59. Abhiman, S. and Sonnhammer, E.L. Large-scale prediction of function shift in protein families with a focus on enzymatic function. *Proteins* 2005, 60(4), 758–768.
60. Abhiman, S. and Sonnhammer, E.L. FunShift: A database of function shift analysis on protein subfamilies. *Nucleic Acids Res.* 2005, 33(Database issue), D197–D200.
61. Eskin, E., Noble, W.S., and Singer, Y. Protein family classification using sparse Markov transducers. *J. Comput. Biol.* 2003, 10(2), 187–213.
62. Harlow, T.J., Gogarten, J.P., and Ragan, M.A. A hybrid clustering approach to recognition of protein families in 114 microbial genomes. *BMC Bioinformatics* 2004, 5, 45.
63. Liu, Y., Engelman, D.M., and Gerstein, M. Genomic analysis of membrane protein families: Abundance and conserved motifs. *Genome Biol.* 2002, %19;3 (10), research0054.
64. Caruso, M.E., Jenna, S., Beaulne, S., Lee, E.H., Bergeron, A., Chauve, C., Roby, P., Rual, J.F., Hill, D.E., Vidal, M., Bosse, R., and Chevet, E. Biochemical clustering of monomeric GTPases of the Ras superfamily. *Mol. Cell Proteomics* 2005, 4(7), 936–944.
65. Leo-Macias, A., Lopez-Romero, P., Lupyan, D., Zerbino, D., and Ortiz, A.R. An analysis of core deformations in protein superfamilies. *Biophys. J.* 2005, 88(2), 1291–1299.
66. Ben Hur, A. and Noble, W.S. Kernel methods for predicting protein-protein interactions. *Bioinformatics* 2005, 21(Suppl 1), i38–i46.
67. Craig, R.A. and Liao, L. Phylogenetic tree information aids supervised learning for predicting protein-protein interaction based on distance matrices. *BMC Bioinformatics* 2007, 8, 6.
68. Mamitsuka, H. Essential latent knowledge for protein-protein interactions: Analysis by an unsupervised learning approach. *IEEE/ACM Trans. Comput. Biol. Bioinform.* 2005, 2(2), 119–130.
69. Yu, H., Paccanaro, A., Trifonov, V., and Gerstein, M. Predicting interactions in protein networks by completing defective cliques. *Bioinformatics* 2006, 22(7), 823–829.
70. La, D. and Livesay, D.R. Predicting functional sites with an automated algorithm suitable for heterogeneous datasets. *BMC Bioinformatics* 2005, 6, 116.
71. Landgraf, R., Xenarios, I., and Eisenberg, D. Three-dimensional cluster analysis identifies interfaces and functional residue clusters in proteins. *J. Mol. Biol.* 2001, 307(5), 1487–1502.
72. Han, L., Cui, J., Lin, H., Ji, Z., Cao, Z., Li, Y., and Chen, Y. Recent progresses in the application of machine learning approach for predicting protein functional class independent of sequence similarity. *Proteomics* 2006, 6(14), 4023–4037.

73. Portugaly, E., Harel, A., Linial, N., and Linial, M. EVEREST: Automatic identification and classification of protein domains in all protein sequences. *BMC Bioinformatics* 2006, 7(1), 277.
74. Sasson, O., Vaaknin, A., Fleischer, H., Portugaly, E., Bilu, Y., Linial, N., and Linial, M. ProtoNet: Hierarchical classification of the protein space. *Nucleic Acids Res.* 2003, 31(1), 348–352.
75. Shen, H.B., Yang, J., Liu, X.J., and Chou, K.C. Using supervised fuzzy clustering to predict protein structural classes. *Biochem. Biophys. Res. Commun.* 2005, 334(2), 577–581.
76. Torshin, I.Y. Clustering amino acid contents of protein domains: Biochemical functions of proteins and implications for origin of biological macromolecules. *Front. Biosci.* 2001, 6, A1–A12.
77. Borodovsky, M. and Peresetsky, A. Deriving non-homogeneous DNA Markov chain models by cluster analysis algorithm minimizing multiple alignment entropy. *Comput. Chem.* 1994, 18(3), 259–267.
78. Plasterer, T.N. PROTEAN. Protein sequence analysis and prediction. *Mol. Biotechnol.* 2000, 16(2), 117–125.
79. Zemla, A., Venclovas, C., Fidelis, K., and Rost, B. A modified definition of Sov, a segment-based measure for protein secondary structure prediction assessment. *Proteins* 1999, 34(2), 220–223.
80. Chou, P.Y. and Fasman, G.D. Prediction of the secondary structure of proteins from their amino acid sequence. *Adv. Enzymol. Relat. Areas Mol. Biol.* 1978, 47, 45–148.
81. Garnier, J., Osguthorpe, D.J., and Robson, B. Analysis of the accuracy and implications of simple methods for predicting the secondary structure of globular proteins. *J. Mol. Biol.* 1978, 120(1), 97–120.
82. Brendel, V., Bucher, P., Nourbakhsh, I.R., Blaisdell, B.E., and Karlin, S. Methods and algorithms for statistical analysis of protein sequences. *Proc. Natl. Acad. Sci. U S A* 1992, 89(6), 2002–2006.
83. Qian, N. and Sejnowski, T.J. Predicting the secondary structure of globular proteins using neural network models. *J. Mol. Biol.* 1988, 202(4), 865–884.
84. Rost, B. and Sander, C. Improved prediction of protein secondary structure by use of sequence profiles and neural networks. *Proc. Natl. Acad. Sci. U S A* 1993, 90(16), 7558–7562.
85. Rost, B. and Sander, C. Prediction of protein secondary structure at better than 70% accuracy. *J. Mol. Biol.* 1993, 232(2), 584–599.
86. Jones, D.T. Protein secondary structure prediction based on position-specific scoring matrices. *J. Mol. Biol.* 1999, 292(2), 195–202.
87. Cuff, J.A. and Barton, G.J. Application of multiple sequence alignment profiles to improve protein secondary structure prediction. *Proteins* 2000, 40(3), 502–511.
88. Petersen, T.N., Lundegaard, C., Nielsen, M., Bohr, H., Bohr, J., Brunak, S., Gippert, G.P., and Lund, O. Prediction of protein secondary structure at 80% accuracy. *Proteins* 2000, 41(1), 17–20.
89. Pollastri, G., Przybylski, D., Rost, B., and Baldi, P. Improving the prediction of protein secondary structure in three and eight classes using recurrent neural networks and profiles. *Proteins* 2002, 47(2), 228–235.
90. Pollastri, G., Baldi, P., Fariselli, P., and Casadio, R. Prediction of coordination number and relative solvent accessibility in proteins. *Proteins* 2002, 47(2), 142–153.
91. Pollastri, G. and McLysaght, A. Porter: A new, accurate server for protein secondary structure prediction. *Bioinformatics* 2005, 21(8), 1719–1720.
92. Lin, K., Simossis, V.A., Taylor, W.R., and Heringa, J. A simple and fast secondary structure prediction method using hidden neural networks. *Bioinformatics* 2005, 21(2), 152–159.
93. Hua, S. and Sun, Z. A novel method of protein secondary structure prediction with high segment overlap measure: Support vector machine approach. *J. Mol. Biol.* 2001, 308(2), 397–407.
94. Ward, J.J., McGuffin, L.J., Buxton, B.F., and Jones, D.T. Secondary structure prediction with support vector machines. *Bioinformatics* 2003, 19(13), 1650–1655.
95. Guo, J., Chen, H., Sun, Z., and Lin, Y. A novel method for protein secondary structure prediction using dual-layer SVM and profiles. *Proteins* 2004, 54(4), 738–743.
96. Kim, H. and Park, H. Protein secondary structure prediction based on an improved support vector machines approach. *Protein Eng.* 2003, 16(8), 553–560.
97. Nguyen, M.N. and Rajapakse, J.C. Multi-class support vector machines for protein secondary structure prediction. *Genome Inform.* 2003, 14, 218–227.
98. Nguyen, M.N. and Rajapakse, J.C. Two-stage support vector machine approach to protein secondary structure prediction. *Pac. Symp. Biocomput.* 2005, 10, 346–357.
99. Karypis, G. YASSPP: Better kernels and coding schemes lead to improvements in protein secondary structure prediction. *Proteins* 2006, 64(3), 575–586.

100. Birzele, F. and Kramer, S. A new representation for protein secondary structure prediction based on frequent patterns. *Bioinformatics* 2006, 22(21), 2628–2634.
101. Yi, T.M. and Lander, E.S. Protein secondary structure prediction using nearest-neighbor methods. *J. Mol. Biol.* 1993, 232(4), 1117–1129.
102. Salamov, A.A. and Solovyev, V.V. Prediction of protein secondary structure by combining nearest-neighbor algorithms and multiple sequence alignments. *J. Mol. Biol.* 1995, 247(1), 11–15.
103. Frishman, D. and Argos, P. Seventy-five percent accuracy in protein secondary structure prediction. *Proteins* 1997, 27(3), 329–335.
104. Frishman, D. and Argos, P. Incorporation of non-local interactions in protein secondary structure prediction from the amino acid sequence. *Protein Eng.* 1996, 9(2), 133–142.
105. King, R.D. and Sternberg, M.J. Identification and application of the concepts important for accurate and reliable protein secondary structure prediction. *Protein Sci.* 1996, 5(11), 2298–2310.
106. Schmidler, S.C., Liu, J.S., and Brutlag, D.L. Bayesian segmentation of protein secondary structure. *J. Comput. Biol.* 2000, 7(1–2), 233–248.
107. Bystroff, C., Thorsson, V., and Baker, D. HMMSTR: A hidden Markov model for local sequence-structure correlations in proteins. *J. Mol. Biol.* 2000, 301(1), 173–190.
108. Asai, K., Hayamizu, S., and Handa, K. Prediction of protein secondary structure by the hidden Markov model. *Comput. Appl. Biosci.* 1993, 9(2), 141–146.
109. Aydin, Z., Altunbasak, Y., and Borodovsky, M. Protein secondary structure prediction for a single-sequence using hidden semi-Markov models. *BMC Bioinformatics* 2006, 7, 178.
110. Ceroni, A., Frasconi, P., Passerini, A., and Vullo, A. A combination of support vector machines and bidirectional recurrent neural networks for protein secondary structure prediction. In *AI*IA 2003: Advances in Artificial Intelligence*, Cappelli, A. and Turini, F. (Eds.), Springer, Berlin/Heidelberg: 2003; pp. 142–153.
111. Selbig, J., Mevissen, T., and Lengauer, T. Decision tree-based formation of consensus protein secondary structure prediction. *Bioinformatics* 1999, 15(12), 1039–1046.
112. King, R.D., Ouali, M., Strong, A.T., Aly, A., Elmaghraby, A., Kantardzic, M., and Page, D. Is it better to combine predictions? *Protein Eng.* 2000, 13(1), 15–19.
113. Albrecht, M., Tosatto, S.C.E., Lengauer, T., and Valle, G. Simple consensus procedures are effective and sufficient in secondary structure prediction. *Protein Eng.* 2003, 16(7), 459–462.
114. Robles, V., Larranaga, P., Pena, J.M., Menasalvas, E., Perez, M.S., Herves, V., and Wasilewska, A. Bayesian network multi-classifiers for protein secondary structure prediction. *Artif. Intell. Med.* 2004, 31(2), 117–136.
115. Lin, H.N., Chang, J.M., Wu, K.P., Sung, T.Y., and Hsu, W.L. HYPROSP II-A knowledge-based hybrid method for protein secondary structure prediction based on local prediction confidence. *Bioinformatics* 2005, 21(15), 3227–3233.
116. Montgomerie, S., Sundararaj, S., Gallin, W., and Wishart, D. Improving the accuracy of protein secondary structure prediction using structural alignment. *BMC Bioinformatics* 2006, 7(1), 301.
117. Adamczak, R., Porollo, A., and Meller, J. Combining prediction of secondary structure and solvent accessibility in proteins. *Proteins* 2005, 59(3), 467–475.
118. Rost, B. Twilight zone of protein sequence alignments. *Protein Eng.* 1999, 12(2), 85–94.
119. Rost, B. and Eyrich, V.A. EVA: Large-scale analysis of secondary structure prediction. *Proteins* 2001, Suppl 5, 192–199.
120. Weston, J. and Watkins, C. Multi-class support vector machines. In *Proceedings of ESANN99*, Verleysen, M. (Ed.), D. Facto Press, Brussels, Belgium: 1999.
121. Crammer, K. and Singer, Y. On the learnability and design of output codes for multiclass problems. *Machine Learning* 2002, 47(2–3), 201–233.
122. Wu, K.P., Lin, H.N., Chang, J.M., Sung, T.Y., and Hsu, W.L. HYPROSP: A hybrid protein secondary structure prediction algorithm–a knowledge-based approach. *Nucleic Acids Res.* 2004, 32(17), 5059–5065.
123. Choi, I.G., Kwon, J., and Kim, S.H. Local feature frequency profile: A method to measure structural similarity in proteins. *Proc. Natl. Acad. Sci. U S A* 2004, 101(11), 3797–3802.
124. Levitt, M. and Gerstein, M. A unified statistical framework for sequence comparison and structure comparison. *Proc. Natl. Acad. Sci. U S A* 1998, 95(11), 5913–5920.

125. Murzin, A.G., Brenner, S.E., Hubbard, T., and Chothia, C. SCOP: A structural classification of proteins database for the investigation of sequences and structures. *J. Mol. Biol.* 1995, 247(4), 536–540.
126. Noble, W.S., Kuang, R., Leslie, C., and Weston, J. Identifying remote protein homologs by network propagation. *FEBS J.* 2005, 272(20), 5119–5128.
127. Dobson, P.D. and Doig, A.J. Predicting enzyme class from protein structure without alignments. *J. Mol. Biol.* 2005, 345(1), 187–199.
128. Ferragina, P., Giancarlo, R., Greco, V., Manzini, G., and Valiente, G. Compression-based classification of biological sequences and structures via the Universal Similarity Metric: Experimental assessment. *BMC Bioinformatics* 2007, 8, 252.
129. Kensche, P.R., van Noort, V., Dutilh, B.E., and Huynen, M.A. Practical and theoretical advances in predicting the function of a protein by its phylogenetic distribution. *J.R. Soc Interface.* 2008, 5(19), 151–170.
130. Brenner, S.E., Chothia, C., and Hubbard, T.J. Assessing sequence comparison methods with reliable structurally identified distant evolutionary relationships. *Proc. Natl. Acad. Sci. U S A* 1998, 95(11), 6073–6078.
131. Pearson, W.R. Comparison of methods for searching protein sequence databases. *Protein Sci.* 1995, 4(6), 1145–1160.
132. Pipenbacher, P., Schliep, A., Schneckener, S., Schonhuth, A., Schomburg, D., and Schrader, R. ProClust: Improved clustering of protein sequences with an extended graph-based approach. *Bioinformatics* 2002, 18(Suppl 2), S182–S191.
133. Friedberg, I. and Godzik, A. Fragnostic: Walking through protein structure space. *Nucleic Acids Res.* 2005, 33(Suppl 2), W249–W251.
134. Kunin, V. and Ouzounis, C. Clustering the annotation space of proteins. *BMC Bioinformatics* 2005, 6(1), 24.
135. Yee, D.P. and Dill, K.A. Families and the structural relatedness among globular proteins. *Protein Sci.* 1993, 2(6), 884–899.
136. Gracy, J. and Argos, P. Automated protein sequence database classification. I. Integration of compositional similarity search, local similarity search, and multiple sequence alignment. *Bioinformatics* 1998, 14(2), 164–173.
137. Lazareva-Ulitsky, B., Diemer, K., and Thomas, P.D. On the quality of tree-based protein classification. *Bioinformatics* 2005, 21(9), 1876–1890.
138. Gerstein, M. Measurement of the effectiveness of transitive sequence comparison, through a third intermediate sequence. *Bioinformatics* 1998, 14(8), 707–714.
139. Yona, G., Linial, N., and Linial, M. ProtoMap: Automatic classification of protein sequences, a hierarchy of protein families, and local maps of the protein space. *Proteins* 1999, 37(3), 360–378.
140. Dubey, A., Hwang, S., Rangel, C., Rasmussen, C.E., Ghahramani, Z., and Wild, D.L. Clustering protein sequence and structure space with infinite Gaussian mixture models. *Pac. Symp. Biocomput.* 2004, 9, 399–410.
141. Mohseni-Zadeh, S., Brezellec, P., and Risler, J.L. Cluster-C, an algorithm for the large-scale clustering of protein sequences based on the extraction of maximal cliques. *Comput. Biol. Chem.* 2004, 28(3), 211–218.
142. Kelil, A., Wang, S., Brzezinski, R., and Fleury, A. CLUSS: Clustering of protein sequences based on a new similarity measure. *BMC Bioinformatics* 2007, 8, 286.
143. Kimura, M. A simple method for estimating evolutionary rates of base substitutions through comparative studies of nucleotide sequences. *J. Mol. Evol.* 1980, 16(2), 111–120.
144. Felsenstein, J. Inferring phylogenies from protein sequences by parsimony, distance, and likelihood methods. *Methods Enzymol.* 1996, 266, 418–427.
145. Felsenstein, J. Taking variation of evolutionary rates between sites into account in inferring phylogenies. *J. Mol. Evol.* 2001, 53(4–5), 447–455.
146. Basu, S., Pan, A., Dutta, C., and Das, J. Chaos game representation of proteins. *J. Mol. Graph. Model.* 1997, 15(5), 279–289.
147. Yu, C.S., Wang, J.Y., Yang, J.M., Lyu, P.C., Lin, C.J., and Hwang, J.K. Fine-grained protein fold assignment by support vector machines using generalized npeptide coding schemes and jury voting from multiple-parameter sets. *Proteins* 2003, 50(4), 531–536.
148. Ding, C.H. and Dubchak, I. Multi-class protein fold recognition using support vector machines and neural networks. *Bioinformatics* 2001, 17(4), 349–358.

149. Dubchak, I., Muchnik, I., Mayor, C., Dralyuk, I., and Kim, S.H. Recognition of a protein fold in the context of the structural classification of proteins (SCOP) classification. *Proteins* 1999, 35(4), 401–407.
150. Hung, L.H. and Samudrala, R. PROTINFO: Secondary and tertiary protein structure prediction. *Nucleic Acids Res.* 2003, 31(13), 3296–3299.
151. Hung, L.H., Ngan, S.C., Liu, T., and Samudrala, R. PROTINFO: New algorithms for enhanced protein structure predictions. *Nucleic Acids Res.* 2005, 33(Web Server issue), W77–W80.
152. Han, S., Lee, B.C., Yu, S.T., Jeong, C.S., Lee, S., and Kim, D. Fold recognition by combining profile-profile alignment and support vector machine. *Bioinformatics* 2005, 21(11), 2667–2673.
153. Park, J., Teichmann, S.A., Hubbard, T., and Chothia, C. Intermediate sequences increase the detection of homology between sequences. *J. Mol. Biol.* 1997, 273(1), 349–354.
154. Liu, Y., Carbonell, J., Weigele, P., and Gopalakrishnan, V. Protein fold recognition using segmentation conditional random fields (SCRFs). *J. Comput. Biol.* 2006, 13(2), 394–406.
155. Bindewald, E., Cestaro, A., Hesser, J., Heiler, M., and Tosatto, S.C. MANIFOLD: Protein fold recognition based on secondary structure, sequence similarity and enzyme classification. *Protein Eng.* 2003, 16 (11), 785–789.
156. Frickey, T. and Lupas, A. CLANS: A Java application for visualizing protein families based on pairwise similarity. *Bioinformatics* 2004, 20(18), 3702–3704.
157. Panchenko, A.R. and Madej, T. Analysis of protein homology by assessing the (dis)similarity in protein loop regions. *Proteins* 2004, 57(3), 539–547.
158. Panchenko, A.R., Wolf, Y.I., Panchenko, L.A., and Madej, T. Evolutionary plasticity of protein families: Coupling between sequence and structure variation. *Proteins* 2005, 61(3), 535–544.
159. Panchenko, A.R. and Madej, T. Structural similarity of loops in protein families: Toward the understanding of protein evolution. *BMC Evol. Biol.* 2005, 5(1), 10.
160. Melvin, I., Ie, E., Kuang, R., Weston, J., Stafford, W.N., and Leslie, C. SVM-Fold: A tool for discriminative multi-class protein fold and superfamily recognition. *BMC Bioinformatics* 2007, 8(Suppl 4), S2.
161. Metfessel, B.A., Saurugger, P.N., Connelly, D.P., and Rich, S.S. Cross-validation of protein structural class prediction using statistical clustering and neural networks. *Protein Sci.* 1993, 2(7), 1171–1182.
162. Gribskov, M., Homyak, M., Edenfield, J., and Eisenberg, D. Profile scanning for three-dimensional structural patterns in protein sequences. *Comput. Appl. Biosci.* 1988, 4(1), 61–66.
163. Sander, C. and Schneider, R. Database of homology-derived protein structures and the structural meaning of sequence alignment. *Proteins* 1991, 9(1), 56–68.
164. Bauer, A. and Bryant, S.H. A measure of success in fold recognition. *Trends Biochem. Sci.* 1997, 22(7), 236–240.
165. Kryshtafovych, A., Fidelis, K., and Moult, J. Progress from CASP6 to CASP7. *Proteins* 2007, 69(Suppl 8), 194–207.
166. Kryshtafovych, A., Prlic, A., Dmytriv, Z., Daniluk, P., Milostan, M., Eyrich, V., Hubbard, T., and Fidelis, K. New tools and expanded data analysis capabilities at the Protein Structure Prediction Center. *Proteins* 2007, 69(Suppl 8), 19–26.
167. Bonneau, R., Strauss, C.E.M., Rohl, C.A., Chivian, D., Bradley, P., Malmstrom, L., Robertson, T., and Baker, D. De novo prediction of three-dimensional structures for major protein families. *J. Mol. Biol.* 2002, 322(1), 65–78.
168. Fang, Q. and Shortle, D. Prediction of protein structure by emphasizing local side-chain/backbone interactions in ensembles of turn fragments. *Proteins* 2003, 53(Suppl 6), 486–490.
169. Kumar, M., Bhasin, M., Natt, N.K., and Raghava, G.P. BhairPred: Prediction of beta-hairpins in a protein from multiple alignment information using ANN and SVM techniques. *Nucleic Acids Res.* 2005, 33(Web Server issue), W154–W159.
170. Fernandez-Fuentes, N., Querol, E., Aviles, F.X., Sternberg, M.J., and Oliva, B. Prediction of the conformation and geometry of loops in globular proteins: Testing ArchDB, a structural classification of loops. *Proteins* 2005, 60(4): 746–757.
171. Wallner, B. and Elofsson, A. Can correct protein models be identified? *Protein Sci.* 2003, 12(5), 1073–1086.
172. Moult, J., Fidelis, K., Kryshtafovych, A., Rost, B., Hubbard, T., and Tramontano, A. Critical assessment of methods of protein structure prediction-Round VII. *Proteins* 2007, 69(Suppl 8), 3–9.

5 Analysis of Coordinated Substitutions in Proteins

David Campo

CONTENTS

> I mean by this expression (Correlation of growth) that the whole organization is so tied together during its growth and development, that when slight variations in any one part occur, and are accumulated through natural selection, other parts become modified.
>
> **Darwin. C, *The Origin of Species* [1]**

5.1 INTRODUCTION

One way of studying protein structure and function is to carry out site-directed mutagenesis where specific residues within a protein are altered, and then to examine the effects of these changes on protein characteristics. Changes in the amino acid (aa) properties (e.g., hydrophobicity, volume, and charge) of the mutated sites can then be correlated with changes in protein characteristics [2]. Another approach is to analyze large families of naturally occurring proteins or protein domains. During divergent evolution, protein sequences change through genetic drift, while the biochemical function of the protein is substantially retained. It is known that the number of sequences exceeds the number of structures by several orders of magnitude and, therefore, the number of three-dimensional protein structures corresponding to a given function is small, from one (like the hand-shaped structure of nucleic acid polymerases) to a small number (for example the four families of endoproteases) [3,4]. The core conformation of homologous proteins persists long after the statistically significant sequence similarities have vanished [5] and this persistence underlies all tools where a function is predicted or a three-dimensional model of a protein is built by extrapolation from an experimental structure of a homologue sequence [6].

By examining patterns of sequence diversity, one can explore how naturally occurring sequence variability and aa properties are important in maintaining protein structure. Analysis of the variation in aa at different sites allows the understanding of the structural–functional role of residues at these positions and to predict protein structure [7]. Some functionally important protein sites are easily detected since they correspond to conserved columns in a multiple-sequence alignment (MSA) but non-conserved sites are also interesting as they may be functionally or structurally important, or possibly key sites of interaction between the protein and its substrate [8]. Experimental and quantitative analyses of proteins often assume that the protein sites are independent, i.e., the presence of a residue at one site is assumed to be independent of residues at other sites. However, the activities and properties of proteins are the result of interactions among their constitutive aa and this leads to the hypothesis that in the course of evolution, substitutions which tend to destabilize a particular structure are probably compensated by other substitutions which confer stability on that structure [9]. Interactions among aa sites include salt bridges between charged residues, hydrogen bonds between electron acceptors and donors, size constraints reflecting structural interactions between large and small side chains, electrostatic interactions, hydrophobic effects, van der Waal's forces, and similar phenomena [2]. It is reasonable to suppose that sites that can compensate for a destabilizing substitution at another site are likely to be close to this site in the three-dimensional structure of the protein. For example, if a salt bond were important to structure and function, a substitution of the positively charged residue with a neutral residue would need to be compensated by a nearby residue substituting from a neutral to a positive residue (Figure 5.1). Similarly, a substitution involving a reduction of volume in the protein core might cause a destabilizing pocket which only one or a few adjacent residues would be capable of filling. Thus, if structural compensation is a general phenomenon, sites which are close together in the three-dimensional structure will tend to evolve in a correlated fashion due to the compensation process [9].

There is experimental evidence indicating that proteins contain pairs of covariant sites, which were found both by analysis of the families of natural proteins with known structures [10–15] and in proteins into which point mutations have been introduced by site-directed mutagenesis [16–18]. In these examples, sites distant in the sequence but near in three-dimensional space in the folded structure have been observed to undergo simultaneous compensatory variation to conserve the overall volume, charge, or hydrophobicity [6]. Several experiments also have provided strong evidence of compensatory mutations in the fast evolution of RNA viruses [15,19–24].

Independent mutations among functionally linked sites would be disadvantageous but simultaneous or sequential compensating mutations may allow the protein to retain function [25]. Furthermore, there are constraints on aa replacements that arise for functional reasons, such as aa bias at recognition sites related to DNA binding in transcriptional regulators. Evolutionarily related sequences should contain the vestiges of these effects in the form of covariant pairs of sites [26] and these interactions can be manifested in covariation between substitutions at pairs of alignment positions in a MSA. The analysis of covariation has been used in protein-engineering approaches [27], sequence-function correlations [2,28], protein structure prediction methods [13,26,29–39], and in finding important motifs in viral proteins [40–43]. However, early studies did not optimally discriminate the three different sources of covariation: (1) chance, (2) common ancestry, and (3) structural or functional constraints. Effectively discriminating among these underlying causes is a difficult task with many statistical and computational difficulties. Improved analyses that discriminate the different sources of covariation confirmed that highly coordinated sites are often functionally related or spatially coupled [2,6,8,22,44,45].

Covariation analyses can also be important in identifying sites that may change the phenotype of a protein, and they could be used as a tentative map for researchers attempting to define functional domains in the protein through mutational analysis. For instance, covariant sites could be used as a

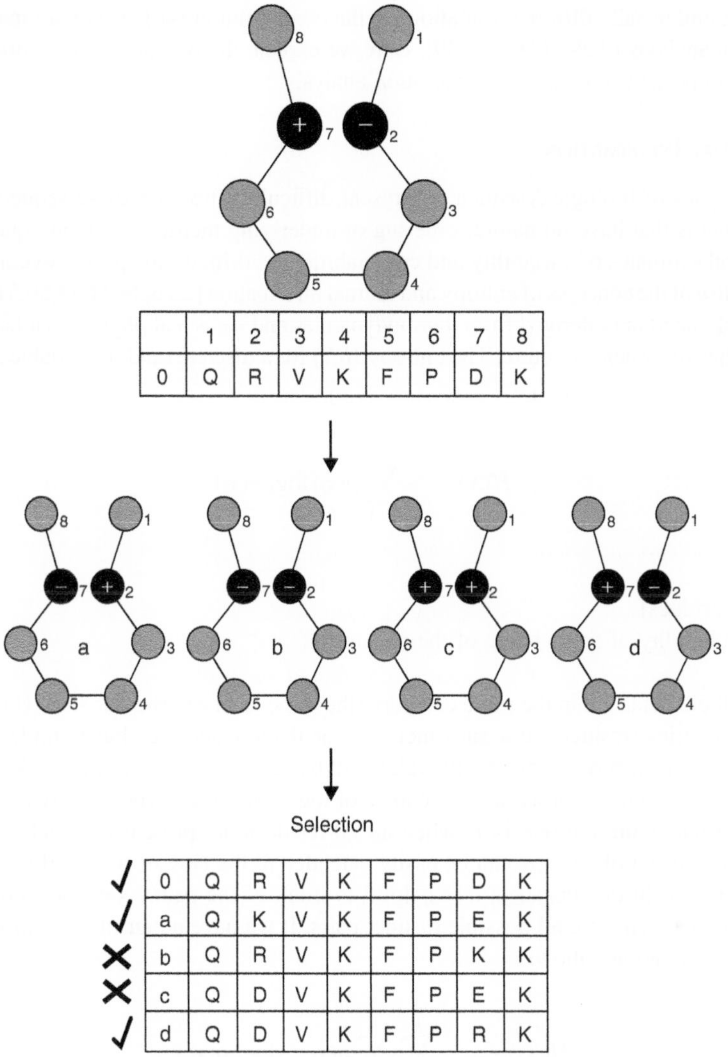

FIGURE 5.1 Schematic representation of coordinated substitutions in a pair of aa sites forming a salt bond in a protein domain. Sequences that contain residues of the same charge at positions 2 and 7 are unstable (b and c) and are eliminated during natural selection. Sequences containing residues of different charges that are stable (0, a, and d) can occur in a multiple-sequence alignment.

guide for reasoned selection of sets of sequences for inclusion in a mixture of peptides for vaccine design. Therefore, by selecting sequences which include pairs of aa that are highly predictive of each other, one may be covering important classes of sequences that are structurally or functionally related. Thus inclusion of peptides with highly covariant aa may be a useful strategy for designing broadly reactive vaccines [41].

5.2 METHODS

Bioinformatics methods for detecting correlated mutations consist of two main steps (1) alignment of homologous sequences and (2) identification of pairs of columns in the alignment in which there is a statistically significant tendency for mutations in one column to be accompanied by corresponding and usually different mutations in the other column [46]. There are many algorithms for covariation analysis [9,29,36,41,47–49]. Here we explain the two principal algorithms and then discuss some important problems of covariation analysis.

5.2.1 Mutual Information

Statistical analyses of biological sequences present difficulties because these sequences are represented by symbols that have no natural ordering or underlying metric [50]. Consequently, conventional statistical estimates of variability and covariability are difficult to apply. Several authors have suggested the use of the concepts of entropy and mutual information [2,26,39,41,44,50,51]. Entropy (H) is a measure of uncertainty derived from thermodynamics and statistical physics that has considerable utility for studies of protein structure. The entropy $H(X)$ for a discrete random variable X is defined as follows:

$$H(X) = -\sum_{i=1}^{k} p(x_i)\log_b p(x_i)$$

where

$k = 20$ (aa residues)
$p(x_i)$ the probability of an aa being of the ith kind

$H = 0$ when all elements are in the same category (the same aa at a particular site). H increases if the number of categories (residues at a site) increases or if the categories have similar probabilities. Thus, the minimum entropy or uncertainty value will be zero when only a single residue occurs at a particular site in all included proteins. The choice of logarithm base b serves to scale the entropy, if $b = k$, then the maximum entropy is 1, when all 20 residues are present in equal frequencies at a given site. The concept of entropy can be easily extended to the case of two random variables with ordered pairs (x_i, y_i). In this instance, it is helpful to think of the pairs as elements of an extended alphabet, whose elements are all possible distinct pairs. If we have a pair of random variables, then pair entropy is defined as follows:

$$H(X,Y) = -\sum_{i=1}^{k}\sum_{j=1}^{l} p(x_i,y_i)\log_b p(x_i,y_i)$$

The relative information content of Y contained in X is termed the mutual information (MI) and is calculated as follows:

$$\mathrm{MI}(X,Y) = H(X) + H(Y) - H(X,Y)$$

In biological sequences, MI describes the extent of association between residues at aa sites X and Y that might arise from evolutionary, functional, or structural constraints [2]. Note that $\mathrm{MI}(X,Y) = \mathrm{MI}(Y,X)$, and if X and Y are independent, then $\mathrm{MI}(X,Y) = 0$, corresponding to the fact that no information is obtained regarding Y by finding out about X. MI is always nonnegative and achieves its maximum value if there is complete covariation. The minimum value of 0 is obtained either when X and Y vary independently or when there is no variation [41]. MI is always lower than the minimum entropy of X and Y and therefore, it is convenient to normalize the MI to compare pairs of different entropy. Martin et al. [8] assessed the performance of various normalizations of MI in enhancing the detection of covariation and found that normalizing MI by the pair entropy optimized the ability to detect coevolving sites over a large range of mutation rates.

5.2.2 Physicochemical Correlation

The information theoretic approach to sequence data has serious shortcomings. For example, it is difficult to describe inverse (negative) relationships among sequence sites, such as those found with compensatory variation associated with aa charge or size. Furthermore, this approach provides little information about the underlying causal complexity of observed covariation [52]. Functionally, significant coordinated substitutions of residues in proteins must result from interactions dependent on the physicochemical property values of the residues [7]. One approach for the analysis of covariation is based on estimation of the correlation coefficient between the values of a physicochemical parameter at a pair of positions of sequence alignment. Let us consider a sample of N aligned sequences of length L. Then, we consider a certain physicochemical aa property f. A value of this property is attributed to every aa in the alignment. As a result, we obtain a matrix whose element f_{ki} is the f value at the ith position of the kth sequence. In the case of evolutionary unrelated sequences, the covariance s_{ij} (if $i \neq j$) and variance (if $i = j$) are equal to

$$S_{ij} = \frac{1}{N-1} \sum_{k=1}^{N} (f_{ki} - \bar{f}_i)(f_{kj} - \bar{f}_j)$$

To estimate the relation between the pair of variables f_i and f_j, the linear correlation coefficient is calculated as

$$r_{ij} = \frac{S_{ij}}{\sqrt{S_{ii} \cdot S_{jj}}}$$

When the correlation coefficient between two sites is negative, an increase in the value of a property at position i will make more likely a substitution at position j that will result in a decrease in the value of the property (a net value compensatory substitution) [53]. When the correlation coefficient is positive, it may be assumed that substitutions are compensatory for the difference between the property values of two residues (the amount of the difference is conserved) [53]. There are other correlation methods that do not assume a linear relationship between variables such as χ^2, Spearman's ρ, and Kendall's τ.

5.2.3 Physicochemical Properties of Amino Acids

A study by Chelvanayagam et al. [6] found that the analysis of covariation involving different physicochemical characteristics improves the number of truly covariant pairs. However, there are many reported aa properties and the selection of the right ones is a difficult choice. Interestingly, Atchley et al. [52] used multivariate statistical analyses on 494 aa properties [54] to produce a small set of highly interpretable numeric patterns of aa variability that can be used in a wide variety of analyses directed toward understanding the evolutionary, structural, and functional aspects of protein variability. Factor analysis was used to produce a subset of numerical descriptors to summarize the set of aa physiochemical properties. Factor analysis simplifies high-dimensional data by generating a smaller number of factors that describe the structure of highly correlated variables. The resultant factors are linear functions of the original data, fewer in number than the original, and reflect clusters of covarying traits that describe the underlying structure of the variables [52].

Factor analysis of the aa attributes resulted in five factors or patterns of highly intercorrelated physiochemical variables, a reduction in dimensionality of two orders of magnitude from the original 494 properties [52] (Table 5.1). POLARF1 reflects polarity and simultaneous covariation in portion of exposed residues versus buried residues, nonbonded energy versus free energy, number of hydrogen bond donors, polarity versus nonpolarity, and hydrophobicity versus hydrophilicity. HELIXF2 is a secondary structure factor. There is an inverse relationship of

TABLE 5.1
Highest Factor Coefficients of AA Properties

POLARF1	
Average nonbonded energy per atom	1.028
Percentage of exposed residues	1.024
Percentage of buried residues	−1.017
Average accessible surface area	1.005
Transfer free energy	−1.003
Residue accessible surface area in folded protein	0.95
Average interactions per side chain atom	−0.928
Average side chain orientation angle	−0.896
Eisenberg hydrophobic index	−0.864
Hydropathy index	−0.856
HELIXF2	
Average relative probability of helix	−1.004
Relative frequency in α-helix	−0.987
α-Helix indices	−0.939
Normalized frequency of coil	0.863
Free energy in α-helical region	0.858
Normalized frequency of turn	0.831
Information measure for loop	0.786
Chou–Fasman parameter of coil conformation	0.78
Helix–coil equilibrium constant	−0.724
Conformational parameter of β-turn	0.693
SIZEF3	
Bulkiness	0.988
Hydrophobicity factor	0.833
Size	0.811
Residue volume	0.794
Average volume of buried residue	0.766
Side chain volume	0.754
Normalized frequency of extended structure	0.706
Molecular weight	0.657
Normalized frequency of left-handed α-helix	−0.641
Normalized frequency of β-sheet, unweighted	0.611
CODONF4	
aa composition of total proteins	0.963
Relative frequency of occurrence	0.931
Number of codons	0.867
aa composition	0.852
Heat capacity	−0.656
Refractivity	−0.621
Average nonbonded energy per residue	−0.507
Molecular weight	−0.504
Conformational parameter of β-turn	−0.439
Normalized frequency of turn	−0.393

TABLE 5.1 (continued)
Highest Factor Coefficients of AA Properties

CHARGEF5

Eisenberg hydrophobic index	−0.864
Number of hydrogen bond donors	0.809
Negative charge	0.451
Positive charge	0.442
Relative mutability	0.337
Isoelectric point	0.224
Number of codons	0.079
Normalized frequency of left-handed α-helix	−0.079
Net charge	0.078
Average nonbonded energy per residue	0.042

Source: From Atchley, W., Zhao, J., Fernandes, A., and Druke, T., *Proc. Natl. Acad. Sci. U S A*, 102, 6395, 2005.

Note: The values of each AAIndex and the source reference can be found at the AAIndex On-Line Database (Kawashima, S. and Kanehisa, M., *Nucleic Acids Res.*, 28, 374, 2000). The names of the factors (e.g., POLARF1) reflect the most important physicochemical property of each factor but the reader must keep in mind that each factor reflects a set of related properties.

relative propensity for various aa in various secondary structural configurations, such as a coil, a turn, or a bend versus the frequency in an α-helix. SIZEF3 relates to molecular size or volume with high factor coefficients for bulkiness, residue volume, average volume of a buried residue, side chain volume, and molecular weight. CODONF4 reflects relative aa composition in various proteins and the number of codons for each aa. These attributes vary inversely with refractivity and heat capacity. CHARGEF5 refers to electrostatic charge with high coefficients on isoelectric point and net charge. Atchley et al. [52] showed how the transformation into one of the five multidimensional factors of physicochemical properties was useful in the analysis of bHLH proteins that bind DNA.

5.2.4 Invariance of Protein Physicochemical Characteristics

An important feature of coordinated substitutions is their additional contribution to the invariance of the integral physicochemical characteristics of a protein, such as the total volume and net charge. Invariance of a physicochemical characteristic may result from the pressure of selection either on the entire protein or on its functionally or structurally significant parts. For example, Afonnikov et al. [36] analyzed the DNA-binding domain of the homeodomain class, finding two conservative physicochemical characteristics preserved due to coordinated substitutions at certain groups of positions in the protein sequence. Integral characteristics of proteins have also been found in Zinc-finger domains, DNAJ domains of heat-shock proteins, DNA-binding domains of CREB, and AP-1 and the Btk PH domain, where information on these characteristics facilitated predictions of their functional motifs [36,53,55]. An integral characteristic (F) of a protein is described as the sum of the values of a physicochemical property at protein positions, with variance D(F) [36]. A permutation procedure can be used to test the hypothesis that the observed sample variance is lower than the expected if correlation between sites is absent.

5.3 SOME PROBLEMS AND THEIR SOLUTIONS

5.3.1 Sequence Conservation

All the methods for detecting correlated mutations are sensitive to the degree of sequence conservation in the alignment [56]. A covariation analysis is supposed to detect how the changes in column i effect column j and, therefore, if there are no changes in a column, the algorithm must choose to report no score, a perfectly high score, or a perfectly low score. Each algorithm for covariation analysis works on certain level of sequence conservation and within that level chooses the residue pairs that truly covary. The MI approach will tend toward a low covariance score for highly conserved pairs of columns in an alignment. The physicochemical correlation approach will tend toward a high covariance score for highly conserved columns. Simulations of protein coevolution have showed that it is very difficult to separate sites which are coevolving from those that are not if either site is highly conserved [8,44]. A first step in the covariation analysis must be the removal of perfectly conserved columns (entropy equal to zero) and depending of the variability of the dataset, a polymorphism cutoff is also recommended. The polymorphism cutoff depends on the diversity of the dataset but protein simulations performed by Martin et al. [8] showed that the ability of MI to filter out false positives is optimized for MSA with mean entropy of 0.3 for alignments of approximately 100 sequences, suggesting a natural entropy cutoff when analyzing real protein MSA.

5.3.2 Sequence Weights

In the analysis of covariation, it is important to include weighting functions that correct for the different numbers of proteins in different branches of an evolutionary tree [6]. It is known that over-representation of some homologous sequences in the sample may cause biases in statistical estimates. In the context of a MSA, this is important because alignments frequently contain very similar (even duplicated) sequences; these can bias the construction of the alignment itself or make some trends (merely due to nonrandom sampling) appear strong [57]. Equally problematic is the low representation of interesting but rare data. Scores averaged over alignment columns are vulnerable to over- or under-representation of certain sequences. A remedy is to assign weights to the sequences in an alignment before calculating any average value. To avoid such biases, different schemes of sequence weighting have been proposed [57–59]. These approaches reduce the weights of over-represented sequences and imply that the distribution of sequences in the sequence space is expected to be homogeneous [36,53].

5.3.3 Statistical Significance

The shape of the distribution of covariation values depends on the particular method used. However, most distributions are skewed, with most pairs having a very low value. To define the pairs of position with a high probability of being structurally or functionally linked, some studies chose the pairs of positions with the highest covariation values, usually using an arbitrary cutoff based on z scores (number of standard deviations from the mean covariation value) [8,26,44]. Recent studies have opted to assess the significance of the covariation values using a permutation procedure. The aa at each site in the sequence alignment are vertically shuffled, creating 10,000 or more random alignments that simulate the distribution of the covariation values under the null hypothesis that substitutions of aa at two sites are statistically independent. If the sequences are effectively unrelated then the pairs of positions with a significant covariation must have structural or functional links. Sequences are unrelated if the relationships by descent have been lost and there is no longer a significant phylogenetic signal or the sequences were obtained by in vitro selection [53,60]. However, in the case of related sequences it is necessary to test whether a correlation value reflects a significant association (possibly due to structural and functional constraints), or, instead, results from evolutionary history and stochastic events [2].

5.3.4 Multiple-Comparisons Problem

The multiple-comparisons problem occurs when one considers a set of statistical inferences simultaneously. In the covariation analysis, there are usually a large number of pairs of sites tested. Technically, the problem of multiple comparisons can be described as the potential increase in false positives that occurs when statistical tests are used repeatedly [61]. If m independent comparisons are performed with a given allowable error (α_i) the experiment-wide significance level α_g is given by

$$\alpha_g = 1 - (1 - \alpha_i)^m$$

To retain the same overall rate of false positives (rather than a higher rate) in a test involving more than one comparison, the standards for each comparison must be more stringent [61]. The Bonferroni correction states that the statistical significance level that should be used for each hypothesis separately is $1/n$ times of what it would be if only one hypothesis were tested [61]. However, many biological applications require a less conservative approach with greater power to detect true positives, at a cost of increasing the likelihood of obtaining false positives. The false discovery rate (FDR) is currently the most popular approach, which controls the expected proportion of false positives instead of the chance of any false positives. Consider testing $H_1, H_2, \ldots, H_m$ based on the corresponding ordered p-values $P_1, P_2, \ldots, P_m$, where $P_1 \leq P_2 \leq, \ldots, \leq P_m$ and H_i denotes the null hypothesis corresponding to P_i. Let k be the largest i for which

$$P_i \leq \frac{i}{m} q$$

Then all $H_i \leq k$ are rejected. This procedure controls the FDR at a proportion q which can be adjusted as low as possible [61].

5.3.5 Background Sequence Covariation

An obvious source of covariation among residues at different sites is common evolutionary history. Felsenstein (1985) showed that related sequences are part of a hierarchically structured phylogeny and, therefore, for statistical purposes, cannot be regarded as being drawn independently from the same distribution. In estimating the significance of the correlation coefficients, homologous sequences cannot be considered statistically independent because they share common evolutionary ancestry. Tree structure generally imparts extreme non-normality in the form of kurtosis to the correlation distribution, and this invalidates significance statistics based on the assumption of the normal distribution [9]. A large number of residue pairs with high correlation are expected simply due to background noise in the presence of phylogenetic structure and all the methods that do not incorporate the effect of this tree structure have many false positives [9]. Recent methods have been developed that address these issues but there is a great lack of agreement between all methods. The following five methods incorporate different solutions to the background sequence covariation problem and have different sensitivity and specificity but a definitive comparison between them is still absent.

1. Removal of multiple covariation [62]. This procedure claims to detect statistical correlations stemming from functional interaction by removing the strong phylogenetic signal that leads to the correlations of each site with many others in the sequence. The method assumes that all sites in a sequence have followed the same phylogeny resulting in a consistent pattern of substitution throughout the sequences thus creating many correlations between sites. Their analysis is biased toward those positions that covary with at most a few others and excludes coevolving groups of positions. However, the amount of coevolution in proteins is unknown and therefore this assumption is highly conservative and lacks experimental or theoretical support.

2. Parametric bootstrap [2,63]. This approach compares the distribution of inter-site mutual information for an alignment of naturally occurring sequences with the distribution of mutual information for artificial sequence data generated using the parametric bootstrap from a random ancestral sequence, a given substitution matrix, and the same tree. Correlated mutations in the set of artificial sequences can arise solely from common ancestry and, therefore, this comparison enables the calculation of the probability that a pair of covarying sites with a certain value of the mutual information statistic did not result from common ancestry [46]. This approach depends crucially on measures of aa similarity or distance. There are many such distance substitution matrices, the most used being those based on the propensity for evolutionary change from one aa to another [64]. However, the exchangeability of aa in a protein context is problematic because phylogenetic relations, mutational bias, and physicochemical effects are difficult to separate [65].
3. Background MI [8,44]. This method makes the assumption that each position in a MSA is affected equally by phylogenetic linkage, and that the majority of positions in the alignment covary only because of linkage. On the basis of these assumptions, each alignment is used as its own null model for the identification of covarying positions. The average MI of all possible pairs and its standard deviation are evaluated, allowing the identification of pairs with high MI values (z scores >4). Gloor et al. [44] used this method to identify nonconserved coevolving sites in MSA from a variety of protein families, finding that coevolving sites in these alignments fall into two general categories. One set is composed of sites that coevolve with only one or two other sites, often displaying direct aa side chain interactions with their coevolving partner. The other set comprises sites that coevolve with many others and are frequently located in regions critical for protein function, such as active sites and surfaces involved in molecular interactions and recognition. Gloor et al. [44] also found that coevolving positions are more likely to change protein function when mutated than are positions showing little coevolution. These results imply that these coevolving positions compose an important subset of the positions in an alignment, and may be as important to the structure and function of the protein family as are highly conserved positions. Interestingly, the analysis of the homeodomain mutations associated with human disease showed that those positions with high levels of covariation are more likely to be associated with a mutant phenotype when mutated [44]. In addition, the *E. coli* ATP synthase e subunit has been extensively mutated in vitro, and Gloor et al. [44] found that positions with high level of covariation are more likely to change the activity of the protein upon in vitro mutagenesis than those with low levels of covariation.
4. Genetic linkage on synonymous (S) and non-synonymous (A) sites [66]. This interesting new method systematically separates the covariation induced by selective interactions between aa from background sequence covariation, using silent (S) versus aa replacing (A) mutations. Covariation between two aa mutations, (A,A), can be affected by selective interactions between aa, whereas covariation within (A,S) pairs or (S,S) pairs cannot. This study performed an analysis of the pol gene in HIV, revealing that (A,A) covariation levels are enormously higher than for either (A,S) or (S,S), and thus cannot be attributed to phylogenetic effects. Inspection of the most prominent (A,A) interactions in the HIV pol gene showed that they are known sites of independently identified drug resistance mutations, and physically cluster around the drug-binding site.
5. Removal of phylogenetic clades [67–69]. To remove the effects of background sequence covariation, their analysis is applied to the complete alignment and to subalignments where specific phylogenetic clades are removed from the tree. Coevolving aa sites that are no longer detected following removal of one of the clades are classified as phylogenetically related sites as they occur in specific branches of the tree. The clades chosen for removal are identified before the covariation analysis is applied and they include

sequences that form a well-defined biological cluster or a cluster with high statistical support [68]. This method was successfully applied to heat-shock protein GroEL, ATPase Hsp90, the Gag protein from HIV-1, and the env gene of the HIV-1 group M subtype, highlighting that almost all detected coevolving sites are functionally or structurally important [68,69]. A possible weakness of this approach is the removal of pairs of positions that are truly linked only in the genomic context of certain sequences that formed a clade due to selective pressure. Some changes could be negatively selected and maintained in the population depending on the particular patterns of epistasis that apply in that genomic context. The positions where these type of changes occur are clade-specific and may define different evolutionary paths of clades. Thus, these covariable changes may still be functionally and structurally important but only in the context of the specific clade.

6. Markov model for sequence coevolution [45]. Recently, Yeang and Haussler [45] proposed a continuous-time Markov process model for sequence coevolution under two hypotheses and testing the most likely for each pair of sites. The null (independent) model hypothesizes that two sites evolve independently. The alternative (coevolutionary) model is obtained from the null model by re-weighting the independent substitution rate matrix to favor double over single changes. This method was applied to all the inter- and intra-domain position pairs in all the known protein domain families in Pfam database [70]. The majority of the inferred coevolving pairs of positions are functionally related or spatially coupled. Many of the coevolving positions are located at functionally important sites of proteins/protein complexes, such as the subunit linkers of superoxide dismutase, the tRNA-binding sites of ribosomes, the DNA-binding region of RNA polymerase, and the active- and ligand-binding sites of various enzymes.

5.4 AVAILABLE SOFTWARE

The following are publicly available programs that are currently used in covariation analysis:

DEPENDENCY [62]. http://www.uhnres.utoronto.ca/tillier/depend2/dependency.html
PCOAT [71]. ftp://iole.swmed.edu/pub/PCOAT/
CRASP [53]. http://wwwmgs.bionet.nsc.ru/mgs/programs/crasp/
CAPS [68]. http://bioinf.gen.tcd.ie/~faresm/software/caps/

5.5 CONCLUSION

During protein evolution, certain substitutions at different sites may occur in a coordinated manner due to interactions between aa residues. The detection of these interactions among separate aa sites is fundamental for understanding protein structure and evolution. Covariation analyses can also be important in identifying sites that may change the phenotype of a protein or be located at functionally important sites of proteins/protein complexes, a useful compass for researchers attempting to define functional domains through experimental analysis. In sequence alignments of homologous proteins, these interactions between sites can be manifested in correlation between substitutions at pairs of alignment positions. However, it is necessary to discriminate the three different sources of covariation: (1) chance, (2) common ancestry, and (3) structural or functional constraints. Effectively discriminating among these underlying causes is a difficult task with many statistical and computational difficulties, which are addressed in very different ways by recent methods. Although there is not a consensus about the best way of discriminating the sources of covariation, all recent methods have confirmed that the detection of coordinated substitutions is a very important tool for protein analyses, predictions of spatial structure, inter-residue contacts, function, and protein–protein interactions.

REFERENCES

1. Darwin, C. *The Origin of Species*, Penguin, Middlesex, United Kingdom, 1859.
2. Atchley, W., Wollenberg, K., Fitch, W., Terhalle, W., and Dress, A. Correlations among amino acid sites in bHLH protein domains: An information theoretic analysis. *Mol Biol Evol* 17, 164–178, 2000.
3. Tolou, H., Nicoli, J., and Chastel, C. Viral evolution and emerging viral infections: What future for the viruses? A theoretical evaluation based on informational spaces and quasispecies. *Virus Genes* 24, 267–274, 2002.
4. Schuster, P. Evolution at molecular resolution. *Nolinear Cooperative Phenomena in Biological Systems*, World Scientific, Singapore, pp. 86–112, 1998.
5. Lesk, A. and Chothia, C. How different amino acid sequences determine similar protein structures: The structure and evolutionary dynamics of the globins. *J Mol Biol* 136, 225–270, 1980.
6. Chelvanayagam, G., Eggenschwiler, A., Knecht, L., Gonnet, G., and Benner, S. An analysis of simultaneous variation in protein structures. *Protein Eng* 10, 307–316, 1997.
7. Tomii, K. and Kanehisa, M. Analysis of amino acid indices and mutation matrices for sequence comparison and structure prediction of proteins. *Protein Eng* 9, 27–36, 1996.
8. Martin, L., Gloor, G., Dunn, S., and Wahl, L. Using information theory to search for co-evolving residues in proteins. *Bioinformatics* 21, 4116–4124, 2005.
9. Pollock, D. and Taylor, W. Effectiveness of correlation analysis in identifying protein residues. *Protein Eng* 10, 647–657, 1997.
10. Chothia, C. and Lesk, A. Evolution of proteins formed by beta-sheets. I. Plastocyanin and azurin. *J Mol Biol* 160, 309–323, 1982.
11. Lesk, A. and Chothia, C. Evolution of proteins formed by beta-sheets. II. The core of the immunoglobulin domains. *J Mol Biol* 160, 325–342, 1982.
12. Oosawa, K. and Simon, M. Analysis of mutations in the transmembrane region of the aspartate chemoreceptor in *Escherichia coli. Proc Natl Acad Sci U S A* 83, 6930–6934, 1986.
13. Altschuh, D., Vernet, T., Berti, P., Moras, D., and Nagai, K. Coordinated amino acid changes in homologous protein families. *Protein Eng* 2, 193–199, 1988.
14. Bordo, D. and Argos, P. Evolution of protein cores. Constraints in point mutations as observed in globin tertiary structures. *J Mol Biol* 211, 975–988, 1990.
15. Mateu, M. and Fersht, A. Mutually compensatory mutations during evolution of the tetramerization domain of tumor suppressor p53 lead to impaired hetero-oligomerization. *Proc Natl Acad Sci U S A* 96, 3595–3599, 1999.
16. Lim, W. and Sauer, R. Alternative packing arrangements in the hydrophobic core of lambda repressor. *Nature* 339, 31–36, 1989.
17. Lim, W., Farruggio, D., and Sauer, R. Structural and energetic consequences of disruptive mutations in a protein core. *Biochemistry* 31, 4324–4333, 1992.
18. Baldwin, E., Hajiseyedjavadi, O., Baase, W., and Matthews, B. The role of backbone flexibility in the accommodation of variants that repack the core of T4 lysozyme. *Science* 262, 1715–1718, 1993.
19. Burch, C. and Chao, L. Epistasis and its relationship to canalization in the RNA virus phi 6. *Genetics* 167, 559–567, 2004.
20. Bonhoeffer, S., Chappey, C., Parkin, N., Whitcomb, J., and Petropoulos, C. Evidence for positive epistasis in HIV-1. *Science* 306, 1547–1550, 2004.
21. Sanjuan, R., Moya, A., and Elena, S. The contribution of epistasis to the architecture of fitness in an RNA virus. *Proc Natl Acad Sci U S A* 101, 15376–15379, 2004.
22. Poon, A. and Chao, L. The rate of compensatory mutation in the DNA bacteriophage phiX174. *Genetics* 170, 989–999, 2005.
23. Mateo, R. and Mateu, M. Deterministic, compensatory mutational events in the capsid of foot-and-mouth disease virus in response to the introduction of mutations found in viruses from persistent infections. *J Virol* 81, 1879–1887, 2007.
24. Garriga, C., Pérez-Elías, M.J., Delgado, R., Ruiz, L., Nájera, R., Pumarola, T., Alonso-Socas, M., García-Bujalance, S., and Menéndez-Arias, L. Mutational patterns and correlated amino acid substitutions in the HIV-1 protease after virological failure to nelfinavir- and lopinavir/ritonavir-based treatments. *J Med Virol* 79, 1617–1628, 2007.

25. Govindarajan, S. et al. Systematic variation of Amino acid substitutions for stringent assessment of pairwise covariation. *J Mol Biol* 328, 1061–1069, 2003.
26. Clarke, N. Covariation of residues in the homeodomain sequence family. *Protein Sci* 4, 2269–2278, 1995.
27. Voigt, C., Mayo, S., Arnold, F., and Wang, Z. Computational method to reduce the search space for directed protein evolution. *Proc Natl Acad Sci U S A* 98, 3778–3783, 2001.
28. Fukami-Kobayashi, K., Schreiber, D., and Benner, S. Detecting compensatory covariation signals in protein evolution using reconstructed ancestral sequences. *J Mol Biol* 319, 729–743, 2002.
29. Göbel, U., Sander, C., Schneider, R., and Valencia, A. Correlated mutations and residue contacts in proteins. *Proteins* 18, 309–317, 1994.
30. Neher, E. How frequent are correlated changes in families of protein sequences? *Proc Natl Acad Sci U S A* 91, 98–102, 1994.
31. Shindyalov, I., kolchanov, N., and Sander, C. Can three dimensional contacts in protein structures be predicted by analysis of correlated mutations? *Protein Eng* 7, 349–358, 1994.
32. Taylor, W. and Hatrick, K. Compensating changes in protein multiple sequence alignments. *Protein Eng* 7, 341–348, 1994.
33. Benner, S., Cannarozzi, G., Gerloff, D., Turcotte, M., and Chelvanayagam, G. Bona fide predictions of protein secondary structure using transparent analyses of multiple sequence alignments. *Chem Rev* 97, 2725–2844, 1997.
34. Nagl, S., Freeman, J., and Smith, T. Evolutionary constraint networks in ligand-binding domains: An information-theoretic approach. *Pac Symp Biocomput*, 90–101, 1999.
35. Larson, S., Di Nardo, A., and Davidson, A. Analysis of covariation in an SH3 domain sequence alignment: Applications in tertiary contact prediction and the design of compensating hydrophobic core substitutions. *J Mol Biol* 303, 433–446, 2000.
36. Afonnikov, D., Oshchepkov, D., and Kolchanov, N. Detection of conserved physico-chemical characteristics of proteins by analyzing clusters of positions with co-ordinated substitutions. *Bioinformatics* 17, 1035–1046, 2001.
37. Nemoto, W., Imai, T., Takahashi, T., Kikuchi, T., and Fujita, N. Detection of pairwise residue proximity by covariation analysis for 3D-structure prediction of G-protein-coupled receptors. *Protein J* 23, 427–435, 2004.
38. Wang, L. Covariation analysis of local amino acid sequences in recurrent protein local structures. *J Bioinform Comput Biol* 3, 1391–1409, 2005.
39. Shackelford, G. and Karplus, K. Contact prediction using mutual information and neural nets. *Proteins* 69, 159–164, 2007.
40. Altschuh, D., Lesk, A., Bloomer, A., and Klug, A. Correlation of co-ordinated amino acid substitutions with function in viruses related to tobacco mosaic virus. *J Mol Biol* 193, 693–707, 1987.
41. Korber, B., Farber, R., Wolpert, D., and Lapedes, A. Covariation of mutations in the V3 loop of human immunodeficiency virus type 1 envelope protein: An information theoretic analysis. *Proc Natl Acad Sci U S A* 90, 7176–7180, 1993.
42. Gilbert, P., Novitsky, V., and Essex, M. Covariability of selected amino acid positions for HIV type 1 subtypes C and B. *AIDS Res Hum Retroviruses* 21, 1016–1030, 2005.
43. Kolli, M., Lastere, S., and Schiffer, C. Co-evolution of nelfinavir-resistant HIV-1 protease and the p1-p6 substrate. *Virology* 347, 405–409, 2006.
44. Gloor, G., Martin, L., Wahl, L., and Dunn, S. Mutual information in protein multiple sequence alignments reveals two classes of coevolving positions. *Biochemistry* 44, 156–165, 2005.
45. Yeang, C. and Haussler, D. Detecting coevolution in and among protein domains. *PLoS Comput Biol* 3, e211, 2007.
46. Noivirt, O., Eisenstein, M., and Horovitz, A. Detection and reduction of evolutionary noise in correlated mutation analysis. *Protein Eng* 18, 247–253, 2005.
47. Olmea, O., Rost, B., and Valencia, A. Effective use of sequence correlation and conservation in fold recognition. *J Mol Biol* 293, 1221–1239, 1999.
48. Lockless, S. and Ranganathan, R. Evolutionarily conserved pathways of energetic connectivity in protein families. *Science* 286, 295–299, 1999.
49. Kass, I. and Horovitz, A. Mapping pathways of allosteric communication in GroEL by analysis of correlated mutations. *Proteins* 48, 611–617, 2002.
50. Atchley, W., Terhalle, W., and Dress, A. Positional dependance, cliques and predictive motifs in the bHLH protein domain. *J Mol Evol* 48, 501–506, 1999.

51. Crooks, G. and Brenner, S. Protein secondary structure: Entropy, correlations and prediction. *Bioinformatics* 20, 1603–1611, 2004.
52. Atchley, W., Zhao, J., Fernandes, A., and Druke, T. Solving the protein sequence metric problem. *Proc Natl Acad Sci U S A* 102, 6395–6400, 2005.
53. Afonnikov, D. and Kolchanov, N. CRASP: A program for analysis of coordinated substitutions in multiple alignments of protein sequences. *Nucleic Acids Res* 32, W64–W68, 2004.
54. Kawashima, S. and Kanehisa, M. AAindex: Amino acid index database. *Nucleic Acids Res* 28, 374, 2000.
55. Shen, B. and Vihinen, M. Conservation and covariance in PH domain sequences: Physicochemical profile and information theoretical analysis of XLA-causing mutations in the Btk PH domain. *Protein Eng Des Sel* 17, 267–276, 2004.
56. Fodor, A. and Aldrich, R. Influence of conservation on calculations of amino acid covariance in multiple sequence alignments. *Proteins* 56, 211–221, 2004.
57. Vingron, M. and Sibbald, P. Weighting in sequence space: A comparison of methods in terms of generalized sequences. *Proc Natl Acad Sci U S A* 90, 8777–8781, 1993.
58. Altschul, S., Carroll, R., and Lipman, D. Weights for data related by a tree. *J Mol Biol* 207, 647–653, 1989.
59. Sibbald, P. and Argos, P. Weighting aligned protein or nucleic acid sequences to correct for unequal representation. *J Mol Biol* 216, 813–818, 1990.
60. Segal, M., Cummings, M., and Hubbard, A. Relating amino acid sequence to phenotype: Analysis of peptide-binding data. *Biometrics* 57, 632–642, 2001.
61. Benjamini, Y. and Hochberg, Y. Controlling the false discovery rate: A practical and powerful approach to multiple testing. *J Royal Stat Soc, Series B* 57, 289–300, 1995.
62. Tillier, E. and Lui, T. Using multiple interdependency to separate functional from phylogenetic correlations in protein alignments. *Bioinformatics* 19, 750–755, 2003.
63. Wollenberg, K. and Atchley, W. Separation of phylogenetic and functional associations in biological sequences by using the parametric bootstrap. *Proc Natl Acad Sci U S A* 97, 3288–3291, 2000.
64. Jones, D., Taylor, W., and Thornton, J. The rapid generation of mutation data matrices from protein sequences. *Comput Appl Biosci* 8, 275–282, 1992.
65. Yampolsky, L. and Stoltzfus, A. The exchangeability of amino acids in proteins. *Genetics* 170, 1459–1472, 2005.
66. Wang, Q. and Lee, C. Distinguishing functional amino acid covariation from background linkage disequilibrium in HIV protease and reverse transcriptase. *Plos ONE* 2, e814, 2007.
67. Fares, M. and Travers, S. A novel method for detecting intramolecular coevolution: Adding a further dimension to selective constraints analysis. *Genetics* 173, 9–23, 2006.
68. Fares, M. and McNally, D. CAPS: Coevolution analysis using protein sequences. *Bioinformatics* 22, 2821–2822, 2006.
69. Travers, S., Tully, D., McCormack, G., and Fares, M. A study of the coevolution patterns operating within the env gene of the HIV-1 group M subtypes. *Mol Biol Evol* 24, 2787–2801, 2007.
70. Bateman, A., Birney, E., Cerruti, L., Durbin, R., Etwiller, L., Eddy, S., Griffiths-Jones, S., Howe, K.L., Marshall, M., and Sonnhammer, E. The Pfam protein families database. *Nucleic Acids Res* 30, 276–280, 2002.
71. Qi, Y. and Grishin, N. PCOAT: Positional correlation analysis using multiple methods. *Bioinformatics* 20, 3697–3699, 2004.

Part II

Protein Engineering Targets: Vaccine

6 Tuning the Immune Response to Our Advantage: Design of Vaccines with Tailored Functions

Till A. Röhn and Martin F. Bachmann

CONTENTS

6.1 INTRODUCTION

One of the most important medical achievements of mankind is the prevention of diseases by vaccination. The average human lifespan has increased by approximately 30 years in the twentieth century and the two most important contributions to this have been sanitation and vaccination [1]. In the past 200 years since Jenner's epochal observation that infection with cowpox, which induces only a mild illness in humans, can prevent and protect against smallpox, vaccination has controlled not only the spread of this disease but also of diphtheria, tetanus, yellow fever, pertussis, *Haemophilus influenza* type B, polio, measles, mumps, rubella, hepatitis A + B, varicella as well as pneumococcal and meningococcal infections. In fact, smallpox could be eradicated by continuous vaccinations in the twentieth century and the WHO expects polio to be erased in the very near future as well [2,3]. Hence, vaccines are considered to be one of the most successful medical interventions against infectious diseases. Despite these accomplishments, however, substantial morbidity and mortality is still caused by pathogens like HIV, HCV, rotavirus, influenza virus, or *M. tuberculosis* and parasite infections like plasmodia, trypanosoma, leishmania, or schistosoma to name just a few.

The goal of any vaccination is the induction of an appropriate and effective immune response but precisely what constitutes an effective immune response is still unclear for many diseases. This is especially true for infectious agents with more complex host–pathogen interactions, or infections in which the pathogen has found ways to undermine, evade, or misdirect the host's immune response. Decades of fundamental research into immunology, molecular cellular biology, and vaccinology have revealed insights into mechanisms of host–pathogen interactions and provided a deeper understanding of the regulatory mechanisms involved in immune responses and of ways how to modulate them [2,4–7]. Through these advances in knowledge it will become possible to improve existing suboptimal vaccines by making them safer and more effective. Moreover, it will hopefully become possible to rationally design new vaccines that evoke the types of immune responses required to prevent and control the devastating chronic diseases for which no effective vaccines exist, yet.

The opportunities for modulation of immune responses broaden the scope of conventional vaccines and offer potential in the prevention and treatment of a far larger diversity of diseases like cancer, allergies, and chronic infections. It may even be possible to develop fundamentally new classes of vaccines that target chronic noninfectious conditions such as autoimmune inflammatory diseases, Alzheimer's disease, hypertension, or addiction to substances of abuse.

6.2 TRADITIONAL VACCINES

The basis for vaccination is the immune system's fundamental property to remember an encounter with a pathogen. As a result, it can mount a faster and stronger response upon subsequent exposure to the same pathogen enabling it to rapidly control infection and eliminate the invader. Hence, a successful vaccine should mimic an infectious agent at the best possible rate to prepare the immune system for the encounter with the real pathogen. The success of this approach, however, depends on decoupling virulence from induction of protective immunity.

An early strategy to artificially induce protective immunity was to deliberately cause a relatively mild infection with the unmodified pathogen, a process termed variolation. Originally this process was applied by Chinese and Indian physicians who inoculated their patients with dried scabs of smallpox victims [8,9]. In the eighteenth century, variolation was introduced in England by Lady Mary Wortley Montagu, the wife of the British ambassador who had lived and learned about variolation in Constantinople. This procedure, however, often resulted in severe smallpox infections and sometimes even caused local outbreaks so that variolation never achieved broad acceptance and certainly does not meet modern criteria of safety.

Jenner's first smallpox vaccine developed for widespread use consisted of a non-virulent bovine analogue of smallpox, which induced an immune response against antigens nearly identical in both poxviruses thus conferring cross-protection without causing disease. Since then, the use of live related or attenuated organisms to induce protective immunity has been adopted to develop vaccines for a number of viral and bacterial pathogens, like tuberculosis, yellow fever, and rabies [10–12]. The concept of attenuation was discovered during Pasteur's work with chicken cholera who observed that virulence of pathogens could be attenuated by exposure to environmental insults such as high temperature, chemicals, and oxygen while maintaining immunogenicity [13]. The induction and selection of mutants with reduced virulence has hence been an important endeavor of vaccine manufacturers. Important ways of generating live attenuated viruses until nowadays have been (1) passage in cell cultures followed by selection of mutants better adapted to replication in the cells used in vitro than in the living host, (2) selection of cold adapted viruses, or (3) cell culture reassortment of attenuated viruses with viruses encoding protective antigens from circulating wild strains [2]. Although empirically developed, these types of vaccines are of great success because the immune system treats live attenuated vaccines essentially in the same way as it treats the infectious pathogens and thus mounts a strong and long-lasting immune response. An important limitation of live organism-based vaccines, however, is safety, as the identification of sufficiently attenuated strains cannot always be warranted. Moreover, live attenuated organisms bear the residual risk of reversion to virulence and transmission to nontarget subjects. Since attenuated pathogens are often released by the infected host, spread to immuno-compromised individuals succumbing to a more severe disease, cannot be controlled. For these reasons, the use of attenuated vaccines comes increasingly under pressure, as evidenced by the fact that the attenuated Sabin polio vaccine has largely been replaced by the inactivated Salk polio vaccine [14].

A second strategy for vaccine development has therefore been the use of inactivated or killed organisms. Also by this approach the immune system is exposed to all the antigens of a pathogen but the pathogen itself is rendered replication deficient and harmless. A number of successful vaccines have been developed based on killed bacteria or inactivated viruses, directed for example against influenza, hepatitis A, or polio (Salk). Vaccines based on killed organisms, however, are generally less immunogenic than their live counterparts. Especially, the induction of cell-mediated immunity, particularly the induction of cytotoxic T-cell responses, is usually suboptimal with inactivated vaccines in comparison to live attenuated ones. This has been observed in vaccines against measles and influenza [15–17]. As inactivated pathogens are unable to replicate, they usually do not induce such long-lasting protective immune responses as live attenuated ones and require booster immunizations to generate memory responses. Vaccines based on killed organisms are therefore often administered several times together with an adjuvant like Alum (aluminum hydroxide gel has been for a long time the only adjuvant approved for use in humans) to increase their potency.

Finally, the third important strategy in traditional vaccine development has been the use of pathogen extracts or subunits as antigens. The first way to do so was the use of inactivated toxins so-called toxoids as vaccine components and was triggered by the comprehension that many bacteria produce toxins which are largely responsible for their pathogenesis. The neutralization of these toxins via vaccine-induced neutralizing antibodies could prevent these diseases. Examples for toxoid-based vaccines have been those against *Clostridium tetani* and *Corynebacterium diphtheriae* developed in the 1920s [18,19]. Later on, components extracted or enriched from bacterial cultures or infected tissues like polysaccharides (meningococci and pneumococci) or proteins (pertussis) were used in subunit vaccines. The great advantage of acellular subunit vaccines is the increased safety and the ability to direct and focus the immune response on certain components of a pathogen that might play a crucial role in pathogenicity or host–pathogen interaction. As with inactivated vaccines, however, increased safety of subunit vaccines comes at the cost of lower immunogenicity and therefore, requires coadministration of adjuvants and repeated booster immunizations (Figure 6.1).

6.3 REQUIREMENTS FOR AN EFFECTIVE VACCINE

During evolution the immune system has developed two main pillars to respond to a microbial infection: The first one is the innate immune response, which provides a first line of defense because it acts within minutes after recognition of molecular patterns found in microbes. The innate immune system possesses a number of means to directly fight a pathogen. Among them is the rapid burst of toxic substances such as NO, radicals, or defensins (toxic peptides) as well as inflammatory chemokines and cytokines that can induce defense mechanisms in the infected cells and lead to the recruitment and activation of further immune cells. The immune mediators released during an innate immune response are especially important also because they condition the immune system for the subsequent development of an adaptive immune response. The major players of the adaptive immune reaction, which develops much slower, within days to weeks, are B- and T-cells. In contrast to the innate immune response, adaptive immunity makes use of selection and clonal expansion to generate antigen-specific cells. While the key function of B-cells is the generation of antigen-specific antibodies, the function of T-cells is more diverse and ranges from lysis of infected cells by cytotoxic T-lymphocytes (CTLs) to the release of effector and regulatory cytokines or activation

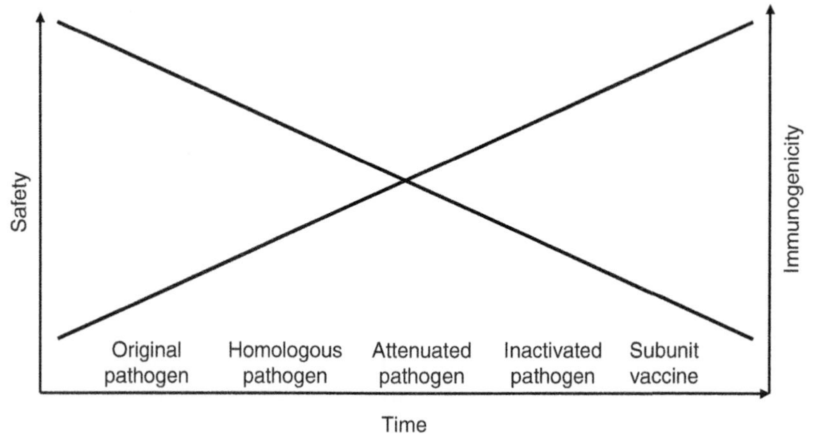

FIGURE 6.1 Since the first attempts of variolation which were often associated with severe infections and local outbreaks, vaccine safety has improved considerably. This is due to the use of attenuated or inactivated pathogens for vaccination and the development of subunit vaccines. Increased safety, however, often comes at the cost of lower immunogenicity so that appropriate adjuvants need to be included to enhance the immune response.

of B-cells by different subsets of T-helper cells (Th). Which type of response is critical to the resolution of an infection varies from pathogen to pathogen. However, while a diverse arsenal of reactions is responsible for resolution of primary infections, antibodies are almost always critical for secondary responses [20]. The primary goal of vaccines, therefore, is normally the induction of antibodies. The hallmark of an adaptive immune response eventually is the development of long-lasting immunological memory crucial to vaccine effectiveness.

Most traditional vaccines have been derived empirically. Hence, the mechanisms underlying the great success of vaccines based on live attenuated organisms remained largely unknown until recently. Increased understanding of the mechanisms involved in the activation of innate immune responses and the regulatory processes influencing the outcome of adaptive immunity, however, started to elucidate some of the parameters important for vaccine development. But what are the key elements required if effective vaccines are to be rationally designed?

First, the right antigen is necessary, against which the adaptive immune response is to be elicited. The selection and proper presentation of the antigen in its correct tertiary structure is of outmost importance especially if conformation-dependent functional antibodies are to be elicited.

Second, immune potentiators (adjuvants) need to be included that sufficiently stimulate the innate immune system, which aids in the generation of robust and long-lasting adaptive immune responses. Third, an appropriate delivery system is of importance to ensure that the antigen is conveyed to the right place, sufficiently stable, present over an extended period and efficiently taken up by antigen presenting cells (APCs). The latter two properties may allow for vaccines to provide rapid and sufficient immune protection against illness and elicit immunological memory for several years or preferably even for a whole lifetime. Ideally, vaccines should induce both B- and T-cell immune responses and condition the immune system in the right way to deal with the targeted pathogen (Figure 6.2).

To be successful in the market, vaccines should furthermore be safe, affordable, and user friendly. Modern criteria of safety imposed on vaccine manufacturers by the regulatory authorities require vaccines first of all to be safe. This is especially true for prophylactic vaccines, which are administered to healthy people but also for therapeutic vaccines given to patients already suffering from diseases such as cancer. Finally, a successful vaccine should also be of low cost, easy to manufacture, biologically stable for an extended period, easy to administer, and should cause few to no side effects in the patients.

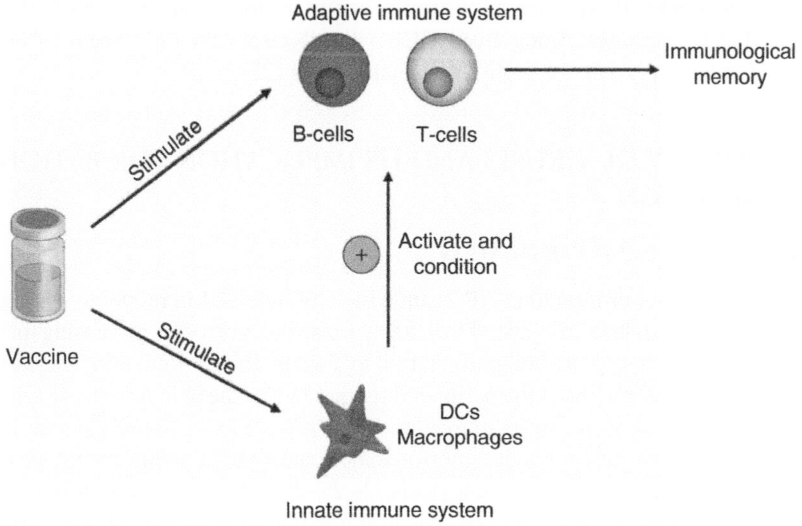

FIGURE 6.2 Ideal vaccines should stimulate both the innate and the adaptive immune system. Stimulation of the release of cytokines and chemokines by activated innate immune cells is crucial for the proper stimulation and conditioning of the adaptive immune response and contributes to the development of immunological memory, the goal of vaccination.

Many of the most successful vaccines were based on attenuated viruses, like polio virus, measles virus, mumps virus, rubella virus, yellow fever virus, and others. Especially, the yellow fever vaccine YF-17D is considered one of the most successful vaccines and has been administered to over 400 million people. It is not only a very potent inducer of cytotoxic T-cell responses but also stimulated neutralizing antibody titers lasting for over 35 years with just one single immunization [21]. In the following, we would like to discuss the unique properties of viruses responsible for their excellent immunogenicity and how these properties could be harnessed for rational vaccine design.

During evolution, the immune system has been constantly exposed to viruses and thus has developed a number of mechanisms to sense and efficiently respond to them. Four key parameters responsible for viral immunogenicity have been identified [22]:

1. Viruses are of particulate nature. APCs and especially dendritic cells (DCs), which are the most important cells for priming of naïve T-cells, and hence induction of adaptive immune responses, endocytose particulate antigens much more efficiently than soluble antigens. Viruses are therefore targeted in vivo to the requisite APCs for optimal antigen-presentation on MHC molecules and induction of adaptive immunity.
2. Viruses are efficiently recognized by the innate immune system through activation of pattern-recognition receptors (PRRs). Engagement of PRRs causes the activation of a number of immune cells and some nonimmune cells, leads to the secretion of cytokines and chemokines, and to the maturation and migration of DCs to secondary lymphoid organs. This immune modulating property of viruses creates an inflammatory environment, which is essential for the induction of strong adaptive immune responses.
3. Viruses exhibit highly repetitive structural motifs. B-cells can be directly activated after recognition of highly repetitive structures on pathogens, which lead to efficient cross-linking of specific immunoglobulins on their surface. This causes vigorous induction of antibody production by B-cells.
4. Viruses replicate in vivo. Hence the immune system is exposed for a relatively long period to the antigen. Antigen load follows characteristic kinetics with peak amounts reached after

a few days to weeks. Persistence of antigen is important for the generation of effector and memory T-cell responses. Long lived $CD8^+$ T-cell responses only seem to efficiently develop if antigen persists for more than about 7 days [23].

6.4 IMMUNOLOGY OF VIRUSES AND ITS IMPLICATION FOR RATIONAL VACCINE DESIGN

6.4.1 Antigen Uptake and Presentation

DCs are the most powerful professional APCs and crucially involved in inducing and orchestrating the adaptive immune response [24–26]. They are specialized not only in sensing pathogens or vaccines but also in antigen capture and presentation to T-cells. Hence, immature DCs are located at strategic anatomical sites throughout the peripheral tissues e.g., in the skin or mucosa where entry of pathogens is most likely to occur. In its immature state, DCs are exceptionally potent in taking up antigens. This can happen either via receptor-mediated endocytosis using receptors recognizing pathogen-associated structures or via phagocytosis [24,27–29]. Phagocytotic uptake of antigens by DCs depends upon a number of antigen-associated properties such as size, shape, surface charge, and hydrophobicity/hydrophilicity [30]. The size of an antigen is a key factor for its efficient uptake by DCs. Large molecular aggregates and particulate antigens have superior properties to bind cellular surfaces than soluble antigens and are therefore more efficiently taken up. In addition, the surfaces of most pathogens, including viruses, have the potential to directly activate components of the complement cascade or bind natural antibodies [31]. This phenomenon leads to opsonization and complement and Fc receptor-mediated enhanced antigen uptake and presentation [32–34]. Hence DCs are able to capably take up particulate antigens with pathogen-like dimensions (10 nm–3 μm), including of course viruses.

Once taken up by DCs, antigens are processed by proteases in endosomal compartments to peptides, which can be presented on MHC class II molecules. DCs in addition also have the property to cross-present endocytosed antigens. This means that antigens are shuttled into the MHC class I processing pathway and are presented on MHC class I molecules. Interestingly, cross-presentation of antigens derived from particulate structures occurs much more efficiently than cross-presentation of soluble antigens and is comparable in its magnitude to presentation on MHC class II molecules [35–37]. Presentation on MHC class I molecules enables priming of CD8 T-cell responses, which are important for the elimination of intracellular pathogens and viruses. The presentation of antigens to CD4 and CD8 T-cells occurs in the T-cell rich zones of secondary lymphoid organs to which DCs migrate after an encounter with a pathogen.

Small particles with the dimensions of viruses are not only taken up and transported by DCs from the site of infection to lymphoid organs but also can efficiently drain and diffuse to the lymph nodes [38]. Here they can directly interact with B-cells and stimulate them. The recognition of antigens in its native conformation by B-cells is important for optimal induction of antibody responses. The superior immunogenicity of particulate antigens, especially those with the dimensions of viruses, has powered the development of vaccine delivery systems that are based on particulate structures. These include mineral salts, oil-in-water emulsions, liposomes, virosomes, ISCOMs, microparticles (composed of biodegradable polymers such as poly(lactide *co*-glycolide), polyanhydrides, hyaluronic acid, poly-L-lysine, chitosan and starch), and virus-like particles (VLPs).

6.4.2 Stimulation of Innate Immunity

To mount an adaptive immune response, the efficient uptake of antigen and its (cross-) presentation on MHC molecules is not sufficient. Indeed, the presentation of antigens on MHC molecules without the concomitant expression of co-stimulatory molecules and release of inflammatory cytokines by DCs results in induction of tolerance rather than immunity. No relevant adaptive immune response will be mounted unless the innate immune system has received an innate signal which leads to the maturation of DCs, upregulation of co-stimulatory molecules, and the release of cytokines. All this serves to induce, condition, and shape the subsequent adaptive immune response. The cells of the innate immune system have a number of detection systems that sense and react to conserved and distinct microbial features, termed pathogen-associated molecular patterns (PAMPs). To be able to sense PAMPs and raise an alarm, the innate immune system has developed a diverse set of PRRs. An important family of PRRs is the toll-like receptors (TLRs).

Mammalian TLRs are a family consisting of at least 12 membrane proteins some of which are expressed on the cell surface, others on membranes of endosomal vesicles or other intracellular organelles [39,40]. TLRs are widely expressed on immune and nonimmune cells, like different subsets of DCs, macrophages, mast cells, natural killer cells, neutrophils, B-cells, endothelial cells, epithelial cells, and fibroblasts and have a broad specificity for conserved molecular patterns like lipids, carbohydrates, peptides, and nucleic-acid structures shared by parasites, bacteria, and viruses. TLRs consist of an extracellular domain of leucine rich repeats, which are directly or indirectly involved in ligand binding and a cytoplasmatic domain. The cytoplasmatic domain contains a signaling part called Toll/interleukin 1 receptor domain (TIR) that interacts with TIR-domain containing adaptor molecules. For most TLRs except for TLR 3 this adaptor molecule is MyD88. For some TLRs e.g., TLR 2 + 4 signals are not only relayed through MyD88 but also through an additional adaptor molecule called TIRAP (TIR-domain containing adapter protein). TLR3 employs TRIF (TIR-domain-containing adaptor protein inducing IFNβ) rather than MyD88 as adaptor molecule. Eventually interaction of TLRs with their ligands leads to nuclear factor κB (NF-κB)-mediated release of cytokines. In addition, signaling through TLRs 3, 4, 7, and 9 can mediate release of type I interferons after phosphorylation and nuclear translocation of interferon-regulatory factor 3 (IRF3) [39,40]. The PAMPs detected by TLRs are very diverse. While TLRs 1, 2, 4, 5, and 6 recognize mainly bacterial products, like LPS, bacterial lipoproteins, flagellin, lipoteichoid acids etc. TLRs 3, 7, 8, and 9 are more specialized in viral detection and recognize mainly nucleic acid structures like viral dsRNA, ssRNA, and CpG oligonucleotides. Engagement of these receptors is a crucial factor in the potent immunogenicity of viruses.

Although TLRs are an important system for microbial sensing they are not the only PRRs. Further cell surface PRRs exist next to TLRs, like mannose receptors, DEC-205, DC-SIGN, dectin-1, etc., commonly referred to as C-type lectin-like receptors (CLRs). Also cytosolic PRRs exist. These cytosolic receptors detect components of microorganisms that reach the cytosol. Two main families of cytosolic receptors exist, the nucleotide-binding oligomerization domain (NOD)-like receptors (NLRs), detecting e.g., bacterial peptidoglycans and the RIG-I-like family of receptors (RLRs) recognizing non-capped RNA [41].

There is growing evidence that TLRs and other PPRs can cooperate in innate immune responses. Different cells of the innate immune system express distinct combinations of PPRs. The integrated sum of signals induced by these different receptors encountered on different cells might contribute to the diverse outcome of immune responses with which the immune system can react to different pathogens and condition the inflamed site for the subsequent adaptive immune response [42]. The adaptive immune system has evolved a number of diverse ways how to respond to the immense number of different pathogens that infect the host. Adaptive immune responses can for example be diverse with respect to the cytokines made by Th-cells or the class of antibodies secreted by B-cells. DCs, which are crucially involved in priming of the T-cell response express a number of different PPRs. It is assumed that different combinations of PPRs stimulated by a pathogen on DCs cause cytokine profiles that together with the strength of the signal received from the T-cell receptor (TCR)–MHC interaction decide whether Th-cells differentiate into the Th1, the Th2, the Th17, or the Treg type. Moreover, DCs exist in multiple subsets that are differentially equipped with distinct sets of PRRs. Myeloid human DCs (mDCs) express a variety of surface TLRs (TLR 1, 2, 5, 6) and endosomal TLRs (TLR 3 and 8). mDCs are therefore able to recognize bacterial, fugal, and viral pathogens and respond to their encounter with maturation, expression of co-stimulatory molecules and secretion of different types of cytokines like IL-12, TNF, and IL-6. Plasmacytoid DCs (pDCs) on the other hand strongly express the endosomal TLR 7 detecting single-stranded RNA and TLR 9 detecting double-stranded DNA. pDCs are therefore specialized in the detection of viral genomes and mediate robust induction of type I interferons like IFN-α [41]. Release of INF-α mediates direct antiviral responses, favors cross-presentation and induction of CTLs, and promotes Th1-type responses. Th1-type immune responses are mainly characterized by the release of IFN-γ which further helps the induction of cell-mediated immunity like CTL induction. A number of TLRs (TLR 3, 4, 5, 8, 11) seem to promote Th1 responses via induction of robust IL-12 responses in DCs. Others like TLR 2 or TLR 2/1 and TLR 2/6 heterodimers favor Th2 responses which are characterized by the release of IL-4, IL-5, IL-10, the induction of IgE antibody responses and eosinophil-mediated responses [43]. Which PPRs or combinations of PPRs promote Th17 responses characterized by the release of pro-inflammatory IL-17A, IL-17F, and TNF or Treg responses characterized by the release of anti-inflammatory IL-10 are still largely unknown.

Adjuvants currently licensed for human use were empirically developed and are essentially limited to aluminum salts (Alum) and the oil-in-water emulsion MF59 [44,45]. Alum is known to favor Th2 immune responses and efficiently enhances antibody responses. Historically, the emphasis of many vaccines has been the induction of humoral immune responses for which Alum is suitable. The need for therapeutic vaccines against chronic infections like HIV, HCV, HSV, tuberculosis, or even cancer, however, requires the elicitation of cell-mediated immunity like CTL responses and therefore requires Th1 immune responses. The growing understanding of innate immune mechanisms and the identification of a number of PRRs which translate microbial signals into potent adaptive immune responses has stimulated high-throughput search for PPR ligands to be used as vaccine adjuvants. As described above the signaling of one or more PRRs on sentinel cells can result in an inflammatory milieu that favors Th1 responses. Combinations of different PRR ligands could be suitable to design adjuvants tailored to the specific requirements of each particular vaccine. Analysis of the highly successful yellow fever vaccine has revealed that it simultaneously activates TLR2, TLR7, TLR8, and TLR9 on different DC subsets [46]. Likely, the combined activation of different TLRs on different DC subsets contributes to the highly efficient and long-lasting immunity of this vaccine. Further insights into the critical parameters of innate immune responses will aid in designing tailor made vaccines that induce the desired strength, duration, and quality of T- and B-cell responses.

6.4.3 Activation of B-Cell Responses

After entering the host, viruses are not only sensed by cells of the innate immune system but also induce an inflammatory response after ligation of PRRs. As mentioned above, they can also directly reach the proximal lymph node and interact with B-cells. Due to their simplicity, viruses usually consist of only a very limited set of structural proteins meaning that they are constrained to use multiple copies of these proteins to assemble. The cores and envelopes of viruses therefore consist of highly repetitive and ordered structures. During evolution the immune system has learned to judge these repetitive viral structures as a molecular danger signal because they are usually not present in the host [47]. Naïve B-cells can be activated by cross-linking of antigen-specific immunoglobulins on their surface without cognate T-cell help. T-cell independent activation of B-cell responses results in rapid generation of high IgM titers [48,49]. Although, these IgM antibodies are usually of low affinity, they have high avidity and can hence often efficiently control viral infection. Thus, the highly repetitive epitopes on the surface of viruses can efficiently cross-link B-cell receptors and induce quick T-cell independent IgM antibody responses. If the antigen is a protein, presentation of peptides on MHC class II and activation of Th-cells will additionally result in a subsequent Th-cell-dependent IgG response in addition to the IgM response. The highly repetitive structure of viruses is therefore an important reason for their superb immunogenicity and the development of vaccine delivery systems based on such highly repetitive structures thus an attractive approach to develop more effective vaccines.

6.5 NOVEL VACCINE TECHNOLOGIES

The development of vaccines has for a long time been on a rather low technological base. Nevertheless novel ideas are beginning to filter through which may aid in designing fundamentally new vaccination approaches. Whole-genome sequencing of pathogens and progresses in molecular biology and bioinformatics have enabled the rapid identification of new antigen candidates without the need for cultivating the pathogen. This approach is called "reverse vaccinology." Open reading frames in whole-genome sequences of pathogens are identified, the genes expressed in *E. coli* and the resulting proteins tested in animal models with regard to their ability to generate functional antibody responses [50]. Reverse vaccinology has already led to libraries of recombinant antigens available for vaccine use. However, the more challenging problem in vaccine development now is (1) to identify the relevant antigens for protection, (2) to present these antigens in the right fashion to the immune system, and (3) to optimize its responses against them. The focus has therefore shifted to translate advances in immunology into the development of novel delivery systems and vaccine adjuvants. An additional focus of interest is the induction of potent cellular immunity, such as CTLs, for the treatment of chronic infections and cancer.

6.5.1 Gene-Based Vaccines

CTL responses are efficiently induced when the antigen is expressed by host cells such as during viral infections because this allows appropriate antigen presentation to CD8 T-cells. Hence, gene-based vaccines have been developed that deliver the antigen DNA into host cells. Gene-based vaccines consist of either plasmid DNA or recombinant viral vectors. DNA vaccines are easy to manufacture because only the sequence of the gene insert needs to be exchanged to create vaccines against different target antigens. DNA vaccines have shown promising results in preclinical studies but successful human clinical trials have remained the exception [51–53]. The delivery of sufficient quantities of DNA to elicit a strong immune response is still a major obstacle in this technology. Also DNA vaccines administered as "naked DNA" in buffer have induced only modest immunity in humans. As an alternative to plasmid DNA, gene-based vaccines can also consist of recombinant viral vectors derived from adenoviruses, poxviruses, or α-viruses (see Chapter 20 in this book). The immunogenicity of viral vector vaccines seems to be superior to DNA vaccines in humans [54]. As with recombinant subunit vaccines, however, coadministration of immune potentiators seems to be essential also for gene-based vaccines to induce sufficient immunity in humans.

6.5.2 Synthetic PRR Ligands as Vaccine Adjuvants

Several immune potentiators that interact with TLRs and enhance vaccine potency are in development. These include small molecular compounds like the imidazoquinolines as TLR 7/8 ligands, nontoxic LPS analogues like synthetic monophosphoryl lipid A (MPL) for TLR4 or unmethylated CpG oligonucleotides (CpG-ODNs) as TLR9 ligands [55–57]. Next to CpG-ODNs also ssRNA derivatives and poly (I:C) analogues are developed as virus related adjuvants [58]. CPG-ODNs are the most advanced viral TLR ligands and have been tested in a number of preclinical models. It turned out that synthetic CpG-ODN-adjuvants are potent activators of DC maturation and induce Th1 and CTL responses [59]. Furthermore, they have a direct stimulatory effect on B-cells which express TLR 9 and induce class switching [60]. The use of CpG-ODNs as vaccine adjuvants is also already undergoing intensive testing in clinical trials for example for hepatitis B vaccines [61,62] and a trivalent killed influenza vaccine [63]. These studies demonstrated enhanced antibody titers, accelerated sero-conversion as well as enhanced responder rates. Also the increased potency of vaccines containing CpG-ODNs as adjuvant inducing CD4 and CD8 T-cell immunity was demonstrated in a clinical trial with melanoma patients [64]. The potency of CpG-ODNs in stimulating Th1-immune responses could even be envisioned to modulate immune responses and redirect for example Th2 allergic responses as shown in clinical trials in combination with recombinant ragweed allergen [65].

Although numerous PAMPs have been identified and synthetic TLR analogues have been produced, their use as adjuvants in vaccines is associated with numerous problematic issues. PRRs are expressed on many immune and nonimmune cells. Hence, the simple mixture of PAMPs with antigens could lead to their systemic exposure causing nonspecific stimulation of the immune system associated with immune-related pathology. This has been demonstrated for CpG-ODNs which upon high dose or chronic systemic exposure in preclinical models cause changes in lymphoid-follicle architecture, suppression of follicular dendritic cells and germinal centre B-lymphocytes, impaired humoral immune responses, and hepatic necrosis [66,67]. Also, splenomegaly in mice has been observed following coadministration of free CpG-ODN with antigens [67].

LPS or endotoxins which consist of a hydrophilic polysaccharide and a lipophilic phospholipid (lipid A) are TLR4 ligands and strong inducers of pro-inflammatory cytokines that may lead to septic shock. Consequently, they cannot be used as vaccine adjuvants. Removal of a phosphate residue, however, yielding MPL, results in a synthetic LPS analogue with lower toxicity but retained adjuvanticity [68]. Synthetic MPLs are promising adjuvants which have been included already in novel vaccines like GlaxoSmithKline's hepatitis B vaccine Fendrix.

Another issue of natural freely administered PAMPs is their problematic pharmacokinetic and pharmacodynamic properties. Viral TLR ligand for example, being highly charged polyions, are strongly bound by serum proteins and unstable in vivo due to the presence of degradative enzymes. Several analogues of TLR ligands have been produced already that overcome these instability problems like CpG-ODNs modified to contain a phosphorothioate backbone as TLR 9 ligand and imidazoquinolines, mentioned above, as TLR 7/8 analogues [68].

An alternative strategy to reduce systemic toxicity and to overcome the pharmacokinetic problems of PPR ligands as adjuvants is to target these immune enhancers directly to APCs and in particular to DCs. One way of doing this is packaging of PPR-ligands into particulate delivery systems. The advantage of such packaging is that the PPR-ligand is not systemically distributed and also protected from degradation. Another advantage of packaging PAMPs into a particulate delivery system is that both the antigen and the adjuvant are co-delivered to the APCs resulting in highly efficient activation of innate and adaptive immunity. The already mentioned microparticles, liposomes, virosomes, ISCOMs, and VLPs have all been used as delivery system for a variety of TLR-ligands [67,69–72]. VLPs are delivery systems that combine a number of advantageous properties. They are discussed in more detail in the next chapter.

6.5.3 Virus-Like Particles

VLPs are a special type of subunit vaccines based on the recombinant expression of virus capsid proteins. The expression of capsid proteins of many viruses leads to the spontaneous assembly into supramolecular, highly repetitive, icosahedral, or rod-like particles similar to the authentic viruses they are derived from but free of viral genetic material [73,74]. Thus VLPs represent a nonreplicating, noninfectious particulate antigen delivery system. VLPs can be extremely diverse in terms of their structure consisting of single or multiple capsid proteins containing lipid envelopes or existing without. The simplest VLPs are non-enveloped and assemble by expression of just one major capsid protein as shown for VLPs derived from papillomaviruses, parvoviruses, polyomaviruses, or different bacteriophages (see Chapters 9 through 11 in this book).

6.5.3.1 Immune Response to VLPs

Accumulated evidence has demonstrated that VLP-based vaccines are very efficient at stimulating both cellular and humoral immune responses [75–77]. Due to the structural similarity of VLPs to the viruses they are derived from, their immunological properties are indeed also very similar. VLPs exhibit numerous features for the potent induction of T-cell responses which makes them attractive vaccine candidates for the treatment of chronic viral infections and cancer. As a result of their particulate nature and dimensions, VLPs are efficiently taken up and processed by DCs for the subsequent presentation to T-cells in lymphoid organs [78]. Antigens delivered to DCs by VLP vaccines are not only presented on MHC class II molecules but also efficiently cross-presented on MHC class I molecules [79]. As discussed above, however, induction of potent effector and long-lasting memory T-cell responses also requires the presentation of antigens together with co-stimulatory molecules and cytokines released by activated DCs. To achieve this, PRRs expressed by DCs need to be stimulated while the antigen is taken up. VLPs are an ideal vaccine delivery system because they can easily co-deliver both antigens and activators of innate immunity. Bacteriophage-derived VLPs recombinantly expressed in *E. coli* for example spontaneously incorporate host cell RNA during their assembly process thus becoming a ready-made delivery system for this normally unstable TLR 7 ligand. Upon uptake by DCs not only the virus capsid protein is processed by DCs but also the TLR 7 ligand in the lumen of the VLPs is set free and binds to the endosomally expressed TLR 7 causing DC maturation [60]. The ability of certain VLPs to assemble and disassemble by changing chemical environment makes it possible to incorporate PAMPs other than host cell RNA. A prominent example for this are CpG-ODN-packaged VLPs which are potent inducers of CTL responses [67,80]. Linking of adjuvant molecules to VLPs is another possibility to co-deliver immune potentiators and was shown to increase immunogenicity [81]. The opportunity to incorporate or attach different types of immune enhancers to VLPs and thereby to co-deliver them to DCs together with the antigen makes it possible to induce different types of T-helper cell responses. Hence, VLPs are not only able to induce potent T-cell responses but to variably modulate them according to the requirements necessary to fight the targeted pathogen.

The excellent immunogenicity of VLPs is to a great extend also derived from its highly ordered and repetitive structure. As described above for viruses, the repetitive display of antigens like capsid proteins of VLPs is a unique way to stimulate B-cell responses. The optimal spacing of antigens for B-cell activation has been found to be 20–25 epitopes spaced by 5–10 nm. This corresponds well to numbers and distances found on many VLPs [82,83] and leads to efficient cross-linking of immunoglobulins on the surface of B-cells. Interaction of B-cells with VLPs, therefore, results in the vigorous induction of antibodies against the highly repetitive and orderly displayed viral capsid proteins. The stimulation of B-cells by VLPs is strong enough to elicit the T-cell independent induction of IgM antibodies. Hence, VLPs can be considered as typical T-cell-independent B-cell antigens. However, as also Th-responses to the VLP capsid proteins are concomitantly induced, efficient class switch from IgM to IgG occurs. PRR ligands incorporated into the lumen of VLPs can also have a direct stimulatory influence on B-cells and influence isotype switching as shown with CpG-ODNs incorporated into VLPs [60]. The stimulation of B-cells by the highly repetitive antigen display on VLPs is so strong that it can even overcome B-cell tolerance. This aspect is discussed later (Figure 6.3).

6.5.3.2 Viral Vaccines Based on VLPs

The excellent immunogenicity of VLPs makes them ideal candidates for the design of vaccines against the virus they are derived from. This is especially of interest, if the VLPs are composed of viral surface proteins which can be recognized by neutralizing antibodies. Examples for such VLPs being marketed or developed as viral vaccines are VLPs derived from hepatitis B virus, papilloma virus, hepatitis C virus, Norwalk virus, HIV (gp120), rotavirus, and parvovirus. For a long time the hepatitis B vaccine Engerix has been the only VLP-based vaccine licensed for human use. In June 2006, however, Merck's quadrivalent prophylactic recombinant VLP vaccine, named Gardasil,

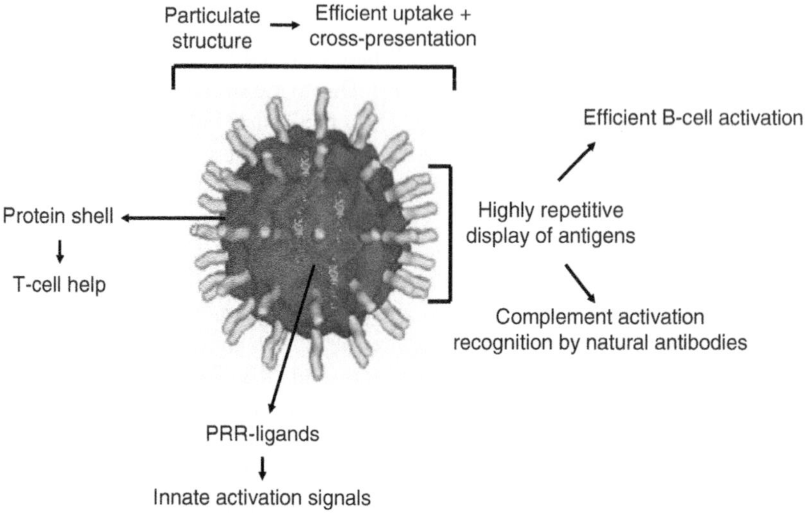

FIGURE 6.3 VLPs are a delivery system that combines a number of properties important for vaccine effectiveness. Due to their particulate structure, VLPs are efficiently taken up by DCs and their antigens are presented on MHC class II and cross-presented on class I molecules. The highly repetitive display of antigens on VLPs leads to the direct activation of B-cells and of components of the complement cascade. Binding of natural antibodies to VLPs can further enhance VLP-uptake by APCs. The protein backbone of VLPs provides activated B-cells with T-cell help and PRR-ligands incorporated in the lumen of VLPs can induce stimulatory signals to the innate immune system.

derived from the L1 major capsid protein of the human papilloma virus (HPV) [84] received approval from the Food and Drug Administration (FDA). It consists of L1 capsid proteins of HPV types 6, 11, 16, and 18. HPV types 16 and 18 have been associated with about 70% of cervical cancers, HPV types 6 and 11 are low-risk types, but are the most frequent cause of genital warts. There are over 100 known types of HPV with 30–40 types that are sexually transmitted (they infect mucosal epithelium of the anogenital region). Infection with high-risk HPV types is present in 99% of cervical cancers. Two of the HPV genes, E6 and E7, are known to inactivate normal tumor suppressor proteins, thereby allowing excessive growth and potential cancer. GlaxoSmithKline has produced and tested a divalent vaccine, which includes VLPs of HPV types 16 and 18, called Cervarix and is approaching FDA approval. Both vaccines are prophylactic, they work by stimulating antibody production that protects the person from infection. The antibodies will coat and neutralize the virus before it infects cells. The vaccines induce antibody responses that are considerably stronger than those observed during normal infection, show 100% responder rates and near 100% protection from the HPV types included in the vaccine [84–86]. HPV-VLPs, however can not only induce HPV antibodies but also elicit potent CTL responses as shown in numerous studies in mice [87,88]. For therapeutic purposes, chimeric HPV-VLPs that contain nonstructural epitopes of the viral oncoproteins E6 and E7 incorporated into the L1 and L2 capsids have been designed to specifically induce CTL responses against these antigens that are expressed and therefore presented on MHC class I molecules of HPV transformed tumor cells. The great success of the papilloma VLP vaccines demonstrates the potential of VLP-based vaccines. Both mentioned papilloma VLP vaccines are expected to reach blockbuster status.

6.5.3.3 VLPs as Carrier and Delivery System for Unrelated Target Antigens

In addition to the straightforward use of VLPs for vaccinations against the virus they are derived from, the efficiency with which they stimulate cellular and humoral immune responses has made them prime candidates as carrier molecules for the delivery of B- and T-cell epitopes from unrelated sources. This greatly extends the boundaries of VLP use. Rational design of VLP-based vaccines can be performed via incorporation of target antigens into the highly ordered particulate structure of VLPs. Integration of antigens into VLPs can be achieved by either genetic fusion or chemical conjugation of unrelated antigens to the surface of VLPs. The coupled antigen now exhibits the same order and repetitiveness as the carrier proteins meaning that it adopts the "viral fingerprint" and is thus able to induce similarly efficient B-cell and T-cell responses. In consequence, any desired antigen may be brought into a highly repetitive array required for optimal B-cell activation. The use of VLPs as carrier and delivery system is especially of interest for antigens that are by itself not or only very weakly immunogenic, like haptens, lipids, carbohydrates, or peptides. Being displayed on VLPs these antigens gain the same immunogenicity as the underlying capsid proteins: they can efficiently activate B-cells, their pharmacokinetics is improved, and their delivery to APCs is enhanced. Furthermore being attached to a protein carrier molecule, the antigens are linked to a source of T-cell help and can be co-delivered with immune stimulators or modulators packaged into the VLP lumen.

Genetic fusion of linear or conformational epitopes to the self-assembly-competent capsid molecules leads to so-called VLP-chimeras. VLP-chimeras present the introduced target antigen on each capsid subunit of the particle and hence display it at a very high density. Numerous VLP-chimeras presenting B- and T-cell epitopes have been designed and tested in preclinical and clinical studies [74,89].

As an alternative to genetic fusion, antigens can be separately expressed and then coupled to the VLP surface [83,90]. This approach increases the chance to incorporate a correctly folded full length target protein as the structure of the protein in not constrained during folding and assembly of the VLP monomers. A further advantage of the coupling approach is that the site of coupling to the VLP can be optimized to achieve maximal exposure of the coupled antigen. For small antigens like peptides the attachment of several antigens per VLP subunit may be possible hence further increasing antigen density. Coupling of antigens to VLP surfaces can result in higher titers than genetic fusion [83,91]. Attachment to VLPs may be done through non-covalent bonds such as biotin–streptavidin interactions [90]. Alternatively, covalent attachment of antigens to VLPs may be achieved via chemical cross-linkers. Numerous approaches to do this exist and depend very much on the functional groups available on the VLP surface and the antigen of choice. One approach that is successfully employed to attach high densities of antigens onto VLPs makes use of heterobifunctional cross-linkers [83,92]. In this particular approach the heterobifunctional cross-linker consists of an amine reactive NHS-ester and a sulfhydryl-reactive maleimide group. Amines exposed on the surface of VLPs are first reacted with the NHS-ester of the cross-linker while preserving the activity of its maleimide group. Following the first reaction, the derivatized VLP is reacted with the target antigen which contains a free SH-group (usually a cysteine) under conditions permitting conjugation of the maleimide group to the sulfhydryl group of the antigen. To ensure that the antigen is coupled to the VLP in a directed and oriented fashion, peptide antigens may be engineered to contain a free cysteine group either at the amino or carboxy terminal end. The capacity of this approach to induce vigorous antibody responses against the immunogen displayed in such a fashion on the VLP surface has been demonstrated for many antigens in mice as well as humans [49,83,93–96].

6.5.3.4 Nonconventional Vaccination Approaches Based on VLP Vaccines

6.5.3.4.1 Vaccination Against Small Molecular Compounds

The ability to couple antigens to the surface of VLPs makes it possible to bring essentially every desired compound into a highly repetitive array. The immune system treats these compounds as if they were viral components and reacts with strong antibody responses. As mentioned above, haptens can be rendered immunogenic via chemical coupling to VLPs. As an example for such an approach, nicotine coupled to VLPs is developed as a vaccine candidate against nicotine addiction. Nicotine, which is a tobacco alkaloid, is the prime substance in cigarettes that drives addiction. During smoking nicotine is taken up through the mucosa into the blood and subsequently enters the brain where it unfolds its addictive virtue. Nicotine itself is not immunogenic. Coupled to VLPs, however, it induces anti-nicotine antibodies in a person. The basis of a nicotine vaccine is to induce strong anti-nicotine antibodies which prevent the alkaloid to enter the brain [97]. As antibody affinities to small molecules are usually low and nicotine concentrations in the blood during smoking high, strong antibody responses have to be induced. A phase IIa clinical trial with a nicotine-VLP vaccine has demonstrated that sufficient antibody titers to sequester the nicotine in the blood could be induced and that smoking cessation was markedly facilitated in individuals that developed strong antibody responses. This example shows the potency of VLPs as carriers for hapten antigens and opens the possibility to develop fundamentally new vaccine approaches to treat diseases that were traditionally not targeted with vaccines.

6.5.3.4.2 Vaccination Against Self-Antigens

A hallmark of the immune system is its ability to distinguish between infectious nonself and noninfectious self. As a consequence, the immune system is able to vigorously attack foreign invaders while sparing the host. Unresponsiveness to self is generally referred to as tolerance and it exists at the T- and B-cell level [98–100]. Self-reactive lymphocytes can be deleted from the repertoire during the induction of central tolerance, which takes place in the thymus for T- and in the bone marrow for B-lymphocytes. The process of central tolerance induction, however, is incomplete so that self-reactive B- and T-lymphocytes can reach the periphery where they are usually unresponsive [101,102]. Interestingly, self-antigens which are attached to VLPs are rendered immunogenic enough to overcome B-cell unresponsiveness. This is firstly because of the highly ordered and repetitive way in which they are presented to B-cells causing strong cross-linking of B-cell receptors and hence their activation. Secondly, the fact that they are attached to a nonself carrier leads to the presentation of nonself peptides on the MHC class II molecules of activated B-cells so that cognate T-cell help can be provided to these B-cells. Thus, the presentation of self-antigens on the surface of VLPs can induce the development of auto-antibodies to these self-molecules.

The use of VLPs to present self-antigens to the immune system offers great potential for treatment of not only cancer but also other chronic diseases like autoimmune diseases or CNS-disorders. For many of these illnesses monoclonal antibodies are under development or already marketed that neutralize the action of important mediators or block receptors. The administration of monoclonal antibodies to treat diseases essentially reflects the process of passive vaccination. The administration of antibodies, however, has a number of disadvantages, such as the very high costs associated with antibody therapies, the need for frequent administration and the development of anti-antibody responses due to the immunogenicity that is observed even with humanized and entirely human antibodies [103]. These are important issues especially when chronic diseases are to be treated which require long-term administration of therapeutic antibodies. Active vaccination against the self-antigens currently targeted with monoclonal antibodies would overcome these disadvantages. Important safety issues such as the reversibility of the antibody response must, however, be carefully assessed, because passive therapy with monoclonal antibodies is more controllable than vaccination which involves active engagement of the host's immune system.

The feasibility of using VLPs conjugated with self-molecules for induction of high titers of IgG auto-antibodies has been demonstrated in numerous preclinical studies and has shown efficacy in treating the targeted diseases. Examples include the induction of auto-antibodies against the inflammatory cytokines TNF for the treatment of rheumatoid arthritis in mice [90,95] and IL-17 for the treatment of rheumatoid arthritis and experimental autoimmune encephalomyelitis (an animal model for multiple sclerosis) in mice [94]. Also auto-antibodies that were raised against RANKL for the treatment of osteoporosis [96], against CCR5 for prevention of HIV infections [104] and against angiotensin II for the treatment of high blood pressure showed efficacy in animal models [93].

A VLP-based vaccine against angiotensin II has already completed a phase II clinical trial and proved to be well tolerated and to reduce blood pressure in patients suffering from hypertension [105].

Vaccination against self-molecules aims to induce a specific antibody response; however, an undesired T-cell response might also be generated, resulting in activated infiltrating T-cells that could potentially cause organ damage. This has been observed in a phase II clinical trial with a vaccine against amyloid β protein (Aβ) which forms the amyloid plaques associated with neuro-degeneration in Alzheimer's disease. The induced antibodies showed efficacy and the disease progressed more slowly in vaccinated patients. However, the trial had to be discontinued due to the development of a sterile meningoencephalitis likely caused by infiltrating Aβ-specific T-cells [106]. The induction of antibody responses by active vaccination against self-molecules therefore aims at avoiding such T-cell responses by using adjuvant free vaccines. The potency of VLPs to break B-cell tolerance and to induce neutralizing antibodies against self-molecules without the addition of further adjuvants might be a way to achieve this. Nevertheless, these safety issues have to be addressed on a case-by-case basis.

6.5.3.4.3 Vaccines for Allergies

Allergic reactions occur when an individual who has developed IgE antibodies after encounter with an innocuous antigen (allergen) is exposed to the same antigen again. The specific IgEs produced in response to the allergen bind to high-affinity IgE receptors, which are localized mostly on mast cells and basophils, and mediate the release of proinflammatory cytokines that recruit eosinophils, basophils, monocytes, neutrophils, and CD4 T-cells which cause the allergic inflammation. The development of IgE responses against allergens occurs when very low doses of the allergen enter the body, usually through mucosal surfaces, and is associated with Th2-type responses. The majority of pathogenic T-cells in allergic reactions are Th2-cells, producing the cytokines IL-4, IL-5, and IL-13. These cytokines regulate IgE synthesis, promote eosinophil differentiation, and trigger the allergic response [107,108]. One common approach to the treatment of allergies is desensitization. The aim of desensitization is to shift the antibody response dominated by IgE to one that is dominated by IgG to prevent IgE-mediated receptor pathways. Patients are desensitized by a number of consecutive injections with escalating doses of allergen over a period of years [109]. This injection schedule is thought to divert the immune response from a Th2 driven one to a Th1 driven one thereby downregulating IgE production and changing the cytokine milieu.

An alternative approach to desensitization is vaccination with allergen-formulations in adjuvant. The aim of vaccination against allergies is also the redirection of Th2-type immune responses towards a Th1-response pattern. These vaccines consist of the allergen as protein or T-cell epitope and a Th1 promoting adjuvant. An adjuvant that has proven especially potent in this context is CpG-ODNs which is a strong Th1 promoting adjuvant [110]. As mentioned above, VLPs are especially suitable to co-deliver CpG-ODNs together with an antigen and might therefore be potent in shifting allergen-specific T-cells from a Th2- to a Th1-type hence suppressing the allergic inflammation. However, once a pathological immune response has been established and the more polarized it has become the more difficult it may be to modulate it by therapeutic vaccination. Therefore, the future of vaccination against allergies might not be to vaccinate in order to reverse an existing pathological immune response but rather to prevent it from happening. Therefore, prophylactic vaccination against common allergens might become as normal as vaccination against bacterial and viral pathogens is already today.

6.6 CONCLUSION

The immune system is an incredibly complex and powerful entity. Its proper function is crucial to protect the organism against foreign invaders and hence to prevent diseases. Its malfunction, however, can actually be the cause of serious diseases as well, like autoimmune diseases or allergies.

Since its commencement over 200 years ago, vaccinology has tried to use the immune system as a tool to fight infectious diseases. The greatest successes of vaccination have been established against pathogens that cause disease in a relatively direct and simple manner. The most important characteristic of these pathogens is that neutralizing antibodies efficiently protect against infection and that induction of such antibodies is relatively straightforward. Use of the immune system against these pathogens has therefore been relatively simple. A great number of challenges, however, remain. Vaccines against chronic infections that have reached epidemic proportions like HIV or HCV are still elusive. Infectious diseases like Malaria or Dengue fever, which represent a serious health problem in many tropical countries, can still not be targeted by vaccination although considerable efforts have been undertaken. And also for the presumably gravest burden of mankind, cancer, no therapeutic vaccines exist yet, although countless attempts have been made.

The increased understanding of the multitude of processes involved in orchestrating an immune response, the discovery of substances that can activate and modulate the immune response and the development of new vaccine delivery systems may give us the opportunity to meet these challenges and have given vaccines a new face. Novel vaccines might not be simply considered as substances that induce a protective immune response against a pathogen anymore. They could rather be considered as substances with which we can tune the immune response to our advantage. Vaccines may be used to activate the right immune cells and to induce the proper inflammatory condition to eliminate chronic infections. They may be used to neutralize, block, or bind self-molecules crucially involved in diseases like cancer, Alzheimer, chronic inflammatory autoimmune diseases, or even hypertension. Vaccines may also be used as immunomodulators to alter the cytokine milieu and the type of antibody responses to reverse allergies.

REFERENCES

1. Cooper, M.R., et al., Medicine at the medical center then and now: One hundred years of progress. *South Med J*, 2002. 95(10): 1113–1121.
2. Plotkin, S.A., Vaccines: Past, present and future. *Nat Med*, 2005. 11(Suppl 4): 5–11.
3. Rappuoli, R., From Pasteur to genomics: Progress and challenges in infectious diseases. *Nat Med*, 2004. 10(11): 1177–1185.
4. Hoebe, K., E. Janssen, and B. Beutler, The interface between innate and adaptive immunity. *Nat Immunol*, 2004. 5(10): 971–974.
5. Levine, M.M. and M.B. Sztein, Vaccine development strategies for improving immunization: The role of modern immunology. *Nat Immunol*, 2004. 5(5): 460–464.
6. O'Hagan, D.T. and N.M. Valiante, Recent advances in the discovery and delivery of vaccine adjuvants. *Nat Rev Drug Discov*, 2003. 2(9): 727–735.
7. Scarselli, M., et al., The impact of genomics on vaccine design. *Trends Biotechnol*, 2005. 23(2): 84–91.
8. Fenner, F., et al., Early efforts at control: Variolation, vaccination, and isolation and quarantine. *History of International Public Health*, F. Fenner, et al. (Eds.), Vol. 6. 1988, Geneva: World Health Organisation. pp. 245–276.
9. Plotkin, S.L. and S.A. Plotkin, A Short History of Vaccination. In, S.A. Plotkin, W.A. Orenstein, and P. Offit (Eds.), *Vaccines*. 4th ed. 2004, Philadelphia: W.B. Saunders. pp. 1–15.
10. Calmette, A., C. Guerin, and M. Breton, Contribution a l'etude da la tuberculose experimental du cobaye (infection et essais da vaccination par la voie digestive). *Ann Inst Pasteur Paris*, 1907. 21: 401–416.
11. Pasteur, L. and C.-E. Chamberland, Sur la vaccination charbonneuse. *C R Acad Sci Paris*, 1881. 92: 1378–1383.
12. Theiler, M. and H.H. Smith, The use of the yellow fever virus by in vitro cultivation for human immunization. *J ExMed*, 1937. 65: 787–800.
13. Pasteur, L., De l'attenuation du virus du cholera des poules. *C R Acad Sci Paris*, 1880. 91: 673–680.
14. Pearce, J.M., Salk and Sabin: Poliomyelitis immunisation. *J Neurol Neurosurg Psychiatry*, 2004. 75(11): 1552.
15. Belshe, R.B., et al., Immunization of infants and young children with live attenuated trivalent cold-recombinant influenza A H1N1, H3N2, and B vaccine. *J Infect Dis*, 1992. 165(4): 727–732.
16. Davies, J.R. and E.A. Grilli, Natural or vaccine-induced antibody as a predictor of immunity in the face of natural challenge with influenza viruses. *Epidemiol Infect*, 1989. 102(2): 325–333.

17. Strebel, P.M., et al., Global measles elimination efforts: The significance of measles elimination in the United States. *J Infect Dis*, 2004. 189(Suppl 1): S251–S257.
18. Ramon, G., Sur le pouvoir floculant et sur les proprietes immunisantes d'une toxine diphterique rendu anatoxique (anatosine). *C R Acad Sci Paris*, 1923. 177: 1338–1340.
19. Ramon, G. and C. Zeller, De la valeur antigenique de l'antioxine tetanique chez l'homme. *C R Acad Sci Paris*, 1926. 182: 245–247.
20. Bachmann, M.F. and M. Kopf, Balancing protective immunity and immunopathology. *Curr Opin Immunol*, 2002. 14(4): 413–419.
21. Pugachev, K.V., F. Guirakhoo, and T. Monath, New developments in flavivirus vaccines with special attention to yellow fever. *Curr Opin Infect Dis*, 2005. 18(5): 387–394.
22. Jennings, G.T. and M.F. Bachmann, Designing recombinant vaccines with viral properties: A rational approach to more effective vaccines. *Curr Mol Med*, 2007. 7(2): 143–155.
23. Bachmann, M.F., et al., Long-lived memory $CD8^+$ T cells are programmed by prolonged antigen exposure and low levels of cellular activation. *Eur J Immunol*, 2006. 36(4): 842–854.
24. Banchereau, J., et al., Immunobiology of dendritic cells. *Annu Rev Immunol*, 2000. 18: 767–811.
25. Janeway, C.A., Jr., and R. Medzhitov, Innate immune recognition. *Annu Rev Immunol*, 2002. 20: 197–216.
26. Shortman, K. and Y.J. Liu, Mouse and human dendritic cell subtypes. *Nat Rev Immunol*, 2002. 2(3): 151–161.
27. Engering, A.J., et al., Mannose receptor mediated antigen uptake and presentation in human dendritic cells. *Adv Exp Med Biol*, 1997. 417: 183–187.
28. Jiang, W., et al., The receptor DEC-205 expressed by dendritic cells and thymic epithelial cells is involved in antigen processing. *Nature*, 1995. 375(6527): 151–155.
29. Reis e Sousa, C. and J.M. Austyn, Phagocytosis of antigens by Langerhans cells. *Adv Exp Med Biol*, 1993. 329: 199–204.
30. Ahsan, F., et al., Targeting to macrophages: Role of physicochemical properties of particulate carriers–liposomes and microspheres–on the phagocytosis by macrophages. *J Control Release*, 2002. 79(1–3): 29–40.
31. Ochsenbein, A.F., et al., Protective T cell-independent antiviral antibody responses are dependent on complement. *J Exp Med*, 1999. 190(8): 1165–1174.
32. Carroll, M.C. and M.B. Fischer, Complement and the immune response. *Curr Opin Immunol*, 1997. 9(1): 64–69.
33. Fanger, N.A., et al., Type I (CD64) and type II (CD32) Fc gamma receptor-mediated phagocytosis by human blood dendritic cells. *J Immunol*, 1996. 157(2): 541–548.
34. Matsushita, M. and T. Fujita, Ficolins and the lectin complement pathway. *Immunol Rev*, 2001. 180: 78–85.
35. Harding, C.V. and R. Song, Phagocytic processing of exogenous particulate antigens by macrophages for presentation by class I MHC molecules. *J Immunol*, 1994. 153(11): 4925–4933.
36. Kovacsovics-Bankowski, M., et al., Efficient major histocompatibility complex class I presentation of exogenous antigen upon phagocytosis by macrophages. *Proc Natl Acad Sci U S A*, 1993. 90(11): 4942–4946.
37. Storni, T. and M.F. Bachmann, Loading of MHC class I and II presentation pathways by exogenous antigens: A quantitative in vivo comparison. *J Immunol*, 2004. 172(10): 6129–6135.
38. Manolova, V., et al., Nanoparticles target distinct dendritic cell populations according to their size. *Eur J Immunol*, 2008. 38(5): 1404–1413.
39. Trinchieri, G. and A. Sher, Cooperation of Toll-like receptor signals in innate immune defence. *Nat Rev Immunol*, 2007. 7(3): 179–190.
40. West, A.P., A.A. Koblansky, and S. Ghosh, Recognition and signaling by toll-like receptors. *Annu Rev Cell Dev Biol*, 2006. 22: 409–437.
41. Creagh, E.M. and L.A. O'Neill, TLRs, NLRs and RLRs: A trinity of pathogen sensors that co-operate in innate immunity. *Trends Immunol*, 2006. 27(8): 352–357.
42. Iwasaki, A. and R. Medzhitov, Toll-like receptor control of the adaptive immune responses. *Nat Immunol*, 2004. 5(10): 987–995.
43. Pulendran, B. and R. Ahmed, Translating innate immunity into immunological memory: Implications for vaccine development. *Cell*, 2006. 124(4): 849–863.

44. Lindblad, E.B., Aluminium compounds for use in vaccines. *Immunol Cell Biol*, 2004. 82(5): 497–505.
45. Ott, G., et al., MF59. Design and evaluation of a safe and potent adjuvant for human vaccines. *Pharm Biotechnol*, 1995. 6: 277–296.
46. Querec, T., et al., Yellow fever vaccine YF-17D activates multiple dendritic cell subsets via TLR2, 7, 8, and 9 to stimulate polyvalent immunity. *J Exp Med*, 2006. 203(2): 413–424.
47. Bachmann, M.F., et al., The influence of antigen organization on B cell responsiveness. *Science*, 1993. 262(5138): 1448–1451.
48. Bachmann, M.F. and R.M. Zinkernagel, Neutralizing antiviral B cell responses. *Annu Rev Immunol*, 1997. 15: 235–270.
49. Maurer, P., et al., A therapeutic vaccine for nicotine dependence: Preclinical efficacy, and phase I safety and immunogenicity. *Eur J Immunol*, 2005. 35(7): 2031–2040.
50. Mora, M., et al., Reverse vaccinology. *Drug Discov Today*, 2003. 8(10): 459–464.
51. Rottinghaus, S.T., et al., Hepatitis B DNA vaccine induces protective antibody responses in human non-responders to conventional vaccination. *Vaccine*, 2003. 21(31): 4604–4608.
52. Ulmer, J.B., B. Wahren, and M.A. Liu, Gene-based vaccines: Recent technical and clinical advances. *Trends Mol Med*, 2006. 12(5): 216–222.
53. Wang, R., et al., Induction of antigen-specific cytotoxic T lymphocytes in humans by a malaria DNA vaccine. *Science*, 1998. 282(5388): 476–480.
54. McConkey, S.J., et al., Enhanced T-cell immunogenicity of plasmid DNA vaccines boosted by recombinant modified vaccinia virus Ankara in humans. *Nat Med*, 2003. 9(6): 729–735.
55. Baldridge, J.R., et al., Taking a Toll on human disease: Toll-like receptor 4 agonists as vaccine adjuvants and monotherapeutic agents. *Expert Opin Biol Ther*, 2004. 4(7): 1129–1138.
56. Cooper, C.L., et al., CPG 7909, an immunostimulatory TLR9 agonist oligodeoxynucleotide, as adjuvant to Engerix-B HBV vaccine in healthy adults: A double-blind phase I/II study. *J Clin Immunol*, 2004. 24 (6): 693–701.
57. Hemmi, H., et al., Small anti-viral compounds activate immune cells via the TLR7 MyD88-dependent signaling pathway. *Nat Immunol*, 2002. 3(2): 196–200.
58. Sloat, B.R. and Z. Cui, Nasal immunization with anthrax protective antigen protein adjuvanted with polyriboinosinic-polyribocytidylic acid induced strong mucosal and systemic immunities. *Pharm Res*, 2006. 23(6): 1217–1226.
59. Krieg, A.M., Therapeutic potential of Toll-like receptor 9 activation. *Nat Rev Drug Discov*, 2006. 5(6): 471–484.
60. Jegerlehner, A., et al., TLR9 signaling in B cells determines class switch recombination to IgG2a. *J Immunol*, 2007. 178(4): 2415–2420.
61. Cooper, C.L., et al., CPG 7909 adjuvant improves hepatitis B virus vaccine seroprotection in antiretroviral-treated HIV-infected adults. *AIDS*, 2005. 19(14): 1473–1479.
62. Halperin, S.A., et al., Comparison of the safety and immunogenicity of hepatitis B virus surface antigen co-administered with an immunostimulatory phosphorothioate oligonucleotide and a licensed hepatitis B vaccine in healthy young adults. *Vaccine*, 2006. 24(1): 20–26.
63. Cooper, C.L., et al., Safety and immunogenicity of CPG 7909 injection as an adjuvant to Fluarix influenza vaccine. *Vaccine*, 2004. 22(23–24): 3136–3143.
64. Speiser, D.E., et al., Rapid and strong human $CD8^+$ T cell responses to vaccination with peptide, IFA, and CpG oligodeoxynucleotide 7909. *J Clin Invest*, 2005. 115(3): 739–746.
65. Simons, F.E., et al., Selective immune redirection in humans with ragweed allergy by injecting Amb a 1 linked to immunostimulatory DNA. *J Allergy Clin Immunol*, 2004. 113(6): 1144–1151.
66. Heikenwalder, M., et al., Lymphoid follicle destruction and immunosuppression after repeated CpG oligodeoxynucleotide administration. *Nat Med*, 2004. 10(2): 187–192.
67. Storni, T., et al., Nonmethylated CG motifs packaged into virus-like particles induce protective cytotoxic T cell responses in the absence of systemic side effects. *J Immunol*, 2004. 172(3): 1777–1785.
68. Marciani, D.J., Vaccine adjuvants: Role and mechanisms of action in vaccine immunogenicity. *Drug Discov Today*, 2003. 8(20): 934–943.
69. Gursel, I., et al., Sterically stabilized cationic liposomes improve the uptake and immunostimulatory activity of CpG oligonucleotides. *J Immunol*, 2001. 167(6): 3324–3328.
70. O'Hagan, D.T. and M. Singh, Microparticles as vaccine adjuvants and delivery systems. *Expert Rev Vaccines*, 2003. 2(2): 269–283.

71. Westwood, A., et al., Immunological responses after immunisation of mice with microparticles containing antigen and single stranded RNA (polyuridylic acid). *Vaccine*, 2006. 24(11): 1736–1743.
72. Zaks, K., et al., Efficient immunization and cross-priming by vaccine adjuvants containing TLR3 or TLR9 agonists complexed to cationic liposomes. *J Immunol*, 2006. 176(12): 7335–7345.
73. Johnson, J.E. and W. Chiu, Structures of virus and virus-like particles. *Curr Opin Struct Biol*, 2000. 10(2): 229–235.
74. Pumpens, P. and E. Grens, HBV core particles as a carrier for B cell/T cell epitopes. *Intervirology*, 2001. 44(2–3): 98–114.
75. Murata, K., et al., Immunization with hepatitis C virus-like particles protects mice from recombinant hepatitis C virus-vaccinia infection. *Proc Natl Acad Sci U S A*, 2003. 100(11): 6753–6758.
76. Paliard, X., et al., Priming of strong, broad, and long-lived HIV type 1 p55gag-specific $CD8^+$ cytotoxic T cells after administration of a virus-like particle vaccine in rhesus macaques. *AIDS Res Hum Retroviruses*, 2000. 16(3): 273–282.
77. Schirmbeck, R., W. Bohm, and J. Reimann, Virus-like particles induce MHC class I-restricted T-cell responses. Lessons learned from the hepatitis B small surface antigen. *Intervirology*, 1996. 39(1–2): 111–119.
78. Beyer, T., et al., Bacterial carriers and virus-like-particles as antigen delivery devices: Role of dendritic cells in antigen presentation. *Curr Drug Targets Infect Disord*, 2001. 1(3): 287–302.
79. Ruedl, C., et al., Cross-presentation of virus-like particles by skin-derived CD8(−) dendritic cells: A dispensable role for TAEur. *J Immunol*, 2002. 32(3): 818–825.
80. Schwarz, K., et al., Efficient homologous prime-boost strategies for T cell vaccination based on virus-like particles. *Eur J Immunol*, 2005. 35(3): 816–821.
81. Kang, S.M. and R.W. Compans, Enhancement of mucosal immunization with virus-like particles of simian immunodeficiency virus. *J Virol*, 2003. 77(6): 3615–3623.
82. Dintzis, R.Z., B. Vogelstein, and H.M. Dintzis, Specific cellular stimulation in the primary immune response: Experimental test of a quantized model. *Proc Natl Acad Sci U S A*, 1982. 79(3): 884–888.
83. Jegerlehner, A., et al., A molecular assembly system that renders antigens of choice highly repetitive for induction of protective B cell responses. *Vaccine*, 2002. 20(25–26): 3104–3112.
84. Koutsky, L.A., et al., A controlled trial of a human papillomavirus type 16 vaccine. *N Engl J Med*, 2002. 347(21): 1645–1651.
85. Evans, T.G., et al., A Phase 1 study of a recombinant viruslike particle vaccine against human papillomavirus type 11 in healthy adult volunteers. *J Infect Dis*, 2001. 183(10): 1485–1493.
86. Harro, C.D., et al., Safety and immunogenicity trial in adult volunteers of a human papillomavirus 16 L1 virus-like particle vaccine. *J Natl Cancer Inst*, 2001. 93(4): 284–292.
87. Gissmann, L., et al., Therapeutic vaccines for human papillomaviruses. *Intervirology*, 2001. 44(2–3): 167–175.
88. Schiller, J.T. and D.R. Lowy, Papillomavirus-like particle vaccines. *J Natl Cancer Inst Monogr*, 2001. 44(2–3): 50–54.
89. Pumpens, P., et al., Hepatitis B virus core particles as epitope carriers. *Intervirology*, 1995. 38(1–2): 63–74.
90. Chackerian, B., D.R. Lowy, and J.T. Schiller, Conjugation of a self-antigen to papillomavirus-like particles allows for efficient induction of protective autoantibodies. *J Clin Invest*, 2001. 108(3): 415–423.
91. Neirynck, S., et al., A universal influenza A vaccine based on the extracellular domain of the M2 protein. *Nat Med*, 1999. 5(10): 1157–1163.
92. Lechner, F., et al., Virus-like particles as a modular system for novel vaccines. *Intervirology*, 2002. 45(4–6): 212–217.
93. Ambuhl, P.M., et al., A vaccine for hypertension based on virus-like particles: Preclinical efficacy and phase I safety and immunogenicity. *J Hypertens*, 2007. 25(1): 63–72.
94. Rohn, T.A., et al., Vaccination against IL-17 suppresses autoimmune arthritis and encephalomyelitis. *Eur J Immunol*, 2006. 36(11): 2857–2867.
95. Spohn, G., et al., A virus-like particle-based vaccine selectively targeting soluble TNF-{alpha} protects from arthritis without inducing reactivation of latent tuberculosis. *J Immunol*, 2007. 178(11): 7450–7457.
96. Spohn, G., et al., Protection against osteoporosis by active immunization with TRANCE/RANKL displayed on virus-like particles. *J Immunol*, 2005. 175(9): 6211–6218.
97. Maurer, P. and M.F. Bachmann, Therapeutic vaccines for nicotine dependence. *Curr Opin Mol Ther*, 2006. 8(1): 11–16.

98. von Boehmer, H. and P. Kisielow, Self-nonself discrimination by T cells. *Science*, 1990. 248(4961): 1369–1373.
99. Ferry, H., et al., B-cell tolerance. *Transplantation*, 2006. 81(3): 308–315.
100. Ohashi, P.S. and A.L. DeFranco, Making and breaking tolerance. *Curr Opin Immunol*, 2002. 14(6): 744–759.
101. Hawiger, D., et al., Dendritic cells induce peripheral T cell unresponsiveness under steady state conditions in vivo. *J Exp Med*, 2001. 194(6): 769–779.
102. Gavin, A., et al., Peripheral B lymphocyte tolerance. *Keio J Med*, 2004. 53(3): 151–158.
103. Kim, S.J., Y. Park, and H.J. Hong, Antibody engineering for the development of therapeutic antibodies. *Mol Cells*, 2005. 20(1): 17–29.
104. Chackerian, B., et al., Induction of autoantibodies to CCR5 in macaques and subsequent effects upon challenge with an R5-tropic simian/human immunodeficiency virus. *J Virol*, 2004. 78(8): 4037–4047.
105. Tissot, A.C., et al., Effect of immunisation against angiotensin II with CYT006-AngQb on ambulatory blood pressure: A double-blind, randomised, placebo-controlled phase IIa study. *Lancet*, 2008. 371(9615): 821–827.
106. Orgogozo, J.M., et al., Subacute meningoencephalitis in a subset of patients with AD after Abeta42 immunization. *Neurology*, 2003. 61(1): 46–54.
107. Barnes, P.J., New directions in allergic diseases: Mechanism-based anti-inflammatory therapies. *J Allergy Clin Immunol*, 2000. 106(1 Pt 1): 5–16.
108. Nickel, R., et al., Chemokines and allergic disease. *J Allergy Clin Immunol*, 1999. 104(4 Pt 1): 723–742.
109. Barnes, P.J., Therapeutic strategies for allergic diseases. *Nature*, 1999. 402(Suppl 6760): B31–B38.
110. Kline, J.N., Effects of CpG DNA on Th1/Th2 balance in asthma. *Curr Top Microbiol Immunol*, 2000. 247: 211–225.

7 Modification of HIV-1 Envelope Glycoprotein to Enhance Immunogenicity

Michael Aidoo, Salvatore Butera, and Dennis Ellenberger

CONTENTS

7.1 INTRODUCTION

Infection by human immunodeficiency virus (HIV) and the disease it causes, acquired immunodeficiency syndrome (AIDS), is clearly one of the most important present-day health threats. Worldwide, according to the UNAIDS/WHO 2006 AIDS epidemic report, it is estimated that HIV/AIDS accounts for 2.9 million deaths and 4.3 million new infections each year (UNIADS/WHO 2006 AIDS Epidemic Update). The immunodeficiency caused by this infection has led to the explosion of another important infectious disease, TB, and the economic costs of the infection to the individual and communities, especially in high prevalent countries is enormous.

The discovery in the mid-1980s of HIV as the causative agent for AIDS prompted tremendous optimism that an efficacious vaccine would soon be developed to protect exposed persons.

After more than 20 years and in spite of enormous resources dedicated to the effort, such a vaccine remains a distant hope. However, an important benefit of the intense effort to produce an HIV vaccine has been the elucidation of the mechanisms of viral pathogenesis, and natural history of infection and disease. While the world waits for a vaccine, the pharmaceutical industry is developing antiretroviral (ARV) drugs that are been used in the fight against HIV/AIDS. These ARVs have prolonged the lives of many HIV-infected persons and drastically reduced perinatal transmission. However, while provision of ARVs to infected persons is critical to HIV/AIDS control, because of several critical limitations, the best preventive tool to combat new HIV infections will be a vaccine.

Early exploratory efforts at HIV vaccine development largely followed classical or historical paradigms. Due to previous experiences with successful antibody-inducing viral vaccines against polio, measles, and hepatitis B, upon the elucidation of the HIV viral structure, initial attempts at developing HIV vaccines focused on inducing antibodies to either critical structures of or the entire envelope protein [1–5]. Despite tremendous effort, virus attenuation and inactivation strategies that have been employed successfully in developing vaccines for other viral infections have thus far not yielded an efficacious HIV vaccine for human use.

As a consequence of the failure of early attempts to develop an HIV vaccine, there was a shift by the HIV vaccine research community away from the classical vaccine paradigm to developing vaccines that induced cell-mediated immune responses, in particular, $CD8^{+}$ cytotoxic T cells (CTL). This bias towards CTL-inducing vaccines was based on many lines of evidence suggesting that CTL immunity played a significant role in virus control throughout the various phases of the infection [6–10]. However, it soon became clear from several preclinical studies of T cell-inducing vaccines that vaccine-elicited CD8 T cells control viremia but do not prevent infection. Thus, optimal HIV vaccine will likely need to induce both neutralizing antibody and T cell immunity, and this realization renewed efforts to develop better and unique vaccine strategies for generating more potent antibody responses.

In this chapter, we focus on efforts to modify the HIV envelope glycoprotein to induce potent neutralizing antibodies that alone or with the help of T cell immunity potentially prevent HIV transmission.

7.2 HIV-1 VIRUS AND HOST CELL INTERACTION

HIV-1 entry into the cell begins with the binding of viral envelope glycoprotein gp120 to CD4 molecules on the target cell surface. Binding of gp120 to CD4 triggers conformational changes in the envelope structure that exposes or forms the co-receptor binding sites. The conformational rearrangements, facilitate the second step of the infection process, which is the binding of the gp120 to the entry co-receptors, mainly the chemokine receptors CCR5 and CXCR4 on the cell surface [11]. Further viral envelope structural changes occur subsequent to viral attachment eventually leading to virion-target cell fusion and the deposition of viral genetic material into the host cell. The exposed surface of HIV-1 gp120 envelope glycoprotein is organized into five variable (V1–V5) and five conserved (C1–C5) regions which fold into a complex secondary structure. These conserved and variable regions combine to form a tertiary structure that is the functional envelope trimer. The initial virus interaction with CD4 on the cell membrane is mediated primarily by discontinuous epitopes consisting of elements of the third (C3) and fourth (C4) conserved regions [12]. Subsequent co-receptor engagement is facilitated by the third variable loop (V3), which also determines CXCR4 or CCR5 utilization and cell tropism [13,14] (Figure 7.1). Sequence analyses of viral isolates revealed that basic amino acids at positions 11 and 25 of V3 can predict co-receptor usage and a switch from CCR5 to CXCR4 usage is accompanied by changes at several amino acid positions in V3 [15].

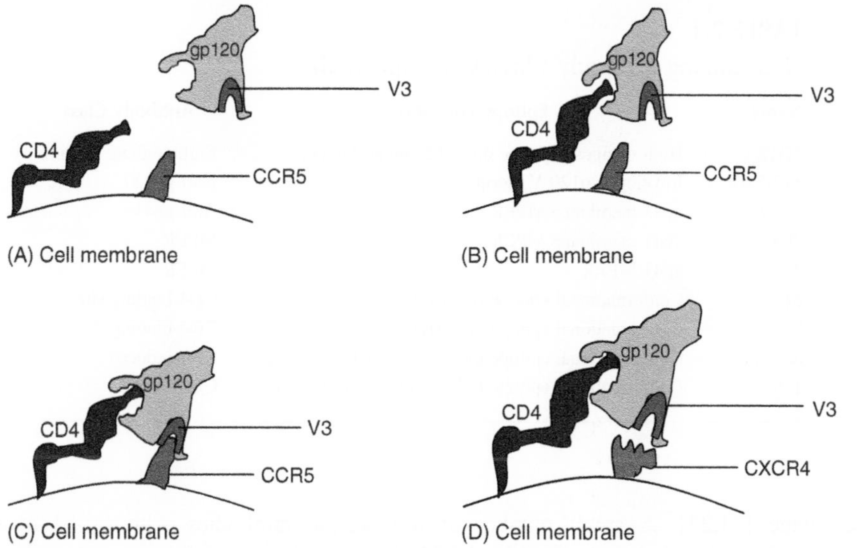

FIGURE 7.1 (A and B) HIV-1 infection begins with the binding of discontinuous epitopes made up of elements from C3 and C4 regions of envelope glycoprotein gp120 to CD4 on the target cell membrane. (C) CD4-binding triggers a conformational change in gp120 that allows for the binding of the V3 region gp120 to entry co-receptors CCR5 or CXCR4. Specific V3 sequences recognize either CCR5 or CXCR4. (D) Co-receptor engagement is aborted or suboptimal when CCR5-specific V3 is exposed to CXCR4.

7.3 INDUCTION AND THE ROLE OF NEUTRALIZING AND NON-NEUTRALIZING ANTIBODIES IN HIV-1 INFECTION

Individuals infected with HIV show strong humoral immune responses. However, most of the antibodies produced by the host immune response are ineffective at neutralizing the contemporary virus. In addition, constant changes in the envelope protein ensure that any neutralizing antibodies produced in response to the evolving virus are soon rendered ineffective. Even so, when effective, HIV antibodies can block viral entry into target cells in vitro. Furthermore, neutralizing antibodies are the only immune molecules which when passively transferred in high concentrations in animal models confer complete protection from simian-human immunodeficiency virus (SHIV) infection [16,17]. It has been suggested that non-neutralizing antibodies induced by soluble highly immunogenic monomeric or non-native forms of the gp120 or gp41 have little or no activity against the relatively poorly immunogenic intact functional envelope trimer thus distracting the immune response away from neutralizing targets [18–20]. This complex pattern of virus–host immune response interaction contrasts with infections by relatively invariant viruses such as measles and chickenpox viruses in which antibodies induced by an initial infection confer life-long protection against subsequent encounters with the virus.

7.4 VIRUS AND VIRAL ENTRY PROCESSES AS TARGETS FOR ANTIBODY-BASED VACCINES

The entire process of viral attachment, from free virus through the CD4 and co-receptor binding steps, provides opportunities for the development of immunogens to prevent infection. Anti-HIV antibodies may bind the free virus prior to cell attachment, preventing contact with the cell by stearic hindrance or may bind to the co-receptor binding sites on the virus-CD4 complex preventing further conformational rearrangements necessary for co-receptor engagement and fusion with host

TABLE 7.1
Neutralizing Antibody Classes and Specificities

Name	Epitope Location	Antibody Class
2G12	High mannose epitope on gp120 outer domain	Carbohydrate specific
447-52D	Envelope gp120 V3 loop	Anti-gp120
4E10	gp41 membrane MPER	Anti-gp41
2F5	gp41 membrane MPER	MPER
Z13	gp41 MPER	MPER
b12	Conformational epitope on gp120	CD4-binding site
b6	Conformational epitope on gp120	CD4-binding site
X5	Conformational epitope induced after CD4 binding	CD4-induced
17b	Conformational epitope induced after CD4 binding	CD4-induced

cell membrane [21,22]. A small number of monoclonal antibodies with broad neutralizing capacity have been isolated from the blood of HIV-infected humans that recognize the virus at various stages of the infection process (Table 7.1). These antibodies can be categorized by the stage of the infection process or site on which they act (1) CD4-binding site antibodies that recognize conformational epitopes on the virus where it binds to CD4 and disrupt viral attachment to CD4 on the cell membrane (e.g., b12), (2) CD4-induced co-receptor binding site antibodies that recognize epitopes that are formed or exposed after CD4 binding (e.g., antibody 17b) [21,22], (3) gp41-binding antibodies that recognize epitopes in the conserved membrane proximal external region (MPER) of gp41 and inhibit fusion (e.g., antibodies 4E10, 2F5, and Z13) [23–25], and (4) glycoprotein-binding antibodies that recognize sugar molecules on the viral envelope and inhibit viral attachment via steric hindrance (e.g., antibody 2G12 [26]. Despite the ability to produce broadly reactive antibodies such as these, complete viral control during natural infection is rare because HIV infections seldom induce high levels of neutralizing antibodies and, further, the virus has adopted mechanisms to evade such antibody recognition.

7.5 MECHANISMS OF HIV-1 ESCAPE FROM ANTIBODY RECOGNITION

The inability of HIV-induced antibodies to adequately control infection has been the subject of intense study, and has revealed many mechanisms that the virus employs to evade immune recognition (reviewed in Refs. [27–29]). Some of these evasion mechanisms are common to many viruses, while others are unique to HIV.

One common mechanism of immune evasion is sequence variation in critical highly immunogenic or immunodominant epitopes of the viral envelope that render neutralizing antibodies ineffective against the variant virus [30–33]. HIV-1 can be categorized into subtypes and circulating recombinant forms (CRFs), largely based on sequence variability within the envelope gene. Currently there are 9 genetic subtypes (designated A–D, F–H, J, and K) and approximately 16 CRFs [34]. This extensive sequence variation is thought to be primarily driven by human antibody and T cell immune pressure on the virus. There are indications that viral sequence diversity is influenced by the MHC makeup of the population within which the virus is being transmitted [35–37].

Another mechanism of immune escape unique to HIV is the use of an array of immunologically inert oligomannose sugars on the exterior envelope to shield conserved epitopes thereby effectively masking these epitopes and making them inaccessible to neutralizing antibodies [38,39], forming what is termed a "silent face." The shielding of conserved epitopes by these sugar molecules maintains these critical envelope epitopes essential for structure and function. There is also variation

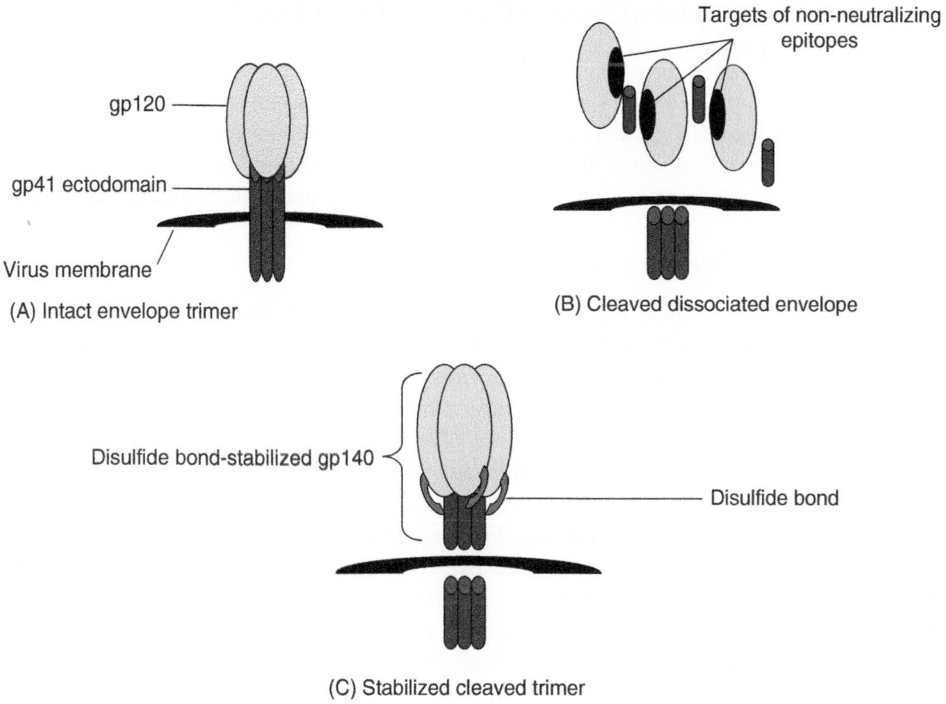

FIGURE 7.2 (A) The functional envelope spike is composed of a trimer of gp120 glycoproteins non-covalently linked to three gp41 transmembrane glycoproteins. (B) The envelope spike dissociates into its monomeric components when cleaved from the viral membrane. Antibodies generated against the cleaved dissociated envelope glycoprotein subunits do not recognize the envelope spike and are therefore non-neutralizing. (C) The cleaved envelope spike can be stabilized by introducing disulfide bond-forming cysteine mutations into the gp120 and gp41 subunits.

in the number and positions of amino acids to which oligomannose sugar molecules can be linked thereby creating several degrees of flexibility in the use of the sugar molecules to protect the virus against neutralization.

The functional HIV-1 envelope glycoprotein is formed as a trimer, with each monomeric unit consisting of the gp120 surface and gp41 transmembrane glycoprotein as a non-covalent heterodimer [22,40–42] (Figure 7.2). The organization of HIV-1 envelope as a trimeric functional unit with non-covalently linked gp120 and gp41 subunits may also serve as an immune evasion mechanism through the conformational shielding of immunogenic epitopes [43]. Furthermore, antibodies directed against the monomeric subunits of the dissociated envelope do not bind the envelope trimer and are non-neutralizing. These non-neutralizing antibodies skew the immune response away from effective neutralization. Evidence of the inability of antibodies to monomeric envelope to neutralize the trimeric envelope was demonstrated in the phase III trial in humans of a soluble monomeric AIDSVAX gp120 vaccine (VaxGen) in which no vaccine efficacy was demonstrated [44]. It appears that, in general, HIV is organized in such a way that conserved epitopes are concealed and not susceptible to neutralization while the exposed variable epitopes have the ability to readily escape neutralization.

Despite these mechanisms by which HIV evade immune recognition, antibodies that can overcome one or more of these evasion mechanisms have been identified [45–51]. Therefore, the challenge of developing an HIV vaccine that stimulates broadly neutralizing antibodies is to manipulate the virus or viral proteins to create an immunogen giving rise to an immune response that can overcome the evasion mechanisms. Many such attempts have been made and the rest of the chapter discusses some of the approaches that have been tested.

7.6 STRATEGIES FOR DESIGNING HIV ENVELOPE-TARGETING IMMUNOGENS

Several strategies to modify the HIV envelope protein have been employed to improve antibody induction and guide the development of vaccine products to overcome viral escape from antibody recognition. These strategies include the development of entry intermediates, engineered envelopes, envelope variable loop deglycosylation and hyperglycosylation, and epitope and trimer mimics.

7.6.1 Strategies to Overcome Genetic Variation

As previously discussed, there is extensive genetic variation within immunogenic epitopes of the HIV envelope protein that is for most part driven by immune pressure. However, antigenic variation at critical epitopes may be at a cost to the virus by, for instance, resulting in viruses that have reduced replication capacity [52,53]. Thus, it benefits the virus to maintain certain critical epitopes despite immune pressure. The existence of such conserved epitopes with limited genetic variation provides opportunities for the development of immunogens that target these epitopes for neutralization.

Consensus, ancestral or center of tree envelope sequences that are evolutionarily closely related to all circulating strains have been constructed to overcome viral genetic diversity (reviewed in Refs. [54,55]). Consensus sequences are constructed by selecting the most common amino acid at each position in the alignment of full-length sequences of a particular subtype or groups of subtypes. On the other hand, ancestral sequences are derived, from phylogenetic tree analyses and predict the sequence of the likely ancestral virus from which a subtype or group of subtypes arose (see Chapter 3). The center of tree sequence is a specific point on the phylogenetic tree that represents a functional sequence with the minimum evolutionary distance between isolates. The aim of all three methods is to obtain a single relatively conserved sequence that best represents viruses across a broad array of known divergence. In studies conducted in small animal models to test consensus HIV-1 subtypes B and C immunogens [56–58], the constructed viruses induced neutralizing antibodies cross-reactive with a panel of diverse viruses and displaying a wider breadth of neutralization than antibodies induced by wild-type virus [57,58]. Cross-reactive T cell responses were also detected, suggesting that these constructs are capable of inducing both cellular and humoral immune responses. The eventual testing of an immunogen based on consensus, ancestral or center of tree sequences in human populations with multiple circulating viruses may provide valuable information on the use of these concepts for HIV vaccine development.

Consensus, ancestral and center of tree sequence construction, while minimizing the distance between variant viruses, is naturally dependent on the viral sequences from which they are derived. Therefore, any biases in virus sampling or rapid changes in the circulating viral pool are likely to adversely impact their design and potential utility. Thus, it is important to consider the possibility of such constraints in the construction of consensus sequences.

7.6.2 Structure Guided Envelope Modifications

7.6.2.1 Stabilization of gp120 + CD4 Complex

When in its trimeric form, the gp120 envelope glycoprotein exhibits flexible conformations varying between the native unbound form, the CD4-bound, and the CD4 co-receptor bound forms [59,60]. Antibodies specific for the various conformations may be needed to overcome viral escape and efficiently block infection. Studying the effect of flexible conformations on viral escape has provided opportunities for developing immunogens that can potentially be used as prophylactic or therapeutic vaccines. The different conformations exhibited by the gp120 glycoprotein subsequent to CD4 binding are critical for viral fusion but are unavailable prior to CD4 binding. Therefore, immunogens based on the stable gp120 + CD4 complexes have been developed to elicit antibodies targeting the various gp120 conformations assumed after CD4 binding [61–63]. The expectation is that such antibodies will recognize crucial co-receptor binding regions and disrupt the interaction of gp120 with entry co-receptors.

Another method of producing HIV immunogens with the capacity to elicit conformational antibodies is the introduction of mutations to stabilize the envelope in the CD4-bound conformation. Information from mutagenesis and x-ray crystallographic studies of the CD4-bound HIV envelope glycoprotein has identified amino acid residues that when altered create a stabilized gp120 in a conformation similar to the CD4-bound conformation which can be used as an immunogen [61,64].

During viral attachment to target cells, conformational changes in gp120 create a cavity within gp120 at the interface of its interaction with CD4. A specific phenylalanine amino acid residue of CD4 occupies the cavity and is essential for gp120–CD4 binding. A single cavity-filling mutation that substitutes serine for tryptophan at amino acid residue 375 (S375W) [64] and a double mutation utilizing S375W and a threonine to serine change at position 257 (T257S) [61] both stabilize gp120 in the CD4-bound conformation with the double mutant conferring a more stable conformation. These mutant gp120s have increased binding capability to soluble CD4 and comparable recognition by a CD4-binding site monoclonal antibody (17b) when compared with the wild-type envelope sequence [61]. Because the stabilized gp120s closely resembled the conformation assumed after CD4 binding, these stabilized mutants were able to bind the CCR5 co-receptor in the absence of CD4 and also bind CD4 with higher affinity than the wild-type protein. Furthermore, in addition to stabilizing gp120, a trimeric assembly of the double mutant demonstrated binding characteristics to the broadly neutralizing antibody IgG1b12 that were similar to that of the trimeric wild-type gp120 [61]. However, despite the relative stability of the gp120 mutants compared to the wild-type envelope, stabilization is incomplete and a small degree of flexibility remains in the double mutant. Because the CD4-bound conformation is incomplete, the double mutant is able to bind CD4-binding site antibodies such as F105.

By studying several cavity-filling mutants with respect to structure and binding characteristics to known neutralizing antibodies, there is now a better understanding of how to develop immunogens that induce antibodies with wider breadth of neutralizing capacity. For example, gp120 stabilized by the double mutations S375W and T257S was tested in rabbits for its ability to induce broadly neutralizing antibodies [61]. As suggested by its relative stability compared to wild-type gp120 in structural studies, the double mutant-stabilized gp120 immunogen induced high levels of binding antibodies and more potent neutralizing antibodies against wild-type subtype B viruses in vitro when compared with the non-mutated gp120 immunogen.

7.6.2.2 Development of Stable Envelope Trimer Mimetics

The functional HIV envelope glycoprotein heterodimer is unstable and easily dissociates into the gp120 and gp41 subunits [65–67] (Figure 7.2A and B). Several immunogens based on monomeric subunits of the envelope trimer have been developed [61,62]. While these products are immunogenic, most of the antibodies elicited have poor neutralizing capacities, often neutralizing only laboratory-adapted isolates or pseudoviruses that are easier to neutralize than primary viruses [68–70]. Furthermore, the antibody responses lack significant neutralization breadth against a panel of diverse isolates to be potentially useful in protecting against infection by many contemporary viruses.

Therefore, an immunogen capable of generating antibodies that bind the functional trimeric structure with significant affinity may actually require preservation of the trimeric structure. Production of an immunogen with an intact trimeric functional HIV envelope unit presents a formidable challenge to vaccine researchers. However, a few methods have been investigated in attempts to develop such an immunogen.

One method of preserving the structure of the Env protein has been by stabilizing the interaction of gp120 and gp41 with paired introduced cysteine residues resulting in a covalent linkage of the two glycoproteins by intermolecular disulfide bonds [65,66,71–73] (Figure 7.2C). By simultaneously replacing amino acids in the C1 and C5 region of gp120 and the disulfide loop region of gp41 with cysteines, Binley et al. generated a folded intermolecular disulfide bond-stabilized gp140 termed SOS gp140. Despite the increased stability of the gp120 and gp41 association, SOS gp140 was still relatively unstable as a trimeric assembly, forming predominantly monomers [74]. Therefore, to improve the stable multimerization of SOS gp140, follow-up studies [75] were conducted in which isoleucine was substituted for proline at position 559 in the gp41 ectodomain. This amino acid substitution resulted in a strengthening of gp41/gp41 interactions and the resulting product, SOSIP gp140, was found to form more stable trimers [75].

For an immunogen to elicit a protective humoral immune response against HIV infection, it would need to generate antibodies capable of neutralizing a wide range of primary viruses preferably at low antibody concentrations. However, in SOSIP gp140 immunogenicity studies in rabbits using various vaccine formulations and immunization regimens, antibody responses neutralized HIV bearing the homologous envelope and sensitive laboratory-adapted virus strains but had limited activity against heterologous primary viruses [73,76]. Further improvements of this promising approach to protein engineering will be necessary.

Another approach to maintaining a stable full length gp160 has been to abolish the proteolytic cleavage site between gp120 and gp41 by amino acid substitutions. Serine to arginine substitutions at positions 508 and 511 [66,67] and replacement of the cleavage sequence REKR with IEGR or other amino acids have been described [72,73]. However, disrupting the cleavage site alone is insufficient to ensure the consistent and stable folding of envelope glycoproteins into the functional trimeric forms. Therefore, further modifications of the envelope glycoprotein have been exploited to obtain well-folded trimers, and in combination with cleavage site disruption, have produced stabilized trimers or other higher oligomer forms.

One such envelope modification is the addition of a GCN4 transcription factor to the C-terminus of the gp41 thereby extending the N-terminal coiled coil region and promoting the formation of stable trimers [66,67,77]. The GCN4 motif normally forms stable homodimers. However, by introducing hydrophobic residues in its heptad repeat region [78] GCN4 can be induced to promote the formation of trimers. Further, by combining intermolecular disulfide bonding, cleavage site disruption [73] and addition of the modified GCN4 trimerization motif to the C-terminus of gp41 [66], a stable gp140 trimer was obtained that was recognized by several neutralizing monoclonal antibodies, suggesting the formation of a proper tertiary structure. It is worth noting that while the GCN4 trimerization motif was needed to form a more stable trimer, a modified envelope with the GCN4 motif but not the cleavage disrupting motif was still susceptible to proteolytic cleavage into monomers [66] suggesting that this form of envelope modification had to be combined with other envelope stabilization methods to achieve optimal envelope immunogens.

7.6.2.3 Structure Guided Immunogens

A small number of polyclonal and monoclonal antibodies with broad neutralization capacities have been obtained from HIV-infected humans [45–51]. Importantly, x-ray crystallography of these broadly neutralizing antibodies bound to gp120 revealed specific epitopes on the envelope spike as specific targets for neutralization [22,79,80]. Furthermore, because these monoclonal antibodies have broad neutralizing capacities, they are thought to bind relatively conserved Env epitopes or conformations. Therefore, studying the dynamics with which these antibodies bind to the viral envelope may provide valuable insight into immunogen design.

A significant step towards developing immunogens to mimic the HIV envelope with neutralization epitopes exposed was recently accomplished using gp120 stabilized in the CD4-bound state [81]. The stabilized gp120 were then tested for binding kinetics to soluble CD4 and the CD4-binding site monoclonal antibody b12 using x-ray crystallographic and structural analyses. Using this approach, a functionally conserved antibody accessible region was identified in envelope glycoprotein that can be targeted by an appropriately designed and delivered immunogen. In studies of an immunogen composed of HIV-1gp120 covalently linked to soluble human CD4 receptor [82,83], antibodies were induced by vaccination of rhesus macaques that neutralized several primary HIV-1 isolates. In subsequent studies, an immunoglobulin fusion protein based on the gp120–CD4 complex inhibited in vitro R5 HIV-1 infection in PBMC and cell-line assays [84]. Protein engineering will be important to further improve these immunogenic platforms.

The knowledge that the most potent antibodies may need to recognize the viral envelope in its functional trimeric form makes a good case for advancing the development of structurally modified immunogens. Such immunogens may be (1) developed as native envelope spike mimics, (2) engineered to better expose neutralizing epitopes, (3) developed as stable CD4-bound or co-receptor bound intermediates, or (4) developed as epitope mimics targeted at specific neutralizing epitopes. Overcoming the structural flexibility of the envelope glycoprotein in its native or CD4-bound state by identifying critical conserved structures or epitopes will be key to such an effort.

7.6.3 Envelope Variable Loop Modifications

7.6.3.1 Envelope Variable Loop Depleted Immunogens

HIV-1 envelope variable loops play a significant role in viral attachment and could be targets for vaccine-elicited immune responses. However, the highly variable nature of these envelope regions makes them unattractive targets for eliciting broadly neutralizing antibodies. Furthermore, envelope variable loops shield more conserved neutralizing epitopes from antibody recognition.

The realization that envelope glycoprotein variable loops are likely obstacles to efficient antibody recognition led to studies aimed at testing the infectivity, immunogenicity, and antibody susceptibility of viruses with various deletions in the envelope variable loops [12,85,86].

In one envelope modification study, a V2-deleted Env mutant [87] demonstrated the importance of this variable loop in HIV evasion from antibody neutralization. Virions bearing V2-deleted Env were significantly more sensitive to neutralization by monoclonal antibodies and other homologous or heterologous HIV-infected patient sera. In other studies, mutagenesis experiments were used to construct envelope sequences with deletions of amino acids critical to loop formation. The resulting mutant envelope sequences lacking the V1 and V2 loop domains (ΔdV1/V2) or the V3 loop domain (ΔV3) were better recognized by conformation-dependent antibodies when compared to the wild-type envelope [85]. However, the improved recognition of mutant Envs over the wild-type Env was not evident when CD4 binding was examined. These experiments suggested that immunogens based on envelope variable loop modification could induce antibodies with neutralizing properties similar to some neutralizing monoclonal antibodies. Therefore, comparative immunogenicity studies of gp140 immunogens based on the wild type (SF162) and V2-deleted mutant (SF162ΔV2) envelopes were conducted in mice and rhesus macaques [88]. The mutant envelope immunogen was consistently better than the wild-type envelope immunogen at eliciting neutralizing antibodies against homologous and heterologous primary viral isolates. Both immunogens elicited similar titers of binding antibodies. The V2 deletion improved the immunogenic properties of the V1, V3, and C5 envelope regions, possibly by exposure of previously hidden or alternate epitopes [89,90]. While this outcome might be limited to the immunogen/virus and animal combination, the study clearly demonstrated that protein engineering of the envelope variable loops remains a valuable tool to be exploited in the search for an HIV vaccine. The immunogenicity of several HIV-1 gp140 immunogens with single or multiple deletions of envelope variable loops have now been tested in rhesus macaques. These additional studies have revealed differential specificities; V1-directed antibodies were associated with neutralization of the homologous virus, while heterologous neutralization of primary viruses was directed at non-V1 epitopes [91]. This outcome suggested that variable loop modified immunogens should be developed to target non-V1 epitopes. However, in a preclinical vaccination and virus challenge study using nonhuman primates, immunization with a V2-deleted SF162 immunogen did not protect against heterologous viral challenge [92].

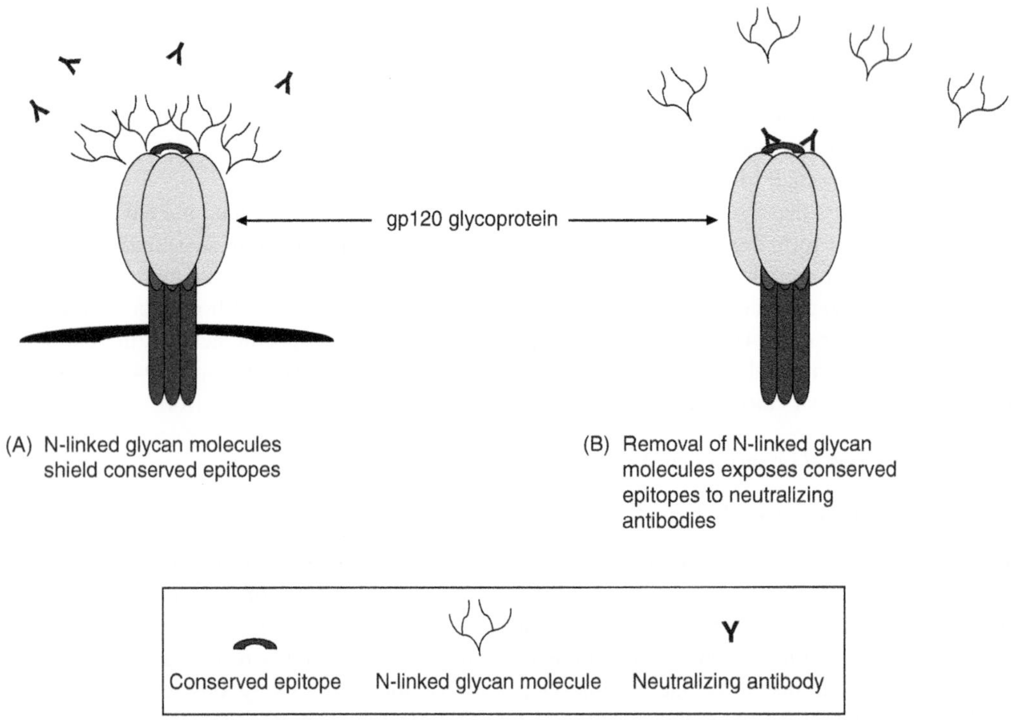

FIGURE 7.3 (A) The HIV-1 envelope is covered by an array of poorly immunogenic N-linked glucose molecules that shield immunogenic epitopes from neutralizing antibodies and thereby present an immunologically silent face to the immune system. (B) Removal of N-linked glucose molecules either by mutagenesis or by removal of envelope variable region loops can expose conserved epitopes to neutralizing antibodies.

7.6.3.2 Envelope Variable Loop Deglycosylation

A significant mechanism by which HIV evades immune recognition is the addition of immunomodulating N-linked carbohydrates on the envelope variable loops to protect immunogenic, conserved epitopes on the envelope glycoprotein (Figure 7.3). Therefore, envelope deglycosylation has been tested as a method to improve the immunogenicity of HIV vaccines. For example, deglycosylation of the SIV envelope glycoprotein was achieved by mutagenesis, replacing the first or third amino acid residue in the 3mer N-glycosylation motif, N-X-T/S, with glutamine or alanine, respectively [93]. Depending on the position of the glycosylation site, the effect of the mutagenesis on virus replication compared to wild-type virus varied from no effect, reduced replication and sometimes either enhanced replication or replication incompetent defective viruses [93,94]. These studies showed that viral envelope glycosylation is critical to viability of the virus particle, proper virus folding, and immunogenicity with the specific effects dependent on the location of the sugar molecules in the envelope protein [93–97]. An immunogen developed by removing certain carbohydrate molecules on the envelope has been reported to induce higher titers and wider breadth of neutralizing antibodies than the wild-type envelope in the rabbit model [98]. In the rhesus macaque model, glycan-deleted SIVmac239 virus was more susceptible to humoral control than the parental virus [96,99,100]. This outcome further suggests that immunogens derived from glycan-deleted viruses may present epitopes to the immune system that, because of glycan molecule masking, are unexposed on the native virus.

7.6.3.3 Manipulation of Envelope Variable Loop Glycosylation

Another method used to develop immunogens based on the HIV-1 glycan shield involves the synthesis of high-mannose oligosaccharides. The monoclonal antibody 2G12 is a potent neutralizing antibody that recognizes carbohydrate molecules on the HIV-1 envelope. To produce carbohydrate specific antibodies that might mimic the interaction of 2G12 with carbohydrate molecules on the HIV-1 envelope, an array of high-mannose oligosaccharides were synthesized. Oligosaccharides likely to be good vaccine candidates were selected based on induction of antibodies with binding characteristics similar to the carbohydrate specific neutralizing antibody 2G12 [101–103].

Engineering of HIV envelope gp120 to create immunogens that involve the use of a combination of alanine substitutions and increased number of N-glycosylation sites in the envelope variable loops and the gp120 core have been tested [104–106]. The rationale for increasing the number of glycosylation sites is to create hyper-glycosylated mutant envelope proteins on which non-neutralizing epitopes are masked by the extra sugar molecules with little or no adverse effect on neutralizing epitopes. The result is a skewing of antibody response away from non-neutralizing epitopes to the neutralizing epitopes.

By testing several mutant monomeric gp120 envelopes with combinations of alanine substitutions and additional N-glycosylation sites, Pantophlet and colleagues [104–106] identified monomeric envelope mutants that bound the neutralizing monoclonal antibodies b12 and 2G12 with affinities similar to or slightly lower than wild-type envelopes while showing little or no binding to a panel of 21 non-neutralizing antibodies. The alanine-substituted hyper-glycosylated mutant envelopes could, therefore, be developed into an immunogen that elicits disproportionately more beneficial neutralizing than non-neutralizing antibodies.

Envelope protein glycosylation is a major mechanism for viral immune evasion and may also be critical in viral particle viability. Therefore, development of immunogens that can overcome the evasion mechanism conferred by these sugar molecules, either by inducing immune responses that recognize the sugar molecules such as the monoclonal antibody 2G12 or that bypass the sugar molecules to recognize neutralizing epitopes will be a major advancement in the search for an HIV vaccine. Glycan shield-recognizing oligosaccharide arrays may also be a positive step in that direction.

7.6.4 Recombinant Proteins

Recombinant DNA and protein technologies are routinely used in vaccine development. For example, the human hepatitis B vaccine was developed by recombinant protein technology [107,108] and is made from purified hepatitis B surface antigen (HBsAg) produced in yeast. For HIV vaccines, recombinant protein technology has been used to enhance envelope expression, produce optimally folded HIV envelopes, and to express a specific or a combination of immunogenic regions [109–114]. Currently, the only candidate HIV vaccine that has been tested in human efficacy trials was a soluble gp120 monomer that was produced as a recombinant protein in chinese hamster ovary (CHO) cells [112]. The HIV gp120 has also been produced in other expression systems such as hepatitis B virus [115–117], vaccinia, and other pox viruses [109,118,119]. Other vaccine vectors including salmonella, BCG [120–123], and yeast [124,125] have also been studied for the expression of HIV envelope.

Poxviruses are vehicles for expressing recombinant HIV proteins (reviewed in Refs. [126,127]). These live viral vectors are attractive because they are stable, can tolerate the insertion of large pieces of foreign genetic material, and efficiently induce both humoral and cell-mediated immune responses to the expressed proteins. The attenuated vaccinia viruses NYVAC, derived from the Copenhagen strain of vaccinia virus, and the modified vaccinia ankara (MVA) strain have been used extensively as immunization vehicles for HIV [128–132]. In addition, canarypox virus (ALVAC), an avian poxvirus with limited replication capacity in mammalian cells has been used to express HIV antigens [128,133,134] and shown to be immunogenic in nonhuman primates. Importantly, these poxvirus recombinants, despite attenuation, retain some replication capacity and continue to express the inserted genes. Thus, the recombinant poxviruses provide the prolonged antigen stimulation required for sustained immune responses.

When used as recombinant live viral vaccines, attenuated vaccinia viruses with HIV envelope inserts stimulate both antibody- and cell-mediated immune responses. Vaccinia recombinants expressing several HIV genes have been developed that are highly immunogenic in preclinical testing [129–131,135]. In other studies, insertion of multiple human, nonhuman primate, and murine MHC restricted epitopes from HIV and plasmodium parasites in a single MVA construct has been described [136,137]. This product was immunogenic in a mouse model and furthermore, the inserted epitopes could be processed in and presented on human cells.

Human adenovirus vectors are frequently used as vaccine platforms to express modified HIV envelopes. The utility of human adenoviruses as vaccine vectors is based on their capacity to produce large amounts of viral progeny, the efficient expression of inserted genes, and lack of integration into host genes. For instance, a recombinant adenovirus serotype 5 (Ad5) virus was engineered to express HIV-1 envelope [110] and upon subsequent immunization of rats resulted in the production of HIV-1 envelope-specific antibodies. Since these initial experiments, several groups now have developed recombinant adenovirus vaccine vectors, including serotypes other than Ad5, that express HIV-1 envelope and other HIV-1 and SIV/SHIV genes [138–143]. Many of these recombinant adenoviruses expressing HIV-1 genes have been tested as vaccines in both animals and humans and found to stimulate both humoral and cell-mediated immune responses [138–141,144–146]. However, the trial of an adenovirus-based HIV vaccine, MRK-Ad5 HIV-1 gag/pol/nef, has recently been discontinued due to evidence after initial data analyses that the vaccine did not protect against infection. The vaccine was moved into phase II B test-of-concept trials after comparative preclinical testing in rhesus macaques showed better virologic control of the infection.

Many HIV vaccine candidates are being developed with recombinant technology and these methods are likely to play a significant role in future vaccines. The attraction of recombinant technology in HIV vaccine development is the flexibility that it offers for the manipulation of immunogens. This flexibility allows for example the easy inclusion of variants of an immunogenic epitope from several viral subtypes or several immunogenic components from a virus into a single vaccine product.

7.6.5 Envelope Proteoliposomes

Solid phase proteoliposomes have been developed as immunogens to mimic the intact functional envelope glycoprotein embedded in a membrane. In this methodology, the cleavage site between gp120 and gp41 is eliminated and the cytoplasmic tail of the envelope glycoprotein deleted to form uncleaved gp160. This uncleaved gp160 is then attached to beads by antibody conjugates and incubated with lipids to create gp160 envelope proteoliposomes (gp160PL) [147]. Immunogenicity studies in rabbits showed the gp160PL to induce higher titers and wider breadth of HIV-1 neutralizing antibodies when compared to monomeric gp120. In addition, unlike monomeric gp120, the breadth of neutralizing immune responses induced by gp160PL increased after boosting even though there was no associated increase in antibody titers [148].

Although gp160PL appears to be more immunogenic than monomeric gp120, it remains to be demonstrated whether it provides significant advantages over other forms of stabilized trimeric envelopes. Comparative preclinical testing of these products will provide a definitive answer.

7.6.6 Virus Inactivation and Attenuation

Some of the most successful human viral vaccines have been formulated simply by chemical, physical, or biological attenuation or inactivation of the etiologic agent (reviewed in Refs. [149,150]), resulting in either killed or live attenuated viral vaccines. Historically, heat inactivation, continuous passage in cell culture, and the use of other biological systems such as eggs to reduce or eliminate virulence have been successful. However, the use of such strategies to develop an HIV vaccine has been unsuccessful, largely due to the heterogeneous nature of HIV envelope glycoprotein. Furthermore, for obvious reasons, the safety of live attenuated HIV-1 in humans remains a concern.

It is believed that the immune response induced by the inactivated intact virus is likely to be better at blocking viral infection than responses induced by immunogens made up of subunits or individual components. However, development of such an immunogen for HIV has proceeded cautiously because of the concern over antibody-induced enhancement of infection in vitro [151–154]. Despite these obstacles encountered in producing an inactivated HIV vaccine, efforts and progress have been made. An immunogen, Remune [155], which is an envelope-depleted and radiation-attenuated virion, was tested as a therapeutic vaccine in HIV-infected persons receiving antiretroviral therapy and reported to improve CD4 T cell recovery and delay progression to AIDS [156,157]. Other forms of viral attenuation such as Aldrithiol-2 (2,2′-dithiodipyridine; AT-2) inactivation that targets zinc finger molecules in the viral nucleocapsid have been reported [158,159]. AT-2 does not act directly on the envelope glycoprotein and HIV and SIV viruses treated with this compound retain the envelope structure and function. Immunization with AT-2 inactivated virus is capable of inducing binding and neutralizing antibodies as well as viral specific cellular responses [160,161].

By first using formaldehyde then heat to inactivate infectious HIV clones, it is possible to generate whole viral immunogens that retain binding to several broadly neutralizing, conformation-dependent monoclonal antibodies [162]. Stabilization of the virion structure by formaldehyde treatment prior to heat inactivation reduced envelope shedding and disruption of the envelope subunits when compared to heat treatment alone. Further, this inactivated virion when used with an appropriate adjuvant, induced antibodies in both mice and rhesus macaques capable of neutralizing heterologous HIV-1 strains [163]. An improvement of this method involved an amino acid substitution (Y706C) in the gp41 endocytosis signal sequence to increase virion envelope incorporation and resulted in the induction of higher neutralizing antibody titers [164]. The ability of the formaldehyde-stabilized, heat-inactivated virion vaccine to prevent viral infection in animal models and humans will be the ultimate test of the utility of this product. Apart from the immunogen preparation, certain elements in its delivery such as adjuvants and vaccination schedule may determine its efficacy.

Another method for HIV inactivation has been described in which the proteins and lipids in the viral envelope are modified by treatment with 1,5-iodonaphthalyzide (INA) followed by UV irradiation [165]. INA binds to viral membrane proteins and eliminates the ability of the virus to fuse with target cell membranes. More importantly, INA treatment does not interfere with viral immunogenicity, retaining viral reactivity to the monoclonal antibodies 2G12, b12, and 4E10. Similar to other forms of inactivated HIV immunogens, the risks of using whole virus as an immunogen remains a significant safety concern.

Of the various methods to generate HIV envelope immunogens discussed in this chapter, inactivated virus vaccines are the most likely to induce both cell-mediated and humoral immunity, a characteristic that may be critical in an efficacious HIV vaccine. However, until the concerns regarding safety are fully addressed, it is unlikely that such a vaccine will progress to clinical trials in humans. Despite this drawback, work continues in this area of vaccine research and it is expected that improvements in the safety aspects of vaccine development will be achieved.

7.7 CONCLUSION

The optimal envelope-based anti-HIV immunogen will induce sufficient titers of long lasting antibodies capable of broad neutralization and blocking HIV transmission in humans. Achieving this level of protection will probably require a combination of technologies that utilize some of the envelope glycoprotein modifications described in this chapter and immunogens that stimulate

TABLE 7.2
Summary and Merits of HIV-1 Envelope Modifications

Modification	Rationale	Advantages	Disadvantages
Consensus and ancestral sequences	Reduce genetic variation between immunogen and circulating viruses by use of a single unique sequence that is closely related to variant viruses	Immunogen best represents the variant viral sequences Maintenance of critical conserved epitopes	Consensus sequence is artificial Consensus sequence changes at constant rate over evolutionary time
Envelope trimer mimetics	Stabilize envelope to form intact undissociated trimer	Stabilized undissociated trimer induces antibodies that best recognizes functional envelope	Antibodies to stabilized trimer may recognize only the homologous virus Non-neutralizing antibodies are generated
Envelope variable loop modification and deglycosylation	Expose immunogenic epitopes by removing immunologically silent sugar molecules	Antibodies to exposed conserved epitopes may recognize viral variants	Loss of reactivity of some antibodies produced in natural infection that can recognize envelopes with intact variable loops
Recombination	Reduce virulence and enhance expression of envelope glycoproteins	Highly immunogenic antigens can be chosen for recombination Augmentation of envelope gene expression Immunogenic antigens or epitopes from variant viruses can be combined in a single immunogen	Preexisting immunity to vector
Inactivation/ attenuation	Reduce virulence while maintaining full antigenic repertoire	Full antigenic complement is presented to immune system Partially replicating immunogen maintains antigenic stimulation	Induction of enhancing antibodies Reversion to virulence

cellular immunity. That there is currently no such vaccine is a demonstration of the challenges that need to be overcome.

Compared to a decade ago, the natural history and pathogenesis of HIV infections are better understood and many modern research tools are now available for vaccine development (Table 7.2). X-ray crystallography has pinpointed targets for neutralizing antibodies by revealing viral epitopes and envelope conformations assumed upon CD4 binding. Furthermore, studies of crystal structures of neutralizing antibody bound to HIV have revealed conformational epitopes that need to be targeted by vaccine-induced antibodies. Structural analyses, together with mutagenesis and protein chemistry have guided the development of immunogens that mimic the functional envelope spike and glycan arrays that have the potential to recognize sugar molecules on the viral envelope. Other tools such as recombinant protein technology provide the ability to manipulate viral proteins for optimal expression. Also, the advent of computer-based bioinformatic tools has made it possible to construct consensus and ancestral sequences that reduce the genetic distances between virus subtypes and are better suited as immunogens to protect against a diverse group of viruses.

Several HIV-1 vaccine candidates are in different stages of development and the International AIDS Vaccine Initiative (IAVI) provides a database of planned and ongoing trials (http://www.iavireport.org/trialsdb/). A cursory look at the database shows a majority of the vaccine candidates have been developed using recombinant technology. Thus, this method of modifying HIV antigens currently appears to be favored. Presently, the only phase III HIV vaccine trial is being conducted in Thailand. The trial is evaluating two recombinant vaccines, an antibody-inducing recombinant gp120 vaccine (AIDSVAX B/E) and a T cell-inducing recombinant canary pox virus vaccine (ALVAC vCP1521) in a prime-boost vaccination strategy. The AIDSVAX B/E is a version of AIDSVAX B/B, the only HIV vaccine candidate that has been tested in human trials but which did not show any efficacy. It is believed that combining this antibody-inducing vaccine with a T cell-inducing vaccine might show efficacy. The outcome of this and other trials may be beneficial in determining which of the modification methods will yield the best product.

REFERENCES

1. Desrosiers, R.C., Wyand, M.S., Kodama, T., Ringler, D.J., Arthur, L.O., Sehgal, P.K., Letvin, N.L., King, N.W., and Daniel, M.D. Vaccine protection against simian immunodeficiency virus infection. *Proc. Natl. Acad. Sci. U S A* 1989, *86*(16), 6353–6357.
2. Murphey-Corb, M., Martin, L.N., vison-Fairburn, B., Montelaro, R.C., Miller, M., West, M., Ohkawa, S., Baskin, G.B., Zhang, J.Y., and Putney, S.D. A formalin-inactivated whole SIV vaccine confers protection in macaques. *Science* 1989, *246*(4935), 1293–1297.
3. Carlson, J.R., McGraw, T.P., Keddie, E., Yee, J.L., Rosenthal, A., Langlois, A.J., Dickover, R., Donovan, R., Luciw, P.A., and Jennings, M.B. Vaccine protection of rhesus macaques against simian immunodeficiency virus infection. *AIDS Res. Hum. Retroviruses* 1990, *6*(11), 1239–1246.
4. Gardner, M.B., Carlson, J.R., Jennings, M., Rosenthal, A., Langlois, A., Haynes, B., Bolognesi, D., and Palker, T.J. SIV vaccine protection of rhesus monkeys. *Biotechnol. Ther.* 1991, *2*(1–2), 9–19.
5. Montefiori, D.C., Graham, B.S., Zhou, J., Zhou, J., Bucco, R.A., Schwartz, D.H., Cavacini, L.A., and Posner, M.R. V3-specific neutralizing antibodies in sera from HIV-1 gp160-immunized volunteers block virus fusion and act synergistically with human monoclonal antibody to the conformation-dependent CD4 binding site of gp120. NIH-NIAID AIDS vaccine clinical trials network. *J. Clin. Invest.* 1993, *92*(2), 840–847.
6. Borrow, P., Lewicki, H., Hahn, B.H., Shaw, G.M., and Oldstone, M.B. Virus-specific CD8 + cytotoxic T-lymphocyte activity associated with control of viremia in primary human immunodeficiency virus type 1 infection. *J. Virol.* 1994, *68*(9), 6103–6110.
7. Koup, R.A., Safrit, J.T., Cao, Y., Andrews, C.A., McLeod, G., Borkowsky, W., Farthing, C., and Ho, D.D. Temporal association of cellular immune responses with the initial control of viremia in primary human immunodeficiency virus type 1 syndrome. *J. Virol.* 1994, *68*(7), 4650–4655.
8. Rowland-Jones, S., Sutton, J., Ariyoshi, K., Dong, T., Gotch, F., McAdam, S., Whitby, D., Sabally, S., Gallimore, A., and Corrah, T. HIV-specific cytotoxic T-cells in HIV-exposed but uninfected Gambian women. *Nat. Med.* 1995, *1*(1), 59–64.
9. Ogg, G.S., Jin, X., Bonhoeffer, S., Dunbar, P.R., Nowak, M.A., Monard, S., Segal, J.P., Cao, Y., Rowland-Jones, S.L., Cerundolo, V., Hurley, A., Markowitz, M., Ho, D.D., Nixon, D.F., and McMichael, A.J. Quantitation of HIV-1-specific cytotoxic T lymphocytes and plasma load of viral RNA. *Science* 1998, *279*(5359), 2103–2106.
10. Schmitz, J.E., Kuroda, M.J., Santra, S., Sasseville, V.G., Simon, M.A., Lifton, M.A., Racz, P., Tenner-Racz, K., Dalesandro, M., Scallon, B.J., Ghrayeb, J., Forman, M.A., Montefiori, D.C., Rieber, E.P., Letvin, N.L., and Reimann, K.A. Control of viremia in simian immunodeficiency virus infection by $CD8^+$ lymphocytes. *Science* 1999, *283*(5403), 857–860.
11. Chen, B., Vogan, E.M., Gong, H., Skehel, J.J., Wiley, D.C., and Harrison, S.C. Structure of an unliganded simian immunodeficiency virus gp120 core. *Nature* 2005, *433*(7028), 834–841.
12. Wyatt, R., Moore, J., Accola, M., Desjardin, E., Robinson, J., and Sodroski, J. Involvement of the V1/V2 variable loop structure in the exposure of human immunodeficiency virus type 1 gp120 epitopes induced by receptor binding. *J. Virol.* 1995, *69*(9), 5723–5733.
13. Hwang, S.S., Boyle, T.J., Lyerly, H.K., and Cullen, B.R. Identification of the envelope V3 loop as the primary determinant of cell tropism in HIV-1. *Science* 1991, *253*(5015), 71–74.

14. Shioda, T., Levy, J.A., and Cheng-Mayer, C. Small amino acid changes in the V3 hypervariable region of gp120 can affect the T-cell-line and macrophage tropism of human immunodeficiency virus type 1. *Proc. Natl. Acad. Sci. U S A* 1992, *89*(20), 9434–9438.
15. Resch, W., Hoffman, N., and Swanstrom, R. Improved success of phenotype prediction of the human immunodeficiency virus type 1 from envelope variable loop 3 sequence using neural networks. *Virology* 2001, *288*(1), 51–62.
16. Shibata, R., Igarashi, T., Haigwood, N., Buckler-White, A., Ogert, R., Ross, W., Willey, R., Cho, M.W., and Martin, M.A. Neutralizing antibody directed against the HIV-1 envelope glycoprotein can completely block HIV-1/SIV chimeric virus infections of macaque monkeys. *Nat. Med.* 1999, *5*(2), 204–210.
17. Mascola, J.R., Stiegler, G., Vancott, T.C., Katinger, H., Carpenter, C.B., Hanson, C.E., Beary, H., Hayes, D., Frankel, S.S., Birx, D.L., and Lewis, M.G. Protection of macaques against vaginal transmission of a pathogenic HIV-1/SIV chimeric virus by passive infusion of neutralizing antibodies. *Nat. Med.* 2000, *6*(2), 207–210.
18. Parren, P.W., Burton, D.R., and Sattentau, Q.J. HIV-1 antibody—debris or virion? *Nat. Med.* 1997, *3*(4), 366–367.
19. Burton, D.R. and Parren, P.W. Vaccines and the induction of functional antibodies: Time to look beyond the molecules of natural infection? *Nat. Med.* 2000, *6*(2), 123–125.
20. Poignard, P., Moulard, M., Golez, E., Vivona, V., Franti, M., Venturini, S., Wang, M., Parren, P.W., and Burton, D.R. Heterogeneity of envelope molecules expressed on primary human immunodeficiency virus type 1 particles as probed by the binding of neutralizing and nonneutralizing antibodies. *J. Virol.* 2003, *77*(1), 353–365.
21. Wyatt, R., Kwong, P.D., Desjardins, E., Sweet, R.W., Robinson, J., Hendrickson, W.A., and Sodroski, J.G. The antigenic structure of the HIV gp120 envelope glycoprotein. *Nature* 1998, *393*(6686), 705–711.
22. Kwong, P.D., Wyatt, R., Robinson, J., Sweet, R.W., Sodroski, J., and Hendrickson, W.A. Structure of an HIV gp120 envelope glycoprotein in complex with the CD4 receptor and a neutralizing human antibody. *Nature* 1998, *393*(6686), 648–659.
23. Purtscher, M., Trkola, A., Gruber, G., Buchacher, A., Predl, R., Steindl, F., Tauer, C., Berger, R., Barrett, N., and Jungbauer, A. A broadly neutralizing human monoclonal antibody against gp41 of human immunodeficiency virus type 1. *AIDS Res. Hum. Retroviruses* 1994, *10*(12), 1651–1658.
24. Zwick, M.B., Labrijn, A.F., Wang, M., Spenlehauer, C., Saphire, E.O., Binley, J.M., Moore, J.P., Stiegler, G., Katinger, H., Burton, D.R., and Parren, P.W. Broadly neutralizing antibodies targeted to the membrane-proximal external region of human immunodeficiency virus type 1 glycoprotein gp41. *J. Virol.* 2001, *75*(22), 10892–10905.
25. Stiegler, G., Kunert, R., Purtscher, M., Wolbank, S., Voglauer, R., Steindl, F., and Katinger, H. A potent cross-clade neutralizing human monoclonal antibody against a novel epitope on gp41 of human immunodeficiency virus type 1. *AIDS Res. Hum. Retroviruses* 2001, *17*(18), 1757–1765.
26. Trkola, A., Purtscher, M., Muster, T., Ballaun, C., Buchacher, A., Sullivan, N., Srinivasan, K., Sodroski, J., Moore, J.P., and Katinger, H. Human monoclonal antibody 2G12 defines a distinctive neutralization epitope on the gp120 glycoprotein of human immunodeficiency virus type 1. *J. Virol.* 1996, *70*(2), 1100–1108.
27. Parren, P.W., Moore, J.P., Burton, D.R., and Sattentau, Q.J. The neutralizing antibody response to HIV-1: Viral evasion and escape from humoral immunity. *AIDS* 1999, *13*(Suppl A), S137–S162.
28. Evans, D.T. and Desrosiers, R.C. Immune evasion strategies of the primate lentiviruses. *Immunol. Rev.* 2001, *183*, 141–158.
29. Haynes, B.F. and Montefiori, D.C. Aiming to induce broadly reactive neutralizing antibody responses with HIV-1 vaccine candidates. *Expert. Rev. Vaccines* 2006, *5*(3), 347–363.
30. Kelly, H.R., Urbanski, M., Burda, S., Zhong, P., Konings, F., Nanfack, J., Tongo, M., Kinge, T., Achkar, J., and Nyambi, P. Neutralizing antibody patterns and viral escape in HIV-1 non-B subtype chronically infected treatment-naive individuals. *Hum. Antibodies* 2005, *14*(3–4), 89–99.
31. Rong, R., Gnanakaran, S., Decker, J.M., Bibollet-Ruche, F., Taylor, J., Sfakianos, J.N., Mokili, J.L., Muldoon, M., Mulenga, J., Allen, S., Hahn, B.H., Shaw, G.M., Blackwell, J.L., Korber, B.T., Hunter, E., and Derdeyn, C.A. Unique mutational patterns in the envelope alpha 2 amphipathic helix and acquisition of length in gp120 hypervariable domains are associated with resistance to autologous neutralization of subtype C human immunodeficiency virus type 1. *J. Virol.* 2007, *81*(11), 5658–5668.

32. Manrique, A., Rusert, P., Joos, B., Fischer, M., Kuster, H., Leemann, C., Niederost, B., Weber, R., Stiegler, G., Katinger, H., Gunthard, H.F., and Trkola, A. In vivo and in vitro escape from neutralizing antibodies 2G12, 2F5 and 4E10. *J. Virol.* 2007, *81*(16), 98793–98808.
33. Sagar, M., Wu, X., Lee, S., and Overbaugh, J. Human immunodeficiency virus type 1 V1-V2 envelope loop sequences expand and add glycosylation sites over the course of infection, and these modifications affect antibody neutralization sensitivity. *J. Virol.* 2006, *80*(19), 9586–9598.
34. McCutchan, F.E. Global epidemiology of HIV. *J. Med. Virol.* 2006, *78*(Suppl 1), S7–S12.
35. Moore, C.B., John, M., James, I.R., Christiansen, F.T., Witt, C.S., and Mallal, S.A. Evidence of HIV-1 adaptation to HLA-restricted immune responses at a population level. *Science* 2002, *296*(5572), 1439–1443.
36. Allen, T.M., Altfeld, M., Geer, S.C., Kalife, E.T., Moore, C., O'sullivan, K.M., Desouza, I., Feeney, M.E., Eldridge, R.L., Maier, E.L., Kaufmann, D.E., Lahaie, M.P., Reyor, L., Tanzi, G., Johnston, M.N., Brander, C., Draenert, R., Rockstroh, J.K., Jessen, H., Rosenberg, E.S., Mallal, S.A., and Walker, B.D. Selective escape from $CD8^+$ T-cell responses represents a major driving force of human immunodeficiency virus type 1 (HIV-1) sequence diversity and reveals constraints on HIV-1 evolution. *J. Virol.* 2005, *79*(21), 13239–13249.
37. Brumme, Z.L., Brumme, C.J., Heckerman, D., Korber, B.T., Daniels, M., Carlson, J., Kadie, C., Bhattacharya, T., Chui, C., Szinger, J., Mo, T., Hogg, R.S., Montaner, J.S., Frahm, N., Brander, C., Walker, B.D., and Harrigan, P.R. Evidence of differential HLA class I-mediated viral evolution in functional and accessory/regulatory genes of HIV-1. *PLoS. Pathog.* 2007, *3*(7), e94.
38. Wei, X., Decker, J.M., Wang, S., Hui, H., Kappes, J.C., Wu, X., Salazar-Gonzalez, J.F., Salazar, M.G., Kilby, J.M., Saag, M.S., Komarova, N.L., Nowak, M.A., Hahn, B.H., Kwong, P.D., and Shaw, G.M. Antibody neutralization and escape by HIV-1. *Nature* 2003, *422*(6929), 307–312.
39. Wu, X., Parast, A.B., Richardson, B.A., Nduati, R., John-Stewart, G., Mbori-Ngacha, D., Rainwater, S.M., and Overbaugh, J. Neutralization escape variants of human immunodeficiency virus type 1 are transmitted from mother to infant. *J. Virol.* 2006, *80*(2), 835–844.
40. Center, R.J., Leapman, R.D., Lebowitz, J., Arthur, L.O., Earl, P.L., and Moss, B. Oligomeric structure of the human immunodeficiency virus type 1 envelope protein on the virion surface. *J. Virol.* 2002, *76*(15), 7863–7867.
41. Zhu, P., Chertova, E., Bess, J., Jr., Lifson, J.D., Arthur, L.O., Liu, J., Taylor, K.A., and Roux, K.H. Electron tomography analysis of envelope glycoprotein trimers on HIV and simian immunodeficiency virus virions. *Proc. Natl. Acad. Sci. U S A* 2003, *100*(26), 15812–15817.
42. Zhu, P., Liu, J., Bess, J., Jr., Chertova, E., Lifson, J.D., Grise, H., Ofek, G.A., Taylor, K.A., and Roux, K.H. Distribution and three-dimensional structure of AIDS virus envelope spikes. *Nature* 2006, *441* (7095), 847–852.
43. Zanetti, G., Briggs, J.A., Grunewald, K., Sattentau, Q.J., and Fuller, S.D. Cryo-electron tomographic structure of an immunodeficiency virus envelope complex in situ. *PLoS. Pathog.* 2006, *2*(8), e83.
44. Pitisuttithum, P., Gilbert, P., Gurwith, M., Heyward, W., Martin, M., van, G.F., Hu, D., Tappero, J.W., and Choopanya, K. Randomized, double-blind, placebo-controlled efficacy trial of a bivalent recombinant glycoprotein 120 HIV-1 vaccine among injection drug users in Bangkok, Thailand. *J. Infect. Dis.* 2006, *194*(12), 1661–1671.
45. Thali, M., Moore, J.P., Furman, C., Charles, M., Ho, D.D., Robinson, J., and Sodroski, J. Characterization of conserved human immunodeficiency virus type 1 gp120 neutralization epitopes exposed upon gp120-CD4 binding. *J. Virol.* 1993, *67*(7), 3978–3988.
46. Roben, P., Moore, J.P., Thali, M., Sodroski, J., Barbas, C.F., III, and Burton, D.R. Recognition properties of a panel of human recombinant Fab fragments to the CD4 binding site of gp120 that show differing abilities to neutralize human immunodeficiency virus type 1. *J. Virol.* 1994, *68*(8), 4821–4828.
47. Binley, J.M., Wrin, T., Korber, B., Zwick, M.B., Wang, M., Chappey, C., Stiegler, G., Kunert, R., Zolla-Pazner, S., Katinger, H., Petropoulos, C.J., and Burton, D.R. Comprehensive cross-clade neutralization analysis of a panel of anti-human immunodeficiency virus type 1 monoclonal antibodies. *J. Virol.* 2004, *78*(23), 13232–13252.
48. Zhang, M.Y., Xiao, X., Sidorov, I.A., Choudhry, V., Cham, F., Zhang, P.F., Bouma, P., Zwick, M., Choudhary, A., Montefiori, D.C., Broder, C.C., Burton, D.R., Quinnan, G.V., Jr., and Dimitrov, D.S. Identification and characterization of a new cross-reactive human immunodeficiency virus type 1-neutralizing human monoclonal antibody. *J. Virol.* 2004, *78*(17), 9233–9242.

49. Trkola, A., Pomales, A.B., Yuan, H., Korber, B., Maddon, P.J., Allaway, G.P., Katinger, H., Barbas, C.F., III; Burton, D.R., and Ho, D.D. Cross-clade neutralization of primary isolates of human immunodeficiency virus type 1 by human monoclonal antibodies and tetrameric CD4-IgG. *J. Virol.* 1995, *69*(11), 6609–6617.
50. Dhillon, A.K., Donners, H., Pantophlet, R., Johnson, W.E., Decker, J.M., Shaw, G.M., Lee, F.H., Richman, D.D., Doms, R.W., Vanham, G., and Burton, D.R. Dissecting the neutralizing antibody specificities of broadly neutralizing sera from human immunodeficiency virus type 1-infected donors. *J. Virol.* 2007, *81*(12), 6548–6562.
51. Choudhry, V., Zhang, M.Y., Sidorov, I.A., Louis, J.M., Harris, I., Dimitrov, A.S., Bouma, P., Cham, F., Choudhary, A., Rybak, S.M., Fouts, T., Montefiori, D.C., Broder, C.C., Quinnan, G.V., Jr., and Dimitrov, D.S. Cross-reactive HIV-1 neutralizing monoclonal antibodies selected by screening of an immune human phage library against an envelope glycoprotein (gp140) isolated from a patient (R2) with broadly HIV-1 neutralizing antibodies. *Virology* 2007, *363*(1), 79–90.
52. Derdeyn, C.A., Decker, J.M., Bibollet-Ruche, F., Mokili, J.L., Muldoon, M., Denham, S.A., Heil, M.L., Kasolo, F., Musonda, R., Hahn, B.H., Shaw, G.M., Korber, B.T., Allen, S., and Hunter, E. Envelope-constrained neutralization-sensitive HIV-1 after heterosexual transmission. *Science* 2004, *303*(5666), 2019–2022.
53. Peut, V. and Kent, S.J. Fitness constraints on immune escape from HIV: Implications of envelope as a target for both HIV-specific T cells and antibody. *Curr. HIV Res.* 2006, *4*(2), 191–197.
54. Gaschen, B., Taylor, J., Yusim, K., Foley, B., Gao, F., Lang, D., Novitsky, V., Haynes, B., Hahn, B.H., Bhattacharya, T., and Korber, B. Diversity considerations in HIV-1 vaccine selection. *Science* 2002, *296*(5577), 2354–2360.
55. Rolland, M., Jensen, M.A., Nickle, D.C., Yan, J., Learn, G.H., Heath, L., Weiner, D., and Mullins, J.I. Reconstruction and function of ancestral center-of-tree human immunodeficiency virus type 1 proteins. *J. Virol.* 2007, *81*(16), 8507–8514.
56. Kothe, D.L., Li, Y., Decker, J.M., Bibollet-Ruche, F., Zammit, K.P., Salazar, M.G., Chen, Y., Weng, Z., Weaver, E.A., Gao, F., Haynes, B.F., Shaw, G.M., Korber, B.T., and Hahn, B.H. Ancestral and consensus envelope immunogens for HIV-1 subtype C. *Virology* 2006, *352*(2), 438–449.
57. Liao, H.X., Sutherland, L.L., Xia, S.M., Brock, M.E., Scearce, R.M., Vanleeuwen, S., Alam, S.M., McAdams, M., Weaver, E.A., Camacho, Z., Ma, B.J., Li, Y., Decker, J.M., Nabel, G.J., Montefiori, D.C., Hahn, B.H., Korber, B.T., Gao, F., and Haynes, B.F. A group M consensus envelope glycoprotein induces antibodies that neutralize subsets of subtype B and C HIV-1 primary viruses. *Virology* 2006, *353*(2), 268–282.
58. Kothe, D.L., Decker, J.M., Li, Y., Weng, Z., Bibollet-Ruche, F., Zammit, K.P., Salazar, M.G., Chen, Y., Salazar-Gonzalez, J.F., Moldoveanu, Z., Mestecky, J., Gao, F., Haynes, B.F., Shaw, G.M., Muldoon, M., Korber, B.T., and Hahn, B.H. Antigenicity and immunogenicity of HIV-1 consensus subtype B envelope glycoproteins. *Virology* 2007, *360*(1), 218–234.
59. Myszka, D.G., Sweet, R.W., Hensley, P., Brigham-Burke, M., Kwong, P.D., Hendrickson, W.A., Wyatt, R., Sodroski, J., and Doyle, M.L. Energetics of the HIV gp120-CD4 binding reaction. *Proc. Natl. Acad. Sci. U S A* 2000, *97*(16), 9026–9031.
60. Rits-Volloch, S., Frey, G., Harrison, S.C., and Chen, B. Restraining the conformation of HIV-1 gp120 by removing a flexible loop. *EMBO J.* 2006, *25*(20), 5026–5035.
61. Dey, B., Pancera, M., Svehla, K., Shu, Y., Xiang, S.H., Vainshtein, J., Li, Y., Sodroski, J., Kwong, P.D., Mascola, J.R., and Wyatt, R. Characterization of human immunodeficiency virus type 1 monomeric and trimeric gp120 glycoproteins stabilized in the CD4-bound state: Antigenicity, biophysics, and immunogenicity. *J. Virol.* 2007, *81*(11), 5579–5593.
62. Crooks, E.T., Moore, P.L., Franti, M., Cayanan, C.S., Zhu, P., Jiang, P., de Vries, R.P., Wiley, C., Zharkikh, I., Schulke, N., Roux, K.H., Montefiori, D.C., Burton, D.R., and Binley, J.M. A comparative immunogenicity study of HIV-1 virus-like particles bearing various forms of envelope proteins, particles bearing no envelope and soluble monomeric gp120. *Virology* 2007, *366*(2), 245–262.
63. Beddows, S., Franti, M., Dey, A.K., Kirschner, M., Iyer, S.P., Fisch, D.C., Ketas, T., Yuste, E., Desrosiers, R.C., Klasse, P.J., Maddon, P.J., Olson, W.C., and Moore, J.P. A comparative immunogenicity study in rabbits of disulfide-stabilized, proteolytically cleaved, soluble trimeric human immunodeficiency virus type 1 gp140, trimeric cleavage-defective gp140 and monomeric gp120. *Virology* 2007, *360*(2), 329–340.

64. Xiang, S.H., Kwong, P.D., Gupta, R., Rizzuto, C.D., Casper, D.J., Wyatt, R., Wang, L., Hendrickson, W.A., Doyle, M.L., and Sodroski, J. Mutagenic stabilization and/or disruption of a CD4-bound state reveals distinct conformations of the human immunodeficiency virus type 1 gp120 envelope glycoprotein. *J. Virol.* 2002, *76*(19), 9888–9899.
65. Yang, X., Farzan, M., Wyatt, R., and Sodroski, J. Characterization of stable, soluble trimers containing complete ectodomains of human immunodeficiency virus type 1 envelope glycoproteins. *J. Virol.* 2000, *74*(12), 5716–5725.
66. Yang, X., Florin, L., Farzan, M., Kolchinsky, P., Kwong, P.D., Sodroski, J., and Wyatt, R. Modifications that stabilize human immunodeficiency virus envelope glycoprotein trimers in solution. *J. Virol.* 2000, *74*(10), 4746–4754.
67. Yang, X., Wyatt, R., and Sodroski, J. Improved elicitation of neutralizing antibodies against primary human immunodeficiency viruses by soluble stabilized envelope glycoprotein trimers. *J. Virol.* 2001, *75*(3), 1165–1171.
68. Moore, J.P. and Ho, D.D. Antibodies to discontinuous or conformationally sensitive epitopes on the gp120 glycoprotein of human immunodeficiency virus type 1 are highly prevalent in sera of infected humans. *J. Virol.* 1993, *67*(2), 863–875.
69. Vancott, T.C., Bethke, F.R., Burke, D.S., Redfield, R.R., and Birx, D.L. Lack of induction of antibodies specific for conserved, discontinuous epitopes of HIV-1 envelope glycoprotein by candidate AIDS vaccines. *J. Immunol.* 1995, *155*(8), 4100–4110.
70. Vancott, T.C., Mascola, J.R., Kaminski, R.W., Kalyanaraman, V., Hallberg, P.L., Burnett, P.R., Ulrich, J.T., Rechtman, D.J., and Birx, D.L. Antibodies with specificity to native gp120 and neutralization activity against primary human immunodeficiency virus type 1 isolates elicited by immunization with oligomeric gp160. *J. Virol.* 1997, *71*(6), 4319–4330.
71. Farzan, M., Choe, H., Desjardins, E., Sun, Y., Kuhn, J., Cao, J., Archambault, D., Kolchinsky, P., Koch, M., Wyatt, R., and Sodroski, J. Stabilization of human immunodeficiency virus type 1 envelope glycoprotein trimers by disulfide bonds introduced into the gp41 glycoprotein ectodomain. *J. Virol.* 1998, *72*(9), 7620–7625.
72. Binley, J.M., Sanders, R.W., Clas, B., Schuelke, N., Master, A., Guo, Y., Kajumo, F., Anselma, D.J., Maddon, P.J., Olson, W.C., and Moore, J.P. A recombinant human immunodeficiency virus type 1 envelope glycoprotein complex stabilized by an intermolecular disulfide bond between the gp120 and gp41 subunits is an antigenic mimic of the trimeric virion-associated structure. *J. Virol.* 2000, *74*(2), 627–643.
73. Beddows, S., Schulke, N., Kirschner, M., Barnes, K., Franti, M., Michael, E., Ketas, T., Sanders, R.W., Maddon, P.J., Olson, W.C., and Moore, J.P. Evaluating the immunogenicity of a disulfide-stabilized, cleaved, trimeric form of the envelope glycoprotein complex of human immunodeficiency virus type 1. *J. Virol.* 2005, *79*(14), 8812–8827.
74. Schulke, N., Vesanen, M.S., Sanders, R.W., Zhu, P., Lu, M., Anselma, D.J., Villa, A.R., Parren, P.W., Binley, J.M., Roux, K.H., Maddon, P.J., Moore, J.P., and Olson, W.C. Oligomeric and conformational properties of a proteolytically mature, disulfide-stabilized human immunodeficiency virus type 1 gp140 envelope glycoprotein. *J. Virol.* 2002, *76*(15), 7760–7776.
75. Sanders, R.W., Vesanen, M., Schuelke, N., Master, A., Schiffner, L., Kalyanaraman, R., Paluch, M., Berkhout, B., Maddon, P.J., Olson, W.C., Lu, M., and Moore, J.P. Stabilization of the soluble, cleaved, trimeric form of the envelope glycoprotein complex of human immunodeficiency virus type 1. *J. Virol.* 2002, *76*(17), 8875–8889.
76. Beddows, S., Kirschner, M., Campbell-Gardener, L., Franti, M., Dey, A.K., Iyer, S.P., Maddon, P.J., Paluch, M., Master, A., Overbaugh, J., VanCott, T., Olson, W.C., and Moore, J.P. Construction and characterization of soluble, cleaved, and stabilized trimeric Env proteins based on HIV type 1 Env subtype A. *AIDS Res. Hum. Retroviruses* 2006, *22*(6), 569–579.
77. Pancera, M., Lebowitz, J., Schon, A., Zhu, P., Freire, E., Kwong, P.D., Roux, K.H., Sodroski, J., and Wyatt, R. Soluble mimetics of human immunodeficiency virus type 1 viral spikes produced by replacement of the native trimerization domain with a heterologous trimerization motif: Characterization and ligand binding analysis. *J. Virol.* 2005, *79*(15), 9954–9969.
78. Harbury, P.B., Zhang, T., Kim, P.S., and Alber, T. A switch between two-, three-, and four-stranded coiled coils in GCN4 leucine zipper mutants. *Science* 1993, *262*(5138), 1401–1407.

79. Saphire, E.O., Parren, P.W., Pantophlet, R., Zwick, M.B., Morris, G.M., Rudd, P.M., Dwek, R.A., Stanfield, R.L., Burton, D.R., and Wilson, I.A. Crystal structure of a neutralizing human IGG against HIV-1: A template for vaccine design. *Science* 2001, *293*(5532), 1155–1159.
80. Ofek, G., Tang, M., Sambor, A., Katinger, H., Mascola, J.R., Wyatt, R., and Kwong, P.D. Structure and mechanistic analysis of the anti-human immunodeficiency virus type 1 antibody 2F5 in complex with its gp41 epitope. *J. Virol.* 2004, *78*(19), 10724–10737.
81. Zhou, T., Xu, L., Dey, B., Hessell, A.J., Van, R.D., Xiang, S.H., Yang, X., Zhang, M.Y., Zwick, M.B., Arthos, J., Burton, D.R., Dimitrov, D.S., Sodroski, J., Wyatt, R., Nabel, G.J., and Kwong, P.D. Structural definition of a conserved neutralization epitope on HIV-1 gp120. *Nature* 2007, *445*(7129), 732–737.
82. Fouts, T.R., Tuskan, R., Godfrey, K., Reitz, M., Hone, D., Lewis, G.K., and DeVico, A.L. Expression and characterization of a single-chain polypeptide analogue of the human immunodeficiency virus type 1 gp120-CD4 receptor complex. *J. Virol.* 2000, *74*(24), 11427–11436.
83. Fouts, T., Godfrey, K., Bobb, K., Montefiori, D., Hanson, C.V., Kalyanaraman, V.S., DeVico, A., and Pal, R. Crosslinked HIV-1 envelope-CD4 receptor complexes elicit broadly cross-reactive neutralizing antibodies in rhesus macaques. *Proc. Natl. Acad. Sci. U S A* 2002, *99*(18), 11842–11847.
84. Vu, J.R., Fouts, T., Bobb, K., Burns, J., McDermott, B., Israel, D.I., Godfrey, K., and DeVico, A. An immunoglobulin fusion protein based on the gp120-CD4 receptor complex potently inhibits human immunodeficiency virus type 1 in vitro. *AIDS Res. Hum. Retroviruses* 2006, *22*(6), 477–490.
85. Wyatt, R., Sullivan, N., Thali, M., Repke, H., Ho, D., Robinson, J., Posner, M., and Sodroski, J. Functional and immunologic characterization of human immunodeficiency virus type 1 envelope glycoproteins containing deletions of the major variable regions. *J. Virol.* 1993, *67*(8), 4557–4565.
86. Cao, J., Sullivan, N., Desjardin, E., Parolin, C., Robinson, J., Wyatt, R., and Sodroski, J. Replication and neutralization of human immunodeficiency virus type 1 lacking the V1 and V2 variable loops of the gp120 envelope glycoprotein. *J. Virol.* 1997, *71*(12), 9808–9812.
87. Stamatatos, L. and Cheng-Mayer, C. An envelope modification that renders a primary, neutralization-resistant clade B human immunodeficiency virus type 1 isolate highly susceptible to neutralization by sera from other clades. *J. Virol.* 1998, *72*(10), 7840–7845.
88. Barnett, S.W., Lu, S., Srivastava, I., Cherpelis, S., Gettie, A., Blanchard, J., Wang, S., Mboudjeka, I., Leung, L., Lian, Y., Fong, A., Buckner, C., Ly, A., Hilt, S., Ulmer, J., Wild, C.T., Mascola, J.R., and Stamatatos, L. The ability of an oligomeric human immunodeficiency virus type 1 (HIV-1) envelope antigen to elicit neutralizing antibodies against primary HIV-1 isolates is improved following partial deletion of the second hypervariable region. *J. Virol.* 2001, *75*(12), 5526–5540.
89. Srivastava, I.K., VanDorsten, K., Vojtech, L., Barnett, S.W., and Stamatatos, L. Changes in the immunogenic properties of soluble gp140 human immunodeficiency virus envelope constructs upon partial deletion of the second hypervariable region. *J. Virol.* 2003, *77*(4), 2310–2320.
90. Ly, A. and Stamatatos, L. V2 loop glycosylation of the human immunodeficiency virus type 1 SF162 envelope facilitates interaction of this protein with CD4 and CCR5 receptors and protects the virus from neutralization by anti-V3 loop and anti-CD4 binding site antibodies. *J. Virol.* 2000, *74*(15), 6769–6776.
91. Derby, N.R., Kraft, Z., Kan, E., Crooks, E.T., Barnett, S.W., Srivastava, I.K., Binley, J.M., and Stamatatos, L. Antibody responses elicited in macaques immunized with human immunodeficiency virus type 1 (HIV-1) SF162-derived gp140 envelope immunogens: Comparison with those elicited during homologous simian/human immunodeficiency virus SHIVSF162P4 and heterologous HIV-1 infection. *J. Virol.* 2006, *80*(17), 8745–8762.
92. Xu, R., Srivastava, I.K., Kuller, L., Zarkikh, I., Kraft, Z., Fagrouch, Z., Letvin, N.L., Heeney, J.L., Barnett, S.W., and Stamatatos, L. Immunization with HIV-1 SF162-derived Envelope gp140 proteins does not protect macaques from heterologous simian-human immunodeficiency virus SHIV89.6P infection. *Virology* 2006, *349*(2), 276–289.
93. Ohgimoto, S., Shioda, T., Mori, K., Nakayama, E.E., Hu, H., and Nagai, Y. Location-specific, unequal contribution of the N glycans in simian immunodeficiency virus gp120 to viral infectivity and removal of multiple glycans without disturbing infectivity. *J. Virol.* 1998, *72*(10), 8365–8370.
94. Reynard, F., Fatmi, A., Verrier, B., and Bedin, F. HIV-1 acute infection env glycomutants designed from 3D model: Effects on processing, antigenicity, and neutralization sensitivity. *Virology* 2004, *324*(1), 90–102.

95. Johnson, W.E., Morgan, J., Reitter, J., Puffer, B.A., Czajak, S., Doms, R.W., and Desrosiers, R.C. A replication-competent, neutralization-sensitive variant of simian immunodeficiency virus lacking 100 amino acids of envelope. *J. Virol.* 2002, *76*(5), 2075–2086.
96. Johnson, W.E., Sanford, H., Schwall, L., Burton, D.R., Parren, P.W., Robinson, J.E., and Desrosiers, R.C. Assorted mutations in the envelope gene of simian immunodeficiency virus lead to loss of neutralization resistance against antibodies representing a broad spectrum of specificities. *J. Virol.* 2003, *77*(18), 9993–10003.
97. Mori, K., Sugimoto, C., Ohgimoto, S., Nakayama, E.E., Shioda, T., Kusagawa, S., Takebe, Y., Kano, M., Matano, T., Yuasa, T., Kitaguchi, D., Miyazawa, M., Takahashi, Y., Yasunami, M., Kimura, A., Yamamoto, N., Suzuki, Y., and Nagai, Y. Influence of glycosylation on the efficacy of an Env-based vaccine against simian immunodeficiency virus SIVmac239 in a macaque AIDS model. *J. Virol.* 2005, *79*(16), 10386–10396.
98. Reynard, F., Willkomm, N., Fatmi, A., Vallon-Eberhard, A., Verrier, B., and Bedin, F. Characterization of the antibody response elicited by HIV-1 Env glycomutants in rabbits. *Vaccine* 2007, *25*(3), 535–546.
99. Reitter, J.N., Means, R.E., and Desrosiers, R.C. A role for carbohydrates in immune evasion in AIDS. *Nat. Med.* 1998, *4*(6), 679–684.
100. Mori, K., Yasutomi, Y., Ohgimoto, S., Nakasone, T., Takamura, S., Shioda, T., and Nagai, Y. Quintuple deglycosylation mutant of simian immunodeficiency virus SIVmac239 in rhesus macaques: Robust primary replication, tightly contained chronic infection, and elicitation of potent immunity against the parental wild-type strain. *J. Virol.* 2001, *75*(9), 4023–4028.
101. Wang, L.X., Ni, J., Singh, S., and Li, H. Binding of high-mannose-type oligosaccharides and synthetic oligomannose clusters to human antibody 2G12: Implications for HIV-1 vaccine design. *Chem. Biol.* 2004, *11*(1), 127–134.
102. Calarese, D.A., Lee, H.K., Huang, C.Y., Best, M.D., Astronomo, R.D., Stanfield, R.L., Katinger, H., Burton, D.R., Wong, C.H., and Wilson, I.A. Dissection of the carbohydrate specificity of the broadly neutralizing anti-HIV-1 antibody 2G12. *Proc. Natl. Acad. Sci. U S A* 2005, *102*(38), 13372–13377.
103. Adams, E.W., Ratner, D.M., Bokesch, H.R., McMahon, J.B., O'Keefe, B.R., and Seeberger, P.H. Oligosaccharide and glycoprotein microarrays as tools in HIV glycobiology; glycan-dependent gp120/protein interactions. *Chem. Biol.* 2004, *11*(6), 875–881.
104. Pantophlet, R., Ollmann, S.E., Poignard, P., Parren, P.W., Wilson, I.A., and Burton, D.R. Fine mapping of the interaction of neutralizing and nonneutralizing monoclonal antibodies with the CD4 binding site of human immunodeficiency virus type 1 gp120. *J. Virol.* 2003, *77*(1), 642–658.
105. Pantophlet, R., Wilson, I.A., and Burton, D.R. Hyperglycosylated mutants of human immunodeficiency virus (HIV) type 1 monomeric gp120 as novel antigens for HIV vaccine design. *J. Virol.* 2003, *77*(10), 5889–5901.
106. Pantophlet, R., Wilson, I.A., and Burton, D.R. Improved design of an antigen with enhanced specificity for the broadly HIV-neutralizing antibody b12. *Protein Eng. Des. Sel.* 2004, *17*(10), 749–758.
107. McAleer, W.J., Buynak, E.B., Maigetter, R.Z., Wampler, D.E., Miller, W.J., and Hilleman, M.R. Human hepatitis B vaccine from recombinant yeast. *Nature* 1984, *307*(5947), 178–180.
108. Hilleman, M.R. Yeast recombinant hepatitis B vaccine. *Infection* 1987, *15*(1), 3–7.
109. Dallo, S., Maa, J.S., Rodriguez, J.R., Rodriguez, D., and Esteban, M. Humoral immune response elicited by highly attenuated variants of vaccinia virus and by an attenuated recombinant expressing HIV-1 envelope protein. *Virology* 1989, *173*(1), 323–329.
110. Dewar, R.L., Natarajan, V., Vasudevachari, M.B., and Salzman, N.P. Synthesis and processing of human immunodeficiency virus type 1 envelope proteins encoded by a recombinant human adenovirus. *J. Virol.* 1989, *63*(1), 129–136.
111. Parry, C., McLain, L., and Dimmock, N.J. Production of long-lived neutralizing antibodies to HIV-1 IIIB in mice with a vaccinia recombinant virus-infected cell vaccine expressing gp160. *AIDS Res. Hum. Retroviruses* 1994, *10*(2), 205–212.
112. Berman, P.W., Huang, W., Riddle, L., Gray, A.M., Wrin, T., Vennari, J., Johnson, A., Klaussen, M., Prashad, H., Kohne, C., deWit, C., and Gregory, T.J. Development of bivalent (B/E) vaccines able to neutralize CCR5-dependent viruses from the United States and Thailand. *Virology* 1999, *265*(1), 1–9.
113. Gomez, C.E. and Esteban, M. Recombinant proteins produced by vaccinia virus vectors can be incorporated within the virion (IMV form) into different compartments. *Arch. Virol.* 2001, *146*(5), 875–892.

114. Gomez-Roman, V.R., Florese, R.H., Peng, B., Montefiori, D.C., Kalyanaraman, V.S., Venzon, D., Srivastava, I., Barnett, S.W., and Robert-Guroff, M. An adenovirus-based HIV subtype B prime/boost vaccine regimen elicits antibodies mediating broad antibody-dependent cellular cytotoxicity against non-subtype B HIV strains. *J. Acquir. Immune. Defic. Syndr.* 2006, *43*(3), 270–277.
115. Berkower, I., Raymond, M., Muller, J., Spadaccini, A., and Aberdeen, A. Assembly, structure, and antigenic properties of virus-like particles rich in HIV-1 envelope gp120. *Virology* 2004, *321*(1), 75–86.
116. von, B.A., Brand, M., Reichhuber, C., Morys-Wortmann, C., Deinhardt, F., and Schodel, F. Principal neutralizing domain of HIV-1 is highly immunogenic when expressed on the surface of hepatitis B core particles. *Vaccine* 1993, *11*(8), 817–824.
117. Eckhart, L., Raffelsberger, W., Ferko, B., Klima, A., Purtscher, M., Katinger, H., and Ruker, F. Immunogenic presentation of a conserved gp41 epitope of human immunodeficiency virus type 1 on recombinant surface antigen of hepatitis B virus. *J. Gen. Virol.* 1996, *77*(Pt 9), 2001–2008.
118. Ourmanov, I., Brown, C.R., Moss, B., Carroll, M., Wyatt, L., Pletneva, L., Goldstein, S., Venzon, D., and Hirsch, V.M. Comparative efficacy of recombinant modified vaccinia virus Ankara expressing simian immunodeficiency virus (SIV) Gag-Pol and/or Env in macaques challenged with pathogenic SIV. *J. Virol.* 2000, *74*(6), 2740–2751.
119. Radaelli, A. and De Giuli, M.C. Expression of HIV-1 envelope gene by recombinant avipox viruses. *Vaccine* 1994, *12*(12), 1101–1109.
120. Fouts, T.R., Tuskan, R.G., Chada, S., Hone, D.M., and Lewis, G.K. Construction and immunogenicity of *Salmonella typhimurium* vaccine vectors that express HIV-1 gp120. *Vaccine* 1995, *13*(17), 1697–1705.
121. Cirillo, J.D., Stover, C.K., Bloom, B.R., Jacobs, W.R., Jr., and Barletta, R.G. Bacterial vaccine vectors and bacillus Calmette-Guerin. *Clin. Infect. Dis.* 1995, *20*(4), 1001–1009.
122. DeVico, A.L., Fouts, T.R., Shata, M.T., Kamin-Lewis, R., Lewis, G.K., and Hone, D.M. Development of an oral prime-boost strategy to elicit broadly neutralizing antibodies against HIV-1. *Vaccine* 2002, *20*(15), 1968–1974.
123. Yu, J.S., Peacock, J.W., Jacobs, W.R., Jr., Frothingham, R., Letvin, N.L., Liao, H.X., and Haynes, B.F. Recombinant *Mycobacterium bovis* Bacillus Calmette-Guerin elicits human immunodeficiency virus type 1 envelope-specific T lymphocytes at mucosal sites. *Clin. Vaccine Immunol.* 2007, *14*(7), 886–893.
124. Griffiths, J.C., Berrie, E.L., Holdsworth, L.N., Moore, J.P., Harris, S.J., Senior, J.M., Kingsman, S.M., Kingsman, A.J., and Adams, S.E. Induction of high-titer neutralizing antibodies, using hybrid human immunodeficiency virus V3-Ty viruslike particles in a clinically relevant adjuvant. *J. Virol.* 1991, *65*(1), 450–456.
125. Layton, G.T., Harris, S.J., Myhan, J., West, D., Gotch, F., Hill-Perkins, M., Cole, J.S., Meyers, N., Woodrow, S., French, T.J., Adams, S.E., and Kingsman, A.J. Induction of single and dual cytotoxic T-lymphocyte responses to viral proteins in mice using recombinant hybrid Ty-virus-like particles. *Immunology* 1996, *87*(2), 171–178.
126. Paoletti, E., Tartaglia, J., and Taylor, J. Safe and effective poxvirus vectors–NYVAC and ALVAC. *Dev. Biol. Stand.* 1994, *82*, 65–69.
127. Perkus, M.E., Tartaglia, J., and Paoletti, E. Poxvirus-based vaccine candidates for cancer, AIDS, and other infectious diseases. *J. Leukoc. Biol.* 1995, *58*(1), 1–13.
128. Cox, W.I., Tartaglia, J., and Paoletti, E. Induction of cytotoxic T lymphocytes by recombinant canarypox (ALVAC) and attenuated vaccinia (NYVAC) viruses expressing the HIV-1 envelope glycoprotein. *Virology* 1993, *195*(2), 845–850.
129. Gomez, C.E., Najera, J.L., Jimenez, V., Bieler, K., Wild, J., Kostic, L., Heidari, S., Chen, M., Frachette, M.J., Pantaleo, G., Wolf, H., Liljestrom, P., Wagner, R., and Esteban, M. Generation and immunogenicity of novel HIV/AIDS vaccine candidates targeting HIV-1 Env/Gag-Pol-Nef antigens of clade C. *Vaccine* 2007, *25*(11), 1969–1992.
130. Gomez, C.E., Najera, J.L., Jimenez, E.P., Jimenez, V., Wagner, R., Graf, M., Frachette, M.J., Liljestrom, P., Pantaleo, G., and Esteban, M. Head-to-head comparison on the immunogenicity of two HIV/AIDS vaccine candidates based on the attenuated poxvirus strains MVA and NYVAC co-expressing in a single locus the HIV-1BX08 gp120 and HIV-1(IIIB) Gag-Pol-Nef proteins of clade B. *Vaccine* 2007, *25*(15), 2863–2885.
131. Smith, J.M., Amara, R.R., McClure, H.M., Patel, M., Sharma, S., Yi, H., Chennareddi, L., Herndon, J.G., Butera, S.T., Heneine, W., Ellenberger, D.L., Parekh, B., Earl, P.L., Wyatt, L.S., Moss, B., and Robinson, H.L. Multiprotein HIV type 1 clade B DNA/MVA vaccine: Construction, safety, and immunogenicity in Macaques. *AIDS Res. Hum. Retroviruses* 2004, *20*(6), 654–665.

132. Ellenberger, D., Otten, R.A., Li, B., Aidoo, M., Rodriguez, I.V., Sariol, C.A., Martinez, M., Monsour, M., Wyatt, L., Hudgens, M.G., Kraiselburd, E., Moss, B., Robinson, H., Folks, T., and Butera, S. HIV-1 DNA/MVA vaccination reduces the per exposure probability of infection during repeated mucosal SHIV challenges. *Virology* 2006, *352*(1), 216–225.
133. Pialoux, G., Excler, J.L., Riviere, Y., Gonzalez-Canali, G., Feuillie, V., Coulaud, P., Gluckman, J.C., Matthews, T.J., Meignier, B., and Kieny, M.P. A prime-boost approach to HIV preventive vaccine using a recombinant canarypox virus expressing glycoprotein 160 (MN) followed by a recombinant glycoprotein 160 (MN/LAI). The AGIS Group, and l'Agence Nationale de Recherche sur le SIDA. *AIDS Res. Hum. Retroviruses* 1995, *11*(3), 373–381.
134. Belshe, R.B., Gorse, G.J., Mulligan, M.J., Evans, T.G., Keefer, M.C., Excler, J.L., Duliege, A.M., Tartaglia, J., Cox, W.I., McNamara, J., Hwang, K.L., Bradney, A., Montefiori, D., and Weinhold, K.J. Induction of immune responses to HIV-1 by canarypox virus (ALVAC) HIV-1 and gp120 SF-2 recombinant vaccines in uninfected volunteers. NIAID AIDS vaccine evaluation group. *AIDS* 1998, *12*(18), 2407–2415.
135. Wyatt, L.S., Earl, P.L., Liu, J.Y., Smith, J.M., Montefiori, D.C., Robinson, H.L., and Moss, B. Multi-protein HIV type 1 clade B DNA and MVA vaccines: Construction, expression, and immunogenicity in rodents of the MVA component. *AIDS Res. Hum. Retroviruses* 2004, *20*(6), 645–653.
136. Hanke, T., Schneider, J., Gilbert, S.C., Hill, A.V., and McMichael, A. DNA multi-CTL epitope vaccines for HIV and *Plasmodium falciparum*: Immunogenicity in mice. *Vaccine* 1998, *16*(4), 426–435.
137. Hanke, T., Blanchard, T.J., Schneider, J., Ogg, G.S., Tan, R., Becker, M., Gilbert, S.C., Hill, A.V., Smith, G.L., and McMichael, A. Immunogenicities of intravenous and intramuscular administrations of modified vaccinia virus Ankara-based multi-CTL epitope vaccine for human immunodeficiency virus type 1 in mice. *J. Gen. Virol.* 1998, *79*(Pt 1), 83–90.
138. Bruce, C.B., Akrigg, A., Sharpe, S.A., Hanke, T., Wilkinson, G.W., and Cranage, M.P. Replication-deficient recombinant adenoviruses expressing the human immunodeficiency virus Env antigen can induce both humoral and CTL immune responses in mice. *J. Gen. Virol.* 1999, *80*(Pt 10), 2621–2628.
139. Shiver, J.W., Fu, T.M., Chen, L., Casimiro, D.R., Davies, M.E., Evans, R.K., Zhang, Z.Q., Simon, A.J., Trigona, W.L., Dubey, S.A., Huang, L., Harris, V.A., Long, R.S., Liang, X., Handt, L., Schleif, W.A., Zhu, L., Freed, D.C., Persaud, N.V., Guan, L., Punt, K.S., Tang, A., Chen, M., Wilson, K.A., Collins, K.B., Heidecker, G.J., Fernandez, V.R., Perry, H.C., Joyce, J.G., Grimm, K.M., Cook, J.C., Keller, P.M., Kresock, D.S., Mach, H., Troutman, R.D., Isopi, L.A., Williams, D.M., Xu, Z., Bohannon, K.E., Volkin, D.B., Montefiori, D.C., Miura, A., Krivulka, G.R., Lifton, M.A., Kuroda, M.J., Schmitz, J.E., Letvin, N.L., Caulfield, M.J., Bett, A.J., Youil, R., Kaslow, D.C., and Emini, E.A. Replication-incompetent adenoviral vaccine vector elicits effective anti-immunodeficiency-virus immunity. *Nature* 2002, *415* (6869), 331–335.
140. Liang, X., Casimiro, D.R., Schleif, W.A., Wang, F., Davies, M.E., Zhang, Z.Q., Fu, T.M., Finnefrock, A.C., Handt, L., Citron, M.P., Heidecker, G., Tang, A., Chen, M., Wilson, K.A., Gabryelski, L., McElhaugh, M., Carella, A., Moyer, C., Huang, L., Vitelli, S., Patel, D., Lin, J., Emini, E.A., and Shiver, J.W. Vectored Gag and Env but not Tat show efficacy against simian-human immunodeficiency virus 89.6P challenge in Mamu-A*01-negative rhesus monkeys. *J. Virol.* 2005, *79*(19), 12321–12331.
141. Xin, K.Q., Jounai, N., Someya, K., Honma, K., Mizuguchi, H., Naganawa, S., Kitamura, K., Hayakawa, T., Saha, S., Takeshita, F., Okuda, K., Honda, M., Klinman, D.M., and Okuda, K. Prime-boost vaccination with plasmid DNA and a chimeric adenovirus type 5 vector with type 35 fiber induces protective immunity against HIV. *Gene Ther.* 2005, *12*(24), 1769–1777.
142. Nanda, A., Lynch, D.M., Goudsmit, J., Lemckert, A.A., Ewald, B.A., Sumida, S.M., Truitt, D.M., Abbink, P., Kishko, M.G., Gorgone, D.A., Lifton, M.A., Shen, L., Carville, A., Mansfield, K.G., Havenga, M.J., and Barouch, D.H. Immunogenicity of recombinant fiber-chimeric adenovirus serotype 35 vector-based vaccines in mice and rhesus monkeys. *J. Virol.* 2005, *79*(22), 14161–14168.
143. Xin, K.Q., Sekimoto, Y., Takahashi, T., Mizuguchi, H., Ichino, M., Yoshida, A., and Okuda, K. Chimeric adenovirus 5/35 vector containing the clade C HIV gag gene induces a cross-reactive immune response against HIV. *Vaccine* 2007, *25*(19), 3809–3815.
144. Santra, S., Seaman, M.S., Xu, L., Barouch, D.H., Lord, C.I., Lifton, M.A., Gorgone, D.A., Beaudry, K.R., Svehla, K., Welcher, B., Chakrabarti, B.K., Huang, Y., Yang, Z.Y., Mascola, J.R., Nabel, G.J., and Letvin, N.L. Replication-defective adenovirus serotype 5 vectors elicit durable cellular and humoral immune responses in nonhuman primates. *J. Virol.* 2005, *79*(10), 6516–6522.

145. Baliga, C.S., van, M.M., Chastain, M., and Sutton, R.E. Vaccination of mice with replication-defective human immunodeficiency virus induces cellular and humoral immunity and protects against vaccinia virus-gag challenge. *Mol. Ther.* 2006, *14*(3), 432–441.
146. Tobery, T.W., Dubey, S.A., Anderson, K., Freed, D.C., Cox, K.S., Lin, J., Prokop, M.T., Sykes, K.J., Mogg, R., Mehrotra, D.V., Fu, T.M., Casimiro, D.R., and Shiver, J.W. A comparison of standard immunogenicity assays for monitoring HIV type 1 gag-specific T cell responses in Ad5 HIV Type 1 gag vaccinated human subjects. *AIDS Res. Hum. Retroviruses* 2006, *22*(11), 1081–1090.
147. Grundner, C., Mirzabekov, T., Sodroski, J., and Wyatt, R. Solid-phase proteoliposomes containing human immunodeficiency virus envelope glycoproteins. *J. Virol.* 2002, *76*(7), 3511–3521.
148. Grundner, C., Li, Y., Louder, M., Mascola, J., Yang, X., Sodroski, J., and Wyatt, R. Analysis of the neutralizing antibody response elicited in rabbits by repeated inoculation with trimeric HIV-1 envelope glycoproteins. *Virology* 2005, *331*(1), 33–46.
149. Stephenson, J.R. Genetically modified viruses: Vaccines by design. *Curr. Pharm. Biotechnol.* 2001, *2*(1), 47–76.
150. Plotkin, S.A. Vaccines, vaccination, and vaccinology. *J. Infect. Dis.* 2003, *187*(9), 1349–1359.
151. Robinson, W.E., Jr., Montefiori, D.C., Mitchell, W.M., Prince, A.M., Alter, H.J., Dreesman, G.R., and Eichberg, J.W. Antibody-dependent enhancement of human immunodeficiency virus type 1 (HIV-1) infection in vitro by serum from HIV-1-infected and passively immunized chimpanzees. *Proc. Natl. Acad. Sci. U S A* 1989, *86*(12), 4710–4714.
152. Robinson, W.E., Jr., Montefiori, D.C., Gillespie, D.H., and Mitchell, W.M. Complement-mediated, antibody-dependent enhancement of HIV-1 infection in vitro is characterized by increased protein and RNA syntheses and infectious virus release. *J. Acquir. Immune. Defic. Syndr.* 1989, *2*(1), 33–42.
153. Mascola, J.R., Mathieson, B.J., Zack, P.M., Walker, M.C., Halstead, S.B., and Burke, D.S. Summary report: Workshop on the potential risks of antibody-dependent enhancement in human HIV vaccine trials. *AIDS Res. Hum. Retroviruses* 1993, *9*(12), 1175–1184.
154. Szabo, J., Prohaszka, Z., Toth, F.D., Gyuris, A., Segesdi, J., Banhegyi, D., Ujhelyi, E., Minarovits, J., and Fust, G. Strong correlation between the complement-mediated antibody-dependent enhancement of HIV-1 infection and plasma viral load. *AIDS* 1999, *13*(14), 1841–1849.
155. Limsuwan, A., Churdboonchart, V., Moss, R.B., Sirawaraporn, W., Sutthent, R., Smutharaks, B., Glidden, D., Trauger, R., Theofan, G., and Carlo, D. Safety and immunogenicity of REMUNE in HIV-infected Thai subjects. *Vaccine* 1998, *16*(2–3), 142–149.
156. Churdboonchart, V., Sakondhavat, C., Kulpradist, S., Na Ayudthya, B.I., Chandeying, V., Rugpao, S., Boonshuyar, C., Sukeepaisarncharoen, W., Sirawaraporn, W., Carlo, D.J., and Moss, R. A double-blind, adjuvant-controlled trial of human immunodeficiency virus type 1 (HIV-1) immunogen (Remune) monotherapy in asymptomatic, HIV-1-infected thai subjects with CD4-cell counts of >300. *Clin. Diagn. Lab Immunol.* 2000, *7*(5), 728–733.
157. Turner, J.L., Kostman, J.R., Aquino, A., Wright, D., Szabo, S., Bidwell, R., Goodgame, J., Daigle, A., Kelley, E., Jensen, F., Duffy, C., Carlo, D., and Moss, R.B. The effects of an HIV-1 immunogen (Remune) on viral load, CD4 cell counts and HIV-specific immunity in a double-blind, randomized, adjuvant-controlled subset study in HIV infected subjects regardless of concomitant antiviral drugs. *HIV Med.* 2001, *2*(2), 68–77.
158. Rossio, J.L., Esser, M.T., Suryanarayana, K., Schneider, D.K., Bess, J.W., Jr., Vasquez, G.M., Wiltrout, T.A., Chertova, E., Grimes, M.K., Sattentau, Q., Arthur, L.O., Henderson, L.E., and Lifson, J.D. Inactivation of human immunodeficiency virus type 1 infectivity with preservation of conformational and functional integrity of virion surface proteins. *J. Virol.* 1998, *72*(10), 7992–8001.
159. Arthur, L.O., Bess, J.W., Jr., Chertova, E.N., Rossio, J.L., Esser, M.T., Benveniste, R.E., Henderson, L.E., and Lifson, J.D. Chemical inactivation of retroviral infectivity by targeting nucleocapsid protein zinc fingers: A candidate SIV vaccine. *AIDS Res. Hum. Retroviruses* 1998, *14*(Suppl 3), S311–S319.
160. Lifson, J.D., Rossio, J.L., Piatak, M., Jr., Bess, J., Jr., Chertova, E., Schneider, D.K., Coalter, V.J., Poore, B., Kiser, R.F., Imming, R.J., Scarzello, A.J., Henderson, L.E., Alvord, W.G., Hirsch, V.M., Benveniste, R.E., and Arthur, L.O. Evaluation of the safety, immunogenicity, and protective efficacy of whole inactivated simian immunodeficiency virus (SIV) vaccines with conformationally and functionally intact envelope glycoproteins. *AIDS Res. Hum. Retroviruses* 2004, *20*(7), 772–787.
161. Frank, I., Santos, J.J., Mehlhop, E., Villamide-Herrera, L., Santisteban, C., Gettie, A., Ignatius, R., Lifson, J.D., and Pope, M. Presentation of exogenous whole inactivated simian immunodeficiency

virus by mature dendritic cells induces $CD4^+$ and $CD8^+$ T-cell responses. *J. Acquir. Immune. Defic. Syndr.* 2003, *34*(1), 7–19.

162. Grovit-Ferbas, K., Hsu, J.F., Ferbas, J., Gudeman, V., and Chen, I.S. Enhanced binding of antibodies to neutralization epitopes following thermal and chemical inactivation of human immunodeficiency virus type 1. *J. Virol.* 2000, *74*(13), 5802–5809.
163. Poon, B., Safrit, J.T., McClure, H., Kitchen, C., Hsu, J.F., Gudeman, V., Petropoulos, C., Wrin, T., Chen, I.S., and Grovit-Ferbas, K. Induction of humoral immune responses following vaccination with envelope-containing, formaldehyde-treated, thermally inactivated human immunodeficiency virus type 1. *J. Virol.* 2005, *79*(8), 4927–4935.
164. Poon, B., Hsu, J.F., Gudeman, V., Chen, I.S., and Grovit-Ferbas, K. Formaldehyde-treated, heat-inactivated virions with increased human immunodeficiency virus type 1 env can be used to induce high-titer neutralizing antibody responses. *J. Virol.* 2005, *79*(16), 10210–10217.
165. Raviv, Y., Viard, M., Bess, J.W., Jr., Chertova, E., and Blumenthal, R. Inactivation of retroviruses with preservation of structural integrity by targeting the hydrophobic domain of the viral envelope. *J. Virol.* 2005, *79*(19), 12394–12400.

8 Engineering Better Influenza Vaccines: Traditional and New Approaches

Peter M. Pushko, Rick A. Bright, Terrence M. Tumpey, and Gale E. Smith

CONTENTS

To add to their troubles both sides were visited by pestilence, a calamity almost heavy enough to turn them from all thoughts of war.

Livius, Titus
The History of Rome. Book XXV, *The Fall of Syracuse*

8.1 INTRODUCTION

8.1.1 Influenza: History and Statistics

Influenza and humans are long-time companions. The disease may have been mentioned by Hippocrates as early as in the 412 B.C. as "the cough of Perinthus" in his work *Of the Epidemics*. The first clear description of influenza dates to 1100s [1,2]. The word "influenza" is originating from the fourteenth century Italy meaning "flow of liquid" or "influence," the latter perhaps suggesting its use by the contemporary scholars to describe the impact of the stars on the appearance of influenza epidemics. In English, influenza is commonly abbreviated to "flu." The synonym grippe is more often used in other languages including French, German, Russian, and Spanish.

Influenza virus is the most frequent cause of acute respiratory illness requiring medical intervention. The causative agent, influenza A virus, was discovered in 1933 [3]. In 1940, influenza B virus that had no antigenic similarity to influenza A has been described [4]. Two years later, in 1942, the first vaccine for influenza A and B viruses was introduced to the U.S. Armed Forces Epidemiological Board and licensed in 1945. Such a rapid introduction of the vaccine reflected the acute need for the preventive measures against the disease. After World War II, the vaccine was also used for civilians and has been significantly improved since then.

In spite of the long history of influenza-related medical research and the fact that effective vaccines exist, influenza remains a major public health problem. Human influenza viruses efficiently infect people and spread from person to person, they affect all age groups, and they can recur in any individual [5,6]. According to the Centers for Disease Control and Prevention (CDC), annual death toll from influenza in the United States reaches 36,000 [7]. The World Health Organization (WHO) estimates that 250,000–500,000 people die annually from influenza worldwide [8], which makes influenza the cause of the greatest number of vaccine-preventable deaths.

Influenza viruses also represent a threat of pandemics, the worst of which has been Spanish influenza, which circled the globe in just 4 months in 1918 and resulted in deaths estimated between 40 million and 100 million worldwide, possibly more than from any other virus on record [9–11]. The United States alone lost 675,000 people to the Spanish influenza, which is more than casualties of World War I, World War II, the Korean War, and the Vietnam War combined. Recently, the 1918 pandemic virus has been reconstructed (Figure 8.1) to study the pathogen and to find effective ways to respond to such extreme pathogenic potential [12]. The second pandemic of the twentieth century, "Asian" H2N2 influenza in 1957–1958 resulted in an estimated one million deaths worldwide and 70,000 deaths in the United States. Finally, in 1968–1969, the last recognized

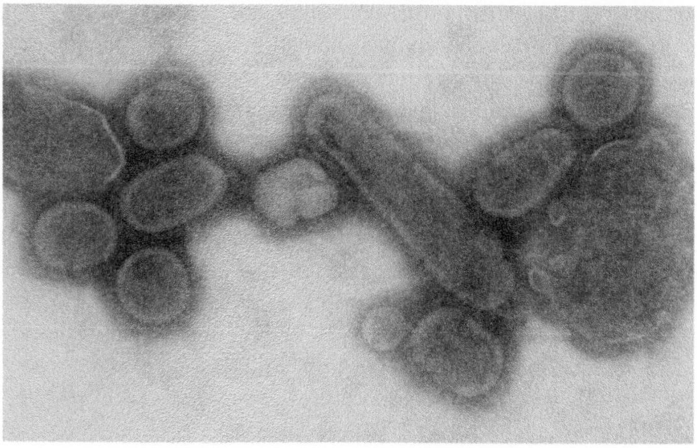

FIGURE 8.1 Influenza A virus particles, by negative staining transmission electron microscopy. This electron micrograph shows recreated 1918 influenza virions that were collected from the supernatant of a 1918-infected MDCK cell culture 18 h after infection. In order to sequester these virions, the MDCK cells were spun down by centrifugation, and the 1918 influenza virus present in the fluid was immediately fixed for negative staining. (Courtesy of the Centers for Disease Control and Prevention, CDC).

pandemic, "Hong Kong" influenza, caused by an H3N2 influenza virus, resulted in approximately 34,000 deaths in the United States. The relatively low death toll in the 1968 pandemic is thought to have been due to the residual cross-reactive immunity to viral hemagglutinin (HA) H3 and especially neuraminidase (NA) N2 proteins from the previous exposures [13].

Currently, the H5N1 strain of avian influenza is considered to be a potential culprit of next influenza pandemic [14]. Since 2003 to May 2008, the number of laboratory-confirmed human cases of avian H5N1 influenza reached 383, from which 241 infections were fatal representing a 63% mortality rate [15]. In addition to H5N1, other strains with pandemic potential exist. For example, in 1999, the avian H9N2 influenza virus was isolated from two children who recovered from influenza-like illnesses in Hong Kong [16]. In 2003, highly pathogenic H7N7 avian influenza virus caused an outbreak in poultry farms in the Netherlands, with 89 human cases and probably human-to-human transmissions [17].

As major human pathogens, influenza viruses remain among the most important targets for prophylactic and therapeutic interventions [13,18,19]. In this review, we summarize the traditional as well as novel approaches used in the development of vaccines for this important pathogen. In some more detail, we describe the development of native virus-like particles (VLPs) as candidate vaccines for influenza, as well as application of protein engineering methods for the development of chimeric VLPs with predicted antigenic characteristics. Most of influenza vaccine research has been done with influenza A virus. However, in most cases, including the existing commercial vaccines, approaches developed for influenza A virus could also be applied successfully to vaccine development for influenza B.

8.1.2 Influenza A, B, C

Influenza viruses belong to the Orthomyxoviridae family [20,21]. There are three types of influenza viruses in the orthomyxovirus family, namely types A, B, and C. Influenza A virus has caused all known influenza pandemics and also causes annual epidemics, being responsible for the majority of severe cases of influenza resulting in hospitalizations or fatalities. Due to these reasons, influenza A virus is a primary vaccine target and an obligatory component of seasonal and pandemic vaccines. The natural reservoirs of influenza A are the aquatic birds of the world, and the virus has been also isolated from a number of other animal hosts including horses, whales, seals, mink, and swine [2]. The pathogenesis of influenza A virus in humans starts usually with infection of upper respiratory mucosal epithelium. The infection then progresses to an acute febrile illness, which is associated with myalgias, headache, cough, rhinitis, and otitis media. Influenza infection is usually self-limited but may progress to pneumonia. Serious complications include encephalopathy, myocarditis, and myositis [22].

In addition to influenza A, influenza B virus is also included in the annual vaccine. Influenza B virus is primarily a human pathogen not associated with an animal reservoir. Influenza B causes similar symptoms and disease and is often clinically indistinguishable from influenza A virus infections. However, the frequency of the severe cases of influenza B infections appears to be significantly lower than that of influenza A and children will usually show symptoms more frequently than adults infected with influenza B virus [23,24].

For influenza C, well-defined outbreaks have rarely been detected in people, and the virus is rarely associated with severe syndromes [25,26]. Influenza C virus has been also isolated from pigs [27]. Most people have antibody to influenza C by early adulthood. Administration of influenza C virus to volunteers induced only mild symptoms [28]. Due to these reasons, influenza C is not included in the current vaccines.

8.1.3 Vaccine-Relevant Antigens

8.1.3.1 Envelope Proteins: Targets for Antiviral Drugs and Vaccines

Influenza A and B virus particles are pleomorphic, mostly spherical in shape and 80–120 nm in diameter (Figure 8.1). Filamentous particles also can be found, with 80–120 nm diameter and up to 2000 nm in length. The virion of influenza A contains a negative-sense, single-stranded genome comprised of eight ribonucleic acid (RNA) segments, numbered 1 through 8 on Figure 8.2A. The outer envelope of the virion particle is made of a lipid bilayer from which protrude glycoprotein spikes of two types, HA, ~14 nm trimer, and NA, ~6 nm tetramer (Figure 8.2). The HA and NA represent two major antigens on the surface of influenza virions. The three-dimensional structures for the HA and NA have been determined using x-ray crystallography (Figure 8.2B), which form the structural basis for rational design and engineering of vaccine-related proteins and antiviral drugs [29,30]. The HA protein is responsible for the attachment of the virus to cell receptors and subsequent fusion of virus envelope with host cell membrane. Human influenza viruses (H1–H3 subtypes) recognize cell surface glycoconjugates containing terminal α-2,6 linked sialyl-galactosyl (α-2,6) sialic acid (SA) moieties that are found on the human respiratory tract epithelium. Conversely, avian influenza viruses preferentially bind SA linked to galactose by an α-2,3 linkage (α-2,3 SA), which is found in high concentrations on the epithelial cells of the intestine of waterfowl and shorebirds. A mutation in the avian virus HA protein resulting in increased α-2,6 SA binding may only require a single amino acid mutation in the avian virus HA protein [29]. HA protein is normally synthesized as HA0 precursor that is cleaved posttranslationally by cellular proteases into HA1 and HA2. Cleavage exposes hydrophobic N-terminus of HA2, which then mediates fusion between viral envelope and endosomal membrane. There is a direct link between cleavage and virulence of avian influenza viruses [31]. Highly pathogenic H5 and H7 viruses contain multiple basic amino acid cleavage site, which can be recognized by furin and PC6 proteases in many host cells and organs that may lead to more efficient spread of the virus and more severe disease in humans and up to 100% mortality in birds.

The NA has a sialidase activity that removes sialic acid from the HA, NA, and host cell surfaces thus facilitating the release of progeny virions from infected cells. In addition, NA may facilitate virus attachment to the epithelial cells by removal of sialic acid from the mucin layer [32]. Depending on antigenic characteristics of the HA and NA, influenza A (but not B) viruses are

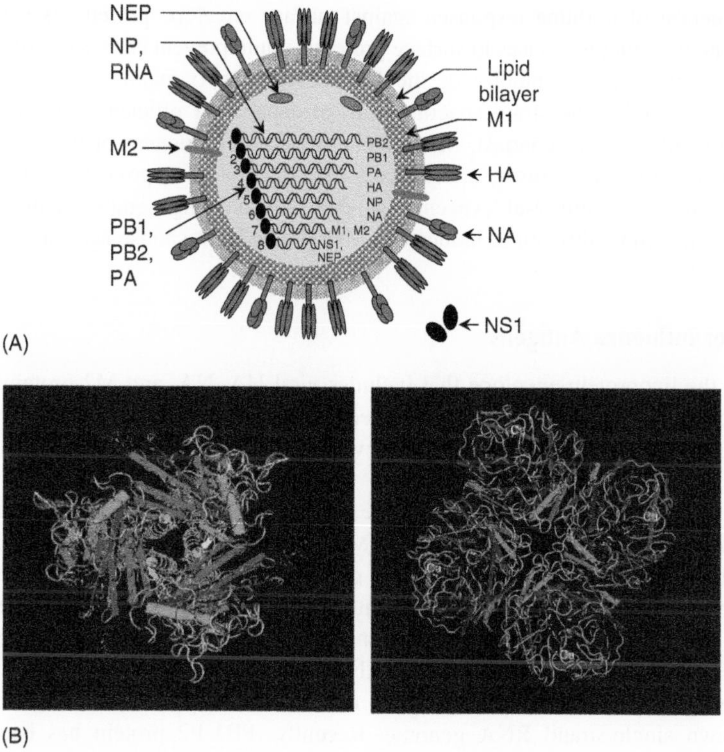

FIGURE 8.2 (A) Structure of influenza A virus particle. Indicated are structural proteins, lipoprotein membrane, and RNA segments. Numbers of RNA segments are also shown, as well as the encoded proteins. NEP, nuclear export protein (formerly NS2); NP, nucleoprotein; RNA, ribonucleic acid; M2, ion channel protein (translated from the same RNA as matrix protein M1); PB1, polymerase basic protein 1; PB2, polymerase basic protein 2; PA, polymerase acid protein; M1, matrix protein 1; HA, hemagglutinin; NA, neuraminidase; NS1, nonstructural protein. (B) **(See color insert following page 142.)** Ribbon representations of three-dimensional structures of the HA trimer (left) and NA tetramer (right). HA (H5 serotype) shown with the ligands as seen on Cn3D image of PDB Id 2IBX at NCBI. (From Yamada, S. et al., *Nature*, 444, 378, 2006.) NA (N1 serotype) monomers contain canonical six-bladed β-propeller structure; also shown are locations of bound ligands and Ca^{++} as seen on Cn3D image of PDB Id 2HTY at NCBI. (From Russell, R.J. et al., *Nature*, 443, 45, 2006.)

divided into antigenic subtypes or serotypes. For influenza A viruses, there are currently 16 different HA serotypes, H1–H16, and 9 different serotypes of NA (N1–N9).

In addition to HA and NA, the outer lipoprotein envelope of the virus particle also contain several molecules of M2, an ion channel protein. The M2 protein is a minor component of the envelope that has been implicated in ion channel activity and efficient uncoating of incoming viruses.

The outer envelope proteins are the attractive targets for vaccines as well as for therapeutic drugs including currently available two types of antivirals that are approved by Food and Drug Administration (FDA). One type of drugs, M2 blockers, such as amantadine and rimantadine, target specifically the M2 protein [33]. However, all influenza B viruses and many influenza A isolates are resistant to M2 blockers [34]. The second type of antivirals, NA inhibitors (such as oseltamivir, also known as Tamiflu), target the NA protein [35]. Drug resistance to the NA-specific antivirals has also been reported [36,37].

Because drug resistant influenza viruses can be found in nature or generated in the process of treatment with antiviral drugs, prophylactic vaccination may be a more effective way to combat influenza. Induction of immune responses against surface envelope proteins is a preferred way, because virus-neutralizing responses to surface proteins would prevent early steps of viral infection. Because of the significant variability of the HA and NA, two major envelope proteins, such responses are expected to be strain specific and not capable of efficient cross-neutralization of many strains of influenza. In contrast, M2 protein, a minor component of the envelope, is well-conserved among influenza viruses, and successful induction of a protective immune response to M2 may result in a "universal" vaccine capable of protecting against many isolates of the virus. However, it is more difficult to induce effective protective response against a minor envelope protein.

8.1.3.2 Inner Influenza Antigens

In addition to the lipoprotein envelope that includes viral HA, NA, and M2 proteins, influenza A virion particle contains an inner core comprised of several internal structural proteins, which potentially can also serve as targets for the development of countermeasures against influenza (Figure 8.2).

The inner side of the envelope is lined by the matrix protein (M1), which is the most abundant structural protein of the influenza A virion. The M1 and M2 proteins are encoded within the same RNA segment 7 (Figure 8.2), from which M2 is generated by using alternatively spliced RNA.

The inner part of the virion contains eight genomic RNA segments complexed with the nucleoprotein (NP) into helical ribonucleoproteins. In addition to NP, three polymerase polypeptides (PB1, PB2, PA) are associated with each RNA segment. Virus RNA synthesis takes place in the cell nucleus, where the virus subverts the cellular transcription machinery to express and replicate its own single-strand RNA genome. Recently, PB1-F2 protein has been discovered, which is also encoded by the PB1 gene.

The nuclear export protein (NEP), previously known as NS2, is also associated with the virion. Finally, NS1 is a cell-associated protein, which is normally not found in the virion particle. NS1 inhibits export of poly-A containing mRNA molecules from the nucleus, which gives preference to viral RNAs in transport and translation. NS1 has also been implicated in induction of apoptosis in influenza-infected cells [38] and in resistance to antiviral interferon (IFN) in highly pathogenic H5N1 influenza [39].

Influenza B virus has an additional protein, NB, which is generated by utilizing an alternative open reading frame from the same RNA segment that encodes the NA. The functions of NB are not completely understood. The NB-deficient live influenza virus could be generated, which was attenuated in mice [40].

Inner virus proteins are normally hidden within the particle and can only become accessible to the host immune effector mechanisms after initiation of infection and virus uncoating or expression within the infected cells.

8.1.4 Immunity to Influenza

8.1.4.1 Innate Immunity

Understanding mechanisms of generating protective immunity is a prerequisite for the development of successful vaccines. The innate immune system is one of the first lines of defense against influenza. The response of the innate immune system to influenza infection involves several mechanisms including virus interaction with toll-like receptors (TLRs), which recognize pathogen-associated molecular patterns and trigger resistance to infection through various pathways such as cytokine cascades. TLR3 on respiratory epithelial cells can activate type 1 IFN production and release of cytokines in response to influenza A virus or double-stranded RNA [41]. TLR7 expressed on dendritic cells and reacts with single-stranded RNA may also be involved in IFN pathway and production of cytokines [42]. Cytokine responses have been induced when volunteers were experimentally infected with influenza and exacerbated cytokine responses have been detected in a limited number of H5N1-infected patients with severe disease [43,44]. IFN appears to contribute to recovery from influenza infection [45]. However, highly pathogenic H5N1 viruses appear to be resistant to the effects of type 1 IFN response when evaluated in a porcine epithelial cell monolayer [39].

Production of IFN-γ was detected in peripheral blood natural killer (NK) cell subsets, as well as in influenza-specific memory CD8 T cells after exposure to influenza virus; IL-2 production by T cells was required for the IFN-g response of NK cells, indicating that memory T cells enhance innate NK-mediated antiviral immunity [46].

The involvement of innate immunity in control of influenza infection has been implicated, when intranasally (i.n.) administered baculovirus induced protection in mice to influenza challenge [47]. It has been shown that baculoviruses may induce maturation of dendritic cells in vivo, production of inflammatory cytokines, and also promote humoral and CD8 T cell adaptive responses against coadministered antigens [48].

8.1.4.2 Adaptive Immunity

During human infection with influenza A virus, serum antibodies to HA, NA, NP, and M1 are generated. High antibody titers to HA and NA envelope glycoproteins typically correlate with a lower attack rate for infection and less severe influenza disease [20]. In contrast, antibody to internal M1 and NP proteins are typically not associated with resistance to infection. Acquired or vaccine-induced immunity to influenza is, therefore, measured as serum IgG antibodies to the HA and NA antigens of the circulating influenza A and B viruses. High levels of hemagglutination-inhibition (HAI) antibody correlate with efficient neutralization of infectivity of the incoming virus before it infects host cells. In contrast, the NA-specific antibody is thought to inhibit virus release from infected cells thus restricting spread of the virus.

While HA-specific antibodies in the serum provide an excellent correlate of protection, the control of disease and clearance of the virus depend on more complex effector mechanisms provided by both humoral and cell-mediated immunity. Immune responses at both mucosal and systemic sites are presumed to be necessary for natural and vaccine-induced protection. Both humoral and cell-mediated responses following influenza infection or vaccination have been studied in detail [49–54]. Antibody and cytotoxic T cell epitopes have been found in all influenza proteins [55]. However, in spite of the fact that antigenic epitopes and immune responses to multiple influenza proteins have been found, their specific roles in protecting human host against infection and re-infection remains to be fully elucidated. Further, the immune mechanisms and relative importance of each component in the exposed host are not completely understood. It has been shown that passive transfer of antibodies can clear influenza A virus infection in mice even in the absence of B or T cells [56]. Little is known about influenza-specific B cells at the single cell level or the homing of effector T or B cells to the respiratory tract.

8.2 CURRENT INFLUENZA VACCINES

Currently, available prophylactic influenza vaccines licensed in the United States are based on viruses grown in embryonated chicken eggs. Seasonal vaccines address the issue of multiple virus strains and significant variability of influenza viruses by including two strains of influenza A virus and one strain of influenza B virus into the vaccine as well as by changing vaccine strains annually to protect against the most current prevalent circulating viruses. The composition of such trivalent seasonal influenza vaccines has been standardized by the WHO and depends on the predicted circulating virus strains [57]. Designed to protect against influenza A and B viruses, trivalent influenza vaccines elicit efficient HA-specific systemic antibodies, which can bind the viruses and inhibit early events in the influenza virus infection. In the United States and Europe, criteria for vaccine immunogenicity, which are based on the induction of virus-neutralizing or HAI antibodies in the serum, have been implemented [58,59]. In 2000, approximately 240 million doses of influenza vaccine have been distributed worldwide [60]. A Federal Medicare rule became effective in 2005 in the United States that requires all long-term care facilities to offer annual vaccination for influenza.

8.2.1 Inactivated Vaccines

8.2.1.1 Whole Virus Vaccines

Historically, the first influenza vaccines were preparations of the whole viruses grown in allantoic fluid of embryonated ages and chemically inactivated, for example by using formalin or β-propiolactone. Such whole virus vaccines were reactogenic because of contamination with egg-derived components. Introduction of zonal centrifugation [61] improved characteristics of the vaccines. Generally, the whole virus vaccines have high degree of immunogenicity, exceeding immunogenicity of the currently used subvirion, or "split" virus, vaccines. However, with the introduction of split virus vaccines, the use of the whole virus vaccines has been essentially discontinued due to the higher rates of adverse reactions [62,63]. The reasons for better immunogenicity of whole virus vaccines remain to be fully elucidated. It has been hypothesized that presentation of influenza antigens in an intact lipid membrane in a whole virus vaccine may result in better processing of antigen, resulting in advantageous epitope presentation and immune responses [64].

Due to a pandemic threat and especially with introduction of egg-independent production methods, the whole virus vaccines may return to the human vaccine market, because higher immunogenicity in a naive population could reduce the vaccine dose needed to provide effective protective immunity and therefore may allow to vaccinate more people. Recently, promising clinical results of safety and immunogenicity for cell-derived, inactivated whole virus vaccine for H5N1 virus have been presented [65].

8.2.1.2 Split Virus Vaccines

A cornerstone of current prophylactic influenza vaccines are inactivated subvirion, or split virus, vaccines. The examples of FDA-approved influenza vaccines for the use in the United States are shown in Table 8.1. The standard split influenza vaccine is a trivalent inactivated virus preparation given by intramuscular (i.m.) injection. Generally, one dose of vaccine for adults contains 45 μg of

TABLE 8.1
Examples of Licensed Human Influenza Vaccines

Vaccine	Manufacturer	Dose	Route	Description	References
Fluzone	Sanofi Pasteur	15 μg each strain	i.m.	Inactivated trivalent	[195]
Fluvirin	Chiron	15 μg each strain	i.m.	Inactivated trivalent	[195]
FluMist	MedImmune	$10^{6.5-7.5}$ $TCID_{50}$ each strain	i.n.	Live attenuated trivalent	[69,70]
Fluarix	GlaxoSmithKline	15 μg each strain	i.m.	Inactivated trivalent	[67,195,196]
FluLaval	ID Biomedical	15 μg each strain	i.m.	Inactivated trivalent	[197]
H5N1	Sanofi Pasteur	90 μg	i.m.	Inactivated monovalent	[68]
Inflexal V[a]	Berna Biotech	15 μg each strain	i.m.	Virosomal trivalent	[198]
Influvac[a]	Solvay Pharmaceuticals	15 μg each strain	i.m.	Subunit trivalent	[131]

[a] Licensed in Europe.

HA, that is, 15 μg for each strain. The methods for vaccine preparation and virus inactivation that results in a split virus were developed in the 1960s [61,66] and are used until today with modifications. Before formulating trivalent vaccine, each of the influenza viruses is produced, inactivated, and purified separately. For example, in the process of manufacturing Fluarix vaccine, virus-containing allantoic fluids are harvested from eggs; each virus is then concentrated and purified by zonal centrifugation using a linear sucrose density gradient solution containing detergent to disrupt the viruses. After dilution, the virus preparation is further purified by diafiltration. Each influenza virus is inactivated by the consecutive effects of sodium deoxycholate and formaldehyde leading to the production of a split virus. Fluarix vaccine is an injectable vaccine indicated for adults 18 years of age and older that underwent rapid licensure process after influenza vaccine shortage occurred in the United States in 2004 [67].

8.2.1.3 Pandemic Vaccine

In 2004, contracts were awarded to Sanofi Pasteur and to Chiron to develop vaccine against the H5N1 avian influenza virus. The first egg-based inactivated split vaccine for H5N1 influenza, which is based on influenza virus strain A/Vietnam/1203/2004 (H5N1, clade 1), has been licensed in the United States in 2007 [68] as an interim pandemic vaccine. Each 1 mL dose is formulated to contain 90 μg of HA. The vaccine is prepared from influenza virus harvested from embryonated chicken eggs and inactivated with formaldehyde. For the production of H5N1 pandemic vaccine, the virus is concentrated and purified in a linear sucrose density gradient using a continuous flow centrifuge. The virus is then chemically disrupted using polyethylene glycol *p*-isooctylphenyl ether (Triton X-100), a nonionic surfactant, producing a split virus, which is further purified by biochemical means.

8.2.2 Live Attenuated Vaccines

In 2003, the first i.n. administered live attenuated influenza vaccine, FluMist, was licensed in the United States (Table 8.1). This live influenza A and B virus vaccine was indicated for healthy individuals ages 5–49 years as an alternative approach to influenza vaccination [69,70]. Initially, to develop such a vaccine, influenza virus was passaged at 25°C in chicken kidney cells and in embryonated eggs that resulted in cold-adapted, highly attenuated virus [71,72]. Live attenuated vaccines are administered as nasal spray and do not require needles for administration. In comparison to inactivated vaccines, which induce good antibody response, live attenuated vaccine appear to induce more balanced responses including mucosal and cellular immunity. A placebo-controlled study in 103 experimentally infected adult patients found that the protective efficacy of FluMist was 85%, compared to 71% with inactivated influenza vaccine [73]. However, as with other live attenuated virus vaccines, FluMist is not administered to individuals with known or suspected immune deficiency diseases including thymic abnormalities, malignancies, or human immunodeficiency virus infection, as well as to those who may be immunosuppressed because of radiation treatment or other immunosuppressive therapies.

8.2.3 Veterinary Vaccines

The number of influenza outbreaks in the poultry sharply increased during the last years. The control of avian influenza in the industry becomes very important for control of potential influenza pandemic. Vaccination can be a powerful tool to prevent outbreaks, along with other measures [74]. Preventive measures should be applied not only to highly pathogenic H5 and H7 viruses with pandemic potential but also to low-pathogenic H5 and H7 viruses, because these may evolve in the poultry into high-pathogenic strains. Veterinary vaccines for influenza are available and have been used successfully in avian influenza control programs. Standard inactivated influenza vaccines are fully or conditionally licensed for parenteral administration and have been successful at providing protection against clinical signs and death among poultry. Avian influenza vaccines generated by reverse genetics [75] or recombinant genetically engineered fowlpox vector vaccines expressing HA have been also successfully tested [76].

8.3 MOLECULAR ENGINEERING OF INFLUENZA VACCINES: KNOWN CHALLENGES

8.3.1 Antigenic Drift and Antigenic Shift

Inactivated and live attenuated trivalent vaccines have generally been found very effective in preventing and limiting the spread of the disease. Lower efficiency of vaccination could be observed during annual epidemics when components of a vaccine did not match well the circulating influenza strains. For example, in 2003 FDA rejected due to safety concerns the use of the most appropriate H3N2 vaccine strain, A/Fujian/411/2002, isolated from the non-approved Madin-Darby canine kidney (MDCK) cells, and instead used the same strain as in the 2002 vaccine formulation [19]. As a result, the H3N2 component of the 2003–2004 influenza virus vaccine did not match circulating strain and showed suboptimal efficacy [77].

Thus, the intrinsic genetic variability of influenza viruses presents the major obstacle for vaccination as well as for vaccine development. Through the genetic processes termed antigenic drift and antigenic shift, influenza virus has the ability to evolve into new variants that can overcome immunity and cause epidemics or pandemics. The emergence of new antigenic variants of influenza viruses through antigenic drift is relatively frequent event, and this phenomenon is responsible for seasonal changes in circulating virus strains and for annual influenza epidemics. Because of antigenic drift, the vaccines for seasonal influenza are reformulated frequently to incorporate viral antigens that match the circulating influenza viruses.

In contrast, antigenic shift is a relatively infrequent event. Unlike antigenic drift, which is caused by mutations within gene segments, the antigenic shift is caused by the reassortment of gene segments between two influenza strains in permissive hosts. Influenza viruses belonging to any of the three different types (influenza A, B, or C) can undergo reassortment, but not between members of different types. The potential for a pandemic arises when such an antigen shift occurs, because most people will not have immunity to such a novel reassortant virus. If such a novel reassortant virus can cause illness and is capable of efficient human-to-human transmission, a pandemic can occur.

8.3.2 Young and Elderly: Suboptimal Responses to Vaccination

Even when vaccine matches circulating viruses, inactivated vaccines tend to have lower efficiency in the young children and especially in elderly [78–80]. Modest responses in the elderly are also observed with live attenuated vaccines [69]. Unfortunately, the young and elderly populations are also more vulnerable to the serious complications of influenza that might result in hospitalization or death [81]. The higher susceptibility of the host to serious influenza in the pediatric and geriatric populations likely reflects diminished capacities of both the innate and adaptive immune systems in the very young and the elderly [82,83]. Young children are often immunologically naive and may also have intrinsic limitations in immune cell functions. Aging of the immune system in the elderly results in weaker responses to both infection and vaccination, despite repeated priming of memory immunity.

8.3.3 Manufacturing in Eggs: Generation of High-Yield Viruses

Preparation of influenza vaccines in eggs is a relatively lengthy process. In a typical chicken embryo operation, one egg is required to produce one 45 μg dose of vaccine protein. However, some influenza viruses do not grow in eggs at efficacies suitable for vaccine manufacturing. In these cases, reassortment with the high-yield influenza virus is done to generate virus that combines the desired antigenic and growth characteristics. Manufacturers receive such seed strains through their government agencies, which generate high-yield viruses using classical reassortment [84] of seasonal strains with the laboratory strain, A/PR/8/34 for influenza A viruses. Such engineered reassortant viruses usually contain six genomic RNA segments from the laboratory strain and the remaining two segments (HA and NA) from the virus of interest. Reassortment is a time-consuming process that involves multiple passaging of seed viruses to allow virus/egg adaptation. For type B viruses, reassortment usually is not used and high-yield seed viruses are obtained by repeated passaging in eggs. It should be noted that during repeated adaptation passages the HA and NA acquire additional mutations, which may cause alteration in their functional and antigenic characteristics [85,86].

Reassortment with the high-yield strains is also used in the process of preparation of live attenuated vaccines. Following reassortment, vaccine seed viruses are generated that contain two genomic segments from the circulating viruses, whereas the remaining six genomic segments are derived from the master seeds of cold-adapted attenuated A and B viruses.

For licensed vaccines, the entire vaccine manufacturing process including generation of seed viruses and all the required tests needs to be completed before annual epidemic begins. As mentioned above, failure to isolate A/Fujian/411/2002 (H3N2) in eggs resulted in its absence from the 2003/2004 vaccine. Furthermore, because of the manufacturing constraints, there is a risk that the demand may outstrip the supply of the vaccine. A significant shortage of influenza vaccine occurred in the United States in 2004. After routine testing required by FDA, Chiron Corporation in the United Kingdom, one of the two suppliers of inactivated influenza vaccine for the United States, found bacterial contamination in a limited number of vaccine lots. As a result, FDA announced that influenza vaccine manufactured by Chiron for the United States market was not safe for use. The remaining vaccine doses were recommended for those at the highest risk of complications from influenza. As a result of the vaccine shortage, FDA had to identify additional sources of vaccine that could be made available under an FDA investigational new drug (IND) application [87]. These efforts resulted in FDA approving INDs that permitted the potential use of vaccines from additional suppliers such as GlaxoSmithKline and Berna Biotech. Dose-sparing studies have been done including intradermal injections and adjuvants to determine if the reduction of vaccine dose may allow adequate protection and increase the number of available vaccine doses [88,89].

8.3.4 Efforts to Improve Current Egg-Based Vaccines

The reduction of vaccine dose might not only increase the number of available vaccine doses but also reduce the reactogenicity of egg-derived vaccines. In spite of the development of purification methods, such as zonal centrifugation and biochemical disruption of viruses, the vaccines derived from eggs remain to be reactogenic, especially in individuals with allergies to egg products.

Attempts have also been made to improve the efficacy and safety of current inactivated vaccines by improving vaccine composition, evaluating adjuvants, and optimizing route of vaccine administration. For example, split vaccine derived from live attenuated recombinant H5N1 influenza virus was tested in mice i.n. with several adjuvants including cholera toxin B subunit containing a trace amount of holotoxin, synthetic double-stranded RNA, or chitin microparticles. Promising effects of these adjuvanted vaccines on IgA and IgG responses, expression of TLR3, and protection from live H5N1 virus challenge have been observed [90]. Further, safety and immunogenicity of split trivalent influenza vaccine formulated with lipid/polysaccharide carrier has been evaluated in a clinical trial as nasal influenza vaccine [91]. Vaccine strains included in this study were A/Johannesburg/82/96 (H1N1), A/Nanchang/933/95 (H3N2), and B/Harbin/07/94. Authors concluded that inactivated nasal influenza vaccine is well-tolerated and immunogenic in healthy adults. In another study, split trivalent vaccines have been evaluated in varying i.m. and i.n. dosages separately and combined [92]. The viruses used were trivalent inactivated vaccine for the 2001–2002 season that included the following influenza virus strains: A/New Caledonia/20/99

(H1N1), A/Panama/2007/99 (H3N2), and B/Victoria/504/2000. Volunteers between the ages of 18 and 45 years received 15, 30, or 60 μg of vaccines by either i.n., i.m., or both routes, 120 μg of vaccine i.m., or placebo. All dosages and routes of vaccine administration were well-tolerated, safe, and induced serum as well as mucosal antibody responses. Overall, high dose of i.m. vaccine with or without i.n. vaccine generated high levels of HAI- and virus-neutralizing serum antibody responses, suggesting the advantage for i.n. vaccine may be limited to induction of nasal secretion antibodies [92].

8.3.5 Vaccine for Pandemic Strains: More Known Challenges

The emergence of pandemic influenza adds additional major challenges for vaccine development [10,93,94]. It is highly unlikely that seasonal influenza vaccines will provide significant levels of protection against potential pandemic strains such as H5N1. Currently available H5N1 inactivated vaccine (Table 8.1) based on clade 1 A/Vietnam/1203/2004 strain may not protect against all potential pandemic viruses including other H5N1 strains, such as clade 2 A/Indonesia/05/2005 virus. In the case of a pandemic, a large number of vaccine doses may be required within a relatively short period. Such a scale-up may be difficult to achieve using current egg-based influenza vaccine manufacturing technology. Also, the virus is expected to be highly lethal to birds including chickens, and the maintenance of a constant supply of embryonated eggs would be difficult in a pandemic. Furthermore, the virus rapidly kills chicken embryos before virus can grow to the yields sufficient for effective production. The latter difficulty can be circumvented by reassortment of pandemic strain virus with the laboratory influenza strains. However, this may result in a new reassortant virus with unpredictable pathogenic and transmission characteristics. Protein engineering methods can be applied to reduce pathogenic potential of such reassortants. For example, deleting of the polybasic cleavage site within the H5 can be used to prevent generation of highly pathogenic virus. In any event, high levels of biosafety containment are needed for the production of pandemic vaccines using live pandemic influenza viruses to protect workers and to prevent the escape of the viruses into the environment.

Further, the risk of incomplete inactivation of highly pathogenic pandemic virus such as H5N1 may be not fully acceptable for human vaccines. Live attenuated vaccines for pandemic influenza viruses also cause concerns, because they may undergo reversion or reassortment with wild-type influenza viruses to regenerate pathogenic or even a new pandemic virus. In addition, live attenuated virus vaccines themselves represent risks for some vaccine recipients, especially for those with immune disorders.

8.4 NOVEL INFLUENZA VACCINES

8.4.1 Requirements for New Vaccines

It is clear that annual influenza epidemics and the probability of an influenza pandemic constitute major public health challenges that require not only timely production of the seasonal vaccines but also, the development of pandemic vaccines and designing strategies for providing large numbers of doses of vaccines that in the case of pandemic, can rapidly induce protective immunity in an immunologically naive population. The current technology for producing influenza virus vaccines involves many steps which are time consuming and require millions of egg embryos. To overcome limitations of current vaccines, the development of egg-independent, cell culture-based influenza vaccines is one of the important priorities. This is being actively pursued with the goal of developing both seasonal and pandemic vaccines that are safe, effective, and simple in manufacturing [94]. In 2005, DHHS awarded a contract for $97 million to Sanofi Pasteur to develop cell culture-based influenza vaccines for the United States.

Other important priorities for improvement of influenza vaccines are the development and application of new technologies such as reverse genetics or protein engineering, which could be used to develop improved vaccines and to shorten the process of preparing current vaccines [19]. For example, application of reverse genetics technology instead of the traditional reassortment technique can facilitate and speed up the manufacturing process for the current inactivated as well as live attenuated influenza vaccines. However, the conventional 12- or 8-plasmid reverse genetics systems have relatively low transfection efficiency of such sets of plasmids, which may impede the rapid generation of vaccine seed viruses. To overcome this potential difficulty, the number of plasmids required to generate influenza virus by reverse genetics have been reduced to only four [95]. The improved system consists of (1) a plasmid that encodes the six gene segments for the internal proteins (PB2, PB1, PA, NP, M, and NS); (2) a second plasmid that encodes the HA and NA segments; (3) a third plasmid that expresses influenza NP protein; and (4) a fourth plasmid expressing PB2, PB1, and PA proteins. These four plasmids are used for generation of influenza virus using transfection of Vero cells. This results in the process that is more efficient than the conventional process involving 12-plasmid systems.

Finally, the novel molecular engineering approaches can be used for the engineering of alternative, conceptually novel, influenza vaccines. Such new influenza vaccines should address the issues that are not completely addressed by currently licensed vaccines, which otherwise have been proven effective and provide a standard for comparison with any new candidate vaccine. The critical parameters for assessment of new influenza vaccines are safety; efficacy in various populations including young, elderly, and the immunologically naive or compromised; speed and ease of production; cost; and acceptance by regulatory agencies and public [19]. Vaccine approaches that have been developed in the past years included recombinant protein subunit vaccines, virosomes, naked DNA vaccines, virus vector vaccines, and other approaches.

Depending on the mechanism of inducing protective immunity, novel influenza vaccines can be divided into two large groups: genetic (or nucleic acid vector-based) vaccines and protein vaccines. In the genetic vaccines, influenza antigen is expressed in the tissues of vaccine recipient from an injected vector construct. In the protein vaccines, antigen is administered directly to vaccine recipient.

8.5 GENETIC VACCINES

8.5.1 DNA Vaccines

DNA vaccination is based upon inoculation of purified DNA expressing immunogen of interest in vivo. DNA vaccines may be especially attractive as "rapid response" vaccines, for example, in the case of a pandemic [96]. There are several methods of administration of DNA vaccines including direct injection into muscle cells, injection of DNA-covered gold particles (gene gun), topical application of DNA, as well as in vivo DNA electroporation. In any case, DNA inoculation results in the production of a protein by host cells, which results in induction of immune response. The immune responses involve both T cell- and antibody-mediated immunity. Studies have also implicated stimulation by DNA of the innate immune system that creates favorable cytokine profiles for Th1 cell-mediated responses [97,98].

Several studies have demonstrated that inoculation of plasmid DNA encoding influenza antigens, particularly HA, elicited specific immune responses and provided protection against influenza virus in preclinical models including mice and ferrets [99–102]. For example, HA- and NA-based DNA vaccines protected mice from live virus challenge [103]. Conversely, M1-, NP-, or NS1-based DNA vaccine provided lower, if any, protection [99,103]. However, protection of mice with NP could be demonstrated using a higher DNA vaccine dose [104]. A DNA vaccine expressing full-length consensus M2 protein induced M2-specific antibody responses and protected mice against lethal influenza virus challenge [105].

Successful DNA vaccination against influenza depends not only on the dose of DNA vaccine or an animal model used but also on many other factors such as immunogen choice and design, method of DNA vaccine administration, and other parameters. For example, immunogenicity of HA-expressing DNA vaccines could be improved by using codon-optimized HA sequences, as shown for H1 or H3 serotypes of human influenza A virus [106]. The authors of this study also engineered two forms of HA antigen, a full-length HA and a secreted form of HA with transmembrane (TM) domain truncated. Both full-length and TM-truncated H3 induced high levels of HAI and neutralizing antibody responses. However, the full-length H1-induced significantly higher HAI and virus-neutralizing antibody responses than did the TM-truncated HA. These data indicated that influenza antigens from different serotypes may have different requirements for the induction of optimal immune responses.

Promising preclinical results achieved with DNA vaccines have generated great interest in developing DNA constructs as a new generation of influenza vaccines with the ability to elicit effective humoral and cellular immune responses. However, DNA vaccines have yet to overcome technical and regulatory hurdles and proceed past phase I/II clinical trials, primarily due to a need to induce more potent immune responses in people [107–109]. Safety issues have been also raised with regard to DNA vaccines, such as development of autoimmunity to DNA or integration of vector into cellular DNA that potentially can lead to insertional mutagenesis, such as activation of oncogenes or inactivation of tumor suppressor genes. FDA and EU developed specific advices with regard to safety testing for DNA vaccines [110,111].

The phase 1 clinical trial of a DNA vaccine developed by the National Institutes of Health and designed to protect against H5N1 influenza infection began in December, 2006 [112].

To improve efficacy of vaccination with DNA vaccines, a prime-boost approach has been often used, in which animals were first vaccinated with DNA and then received booster inoculation with protein. For example, this has been shown recently with DNA/protein prime-boost involving H7 and M1 proteins [113]. Prime-boost DNA/protein vaccination appeared to be more advantageous compared to DNA/DNA or protein/protein vaccinations [113]. Another strategy of improving efficacy of DNA vaccines may be primary vaccination with DNA and a boost with virus vector vaccine. Thus, enhanced vaccine protection was afforded if a recombinant adenoviral boost immunization to NP was included as a DNA prime and boost regime [114].

8.5.2 Virus Vector Vaccines

Virus vector vaccines represent another strategy for the development of influenza vaccines including those caused by avian strains with pandemic potential. Unlike DNA vaccines that involve direct inoculation of "naked" nucleic acid, virus vector vaccines are based upon inoculation of virus vector. The latter functions as an efficient vehicle to deliver the viral nucleic acid (DNA or RNA) into host cells. The viral nucleic acid is configured to express influenza vaccine-related immunogen of interest in vivo. Similarly to the DNA vaccines, inoculation of virus vector results in the production of a vaccine by host cells in vivo, which results in the generation of immune response including both cellular and humoral immunity. Generally, virus vector can be either propagation-competent (live virus) or propagation-incompetent. In the latter case, virus vector is capable of infecting cell and expressing antigen of interest, but it cannot spread to other cells from the initially infected cell [115,116].

Propagation-competent, live recombinant fowlpox-based vaccines for avian influenza have been developed and used in chickens [76,117]. A fowlpox vaccine with an H5 gene insert protected chickens against clinical signs and death following challenge by nine different H5 avian influenza viruses, which were isolated from different continents over a 38 year period and had 87%–100% sequence similarity with the H5 within the vaccine [76]. Fowlpox-based vaccine expressing H5 gene was granted a license in the United States for emergency use in 1989 and full registration in Mexico, Guatemala, and El Salvador. This vaccine is administered to 1 day old chickens and has been also found immunogenic in cats suggesting possibility for use in mammals [117].

Recently, candidate vaccines expressing H5 gene from either A/Hong Kong/156/97 or A/Vietnam/1194/04 have also been developed from highly attenuated strain MVA of vaccinia virus. The vaccines were tested in C57BL/6J mice, and the data suggested that recombinant MVA expressing the H5 of influenza virus A/Vietnam/1194/04 is a promising candidate for the induction of protective immunity against various H5N1 influenza strains [118].

Propagation-incompetent virus vectors have been also developed as influenza vaccine candidates. Alphavirus replicon vectors expressing HA from A/Puerto Rico/8/34 virus induced antibody response and protected BALB/c mice from challenge with homologous influenza virus [115]. A single dose of alphavirus replicon particles expressing HA from A/Hong Kong/156/97 (H5N1) completely protected 2 week old chickens from infection with lethal parent virus [119]. A phase I clinical trial involving alphavirus replicon vaccine for influenza expressing HA protein from H3N2 influenza A/Wyoming/03/2003 is currently recruiting patients.

A propagation-incompetent, human adenoviral vector containing H5 influenza gene (HAd-H5), induced in BALB/c mice both humoral and cell-mediated immune responses against avian H5N1 influenza viruses isolated from people. Vaccination of mice with HAd-H5 provided effective protection from H5N1 infection using antigenically distinct strains of H5N1 influenza viruses [116]. Another propagation-incompetent adenoviral vector expressing A/Vietnam/1203/2004 (H5N1) HA was also used to vaccinate BALB/c mice, which were later exposed to a lethal i.n. dose of homologous H5N1 virus. Mice vaccinated with full-length HA were fully protected from challenge. A single subcutaneous vaccination also protected chickens from a lethal i.n. challenge with live influenza virus [120].

Safety concerns are among the major issues that have been raised for the virus vector vaccines. Propagation-competent vectors, such as poxviruses, may be not safe for the use in people, especially in immunocompromised individuals. Disseminated vaccinia was detected in a military recruit with human immunodeficiency virus (HIV) disease [121]. Propagation-incompetent vectors appear to be safer; however, live virus still may be rescued at low frequency as a result of recombination during virus vector preparation [115]. Additional limitation for the use of viral vectors is preexisting immunity, such as in the case of adenovirus vectors. Because large part of the population has antibody to adenovirus, this may limit the ability of the vector to generate effective immune response. Inoculation with the virus vector may also induce strong anti-vector immunity, which can preclude effective booster vaccinations. To address such a limitation, a prime-boost approach is used, in which primary vaccination is carried out using DNA vector, whereas the booster injection is provided by virus vector. For example, mice primed with M2-DNA and then boosted with recombinant adenovirus expressing M2 (M2-Ad) had enhanced antibody responses that cross-reacted with human and avian M2 sequences and produced T-cell responses [105]. This M2 prime-boost vaccination conferred broad protection against challenge with lethal influenza A, including an H5N1 strain.

8.6 PROTEIN VACCINES

8.6.1 Peptide and Subunit Vaccines: Role of Adjuvants

Peptides and subunit proteins are attractive vaccine candidates because they are safe, not reactogenic, and they can be generated and purified in large quantities. Many antibody and T cell epitopes for influenza virus have been reported in the literature, including a number of protective epitopes against influenza [55]. Methods for fine mapping of viral epitopes have also been developed [122–127]. Potentially, information regarding protective epitopes can be used for the development of epitope-based peptide vaccines. However, despite high safety profile of synthetic peptides, attempts to develop influenza vaccines based on a limited number of peptides face several problems including HLA polymorphism, limitations in the immunogenicity of peptide-based immunogens, and the high mutation rate of influenza viruses. Further, most epitopes are conformation-dependent. The majority of identified influenza epitopes are located within the HA and NP proteins [55]. Peptides derived from the conserved influenza proteins may offer some advantages as a basis for the development of potential peptide-based vaccines. For example, M2 protein is highly conserved across influenza A subtypes. The efficacy of M2-based peptide vaccine has been evaluated in mice [105]. Animals were vaccinated with M2 peptide of a widely shared consensus sequence. Vaccination induced serum antibodies that cross-reacted with divergent M2 peptide from an H5N1 subtype [105]. However, epitope vaccines appear to hold even more promise when they are administered not as soluble synthetic peptides but in conjunction with other immunologically active components such as adjuvants or engineered VLPs (Section 8.6.3). For example, M2 protein incorporated into liposomal formulations containing a lipid adjuvant, MPL were shown to enhance immune response to M2 protein and elicit protection against lethal homologous challenge [128].

Subunit vaccines comprising purified influenza proteins are also well-tolerated clinically [129]. A trivalent, baculovirus-generated, recombinant HA0 (rHA0) vaccine was found safe and immunogenic in a healthy adult population, and inclusion of a NA protein did not appear to be required for protection [130]. Participants of this clinical study received (1) a single injection of saline placebo or (2) 75 μg of an rHA0 vaccine containing 15 μg of HA from influenza A/New Caledonia/20/99 (H1N1) and influenza B/Jiangsu/10/03 virus and 45 μg of HA from influenza A/Wyoming/3/03 (H3N2) virus or (3) 135 μg of rHA0 containing 45 μg of HA each from all 3 components. HAI antibody responses to the H1 component were seen in 51% of 75 μg vaccine, and 67% of 135 μg vaccine recipients, while responses to B were seen in 65% of 75 μg vaccine, and 92% of 135 μg vaccine recipients [130].

Commercial trivalent subunit vaccines have been used for vaccination of people, such as Influvac in the Netherlands (Table 8.1) that contains 15 μg of HA for each strain [131]. The safety and immunogenicity of such commercial trivalent subunit influenza vaccine was compared to an experimental virosome-formulated influenza vaccine in elderly patients [132]. Virosome vaccine was generated by incorporating egg-purified HA into the membrane of phosphatidylcholine liposomes. Both vaccines elicited anti-HA antibody titer to all three vaccine components in 1 month after immunization. However, significantly more patients vaccinated with the virosome vaccine mounted a more than fourfold higher response to influenza as compared with those who received subunit vaccine. Approximately, 68% of patients immunized with the virosome vaccine attained protective levels of antibody to all three vaccine components versus 38% for the subunit vaccine [132].

In general, most of the peptide and purified subunit proteins were found to be relatively weak immunogens. Efficacy of subunit vaccines could be considerably improved by using adjuvants such as squalene, which represents an unsaturated aliphatic hydrocarbon (MF59), or by using liposomes [133,134]. A number of adjuvants for infectious diseases including influenza have been reviewed elsewhere [135]. Liposome-based virosome vaccines have been commercially available for a number of years, as described in Section 8.6.2. Immunogenicity study with MF59 adjuvant and Chiron's Fluad subunit vaccine demonstrated that a consistently higher immune response is observed in MF59-adjuvanted subunit vaccine as compared to non-adjuvanted subunit and split influenza vaccines [133]. MF59-adjuvanted vaccine was clinically well tolerated, also after re-immunization in subsequent influenza seasons. An MF59-adjuvanted inactivated influenza vaccine containing A/Panama/2007/99 (H3N2) induced broader serological protection against heterovariant influenza virus strain A/Fujian/411/02 (H3N2) than a subunit and a split influenza vaccines [136]. The results showed that, while less than 80% of elder people vaccinated with conventional vaccines had protective levels of antibodies against the A/Fujian/2002 heterovariant strain, those vaccinated with the MF59-adjuvanted vaccine had protective levels of antibodies in more than 98% of the cases. Importantly, it has been also shown that vaccination with a subunit influenza vaccine with the MF59 adjuvant neither induced anti-squalene antibodies nor enhanced preexisting anti-squalene antibody titers [137]. This is important, because presence of anti-squalene antibodies in some recipients of squalene-contaminated anthrax vaccine has been linked to the Gulf War syndrome [138].

Other adjuvant systems are also being evaluated in the context of peptides, subunit proteins, as well as of inactivated influenza vaccines including adjuvants based on CD40 monoclonal antibody [139,140]. CD40 is a co-stimulatory receptor on B lymphocytes, and signaling through CD40 molecules greatly enhances lymphocyte activation in the presence of antigen receptor stimulation. Authors engineered conjugates of CD40 monoclonal antibody with three potential influenza vaccines: a peptide-based vaccine containing T- and B-cell epitopes from virus HA, a killed whole virus vaccine, and a commercial split virus vaccine. CD40 mAb conjugates in each case were found to be more immunogenic [140]. Inoculation of mice i.n. with an anti-CD40 monoclonal antibody and NP366–374 peptide, corresponding to a CTL epitope on NP, encapsulated in liposome resulted in induction of protective CTL responses against influenza A virus [141].

8.6.2 Virosomal Vaccines

The combination of a liposome with subunits of the influenza virus is called virosome. The viral antigens HA and NA can be anchored into the liposome lipid layer, thus resembling envelope of a natural influenza virus particle. Commercial virosomal influenza vaccines are available in Europe such as Inflexal V (Table 8.1) licensed for all age groups (up from 6 months). Inflexal V represents an inactivated vaccine which has virosomes in its formulation acting as carrier/adjuvant and it is composed by highly purified surface antigens of influenza virus strains A and B, propagated in embryonated chicken eggs and inactivated with β-propiolactone. The vaccine's antigen composition follows annual WHO recommendations and contains 15 μg of HA for each recommended strain. Inflexal V was originally introduced in 1997 and is registered now in more than 40 countries. The vaccine has very good safety record after approximately 30 million doses administered so far. Neither formaldehyde nor thiomersal (thimerosal), an organomercury preservative, is contained within the vaccine. The manufacturing process allows residual quantities of antibiotics, detergent, and chicken proteins to be minimal compared with other influenza vaccines. Another virosomal influenza vaccine, Invivac, has been also evaluated clinically and has been on the market since 2004 although discontinued in the United Kingdom in December 2006. Virosomal vaccines mostly target the elderly, as well as other populations with impaired immune responses to conventional influenza vaccines [132,142,143].

Virosomal nasal vaccine, NasalFlu, containing strong mucosal adjuvant, *E. coli* heat-labile enterotoxin (LT), also has been licensed in Switzerland but later withdrawn due to increased number of cases of Bell's palsy in association with vaccination [144]. Unfortunately, LT appears to be an important component of the vaccine, as phase I clinical study showed that the use of LT as a mucosal adjuvant is necessary to obtain a humoral immune response comparable to that with parenteral vaccination [145].

8.6.3 Virus-Like Particles as Influenza Vaccines

During the course of infection in virus-infected host cells, structural proteins, and genomic nucleic acids of influenza virus assemble into progeny virion particles, which are released from infected cells. Recently, it has been shown that influenza structural proteins maintain this intrinsic ability to self-assemble into influenza VLPs, following expression of the structural genes in cell culture systems [146–148]. The size and morphology of such self-assembled influenza VLPs resemble those of influenza virions. However, VLPs are noninfectious, because in the cell culture expression systems, viral proteins self-assemble in the absence of the viral genetic material.

For many other viruses, VLPs have been also generated that closely resembled structures formed by their counterpart viruses, including hepatitis B viruses (HBV) [149,150], human papilloma virus (HPV) [151,152], severe acute respiratory syndrome (SARS) coronavirus [153], Ebola and Marburg filoviruses [154], and other viruses [155]. Such noninfectious VLPs were found to be promising candidates for the production of vaccines against many diseases because highly repetitive structure of VLPs and high density display of epitopes is very effective in eliciting strong immune responses. VLPs are often morphologically and antigenically indistinguishable from the respective viruses, and the epitopes within VLPs preserve native conformation found in the viruses. Further, the size and particulate nature of VLPs may facilitate their uptake by dendritic cells, which is advantageous for the development of effective immunity [156].

VLP-based vaccines for influenza can be divided into native VLPs and chimeric VLPs, depending on the approach used to generate VLPs and the way of inducing immunity against influenza virus.

8.6.3.1 Native Influenza VLPs

Native VLPs are formed by self-assembly of native, full-length structural proteins. In these cases, VLPs are generated that have morphological and antigenic characteristics similar, if not identical, to naturally occurring authentic viral and subviral particles. Because native VLPs are antigenically equivalent to the wild-type viruses, when administered to a susceptible host, such VLPs induce the immune responses that in many aspects resemble those induced by the respective cognate viruses.

In many cases, native VLPs have already proven successful as vaccines against the respective virus infections. For example, small envelope protein, HBsAg, of HBV forms in yeast or mammalian cells 22 nm subviral particles that are essentially identical to a natural product of HBV infection, found in patient blood at levels greater than the virion itself. Both plasma-derived and recombinant 22 nm particles provided the successful HBV vaccines. Similarly, expression of the L1 protein of HPV6 and HPV16 in cultured cells have lead to the assembly of VLPs [151,152] that are similar to the virus particles formed during papillomavirus replication, although the natural particles also contain the L2 protein. Gardasil, a recombinant quadrivalent VLP vaccine based on the L1 proteins of HPV (types 6, 11, 16, and 18) have been recently licensed for the commercial use as HPV vaccine. The licensure of the HPV VLP vaccine will undoubtedly provide an impetus for further development of VLP-based vaccines for other viruses.

The expression of native VLPs for influenza virus poses some challenges. Unlike HBV or HPV viruses, influenza virus particle consists of multiple structural proteins. The protein–protein interactions and the roles of each protein, lipids, and host cell factors in the assembly and morphogenesis of VLPs are not fully understood. To achieve correct assembly of native influenza VLPs, proper interactions should take place between the inner influenza proteins, between the surface protein subunits within the lipid bilayer, and between the surface envelope and the inner influenza proteins.

Introduction of the efficient protein expression systems as well of the reverse genetics methods have been critical for efficient co-expression of influenza proteins including VLPs [157–161]. The examples of native influenza VLPs that have been reported in the literature are shown in Table 8.2. In the initial experiments, to generate influenza VLPs, all 10 proteins have been expressed in cultured mammalian cells. For example, Mena et al. used COS-1 cells, which were initially infected with recombinant vaccinia virus expressing the T7 RNA polymerase, and then transfected with plasmids to express all 10 influenza proteins [162]. Influenza VLPs resembling wild-type influenza virus were found in the culture medium. Further, these VLPs were also capable of encapsidating foreign gene and transferring it to fresh MDCK cells, in which expression of foreign gene could be detected.

Watanabe et al. generated influenza VLPs using nine proteins in transfected 293T cells [148]. NS2-knockout VLPs were injected into mice, which later were challenged with antigenically homologous influenza virus. Protective effect of such influenza VLPs has been detected in these experiments, demonstrating the potential of such non-replicating VLPs as a vaccine approach.

By using baculovirus expression system, influenza VLPs have been generated in insect Sf9 cells by using four structural proteins, HA, NA, M1, and M2 [146]. Furthermore, efficient formation of

TABLE 8.2
Examples of Native Influenza VLP Vaccines

Strain	Genes	Size (nm)	Expression	Testing, Model	References
H3N2	PB2, PB1, PA, NP, HA, NA, M1, M2, NS1, NS2	80–120	COS1	nt	[162,171]
H1N1	PB2, PB1, PA, NP, NA, NA, M1, M2, NS2	nt	293T	Mice	[148]
H3N2	HA, NA, M1, M2	~100	Sf9	nt	[146]
H9N2	HA, NA, M1	80–120	Sf9	Mice, rats, ferrets	[134,147]
H3N2	HA, M1	nt	Sf9	Mice	[166]
H1N1	HA, M1	80–120	Sf9	Mice	[167]
H3N2	HA, NA, M1	nt	Sf9	Mice, ferrets	[163]
H5N1	HA, NA, M1	nt	Sf9	Mice	[164]
H3N2	HA, NA	100	293T	nt	[168]

Note: nt, not tested.

influenza VLPs have been achieved in Sf9 cells following expression of only HA, NA, and M1 proteins from a single baculovirus construct [134,147,163]. For example, to generate H9N2 influenza VLPs in Sf9 cells, the HA, NA, and M1 genes were derived from influenza A/Hong Kong/1073/99 (H9N2) virus and introduced into recombinant baculovirus, each gene within its own expression cassette, which included a polyhedrin promoter and transcription termination sequences (Figure 8.3). As expected, the HA was expressed as HA0 in Sf9 cells, and no significant processing into HA1 and HA2 was observed. Influenza VLPs were purified from culture media by sucrose gradient centrifugation and presence of HA, NA, and M1 in VLPs was confirmed by SDS-PAGE, Western blot, hemagglutination assay, and NA enzyme activity assay. Electron microscopic examination of negatively stained samples revealed the presence of H9N2 VLPs with a diameter of approximately 80–120 nm, which showed surface spikes, characteristic of influenza HA protein on virions. We also observed that VLPs were associated frequently as groups (Figure 8.4).

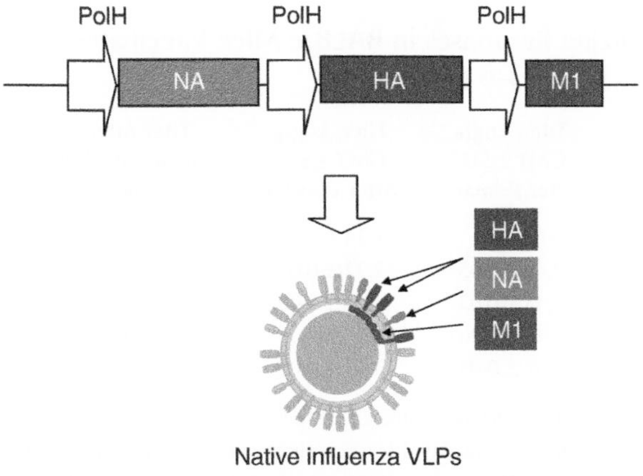

FIGURE 8.3 Baculovirus transfer vector for co-expression of influenza proteins and production of influenza native VLPs. Indicated are the polyhedrin promoter (PolH) and influenza genes. Positions of HA, NA, and M1 proteins on the surface of influenza VLPs are also indicated.

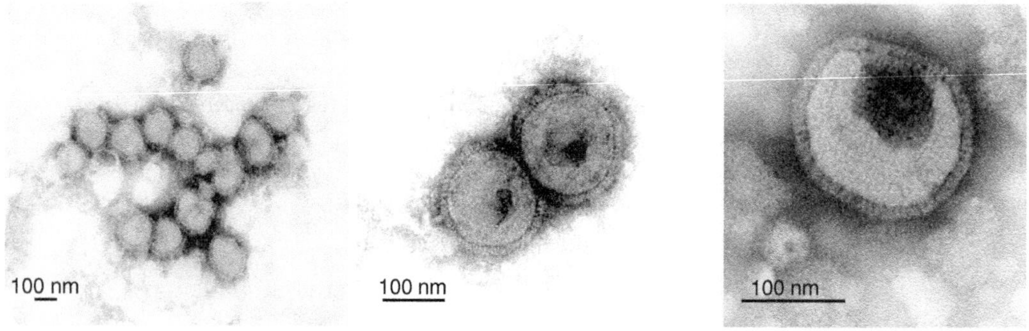

FIGURE 8.4 Negative staining electron microscopy of H9N2 influenza native VLPs. Influenza VLPs were generated from influenza A/Hong Kong/1073/99 (H9N2) HA, NA, and M1 proteins, as shown in Figure 8.3. Bars represent 100 nm. For electron microscopy, VLPs were adsorbed on freshly discharged plastic/carbon-coated grids and stained with 2% sodium phosphotungstate, pH 6.5. Stained VLPs were observed by transmission electron microscope at magnifications ranging from 6,000 to 100,000 x.

Such three-protein VLPs comprised of HA, NA, and M1-induced robust immune responses in preclinical studies [134,147,163]. Furthermore, in the recent study, we compared in BALB/c mice immunogenicity and protective capacity of baculovirus-generated VLPs, subunit HA antigen, and liposome-adjuvanted VLP and HA antigens as influenza vaccines. The following vaccine candidates were prepared (1) H9N2 influenza VLPs, (2) subunit recombinant HA0 protein (rH9), (3) rH9 combined with a nonphospholipid liposome adjuvant (Novasomes), and (4) H9N2 VLPs combined with Novasomes [134]. Animals were vaccinated twice i.m. with 10 μg doses of each influenza vaccine. Following a single inoculation, virus-neutralizing serum antibody was detected in all animals except mock controls, and booster inoculation increased the titers (Table 8.3). Novasomes increased immunogenicity of both VLP and rH9 antigens. This has been confirmed by ELISA as well as HAI assays. Vaccinated animals were challenged i.n. with A/Hong Kong/1073/99 virus, and we determined morbidity (measured by weight loss) as well as virus replication in the challenged mice. Average weight loss of up to 17% was observed in non-vaccinated control animals

TABLE 8.3
Virus Neutralizing Responses in BALB/c Mice Vaccinated with H9N2 VLPs and H9 Protein

Vaccine[a]	Titer, Log_{10} GMT ± SD, After Primary	Titer, Log_{10} GMT ± SD, After Booster	Titer After Primary, Pooled Serum	Titer After Booster, Pooled Serum
PBS Placebo	ND	1.14 ± 0.44	10	10
VLP	2.59 ± 0.05	2.72 ± 0.05	390	549
VLP+Novasomes	2.48 ± 0.03	3.11 ± 0.05	265	1338
rH9	2.44 ± 0.06	2.61 ± 0.05	294	378
rH9+Novasomes	2.82 ± 0.03	3.63 ± 0.02	639	4358

[a] BALB/c mice received vaccine inoculations or PBS as placebo on days 0 and 28. Virus-neutralizing (VN) titers were determined on day 14 after primary inoculation and on day 42 after booster inoculation. Influenza A/Hong Kong/1073/99 (H9N2) virus (100 $TCID_{50}$) was used for determination of VN titer. All sera were diluted 1:10 prior the assay. A titer of 10 indicates that the serum was negative for VN antibodies.

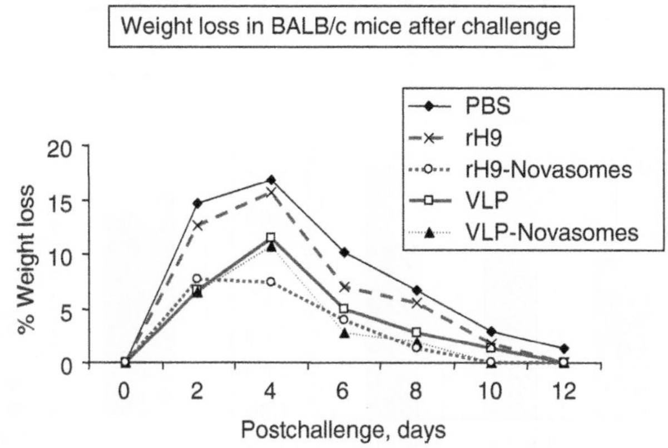

FIGURE 8.5 Weight loss in BALB/c mice vaccinated with H9N2 VLPs or rH9 antigens. Animals were vaccinated i.m. on days 0 and 21 with VLPs or rH9 antigen with or without Novasome adjuvant as indicated. Negative control animal were inoculated with PBS. Animals were challenged i.n. with 10^6 EID_{50} of influenza A/Hong Kong/1073/99 (H9N2) virus.

as well as in the animals vaccinated with rH9 subunit recombinant vaccine without adjuvant (Figure 8.5). However, animals vaccinated with VLPs with or without Novasomes showed only approximately 10% weight loss demonstrating protective effect of VLPs. High degree of protection was also observed in the animals, which received rH9 protein with Novasomes. Protection correlated with the virus-neutralizing titers (Table 8.3). Active replication of influenza virus was detected in lungs as well as in upper respiratory tissues of negative control animals at days 3 and 5 postchallenge (Figure 8.6). In contrast, a significant ($p < .05$) reduction of virus titers in the lung and nose tissues were observed among all vaccinated animals at day 3 postchallenge, as compared to non-vaccinated controls. Among vaccinated groups, the lowest protection was detected in animals vaccinated with subunit rH9 protein without adjuvant, which correlated with weight loss data. No virus was detected on day 5 postchallenge in nose tissues of all vaccinated animals. Taken

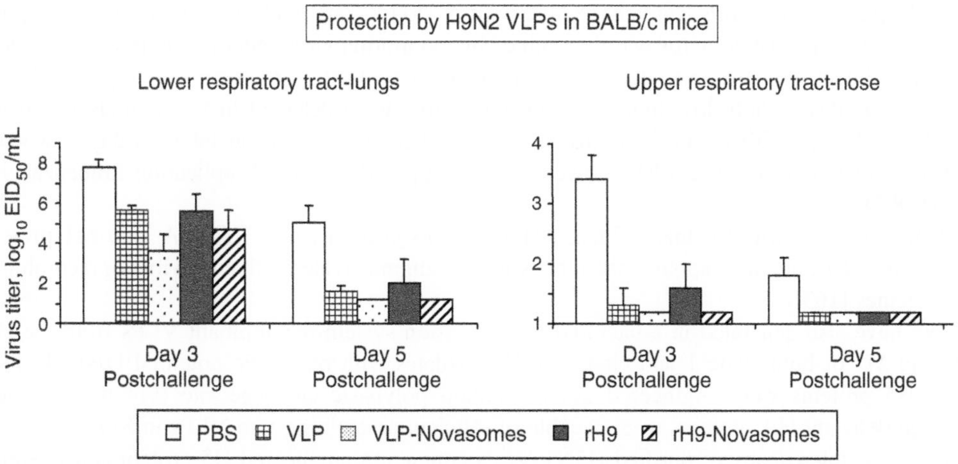

FIGURE 8.6 Titers of replicating influenza A/Hong Kong/1073/99 virus in lungs and nose tissues of vaccinated versus non-vaccinated BALB/c mice. Animals were vaccinated i.m. on days 0 and 21 with VLPs or rH9 antigen with or without Novasome adjuvant as indicated. Negative control animal were inoculated with PBS. Animals were challenged i.n. with 10^6 EID_{50} of influenza A/Hong Kong/1073/99 (H9N2) virus.

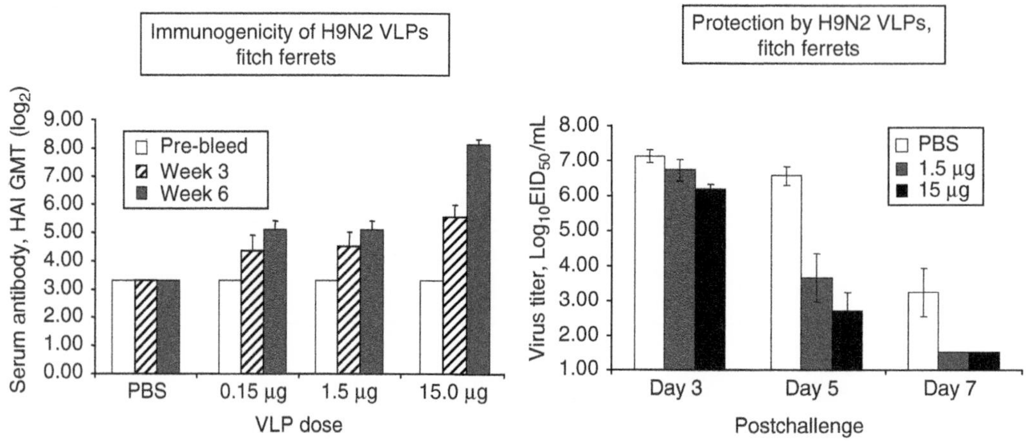

FIGURE 8.7 Immunogenicity and protective effect of H9N2 VLPs in Fitch ferrets. Left panel: Induction of serum HAI antibody in ferrets vaccinated with escalating doses of influenza H9N2 VLPs. Bars indicate standard deviation. GMT, geometric mean titer. Right panel: Protection of VLP-vaccinated ferrets against challenge with influenza A/Hong Kong/1073/99 (H9N2) virus. Shown are the titers of replicating virus in nose tissues of VLP-vaccinated (filled bars) and non-vaccinated (open bars) animals. Animals were challenged i.n. with 10^6 EID_{50} of influenza A/Hong Kong/1073/99 (H9N2) virus. Detection limit was 1.2 log_{10} EID_{50}/ml of virus. Also shown are standard deviation values.

together, these results suggest that VLPs have advantage over recombinant subunit rH9 in terms of both immunogenicity and protection. Interestingly enough, addition of liposome adjuvant significantly improved characteristics of rH9 as a vaccine. However, safety and long-term effects of the Novasomes as adjuvant require further studies [134].

Protective capacity of H9N2 influenza VLPs was also confirmed in ferrets [134]. Ferrets are considered to be the most suitable animal model for preclinical evaluation of human influenza viruses. Ferrets received primary and booster vaccinations with 0.15, 1.5, or 15 μg of HA antigen within the H9N2 VLPs. Control animals received PBS only. As little as 0.15 μg of VLPs induced detectable HAI antibody responses (Figure 8.7). Following vaccinations, animals that received 1.5 or 15 μg doses of VLPs were challenged with A/Hong Kong/1073/99 (H9N2) virus. Replication of the challenge H9N2 virus was determined by monitoring virus shedding in nasal washes after challenge. Infectious virus was detected in nasal washes of all animals on day 3 postchallenge. However, on day 5, only low titers of replicating virus were detected in the animals vaccinated with 1.5 or 15 μg of VLPs. In these groups, virus replication was undetectable on day 7, whereas control animals that received PBS showed over 3 log_{10} EID_{50}/mL of replicating influenza virus (Figure 8.7).

It has been shown also that influenza H3N2 VLPs comprised of HA, NA, and M1 proteins induced broader immune responses than the whole virion inactivated influenza virus or recombinant HA vaccines [163].

We have also generated in a baculovirus expression system recombinant VLPs from the HA, NA, and M1 of both clade 1 and clade 2 H5N1 isolates with pandemic potential [164]. In these cases, H5 proteins were engineered not to contain polybasic cleavage site (Figure 8.8). Such cleavage-deficient HA proteins were capable of efficient assembling into VLPs in Sf9 insect cells. Cryoelectron microscopy of purified H5N1 VLPs containing engineered H5 protein is also shown on Figure 8.8. VLP vaccines were purified and administered to mice in either a one-dose or two-dose regimen and the immune responses were compared to those induced by recombinant HA (rH5). Mice vaccinated with VLPs were protected against challenge regardless if the H5N1 clade was homologous or heterologous to the vaccine. However, rH5-vaccinated mice had significant

```
MEKIVLLLAI VSLVKSDQIC IGYHANNSTE QVDTIMEKNV TVTHAQDILE KTHNGKLCDL
DGVKPLILRD CSVAGWLLGN PMCDEFINVP EWSYIVEKAN PTNDLCYPGS FNDYEELKHL
LSRINHFEKI QIIPKSSWSD HEASSGVSSA CPYLGSPSFF RNVVWLIKKN STYPTIKKSY
NNTNQEDLLV LWGIHHPNDA AEQTRLYQNP TTYISIGTST LNQRLVPKIA TRSKVNGQSG
RMEFFWTILK PNDAINFESN GNFIAPEYAY KIVKKGDSAI MKSELEYGNC NTKCQTPMGA
INSSMPFHNI HPLTIGECPK YVKSNRLVLA TGLRNSPQRE S----RGLFG AIAGFIEGGW
QGMVDGWYGY HHSNEQGSGY AADKESTQKA IDGVTNKVNS IIDKMNTQFE AVGREFNNLE
RRIENLNKKM EDGFLDVWTY NAELLVLMEN ERTLDFHDSN VKNLYDKVRL QLRDNAKELG
NGCFEFYHKC DNECMESIRN GTYNYPQYSE EARLKREEIS GVKLESIGTY QILSIYSTVA
SSLALAIMMA GLSLWMCSNG SLQCRICI*
```

(A)

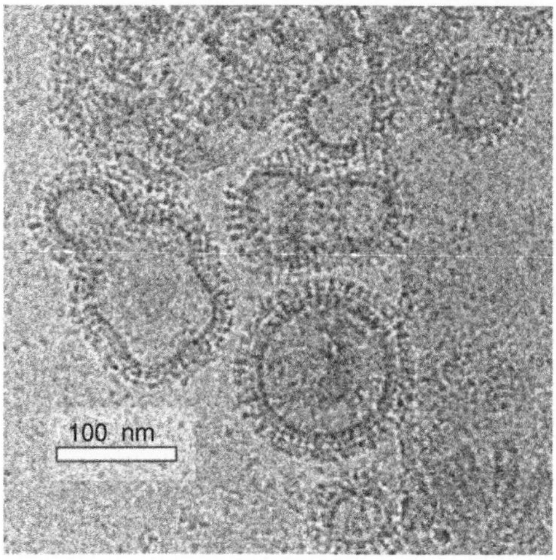

(B)

FIGURE 8.8 (A) Amino acid sequence of engineered H5 protein (HA0 form) with the deleted polybasic cleavage site (shadowed) between HA1 and H2 domains. Also shown are predicted signal peptide (underlined) and transmembrane domain (highlighted). (B) Cryoelectron electron microscopy of H5N1 influenza VLPs containing engineered H5 protein. Native influenza VLPs were generated from influenza A/Indonesia/5/05 (H5N1) HA, NA, and M1 proteins. The HA protein was engineered to lack the multibasic cleavage site. Bar represents 100 nm.

weight loss and death following challenge with the heterologous clade virus. The association rate of antibody binding to HA correlated with protection and was enhanced using H5N1 VLPs, particularly when delivered intranasally, compared to rH5 vaccines. The results showed that native H5N1 VLPs are effective influenza vaccine immunogens that elicit cross-clade protective immune responses to emerging H5N1 influenza isolates [164].

Three-protein VLPs comprised of HA, NA, and M1, were also generated in COS-1 cells transfected with expression plasmids [165]. Further, influenza VLPs have also been successfully generated in Sf9 cells from the HA and M1 proteins only (Table 8.2), and such two-protein VLPs also induced detectable protective immune responses in mice [166,167].

In the experiments that involved Sf9 cells and recombinant baculoviruses, two strategies have been used for expression of influenza VLPs. In one strategy, influenza proteins were co-expressed from a single recombinant baculovirus [134,146,147,163]. The other strategy involved coinfection of the Sf9 cells with the two recombinant baculoviruses, one expressing the HA protein, whereas the other recombinant baculovirus expressed the M1 protein [166,167].

Most of the approaches for expressing influenza VLPs included M1 protein, because M1 is the most abundant protein in the virion and because it has been shown to be the driving force of influenza virus budding [165]. However, recently, two-protein influenza VLPs were also generated that consisted only from HA and NA (Table 8.2) and did not include M1 [168]. HA protein, when expressed in noncytotoxic mammalian cell culture and treated with exogenous NA or co-expressed with viral NA, could be released from cells independently of M1. Incorporation of M1 into VLPs required HA expression, although when M1 was omitted from the VLPs, particles with morphologies similar to those of wild-type viruses were also observed.

Taken together, the data demonstrated that native influenza VLPs can be generated in eukaryotic cell culture expression systems. Furthermore, in several preclinical studies that involved various animal models, native influenza VLP vaccines have shown excellent characteristics as influenza vaccines. Influenza VLPs induced broad and robust immune responses as well as efficient protection against homologous and heterologous live influenza virus challenges. Novavax, Inc. recently announced initiation of the first clinical studies involving native influenza pandemic H5N1, clade 2 VLP vaccine [169].

The expression of native influenza VLPs also contributes to better understanding of influenza virus assembly and morphogenesis. Virion particle contains nine structural proteins. Studies with VLPs demonstrated that expression of M1 alone forms VLPs and that the expression of M1 can drive HA and NA into budding VLPs that are released into culture medium [146,165]. However, an alternative influenza assembly model suggests that not the M1, but envelope glycoproteins control influenza virus budding by sorting to lipid raft microdomains and recruiting the internal viral core components. Recent demonstration of VLPs comprised of HA and NA is consistent with the latter model [168].

VLPs comprised of only two proteins, HA and M1, morphologically resemble wild-type influenza virions including characteristic surface spikes, and are capable of inducing protection against lethal challenge with influenza in BALB/c mice [166,167]. Presence of additional influenza proteins, such as NA, M1, and M2, in the VLPs [146,147,163] may provide additional benefits including inducing broader immune responses, especially in outbred populations, because the NA, M1, and M2 proteins contain additional antigenic epitopes [55]. However, co-expression of several proteins in the cells during manufacturing process may reduce the overall levels of expression of VLPs due to promoter dilution effect or because of excessive metabolic burden [170]. Furthermore, over-expression of M2 may also dramatically decrease the yields of VLPs [171]. The authors hypothesized that over-expression of M2, an ion channel protein, inhibited intracellular transport and drastically reduced accumulation of co-expressed HA and hence reduced the number of VLPs. It appears that the optimal composition of native VLPs still needs to be determined, which would ensure efficient protective characteristics and high levels of production of VLP-based influenza vaccines. Purification of VLPs may represent another potential challenge. Also, it remains to be determined if any molecules such as nucleic acids or host cell proteins are incorporated into the influenza VLPs and whether these play a role in the assembly or whether presence of these molecules may affect the use of VLPs as vaccines from the safety/regulatory points of view.

8.6.3.2 Protein Engineering of Chimeric VLPs Containing Influenza Epitopes

The structural components of native VLPs derived from many viruses have proven amenable to the insertion of foreign antigenic sequences, allowing protein engineering of "chimeric" VLPs that expose foreign antigens on their surface. In other words, many VLPs can be exploited as "carriers" or "platforms" for the presentation of foreign epitopes or targeting molecules on chimeric VLPs [149,150,172–174]. The development of such chimeric VLPs on the basis of VLP carriers is described in detail in the Chapter 9. This method can significantly improve immunogenicity of peptide epitopes. In some cases, foreign peptide epitopes or the entire proteins, or even nonprotein antigens may be chemically conjugated to preformed VLPs [174]. Alternatively, foreign antigen may be exposed on the surface of VLPs via modification of the VLP gene sequences, such that fusion proteins comprised of VLP protein and foreign antigen are assembled into VLPs during de novo synthesis [175]. In this case, it is important that foreign epitope is inserted so that it does not interfere with the assembly process of the subunits of carrier VLPs [176]. Methods for mapping protein domains that are exposed on the surface or internalized within VLPs, have been developed [173,177,178]. Also, three-dimensional structures for many viruses have been determined by cryoelectron microscopy, x-ray crystallography, or a combination of both methods, which greatly facilitates rational design of the proteins that assemble into chimeric VLPs [179–182].

Because chimeric VLPs consist of a core platform VLP as well as of the antigenic epitopes of interest, administration of such chimeric VLPs in vivo usually induces responses to the antigenic epitopes of interest as well as to the epitopes within carrier VLPs.

VLPs derived from several viruses have been configured to carry and display influenza antigenic epitopes, especially fragments derived from M2 protein. Extracellular part of M2 protein, ~23 aa residues, is highly conserved in known human influenza A strains. Therefore, M2-based vaccine may be able to protect from all human influenza A strains that could result in a "universal" influenza vaccine and a major improvement over currently used vaccines.

As mentioned above, foreign epitopes can be fused to carrier VLP through either chemical conjugation or genetic fusion. An example of conjugated vaccine is represented by HPV VLPs, to which extracellular domain of influenza M2 protein was chemically conjugated. Conjugates comprised approximately 4,000 copies of the antigenic peptide per VLP. Such chimeric M2-HPV VLPs were in average larger in size as compared to HPV VLP carrier alone. Such conjugate vaccine has been formulated with adjuvant and administered to mice, in which it induced immune responses and protection against lethal challenge with influenza virus [174].

Genetic fusion of influenza M2 protein sequences with the core gene (HBcAg) of HBV provides an example of an alternative approach to generate chimeric M2-HBcAg VLPs. The M2 gene fragment corresponding to extracellular domain was fused to the HBcAg gene to create fusion gene coding for M2-HBc VLPs. Administration of M2-HBc VLPs purified from *E. coli* to mice provided up to 100% protection against a lethal virus challenge. The protection appeared to be mediated by antibodies, as it was transferable using serum [175]. However, it appears that such M2-specific antibodies neither bind efficiently to the free influenza virus nor neutralize virus infection, but bind to M2 protein expressed on the surface of virus-infected cells. Therefore, the M2 antiserum does not prevent infection but only reduces disease at low challenge doses. At higher challenge doses, the M2 antiserum fails to protect mice [183]. Attempts to improve vaccine characteristics of chimeric M2-HBc VLPs were carried out by optimizing the antigen design by either displaying more copies of the M2 polypeptide, or by inserting the M2 polypeptide into different sites of the HBcAg carrier, the amino-terminus or the immunodominant loop. Such engineered variants showed promise after i.n. coadministration with cholera toxin A1, a powerful vaccine adjuvant [184,185].

Further, characteristics of chemically conjugated chimeric M2-HBc VLPs were compared to those of genetically fused M2-HBc VLPs. The conjugated M2 appeared to be better accessible on the surface of HBc VLPs and induced stronger immune response than genetically fused M2 [186].

In another study, chimeric VLPs were generated, which were formed by using retrovirus Gag protein core and the complete influenza HA, NA, and M2 proteins [187]. By using this approach, influenza VLPs were generated that contained retroviral core and the outer surface that mimicked properties of the influenza viral surface of two highly pathogenic influenza viruses of either H7N1 or H5N1 antigenic subtype. Such chimeric VLPs induced high-titer neutralizing antibodies in mice [187]. The phenomenon of generating influenza VLPs using core structure from unrelated viruses deserves further study and may provide important information regarding mechanism of influenza particle assembly and the specificity of interactions between the core and envelope proteins in such chimeric VLPs. Recently, it has been shown that VLPs of brome mosaic virus, a non-enveloped icosahedral virus, can be generated that contain spherical gold nanoparticle cores. Interestingly enough, variation of the gold core diameter provided control over the VLP structure, as the number of subunits required for a complete particle increased with the core diameter [188]. This technology could also be useful for studying, and possibly controlling, the assembly process of influenza VLPs.

Thus, many types of chimeric VLPs expressing influenza antigens have been generated using protein engineering methods and showed promise as vaccines in preclinical models. However, the limited number and the relatively small size of influenza antigens that can be incorporated into some VLPs may limit the practical utility of many chimeric influenza VLP approaches, while manufacturing and regulatory considerations may also prove to be a significant barrier to vaccine development. The first clinical study for a M2-based universal influenza vaccine has been announced [189].

8.6.3.3 Chimeric Influenza VLPs as a Carrier of Foreign Epitopes

The available data suggest that not only influenza epitopes can be displayed on the surface of carrier VLPs derived from unrelated viruses but also influenza VLPs themselves may also be engineered as a carrier for foreign antigenic epitopes. For example, it has been shown that influenza VLPs are capable of incorporating vesicular stomatitis virus (VSV) G protein [146]. In another study, *Bacillus anthracis* protective antigen (BPA) fragments, 90 or 140 aa in length, were inserted at the C-terminal flank of the HA signal peptide and expressed as the HA1 subunit. The chimeric proteins could be cleaved into the HA1 and HA2 subunits by trypsin and could be also incorporated into recombinant influenza viruses suggesting that viral envelope can tolerate foreign inserts without precluding assembly. The inserted BPA domains were maintained in the HA gene segments following several passages in MDCK cells or embryonated chicken eggs. Immunization of mice with either recombinant viruses or with DNA plasmids that express the chimeric BPA/HA proteins induced antibody responses against both the HA and BPA components of the protein [190]. Although VLPs were not generated in this study, these experiments suggests that similar modifications of HA protein with foreign epitopes may be compatible with the formation of chimeric influenza VLPs.

8.7 INFLUENZA AS A GENETIC VIRUS VECTOR

Using molecular engineering technologies, a concept of using influenza as a genetic vector has been applied to the development of vaccine candidates for several pathogens [191,192]. Both propagation-competent and propagation-incompetent influenza virus vectors have been described. Strategies for the construction of propagation-competent influenza vectors included the insertion of foreign antigenic epitopes into influenza virus glycoproteins [190], rescue of bicistronic genes into infectious viruses, and the expression of polyproteins. Influenza virus vectors have been obtained, which express both B- and T-cell epitopes from different pathogens. These constructs have been shown to induce in vaccinated animals systemic and local antibody responses, and cytotoxic T-cell responses against the expressed antigenic epitopes [191].

For example, vaccination of mice with recombinant influenza and vaccinia viruses expressing antigens from *Plasmodium yoelii* resulted in a significant protective immune response against malaria in this model. Mice immunized with recombinant influenza viruses expressing human immunodeficiency virus (HIV) epitopes generated long-lasting HIV-specific serum antibody response as well as secretory IgA in the nasal, vaginal, and intestinal mucosa [192].

Evaluation of propagation-incompetent influenza vectors has been also initiated. As mentioned in the previous chapters, in the COS-1 cells expressing all 10 influenza virus-encoded proteins, the transfected CAT RNA could be rescued into influenza VLPs that were budded into the supernatant fluids [162]. The released VLPs not only resembled influenza virions but also transferred the encapsidated CAT RNA to MDCK cell cultures. Such VLPs required trypsin treatment to deliver the RNA to fresh cells and could be neutralized by a monoclonal antibody specific for the influenza A virus HA. These data indicated that influenza VLPs are capable of encapsidating a synthetic virus-like RNA, which can be delivered to fresh cells for expression of foreign gene of interest. For other virus vectors, it has been shown that such "vector VLPs" encapsidate nucleic acids by utilizing the ability of viral structural proteins to recognize specific encapsidation signals within the nucleic acid sequences. As with other propagation-incompetent virus vectors, the advantage of such vectors is that when injected in a susceptible host, they can generate immune response almost exclusively against the foreign antigen expressed from the vector, whereas low, if any, immune response is generated against the structural proteins of the vector itself [115]. However, more experiments are needed to demonstrate the potential of influenza VLPs as a vaccine vector system.

8.8 CONCLUSION

Every year, influenza virus causes up to 500,000 of vaccine-preventable deaths in the world [8]. Furthermore, a constant menacing threat of a world pandemic has a potential to cause even a larger numbers of deaths and illnesses over a short period than any other known health threat. Because vaccination is a powerful and cost-effective countermeasure to the threat of either seasonal or pandemic outbreaks of influenza, vaccines against influenza viruses have been developed and licensed during the last decades in the United States, Europe, and other countries. The existing vaccines have reduced significantly the negative impact of influenza disease on public health and world economies.

However, several important issues still remain to be addressed to improve influenza vaccines. As discussed in detail in the previous chapters, these issues include suboptimal efficacy of the existing vaccines in the young and elderly patients, reactogenicity in some people, and the difficulty to accommodate rapid changes in the vaccine composition in response to, and in order to match, the changing circulating influenza virus strains. Many of these issues stem from using relatively inefficient, chicken egg-based technology for the production of the majority of licensed vaccines. Potentially, limitations in vaccine efficacy, safety, and manufacturing may cause, and have caused in the past, suboptimal vaccine efficacy and reduced protection of the population, as well as delays in supply, and even shortages of the vaccines. Such limitations may become insurmountable in the case of a pandemic, which can result in significant loss of human lives and massive economic damage.

Significant efforts have been undertaken by both vaccine manufacturers as well as research community to address these important issues. Among strategies for improvement of existing vaccines are the attempts to adopt cultured cells instead of chicken eggs for vaccine manufacturing, which could bring significant improvements to both the vaccine production and vaccine safety. Further, faster and newer methods of generating vaccine seed viruses are being implemented such as using the reverse genetics technology. In addition, steps are being taken for improvements in virus growth, virus inactivation, and vaccine purification methods. Alternative vaccine inoculation schedules and advanced adjuvants are also being evaluated to improve efficacy and safety of influenza vaccines by possibly using smaller vaccine doses. A promising way of improving vaccination against a respiratory virus such as influenza could be mucosal, such as i.n., vaccine administration. Inactivated influenza vaccines are currently administered i.m. to patients. In general, i.m. vaccinations result in the elicitation of effective virus-neutralizing serum antibody response and effective protection against potentially severe or even fatal consequences. Live attenuated FluMist vaccine is administered i.n. and appears to induce a more balanced immune response. The advantage of i.n. administration could be the elicitation of effective mucosal IgA that may be able to bind and stop the incoming virus in the upper nasal pharyngeal area and thus prevent its passage to the lower respiratory tract.

In addition to improvement of current vaccines, new vaccine candidates are also being actively developed using molecular technologies including protein engineering. Novel vaccine candidates encompass every approach available today, ranging from synthetic peptides to the whole inactivated viruses as well as to recently developed influenza VLPs, which mimic whole influenza viruses. No stone is left unturned in influenza vaccine research, and virtually every new concept or technology is being applied or adapted for influenza research or developed specifically for influenza research and vaccine development. The impact of influenza virus and influenza disease on public health and economy as well as the significant market for influenza vaccines provide an impetus to these efforts.

This concerted research resulted in a considerable progress in our understanding of influenza virus and the disease as well as of influenza vaccines. We can use this knowledge to further advance influenza research and to create vaccines that are safer and even more effective than the ones we are currently using. Importantly, the requirements for inducing effective protective immunity as well as critically important protective antigens have been identified during the past years of influenza vaccine research. The available data strongly suggest that HA, the major surface envelope glycoprotein, is the most important antigen in the currently used influenza vaccines. Presence of functionally active HA is the standard requirement for all currently used inactivated influenza vaccines, because eliciting of virus-neutralization response correlates with the presence of functionally active HA in the vaccine. It appears that other antigenic components of the vaccine are of lesser clinical importance for protection. For example, NA, another influenza surface antigen, appears not to significantly affect protection [130], although in other studies, some protective effect of NA has been implicated retrospectively [13]. The protective potential of NA as well as of other influenza antigens has also been confirmed in numerous preclinical studies.

The available comparative studies of different inactivated vaccines have demonstrated that different vaccines have varying immunogenic characteristics. For example, the whole virus vaccines appear to have higher immunogenicity when compared to split vaccines [64], and virosome vaccines appear to have higher immunogenicity when compared to subunit vaccine [132]. It is plausible to suggest that the higher level of immunogenicity depends on the amount of correctly folded and antigenically active HA. Correct conformation of HA within a lipid membrane is also required for fusion activity [193]. Functional activity of HA, such as the ability to agglutinate erythrocytes, also depends on the correct folding of the HA protein. The higher is the chance of correctly folded HA in vaccine, the higher is the immunogenicity and protective capacity of HA. Our data demonstrated that baculovirus-generated influenza VLPs have higher immunogenicity profile in mice as compared to baculovirus-generated subunit recombinant HA [134]. Interestingly enough, immunogenicity of the subunit HA could be drastically improved by adding Novasomes, which represented non-phospolipid liposomes [134] Further, immunogenicity of split vaccines can also be significantly improved by adding lipid-containing liposomes or other lipid-based adjuvants, such as MF59, an oil-in-water emulsion containing unsaturated aliphatic hydrocarbon squalene [133]. Virosome vaccines representing liposome-embedded subunits of HA represent an example of highly immunogenic influenza vaccines. Thus, highly immunogenic influenza vaccines including whole-virus vaccines, VLPs, and virosomes have lipoprotein membrane or other lipid components as an integral part of the vaccine, suggesting that lipids play an important role in the correct conformation and functional activity of HA. Presence of lipid components appears to result in higher overall immunogenicity and protective efficacy of the vaccines. The reasons for such importance of lipid components in the correct conformation of HA probably stem from the intimate involvement of lipid-containing cellular membranes in the life cycle of influenza virus including budding of progeny particles from the infected cells. In the influenza virus, HA and NA proteins are embedded in the outer envelope, which is derived from cell plasma membrane during budding. The plasma membrane is not homogeneous, and the lipids of the plasma membrane are distributed non-randomly and can self-associate and organize into a liquid ordered phase. Sphingolipids and cholesterol contribute to cellular membrane organization by packing densely to form microdomains in the plasma membrane commonly called lipid rafts. These specialized membrane regions are believed to be involved in the budding of many enveloped viruses including influenza virus. Influenza virus HA and NA are known to associate with lipid rafts, and clustering of HA within the rafts is an intrinsic property of the HA protein. NA is also found sequestered within the same microdomains as HA, whereas the M2 ion channel protein does not concentrate within the raft-like microdomains. Depletion of cholesterol from cells decreased the diameter of the HA clusters [194].

Such intimate involvement of plasma membrane in the distribution of HA in the membrane may also trigger the HA folding or stabilize correct conformation of functionally active HA. Alternatively, lipid membrane may contribute to the advantageous presentation of influenza antigens, which can result in better antigen processing [64]. Absence of lipid component in a vaccine, such as in the case of subunit vaccines, results in significant losses of immunogenic activity and require higher doses to induce protection. However, immunogenicity of the subunit vaccines can be restored by adding lipids, as in the case of liposomes/virosomes. Taken together, these observations strongly support the idea of important role of lipids for optimal immunogenicity of influenza vaccines. This phenomenon requires further studies and comparison of the activity of HA protein as well as of the protein and lipid content in the various vaccines. Presence of the lipids may also be important for optimal immunogenicity of influenza vaccines generated using protein engineering methods, such as chimeric VLPs. It is clear that despite tremendous progress in our understanding of influenza virus and the disease, there are still many questions left unanswered and research remains to be done in order to better understand the virus and the disease and to develop better vaccines for this important human pathogen.

ACKNOWLEDGMENTS

We thank Kutub Mahmood, Louis Potash, Paul Pumpens, and Robin Robinson for reading the manuscript and stimulating discussions, and Terje Dokland for cryoelectron microscopy.

REFERENCES

1. Potter, C.W., Chronicle of influenza pandemics. *Textbook of Influenza*, R.G. Webster, K.G. Nicholson, and A.J. Hay (Eds.), 1998, Oxford, Blackwell Science, pp. 3–18.
2. Webby, R.J. and R.G. Webster, Emergence of influenza A viruses. *Philos Trans R Soc Lond B Biol Sci*, 2001, 356(1416): 1817–1828.
3. Smith, W., C.H. Andrewes, and P.P. Laidlaw, A virus obtained from influenza patients. *Lancet*, 1933, 2: 66–68.
4. Horsfall, F.L.J., E.H. Lennette, E.R. Rickard, C.H. Andrewes, W. Smith, and C.H. Stuart-Harris, The nomenclature of influenza. *Lancet*, 1940, ii: 413.
5. Cox, N.J. and K. Subbarao, Influenza. *Lancet*, 1999, 354(9186): 1277–1282.
6. Brankston, G., et al., Transmission of influenza A in human beings. *Lancet Infect Dis*, 2007, 7(4): 257–265.
7. CDC, http://www.cdc.gov/flu/keyfacts.htm 2007, Centers for Disease Control and Prevention.
8. WHO, http://www.who.int/mediacentre/factsheets/2003/fs211/en/print.html:2004. World Health Organization, Influenza Fact Sheet N211.
9. Taubenberger, J.K., et al., Integrating historical, clinical and molecular genetic data in order to explain the origin and virulence of the 1918 Spanish influenza virus. *Philos Trans R Soc Lond B Biol Sci*, 2001, 356 (1416): 1829–1839.
10. de Jong, M.D. and T.T. Hien, Avian influenza A (H5N1). *J Clin Virol*, 2006, 35(1): 2–13.
11. Johnson, N.P. and J. Mueller, Updating the accounts: Global mortality of the 1918–1920 Spanish influenza pandemic. *Bull Hist Med*, 2002, 76(1): 105–115.
12. Tumpey, T.M., et al., Characterization of the reconstructed 1918 Spanish influenza pandemic virus. *Science*, 2005, 310(5745): 77–80.
13. Lipatov, A.S., et al., Influenza: Emergence and control. *J Virol*, 2004, 78(17): 8951–8959.
14. CDC, Update: Influenza activity-United States and Worldwide, 2005–06 season, and composition of the 2006–07 influenza vaccine. 2006, 55: 648–653.
15. WHO, http://www.who.int/csr/disease/avian_influenza/country/cases_table_2008_05_28/en/print.html. 2008, World Health Organization.
16. Peiris, M., et al., Human infection with influenza H9N2. *Lancet*, 1999, 354(9182): 916–917.
17. Koopmans, M., et al., Transmission of H7N7 avian influenza A virus to human beings during a large outbreak in commercial poultry farms in the Netherlands. *Lancet*, 2004, 363(9409): 587–593.

18. Palese, P., Influenza: Old and new threats. *Nat Med*, 2004, 10(Suppl 12): S82–S87.
19. Palese, P., Making better influenza virus vaccines? *Emerg Infect Dis*, 2006, 12(1): 61–65.
20. Wright, P., Orthomyxoviruses. *Fields Virology*, D.M. Knipe and Howley, P. (Eds.), 2002: Wiliams and Wilkins, Lippincott, pp. 1533–1579.
21. Palese, P.S. and M.L. Shaw, Orthomyxoviridae: The viruses and their replication, In: *Fields Virology, Fifth Edition*, D.M. Knipe and P.M. Howley, (Eds.), 2007, Lippincott Williams & Wilkins: Philadelphia.
22. Arvin, A.M. and H.B. Greenberg, New viral vaccines. *Virology*, 2006, 344(1): 240–249.
23. Glezen, W.P., et al., Epidemiologic observations of influenza B virus infections in Houston, Texas. 1976–1977. *Am J Epidemiol*, 1980, 111(1): 13–22.
24. Hite, L.K., et al., Medically attended pediatric influenza during the resurgence of the Victoria lineage of influenza B virus. *Int J Infect Dis*, 2007, 11(1): 40–47.
25. Moriuchi, H., et al., Community-acquired influenza C virus infection in children. *J Pediatr*, 1991, 118(2): 235–238.
26. Manuguerra, J.C., C. Hannoun, and M. Aymard, Influenza C virus infection in France. *J Infect*, 1992, 24 (1): 91–99.
27. Kimura, H., et al., Interspecies transmission of influenza C virus between humans and pigs. *Virus Res*, 1997, 48(1): 71–79.
28. Joosting, A.C., et al., Production of common colds in human volunteers by influenza C virus. *Br Med J*, 1968, 4(5624): 153–154.
29. Yamada, S., et al., Haemagglutinin mutations responsible for the binding of H5N1 influenza A viruses to human-type receptors. *Nature*, 2006, 444(7117): 378–382.
30. Russell, R.J., et al., The structure of H5N1 avian influenza neuraminidase suggests new opportunities for drug design. *Nature*, 2006, 443(7107): 45–49.
31. Horimoto, T. and Y. Kawaoka, Pandemic threat posed by avian influenza A viruses. *Clin Microbiol Rev*, 2001, 14(1): 129–149.
32. Palese, P., et al., Characterization of temperature sensitive influenza virus mutants defective in neuraminidase. *Virology*, 1974, 61(2): 397–410.
33. Hay, A.J., et al., The molecular basis of the specific anti-influenza action of amantadine. *Embo J*, 1985, 4(11): 3021–3024.
34. Monto, A.S. and N.H. Arden, Implications of viral resistance to amantadine in control of influenza A. *Clin Infect Dis*, 1992, 15(2): 362–367; discussion 368–369.
35. Palese, P. and R.W. Compans, Inhibition of influenza virus replication in tissue culture by 2-deoxy-2,3-dehydro-N-trifluoroacetylneuraminic acid (FANA): Mechanism of action. *J Gen Virol*, 1976, 33(1): 159–163.
36. Kiso, M., et al., Resistant influenza A viruses in children treated with oseltamivir: Descriptive study. *Lancet*, 2004, 364(9436): 759–765.
37. Le, Q.M., et al., Avian flu: Isolation of drug-resistant H5N1 virus. *Nature*, 2005, 437(7062): 1108.
38. Schultz-Cherry, S., et al., Influenza virus ns1 protein induces apoptosis in cultured cells. *J Virol*, 2001, 75(17): 7875–7881.
39. Seo, S.H., E. Hoffmann, and R.G. Webster, Lethal H5N1 influenza viruses escape host anti-viral cytokine responses. *Nat Med*, 2002, 8(9): 950–954.
40. Hatta, M. and Y. Kawaoka, The NB protein of influenza B virus is not necessary for virus replication in vitro. *J Virol*, 2003, 77(10): 6050–6054.
41. Guillot, L., et al., Involvement of toll-like receptor 3 in the immune response of lung epithelial cells to double-stranded RNA and influenza A virus. *J Biol Chem*, 2005, 280(7): 5571–5580.
42. Barchet, W., et al., Dendritic cells respond to influenza virus through TLR7- and PKR-independent pathways. *Eur J Immunol*, 2005, 35(1): 236–242.
43. Hayden, F.G., et al., Local and systemic cytokine responses during experimental human influenza A virus infection. Relation to symptom formation and host defense. *J Clin Invest*, 1998, 101(3): 643–649.
44. de Jong, M.D., et al., Fatal outcome of human influenza A (H5N1) is associated with high viral load and hypercytokinemia. *Nat Med*, 2006, 12(10): 1203–1207.
45. Garcia-Sastre, A., et al., The role of interferon in influenza virus tissue tropism. *J Virol*, 1998, 72(11): 8550–8558.
46. He, X.S., et al., T cell-dependent production of IFN-gamma by NK cells in response to influenza A virus. *J Clin Invest*, 2004, 114(12): 1812–1819.

47. Abe, T., et al., Baculovirus induces an innate immune response and confers protection from lethal influenza virus infection in mice. *J Immunol*, 2003, 171(3): 1133–1139.
48. Hervas-Stubbs, S., et al., Insect baculoviruses strongly potentiate adaptive immune responses by inducing type I IFN. *J Immunol*, 2007, 178(4): 2361–2369.
49. Brown, D.M., E. Roman, and S.L. Swain, CD4 T cell responses to influenza infection. *Semin Immunol*, 2004, 16(3): 171–177.
50. Powell, T.J., et al., $CD8^+$ T cells responding to influenza infection reach and persist at higher numbers than $CD4^+$ T cells independently of precursor frequency. *Clin Immunol*, 2004, 113(1): 89–100.
51. Thomas, P.G., et al., Cell-mediated protection in influenza infection. *Emerg Infect Dis*, 2006, 12(1): 48–54.
52. He, X.S., et al., Analysis of the frequencies and of the memory T cell phenotypes of human $CD8^+$ T cells specific for influenza A viruses. *J Infect Dis*, 2003, 187(7): 1075–1084.
53. Sasaki, S., et al., Comparison of the influenza virus-specific effector and memory B-cell responses to immunization of children and adults with live attenuated or inactivated influenza virus vaccines. *J Virol*, 2007, 81(1): 215–228.
54. Zeman, A.M., et al., Humoral and cellular immune responses in children given annual immunization with trivalent inactivated influenza vaccine. *Pediatr Infect Dis J*, 2007, 26(2): 107–115.
55. Bui, H.H., et al., Ab and T cell epitopes of influenza A virus, knowledge and opportunities. *Proc Natl Acad Sci U S A*, 2007, 104(1): 246–251.
56. Scherle, P.A., G. Palladino, and W. Gerhard, Mice can recover from pulmonary influenza virus infection in the absence of class I-restricted cytotoxic T cells. *J Immunol*, 1992, 148(1): 212–217.
57. Harper, S.A., et al., Prevention and control of influenza. Recommendations of the Advisory Committee on Immunization Practices (ACIP). *MMWR Recomm Rep*, 2005, 54(RR-8): 1–40.
58. Wood, J.M., Standardization of inactivated influenza vaccines. *Textbook of Influenza*, R.G. Webster, K.G. Nicholson, and A.J. Hay (Eds.), 1998, Oxford, Blackwell Science, pp. 333–345.
59. Wood, J.M. and R.A. Levandowski, The influenza vaccine licensing process. *Vaccine*, 2003, 21(16): 1786–1788.
60. van Essen, G.A., et al., Influenza vaccination in 2000: Recommendations and vaccine use in 50 developed and rapidly developing countries. *Vaccine*, 2003, 21(16): 1780–1785.
61. Gerin, J.L. and N.G. Anderson, Purification of influenza virus in the K-II zonal centrifuge. *Nature*, 1969, 221(5187): 1255–1256.
62. Barry, D.W., et al., Comparative trial of influenza vaccines. I. Immunogenicity of whole virus and split product vaccines in man. *Am J Epidemiol*, 1976, 104(1): 34–46.
63. Gross, P.A., et al., A controlled double-blind comparison of reactogenicity, immunogenicity, and protective efficacy of whole-virus and split-product influenza vaccines in children. *J Infect Dis*, 1977, 136(5): 623–632.
64. Hovden, A.O., R.J. Cox, and L.R. Haaheim, Whole influenza virus vaccine is more immunogenic than split influenza virus vaccine and induces primarily an IgG2a response in BALB/c mice. *Scand J Immunol*, 2005, 62(1): 36–44.
65. Müller, M., et al. Safety and immunogenicity of a cell-culture (Vero) derived whole virus H5N1 vaccine. In: *International Meeting on Emerging Diseases and Surveillance*, 2007, Vienna, Austria.
66. Davenport, F.M., et al., Comparisons of serologic and febrile responses in humans to vaccination with influenza A viruses or their hemagglutinins. *J Lab Clin Med*, 1964, 63: 5–13.
67. Treanor, J.J., et al., Rapid licensure of a new, inactivated influenza vaccine in the United States. *Hum Vaccin*, 2005, 1(6): 239–244.
68. FDA_News. 2007. Available from: http://www.fda.gov/bbs/topics/NEWS/2007/NEW01611.html
69. Cox, R.J., K.A. Brokstad, and P. Ogra, Influenza virus: Immunity and vaccination strategies. Comparison of the immune response to inactivated and live, attenuated influenza vaccines. *Scand J Immunol*, 2004, 59(1): 1–15.
70. Belshe, R., et al., Safety, immunogenicity and efficacy of intranasal, live attenuated influenza vaccine. *Expert Rev Vaccines*, 2004, 3(6): 643–654.
71. Maassab, H.F., C.A. Heilman, and M.L. Herlocher, Cold-adapted influenza viruses for use as live vaccines for man. *Adv Biotechnol Processes*, 1990, 14: 203–242.
72. Murphy, B.R. and K. Coelingh, Principles underlying the development and use of live attenuated cold-adapted influenza A and B virus vaccines. *Viral Immunol*, 2002, 15(2): 295–323.

73. Treanor, J.J., et al., Evaluation of trivalent, live, cold-adapted (CAIV-T) and inactivated (TIV) influenza vaccines in prevention of virus infection and illness following challenge of adults with wild-type influenza A (H1N1), A (H3N2), and B viruses. *Vaccine*, 1999, 18(9–10): 899–906.
74. Capua, I. and S. Marangon, Control of avian influenza in poultry. *Emerg Infect Dis*, 2006, 12(9): 1319–1324.
75. Tian, G., et al., Protective efficacy in chickens, geese and ducks of an H5N1-inactivated vaccine developed by reverse genetics. *Virology*, 2005, 341(1): 153–162.
76. Swayne, D.E., et al., Protection against diverse highly pathogenic H5 avian influenza viruses in chickens immunized with a recombinant fowlpox vaccine containing an H5 avian influenza hemagglutinin gene insert. *Vaccine*, 2000, 18(11–12): 1088–1095.
77. CDC, Update: Influenza activity United States, 2004–2005 season. *MMWR Morb Mortal Wkly Rep*, 2005, 54(8): 193–196.
78. Goodwin, K., C. Viboud, and L. Simonsen, Antibody response to influenza vaccination in the elderly: A quantitative review. *Vaccine*, 2006, 24(8): 1159–1169.
79. Wright, P.F., The use of inactivated influenza vaccine in children. *Semin Pediatr Infect Dis*, 2006, 17(4): 200–205.
80. Beyer, W.E., et al., Antibody induction by influenza vaccines in the elderly: A review of the literature. *Vaccine*, 1989, 7(5): 385–394.
81. Thompson, W.W., et al., Mortality associated with influenza and respiratory syncytial virus in the United States. *Jama*, 2003, 289(2): 179–186.
82. Katz, J.M., et al., Immunity to influenza: The challenges of protecting an aging population. *Immunol Res*, 2004, 29(1–3): 113–124.
83. Munoz, F.M., Influenza virus infection in infancy and early childhood. *Paediatr Respir Rev*, 2003, 4(2): 99–104.
84. Kilbourne, E.D., Future influenza vaccines and the use of genetic recombinants. *Bull World Health Organ*, 1969, 41(3): 643–645.
85. Lugovtsev, V.Y., G.M. Vodeiko, and R.A. Levandowski, Mutational pattern of influenza B viruses adapted to high growth replication in embryonated eggs. *Virus Res*, 2005, 109(2): 149–157.
86. Widjaja, L., et al., Molecular changes associated with adaptation of human influenza A virus in embryonated chicken eggs. *Virology*, 2006, 350(1): 137–145.
87. FDA. 2005. Available from: http://www.fda.gov/ola/2005/influenza0210.html
88. Auewarakul, P., et al., Antibody responses after dose-sparing intradermal influenza vaccination. *Vaccine*, 2007, 25(4): 659–663.
89. Sambhara, S., et al., Heterosubtypic immunity against human influenza A viruses, including recently emerged avian H5 and H9 viruses, induced by FLU-ISCOM vaccine in mice requires both cytotoxic T-lymphocyte and macrophage function. *Cell Immunol*, 2001, 211(2): 143–153.
90. Asahi-Ozaki, Y., et al., Intranasal administration of adjuvant-combined recombinant influenza virus HA vaccine protects mice from the lethal H5N1 virus infection. *Microbes Infect*, 2006, 8(12–13): 2706–2714.
91. Halperin, S.A., et al., Phase I, randomized, controlled trial to study the reactogenicity and immunogenicity of a nasal, inactivated trivalent influenza virus vaccine in healthy adults. *Hum Vaccin*, 2005, 1(1): 37–42.
92. Atmar, R.L., et al., A dose-response evaluation of inactivated influenza vaccine given intranasally and intramuscularly to healthy young adults. *Vaccine*, 2007, 25(29): 5367–5373.
93. Subbarao, K. and T. Joseph, Scientific barriers to developing vaccines against avian influenza viruses. *Nat Rev Immunol*, 2007, 7(4): 267–278.
94. Horimoto, T. and Y. Kawaoka, Strategies for developing vaccines against H5N1 influenza A viruses. *Trends Mol Med*, 2006, 12(11): 506–514.
95. Neumann, G., et al., An improved reverse genetics system for influenza A virus generation and its implications for vaccine production. *Proc Natl Acad Sci U S A*, 2005, 102(46): 16825–16829.
96. Forde, G.M., Rapid-response vaccines—Does DNA offer a solution? *Nat Biotechnol*, 2005, 23(9): 1059–1062.
97. Sato, Y., et al., Immunostimulatory DNA sequences necessary for effective intradermal gene immunization. *Science*, 1996, 273(5273): 352–354.
98. Klinman, D.M., G. Yamshchikov, and Y. Ishigatsubo, Contribution of CpG motifs to the immunogenicity of DNA vaccines. *J Immunol*, 1997, 158(8): 3635–3639.

99. Robinson, H.L., et al., DNA immunization for influenza virus: Studies using hemagglutinin- and nucleoprotein-expressing DNAs. *J Infect Dis*, 1997, 176(Suppl 1): S50–S55.
100. Robinson, H.L., L.A. Hunt, and R.G. Webster, Protection against a lethal influenza virus challenge by immunization with a haemagglutinin-expressing plasmid DNA. *Vaccine*, 1993, 11(9): 957–960.
101. Webster, R.G., et al., Protection of ferrets against influenza challenge with a DNA vaccine to the haemagglutinin. *Vaccine*, 1994, 12(16): 1495–1498.
102. Kodihalli, S., et al., DNA vaccine encoding hemagglutinin provides protective immunity against H5N1 influenza virus infection in mice. *J Virol*, 1999, 73(3): 2094–2098.
103. Chen, Z., et al., Comparison of the ability of viral protein-expressing plasmid DNAs to protect against influenza. *Vaccine*, 1998, 16(16): 1544–1549.
104. Ulmer, J.B., et al., Heterologous protection against influenza by injection of DNA encoding a viral protein. *Science*, 1993, 259(5102): 1745–1749.
105. Tompkins, S.M., et al., Matrix protein 2 vaccination and protection against influenza viruses, including subtype H5N1. *Emerg Infect Dis*, 2007, 13(3): 426–435.
106. Wang, S., et al., Hemagglutinin (HA) proteins from H1 and H3 serotypes of influenza A viruses require different antigen designs for the induction of optimal protective antibody responses as studied by codon-optimized HA DNA vaccines. *J Virol*, 2006, 80(23): 11628–11637.
107. Laddy, D.J. and D.B. Weiner, From plasmids to protection: A review of DNA vaccines against infectious diseases. *Int Rev Immunol*, 2006, 25(3–4): 99–123.
108. Donnelly, J., K. Berry, and J.B. Ulmer, Technical and regulatory hurdles for DNA vaccines. *Int J Parasitol*, 2003, 33(5–6): 457–467.
109. Ulmer, J.B., U. Valley, and R. Rappuoli, Vaccine manufacturing: Challenges and solutions. *Nat Biotechnol*, 2006, 24(11): 1377–1383.
110. Robertson, J.S. and K. Cichutek, European Union guidance on the quality, safety and efficacy of DNA vaccines and regulatory requirements. *Dev Biol (Basel)*, 2000, 104: 53–56.
111. Smith, H.A. and D.M. Klinman, The regulation of DNA vaccines. *Curr Opin Biotechnol*, 2001, 12(3): 299–303.
112. NIH. 2007. Available from: http://www.nih.gov/news/pr/jan2007/niaid-02.htm
113. Le Gall-Recule, G., et al., Importance of a prime-boost DNA/protein vaccination to protect chickens against low-pathogenic H7 avian influenza infection. *Avian Dis*, 2007, 51(Suppl 1): 490–494.
114. Epstein, S.L., et al., Protection against multiple influenza A subtypes by vaccination with highly conserved nucleoprotein. *Vaccine*, 2005, 23(46–47): 5404–5410.
115. Pushko, P., et al., Replicon-helper systems from attenuated Venezuelan equine encephalitis virus: Expression of heterologous genes in vitro and immunization against heterologous pathogens in vivo. *Virology*, 1997, 239(2): 389–401.
116. Hoelscher, M.A., et al., Development of adenoviral-vector-based pandemic influenza vaccine against antigenically distinct human H5N1 strains in mice. *Lancet*, 2006, 367(9509): 475–481.
117. Bublot, M., et al., Development and use of fowlpox vectored vaccines for avian influenza. *Ann N Y Acad Sci*, 2006, 1081: 193–201.
118. Kreijtz, J.H., et al., Recombinant modified vaccinia virus Ankara-based vaccine induces protective immunity in mice against infection with influenza virus H5N1. *J Infect Dis*, 2007, 195(11): 1598–1606.
119. Schultz-Cherry, S., et al., Influenza virus (A/HK/156/97) hemagglutinin expressed by an alphavirus replicon system protects chickens against lethal infection with Hong Kong-origin H5N1 viruses. *Virology*, 2000, 278(1): 55–59.
120. Gao, W., et al., Protection of mice and poultry from lethal H5N1 avian influenza virus through adenovirus-based immunization. *J Virol*, 2006, 80(4): 1959–1964.
121. Redfield, R.R., et al., Disseminated vaccinia in a military recruit with human immunodeficiency virus (HIV) disease. *N Engl J Med*, 1987, 316(11): 673–676.
122. Meisel, H., et al., Fine mapping and functional characterization of two immuno-dominant regions from the preS2 sequence of hepatitis B virus. *Intervirology*, 1994, 37(6): 330–339.
123. Bichko, V., et al., Epitopes recognized by antibodies to denatured core protein of hepatitis B virus. *Mol Immunol*, 1993, 30(3): 221–231.
124. Sominskaya, I., et al., Tetrapeptide QDPR is a minimal immunodominant epitope within the preS2 domain of hepatitis B virus. *Immunol Lett*, 1992, 33(2): 169–172.

125. Sominskaya, I., et al., Determination of the minimal length of preS1 epitope recognized by a monoclonal antibody which inhibits attachment of hepatitis B virus to hepatocytes. *Med Microbiol Immunol*, 1992, 181(4): 215–226.
126. Ulrich, R., et al., Precise localization of the epitope of major BLV envelope protein. *Acta Virol*, 1991, 35(3): 302.
127. Levy, R., et al., Fine and domain-level epitope mapping of botulinum neurotoxin type A neutralizing antibodies by yeast surface display. *J Mol Biol*, 2007, 365(1): 196–210.
128. Ernst, W.A., et al., Protection against H1, H5, H6 and H9 influenza A infection with liposomal matrix 2 epitope vaccines. *Vaccine*, 2006, 24(24): 5158–5168.
129. Nicholson, K.G., et al., Clinical studies of monovalent inactivated whole virus and subunit A/USSR/77 (H1N1) vaccine: Serological responses and clinical reactions. *J Biol Stand*, 1979, 7(2): 123–136.
130. Treanor, J.J., et al., Safety and immunogenicity of a baculovirus-expressed hemagglutinin influenza vaccine: A randomized controlled trial. *JAMA*, 2007, 297(14): 1577–1582.
131. Brands, R., et al., Influvac: A safe Madin Darby Canine Kidney (MDCK) cell culture-based influenza vaccine. *Dev Biol Stand*, 1999, 98: 93–100; discussion 111.
132. Conne, P., et al., Immunogenicity of trivalent subunit versus virosome-formulated influenza vaccines in geriatric patients. *Vaccine*, 1997, 15(15): 1675–1679.
133. Podda, A., The adjuvanted influenza vaccines with novel adjuvants: Experience with the MF59-adjuvanted vaccine. *Vaccine*, 2001, 19(17–19): 2673–2680.
134. Pushko, P., et al., Evaluation of influenza virus-like particles and Novasome adjuvant as candidate vaccine for avian influenza. *Vaccine*, 2007, 25(21): 4283–4290.
135. O'Hagan, D.T., M.L. MacKichan, and M. Singh, Recent developments in adjuvants for vaccines against infectious diseases. *Biomol Eng*, 2001, 18(3): 69–85.
136. Del Giudice, G., et al., An MF59-adjuvanted inactivated influenza vaccine containing A/Panama/1999 (H3N2) induced broader serological protection against heterovariant influenza virus strain A/Fujian/2002 than a subunit and a split influenza vaccine. *Vaccine*, 2006, 24(16): 3063–3065.
137. Del Giudice, G., et al., Vaccines with the MF59 adjuvant do not stimulate antibody responses against squalene. *Clin Vaccine Immunol*, 2006, 13(9): 1010–1013.
138. Asa, P.B., R.B. Wilson, and R.F. Garry, Antibodies to squalene in recipients of anthrax vaccine. *Exp Mol Pathol*, 2002, 73(1): 19–27.
139. Barr, T., et al., Antibodies against cell surface antigens as very potent immunological adjuvants. *Vaccine*, 2006, 24(Suppl 2): S2–20–1.
140. Hatzifoti, C. and A.W. Heath, CD40-mediated enhancement of immune responses against three forms of influenza vaccine. *Immunology*, 2007, 122(1): 98–106.
141. Ninomiya, A., et al., Intranasal administration of a synthetic peptide vaccine encapsulated in liposome together with an anti-CD40 antibody induces protective immunity against influenza A virus in mice. *Vaccine*, 2002, 20(25–26): 3123–3129.
142. Glück, R., et al., Immunogenicity of new virosome influenza vaccine in elderly people. *Lancet*, 1994, 344 (8916): 160–163.
143. de Bruijn, I.A., et al., Clinical experience with inactivated, virosomal influenza vaccine. *Vaccine*, 2005, 23 (Suppl 1): S39–S49.
144. Mutsch, M., et al., Use of the inactivated intranasal influenza vaccine and the risk of Bell's palsy in Switzerland. *N Engl J Med*, 2004, 350(9): 896–903.
145. Glück, U., J.O. Gebbers, and R. Glück, Phase 1 evaluation of intranasal virosomal influenza vaccine with and without *Escherichia coli* heat-labile toxin in adult volunteers. *J Virol*, 1999, 73(9): 7780–7786.
146. Latham, T. and J.M. Galarza, Formation of wild-type and chimeric influenza virus-like particles following simultaneous expression of only four structural proteins. *J Virol*, 2001, 75(13): 6154–6165.
147. Pushko, P., et al., Influenza virus-like particles comprised of the HA, NA, and M1 proteins of H9N2 influenza virus induce protective immune responses in BALB/c mice. *Vaccine*, 2005, 23(50): 5751–5759.
148. Watanabe, T., et al., Immunogenicity and protective efficacy of replication-incompetent influenza virus-like particles. *J Virol*, 2002, 76(2): 767–773.
149. Pumpens, P. and E. Grens, HBV core particles as a carrier for B cell/T cell epitopes. *Intervirology*, 2001, 44(2–3): 98–114.
150. Borisova, G.P., et al., Recombinant core particles of hepatitis B virus exposing foreign antigenic determinants on their surface. *FEBS Lett*, 1989, 259(1): 121–124.

151. Sasagawa, T., et al., Synthesis and assembly of virus-like particles of human papillomaviruses type 6 and type 16 in fission yeast *Schizosaccharomyces pombe. Virology*, 1995, 206(1): 126–135.
152. Kirnbauer, R., Papillomavirus-like particles for serology and vaccine development. *Intervirology*, 1996, 39(1–2): 54–61.
153. Mortola, E. and P. Roy, Efficient assembly and release of SARS coronavirus-like particles by a heterologous expression system. *FEBS Lett*, 2004, 576(1–2): 174–178.
154. Bosio, C.M., et al., Ebola and Marburg virus-like particles activate human myeloid dendritic cells. *Virology*, 2004, 326(2): 280–287.
155. Noad, R. and P. Roy, Virus-like particles as immunogens. *Trends Microbiol*, 2003, 11(9): 438–444.
156. Fifis, T., et al., Size-dependent immunogenicity: Therapeutic and protective properties of nano-vaccines against tumors. *J Immunol*, 2004, 173(5): 3148–3154.
157. Luytjes, W., et al., Amplification, expression, and packaging of foreign gene by influenza virus. *Cell*, 1989, 59(6): 1107–1113.
158. Garcia-Sastre, A. and P. Palese, Genetic manipulation of negative-strand RNA virus genomes. *Annu Rev Microbiol*, 1993, 47: 765–790.
159. Neumann, G. and Y. Kawaoka, Reverse genetics of influenza virus. *Virology*, 2001, 287(2): 243–250.
160. Schickli, J.H., et al., Plasmid-only rescue of influenza A virus vaccine candidates. *Philos Trans R Soc Lond B Biol Sci*, 2001, 356(1416): 1965–1973.
161. St Angelo, C., et al., Two of the three influenza viral polymerase proteins expressed by using baculovirus vectors form a complex in insect cells. *J Virol*, 1987, 61(2): 361–365.
162. Mena, I., et al., Rescue of a synthetic chloramphenicol acetyltransferase RNA into influenza virus-like particles obtained from recombinant plasmids. *J Virol*, 1996, 70(8): 5016–5024.
163. Bright, R.A., et al., Influenza virus-like particles elicit broader immune responses than whole virion inactivated influenza virus or recombinant hemagglutinin. *Vaccine*, 2007, 25(19): 3871–3878.
164. Bright, R.A., et al., Cross-clade protective immune responses to influenza viruses with H5N1 HA and NA elicited by an influenza virus-like particle. PLoS ONE, 2008, 3(1): e1501.
165. Gomez-Puertas, P., et al., Influenza virus matrix protein is the major driving force in virus budding. *J Virol*, 2000, 74(24): 11538–11547.
166. Galarza, J.M., T. Latham, and A. Cupo, Virus-like particle (VLP) vaccine conferred complete protection against a lethal influenza virus challenge. *Viral Immunol*, 2005, 18(1): 244–251.
167. Quan, F.S., et al., Virus-like particle vaccine induces protective immunity against homologous and heterologous strains of influenza virus. *J Virol*, 2007, 81(7): 3514–3524.
168. Chen, B.J., et al., Influenza virus hemagglutinin and neuraminidase, but not the matrix protein, are required for assembly and budding of plasmid-derived virus-like particles. *J Virol*, 2007, 81(13): 7111–7123.
169. Novavax, Inc. Novavax begins human clinical testing of novel pandemic flu vaccine. 2007. Available from: http://www.novavax.com/download/File/Clinical_Trial_FO.pdf
170. Roldao, A., et al., Modeling rotavirus-like particles production in a baculovirus expression vector system: Infection kinetics, baculovirus DNA replication, mRNA synthesis and protein production. *J Biotechnol*, 2007, 128(4): 875–894.
171. Gomez-Puertas, P., et al., Efficient formation of influenza virus-like particles: Dependence on the expression levels of viral proteins. *J Gen Virol*, 1999, 80(Pt 7): 1635–1645.
172. Pumpens, P., et al., Evaluation of HBs, HBc, and frCP virus-like particles for expression of human papillomavirus 16 E7 oncoprotein epitopes. *Intervirology*, 2002, 45(1): 24–32.
173. Pushko, P., et al., Analysis of RNA phage fr coat protein assembly by insertion, deletion and substitution mutagenesis. *Protein Eng*, 1993, 6(8): 883–891.
174. Ionescu, R.M., et al., Pharmaceutical and immunological evaluation of human papillomavirus viruslike particle as an antigen carrier. *J Pharm Sci*, 2006, 95(1): 70–79.
175. Neirynck, S., et al., A universal influenza A vaccine based on the extracellular domain of the M2 protein. *Nat Med*, 1999, 5(10): 1157–1163.
176. Pumpens, P. and Grens, E., Artificial genes for chimeric virus-like particles., In: *Artificial DNA*, Y.E. Khudyakov and Fields, H.A., (Eds.), 2003, CRC Press: New York, pp. 249–327.
177. Pushko, P., et al., Identification of hepatitis B virus core protein regions exposed or internalized at the surface of HBcAg particles by scanning with monoclonal antibodies. *Virology*, 1994, 202(2): 912–920.

178. Luo, L., et al., Mapping of functional domains for HIV-2 gag assembly into virus-like particles. *Virology*, 1994, 205(2): 496–502.
179. Rossmann, M.G., et al., Combining X-ray crystallography and electron microscopy. *Structure*, 2005, 13(3): 355–362.
180. Zhang, Y., et al., Structure of immature West Nile virus. *J Virol*, 2007, 81(11): 6141–6145.
181. Prasad, B.V., et al., X-ray crystallographic structure of the Norwalk virus capsid. *Science*, 1999, 286 (5438): 287–290.
182. Crowther, R.A., et al., Three-dimensional structure of hepatitis B virus core particles determined by electron cryomicroscopy. *Cell*, 1994, 77(6): 943–950.
183. Jegerlehner, A., et al., Influenza A vaccine based on the extracellular domain of M2: Weak protection mediated via antibody-dependent NK cell activity. *J Immunol*, 2004, 172(9): 5598–5605.
184. De Filette, M., et al., Improved design and intranasal delivery of an M2e-based human influenza A vaccine. *Vaccine*, 2006, 24(44–46): 6597–6601.
185. Fiers, W., et al., A universal human influenza A vaccine. *Virus Res*, 2004, 103(1–2): 173–176.
186. Jegerlehner, A., et al., A molecular assembly system that renders antigens of choice highly repetitive for induction of protective B cell responses. *Vaccine*, 2002, 20(25–26): 3104–3112.
187. Szecsi, J., et al., Induction of neutralising antibodies by virus-like particles harbouring surface proteins from highly pathogenic H5N1 and H7N1 influenza viruses. *Virol J*, 2006, 3: 70.
188. Sun, J., et al., Core-controlled polymorphism in virus-like particles. *Proc Natl Acad Sci U S A*, 2007, 104 (4): 1354–1359.
189. Available from: http://www.medicalnewstoday.com/articles/77166.php
190. Li, Z.N., et al., Chimeric influenza virus hemagglutinin proteins containing large domains of the Bacillus anthracis protective antigen: Protein characterization, incorporation into infectious influenza viruses, and antigenicity. *J Virol*, 2005, 79(15): 10003–10012.
191. Garcia-Sastre, A. and P. Palese, Influenza virus vectors. *Biologicals*, 1995, 23(2): 171–178.
192. Palese, P., et al., Development of novel influenza virus vaccines and vectors. *J Infect Dis*, 1997, 176 (Suppl 1): S45–S49.
193. Lai, A.L., et al., Fusion peptide of influenza hemagglutinin requires a fixed angle boomerang structure for activity. *J Biol Chem*, 2006, 281(9): 5760–5770.
194. Leser, G.P. and R.A. Lamb, Influenza virus assembly and budding in raft-derived microdomains: A quantitative analysis of the surface distribution of HA, NA and M2 proteins. *Virology*, 2005, 342(2): 215–227.
195. CDC, Updated interim influenza vaccination recommendations—2004–2005 influenza season. *MMWR Morb Mortal Wkly Rep*, 2004, 53(50): 1183–1184.
196. Heckler, R., et al., Cross-protection against homologous drift variants of influenza A and B after vaccination with split vaccine. *Intervirology*, 2007, 50(1): 58–62.
197. FDA, Additional flu vaccine. *FDA Consum*, 2006, 40(6): 4.
198. Salleras, L., et al., Effectiveness of virosomal subunit influenza vaccine in preventing influenza-related illnesses and its social and economic consequences in children aged 3–14 years: A prospective cohort study. *Vaccine*, 2006, 24(44–46): 6638–6642.

9 Construction of Novel Vaccines on the Basis of Virus-Like Particles: Hepatitis B Virus Proteins as Vaccine Carriers

Paul Pumpens, Rainer G. Ulrich, Kestutis Sasnauskas, Andris Kazaks, Velta Ose, and Elmars Grens

CONTENTS

9.1 INTRODUCTION

Food and Drug Administration's (FDA)-approved vaccines against hepatitis B and human papilloma viruses represent genetically engineered virus-like particles (VLPs) generated in heterologous expression systems. VLP-based vaccines are also being developed against malaria, HIV/AIDS, hepatitis C, human and avian influenza, as well as against other diseases. Moreover, VLPs from almost all classes of viruses are being evaluated now or have just been adopted to use as a scaffold for presentation of foreign immunological epitopes on the surface of chimeric VLPs. Two major strategies have been used to present foreign protein epitopes on the surface of VLPs. They are (1) classical gene fusion techniques and (2) chemical coupling of epitope peptides to the VLP surface. VLP technologies possess obvious advantages for generation of safe and efficacious vaccines. First, the repetitive antigenic structure of VLPs makes them highly immunogenic. Second, "classical" VLPs are lacking viral genomes or genes and are noninfectious, although they are mimicking infectious viruses in their structural and immunological features. Third, VLPs are generated by highly efficient heterologous expression of the cloned viral structural genes with subsequent quantitative *in vivo* or *in vitro* self-assembly of their products. Fourth, VLPs can be obtained by simple and efficient purification procedures. A broad range of viral structural proteins is able to form "autologous" VLPs consisting solely of structural protein(s) of the target virus. Many of them have been tested successfully for the construction of chimeric VLPs retaining their VLP-forming ability, but carrying foreign epitopes. VLP technologies allow the generation of (1) uniform chimeric VLPs consisting of identical fusion protein subunits, (2) mosaic VLPs consisting of carrier and fusion protein subunits, and (3) pseudotyped VLPs consisting of nonfused autologous and foreign proteins. VLPs can be used for a broad range of applications, but first of all for vaccine development. In this respect, generation of vaccines against hepatitis B virus (HBV) and hepatitis C virus (HCV) infections is of special interest.

9.2 CONCEPT OF THE VLP APPROACH

9.2.1 Recombinant and Chimeric VLPs

The development of genetic engineering techniques in the 1970s offered a broad range of applications, which were immediately followed by the expression of viral and nonviral genes in efficient heterologous expression systems, first of all, in bacteria and yeast. Special attention has been devoted to the synthesis of viral structural proteins as constituents of viral capsids and envelopes with their subsequent spontaneous self-assembly into correctly organized capsid-like particles (CLPs) and VLPs. We use the term VLPs exclusively for recombinant structures, although this term is sometimes also used to designate structures of unknown nature and origin observed in tissue samples by electron microscopy. Whereas for enveloped viruses both types of recombinant structures can be generated, for naked, i.e., nonenveloped viruses, the terms CLPs and VLPs describe the same structure. For reasons of simplicity, we are using the term VLPs in a broader sense to describe capsid and envelope structures that resemble those of the infectious virus particles. The appropriate research branch of the molecular biotechnology was named VLP biotechnology (for a review, see Ref. [1]). Due to their symmetric three-dimensional organization, recombinant VLPs appear as the most immunogenic viral antigens. Therefore, they have attracted and continue to attract attention as the most promising antiviral vaccine candidates. It is not surprising that the first recombinant vaccine approved for human use in 1986 was based on the hepatitis B surface (HBs) antigen, which formed 22 nm-like VLPs in yeast. These yeast-derived HBs particles are very similar to natural 22 nm HBs particles from human blood.

At the same time, in addition to recombinant VLPs, the idea of the design of chimeric VLPs, i.e., recombinant VLPs carrying foreign immunological epitopes, has become highly attractive (for the first review articles, see Refs. [2,3]; more recent and specific reviews are referenced in the appropriate sections). The recombinant VLPs built symmetrically from hundreds of protein molecules of one or more types seemed to be the most efficient molecular carriers for a regular arrangement of epitope chains on desired positions of their outer surface. The regular repetitive pattern and correct conformation of inserted epitopes were the major factors, which encouraged persistent work on the generation of highly immunogenic foreign epitope-carrying chimeric VLPs. The chimeric VLP approach opened thereby the way to the VLP protein engineering with intent to construct (1) high-precision epitope-based and (2) multi-targeted vaccines.

9.2.2 VLP Protein Engineering

In fact, the idea of protein engineering and namely of the designed genetic engineering-based reconstruction of natural proteins has been started in the early 1980s by Alan R. Fersht and his colleagues. This team conceived the basic idea of mutational intervention into the structure of proteins, based on spatial knowledge and oriented toward the creation of artificially improved proteins with desired functions (for a review, see Ref. [4]).

In contrast to monomeric and oligomeric protein carriers, chimeric VLP structures are able to provide a high density of symmetrically exposed foreign oligopeptides on the particle, which is a prerequisite for the high immunological activity of the VLPs and their putative use as highly efficient vaccines. The interest in such rational-based manipulations on chimeric VLPs was reinforced by the simultaneous development of structural knowledge about viral capsids and envelopes by the use of high-resolution techniques. X-ray crystallography and electron cryomicroscopy revealed high-resolution structures of such favorite VLP carriers based on RNA phages [5–13], tobacco mosaic virus [14,15], poliovirus [16], rhinovirus [17], bluetongue virus [18], hepatitis B virus [5,13,19,20], and many others (Chapters 6,8,10–12).

Concurrent with the VLP protein engineering, the early 1980s were a time when the significance of specific epitopes as antigen regions, which are necessary and sufficient for (1) the induction of antigen-specific immune responses and (2) the recognition of antigens by monoclonal or polyclonal antibodies, gained general acceptance, and many epitope sequences were mapped. Fine mapping of epitopes resulted, first, in the determination of an epitope sequence, which could be defined as the shortest stretch within a protein molecule capable of binding to antibodies. The next level of recognition within the epitope can be defined as an antigenic determinant, namely, the exact side groups of amino acid (aa) residues involved in epitope recognition by the paratope sequences of antibodies. The mapping techniques based, in general, on the examination of libraries of short overlapping synthetic peptides or fusion proteins (for the earliest mapping protocols see [21–23]) revealed the existence of linear and conformational forms of epitopes. The term linear epitope defines that the antigenic determinants are localized in a short stretch of aa residues (usually from 4 to 10 aa in length). Their recognition by antibodies is simply determined by the aa sequence deprived of specific folding. Linear epitopes can be identified using short synthetic peptides and SDS-denatured fusion proteins. By contrast, conformational epitopes presuppose dispersion of antigenic determinants in more distant polypeptide stretches or dependence on a specific protein folding. Gene engineering offered experimental resources to construct DNA copies encoding both linear and conformational epitopes, to define their length and composition at a single aa resolution, to combine them in a different order, and to introduce them into carrier genes, i.e., genes encoding self-assembly competent polypeptides. Conformational epitopes may have a discontinuous character. The antigenic determinants forming the discontinuous epitope are located in protein stretches, which are brought together sterically, but are separated in the linear protein sequence.

Summarizing, the development of protein structure determination and prediction techniques as well as the success in structural investigations of putative VLP carrier moieties and insertion-intended sequences completed the novel specific branch of molecular biology and biotechnology. We can define the VLP protein engineering as a discipline working on knowledge-based approaches to the theoretical (*in silico*) design and experimental (*in vivo* and *in vitro*) construction of recombinant genomes and genes that may enable the efficient synthesis and correct self-assembly of chimeric VLPs with programmed structural and functional properties.

9.2.3 VLP Types and Strategies for VLP Protein Engineering

An overview of general VLP types is shown in Figure 9.1. From the point of view of the infectivity, the VLP approach can be realized in two general ways (Figure 9.1A): first, as noninfectious VLPs, which are the result of the expression of one or more isolated viral structural genes in a suitable prokaryotic or eukaryotic expression system. After self-assembly *in vitro* or *in vivo* and subsequent purification, such VLPs do not contain any genomic material capable of productive infection or replication. The second variant is represented by infectious VLPs, which could be further divided into two categories: (1) replication competent, which carry full-length infectious viral genomes with hybrid genes, allowing the production of chimeric virus progeny, and (2) replication noncompetent, which are infectious but give rise to chimeric progeny only in special replication conditions, e.g., in the presence of special helpers. In this review, we will concentrate exclusively on noninfectious VLPs.

Two general strategies are used to present foreign protein epitopes on the surface of VLPs (Figure 9.1B): (1) gene fusion techniques and (2) chemical coupling of epitope peptides to the VLP surface. As to the latter, besides the classical chemical cross linking, sophisticated procedures have been developed for a site-specific attachment of a foreign antigen to the VLP carrier (Chapter 6). Here, we review only VLPs generated by the expression of fusion genes.

Structural genes allowing the generation of VLPs are obtained usually from viral genomes by PCR amplification using specific synthetic oligonucleotides as primers. Individual VLP genes could also be generated by chemical synthesis. The first fully synthetic genes are those encoding HBV core (HBc) [24] and tobacco mosaic virus coat [25]. DNA copies of foreign oligopeptide sequences can also be prepared by chemical synthesis, as well as by PCR amplification with the use of specific synthetic primers. These epitope-encoding DNA copies must be adjusted to specific insertion sites in the VLP carrier genes, which are selected and proven theoretically and experimentally for each VLP candidate.

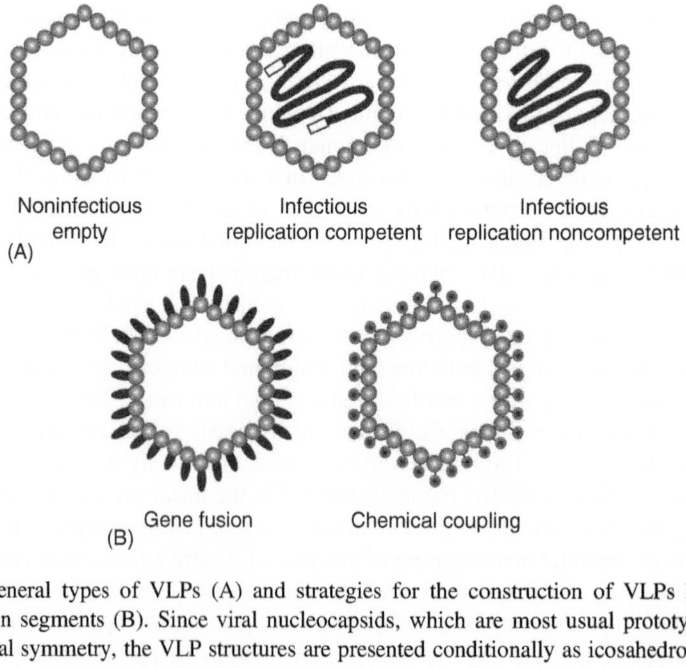

FIGURE 9.1 General types of VLPs (A) and strategies for the construction of VLPs harboring foreign epitopes or protein segments (B). Since viral nucleocapsids, which are most usual prototypes of the VLPs, possess icosahedral symmetry, the VLP structures are presented conditionally as icosahedrons.

One of the most crucial steps for successful implementation of the chimeric VLP production consists in the choice of expression system among prokaryotic bacterial cells and eukaryotic yeast, plant, insect, and mammalian cells, as well as whole eukaryotic organisms (transgenic animals, plants). The choice of the expression system determines the nature of the required regulatory elements, which are added to genomes or genes to be expressed. These regulatory elements (promoters, ribosome binding sites, transcription terminators, etc.) are prepared most often by chemical synthesis and comprise, after their introduction into the appropriate positions within the expression cassette, the essential parts of sophisticated vectors responsible for high-level production of VLPs.

9.2.4 Structural Classification of the Putative VLPs

Possible structural variants of the VLPs are depicted in Figure 9.2. The simplest variant of the VLP protein engineering is represented by the self-assembly of a single viral structural protein into single-shelled VLPs. It is performed by the heterologous or sometimes homologous (for example, in the case of production of RNA phage capsids in bacteria, or coats and capsids of animal and human viruses in eukaryotic cell lines) expression of a single viral structural gene, product of which is able to form spontaneously VLPs *in vivo* or at least *in vitro*, after purification in a nonassembled form. We designated this type of VLPs consisting of only one sort of viral structural protein as uniform VLPs.

More complicated complex VLPs are generated by the co-expression of different genes, products of which are involved in the VLP structure. For example, complex VLPs are represented, first, by double-shelled VLPs of the bluetongue virus (BTV). They appear after co-expression of BTV capsid genes, responsible for the formation of double-shelled BTV capsids [26–28]. Another prototype example of complex VLPs are enveloped VLPs obtained by the simultaneous synthesis and co-assembly of two different viral protein species, i.e., capsid and envelope proteins, after the

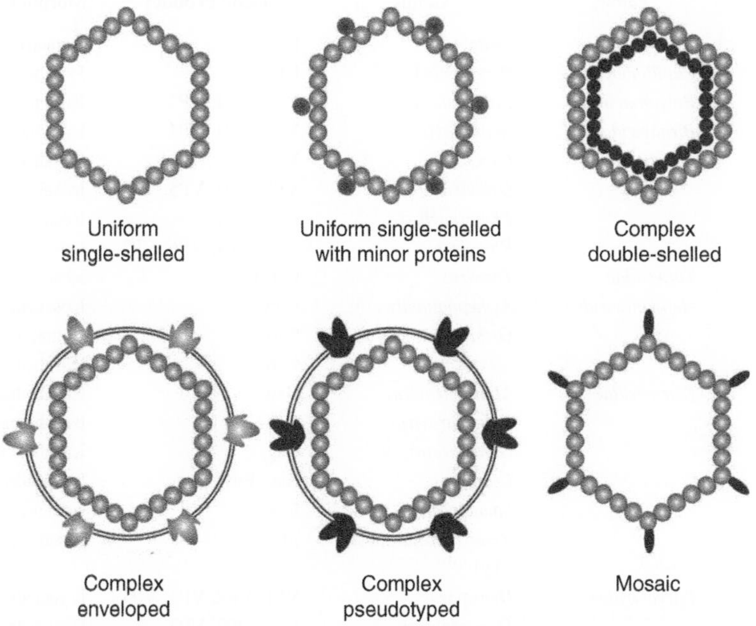

FIGURE 9.2 Structural variants of VLPs. All structural units of both uniform and complex VLPs including not only capsids, but also envelopes can be chimeric. In addition to capsid icosahedrons, envelopes are presented conditionally as phospholipid bilayers with immersed envelope proteins.

co-expression of the corresponding genes. This approach is a widely used strategy for protein engineering of retrovirus-derived VLPs, especially, those based on the human immunodeficiency virus (HIV), by the coproduction of the Gag and Env proteins [29]. For generation of avian influenza VLPs, three proteins have been used, HA, NA, and M1 [30,31]. Co-expression of structural *gag* and *env* genes of different retroviruses, as well as of capsid and envelope genes of unrelated viruses results in the formation of pseudotyped VLPs.

The term mosaic VLPs has been introduced to describe VLPs consisting of different sorts of subunits, one of which is a carrier protein itself as a helper for self-assembly, but others are carrier-foreign protein fusions. For the generation of mosaic VLPs at least two different approaches have been followed: (1) co-expression of two different genes responsible for the synthesis of both helper and fusion protein from the same or from two different plasmids with different selection markers or (2) synthesis of both fusion and helper proteins from the same gene mediated by stop codon readthrough, translational re-initiation, or ribosomal frameshifting [29,32,33].

9.3 ORIGIN OF THE VLP CANDIDATES

9.3.1 VLP Prototypes Originating from Taxonomically Different Viruses

In the last 5 years after publication of our previous review on VLP structures [1], the list of available VLP carrier candidates has been markedly extended. Table 9.1 presents up to date status of the putative VLP models. Table 9.1 shows recombinant "native" VLP candidates, namely, those, which have been generated in different expression systems without any foreign insertions into the genes

TABLE 9.1
Summary of Viral Structural Proteins Allowing the Formation of VLPs after Expression of Their Corresponding Cloned Genes

Genome Type	Family	Genus	Gene Product	Morphology of the VLPs
		Deltavirus	L-HDAg	Isometric
dsDNA, no RNA	*Papillomaviridae*	*Papillomavirus*	L1, L2	Icosahedral, $T = 7d$, 1
stage	*Polyomaviridae*	*Polyomavirus*	VP1, VP2, VP3	Icosahedral, $T = 7d$, 71
dsRNA	*Birnaviridae*	*Avibirnavirus*	VP2, VP3, VP4	Icosahedral, $T = 13$, 1
	Reoviridae	*Cypovirus*	VP1	Icosahedral
		Orbivirus	VP2, VP3, VP5, VP7	Icosahedral, $T = 13l$, 1
		Phytoreovirus	P3, P7, P8	Icosahedral, $T = 13l$, 1
		Rotavirus	VP2, VP4, VP6, VP7	Icosahedral, $T = 13$?
	Totiviridae	*Totivirus*	Capsid	Icosahedral, $T = 1$?
Retroid	*Hepadnaviridae*	*Avihepadnavirus*	Core	Icosahedral, $T = 3$, 4
		Orthohepadnavirus	Surface	Isometric
			Core	Icosahedral, $T = 3$, 4
	Retroviridae	*Alpharetrovirus*	Gag	Icosahedral ?
		Deltaretrovirus	Gag	Icosahedral ?
		Gammaretrovirus	Gag	Icosahedral ?
		Lentivirus	Gag, Env	Icosahedral ?
		Spumavirus	Env	Isometric
		Yeast retrotransposon Ty1	p1	Icosahedral, $T = 3$, 4, 7, 9
ssDNA	*Parvoviridae*	*Densovirus*	VP1, VP2, VP3, VP4	Icosahedral
		Dependovirus	VP1, VP2, VP3	Icosahedral, $T = 1$
		Erythrovirus	VP1, VP2	Icosahedral, $T = 1$
		Parvovirus	VP2	Icosahedral, $T = 1$

TABLE 9.1 (continued)
Summary of Viral Structural Proteins Allowing the Formation of VLPs after Expression of Their Corresponding Cloned Genes

Genome Type	Family	Genus	Gene Product	Morphology of the VLPs
ssRNA of negative polarity	*Arenaviridae*	*Arenavirus*	Z	Pleomorphic
	Bunyaviridae	*Hantavirus*	NC, G1, G2	Pleomorphic
	Filoviridae	*Ebola-like viruses*	GP, VP40	Filamentous
		Marburg-like viruses		
	Paramyxoviridae	*Morbillivirus*	M, N, F, HN	Isometric
		Pneumovirus		
		Respirovirus		
		Rubulavirus		
	Orthomyxoviridae	*Influenzavirus A*	PB1, PB2, PA, HA, NA, NP, M1, M2, NS2	Pleomorphic filamentous
		Thogotovirus	PB1, PB2, PA, NP, M, GP	Pleomorphic
ssRNA of positive polarity, no DNA stage	*Astroviridae*	*Mamastrovirus*	ORF2	Isometric
	Bromoviridae	*Alfamovirus*	Coat	Icosahedral, $T=1$
		Bromovirus	Coat	Icosahedral, $T=3$
	Caliciviridae	*Lagovirus*	Capsid	Icosahedral
		Norovirus	Capsid	Icosahedral, $T=3$
		Sapovirus	Capsid	Icosahedral
		Vesivirus	Capsid	Icosahedral
	Comoviridae	*Comovirus*	L, S	Icosahedral, $T=$p3
		Nepovirus	Capsid	Icosahedral, $T=$p3
	Flaviviridae	*Flavivirus*	C, prM, E	Isometric
		Hepacivirus	Core, E1, E2	Isometric
	Hepeviridae	*Hepatitis E virus*	Capsid	Icosahedral, $T=1$
	Leviviridae	*Allolevivirus*	Coat	Icosahedral, $T=3$
		Levivirus	Coat	Icosahedral, $T=3$
	Luteoviridae	*Luteovirus*	Coat	Icosahedral, $T=3$
	Nidovirales	*Coronaviridae*	S, M, E	Isometric
	Nodaviridae	*Alphanodavirus*	Coat	Icosahedral, $T=3$
		Betanodavirus	Capsid	Icosahedral, $T=3$
		Pecluvirus	Coat	Rod
	Picornaviridae	*Unclassified*	VP2	Icosahedral
	Potyviridae	*Potyvirus*	Coat	Rod
		Sobemovirus	Coat	Icosahedral, $T=3$
	Tetraviridae	*Betatetravirus*	Capsid	Icosahedral, $T=4$
		Omegatetravirus	Capsid	Icosahedral, $T=4$
		Tobamovirus	Coat	Rod
	Togaviridae	*Rubivirus*	C, E1, E2	Isometric
	Tymoviridae	*Tymovirus*	Coat	Icosahedral, $T=3$

Source: From NCBI, http://www.ncbi.nlm.nih.gov.
Note: The VLP carriers are grouped in taxonomic order according to the virus taxonomy.

encoding proteins participating in VLP formation. The latter are grouped in taxonomic order in accordance with the virus taxonomy given by the NCBI and available at the http://www.ncbi.nlm.nih.gov. The putative VLPs represent all structural classes of viruses based on nucleic acid content and polarity of the viral nucleic acid, i.e., dsDNA, dsRNA, "retroid," ssDNA, and ssRNA of both negative and positive polarity. The parasitic deltavirus, standing aside of these structural classes, also serves as a source for generation of recombinant VLPs. It is obvious that the largest number of VLPs is evolving from the virus class with single-stranded RNA genomes of positive polarity, lacking a DNA intermediate stage. Numerous families and genera of this most diverse structural class of viruses have been involved in the construction of differently sized and structured VLPs. As to organization and symmetry, this group of VLPs is represented by the structures of two different symmetries: icosahedral (prevalent number of prototypes, in total 18) and rod-like (only three prototypes), as well as by isometric VLPs without any distinct symmetry (three prototypes). The virus class with single-stranded RNA genomes of positive polarity is represented by such successful vaccine candidate VLPs as those based on the *Leviviridae* family (Chapter 6), and widely used rod-like VLPs based on the *Potyviridae* family and *Tobamovirus* genus. In the context of the present review, the *Flaviviridae* family is most exciting, since it contains HCV, vaccine against which is one of the most relevant keynotes of the present-day healthcare. Another important subject of the present review, HBV, which presents such widely used VLP carriers as core, or HBc, and surface, or HBs, antigens, belongs to the *Hepadnaviridae* family in the Retroid-like virus group. To the same class of viruses, but to other families belong retroviruses, which gave raise to the class of pseudotyped VLPs, and retrovirus-like yeast particles, which played an important role in the development of the VLP ideology.

The generation of VLPs originating from certain viruses has been motivated strongly by the lack of efficient cell culture propagation systems, like HBV, HCV, human papilloma virus (HPV) (Chapter 11), Norwalk virus, and others. An additional important reason for these efforts is the possibility to substitute infectious viruses requiring high-level biosafety containment laboratories for their handling, like Ebola virus, by VLPs as efficient surrogates for virus diagnostics, vaccine development, and basic research applications. Moreover, the most critical challenge nowadays consists in the substitution of live vaccines, for example, against influenza virus (Chapter 8) by the appropriate highly efficient VLPs.

Table 9.2 presents a list of VLP prototypes applied for the construction of chimeric particles, uniform or mosaic, containing foreign insertions, and obtained after expression of the appropriate fused VLP-forming protein genes. This list also shows a broad coverage of all virus classes, with exception of the deltavirus and negative-stranded ssRNA viruses. Again, the *Leviviridae* and *Hepadnaviridae* families present the most widespread and well-studied VLP prototypes for the generation of chimeric VLP candidates.

9.3.2 First Generation of VLP Candidates

The earliest candidates for VLP protein engineering appeared in the middle of 1980s. First, they were derived from the rod-shaped representatives of *Inoviridae* family (filamentous bacteriophage f1) [34] and *Tobamovirus* genus (tobacco mosaic virus, TMV) [25]. Second, the HBs protein, an outer envelope of the HBV, was used to expose foreign sequences on the so-called 22 nm particles formed by them [35,36]. Next, icosahedral particles have been suggested: HBc [2,34,37,38], capsids of picornaviruses [39] and RNA phages [40], and the yeast retrotransposon Ty1 [41].

The first attempts were extremely fruitful not only from the viewpoint of two different VLP candidate symmetries (icosahedral and rod-like) involved, but also in laying down the course of development of the two main directions of VLP engineering based on infectious and noninfectious structures. The latter were represented by VLPs derived from HBs and HBc, coats of TMV, RNA phages, and yeast retrotransposon Ty1, and formed the basis of a long list of noninfectious VLP models, whereas chimeric derivatives of phage f1 and poliovirus initiated the line of infectious replication-competent VLPs. The development of phage f1 VLPs was the starting point for the construction of DNA phage-display libraries. Additional examples of favorite replication-competent VLPs are based on poliovirus, rhinovirus, and plant (tobacco and cowpea mosaic) viruses. In the present review, we have restricted ourselves to noninfectious VLP carriers based on the structural HBV and HCV proteins.

TABLE 9.2
Summary of Viral Structural Proteins Applied for the Construction of Chimeric and Mosaic VLPs

	Viral Origin of the VLP Carrier Protein		
Genome Type	**Family**	**Species**	**Target Gene**
dsDNA, no intermediate RNA stage	*Papillomaviridae*	Bovine papillomavirus 1	L1
		Human papillomavirus 16	L1
			L2
	Polyomaviridae	Murine polyomavirus	VP1, VP2
		B-cell lymphotropic polyomavirus	VP1, VP2
		BK polyomavirus	
		Hamster polyomavirus	
		JC polyomavirus	
		Simian virus 40	
dsRNA	*Birnaviridae*	Infectious bursal disease virus	VP2
	Reoviridae	Bluetongue virus	VP3
			VP7
dsDNA and RNA intermediate (Retroid)	*Hepadnaviridae*	Hepatitis B virus	Surface
			Core
	Retroviridae	Human immunodeficiency virus 1, 2	Gag
		Yeast retrotransposon Ty1	p1
ssDNA	*Parvoviridae*	Human parvovirus B19	VP1
			VP2
		Porcine parvovirus	VP2
ssRNA of positive polarity, no intermediate DNA stage	*Bromoviridae*	Alfalfa mosaic virus	Coat
	Caliciviridae	Rabbit hemorrhagic disease virus	VP60
	Hepeviridae	Hepatitis E virus	Capsid
	Leviviridae	RNA phage Qβ	Coat
		RNA phages MS2, fr, GA, PP7	Coat
	Nodaviridae	Flock house virus	Coat
	Potyviridae	Johnsongrass mosaic virus	Coat
	Tobamovirus genus	Tobacco mosaic virus	Coat

9.3.3 Structural HBV Proteins as the First VLP Candidates

HBV proteins, outer HBs and inner HBc, are very attractive to start the VLP story, since they are the earliest and most well studied, and favorite VLP prototypes. History of their investigation started in the early 1980s, after the initial cloning of the HBV genome in 1978, which allowed generating these two first classical noninfectious VLP carrier candidates. The outer envelope-forming HBs and inner nucleocapsid-forming HBc are encoded by the HBV genes S and C, respectively. The HBs and HBc carriers gave onset to two basic classes of future VLPs: (1) asymmetrical, so-called isometric, spherical pleomorphic lipid-containing envelopes and (2) highly symmetrical particles (icosahedral in the case of the HBc and many other cases, less common—rod-symmetrical). From the point of view of the involvement of both HBs and HBc in the construction of the actual chimeric vaccines, they have become the most favorite and applied model species. A high level of synthesis and an ability to correct self-assembly in heterologous expression systems were the decisive factors that forced practical success of these VLP carriers. While HBs envelopes were capable of self-assembly in all the systems studied with the exception of bacteria, HBc formed native-like capsids in the latter as well.

9.4 HEPATITIS B SURFACE PROTEIN AS VACCINE CARRIER

9.4.1 Fine Structure

HBV large (L), middle (M), and short (S) surface proteins are encoded by a single open reading frame S encoding a total of 389 (almost all genotypes) or 400 (genotype A) aa in length. All HBs proteins contain the essential S sequence of 226 aa, whereas the M protein has a 55-aa preS2 extension at the N-terminus, and the L protein has a preS1 N-terminal extension of 108 or 119 aa (depending on the HBV genotypes), in addition to the preS2 [42]. The additional N-terminal 11 aa of the preS1 of the HBV genotype A seem to be nonessential for any structural and immunological features of the HBs proteins. The proportion of S to M to L (S:M:L) proteins in the HBV particle envelope is about 7:2:1 [43].

For use as a molecular carrier HBs has the unique property of self-assembling with cellular lipids into noninfectious empty envelope particles, spherical or tubular, of about 22 nm in diameter, which are secreted in large excess from hepatocytes during hepatitis B infection. Major structural and immunological features of the HBs monomeric subunit are shown in Figure 9.3. A linear presentation of the HBs protein with location of the putative α-helices and major B-cell and CTL epitopes including major immunodominant B-cell epitope "a" (Figure 9.3A) is illustrated with localization of the same crucial elements on the putative three-dimensional map of the HBs monomer (Figure 9.3B).

To simulate the possible three-dimensional structure of the HBs monomers and 22 nm particles, many alternative mathematical and experimental approaches have been employed. First of all, computer-aided topological models have been constructed [44–46]. A summarizing picture of such modeling is presented in Figure 9.3B. These models were confirmed by the identification of membrane-associated regions protected against proteolysis [47], by circular dichroism studies [48], and small-angle neutron scattering [49]. By the latter method, recombinant yeast HBs protein was characterized as a spherical particle with a 29 nm diameter, where lipids and carbohydrates form a spherical core with a 24 nm diameter, and the S protein on the surface. Examination of yeast-derived HBs by high-resolution negative staining electron microscopy (EM) revealed a spherical to slightly ovoid character of particles with an average diameter of 27.5 nm, consisting of 4 nm subunits with a minute central pore. Ice-embedding EM gave a diameter value of 23.7 nm and a 7–8 nm thick cortex surrounding an electron translucent core [50] that corresponds well to the small-angle neutron scattering data. Human HBs particles, examined by the same EM methods, were found to be smaller in size [50]. Briefly, according to the proposed folding of L, M, and S proteins, they consist of four membrane-spanning helices that are assembled into a highly hydrophobic complex where access to the water environment is not allowed (Figure 9.3B). The proteins project their N and C termini (including preS1 and preS2 sequences of L and M proteins) as well as the second hydrophilic region of protein S, bearing the major immunodominant B-cell epitope "a" (aa residues 111–149 of the protein S corresponding to aa 274–312 of the whole sequence in Figure 9.3A) to the outer surface of HBs particles.

The HBs epitope "a" of the three surface proteins L, M, and S is of special importance since it induces virus-neutralizing humoral responses and confers the protection as demonstrated for HBV vaccines. The exact three-dimensional structure of the determinant "a," which still remains unknown, is crucial since the vast majority of induced anti-HBs antibodies recognize conformational epitopes in this region. Mutations in the "a" epitope may lead to vaccine and diagnostic failure due to HBV escape variants.

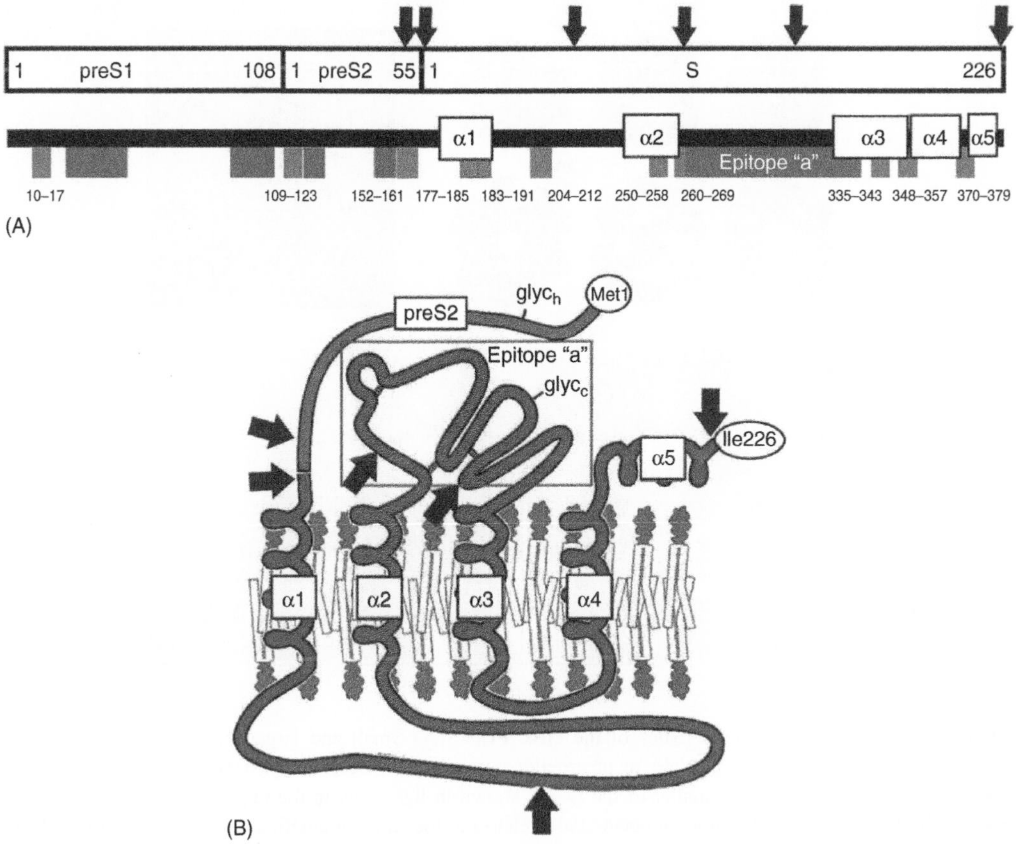

FIGURE 9.3 General structural and immunologically relevant motifs of the HBs protein. (A) A linear presentation of the S protein with its preS1 and preS2 extensions. Location of the transmembrane α-helices and major immunodominant epitopes (B cell, dark grey; CTL, light grey) is presented. Numbers represent aa positions of the CTL epitopes. (B) Putative topological model of the preS2-S protein, the lipid bilayer of the ER membrane is symbolized by the lipid molecules. The top of the figure corresponds to the lumen of the ER or, after multimerization and budding of the HBs protein, to the surface of the particle. The bottom of the figure corresponds to the cytosol or to the interior of the particle. Major immunodominant epitope "a" is boxed. Cysteine bridges and positions of complex biantennary glycoside ($glyc_c$) within the "a" epitope and hybrid-type glycoside ($glyc_h$) on the preS2 segment are shown. Major targets for foreign insertions are depicted by blue arrows on the linear and three-dimensional presentations of the HBs protein. ((B) Courtesy of Bruce Boschek and Wolfram H. Gerlich.)

Recent cryo-EM reconstruction of spherical HBs particles at approximately 12 Å resolution [51] is shown in Figure 9.4A and B. This revolutionary finding proposes that the HBs particles possess octahedral symmetry. According to the cryo-EM reconstruction, the HBs particles are built of the HBs dimers as building blocks. Figure 9.4B and C presents typical EM images of spherical 22 nm particles from human blood and yeast-derived HBs particles, respectively.

In contrast to viral particles, which contain the L and M proteins, the spherical 22 nm particles from human blood consist mainly of the S protein. Eukaryotic cell lines showed the capability of producing similar 22 nm particles after expression of S and M, but not L, genes within the appropriate expression cassettes (for review see Ref. [1]).

The S gene of HBV was one of the first viral genes expressed in bacteria [52–55]. Despite serious efforts mounted to generate "bacterial" HBs, self-assembly of the latter failed, possibly because specific folding and budding mechanisms are missing in bacteria [56–59]. Expression of the

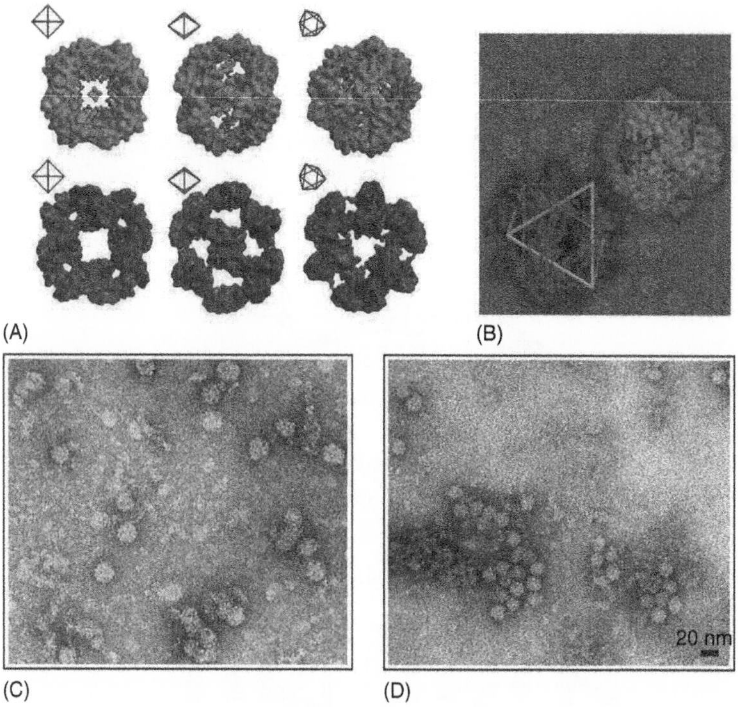

FIGURE 9.4 Structural characteristics of the HBs VLPs. (A) Small and large S particle reconstructions. Surface-rendered views of S particle reconstructions. S particles occur in two sizes, brought about by conformational switching. The smaller of the two is shown in light grey in the upper row, the larger in dark grey in the lower row. Insets show an octahedral skeleton in the same orientations at the actual structures, to indicate the relationship of the dispositons shown for the density maps to the symmetry of the actual object. (B) Views of the small (light grey) and large (dark grey) S particle structures, viewed along their threefold axes. An octahedral skeleton is shown within each semitransparent rendered surface. The background shows a typical cryo-EM field of view such as that which afforded the data used to generate these three-dimensional maps. (Courtesy of Robert Gilbert and David J. Rowlands.) (C) Electron microscopy of the HBs particles from blood of a human carrier. (D) Electron microscopy of the HBs particles from *S. cerevisiae* yeast. Scale bar, 20 Å.

HBs proteins in mammalian cells has been used to prove the usefulness of powerful eukaryotic vector systems based on regulatory elements taken from SV40, retro-, vaccinia, herpes simplex, papilloma, adeno-, varicella zoster, and other viruses. Expression of the S gene was achieved in transgenic mice as well. Reasonable levels of HBs synthesis have also been documented in baculovirus vector-driven insect cells, silkworm larvae, and transgenic plants such as tobacco, potato, lettuce, and lupin (for review see Ref. [1]). However, results of the greatest practical utility were achieved by the successful expression of S, M, and L genes in yeast, where the expression products self-assembled into particles with a widely estimated diameter of 20 ± 4 nm [60–62]. It is worth mentioning that in yeast, in contrast to eukaryotic cell lines, expression of the L gene may ensure formation of spherical particles exclusively consisting of the entire L protein [63]. Such particles contain from about 110 HBs molecules and demonstrate a broad range of diameter between 50 and 500 nm.

Although our knowledge of the folding and molecular architecture of the carrier particles is insufficient for the reliable prediction of preferential insertion sites, currently available data provide some hints for replacing surface-exposed protein domains by foreign sequences without destabilizing the self-assembly of the HBs particles. Direct mutational analysis of the HBs proteins revealed domains indispensable for self-assembly [64,65]. The role of intra- and intermolecular disulfide bonds within the hydrophilic regions of S protein, which are also responsible for its antigenic properties, was tested experimentally by site-directed *in vitro* mutagenesis substituting alanine for cysteine [48,66,67]. Moreover, mutagenesis of the HBs gene led to the important conclusion that mutated, assembly noncompetent polypeptides can be rescued into mosaic particles in the presence of native helpers [64].

The most preferable sites for foreign insertions are localized in preS2 and S parts of the HBs protein and are depicted in Figure 9.3, both on linear (Figure 9.3A) and putative three-dimensional (Figure 9.3B) maps.

9.4.2 Approved S-Based Hepatitis B Vaccines

The most prominent success of VLP protein engineering was reached by the construction of widely approved human hepatitis B vaccines on the basis of recombinant HBs protein. Yeast-derived HBs vaccine remained for long time not only the first but also the only available gene-engineered vaccine in the world.

Although human-plasma-derived and recombinant yeast-derived HBs proteins possess many similar properties, it is necessary to keep in mind that differences do exist. Besides glycosylation, which is different in yeast, such differences are clearly seen not only in the size of particles, but also in both epitope presentation and anti-HBs-binding properties [68].

However, this pioneering development of the HBs protein in yeast resulted, first, in construction of two widely available safe human vaccine products, which are used since 1987 and are still leading in the hepatitis B vaccine market: Engerix-B (developed by SmithKline Beecham, now produced by GlaxoSmithKline Biologicals, Belgium) and H-B-Vax II or Recombivax (Merck, Sharp & Dohme, United States), both of which are based on nonglycosylated protein S-derived particles and produced in *S. cerevisiae* yeast.

The next popular *S. cerevisiae* S-antigen based vaccine of this type is Euvax B, which is developed by the Aventis Pasteur (United States) and produced by the LG Life Sciences (South Korea). These vaccines undergo permanent improvements including combination with other vaccines (hepatitis A, diphtheria, tetanus, acellular pertussis, inactivated poliovirus, *Haemophilus influenzae* type b) into multi-targeted formulations (for more details see [69]). Recently, a new hepatitis B vaccine FENDrix was developed by the GlaxoSmithKline Biologicals and licensed in Europe [70]. The FENDrix vaccine contains the same *S. cerevisiae*-produced protein S, but it is formulated with a novel adjuvant system: aluminum phosphate and 3-O-desacyl-4-monophosphoryl lipid A. It possesses superior immunogenicity and induces higher antibody concentrations that reach protective levels in a faster fashion and retain for a longer period. The FENDrix vaccine is intended for use in adults with renal insufficiency including prehemodialysis and hemodialysis patients [71].

Some other registered and commercially successful hepatitis B vaccines using yeast-derived protein S are produced by original technologies based on the employment of highly productive species of yeast other than *S. cerevisiae*. First, methylotrophic yeast *Hansenula polymorpha* is used for the production of the S protein. Brazilian vaccine ButaNG was developed using *H. polymorpha* by the N.G. Biotecnologia Ltda and produced by the Instituto Butantan [72,73]. Hepavax-Gene vaccine was also developed using *H. polymorpha* by the Rhein Biotech (Germany), now belonging to the Dynavax company (United States), and produced by the Korean Green Cross. A new low-cost *H. polymorpha*-produced vaccine GeneVac B is registered in India [74]. Second, *Pichia pastoris* yeast is used for another set of hepatitis B vaccines. They are Cuban S-based vaccine Heberbiovac HB elaborated by the Centro de Ingenieria Genetica Y Biotecnologia and produced by the Heberbiotec SA (Cuba) company [75] and its analogue, which is produced in India under the trademark Enivac HB [76]. A low-cost *P. pastoris*-produced vaccine Shanvac B is registered in India [77].

The pioneering character of the HBs-based developments seems to be very close to the next revolutionary achievement, namely, to one of the first practical employments of the immunostimulatory sequences (ISS) targeting toll-like receptors (TLRs) [78,79]. Heplisav, a hepatitis B prophylactic and therapeutic vaccine, is developed by the California-based Dynavax company with a strong emphasize on the recently acquired Rhein Biotech company and is currently in phase III trials [80]. Under the terms of the agreement with the Coley Pharmaceutical, the Dynavax receives a nonexclusive license under Coley's ISS oligonucleotide patent estate for the commercialization of the Heplisav vaccine. The principle behind Coley's ISS drug targeting TLR9 is to use synthetic forms of cytosine–phosphate–guanine (CpG) to bind and activate the receptor. In turn, that especially triggers a T-helper 1 (Th1) immune response resulting in the activation of T cells (Chapter 6). Heplisav combines ISS with the HBs and is designed to significantly enhance the level, speed, and longevity of protection against hepatitis B. In several clinical studies, according to Coley, Heplisav has been shown to induce a protective level of antibodies against hepatitis B faster and with fewer doses than conventional hepatitis B vaccines. Additionally, Heplisav has provided 100% seroprotection in all subjects who have received the full regimen, including those who belong to so-called difficult-to-immunize patients group.

Ability of added CpG as an adjuvant to enhance immunogenicity of Engerix-B vaccine has been approved before in a phase I/II clinical study for the first clinical evaluation of the safety, tolerability, and immunogenicity of CpG when added to a commercial HBV vaccine [81–83]. Strikingly, most CpG Engerix-B-vaccinated subjects developed protective levels of anti-HBs IgG within just 2 weeks after application of the priming vaccine dose. A trend towards higher rates of specific cytotoxic T lymphocyte CTL responses has been also reported.

In preclinical animal studies, many other immunostimulatory adjuvants were suggested to improve the efficacy of the S protein-based vaccination, firstly, to improve cytotoxic response: PLGA (poly(D,L-lactic-*co*-glycolic acid)) microspheres [84,85], muramyl dipeptide C, or MDP-C [86], and so-called nanodecoy systems, i.e., spherical ceramic cores formed by self-assembling hydroxyapatite and cellobiose [87].

9.4.3 preS-Containing Hepatitis B Vaccines

Many attempts have been undertaken to improve the hepatitis B vaccine by fusion of the preS elements to the HBs. First attempts to enrich putative HBs vaccines with the preS2 [88] and preS1 [89] peptides were realized by covalent binding of the appropriate synthetic peptides to HBs. Then a chimeric protein, which carries the preS1 region aa 21–47 at the C-terminus of the S protein (at aa position 223), was expressed by a recombinant vaccinia virus and secreted from several mammalian cell lines [90].

HBV envelope was rearranged also by fusing part or all of the preS1 region to either the N- or C-terminus of the S protein [91]. Fusion of the first 42 residues of preS1 to either site allowed efficient secretion of the modified particles and rendered the linked sequence accessible at the surface of the particle. In opposite, fusion of the preS1 sequences to the C-terminus of the M protein completely blocked secretion. Although all these particles displayed preS1, preS2, and S-protein antigenicity, high titers of preS1- and preS2-, but not S-specific, antibodies were induced. Another group constructed three fusion proteins containing preS2 (aa 120–146) and preS1 (aa 21–47) epitopes at the N-terminus and truncated C-terminus of the S protein and expressed them using a recombinant vaccinia virus system [92]. Fusion proteins were efficiently secreted in particulate form, which displayed S, preS1, and preS2 antigenicity and elicited strong antibody responses against S, preS1, and preS2 in mice.

GenHevac B is the first approved preS-, namely pre-S2-containing hepatitis B vaccine. It was generated in 1993 by the Pasteur Mérieux Connaught (France) via expression of full-length nonreconstructed M and S genes. GenVac B is produced in mammalian cells and contains both M and S proteins. This vaccine showed good efficacy for therapeutic vaccination in chronic hepatitis B patients, with or without combination with interferon-α [93].

Next commercial preS-containing vaccine is the Bio-Hep-B vaccine, which is produced by Biotechnology General Ltd (Israel) via expression of M and L genes in mammalian cells. It contains therefore not only the M and S, but also some L protein molecules. The Bio-Hep-B vaccine showed improved efficacy in a specific group of patients with end-stage renal disease (ESRD), who were badly responding to classical S vaccines [94]. The Bio-Hep-B vaccine was highly efficient in newborns after only two injections and elicited multivalent response against L, M, and S proteins [95].

Sci-B-Vac vaccine was developed by the SciGene company (Hong Kong) [96]. It is produced in mammalian cell lines and contains major S and minor M and L proteins, similar to the Bio-Hep-B vaccine. It showed high therapeutic efficacy when applied to chronic hepatitis B patients after liver transplantation receiving additional lamivudine prophylaxis. The effect is due to its high and multivalent (S, preS1, preS2) immunogenicity, which is a key factor to prevent hepatitis B recurrence due to emergence of S escape mutants [97].

A very similar Hepimmune vaccine was developed by the Berna Biotech company (Switzerland). It is also produced in mammalian cells and contains S, M, and L proteins. A trial test in non- and low responders demonstrated enhanced antibody responses to the Hepimmune vaccine in comparison to conventional S-based vaccines [98]. It appears also to be a good candidate for short-time immunization protocols, which are developed for vaccination of living liver donors before transplantation of the liver to chronic hepatitis B patients [99]. The time to achieve hepatitis B immunity in those liver donors is usually short (1–2 months). For this reason, the authors established a short-time immunization protocol (four injections in 2 weeks intervals) with the Hepimmune and showed efficient cellular and humoral immune responses.

Further development of the preS inclusion strategy by the Medeva plc company led to the generation of an efficient preS1- and preS2-containing vaccine Hepagene (or Hepacare, or Hep B-3), which is the first vaccine of this type to have a pan-European license. Unlike preS-containing vaccines described above, the Hepagene vaccine was generated by selection of desired preS epitopes and their combination within the recombinant genes. The Hepagene vaccine contains the preS1 sequence spanning aa 20–47, the complete preS2, and the complete S region of two different subtypes, *adw* and *ayw*. It is produced in the mouse C127I clonal cell line after transfection of the cells with genes encoding the three antigens. Because these genes are expressed in mammalian cells, the Hepagene components are glycosylated and therefore resemble the native viral proteins more closely. The Hepagene vaccine was shown to stimulate stronger and more rapid cellular and humoral immune responses than S-based vaccines and to circumvent anti-HBs nonresponsiveness [100–104]. Clinical trials in both Europe and the United States have clearly demonstrated that the Hepagene vaccine is highly immunogenic for both B- and T cells and induces higher anti-S "a" determinant antibody titers than S-based vaccines [105]. After joining of the Medeva to the Celltech company, the Hepagene vaccine is further evaluated by Celltech and PowderJect companies utilizing PowderJect's needleless injection technology [106].

An yeast-derived HBV vaccine containing preS1 (aa 12–52) and preS2 (aa 133–145) sequences within a rearranged L* protein, in addition to the S protein, was also constructed [107,108]. The vaccine particles consisted of S and L* at a ratio of 7:3.

9.4.4 DNA Vaccines against Hepatitis B

In line with its pioneering traditions, HBs served as one of the first DNA vaccine models [109–115], which is based on a simple conversion of DNA expression vectors for HBs-based VLPs to the corresponding DNA vaccines. Although no DNA vaccine candidate is approved yet, DNA-based vaccination was demonstrated to improve anti-HBs immune responses in animal systems. In DNA vaccination, HBs proteins are expressed in *in vivo* transfected cells of the vaccine recipients in their native conformation with correct posttranslational modifications from antigen-encoding expression plasmid DNA. This ensures the integrity of antibody-defined epitopes and supports the generation of protective virus-neutralizing antibodies. Furthermore, DNA vaccination is an exceptionally potent strategy to stimulate CTL responses because antigenic peptides are efficiently generated by endogenous processing of intracellular protein antigens and thereby presented by the MHC class I pathway. These key features make DNA-based immunization an attractive strategy not only for prophylactic, but also for therapeutic vaccination against hepatitis B [116–119].

In fact, the problem of therapeutic vaccination against chronic hepatitis B remains highly important. Despite the availability of effective hepatitis B vaccines for many years, over 370 million people worldwide are persistently infected with HBV. Since viral persistence is related to poor HBV-specific T-cell responses, DNA vaccination as a strong T-cell inducer is thought to be an adequate measure.

A first phase I clinical trial was performed in chronic HBV carriers to investigate whether HBs DNA vaccination could restore T-cell responsiveness [120]. This study provided evidence that DNA vaccination can activate T-cell responses in some chronic HBV carriers who do not respond to current antiviral therapies. A subsequent phase I clinical trial confirmed the safety and immunological efficacy of HBs DNA vaccination and demonstrated restoration or activation of T-cell responses in chronic HBV carriers [121]. Safety and efficacy of the first powdered HBs DNA vaccine was demonstrated in a clinical study [122]. This clinical study evaluated for the first time the performance of a single-use disposable, commercial prototype device for particle-mediated epidermal delivery of DNA vaccine. However, patients were previously immunized with a licensed protein-based HBV vaccine, but received a single boost vaccination of the DNA vaccine.

Although the initial expectations of the performance of DNA vaccination in chronic hepatitis B patients were optimistic, a therapeutic HBs DNA immunization approach in a chimpanzee chronic HBV carrier with a high viral load failed to control HBV viremia [123].

The obvious difficulties in DNA vaccination pressed investigators to apply many advanced immunization strategies, with the use of different kinds of adjuvants and prime/boost regimens. Thus, primary immunization with a DNA construct encoding interleukin (IL)-12 and boosting with canarypox vectors expressing all HBs genes 6 months thereafter was tried, but did not lead to reduction in viral load in the chimpanzee model [123].

Promising results were obtained in a murine model by vaccine regimens using combinations of plasmid DNA encoding the gene M, poxvirus (modified vaccinia virus Ankara, MVA) encoding the same gene, and HBs protein (Engerix-B vaccine) to induce strong HBs-specific cellular and humoral responses [124]. Boosting of the humoral immune response to HBs DNA vaccine was achieved by coadministration of prothymosin α as an adjuvant [125].

Recently, an optimization of the HBs DNA vaccine plasmid vector was tried by modification of the content of immunostimulatory CpG motifs [126]. However, two doses of DNA vaccine did not generate any detectable anti-HBs antibody response in either of two chimpanzees in this study.

Progressive oral and intranasal immunization schemes are under investigation for putative HBs DNA vaccines. Single-dose oral immunization with biodegradable PLGA microparticles carrying HBs-encoding DNA led to the induction of a long-lasting and stable mucosal and systemic immune response in mice [127]. In contrast, naked DNA vaccines given by intramuscular injection induced only systemic cellular and humoral responses to HBs, which were much lower than the responses elicited by oral administration of DNA encapsulated in PLGA microparticles at equivalent doses. Intranasal HBs DNA vaccination using a cationic emulsion as a mucosal gene carrier dramatically enhanced S gene expression in both nasal tissue and lung and increased the cellular and humoral response in mice [128]. In contrast, very weak humoral and cellular immunities were observed following immunization with naked DNA.

Recently, a modified HBs with a preS1 peptide (aa 21–47) fused to the C terminus (at aa 223) of S protein was shown to induce strong humoral and cytotoxic T-cell responses in transgenic mice after a single DNA injection [129]. In that way, the authors tried to model the vaccination of chronic carriers.

9.4.5 Plant-Derived HBs Protein as an Edible Vaccine Candidate

Plant-derived protein S appeared in the next pioneering step as a real edible vaccine candidate. This plant-derived antigen showed a strong immunogenicity in mice [130,131] and humans [132]. Vegetable and fruit crops would be ideal host systems for the production of an oral hepatitis B vaccine. Considerable levels of the protein S production are achieved currently in banana, carrots, and cherry tomatilla. However, nonfood/feed crops such as tobacco, potato (uncooked), lettuce, and lupin, as well as *in vitro* systems such as plant cell cultures and hairy roots have many advantages, which may be exploited for the future production of hepatitis B vaccines (for a recent review see Ref. [133], and Chapters 18 and 19).

For the first time, a double-blind placebo-controlled clinical trial was performed on HBs protein produced in potatoes and delivered orally to previously vaccinated individuals [134]. The potatoes accumulated HBs protein at approximately 8.5 μg/g of potato tuber, and doses of 100 g of tuber were administered by ingestion. After volunteers ate uncooked potatoes, serum anti-HBs titers increased definitely, and the first plant-derived orally delivered hepatitis B vaccine was suggested by the authors as a viable component of a global immunization program.

A candidate of an edible S-based hepatitis B vaccine was produced in tomato fruits [135–137]. More, the authors intended to generate a bivalent HIV and hepatitis B vaccine [136,137]. However, this vaccine candidate was tested only by feeding laboratory mice yet.

Very recently, the L protein gene was expressed in tomato fruits resulting in the formation of the appropriate HBs VLPs [138].

Interesting alternative variant of an edible vaccine does not use VLPs, but is based on encapsulation of a HBs peptide representing residues 127–145 of the immunodominant epitope "a" in PLGA microparticles [139].

Unfortunately, all published data related to the edible vaccine candidates are missing the HBs VLPs as a reference control. Highly purified HBs protein (bulk of the ButaNG vaccine described above), as well as whole yeast cells containing HBs protein (biomass after fermentation at the ButaNG manufacturing) did not elicit any detectable anti-HBs antibodies in mice multi-fed with mentioned preparations [140].

9.4.6 Vaccine Candidates for Other Diseases

Despite differences among human- and yeast-derived HBs, the high-level expression in yeast stimulated numerous attempts to employ the HBs VLPs for protein engineering studies. Chronologically, the first example of chimeric HBs-based VLPs was the insertion of a long segment from herpes simplex virus type 1 glycoprotein gD into the C-terminal part of the preS2 sequence and the expression of such chimeric gene in yeast [36].

At virtually the same time, a VP1 segment spanning aa 93–103, which included the linear part of the neutralization epitope from both poliovirus types 1 and 2 (PV-1 and PV-2), was inserted into positions 50 and 113 of the two hydrophilic domains of the S protein [35,141–144]. Recombinant genes were expressed in mouse L cells from the SV-40 early promoter and enabled secretion of chimeric 22 nm particles. Yields of secreted 22 nm particles were dependent on the site of insertion. Cotransfection with different plasmids carrying either modified or unmodified S genes led to the formation of mosaic particles presenting HBs-polioVP1 fusion polypeptides on the helper HBs. Surprisingly, mosaic particles induced much higher titers of neutralizing antibodies to poliovirus than did the chimeric ones. This study thus initiated the promising idea of designing multivalent particles carrying various peptide sequences or presenting several heterologous epitopes of interest on the surface of the same VLP carrier molecule.

Next, chimeric HBsAg derivatives were generated harboring sequences from HIV. Because of an overlap between populations at risk for HBV and HIV infections, this stimulated the sound hope to develop a bivalent hepatitis B/HIV vaccine [145–147]. First, an 84-aa-residue-long gp120 fragment of HIV-1 was inserted into the preS2 part of the protein M and the fusion protein was produced in eukaryotic cells [147]. Immunization with these chimeric particles allowed not only for the generation of a humoral response in rabbits [147], but also the induction of neutralizing antibodies and proliferative T-cell and CTL responses in rhesus monkeys to both parts of the hybrid particle, i.e., HIV and HBs protein [148,149].

The same laboratory attempted to repeat this work with the V2 region of the simian immunodeficiency virus (SIV) gp140 protein [150]. Despite having consistent SIV-neutralizing antibody titers, vaccinated macaques were not protected against a homologous SIV challenge. Further, HBs served as a carrier for selected epitopes of HIV-1 that were shown to elicit neutralizing antibodies. Chimeras harboring the 6-aa sequence ELDKWA from the gp41 protein as an internal fusion induced antibodies in mice, which recognized HIV-1 gp160, but failed to neutralize HIV-1 *in vitro* [145]. Much attention was given to the V3 epitope of the gp120, which not only contained T-, CTL-, and B-cell epitopes including those inducing neutralizing antibodies, but also participated in coreceptor interaction and guided cell tropism of the virus. Initially, the V3 sequence of HIV-1 subtype MN was introduced into position 139 of the protein S and produced in yeast [151]. Further, it was inserted into the protein M [152], as well as at the N- and C-terminal positions of the protein S [153]. In mice these DNA vaccines elicited rapid and strong B-cell and CTL responses.

Recently, a putative bivalent HBV–HIV vaccine on the basis of the HBs carrying Gag and Env envelopes has been achieved in tomato plants [137]. Mice fed with the tomato fruits demonstrated HBV- and HIV-specific antibodies in the serum.

Development of HBs-based malaria vaccine candidates was also reported. Vaccine candidates have been constructed by using epitopes from three different human malaria parasite *Plasmodium falciparum* markers. First, computer-predicted B- and T-cell epitopes from the major merozoite surface antigen gp190 sequence (up to 61 aa residues in length) were introduced by replacing the proper antigenic determinants of the protein S, with further successful production of chimeric particles in mammalian cells via vaccinia virus expression [154]. Then, chimeric particles carrying 16 repeats of tetrapeptide NANP of the circumsporozoite (CS) protein of *P. falciparum* were produced by yeast cells and tested successfully on experimentally vaccinated volunteers [155]. As a further vaccine candidate, a sexual stage/sporozoite-specific antigen Pfs16 was fused to the HBs protein and synthesized in yeast cells in the form of mosaic particles in the presence of the helper HBs [156]. The most promising malaria vaccine in our days—RTS,S—contains 19 NANP repeats and the carboxy terminus (aa 210–398) of the CS antigen within the HBs–CS fusion co-expressed with the wild-type protein S as a helper [157–159]. According to recent phase I and IIa trials in children and adults, the RTS,S vaccine was found to be safe and well tolerated [160–163]. The merozoite surface protein 1 gene of *P. vivax* was also suggested as a vaccine candidate in combination with HBs as a carrier [164].

The idea of replacing the major immunodominant region of the S protein (aa 139–172) for the foreign epitopes was also exploited for the presentation of a segment of the human papillomavirus oncoprotein E7 spanning aa 35–98 aa [165].

Future improvement of the HBs as an epitope carrier may also be achieved by the addition of, for example, a Th epitope derived from tetanus toxoid [166]. Such chimeric 22 nm particles produced in a recombinant adenovirus expression system showed a several-fold enhancement of the anti-HBs response in mice, relative to native HBsAg, and were suggested for further exploitation as a new class of highly immunogenic HBs carriers.

In the case of HCV sequences on the HBs carrier, both DNA-based immunization and vaccinia virus expression approaches have been tested first. In the plasmid DNA vector, the 58-aa-long N-terminal fragment of the HCV nucleocapsid was fused to the M or S proteins [167]. In the vaccinia virus expression system, the five hydrophilic domains of the HCV E2 envelope as well as the hypervariable region (HVR) of the E2 protein were presented on the HBsAg surface [168].

In another study, chimeric HBs–HVR VLPs carrying the HVR domain within the "a" determinant region of HBs elicited HVR1-specific antibodies in mice [169]. Interestingly, the HBs–HVR1 VLPs are able to induce a primary immune response to HVR1 in anti-HBs positive mice and, hence, they may be used successfully as a vaccine in HBV carriers and vaccinees.

9.5 HEPATITIS B CORE PROTEIN AS VACCINE CARRIER

9.5.1 Fine Structure

In contrast to the HBs, employment of the HBc as a VLP carrier has been reviewed extensively [1,170–177]. The HBc was first reported as a promising VLP carrier in 1986 [38] and published in 1987 [2,37] essentially at the same time as the use of HBs particles for the construction of chimeric VLPs. Unlike the HBs particle, which is provided with octahedral symmetry (see Section 9.4.1), the HBc particle represents an icosahedron. The multifunctional character of the HBc in the virus replication and assembly seems to be the basis for its unusual flexibility for foreign insertions. The HBc protein, fully or in part, is present in at least four different functionally relevant HBV polypeptides: p25 (preC protein), p22 (N-terminally cleaved form of the p25), p21 (HBc monomer as such), and p17 (HBe antigen, N- and C-terminally cleaved form of the p25) [178]. From these polypeptides, only the HBc is able to self-assemble and serves therefore as a target for the VLP engineering. Besides capsid building, the p21 protein participates in the viral life cycle and its regulation, including synthesis of dsDNA (as a cofactor of the viral reverse transcriptase-DNA polymerase (Pol)), viral maturation, recognition of viral envelope proteins, and budding from the cell [179]. The HBc can recognize specific sites of the envelope proteins S and L, in order to build up the HBs envelope on the HBc capsid [51,180,181].

In many ways HBc holds a unique position among other VLP carriers because of its high-level synthesis and efficient self-assembly in virtually all known homologous and heterologous expression systems, including bacteria, in contrast to the HBs gene expression. Correct folding of the HBc monomer and formation of authentic HBc particles have been documented in various mammalian cell cultures driven by different expression systems including retro-, adeno-, and vaccinia virus vectors. The HBc self-assembled well in frog *Xenopus* oocytes, insect *Spodoptera* cells, yeast *S. cerevisiae*, plant *Nicotiana tabacum*, and bacteria *Escherichia coli, Bacillus subtilis*, *Salmonella*, and *Acetobacter* species (for review see Ref. [1]). After publishing the latter review, highly efficient synthesis of the correctly folded HBc VLPs has been shown in *P. pastoris* yeast [182–184] and by an alphavirus-driven expression system [185].

Recently, efficient HBc expression and self-assembly was achieved in tobacco and cowpea plants by constructs based on either potato virus X (PVX) or cowpea mosaic virus (CPMV) vectors, respectively [186]. Since the constructs were able to infect efficiently the above-mentioned hosts, remained genetically stable during infection, and allowed purification of the HBc, the study provided a practical method for cheap and rapid production of assembled HBc particles in plants. Another rapid high-level HBc production model was achieved in *Nicotiana benthamiana* leaves with levels of up to 7.14% of total soluble protein or 2.38 mg HBc per gram of fresh weight at 7 days after infection [187].

General structural and major immunological features of the HBc protein are shown in Figure 9.5. A linear presentation of the HBc with location of the major immunodominant region (MIR), a border of the self-assembly competence, α-helix structural elements, and major immunodominant B-cell and CTL epitopes (Figure 9.5A) is illustrated with presentation of the same crucial elements on a three-dimensional map of the HBc monomeric subunit (Figure 9.5B). The tetrameric asymmetric unit of the HBc particles is shown in Figure 9.5C.

The major HBc B-cell epitopes (c and e1) are localized within the MIR, on the tip of the spike, around the most protruding region between aa 78 and 82. The next important epitope, e2, is localized on the other surface-exposed region of the HBc particle, around aa position 130.

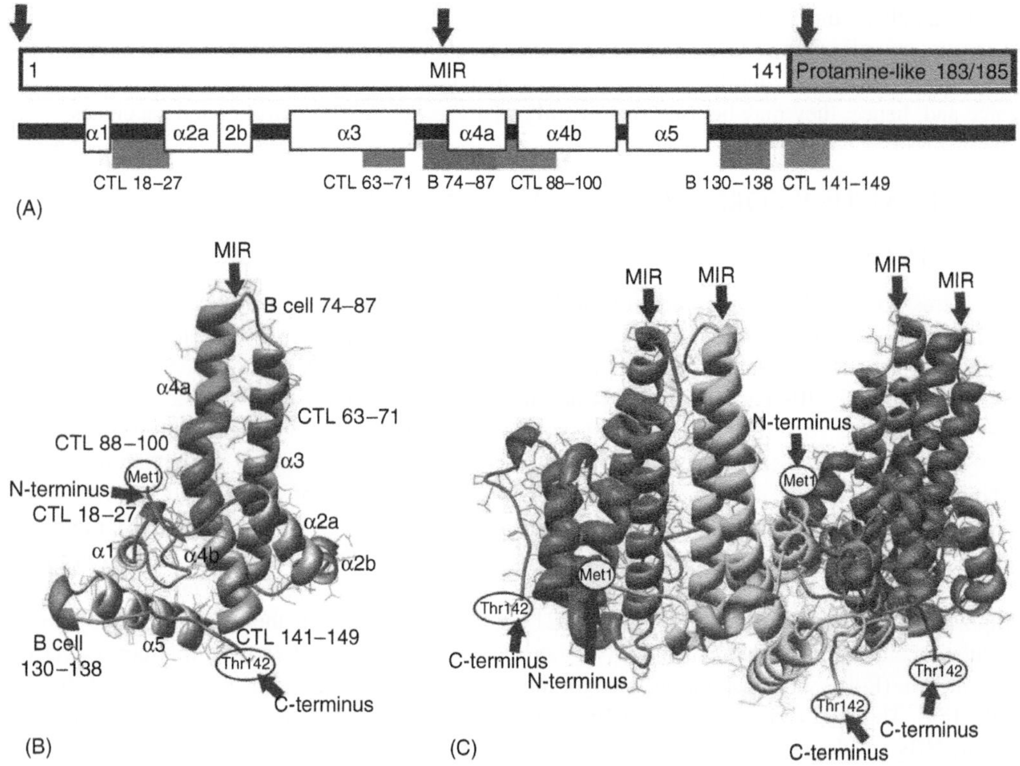

FIGURE 9.5 (See color insert following page 142.) General structural and immunological features of the HBc protein. (A) A linear presentation of the HBc with localization of the major immunodominant region (MIR), border of the self-assembly competence before the protamine-like C-terminal region. Location of the α-helices and major immunodominant epitopes (B cell, blue; CTL, orange) is presented. For details of the immunological epitopes, see reviews of Refs. [174,175]. (B) A three-dimensional presentation of the HBc chain A according to the x-ray structure [13] with localization of α-helices, immunodominant epitopes (the same colors as in A). (C) Tetrameric asymmetric unit of the HBc particles. Chains are colored as follows: A, orange red; B, gold; C, green; D, blue. Major targets for foreign insertions are depicted by blue arrows on the linear and three-dimensional presentations of the HBc protein. When located on the forefont of the map, the N- and C-terminal amino acid residues are deciphered.

The human CTL epitope at aa 18–27 deserves special notice because it is regarded as a promising HBV vaccine candidate beyond the whole HBc VLPs (see Section 9.5.3).

Full three-dimensional organization of the HBc is presented in Figure 9.6, where x-ray maps (Figure 9.6A) are supplemented with electron micrographs of the natural HBc within the HBV virions (Figure 9.6B), recombinant *E. coli*-expressed HBc of full-length (Figure 9.6C), and without C-terminus (Figure 9.6D) and its chimeric derivatives with long foreign insertions at the C-terminus (Figure 9.6E) and within the MIR (Figure 9.6F). Figure 9.6G presents specific immunogold mapping of foreign insertions at the MIR on the outer surface of the chimeric HBc.

The fine structure of HBc particles was revealed first by electron cryomicroscopy and image reconstruction at 7.4 Å resolution [5,19,20]. High-resolution three-dimensional HBc structure was obtained by x-ray crystallography at 3.3 Å resolution [13]. Organization of HBc particles was found largely α-helical and quite different from previously known viral capsid proteins with β-sheet jelly-roll packings [13,20]. In brief, the HBc monomer fold is stabilized by a hydrophobic core that is highly conserved among human HBV variants. Association of two amphipathic α-helical hairpins results in the formation of a dimer with a four-helix bundle as the major central feature (Figure 9.3).

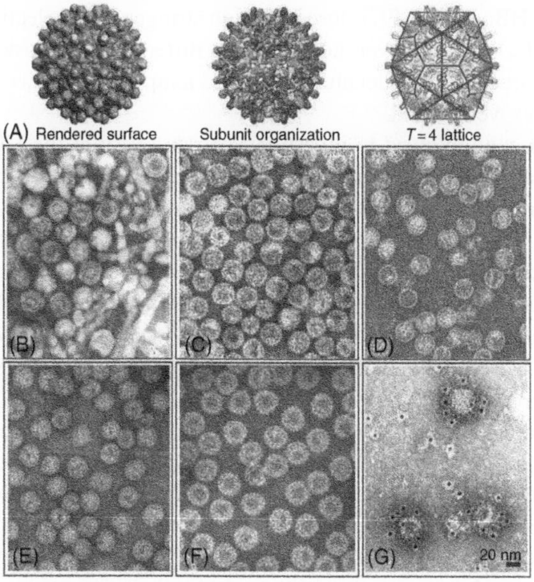

FIGURE 9.6 (See color insert following page 142.) Structural characteristics of the VLPs on the basis of the HBc particles. (A) X-ray structure of the HBc particle (From Wynne, S.A., Crowther, R.A., and Leslie, A.G., *Mol. Cell*, 3, 771, 1999.) according to the VIPER presentation directed by Vijay Reddy (From Reddy, V.S. et al., *J. Virol.*, 75, 11943, 2001.). (B) HBV preparation from the blood of an infected patient, envelopes and internal cores of HBV virions (so-called Dane particles) are visible, as well as filamentous forms of the 22 nm HBs particles, (C–G), electron microscopy of the *E. coli*-expressed HBc VLPs: (C) full-length HBc particles without any insertions; (D) C-terminally truncated HBc particles without any insertions; (E) C-terminally truncated HBc particles with long C-terminal insertion; (F) full-length HBc particles with long internal (into the MIR) insertion; (G) immunogold mapping of the C-terminally truncated HBc particles with internal insertion (into the MIR) with specific monoclonal antibodies against inserted epitope.

The dimers are able to assemble into two types of particles, large and small, that are 34 and 30 nm in diameter and correspond to the triangulation number $T=4$ and $T=3$ packings, containing 240 and 180 HBc molecules, respectively. The MIR with the central aa positions 76–81 is located at the tips of the α-helical hairpins that form spikes on the capsid surface. It is significant that the existence of two size classes of HBV cores, with predominantly larger and less frequently occurring smaller particles, was detected earlier using particles from HBV-infected human liver, by either electron microscopy of a negatively stained specimen [188] or by gel filtration [189]. In addition to the highly exposed MIR, the region between aa 127 and 133 is the next exposed and accessible epitope on the particle surface. This region is located at the end of the C-terminal α-helix and forms small protrusions on the surface of the HBc particle.

Using electron cryomicroscopy, most intimate mechanisms of the HBc maturation process were studied [190]. Comparing the structures of the bacterially expressed RNA-containing HBc particles and the mature DNA-containing HBc particles extracted from virions, the authors found significant differences in the structure between the RNA- and DNA-containing cores. One such difference is in a hydrophobic pocket, formed largely from residues that, upon mutation, lead to abnormal secretion. Since the envelopment of the HBc must not happen before reverse transcription is completed, a change in capsid structure may signal maturation. Therefore, during the infective cycle, the HBc assembles in an immature state around a complex of viral pregenomic RNA and polymerase. After reverse transcription with concomitant degradation of the RNA, the now mature core buds through a cellular membrane containing the surface proteins to become enveloped [190].

Using a highly efficient electron microscopy approach, so-called immature secretion phenomenon of the HBc mutant I97L harboring an isoleucine to leucine exchange at aa residue 97 was investigated [191]. The HBc mutant I97L loses the high stringency of selectivity in genome maturity during virion export [192,193]. However, no significant differences in capsid stability between wild-type and mutant I97L particles under denaturing pH and temperature, as well as in the ratio between $T=3$ and $T=4$ particles were found.

Electron cryomicroscopy allowed to investigate the diversity of the HBc epitopes and to map discontinuous conformational epitopes on the MIR [194] and at the interface between two adjacent subunits [195]. The epitopes were mapped by labeling the HBc particles with monoclonal Fabs at approximately 11 and 10 Å resolution, respectively. Very important, a conformational, but continuous epitope consisting of a loop-helix motif (residues 77–87) on one of the two polypeptide chains in the spike was mapped within the MIR [196]. Another mapped MIR epitope, like most conformational epitopes, is discontinuous, consisting of a loop on one polypeptide chain (residues 74–78) combined with a loop-helix element (residues 78–83) on the other [196].

After electron microscopy and x-ray crystallography, HBc particles were subjected to atomic force microscopy investigations [197]. Interestingly, it was the first time, when the HBc VLPs produced in *S. cerevisiae* yeast were investigated by highly sensitive structural methods. It was found both by electron and atomic force microscopy that also in yeast the HBc monomers could self-assemble into two size classes of HBc particles. The mean diameters of the latter were determined as 30.1 and 31.3 nm for larger particles and 21.5 and 22.5 nm for smaller ones by electron and atomic force microscopy, respectively.

Of particular structural value was the demonstration of the dispensability of the C-terminal protamine-like arginine-rich domain of the HBc protein for its self-assembly capabilities in the so-called HBcΔ particles [198–200]. The HBcΔ particles formed by C-terminally truncated polypeptides are practically indistinguishable from the HBc particles of the same $T=3$ or $T=4$ symmetry, but formed by full-length HBc polypeptides, as shown by electron cryomicroscopy [20]. Unlike the full-length HBc particles, HBcΔ particles are unable to encapsidate nucleic acid and appear as empty shells. Using negative staining and electron microscopy, HBc particles appear as "thick-walled" spherical particles with little interior space, whereas HBcΔ particles appear as "thin-walled" spherical particles with a much larger inner space [191]. The unusual molecular flexibility of the C-terminal protamine-like domain has been revealed by structural analysis of HBc particles using NMR spectroscopy [201]. The C-terminal limit for self-assembly of HBcΔ particles was mapped experimentally between aa residues 139 and 144 [200,202,203]. According to the most recent data [204], it maps at position 140 (the HBc1–140 is the last molecule retaining self-assembly capacity) and gives predominantly the $T=3$ isomorph and a proportion of $T=4$ isomorph of approximately 18%. The relative proportion of $T=4$ to $T=3$ capsids increases with the length of the HBc polypeptide. The empty HBcΔ particles seem to be very promising candidates for further vaccine and gene therapy applications.

The peptide 141-STLPETTVV-149, linking the shell-forming HBc "core" domain and the nucleic acid-binding protamine-like domain, has a morphogenic role [205]. The linker peptide is attached to the HBc inner surface as a hinged strut, forming a mobile array, an arrangement with implications for morphogenesis and management of encapsidated nucleic acid. The peptide was found to be necessary for the assembly of protamine domain-containing capsids, although its size-determining effect tolerates some modifications. Although largely invisible in a capsid x-ray structure, the linker peptide was visualized by electron cryomicroscopy difference imaging. Interestingly, a closely sequence-similar peptide in cellobiose dehydrogenase, which has an extended conformation, offers a plausible prototype for knowledge-based protein engineering [205].

As to formation of the HBc dimers as basic units of the HBc folding, it was shown that authentic HBc monomers may be involved in heterologous dimers with HBc mutant monomers lacking aa stretches 86–93 and even aa 77–93 at the MIR with further formation of mosaic particles [206].

It is obvious that *in vitro* self-assembly of the HBc protein is highly dependent on protein and NaCl concentration. However, it was demonstrated recently that micromolar concentrations of Zn^{2+}

are sufficient to initiate assembly of HBc protein, whereas other mono- and divalent cations elicited assembly only at millimolar concentrations, similar to those required for NaCl-induced assembly [207]. Probably, zinc ions are not the true *in vivo* activators of the HBc self-assembly, but they may provide a model for assembly regulation.

Summarizing, a unique three-dimensional folding of the HBc protein, its exceptionally high immunogenicity, presence of encapsidated RNA and DNA, variants, and polymorphisms of the HBc protein are summarized in excellent recent reviews [208–210].

9.5.2 Optimal Sites for the Foreign Insertions

The HBc protein accepts foreign insertions and retains self-assembly competence, when they are directed to the N- and C-termini and to the MIR, namely, to the tips of the HBc spikes. Historically, N-terminal insertions were the first ones in which chimeric HBc particles carrying the VP1 epitope aa 141–160 of foot and mouth disease virus (FMDV) were demonstrated using the vaccinia virus expression system [37,38], yeast [211], and bacteria [212].

The N-terminus of the HBc protein was used also as an insertion target for relatively short epitopes from the HBV (preS); retroviruses: HIV-1 (gp120, p24, gp41, p34 Pol, p17 Gag), SIV (Env), and feline leukemia virus (gp70); human cytomegalovirus (gp58), human rhinovirus type 2 (VP2), poliovirus type 1 (VP1) (for more details see Refs. [1,175]).

Remarkable breakthrough in the application of the HBc VLPs for vaccine development was also based on a construct harboring the insert at the N-terminal site of HBc. *E. coli*-expressed chimeric particles carrying 23 aa of the extracellular domain of influenza A minor protein M2 provided up to 100% protection against a lethal virus challenge in mice after intraperitoneal or intranasal administration [213].

Fusion of 45 N-terminal aa of the *Puumala hantavirus* nucleocapsid protein to the N-terminus of the HBcΔ allowed for the formation of chimeric VLPs, which induced a strong antibody response and some protection in the bank vole model [214]. However, the addition of 120 N-terminal aa of the hantavirus nucleocapsid protein to the N-terminus of the HBcΔ prevented self-assembly, in contrast to their insertion into the MIR [215].

In general, accessibility of the foreign insertions on the HBc particle surface is good. The maximum length of N-terminal insertions is achieved in following models: 57 aa of the total insertion with the 45 N-terminal aa of the hantavirus nucleocapsid [214], 52 aa of the HIV gp120 [216], and 50 aa of the human chorionic gonadotropin [211].

Regarding the C-terminal insertions, HBc positions 144, 149, and 156, and, recently, 163 and 167 [182] were used most frequently as target sites for fusions of foreign epitopes (for more details, see Refs. [1,175]). The C-terminal modifications involved two types of vectors, encoding either full-length or C-terminally truncated HBc. Despite the fact that capsids formed by the C-terminally truncated HBc derivatives (HBcΔ) are usually less stable than the capsids formed by full-length HBc proteins, high-level synthesis in bacteria and dissociation/reassociation capabilities of the HBcΔ are important advantages.

Full-length HBc vectors were used for insertion of epitopes from HBV (up to 55 aa from the preS); retroviruses: HIV-1 (about 50 aa of the gp41), SIV (20 aa fragments of the Env), and bovine leukemia virus (BLV) (about 50 aa of the Env gp51); murine cytomegalovirus (a 9-aa-long immediate early CTL epitope from the pp89). HBcΔ vectors were used for the expression of epitopes from HBV (preS1); retroviruses: HIV-1 (gp120, gp41, Gag and Nef), BLV (gp51); human cytomegalovirus (gp58), FMDV (VP1), hantavirus (45 aa of the nucleocapsid), HCV (core, NS3), hepatitis E virus capsid, and bacteria like *Porphyromonas gingivalis* (a 47-aa residues epitope) and *Bordetella pertussis* (a 30-aa epitope of outer membrane protein P69) (for more details, see references [1,175]. A 17 kDa nuclease was packaged into the interior of HBc capsids after fusion to the HBc position 155 [217]. The packaged nuclease retained enzymatic activity, and the chimeric protein was able to form mosaic particles with the wild-type HBc protein.

The capacity of the C-terminal vectors usually exceeded 100 aa residues, depending on the structure of insertion. In some cases, the C-terminal vectors demonstrated an extraordinary insertion capacity: Three copies of the HCV core of 559 aa in length did not prevent self-assembly of chimeras, and even four HCV core copies of 741 aa allowed for some production and self-assembly of chimeric VLPs [218]. In some cases, the C-terminally inserted sequences are exposed, at least partially, at the surface of the HBc particle.

The most interesting and promising site for foreign insertions is the MIR, located at the tips of the spikes of the HBc molecule, which allow full exposure of the inserted peptide on the VLP surface. The MIR insertion site was used for the introduction of epitopes from HBV (up to 27-aa epitopes of the preS, 39 aa of the HBs domain "a," positions 111–149); retroviruses HIV-1 (V3 loop of the gp120) and SIV (Env); the human rhinovirus type 2 (18 aa of the VP2); HPV (E7); FMDV (an RGD-containing epitope from the VP1 protein); hepatitis E virus (42 aa of the capsid); malaria agents *P. falciparum* and of two rodent malaria agents, *P. berghei* and *P. yoelli* (CS protein repeat epitopes); and a cattle theileriosis agent *Theileria annulata* (C-terminal segment (SR1) of SPAG-1, a sporozoite surface antigen) [1,175]. The MIR vectors demonstrated unexpectedly high capacity for the size of foreign insertions. The entire 120-aa-long immunoprotective region of the hantavirus nucleocapsid protein and moreover copies of this region from two different hantavirus species were inserted into the MIR of the HBcΔ resulting in VLP formation [215]. Green fluorescent protein (GFP) of 238 aa was also natively displayed on the surface of the HBc particles [219].

The MIR vectors have been used for the construction of the first multivalent particles, i.e., for the simultaneous insertion of different foreign sequences from the HBV preS1 and preS2 regions into the MIR and N-terminus [220] or into the MIR and C-terminus [221] or from the HBs and preS2 into the MIR and C-terminus of the HBc protein [222]. Later, multivalent HBcΔ particles carrying different hantavirus nucleocapsid protein epitopes at the MIR and C terminus were constructed [223].

The first mosaic HBc particles carrying both chimeric and wild-type HBc monomers were also constructed on the basis of full-length HBc vectors for internal insertions. In this case, an epitope of 8 aa from the Venezuelan equine encephalomyelitis virus E2 protein was inserted into position 81 of the HBc molecule and co-expressed with a HBc helper [224]. Further development of the technology of mosaic particles employed a new strategy for constructing mosaic particles. It was based on the introduction of a linker containing translational stop codons (UGA or UAG) between sequences encoding a C-terminally truncated HBcΔ and a foreign protein sequence. Expression of such a recombinant gene in an *E. coli* suppressor strain leads to the simultaneous synthesis of both HBcΔ as a helper moiety and a read-through fusion protein containing a foreign sequence [32,215,223]. This technology allowed for the incorporation into and presentation onto mosaic particles of segments of hantavirus nucleocapsid protein that were 45 [214], 114 [223], 120 [215], and even 213 [225] aa in length, although nonmosaic HBcΔ carrying the larger hantavirus segments at the C terminus were unable to self-assemble.

Prediction of the possible self-assembly competence of the HBc after foreign insertions, to self-assemble or not to self-assemble, remains the central question of the VLP protein engineering. Unfortunately, so far there is no valid algorithm for *in silico* prediction of the assembly competence of HBc fusion proteins, and it remains to prove the self-assembly of each construct experimentally by the trial-and-error method. However, at least aa residues with high hydrophobicity, large volume, and high beta-strand index were discussed as potential reasons for lacking self-assembly competence of chimeric HBc monomers [226]. In addition, a comparison of the assembly competence of 45, 80, and 120 N-terminal segments of hantavirus nucleocapsid proteins was discussed to be based perhaps on the structure of the sequence; the 45 and 120 aa-long segments might be assembly competence due to a flexible structure at the linker regions to HBc, whereas the 80 aa-long nonassembly competent segment linker region is not so flexible [227]. A pioneering idea for structure probing and self-assembly control of the HBc VLPs was proposed by Vogel et al. [228]. Nevertheless, it remains very difficult to provide users with reliable recommendations how to obtain correct folding and thereby self-assembly competence of chimeric HBc proteins.

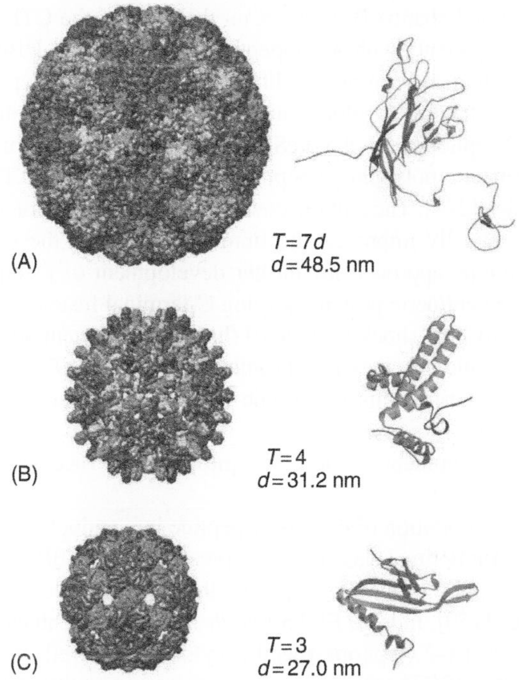

FIGURE 9.7 (See color insert following page 142.) Comparative size and structure of the most popular VLP carriers: (A) Mouse polyomavirus; (B) HBc; and (C) RNA phage Qβ. Subunit organization (left) and subunit folds (right) taken from the VIPER program are presented (From Reddy, V.S. et al., *J. Virol.*, 75, 11943, 2001.). The VLP representations are scaled according to their relative sizes, while the subunits are not. The triangulation (*T*) numbers and average diameters of the VLPs are given in the middle.

Figure 9.7 demonstrates the structure of the HBc VLP, its triangulation number, and average diameter, as well as a structural map of the HBc monomer in comparison with the same data for two popular and widely used VLP carriers: mouse polyomavirus and RNA phage Qβ. The VLP maps are scaled according to their relative sizes, while the subunits are not. Therefore, the HBc demonstrates an intermediate size among other successful VLP carriers and absolutely different organization of the capsid (triangulation number) and monomer (α-helical organization).

9.5.3 Hepatitis B Vaccine Candidates

Among the HBV polypeptides, HBc induces the strongest B-cell, T-cell, and CTL response (for review see Ref. [229]). The HBV-infected patients develop a strong and long-lasting humoral anti-HBc response, in contrast to the low anti-HBs response [230]. In addition, HBc is known to function as both a T-cell-dependent and T-cell-independent antigen [231]. The same is true also for chimeric derivatives of the HBc VLPs [232]. Following immunization, it primes preferentially Th1 cells, does not require an adjuvant [233], and is able to trigger an anti-HBs response [234]. Enhanced immunogenicity of the HBc can be explained by its ability to be presented by B cells as the primary antigen to T cells in mice [235–237]. HBc elicits a strong CTL response during HBV infection [238] and this response is maintained for decades following clinical recovery, apparently keeping the virus under control [239].

In spite of its nature as an internal antigen, HBc particles were found in the middle 1980s to be able to provide protection against HBV infection in chimpanzees [240–242]. Attempts to include HBc into HBV vaccines by genetic immunization [243–245] or as a CTL epitope vaccine (aa 18–27) [246] have been published.

Possibility of therapeutic hepatitis B vaccines on the basis of the CTL epitope 18–27 deserves noticing. Immunization of patients with a lipopeptide containing the HBc 18–27 epitope elicited specific CTL responses in humans expressing different HLA-A2 supertype molecules and demonstrated potential usefulness for the development of a vaccine [247]. Further attempts in this field tried to combine the 18–27 epitope with the preS2 B-cell epitope and a common T-helper epitope of tetanus toxoid. This therapeutic polyepitope peptide was able to induce CTL-mediated cytotoxicity in HLA-A2 transgenic mice [248]. The authors concluded that introduction of the universal T-helper and B-cell epitopes dramatically improved the immunogenicity of the CTL epitope *in vivo* and recommended the polyepitope approach for further development of therapeutic peptide vaccines. Yeast *P. pastoris*-expressed chimeric protein carrying C-terminal fusion of the 18–27 epitope to the *Mycobacterium tuberculosis* heat-shock protein 70 (hsp70) as a potent adjuvant was suggested for the development of therapeutic vaccines for chronic hepatitis B [249]. Such chimeric protein was able to activate *in vitro* dendritic cells in chronic hepatitis B patients and healthy controls and generate a HBV-specific CTL response.

A next stage of the development of the CTL-epitope-based vaccines is connected with DNA vaccination.

In order to promote transportation of the 18–27 peptide into endoplasmic reticulum (ER) to bind to MHC class I molecules for optimal class I antigen presentation, an ER targeting sequence (ERTS) was fused with the 18–27 epitope-encoding gene and the resulting recombinant gene was delivered as a DNA vaccine to mice [250]. Indeed, ERTS fusion significantly enhanced specific $CD8^+$ T cell responses in terms of the 18–27 epitope-specific cytolysis as well as IFN-γ secretion. More advanced single-chain trimer (SCT) technique was used to present the epitope 18–27 (and another CTL epitope 107–115) after corresponding DNA vaccination [251]. The study showed that the SCT presentation of the HBc epitopes ensured functional and stable epitope-specific $CD8^+$ T-cell responses in HLA-A2.1/Kb transgenic mice and might be useful for the control of HBV infection in HLA-A2-positive HBV carriers.

Obvious difficulties in the development of vaccination approaches based on isolated epitopes and polyepitope peptides forced the use of the HBc VLPs as a putative component of hepatitis B vaccines. Moreover, 20 years after first enthusiastic works of the HBc vaccination pioneers [240–242], the interest in the HBc as a vaccine component is again rising. Now, the Rhein Biotech company is trying to introduce the *E. coli*-produced HBcΔ into putative therapeutic hepatitis B vaccines.

Recent demonstration of high mucosal immunogenicity including cellular response to HBc particles after nasal administration may pave the way to further vaccine developments based on HBc [252–254]. It is important to note that also plant-derived HBc particles demonstrated in mice high mucosal immunogenicity after oral and nasal administration [187].

As in the case of HBs, novel putative HBc vaccines were constructed on the basis of plasmid DNA. DNA vaccination via a gene gun elicited a specific anti-HBc response and induced a HBc-specific CTL response in immunized Balb/c mice [255]. Specific T-cell response to the HBc DNA vaccine was enhanced by addition of the DNA fragment encoding the IL-1β polypeptide fragment spanning aa 163–171, which might represent a good candidate for an adjuvant of DNA vaccines [256].

In addition to the development of optimized administration types and routes, a novel system of antigen displaying has been adopted for HBc VLPs. Empty bacterial cell envelopes, i.e., *E. coli* ghosts carrying HBcΔ VLPs anchored either in the inner or the outer membrane, induced significant immune responses by subcutaneous immunization of mice [257].

However, from the very beginning, the idea of combining the most relevant epitopes from other HBV proteins on the HBc carrier seemed the most promising way to get a highly efficient HBV vaccine. It was not surprising that combinations of the HBc carrier with the preS and S at the MIR, and, possibly, HBV polymerase epitopes at the C-terminus were regarded as the most prospective HBV vaccine in the future. Some attempts have been made in this direction. Fifteen years ago, the

British company Medeva started a *Hepacore* project, which aimed at the construction of therapeutic vaccines on the basis of chimeric HBc derivatives. A healthy volunteer study using chimeric HBc particles containing the preS1 sequence 20–47 inserted into the MIR has been planned for 2000, but never occurred after joining the Medeva to the Celltech company.

Since previous review in 2002 [1], a set of novel preS-carrying HBc VLPs have been constructed: the HBcΔ (aa 1–144) bearing the preS1 (aa 1–42) [258], HBc exposing one to three tandem copies of the preS1 (aa 21–47) at the MIR [259], and numerous variants of the preS1 (aa 21–47) insertions into the MIR with and without internal MIR deletions in HBc and HBcΔ VLPs [260,261]. Moreover, practically full-length preS1, with deletion of the inner hydrophobic membrane-targeting fragment, was exposed on the surface of the HBc and HBcΔ VLPs [260,261]. A strong therapeutic potential was shown with the HBcΔ (aa 1–155) variant bearing the C-terminally fused preS1 (aa 3–55) peptide, since immunization of transgenic mice with the chimeric VLPs reduced the titer of HBs antigen and HBV DNA in their sera [262]. The preS1 peptide added at the C-terminus was found surface accessible.

In addition, numerous variants of mosaic HBc particles carrying complete preS sequence at the MIR have been constructed with and without MIR deletions [263].

Expression of recombinant HBc-encoding genes in *Salmonella typhimurium* promised the generation of oral hepatitis B vaccines on the basis of live, avirulent strains of *Salmonella* species [264,265]. Thus, efficacy of a single oral immunization of BALB/c mice with a recombinant *S. typhimurium* carrying HBc-preS chimera has been shown [266]. In this case, HBc-preS chimera contained the preS1 epitope aa 27–53 inserted between aa positions 75 and 81 of the HBc protein, and the preS2 epitope aa 133–143 fused to position 156 of the HBc protein. However, volunteers that received the oral *Salmonella* HBc-preS vaccine failed to develop humoral and cellular responses to hepatitis B antigens [267].

Nevertheless, this vaccination approach remains worth further developing. In particular, nasal vaccination of mice with recombinant attenuated strains of *S. typhimurium* was found more efficient at inducing antibody responses than oral vaccination [268]. Recently, *Salmonella* vehicles for delivering HBc particles were optimized by chromosomal integration of the target gene and employment of promoters inducible *in vivo*, and therefore advanced candidate live oral vaccines were suggested [269].

A promising *Salmonella* expression variant of the HBc-derived chimeras carrying HBs fragments was achieved with an internally inserted "a" epitope of the HBs [270,271]. A single rectal immunization of mice with this HBc-HBs recombinant induced humoral and cellular immune responses to HBc and HBs as well as the generation of a specific mucosal immunity [270].

Very recently, a novel attenuated strain of bacteria, *Mycobacterium smegmatis* was used for expression of HBcΔ (aa 1–155) bearing preS1 (aa 1–55) fragment as a live vaccine candidate [272]. As a result, vaccination with live recombinant *M. smegmatis* induced a stronger cellular immune response and a longer persistence of the humoral immune response than that induced with the DNA vaccination using a corresponding chimeric construct.

9.5.4 Vaccine Candidates against Other Diseases

HBc has been tried as a carrier moiety for the development of many vaccine candidates. First, the ability of HBc particles carrying the N-terminal insertion of the FMDV major VP1 epitope (aa 141–160) for the induction of FMDV-neutralizing antibodies was demonstrated [37]. Next, an RGD-containing epitope from the FMDV VP1 protein was exposed within the MIR, and the HBc-RGD chimeric particles not only elicited high levels of FMDV-neutralizing antibodies in guinea pigs but mediated by the RGD motif insertion also bound specifically to cultured eukaryotic cells and to purified integrins [273].

Next generation of the HBc-FMDV vaccine candidates presented a combination of the FMDV epitopes VP1 (141–160)-VP4 (21–40)-VP1 (141–160) on the MIR with deletions of the intrinsic

HBc stretches: aa 75–78, aa 75–80, and aa 75–82 [274]. The first plant-derived HBc-FMDV vaccine candidate with the FMDV VP1 (140–160) G-H loop epitope being inserted at the HBc MIR was produced in transgenic tobacco and induced a strong protection in immunized mice against FMDV challenge [275].

An attempt to construct a contraceptive vaccine candidate by N-terminal insertion of the fragment of the human chorionic gonadotropin has been described [211]. From other early attempts, C-terminal addition of the CTL epitope from the murine cytomegalovirus pp89 protein ensured a protective immune response against lethal MCMV infection [276].

MIR insertions into the HBc particles were thoroughly investigated for the possible construction of vaccines against infectious diseases caused by intracellular parasites. The first vaccines tested were malaria vaccines, in which chimeric HBc-CS particles carrying circumsporozoite (CS) protein repeat epitopes of two rodent malaria agents, *P. berghei* and *P. yoelli*, were expressed in *S. typhimurium* and elicited high levels of antisporozoite antibodies and protection in mice [277,278]. However, similar hybrid HBc particles containing *P. falciparum* $(NANP)_4$ repeat elicited antibody titers that were lower in orders of magnitude and were poorly reactive with *P. falciparum* sporozoites [277]. An optimal *P. falciparum* CS-HBc immunogen should elicit not only high levels of antisporozoite antibodies, but also parasite-specific T cells to target both extracellular and intracellular parasite stages. To improve efficiency of the HBc-CS vaccine, two T-cell epitopes from the CS stage protein have been added. A series of recombinant HBc-CS proteins containing the parasite-derived B-, T1-, and T*-cell epitopes in various combinations and configurations were assayed for immunogenicity in mice [279,280]. One particle, ICC-1132, contained the T1 and B repeat epitopes inserted at the HBc MIR and the universal T-cell epitope T* fused to the C-terminus of the truncated HBc protein. In preparation for a phase I clinical trial, the ICC-1132 vaccine candidate elicited potent anamnestic antibody responses in mice primed with *P. falciparum* sporozoites, suggesting potential efficacy in enhancing the sporozoite-primed immune responses of individuals living in areas where malaria is endemic [281]. A phase I trial of the ICC-1132, which received a name Malarivax, supported it as a promising malaria vaccine candidate for further clinical testing using more potent adjuvant formulations than the alum adjuvant used in this initial study [282]. In order to improve immunogenicity of the vaccine candidate, a more potent water-to-oil adjuvant has been tried. First phase I trial in the United Kingdom used a formulation of the ICC-1132 in a Seppic ISA720 adjuvant (France) and showed induction of anti-NANP antibodies and a modest T-cell response, but no evidence of protection from experimental challenge with *P. falciparum* sporozoites [283]. Second phase I trial in Germany with formulation of the ICC-1132 vaccine candidate in the same ISA720 adjuvant demonstrated safety, tolerability, and efficacy of the vaccine to induce cellular and humoral responses; however, no experimental challenge was performed [284]. In preclinical studies in the Netherlands, rhesus macaques were immunized with the ICC-1132 vaccine candidate in alum or ISA720 adjuvants, and recommendations for vaccine and adjuvant dosage were elaborated [285]. Therefore, the future of this prospective malaria vaccine candidate remains to be elucidated.

In a similar approach, but with animal vaccine, a C-terminal segment (SR1) of SPAG-1, a sporozoite surface antigen of *T. annulata*, an infectious agent of cattle theileriosis, has been expressed in a MIR-insertion construct [286]. The chimeric HBc-SR1 particles not only induced high titers of neutralizing antibodies and a significant T-cell response, but they also showed some evidence of protection against a sporozoite challenge in cattle [286], which allowed for their recommendation to be included in a future multicomponent vaccine for cattle [287].

Serious attempts to construct a therapeutic vaccine against HPV16-associated anogenital cancer was undertaken using the MIR insertions of B-, T-, and CTL epitopes from the E7 oncoprotein of HPV16 into HBc [288–290]. Humoral and T-cell proliferative responses to the chimeras were elicited successfully [290], also in the case of *Salmonella*-driven expression [288], but the appropriate chimeric particles failed to prime E7-directed CTL responses in mice [289].

HBc VLPs played a role in the development of hantavirus vaccine candidates. Chimeric HBcΔ carrying the 45 N-terminal aa of the Puumala hantavirus, strain CG18–20, nucleocapsid protein inserted internally induced a protective immunity in 80% of the immunized bank voles [214,291]. These investigations resulted also in the identification of a second protective region in the nucleocapsid protein (aa 75–119). A combination of these two protective regions in a single HBcΔ molecule (aa 1–45 as MIR insertion; aa 75–119 at the C-terminal position 144) were demonstrated to induce a protective immunity in the bank vole model [223]. Interestingly, the 45 N-terminal segment of another Puumala virus strain (Vranica/Hällnäs) presented on HBcΔ VLPs induced only a very low level of protection, which could be strongly improved by insertion of a 120 aa-long segment of the same nucleocapsid protein [214,292]. These investigations were based on the surprising observation of a high insertion capacity of the MIR of HBcΔ tolerating internal insertions of the N-terminal 120 aa of the nucleocapsid protein from three different hantavirus species: Puumala, Dobrava-Belgrade, and Hantaan viruses [293]. Because of this unique appearance, chimeric particles evoked high structural interest and were subjected to thorough electron cryomicroscopy investigation. It was found that chimeric particles presented the inserted hantavirus nucleocapsid fragment, at least in part, on their surface. A map computed from $T = 3$ shells, which represented 90% of particle population, showed additional density splaying out from the tips of the spikes producing the effect of an extra shell of density at an outer radius compared with wild-type HBcΔ particles. The inserted hantavirus nucleocapsid protein segment appeared flexibly linked to the HBc spikes and only partially icosahedrally ordered. Immunization of mice with chimeric particles induced a high-titered and highly cross-reactive hantavirus nucleocapsid-specific antibody response. Even without adjuvant, these different chimeric particles induced high-titered and strongly cross-reactive hantavirus nucleocapsid-specific antibody responses in mice [294]. It is important to note that these investigations demonstrated that preexisting anti-HBc antibodies did not abrogate the induction of the hantavirus nucleocapsid protein-specific immune responses. This might be mediated by masking of the MIR by the 120 aa-long insertion; the effect of preexisting anti-HBc antibodies on the induction of immune responses against chimeric HBc particles harboring shorter insertions remains to be proven.

HBc played an important role in the development of the design of influenza vaccines (for more details and discussion see Chapter 8). In the initial study, chimeric HBc particles carrying N-terminally 23 aa of the extracellular domain of the minor protein M2 of the influenza A virus were found providing up to 100% protection against a lethal virus challenge in mice after intraperitoneal or intranasal administration [213]. The advantage of the external domain, M2e, as an antigen is the conservation of its sequence that has hardly changed since the first influenza virus was isolated in 1933, despite numerous epidemics and several pandemics. Since the external domain, M2e, is only 23 aa long, it is weakly immunogenic, but presentation on the *E. coli*-expressed HBc VLPs, N-terminally and/or internally at the MIR, enhanced its ability to induce M2-specific antibodies [295]. Several adjuvants were tested in conjunction with intraperitoneal vaccine administration, while the nontoxic enterotoxin mutant LT (R192G) was used for intranasal vaccination. Appropriate combinations of vaccine construct and this adjuvant were reported to induce high anti-M2e IgG2a serum titers (above 10,000), and to provide complete protection in mice [295]. Chimeric HBc-M2 VLP constructs were further optimized by enhancing the number of N-terminally fused M2e copies from one to three [296]. Such optimized HBc-M2e vaccine induced an anti-M2e antibody titer even higher than that of anti-HBc. Further development of the vaccine candidate was connected with its design, i.e., with the improvement of administration routes and adjuvants. Intranasal immunization in combination with the cholera toxin and *Staphylococcus aureus* protein A-derived CTA1-DD adjuvant was shown to provide full protection from a potentially lethal infection in mice [297]. Combination of the most immunogenic vaccine candidate with the selected intranasal immunization route also ensured full protectivity in mice [298].

An alternative attempt to construct a HBc-M2 VLP-based vaccine was performed by chemical coupling of the M2e peptide to the surface of HBc VLPs [299]. The obtained data were in strong contrast to investigations using chimeric HBc-M2 VLPs. The anti-M2e antibodies induced by HBc-M2e vaccination were not virus-neutralizing, and the observed level of protection was much weaker protection than that achieved by immunization with classical inactivated viral preparations.

Continuating efforts in the HIV vaccine candidate development, some pioneering approaches appeared during recent years. First, HBc VLPs carrying HIV-1 Env V3 epitope were combined with so-called hemagglutinating virus of Japan (HVJ) protein incorporated into an anionic liposome and provided a potent T-cell response in mice [300]. Second, a DNA vaccination strategy was tested, where one or more HIV or SIV CTL epitopes were fused to the HBc carrier [301]. Immunization of mice with a HBc-HIV epitope DNA vaccine induced CTL responses that significantly exceeded levels induced with DNA encoding either the whole HIV antigen or the epitope alone. In rhesus macaques, a multi-epitope hybrid HBc-SIV DNA vaccine induced CTL responses to 13 different epitopes, including 3 epitopes that were previously not detected in SIV-infected macaques [301]. These novel data clearly demonstrate that immunization with hybrid HBc-epitope DNA vaccines appears as a promising strategy in a further vaccine development.

Novel ambitious project was developed for the display of HCV B- and T-cell epitopes on HBc VLPs [302]. C-terminal fusion of HCV core (98 aa) and NS3 (155 aa) protein fragments to the HBcΔ carrier ensured self-assembly of chimeric HBcΔ-HCV monomers. However, chimeric VLPs induced rather low humoral and cellular responses to HCV epitopes. Another group inserted a HCV CTL epitope from the core region (aa 35–44) into the MIR of the HBc and showed strong cellular response after DNA immunization in mice [303].

Phage-display-selected epitopes mimicking hepatitis E virus-neutralizing epitopes were exposed at the MIR of the HBc, and corresponding VLPs were obtained [304]. This is a first attempt to mimic conformational epitopes on the HBc surface. Since selected epitopes are very short, 7 aa in length, it would be very interesting to get information on their immunogenicity.

It is worth to mention some attempts, which have been performed to construct antibacterial vaccines on the basis of the HBc particles. Although the C-terminal additions have not met with success in terms of inducing high-titer antibodies, a conserved sequence of the 47-aa residues, multiple copies of which are present in several proteins of *Porphyromonas gingivalis*, was added to the C-terminus of the HBc protein [305]. Although in this case the chimeric particles purified from *E. coli* were recognized by the host's immune system and induced specific antibodies, they did not protect mice against bacterial challenge. By comparison, the N-terminal insertion of a 30-aa epitope of outer membrane protein P.69 (pertactin) from another bacterium *B. pertussis*, prevented the growth of *B. pertussis* in the lungs of infected mice, and protection correlated with high titers of anti-P.69 antibodies [306].

An interesting approach is connected with the putative HBc-based DNA vaccines for animal use. Three promising CTL epitopes for designing a vaccine against *Taenia solium* (pork tapeworm) cysticercosis were inserted at the MIR of HBcΔ, and one epitope was fused to the C-terminus of the same carrier molecule [307]. After immunization of pigs with the corresponding DNA, the relative protective rate against infectious *T. solium* eggs was 83%.

Now, let us say a few words about the role of immunostimulatory CpGs in the development of HBc VLPs. It was known before that the DNA rich in nonmethylated CpG motifs greatly facilitate induction of immune responses against coadministered antigens. As stated above, the role of CpG oligonucleotides in triggering a strong T-helper and CTL response is crucial and they appear therefore among the most promising adjuvants known to date (Chapter 6). Many attempts were made to combine HBc VLPs with CpGs. The pioneering work in this field postulated, first, that packaging CpGs into HBc and RNA phage Qβ VLPs is a simple and attractive way to reduce two drawbacks of CpGs as adjuvants: unfavorable pharmacokinetics and exhibition of systemic side effects, i.e., splenomegaly [308]. Second, VLPs enhance stability of CpGs protecting them from DNase I digestion. Third, the most important fact is that CpGs ensures a strong response against CTL epitopes inserted into VLPs.

Recently, a novel vaccine approach based on HBc-VLP-pulsed dendritic cells was developed [309]. This approach is based on the finding that immature murine bone marrow-derived dendritic cells could capture HBc-VLPs or CpG-loaded HBc-VLPs efficiently and present the antigen to syngeneic murine spleen T cells *in vitro*. Immunization with dendritic cells showed that compared to

VLP-pulsed dendritic cells, CpG-loaded VLP-pulsed dendritic cells elicit stronger T-cell responses *in vivo*. In a tumor therapy model, the growth of established tumors was significantly inhibited by single immunization of dendritic cells pulsed with CpG-loaded HBc-VLPs, resulting in a significantly longer survival of immunized animals and strikingly high frequencies of protective CTLs. This is a step toward a novel field of the development of effective T-cell-based vaccines.

Another approach to the inclusion of CpGs into putative HBc vaccine candidates is the use of DNA vaccine vectors containing CpG motifs. First attempt was made in a rabbit model of atherosclerosis. Rabbits were intramuscularly immunized with DNA vaccine encoding a 26-aa B-cell epitope of the cholesteryl ester transfer protein (CETP) inserted into HBc [310]. After anti-CETP antibodies were successfully produced, rabbits were fed with a high-cholesterol diet for 15 weeks. It was found that immunized rabbits had a significant increase in HDL fraction of plasma cholesterol and decrease in LDL. More importantly, DNA immunization markedly reduced the average percentage of aortic lesions in the entire aorta area. Thus, the HBc-CETP DNA vaccine containing CpG motifs could significantly inhibit the progression of atherosclerosis and can be potentially developed into a suitable DNA vaccine against atherosclerosis.

9.6 HEPATITIS B VLPS IN THE FUTURE

It has been more than 20 years since HBs and HBc proteins attracted so much attention as effective self-assembling epitope carriers for poorly immunogenic epitopes. With time, optimistic expectations began dissipating, though. Some researchers started discussing a shift in a focus of research from human to rodent hepadnovirus core proteins. Such a consideration was spurred by some perceived disadvantages associated with the use of human HBc for vaccine design [311]. Despite of this, the major concepts and ideas for the use of HBs and HBc have survived and are still being evaluated. For example, many HBs- and HBc-based vaccine candidates against malaria, influenza, HCV, and other human and animal diseases still remain to be the focus of many research programs.

One of the most attractive approaches for the future development of HBV-based VLP vaccines could be a simultaneous exploitation of advantages of both chimeric HBs and HBc proteins while minimizing their drawbacks. The easiest way to exploit such a combination of HBs and HBc is through the use of DNA vaccination, which allows for a simultaneous expression and unlimited modifications of both genes. Active research efforts in this direction are ongoing. The HBs and HBc genes have been recently coexpressed using a synthetic bidirectional promoter under the control of the tetracycline-inactivated transactivator. A genetic construct containing this expression unit was transfected into mammalian cells and was also injected into mice [312]. Coadminstration of both genes as DNA vaccine to mice induced an immune response against both antigens. No interference between immunoresponses to both antigens were observed [313]. The first trial of therapeutic immunization with a DNA vaccine expressing both HBs and HBc genes in chronically infected chimpanzees was successful in an animal having a relatively low viral load, but was less successful in an animal with a high virus load [123].

Another approach is associated with the selection of alternative administration routes such as a nasal immunization regimen, which was recently used for a novel therapeutic HBV vaccine based on the combination of the HBs and HBc proteins [314]. An obvious advantage of this strategy is activation of mucosal T cells in order to overcome the systemic immune downregulation due to HBV infection. A special study showed that two different HBc variants do not differ in their adjuvant effect on the HBs protein [253,254]. Phase I clinical trial in healthy adults of the nasal vaccine candidate (NASVAC) containing both HBs and HBc antigens confirmed its safety, tolerability, and immunogenicity [315]. It is necessary to emphasize that this study is the first demonstration of safety and immunogenicity for a nasal vaccine candidate comprising both HBs and HBc antigens.

The next logical step in the direction of the development of DNA vaccines as a special case of DNA protein engineering is the use of chimeric HBs and HBc derivatives. An obvious advantage of

DNA vaccination is the elimination of always critical self-assembly and low-production barriers typical for engineering at the protein level.

Besides being effective protein carriers, HBs and HBc proteins themselves contain antigenic epitopes, which alone or in combination with other viral epitopes can serve as attractive candidates for insertion into other VLP models. For example, a set of selected B-cell epitopes derived from the HBV S protein (aa 124–147) and HCV core (aa 2–21 and 22–40) and E1 (aa 315–328) proteins were displayed in five different sites of the flock house virus capsid protein and expressed in *E. coli* [316]. Similarly, an epitope from HBV preS1 was presented on the surface of yeast-expressed hamster polyomavirus-derived VLPs [317].

Further advances in this research field may come from the ever surprising discoveries in HBV molecular biology. One of such discoveries is the identification of a novel HBV splice-generated protein (HBSP) in the liver biopsy specimens from patients with chronic active hepatitis B [318]. This study showed for the first time that the HBSP generated *in vivo* from an alternative reading frame within the HBV genome activates T-cell responses in HBV-infected patients. Immunization of HLA-transgenic mice with the corresponding HBSP DNA induced a specific T-cell response. Given that hepatitis B is an immune response-mediated disease, the detection of T-cell responses directed against HBSP in patients with chronic hepatitis B suggests a potential role for this protein in liver disease progression [318].

ACKNOWLEDGMENTS

We thank R. Anthony Crowther (Cambridge) for the generous gift of the HBc images, Robert Gilbert and David J. Rowlands for the three-dimensional HBs maps, Wolfram H. Gerlich and Bruce Boschek (Giessen) for the HBs image, Vijay Reddy (San Diego) for the three-dimensional images of VLPs, and Peter Pushko (Frederick) for constant support. We are grateful to Nikolai Granovski (Sao Paulo) and Yury Khudyakov (Atlanta) for critical reading of the manuscript, helpful discussion, and provision of unpublished data.

REFERENCES

1. Pumpens, P. and Grens, E., Artificial genes for chimeric virus-like particles, In: *Artificial DNA: Methods and Applications*, Khudyakov, Y.E. and Fields, H.A. (Eds.), CRC Press LLC, Boca Raton, Florida, 2002, Chapter 8.
2. Borisova, G.P. et al., Recombinant capsid structures for exposure of protein antigenic epitopes, *Mol. Gen. (Life Sci. Adv.)*, 6, 169, 1987.
3. Gren, E.J. and Pumpen, P.P., Recombinant virus capsids as a new generation of immunogenic proteins and vaccines, *J. All-Union Mendeleyev's Chem. Soc. (in Russian)*, 33, 531, 1988.
4. Fersht, A. and Winter, G., Protein engineering, *Trends. Biochem. Sci.*, 17, 292, 1992.
5. Böttcher, B., Wynne, S.A., and Crowther, R.A., Determination of the fold of the core protein of hepatitis B virus by electron cryomicroscopy, *Nature*, 386, 88, 1997.
6. Crowther, R.A., Amos, L.A., and Finch, J.T., Three-dimensional image reconstructions of bacteriophages R17 and f2, *J. Mol. Biol.*, 98, 631, 1975.
7. Golmohammadi, R. et al., The refined structure of bacteriophage MS2 at 2.8 Å resolution, *J. Mol. Biol.*, 234, 620, 1993.
8. Golmohammadi, R. et al., The crystal structure of bacteriophage Qbeta at 3.5 Å resolution, *Structure*, 4, 543, 1996.
9. Liljas, L. et al., Crystal structure of bacteriophage fr capsids at 3.5 Å resolution, *J. Mol. Biol.*, 244, 279, 1994.
10. Tars, K. et al., The three-dimensional structure of bacteriophage PP7 from *Pseudomonas aeruginosa* at 3.7-Å resolution, *Virology*, 272, 331, 2000.
11. Tars, K. et al., The crystal structure of bacteriophage GA and a comparison of bacteriophages belonging to the major groups of *Escherichia coli* leviviruses, *J. Mol. Biol.*, 271, 759, 1997.
12. Valegård, K. et al., The three-dimensional structure of the bacterial virus MS2, *Nature*, 345, 36, 1990.

13. Wynne, S.A., Crowther, R.A., and Leslie, A.G., The crystal structure of the human hepatitis B virus capsid, *Mol. Cell*, 3, 771, 1999.
14. Namba, K. and Stubbs, G., Structure of tobacco mosaic virus at 3.6 Å resolution: Implications for assembly, *Science*, 231, 1401, 1986.
15. Namba, K., Pattanayek, R., and Stubbs, G., Visualization of protein-nucleic acid interactions in a virus. Refined structure of intact tobacco mosaic virus at 2.9 Å resolution by X-ray fiber diffraction, *J. Mol. Biol.*, 208, 307, 1989.
16. Hogle, J.M., Chow, M., and Filman, D.J., Three-dimensional structure of poliovirus at 2.9 Å resolution, *Science*, 229, 1358, 1985.
17. Rossmann, M.G. et al., Structure of a human common cold virus and functional relationship to other picornaviruses, *Nature*, 317, 145, 1985.
18. Grimes, J.M. et al., An atomic model of the outer layer of the bluetongue virus core derived from X-ray crystallography and electron cryomicroscopy, *Structure*, 5, 885, 1997.
19. Conway, J.F. et al., Visualization of a 4-helix bundle in the hepatitis B virus capsid by cryo-electron microscopy, *Nature*, 386, 91, 1997.
20. Crowther, R.A. et al., Three-dimensional structure of hepatitis B virus core particles determined by electron cryomicroscopy, *Cell*, 77, 943, 1994.
21. Cabilly, S., Combinatorial peptide library protocols, *Methods in Molecular Biology, Vol. 87*, Humana Press, Totowa, New Jersey, 1998, p. 328.
22. Morris, G.E., Epitope mapping protocols, *Methods in Molecular Biology, Vol. 66*, Humana Press, Totowa, New Jersey, 1996, p. 432.
23. Sominskaya, I. et al., Determination of the minimal length of preS1 epitope recognized by a monoclonal antibody which inhibits attachment of hepatitis B virus to hepatocytes, *Med. Microbiol. Immunol. (Berl.)*, 181, 215, 1992.
24. Nassal, M., Total chemical synthesis of a gene for hepatitis B virus core protein and its functional characterization, *Gene*, 66, 279, 1988.
25. Haynes, J.R. et al., Development of a genetically-engineered, candidate polio vaccine employing the self-assembling properties of tobacco mosaic virus coat protein, *Biotechnology*, 4, 637, 1986.
26. Noad, R. and Roy, P., Virus-like particles as immunogens, *Trends Microbiol.*, 11, 438, 2003.
27. Roy, P., Genetically engineered particulate virus-like structures and their use as vaccine delivery systems, *Intervirology*, 39, 62, 1996.
28. Roy, P. and Noad, R., Bluetongue virus assembly and morphogenesis, *Curr. Top. Microbiol. Immunol.*, 309, 87, 2006.
29. Tobin, G.J., Nagashima, K., and Gonda, M.A., Immunologic and ultrastructural characterization of HIV pseudovirions containing Gag and Env precursor proteins engineered in insect cells, *Methods*, 10, 208, 1996.
30. Pushko, P. et al., Influenza virus-like particles comprised of the HA, NA, and M1 proteins of H9N2 influenza virus induce protective immune responses in BALB/c mice, *Vaccine*, 23, 5751, 2005.
31. Pushko, P. et al., Evaluation of influenza virus-like particles and Novasome adjuvant as candidate vaccine for avian influenza, *Vaccine*, 25, 4283, 2007.
32. Koletzki, D. et al., Mosaic hepatitis B virus core particles allow insertion of extended foreign protein segments, *J. Gen. Virol.*, 78, 2049, 1997.
33. Kozlovska, T.M. et al., RNA phage Q beta coat protein as a carrier for foreign epitopes, *Intervirology*, 39, 9, 1996.
34. Smith, G.P., Filamentous fusion phage: Novel expression vectors that display cloned antigens on the virion surface, *Science*, 228, 1315, 1985.
35. Delpeyroux, F. et al., A poliovirus neutralization epitope expressed on hybrid hepatitis B surface antigen particles, *Science*, 233, 472, 1986.
36. Valenzuela, P. et al., Antigen engineering in yeast: Synthesis and assembly of hybrid hepatitis B surface antigen—herpes simplex 1 gD particles, *Biotechnology*, 3, 323, 1985.
37. Clarke, B.E. et al., Improved immunogenicity of a peptide epitope after fusion to hepatitis B core protein, *Nature*, 330, 381, 1987.
38. Newton, S.E. et al., New approaches to FMDV antigen presentation using vaccinia virus, In: *Vaccines 87. Modern Approaches to New Vaccines: Prevention of AIDS and Other Viral, Bacterial, and Parasitic Diseases*. Chanock, R.M., Lerner, R.A., Brown, F., and Ginsberg, H. (Eds). Cold Spring Harbor, New York. Cold Spring Harbor Laboratory Press, 12, 1987.

39. Burke, K.L. et al., Antigen chimaeras of poliovirus as potential new vaccines, *Nature*, 332, 81, 1988.
40. Kozlovskaya, T.M. et al., Genetically engineered mutants of the envelope protein of the RNA-containing bacteriophage fr (in Russian), *Mol. Biol. (Mosk.)*, 22, 731, 1988.
41. Adams, S.E. et al., The expression of hybrid HIV: Ty virus-like particles in yeast, *Nature*, 329, 68, 1987.
42. Heermann, K.H. and Gerlich, W.H., Surface proteins of hepatitis B viruses, In: *Molecular Biology of the Hepatitis B Virus*, McLachlan, A. (Ed.), CRC Press, Boca Raton, Florida, 1991, pp. 109–143.
43. Heermann, K.H. et al., Large surface proteins of hepatitis B virus containing the pre-s sequence, *J. Virol.*, 52, 396, 1984.
44. Berting, A. et al., Computer-aided studies on the spatial structure of the small hepatitis B surface protein, *Intervirology*, 38, 8, 1995.
45. Sonveaux, N., Ruysschaert, J.M., and Brasseur, R., Proposition of a three-dimensional representation of the constitutive protein of the hepatitis B surface antigen particles, *J. Protein Chem.*, 14, 477, 1995.
46. Stirk, H.J., Thornton, J.M., and Howard, C.R., A topological model for hepatitis B surface antigen, *Intervirology*, 33, 148, 1992.
47. Sonveaux, N. et al., The topology of the S protein in the yeast-derived hepatitis B surface antigen particles, *J. Biol. Chem.*, 269, 25637, 1994.
48. Antoni, B.A. et al., Site-directed mutagenesis of cysteine residues of hepatitis B surface antigen. Analysis of two single mutants and the double mutant, *Eur. J. Biochem.*, 222, 121, 1994.
49. Sato, M. et al., Peripherally biased distribution of antigen proteins on the recombinant yeast-derived human hepatitis B virus surface antigen vaccine particle: Structural characteristics revealed by small-angle neutron scattering using the contrast variation method, *J. Biochem. (Tokyo)*, 118, 1297, 1995.
50. Yamaguchi, M. et al., Fine structure of hepatitis B virus surface antigen produced by recombinant yeast: Comparison with HBsAg of human origin, *FEMS Microbiol. Lett.*, 165, 363, 1998.
51. Gilbert, R.J. et al., Hepatitis B small surface antigen particles are octahedral, *Proc. Natl. Acad. Sci. U S A*, 102, 14783, 2005.
52. Burrell, C.J. et al., Expression in *Escherichia coli* of hepatitis B virus DNA sequences cloned in plasmid pBR322, *Nature*, 279, 43, 1979.
53. Charnay, P. et al., Biosynthesis of hepatitis B virus surface antigen in *Escherichia coli*, *Nature*, 286, 893, 1980.
54. Edman, J.C. et al., Synthesis of hepatitis B surface and core antigens in *E. coli*, *Nature*, 291, 503, 1981.
55. Mackay, P. et al., Production of immunologically active surface antigens of hepatitis B virus by *Escherichia coli*, *Proc. Natl. Acad. Sci. U S A*, 78, 4510, 1981.
56. Fujisawa, Y. et al., Direct expression of hepatitis B surface antigen gene in *E. coli*, *Nucleic. Acids. Res.*, 11, 3581, 1983.
57. Pumpen, P. et al., Expression of hepatitis B virus surface antigen gene in *Escherichia coli*, *Gene*, 30, 201, 1984.
58. Pumpen, P.P. et al., Synthesis of the surface antigen of the hepatitis B virus in *Escherichia coli* (in Russian), *Dokl. Akad. Nauk. SSSR.*, 271, 230, 1983.
59. Smirnov, V.D. et al., Synthesis and expression of the DNA fragment coding the antigenic determinant of the surface antigen protein of hepatitis B virus (in Russian), *Bioorg. Khim.*, 9, 1388, 1983.
60. Dehoux, P. et al., Expression of the hepatitis B virus large envelope protein in *Saccharomyces cerevisiae*, *Gene*, 48, 155, 1986.
61. Itoh, Y., Hayakawa, T., and Fujisawa, Y., Expression of hepatitis B virus surface antigen P31 gene in yeast, *Biochem. Biophys. Res. Commun.*, 138, 268, 1986.
62. Valenzuela, P. et al., Synthesis and assembly of hepatitis B virus surface antigen particles in yeast, *Nature*, 298, 347, 1982.
63. Yamada, T. et al., Physicochemical and immunological characterization of hepatitis B virus envelope particles exclusively consisting of the entire L (pre-S1 + pre-S2 + S) protein, *Vaccine*, 19, 3154, 2001.
64. Bruss, V. and Ganem, D., Mutational analysis of hepatitis B surface antigen particle assembly and secretion, *J. Virol.*, 65, 3813, 1991.
65. Prange, R. et al., Mutational analysis of HBsAg assembly, *Intervirology*, 38, 16, 1995.
66. Mangold, C.M. et al., Secretion and antigenicity of hepatitis B virus small envelope proteins lacking cysteines in the major antigenic region, *Virology*, 211, 535, 1995.
67. Mangold, C.M. et al., Analysis of intermolecular disulfide bonds and free sulfhydryl groups in hepatitis B surface antigen particles, *Arch. Virol.*, 142, 2257, 1997.

68. Heijtink, R.A. et al., Anti-HBs characteristics after hepatitis B immunisation with plasma-derived and recombinant DNA-derived vaccines, *Vaccine*, 18, 1531, 2000.
69. Hepatitis B. World Health Organization. Document [WHO/CDS/CSR/LYO/2002.2:Hepatitis B], 2002, pp. 1–76.
70. Kundi, M., New hepatitis B vaccine formulated with an improved adjuvant system, *Expert. Rev. Vaccines.*, 6, 133, 2007.
71. Nevens, F. et al., Immunogenicity and safety of an experimental adjuvanted hepatitis B candidate vaccine in liver transplant patients, *Liver Transpl.*, 12, 1489, 2006.
72. Costa, A.A. et al., Preliminary report of the use on adults of a recombinant yeast-derived hepatitis B vaccine manufactured by Instituto Butantan, *Rev. Inst. Med. Trop. Sao Paulo*, 39, 39, 1997.
73. Ioshimoto, L.M. et al., Safety and immunogenicity of hepatitis B vaccine ButaNG in adults, *Rev. Inst. Med. Trop. Sao Paulo*, 41, 191, 1999.
74. Kulkarni, P.S. et al., Immunogenicity of a new, low-cost recombinant hepatitis B vaccine derived from Hansenula polymorpha in adults, *Vaccine*, 24, 3457, 2006.
75. Galban, G.E. et al., Field trial of the Cuban recombinant vaccine against hepatitis B (Heberbiovac HB). Study in newborn infants born to AgsHB + mothers, *Rev. Cubana Med. Trop.*, 44, 149, 1992.
76. Kaur, H. and Mani, A., Seroprotection following Enivac-HB, a recombinant hepatitis B vaccine, *Indian J. Gastroenterol.*, 19, 41, 2000.
77. Abraham, P. et al., Evaluation of a new recombinant DNA hepatitis B vaccine (Shanvac-B), *Vaccine*, 17, 1125, 1999.
78. Schmidt, C., Toll-like receptor therapies compete to reduce side effects, *Nat. Biotechnol.*, 24, 230, 2006.
79. Sung, J.J. and Lik-Yuen, H., HBV-ISS (Dynavax), *Curr. Opin. Mol. Ther.*, 8, 150, 2006.
80. Malhame, M., Dynavax Completes Enrollment of HEPLISAV(TM) Phase 3, *Dynavax Technologies Corporation*, 2007.
81. Cooper, C.L. et al., CPG 7909, an immunostimulatory TLR9 agonist oligodeoxynucleotide, as adjuvant to Engerix-B HBV vaccine in healthy adults: A double-blind phase I/II study, *J. Clin. Immunol.*, 24, 693, 2004.
82. Cooper, C.L. et al., CPG 7909 adjuvant improves hepatitis B virus vaccine seroprotection in antiretroviral-treated HIV-infected adults, *AIDS*, 19, 1473, 2005.
83. Siegrist, C.A. et al., Co-administration of CpG oligonucleotides enhances the late affinity maturation process of human anti-hepatitis B vaccine response, *Vaccine*, 23, 615, 2004.
84. Feng, L. et al., Pharmaceutical and immunological evaluation of a single-dose hepatitis B vaccine using PLGA microspheres, *J. Control Release*, 112, 35, 2006.
85. Jaganathan, K.S. et al., Development of a single-dose stabilized poly(D,L-lactic-co-glycolic acid) microspheres-based vaccine against hepatitis B, *J. Pharm. Pharmacol.*, 56, 1243, 2004.
86. Yang, H.Z. et al., A novel immunostimulator, N-[alpha-O-benzyl-N-(acetylmuramyl)-L-alanyl-D-isoglutaminyl]-N6-trans-(m-nitrocinnamoyl)-L-lysine, and its adjuvancy on the hepatitis B surface antigen, *J. Med. Chem.*, 48, 5112, 2005.
87. Goyal, A.K. et al., Nanodecoy system: A novel approach to design hepatitis B vaccine for immunopotentiation, *Int. J. Pharm.*, 309, 227, 2006.
88. Machida, A. et al., A synthetic peptide coded for by the pre-S2 region of hepatitis B virus for adding immunogenicity to small spherical particles made of the product of the S gene, *Mol. Immunol.*, 24, 523, 1987.
89. Neurath, A.R., Strick, N., and Girard, M., Hepatitis B virus surface antigen (HBsAg) as carrier for synthetic peptides having an attached hydrophobic tail, *Mol. Immunol.*, 26, 53, 1989.
90. Xu, X. et al., A modified hepatitis B virus surface antigen with the receptor-binding site for hepatocytes at its C terminus: Expression, antigenicity and immunogenicity, *J. Gen. Virol.*, 75, 3673, 1994.
91. Prange, R. et al., Properties of modified hepatitis B virus surface antigen particles carrying preS epitopes, *J. Gen. Virol.*, 76, 2131, 1995.
92. Hui, J. et al., Expression and characterization of chimeric hepatitis B surface antigen particles carrying preS epitopes, *J. Biotechnol.*, 72, 49, 1999.
93. Senturk, H. et al., Therapeutic vaccination in chronic hepatitis B, *J. Gastroenterol. Hepatol.*, 17, 72, 2002.
94. Weinstein, T. et al., Improved immunogenicity of a novel third-generation recombinant hepatitis B vaccine in patients with end-stage renal disease, *Nephron Clin. Pract.*, 97, c67, 2004.
95. Madalinski, K. et al., Presence of anti-preS1, anti-preS2, and anti-HBs antibodies in newborns immunized with Bio-Hep-B vaccine, *Med. Sci. Monit.*, 10, I10, 2004.

96. Yap, I., Guan, R., and Chan, S.H., Recombinant DNA hepatitis B vaccine containing Pre-S components of the HBV coat protein—a preliminary study on immunogenicity, *Vaccine*, 10, 439, 1992.
97. Lo, C.M. et al., Efficacy of a pre-S containing vaccine in patients receiving lamivudine prophylaxis after liver transplantation for chronic hepatitis B, *Am. J. Transplant.*, 7, 434, 2007.
98. Rendi-Wagner, P. et al., Comparative immunogenicity of a PreS/S hepatitis B vaccine in non- and low responders to conventional vaccine, *Vaccine*, 24, 2781, 2006.
99. Schumann, A. et al., Cellular and humoral immune response to a third generation hepatitis B vaccine, *J. Viral Hepat.*, 14, 592, 2007.
100. Jones, C.D. et al., Characterization of the T- and B-cell immune response to a new recombinant pre-S1, pre-S2 and SHBs antigen containing hepatitis B vaccine (Hepagene); evidence for superior anti-SHBs antibody induction in responder mice, *J. Viral. Hepat.*, 5 Suppl 2, 5, 1998.
101. Jones, C.D. et al., T-cell and antibody response characterisation of a new recombinant pre-S1, pre-S2 and SHBs antigen-containing hepatitis B vaccine; demonstration of superior anti-SHBs antibody induction in responder mice, *Vaccine*, 17, 2528, 1999.
102. McDermott, A.B. et al., Hepatitis B third-generation vaccines: Improved response and conventional vaccine non-response—evidence for genetic basis in humans, *J. Viral. Hepat.*, 5 Suppl 2, 9, 1998.
103. Pride, M.W. et al., Evaluation of B and T-cell responses in chimpanzees immunized with Hepagene, a hepatitis B vaccine containing pre-S1, pre-S2 gene products, *Vaccine*, 16, 543, 1998.
104. Waters, J.A. et al., A study of the antigenicity and immunogenicity of a new hepatitis B vaccine using a panel of monoclonal antibodies, *J. Med. Virol.*, 54, 1, 1998.
105. Page, M., Jones, C.D., and Bailey, C., A novel, recombinant triple antigen hepatitis B vaccine (Hepacare), *Intervirology*, 44, 88, 2001.
106. Jones, T., Hepagene (PowderJect), *Curr. Opin. Investig. Drugs*, 3, 987, 2002.
107. Leroux-Roels, G. et al., Hepatitis B vaccine containing surface antigen and selected preS1 and preS2 sequences. 2. Immunogenicity in poor responders to hepatitis B vaccines, *Vaccine*, 15, 1732, 1997.
108. Leroux-Roels, G. et al., Hepatitis B vaccine containing surface antigen and selected preS1 and preS2 sequences. 1. Safety and immunogenicity in young, healthy adults, *Vaccine*, 15, 1724, 1997.
109. Davis, H.L., Michel, M.L., and Whalen, R.G., DNA-based immunization induces continuous secretion of hepatitis B surface antigen and high levels of circulating antibody, *Hum. Mol. Genet.*, 2, 1847, 1993.
110. Davis, H.L. et al., Direct gene transfer in skeletal muscle: Plasmid DNA-based immunization against the hepatitis B virus surface antigen, *Vaccine*, 12, 1503, 1994.
111. Davis, H.L., Michel, M.L., and Whalen, R.G., Use of plasmid DNA for direct gene transfer and immunization, *Ann. N.Y. Acad. Sci.*, 772, 21, 1995.
112. Davis, H.L. et al., DNA-mediated immunization to hepatitis B surface antigen: Longevity of primary response and effect of boost, *Vaccine*, 14, 910, 1996.
113. Davis, H.L. et al., DNA-based immunization against hepatitis B surface antigen (HBsAg) in normal and HBsAg-transgenic mice, *Vaccine*, 15, 849, 1997.
114. Michel, M.L. et al., DNA-mediated immunization to the hepatitis B surface antigen in mice: Aspects of the humoral response mimic hepatitis B viral infection in humans, *Proc. Natl. Acad. Sci. U S A*, 92, 5307, 1995.
115. Schirmbeck, R. et al., Nucleic acid vaccination primes hepatitis B virus surface antigen-specific cytotoxic T lymphocytes in nonresponder mice, *J. Virol.*, 69, 5929, 1995.
116. Geissler, M. et al., Induction of anti-hepatitis B virus immune responses through DNA immunization, *Methods Mol. Med.*, 96, 43, 2004.
117. Michel, M.L. and Loirat, D., DNA vaccines for prophylactic or therapeutic immunization against hepatitis B, *Intervirology*, 44, 78, 2001.
118. Michel, M.L. and Mancini-Bourgine, M., Therapeutic vaccination against chronic hepatitis B virus infection, *J. Clin. Virol.*, 34 Suppl 1, S108, 2005.
119. Schirmbeck, R. and Reimann, J., Revealing the potential of DNA-based vaccination: Lessons learned from the hepatitis B virus surface antigen, *Biol. Chem.*, 382, 543, 2001.
120. Mancini-Bourgine, M. et al., Induction or expansion of T-cell responses by a hepatitis B DNA vaccine administered to chronic HBV carriers, *Hepatology*, 40, 874, 2004.
121. Mancini-Bourgine, M. et al., Immunogenicity of a hepatitis B DNA vaccine administered to chronic HBV carriers, *Vaccine*, 24, 4482, 2006.
122. Roberts, L.K. et al., Clinical safety and efficacy of a powdered Hepatitis B nucleic acid vaccine delivered to the epidermis by a commercial prototype device, *Vaccine*, 23, 4867, 2005.

123. Shata, M.T. et al., Attempted therapeutic immunization in a chimpanzee chronic HBV carrier with a high viral load, *J. Med. Primatol.*, 35, 165, 2006.
124. Hutchings, C.L. et al., Novel protein and poxvirus-based vaccine combinations for simultaneous induction of humoral and cell-mediated immunity, *J. Immunol.*, 175, 599, 2005.
125. Jin, Y. et al., Boosting immune response to hepatitis B DNA vaccine by coadministration of prothymosin alpha-expressing plasmid, *Clin. Diagn. Lab Immunol.*, 12, 1364, 2005.
126. Payette, P.J. et al., Testing of CpG-optimized protein and DNA vaccines against the hepatitis B virus in chimpanzees for immunogenicity and protection from challenge, *Intervirology*, 49, 144, 2006.
127. He, X.W. et al., Induction of mucosal and systemic immune response by single-dose oral immunization with biodegradable microparticles containing DNA encoding HBsAg, *J. Gen. Virol.*, 86, 601, 2005.
128. Kim, T.W. et al., Induction of immunity against hepatitis B virus surface antigen by intranasal DNA vaccination using a cationic emulsion as a mucosal gene carrier, *Mol. Cells*, 22, 175, 2006.
129. Hui, J. et al., Immunization with a plasmid encoding a modified hepatitis B surface antigen carrying the receptor binding site for hepatocytes, *Vaccine*, 17, 1711, 1999.
130. Kong, Q. et al., Oral immunization with hepatitis B surface antigen expressed in transgenic plants, *Proc. Natl. Acad. Sci. U S A*, 98, 11539, 2001.
131. Richter, L.J. et al., Production of hepatitis B surface antigen in transgenic plants for oral immunization, *Nat. Biotechnol.*, 18, 1167, 2000.
132. Kapusta, J. et al., Oral immunization of human with transgenic lettuce expressing hepatitis B surface antigen, *Adv. Exp. Med. Biol.*, 495, 299, 2001.
133. Kumar, G.B., Ganapathi, T.R., and Bapat, V.A., Production of hepatitis B surface antigen in recombinant plant systems: An update, *Biotechnol. Prog.*, 23, 532, 2007.
134. Thanavala, Y. et al., Immunogenicity in humans of an edible vaccine for hepatitis B, *Proc. Natl. Acad. Sci. U S A*, 102, 3378, 2005.
135. Shchelkunov, S.N. et al., The obtaining of transgenic tomato plant producing chimerical proteins TBI-HBsAg, *Dokl. Biochem. Biophys.*, 396, 139, 2004.
136. Shchelkunov, S.N. et al., Study of immunogenic properties of the candidate edible vaccine against human immunodeficiency and hepatitis B viruses based on transgenic tomato fruits, *Dokl. Biochem. Biophys.*, 401, 167, 2005.
137. Shchelkunov, S.N. et al., Immunogenicity of a novel, bivalent, plant-based oral vaccine against hepatitis B and human immunodeficiency viruses, *Biotechnol. Lett.*, 28, 959, 2006.
138. Lou, X.M. et al., Expression of the human hepatitis B virus large surface antigen gene in transgenic tomato plants, *Clin. Vaccine Immunol.*, 14, 464, 2007.
139. Rajkannan, R. et al., Development of hepatitis B oral vaccine using B-cell epitope loaded PLG microparticles, *Vaccine*, 24, 5149, 2006.
140. Granovski, N., personal communication, 2007.
141. Delpeyroux, F. et al., Insertions in the hepatitis B surface antigen. Effect on assembly and secretion of 22 nm particles from mammalian cells, *J. Mol. Biol.*, 195, 343, 1987.
142. Delpeyroux, F. et al., Construction and characterization of hybrid hepatitis B antigen particles carrying a poliovirus immunogen, *Biochimie*, 70, 1065, 1988.
143. Delpeyroux, F. et al., Presentation and immunogenicity of the hepatitis B surface antigen and a poliovirus neutralization antigen on mixed empty envelope particles, *J. Virol.*, 62, 1836, 1988.
144. Delpeyroux, F. et al., Structural factors modulate the activity of antigenic poliovirus sequences expressed on hybrid hepatitis B surface antigen particles, *J. Virol.*, 64, 6090, 1990.
145. Eckhart, L. et al., Immunogenic presentation of a conserved gp41 epitope of human immunodeficiency virus type 1 on recombinant surface antigen of hepatitis B virus, *J. Gen. Virol.*, 77, 2001, 1996.
146. Filatov, F.P. et al., Recombinant surface proteins of the hepatitis B virus, exhibiting the immunodominant membrane protein of the HIV-1 virus (in Russian), *Dokl. Akad. Nauk. SSSR*, 327, 172, 1992.
147. Michel, M.L. et al., Induction of anti-human immunodeficiency virus (HIV) neutralizing antibodies in rabbits immunized with recombinant HIV–hepatitis B surface antigen particles, *Proc. Natl. Acad. Sci. U S A*, 85, 7957, 1988.
148. Michel, M.L. et al., T- and B-lymphocyte responses to human immunodeficiency virus (HIV) type 1 in macaques immunized with hybrid HIV/hepatitis B surface antigen particles, *J. Virol.*, 64, 2452, 1990.

149. Schlienger, K. et al., Human immunodeficiency virus type 1 major neutralizing determinant exposed on hepatitis B surface antigen particles is highly immunogenic in primates, *J. Virol.*, 66, 2570, 1992.
150. Schlienger, K. et al., Vaccine-induced neutralizing antibodies directed in part to the simian immunodeficiency virus (SIV) V2 domain were unable to protect rhesus monkeys from SIV experimental challenge, *J. Virol.*, 68, 6578, 1994.
151. Sasnauskas, K., unpublished data, 1995.
152. Fomsgaard, A. et al., Improved humoral and cellular immune responses against the gp120 V3 loop of HIV-1 following genetic immunization with a chimeric DNA vaccine encoding the V3 inserted into the hepatitis B surface antigen, *Scand. J. Immunol.*, 47, 289, 1998.
153. Bryder, K. et al., Improved immunogenicity of HIV-1 epitopes in HBsAg chimeric DNA vaccine plasmids by structural mutations of HBsAg, *DNA Cell Biol.*, 18, 219, 1999.
154. von Brunn, A. et al., Epitopes of the human malaria parasite *P. falciparum* carried on the surface of HBsAg particles elicit an immune response against the parasite, *Vaccine*, 9, 477, 1991.
155. Vreden, S.G. et al., Phase I clinical trial of a recombinant malaria vaccine consisting of the circumsporozoite repeat region of *Plasmodium falciparum* coupled to hepatitis B surface antigen, *Am. J. Trop. Med. Hyg.*, 45, 533, 1991.
156. Moelans, I.I. et al., Induction of *Plasmodium falciparum* sporozoite-neutralizing antibodies upon vaccination with recombinant Pfs16 vaccinia virus and/or recombinant Pfs16 protein produced in yeast, *Mol. Biochem. Parasitol.*, 72, 179, 1995.
157. Gordon, D.M. et al., Safety, immunogenicity, and efficacy of a recombinantly produced *Plasmodium falciparum* circumsporozoite protein-hepatitis B surface antigen subunit vaccine, *J. Infect. Dis.*, 171, 1576, 1995.
158. Kester, K.E. et al., Efficacy of recombinant circumsporozoite protein vaccine regimens against experimental *Plasmodium falciparum* malaria, *J. Infect. Dis.*, 183, 640, 2001.
159. Stoute, J.A. et al., A preliminary evaluation of a recombinant circumsporozoite protein vaccine against *Plasmodium falciparum* malaria. RTS, S Malaria Vaccine Evaluation Group, *N. Engl. J. Med.*, 336, 86, 1997.
160. Kester, K.E. et al., A phase I/IIa safety, immunogenicity, and efficacy bridging randomized study of a two-dose regimen of liquid and lyophilized formulations of the candidate malaria vaccine RTS,S/AS02A in malaria-naive adults, *Vaccine*, 25, 5359, 2007.
161. Macete, E. et al., Safety and immunogenicity of the RTS,S/AS02A candidate malaria vaccine in children aged 1–4 in Mozambique, *Trop. Med. Int. Health*, 12, 37, 2007.
162. Macete, E.V. et al., Evaluation of two formulations of adjuvanted RTS, S malaria vaccine in children aged 3 to 5 years living in a malaria-endemic region of Mozambique: A Phase I/IIb randomized double-blind bridging trial, *Trials*, 8, 11, 2007.
163. Stoute, J.A. et al., Phase 1 randomized double-blind safety and immunogenicity trial of *Plasmodium falciparum* malaria merozoite surface protein FMP1 vaccine, adjuvanted with AS02A, in adults in western Kenya, *Vaccine*, 25, 176, 2007.
164. de Oliveira, C.I. et al., Antigenic properties of the merozoite surface protein 1 gene of *Plasmodium vivax*, *Vaccine*, 17, 2959, 1999.
165. Pumpens, P. et al., Evaluation of HBs, HBc, and frCP virus-like particles for expression of human papillomavirus 16 E7 oncoprotein epitopes, *Intervirology*, 45, 24, 2002.
166. Chengalvala, M.V. et al., Enhanced immunogenicity of hepatitis B surface antigen by insertion of a helper T cell epitope from tetanus toxoid, *Vaccine*, 17, 1035, 1999.
167. Major, M.E. et al., DNA-based immunization with chimeric vectors for the induction of immune responses against the hepatitis C virus nucleocapsid, *J. Virol.*, 69, 5798, 1995.
168. Lee, I.H., Kim, C.H., and Ryu, W.S., Presentation of the hydrophilic domains of hepatitis C viral E2 envelope glycoprotein on hepatitis B surface antigen particles, *J. Med. Virol.*, 50, 145, 1996.
169. Netter, H.J. et al., Immunogenicity of recombinant HBsAg/HCV particles in mice pre-immunised with hepatitis B virus-specific vaccine, *Vaccine*, 21, 2692, 2003.
170. Karpenko, L.I. et al., Analysis of foreign epitopes inserted in HBcAG. Possible routes for solving the problem of chimeric core particle self assembly (in Russian), *Mol. Biol. (Mosk.)*, 34, 223, 2000.
171. Milich, D.R. et al., The hepatitis nucleocapsid as a vaccine carrier moiety, *Ann. N.Y. Acad. Sci.*, 754, 187, 1995.
172. Murray, K. and Shiau, A.L., The core antigen of hepatitis B virus as a carrier for immunogenic peptides, *Biol. Chem.*, 380, 277, 1999.

173. Pumpens, P. et al., Hepatitis B virus core particles as epitope carriers, *Intervirology*, 38, 63, 1995.
174. Pumpens, P. and Grens, E., Hepatitis B core particles as a universal display model: A structure-function basis for development, *FEBS Lett.*, 442, 1, 1999.
175. Pumpens, P. and Grens, E., HBV core particles as a carrier for B cell/T cell epitopes, *Intervirology*, 44, 98, 2001.
176. Schödel, F. et al., Hybrid hepatitis B virus core antigen as a vaccine carrier moiety: I. presentation of foreign epitopes, *J. Biotechnol.*, 44, 91, 1996.
177. Ulrich, R. et al., Core particles of hepatitis B virus as carrier for foreign epitopes, *Adv. Virus Res.*, 50, 141, 1998.
178. Scaglioni, P.P., Melegari, M., and Wands, J.R., Posttranscriptional regulation of hepatitis B virus replication by the precore protein, *J. Virol.*, 71, 345, 1997.
179. Kann, M. and Gerlich, W.H., Replication of hepatitis B virus, In: *Molecular Medicine of Viral Hepatitis*, Harrison, T.J. and Zuckerman, A.J. (Eds.), John Wiley & Sons, New York, 1997, pp. 63–87.
180. Dyson, M.R. and Murray, K., Selection of peptide inhibitors of interactions involved in complex protein assemblies: Association of the core and surface antigens of hepatitis B virus, *Proc. Natl. Acad. Sci. U S A*, 92, 2194, 1995.
181. Poisson, F. et al., Both pre-S1 and S domains of hepatitis B virus envelope proteins interact with the core particle, *Virology*, 228, 115, 1997.
182. Kazaks, A., unpublished data, 2007.
183. Rolland, D. et al., Purification of recombinant HBc antigen expressed in *Escherichia coli* and *Pichia pastoris*: Comparison of size-exclusion chromatography and ultracentrifugation, *J. Chromatogr. B Biomed. Sci. Appl.*, 753, 51, 2001.
184. Watelet, B. et al., Characterization and diagnostic potential of hepatitis B virus nucleocapsid expressed in *E. coli* and *P. pastoris*, *J. Virol. Methods*, 99, 99, 2002.
185. Braun, S. et al., Proteasomal degradation of core protein variants from chronic hepatitis B patients, *J. Med. Virol.*, 79, 1312, 2007.
186. Mechtcheriakova, I.A. et al., The use of viral vectors to produce hepatitis B virus core particles in plants, *J. Virol. Methods*, 131, 10, 2006.
187. Huang, Z. et al., Rapid, high-level production of hepatitis B core antigen in plant leaf and its immunogenicity in mice, *Vaccine*, 24, 2506, 2006.
188. Cohen, B.J. and Richmond, J.E., Electron microscopy of hepatitis B core antigen synthesized in *E. coli*, *Nature*, 296, 677, 1982.
189. Gerlich, W.H. et al., Specificity and localization of the hepatitis B virus-associated protein kinase, *J. Virol.*, 42, 761, 1982.
190. Roseman, A.M. et al., A structural model for maturation of the hepatitis B virus core, *Proc. Natl. Acad. Sci. U S A*, 102, 15821, 2005.
191. Newman, M. et al., Stability and morphology comparisons of self-assembled virus-like particles from wild-type and mutant human hepatitis B virus capsid proteins, *J. Virol.*, 77, 12950, 2003.
192. Yuan, T.T. et al., The mechanism of an immature secretion phenotype of a highly frequent naturally occurring missense mutation at codon 97 of human hepatitis B virus core antigen, *J. Virol.*, 73, 5731, 1999.
193. Yuan, T.T., Tai, P.C., and Shih, C., Subtype-independent immature secretion and subtype-dependent replication deficiency of a highly frequent, naturally occurring mutation of human hepatitis B virus core antigen, *J. Virol.*, 73, 10122, 1999.
194. Belnap, D.M. et al., Diversity of core antigen epitopes of hepatitis B virus, *Proc. Natl. Acad. Sci. U S A*, 100, 10884, 2003.
195. Conway, J.F. et al., Characterization of a conformational epitope on hepatitis B virus core antigen and quasiequivalent variations in antibody binding, *J. Virol.*, 77, 6466, 2003.
196. Harris, A. et al., Epitope diversity of hepatitis B virus capsids: Quasi-equivalent variations in spike epitopes and binding of different antibodies to the same epitope, *J. Mol. Biol.*, 355, 562, 2006.
197. Chen, H. et al., Purification of the recombinant hepatitis B virus core antigen (rHBcAg) produced in the yeast *Saccharomyces cerevisiae* and comparative observation of its particles by transmission electron microscopy (TEM) and atomic force microscopy (AFM), *Micron*, 35, 311, 2004.
198. Borisova, G.P. et al., Genetically engineered mutants of the core antigen of the human hepatitis B virus preserving the ability for native self-assembly (in Russian), *Dokl. Akad. Nauk. SSSR.*, 298, 1474, 1988.
199. Gallina, A. et al., A recombinant hepatitis B core antigen polypeptide with the protamine-like domain deleted self-assembles into capsid particles but fails to bind nucleic acids, *J. Virol.*, 63, 4645, 1989.

200. Inada, T. et al., Synthesis of hepatitis B virus e antigen in *E. coli*, *Virus Res.*, 14, 27, 1989.
201. Bundule, M.A. et al., C-terminal polyarginine tract of hepatitis B core antigen is located on the outer capsid surface (in Russian), *Dokl. Akad. Nauk. SSSR.*, 312, 993, 1990.
202. Birnbaum, F. and Nassal, M., Hepatitis B virus nucleocapsid assembly: Primary structure requirements in the core protein, *J. Virol.*, 64, 3319, 1990.
203. Seifer, M. and Standring, D.N., Assembly and antigenicity of hepatitis B virus core particles, *Intervirology*, 38, 47, 1995.
204. Zlotnick, A. et al., Dimorphism of hepatitis B virus capsids is strongly influenced by the C-terminus of the capsid protein, *Biochemistry*, 35, 7412, 1996.
205. Watts, N.R. et al., The morphogenic linker peptide of HBV capsid protein forms a mobile array on the interior surface, *EMBO J.*, 21, 876, 2002.
206. Kazaks, A. et al., Mosaic particles formed by wild-type hepatitis B virus core protein and its deletion variants consist of both homo- and heterodimers, *FEBS Lett.*, 549, 157, 2003.
207. Stray, S.J., Ceres, P., and Zlotnick, A., Zinc ions trigger conformational change and oligomerization of hepatitis B virus capsid protein, *Biochemistry*, 43, 9989, 2004.
208. Chain, B.M. and Myers, R., Variability and conservation in hepatitis B virus core protein, *BMC Microbiol.*, 5, 33, 2005.
209. Steven, A.C. et al., Structure, assembly, and antigenicity of hepatitis B virus capsid proteins, *Adv. Virus Res.*, 64, 125, 2005.
210. Vanlandschoot, P., Cao, T., and Leroux-Roels, G., The nucleocapsid of the hepatitis B virus: A remarkable immunogenic structure, *Antiviral Res.*, 60, 67, 2003.
211. Beesley, K.M. et al., Expression in yeast of amino-terminal peptide fusions to hepatitis B core antigen and their immunological properties, *Biotechnology (N. Y.)*, 8, 644, 1990.
212. Clarke, B.E. et al., Presentation and immunogenicity of viral epitopes on the surface of hybrid hepatitis B virus core particles produced in bacteria, *J. Gen. Virol.*, 71, 1109, 1990.
213. Neirynck, S. et al., A universal influenza A vaccine based on the extracellular domain of the M2 protein, *Nat. Med.*, 5, 1157, 1999.
214. Koletzki, D. et al., Puumala (PUU) hantavirus strain differences and insertion positions in the hepatitis B virus core antigen influence B-cell immunogenicity and protective potential of core-derived particles, *Virology*, 276, 364, 2000.
215. Koletzki, D. et al., HBV core particles allow the insertion and surface exposure of the entire potentially protective region of Puumala hantavirus nucleocapsid protein, *Biol. Chem.*, 380, 325, 1999.
216. Moriarty, A.M. et al., Expression of HIV Gag and Env B-cell epitopes on the surface of HBV core particles and analysis of the immune responses generated to those epitopes, In: *Vaccines 90*, Brown, F. et al. (Eds.), Cold Spring Harbor Laboratory, Cold Spring Harbor, New York, 1990, pp. 225–229.
217. Beterams, G., Böttcher, B., and Nassal, M., Packaging of up to 240 subunits of a 17 kDa nuclease into the interior of recombinant hepatitis B virus capsids, *FEBS Lett.*, 481, 169, 2000.
218. Yoshikawa, A. et al., Chimeric hepatitis B virus core particles with parts or copies of the hepatitis C virus core protein, *J. Virol.*, 67, 6064, 1993.
219. Kratz, P.A., Böttcher, B., and Nassal, M., Native display of complete foreign protein domains on the surface of hepatitis B virus capsids, *Proc. Natl. Acad. Sci. U S A*, 96, 1915, 1999.
220. Makeeva, I.V. et al., Heterologous epitopes in the central part of the hepatitis B virus core protein (in Russian), *Mol. Biol. (Mosk.)*, 29, 211, 1995.
221. Schödel, F. et al., The position of heterologous epitopes inserted in hepatitis B virus core particles determines their immunogenicity, *J. Virol.*, 66, 106, 1992.
222. Borisova, G. et al., Spatial structure and insertion capacity of immunodominant region of hepatitis B core antigen, *Intervirology*, 39, 16, 1996.
223. Ulrich, R. et al., New chimaeric hepatitis B virus core particles carrying hantavirus (serotype Puumala) epitopes: Immunogenicity and protection against virus challenge, *J. Biotechnol.*, 73, 141, 1999.
224. Loktev, V.B. et al., Design of immunogens as components of a new generation of molecular vaccines, *J. Biotechnol.*, 44, 129, 1996.
225. Kazaks, A. et al., Stop codon insertion restores the particle formation ability of hepatitis B virus core-hantavirus nucleocapsid protein fusions, *Intervirology*, 45, 340, 2002.
226. Karpenko, L.I. et al., Insertion of foreign epitopes in HBcAg: How to make the chimeric particle assemble, *Amino Acids*, 18, 329, 2000.

227. Koletzki, D., Untersuchungen Zum Insertions- und Assemblierungsverhalten chimärer HBV-Core-Proteins. PhD thesis, Humboldt-Universität Berlin, 1998.
228. Vogel, M. et al., *In vitro* assembly of mosaic hepatitis B virus capsid-like particles (CLPs): Rescue into CLPs of assembly-deficient core protein fusions and FRET-suited CLPs, *FEBS Lett.*, 579, 5211, 2005.
229. Chisari, F.V. and Ferrari, C., Hepatitis B virus immunopathogenesis, *Annu. Rev. Immunol.*, 13, 29, 1995.
230. Hoofnagle, J.H., Gerety, R.J., and Barker, L.F., Antibody to hepatitis-B-virus core in man, *Lancet*, 2, 869, 1973.
231. Milich, D.R. and McLachlan, A., The nucleocapsid of hepatitis B virus is both a T-cell-independent and a T-cell-dependent antigen, *Science*, 234, 1398, 1986.
232. Fehr, T. et al., T cell-independent type I antibody response against B cell epitopes expressed repetitively on recombinant virus particles, *Proc. Natl. Acad. Sci. U S A*, 95, 9477, 1998.
233. Milich, D.R. et al., The hepatitis B virus core and e antigens elicit different Th cell subsets: Antigen structure can affect Th cell phenotype, *J. Virol.*, 71, 2192, 1997.
234. Milich, D.R. et al., Antibody production to the nucleocapsid and envelope of the hepatitis B virus primed by a single synthetic T cell site, *Nature*, 329, 547, 1987.
235. Milich, D.R. et al., Role of B cells in antigen presentation of the hepatitis B core, *Proc. Natl. Acad. Sci. U S A*, 94, 14648, 1997.
236. Cao, T. et al., Hepatitis B virus core antigen binds and activates naive human B cells *in vivo*: Studies with a human PBL-NOD/SCID mouse model, *J. Virol.*, 75, 6359, 2001.
237. Lazdina, U. et al., Priming of cytotoxic T cell responses to exogenous hepatitis B virus core antigen is B cell dependent, *J. Gen. Virol.*, 84, 139, 2003.
238. Mondelli, M. et al., Specificity of T lymphocyte cytotoxicity to autologous hepatocytes in chronic hepatitis B virus infection: Evidence that T cells are directed against HBV core antigen expressed on hepatocytes, *J. Immunol.*, 129, 2773, 1982.
239. Rehermann, B. et al., The hepatitis B virus persists for decades after patients' recovery from acute viral hepatitis despite active maintenance of a cytotoxic T-lymphocyte response, *Nat. Med.*, 2, 1104, 1996.
240. Iwarson, S. et al., Protection against hepatitis B virus infection by immunization with hepatitis B core antigen, *Gastroenterology*, 88, 763, 1985.
241. Murray, K. et al., Hepatitis B virus antigens made in microbial cells immunise against viral infection, *EMBO J.*, 3, 645, 1984.
242. Murray, K. et al., Protective immunisation against hepatitis B with an internal antigen of the virus, *J. Med. Virol.*, 23, 101, 1987.
243. Sällberg, M. et al., Genetic immunization of chimpanzees chronically infected with the hepatitis B virus, using a recombinant retroviral vector encoding the hepatitis B virus core antigen, *Hum. Gene Ther.*, 9, 1719, 1998.
244. Townsend, K. et al., Characterization of $CD8^+$ cytotoxic T-lymphocyte responses after genetic immunization with retrovirus vectors expressing different forms of the hepatitis B virus core and e antigens, *J. Virol.*, 71, 3365, 1997.
245. Wild, J. et al., Polyvalent vaccination against hepatitis B surface and core antigen using a dicistronic expression plasmid, *Vaccine*, 16, 353, 1998.
246. Heathcote, J. et al., A pilot study of the CY-1899 T-cell vaccine in subjects chronically infected with hepatitis B virus. The CY1899 T Cell Vaccine Study Group, *Hepatology*, 30, 531, 1999.
247. Livingston, B.D. et al., Immunization with the HBV core 18–27 epitope elicits CTL responses in humans expressing different HLA-A2 supertype molecules, *Hum. Immunol.*, 60, 1013, 1999.
248. Shi, T.D. et al., Therapeutic polypeptides based on HBcAg (18–27) CTL epitope can induce antigen-specific CD(8)(+) CTL-mediated cytotoxicity in HLA-A2 transgenic mice, *World J. Gastroenterol.*, 10, 1222, 2004.
249. Peng, M. et al., Novel vaccines for the treatment of chronic HBV infection based on mycobacterial heat shock protein 70, *Vaccine*, 24, 887, 2006.
250. Xu, W. et al., Endoplasmic reticulum targeting sequence enhances HBV-specific cytotoxic T lymphocytes induced by a CTL epitope-based DNA vaccine, *Virology*, 334, 255, 2005.
251. Zhang, Y. et al., Hepatitis B virus core antigen epitopes presented by HLA-A2 single-chain trimers induce functional epitope-specific $CD8^+$ T-cell responses in HLA-A2.1/Kb transgenic mice, *Immunology*, 121, 105, 2007.
252. Lobaina, Y. et al., Mucosal immunogenicity of the hepatitis B core antigen, *Biochem. Biophys. Res. Commun.*, 300, 745, 2003.

253. Lobaina, Y. et al., Immunological characterization of two hepatitis B core antigen variants and their immunoenhancing effect on co-delivered hepatitis B surface antigen, *Mol. Immunol.*, 42, 289, 2005.
254. Lobaina, Y. et al., Comparative study of the immunogenicity and immunoenhancing effects of two hepatitis B core antigen variants in mice by nasal administration, *Vaccine*, 24 Suppl 2, S2, 2006.
255. Xing, Y.P. et al., Novel DNA vaccine based on hepatitis B virus core gene induces specific immune responses in Balb/c mice, *World J. Gastroenterol.*, 11, 4583, 2005.
256. Shao, H.J. et al., Enhancement of immune responses to the hepatitis B virus core protein through DNA vaccines with a DNA fragment encoding human IL-1beta 163–171 peptide, *Acta Virol.*, 47, 217, 2003.
257. Jechlinger, W. et al., Comparative immunogenicity of the hepatitis B virus core 149 antigen displayed on the inner and outer membrane of bacterial ghosts, *Vaccine*, 23, 3609, 2005.
258. Zhao, Y. and Zhan, M., The coexpression of the preS1 (1–42) and the core (1–144) antigen of HBV in *E. coli*, *Chin Med. Sci. J.*, 17, 68, 2002.
259. Yang, H.J. et al., Expression and immunoactivity of chimeric particulate antigens of receptor binding site-core antigen of hepatitis B virus, *World J. Gastroenterol.*, 11, 492, 2005.
260. Sominskaya, I. et al., unpublished data, 2000.
261. Sominskaya, I. et al., HBc-preS1 VLPs as potential vaccine candidates, In: *The Molecular Biology of Hepatitis B Viruses*, Bertoletti, A. and Tavis, J.E. (Eds.), Centro Congressi Giovanni XXIII, Bergamo, Italy, Abstracts, 2003, p. 190.
262. Chen, X. et al., Recombinant hepatitis B core antigen carrying preS1 epitopes induce immune response against chronic HBV infection, *Vaccine*, 22, 439, 2004.
263. Kazaks, A. et al., Mosaic hepatitis B virus core particles presenting the complete preS sequence of the viral envelope on their surface, *J. Gen. Virol.*, 85, 2665, 2004.
264. Schödel, F., Milich, D.R., and Will, H., Hepatitis B virus nucleocapsid/pre-S2 fusion proteins expressed in attenuated *Salmonella* for oral vaccination, *J. Immunol.*, 145, 4317, 1990.
265. Schödel, F. et al., A virulent *Salmonella* expressing hybrid hepatitis B virus core/pre-S genes for oral vaccination, *Vaccine*, 11, 143, 1993.
266. Schödel, F. et al., Hybrid hepatitis B virus core-pre-S proteins synthesized in avirulent *Salmonella typhimurium* and *Salmonella typhi* for oral vaccination, *Infect. Immun.*, 62, 1669, 1994.
267. Tacket, C.O. et al., Safety and immunogenicity in humans of an attenuated *Salmonella typhi* vaccine vector strain expressing plasmid-encoded hepatitis B antigens stabilized by the Asd-balanced lethal vector system, *Infect. Immun.*, 65, 3381, 1997.
268. Nardelli-Haefliger, D. et al., Nasal vaccination with attenuated *Salmonella typhimurium* strains expressing the hepatitis B nucleocapsid: Dose response analysis, *Vaccine*, 19, 2854, 2001.
269. Stratford, R. et al., Optimization of *Salmonella enterica* serovar *typhi* ΔaroC ΔssaV derivatives as vehicles for delivering heterologous antigens by chromosomal integration and *in vivo* inducible promoters, *Infect. Immun.*, 73, 362, 2005.
270. Karpenko, L.I. et al., Isolation and study of recombinant strains of *Salmonella typhimurium* SL 7207, producing HBcAg and HBcAg-HBs (in Russian), *Vopr. Virusol.*, 45, 10, 2000.
271. Karpenko, L.I. and Il'ichev, A.A., Chimeric hepatitis B core antigen particles as a presentation system of foreign protein epitopes (in Russian), *Vestn. Ross. Akad. Med. Nauk.*, 3, 6, 1998.
272. Yue, Q. et al., Immune responses to recombinant *Mycobacterium smegmatis* expressing fused core protein and preS1 peptide of hepatitis B virus in mice, *J. Virol. Methods*, 141, 41, 2007.
273. Chambers, M.A. et al., Chimeric hepatitis B virus core particles as probes for studying peptide-integrin interactions, *J. Virol.*, 70, 4045, 1996.
274. Zhang, Y.L. et al., Enhanced immunogenicity of modified hepatitis B virus core particle fused with multiepitopes of foot-and-mouth disease virus, *Scand. J. Immunol.*, 65, 320, 2007.
275. Huang, Y. et al., Immunogenicity of the epitope of the foot-and-mouth disease virus fused with a hepatitis B core protein as expressed in transgenic tobacco, *Viral Immunol.*, 18, 668, 2005.
276. Del Val, M. et al., Protection against lethal cytomegalovirus infection by a recombinant vaccine containing a single nonameric T-cell epitope, *J. Virol.*, 65, 3641, 1991.
277. Schödel, F. et al., Immunity to malaria elicited by hybrid hepatitis B virus core particles carrying circumsporozoite protein epitopes, *J. Exp. Med.*, 180, 1037, 1994.
278. Schödel, F. et al., Immunization with hybrid hepatitis B virus core particles carrying circumsporozoite antigen epitopes protects mice against *Plasmodium yoelii* challenge, *Behring Inst. Mitt.*, 98, 114, 1997.

279. Milich, D.R. et al., Conversion of poorly immunogenic malaria repeat sequences into a highly immunogenic vaccine candidate, *Vaccine*, 20, 771, 2001.
280. Sällberg, M. et al., A malaria vaccine candidate based on a hepatitis B virus core platform, *Intervirology*, 45, 350, 2002.
281. Birkett, A. et al., A modified hepatitis B virus core particle containing multiple epitopes of the *Plasmodium falciparum* circumsporozoite protein provides a highly immunogenic malaria vaccine in preclinical analyses in rodent and primate hosts, *Infect. Immun.*, 70, 6860, 2002.
282. Nardin, E.H. et al., Phase I testing of a malaria vaccine composed of hepatitis B virus core particles expressing *Plasmodium falciparum* circumsporozoite epitopes, *Infect. Immun.*, 72, 6519, 2004.
283. Walther, M. et al., Safety, immunogenicity and efficacy of a pre-erythrocytic malaria candidate vaccine, ICC-1132 formulated in Seppic ISA 720, *Vaccine*, 23, 857, 2005.
284. Oliveira, G.A. et al., Safety and enhanced immunogenicity of a hepatitis B core particle *Plasmodium falciparum* malaria vaccine formulated in adjuvant Montanide ISA 720 in a phase I trial, *Infect. Immun.*, 73, 3587, 2005.
285. Langermans, J.A. et al., Effect of adjuvant on reactogenicity and long-term immunogenicity of the malaria vaccine ICC-1132 in macaques, *Vaccine*, 23, 4935, 2005.
286. Boulter, N.R. et al., *Theileria annulata* sporozoite antigen fused to hepatitis B core antigen used in a vaccination trial, *Vaccine*, 13, 1152, 1995.
287. Boulter, N. et al., Evaluation of recombinant sporozoite antigen SPAG-1 as a vaccine candidate against *Theileria annulata* by the use of different delivery systems, *Trop. Med. Int. Health*, 4, A71, 1999.
288. Londono, L.P. et al., Immunisation of mice using *Salmonella typhimurium* expressing human papillomavirus type 16 E7 epitopes inserted into hepatitis B virus core antigen, *Vaccine*, 14, 545, 1996.
289. Street, M. et al., Differences in the effectiveness of delivery of B- and CTL-epitopes incorporated into the hepatitis B core antigen (HBcAg) c/e1-region, *Arch. Virol.*, 144, 1323, 1999.
290. Tindle, R.W. et al., Chimeric hepatitis B core antigen particles containing B- and Th-epitopes of human papillomavirus type 16 E7 protein induce specific antibody and T-helper responses in immunised mice, *Virology*, 200, 547, 1994.
291. Ulrich, R. et al., Chimaeric HBV core particles carrying a defined segment of Puumala hantavirus nucleocapsid protein evoke protective immunity in an animal model, *Vaccine*, 16, 272, 1998.
292. Kruger, D.H., Ulrich, R., and Lundkvist, A., Hantavirus infections and their prevention, *Microbes Infect.*, 3, 1129, 2001.
293. Geldmacher, A. et al., An amino-terminal segment of hantavirus nucleocapsid protein presented on hepatitis B virus core particles induces a strong and highly cross-reactive antibody response in mice, *Virology*, 323, 108, 2004.
294. Geldmacher, A. et al., A hantavirus nucleocapsid protein segment exposed on hepatitis B virus core particles is highly immunogenic in mice when applied without adjuvants or in the presence of pre-existing anti-core antibodies, *Vaccine*, 23, 3973, 2005.
295. Fiers, W. et al., A "universal" human influenza A vaccine, *Virus Res.*, 103, 173, 2004.
296. De Filette, M. et al., Universal influenza A vaccine: Optimization of M2-based constructs, *Virology*, 337, 149, 2005.
297. De Filette, M. et al., The universal influenza vaccine M2e-HBc administered intranasally in combination with the adjuvant CTA1-DD provides complete protection, *Vaccine*, 24, 544, 2006.
298. De Filette, M. et al., Improved design and intranasal delivery of an M2e-based human influenza A vaccine, *Vaccine*, 24, 6597, 2006.
299. Jegerlehner, A. et al., Influenza A vaccine based on the extracellular domain of M2: weak protection mediated via antibody-dependent NK cell activity, *J. Immunol.*, 172, 5598, 2004.
300. Takeda, S. et al., Hemagglutinating virus of Japan protein is efficient for induction of CD4 + T-cell response by a hepatitis B core particle-based HIV vaccine, *Clin. Immunol.*, 112, 92, 2004.
301. Fuller, D.H. et al., Immunogenicity of hybrid DNA vaccines expressing hepatitis B core particles carrying human and simian immunodeficiency virus epitopes in mice and rhesus macaques, *Virology*, 364, 245, 2007.
302. Mihailova, M. et al., Recombinant virus-like particles as a carrier of B- and T-cell epitopes of hepatitis C virus (HCV), *Vaccine*, 24, 4369, 2006.
303. Chen, J.Y. and Li, F., Development of hepatitis C virus vaccine using hepatitis B core antigen as immunocarrier, *World J. Gastroenterol.*, 12, 7774, 2006.

304. Gu, Y. et al., Selection of a peptide mimicking neutralization epitope of hepatitis E virus with phage peptide display technology, *World J. Gastroenterol.*, 10, 1583, 2004.
305. Dawson, J.A. and Macrina, F.L., Construction and immunologic evaluation of a *Porphyromonas* gingivalis subsequence peptide fused to hepatitis B virus core antigen, *FEMS Microbiol. Lett.*, 175, 119, 1999.
306. Charles, I.G. et al., Identification and characterization of a protective immunodominant B cell epitope of pertactin (P.69) from *Bordetella pertussis*, *Eur. J. Immunol.*, 21, 1147, 1991.
307. Wu, L. et al., DNA vaccine against *Taenia solium* cysticercosis expressed as a modified hepatitis B virus core particle containing three epitopes shared by *Taenia crassiceps* and *Taenia solium*, *J. Nanosci. Nanotechnol.*, 5, 1204, 2005.
308. Storni, T. et al., Nonmethylated CG motifs packaged into virus-like particles induce protective cytotoxic T cell responses in the absence of systemic side effects, *J. Immunol.*, 172, 1777, 2004.
309. Song, S. et al., Augmented induction of CD8(+) cytotoxic T-cell response and antitumor effect by DCs pulsed with virus-like particles packaging with CpG, *Cancer Lett.*, 2007 July 24; [Epub ahead of print].
310. Mao, D. et al., Intramuscular immunization with a DNA vaccine encoding a 26-amino acid CETP epitope displayed by HBc protein and containing CpG DNA inhibits atherosclerosis in a rabbit model of atherosclerosis, *Vaccine*, 24, 4942, 2006.
311. Billaud, J.N. et al., Advantages to the use of rodent hepadnavirus core proteins as vaccine platforms, *Vaccine*, 25, 1593, 2007.
312. Kwissa, M. et al., Polyvalent DNA vaccines with bidirectional promoters, *J. Mol. Med.*, 78, 495, 2000.
313. Musacchio, A. et al., Multivalent DNA-based immunization against hepatitis B virus with plasmids encoding surface and core antigens, *Biochem. Biophys. Res. Commun.*, 282, 442, 2001.
314. Aguilar, J.C. et al., Development of a nasal vaccine for chronic hepatitis B infection that uses the ability of hepatitis B core antigen to stimulate a strong Th1 response against hepatitis B surface antigen, *Immunol. Cell Biol.*, 82, 539, 2004.
315. Betancourt, A.A. et al., Phase I clinical trial in healthy adults of a nasal vaccine candidate containing recombinant hepatitis B surface and core antigens, *Int. J. Infect. Dis.*, 11, 394, 2007.
316. Xiong, X.Y., Liu, X., and Chen, Y.D., Expression and immunoreactivity of HCV/HBV epitopes, *World J. Gastroenterol.*, 11, 6440, 2005.
317. Gedvilaite, A. et al., Formation of immunogenic virus-like particles by inserting epitopes into surface-exposed regions of hamster polyomavirus major capsid protein, *Virology*, 273, 21, 2000.
318. Mancini-Bourgine, M. et al., Hepatitis B virus splice-generated protein induces T-cell responses in HLA-transgenic mice and hepatitis B virus-infected patients, *J. Virol.*, 81, 4963, 2007.
319. Reddy, V.S. et al., Virus Particle Explorer (VIPER), a website for virus capsid structures and their computational analyses, *J. Virol.*, 75, 11943, 2001.

10 Polyomavirus-Derived Virus-Like Particles

Rainer G. Ulrich, Alma Gedvilaite, Tatyana Voronkova, Andris Kazaks, Kestutis Sasnauskas, and Reimar Johne

CONTENTS

10.1 POLYOMAVIRUSES: INTRODUCTION AND TAXONOMY

Initially, polyomaviruses had been grouped together with papillomaviruses into the family *Papovaviridae* due to similarities of both virus groups regarding virion morphology and genome composition. However, as noticeable differences between both virus groups in genome organization and biological properties became evident, a novel family *Polyomaviridae* with the genus *Polyomavirus* was created in 2000, in which all of the polyomavirus species have been placed (Hou et al., 2005).

Until now, a total of 19 different polyomaviruses have been described infecting a broad range of mammalian and avian species (Johne and Müller, 2007). Table 10.1 summarizes biological and structural properties of the most relevant representatives. Because of their small genome size and similarities in genome structure, polyomaviruses have been intensively studied as models for eukaryotic gene organization, gene expression, and genome replication. The monkey polyomavirus simian virus 40 (SV-40) and the murine polyomavirus (MPyV) were among the first viruses whose genomes had been totally sequenced. Fundamental insights into cellular biology such as the discovery of alternative splicing and the abundant tumor suppressor p53 have been gained through the study of SV-40. The prominent property of SV-40 and MPyV to cause tumor growth after inoculation into newborn rodents not only led to intensive studies on pathways regulating the cell cycle but also raised concerns about the involvement of polyomaviruses in human tumor disease. Particularly, the role of SV-40 as a causative tumor-inducing agent in persons that were immunized in the 1950s and 1960s with poliomyelitis vaccine charges, which were contaminated with SV-40, is still a matter of debate (Garcea and Imperiale, 2003).

The human JC polyomavirus (JCPyV) and BK polyomavirus (BKPyV) are broadly distributed among the human population and usually cause subclinical persistent infections. However, disease

TABLE 10.1
Summary of Polyomaviruses, Their Structural Proteins, and Formation of Autologous VLPs in Different Expression Systems

Polyomavirus (Abbreviation)	Natural Host	Disease in the Natural Host	Late Structural Proteins (Amino acids)			Formation of Autologous VLPs after Expression of Structural Protein(s) in			
			VP1	VP2	VP3	Bacteria (Dialysis Against Ca^{2+})	Yeast	Insect Cells	Mammalian Cells
BK polyomavirus (BKPyV)	Human	Post-transplantation polyomavirus-associated nephropathy	362	351	232	nd	VP1: + (Hale et al., 2002)	VP1: + (Viscidi and Clayman, 2006)	nd
JC polyomavirus (JCPyV)	Human	Progressive multifocal leucoencephalopathy in AIDS patients	354	344	225	nd	VP1: + (Chen et al., 2001)	VP1: + (Goldmann et al., 1999)	nd
KI polyomavirus (KIPyV)	Human	May be associated with respiratory disease	378	400	257	nd	nd	nd	nd
WU polyomavirus (WU)	Human	May be associated with respiratory disease	369	415	272	nd	nd	nd	nd
Merkel cell polyomavirus (McPyv)	Human	May be associated with Merkel cell carcinoma	423	241	196	nd	nd	nd	nd
Simian virus 40 (SV-40)	Monkey	Not known	364	352	234	VP1: + (Wrobel et al., 2000)	VP1: + (Sasnauskas et al., 2002)	VP1: +, VP1−3: + (Kosukegawa et al., 1996)	VP1−3: + [Cos-7 cells] (Asano et al., 1985)
B-lymphotropic polyomavirus (LPyV)	Monkey	Not known	368	356	237	nd	VP1: + (A. Gedvilaite, unpublished data)	VP1: + (Pawlita, et al., 1996)	nd
Hamster polyomavirus (HaPyV)	Hamster	Epithelioma and lymphoma	384	345	221	VP1: + (Voronkova et al., 2007)	VP1:+ (Sasnauskas et al., 1999)	VP1:+ (Voronkova et al., 2007)	nd

Murine polyomavirus (MPyV)	Mouse	Not known	384	319	204	VP1: + (Salunke et al., 1985)	VP1: + (Sasnauskas et al., 2002)	VP1: + (Soeda et al., 1998), VP1–3: + (An et al., 1999a)	nd
Murine pneumotropic virus (MPtV)	Mouse	Respiratory disease in suckling mice	373	341	222	nd	nd	VP1: + (Tegerstedt et al., 2003)	nd
Avian polyomavirus (APV) [Budgerigar fledgling disease polyomavirus (BFPyV)]	Parrot (and other birds)	Budgerigar fledgling disease	343 VP4: 176	341	235	VP1: + (Rodgers et al., 1994)	VP1: + (Sasnauskas et al., 2002)	VP1: – VP1–3: – (An et al., 1999b)	VP1: – VP1, 3: + VP1, 3, 4: + [chicken fibroblasts] (Johne and Müller, 2004)
Goose hemorrhagic polyomavirus (GHPV)	Goose	Hemorrhagic nephritis and enteritis of geese	353 ORF-X: 169	326	217	nd	VP1: (+)* VP1, 2: + (Zielonka et al., 2006)	VP1 : + (Zielonka et al., 2006)	nd
Finch polyomavirus (FPyV)	Bullfinch	Increased mortality of fledglings, dermatitis	358 ORF-X: 205	354	244	nd	VP1: + (Zielonka et al., unpublished data)	nd	nd
Crow polyomavirus (CPyV)	Jackdaw	May be associated with increased mortality	353 ORF-X: 150	333	227	nd	VP1: + (Zielonka et al., unpublished data)	nd	nd

Note: + VLP formation observed; – no VLPs observed; * small (20 nm) VLP only; nd, not done.

may occur due to severe immunosuppression. JCPyV is the etiologic agent of progressive multifocal leucoencephalopathy (PML; Padgett et al., 1971), a fatal demyelinating brain disease in AIDS patients. BKPyV causes the polyomavirus-associated nephropathy, which is a severe complication after renal transplantation with increasing importance (Hirsch, 2002). The etiologic role of the two recently discovered human KI and WU polyomaviruses (Allander et al., 2007; Gaynor et al., 2007) in respiratory disease is not clear until now. Very recently, a novel polyomavirus designated as Merkel cell polyomavirus (McPyv) has been detected in samples of Merkel cell carcinoma, which is a rare but aggressive human skin cancer (Feng et al., 2008).

An additional primate polyomavirus worth mentioning is the B-lymphotropic polyomavirus (LPyV) showing a marked tropism to lymphoblastoid cells (zur Hausen and Gissmann, 1979), which is in contrast to the epithelial (kidney) cell tropism of most polyomaviruses. Additional rodent polyomaviruses are the hamster polyomavirus (HaPyV) and the murine pneumotropic virus (MPtV) which cause skin tumors in the Syrian hamster (Graffi et al., 1968) and pneumonia in suckling mice (Greenlee, 1979), respectively.

In contrast to their mammalian relatives, the polyomaviruses of birds are the causative agents of acute and chronic inflammatory diseases with high mortality rates whereas tumor induction has never been observed (Johne and Müller, 2007). The budgerigar fledgling disease polyomavirus (BFPyV) causes a devastating disease with high mortality rates in young budgerigars. As BFPyV also causes acute disease in other parrots as well as chronic feather disorders in a variety of bird species, the designation avian polyomavirus (APV) has been suggested for this virus, which is now widely used (Johne and Müller, 1998; 2007). The recently discovered goose hemorrhagic polyomavirus (GHPV) is the etiological agent of hemorrhagic nephritis and enteritis of geese (Guerin et al., 2000), which is a disease of young goslings with mortality rates up to 60%. GHPV and probably most of the polyomaviruses of birds (with the exception of APV) cannot be efficiently propagated in cell culture. Recently, the multiple-primed rolling-circle amplification technique (RCA, Figure 10.1)

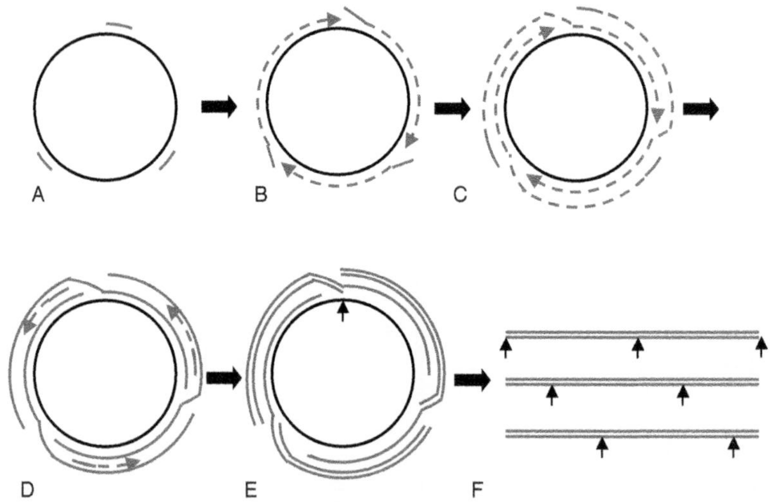

FIGURE 10.1 Principle of multiple-primed rolling-circle amplification technique (RCA). Random hexamer primers are used for annealing (A) followed by isothermal primer extension using the phi29 DNA polymerase (B). When the polymerase reaches a site where a primer has been already bound, it displaces the respective strand and DNA synthesis continues (B, C). By this, a long concatemeric DNA is produced in the case of a circular template. During first-strand synthesis, random hexamer primers anneal to the newly synthesized strand (D) followed by extension (E) leading to products consisting of linear and mainly double-stranded DNA (F). By the use of a single-cutting restriction enzyme (arrow), genome-length units can be generated from the products and cloned for subsequent analysis.

has been applied for amplifying and cloning of the whole genomes of two novel polyomaviruses (finch polyomavirus, FPyV, and crow polyomavirus, CPyV) directly from infected organ samples without the need of cell culture propagation (Johne et al., 2006).

10.2 STRUCTURE AND GENOME ORGANIZATION OF POLYOMAVIRUSES

Polyomaviruses are nonenveloped viruses with an icosahedral capsid, approximately 45 nm in diameter. The crystal structures of the virions of SV-40 (Liddington et al., 1991) and MPyV (Stehle et al., 1994) have been resolved. The capsid mainly consists of 72 pentamers of the viral protein 1 (VP1) which are arranged in a $T=7$ symmetry. At the inner surface of each VP1 pentamer, one molecule of the minor capsid proteins, either VP2 or VP3, is situated. Within the capsid, the double-stranded circular DNA genome, approximately 5 kbp in length, is associated with cell-derived histones resulting in a chromosome-like genome structure.

All polyomaviruses possess a very similar genome organization. The genome is transcribed bidirectionally from a noncoding regulatory region for the expression of early and late genes. The early genome region encodes a large and a small tumor (T) antigen which are expressed by partial or alternative splicing of the mRNA. They are involved in viral genome replication and malignant transformation of cells *in vitro* and *in vivo*. By alternative splicing, MPyV and HaPyV encode an additional transforming protein designated as middle T antigen. The late region encodes the structural proteins VP1, VP2, and VP3, which are translated from partially overlapping open reading frames (ORFs) (Figure 10.2A and Table 10.1). VP1 represents the major capsid protein that interacts with the cellular receptor molecule. The minor capsid proteins VP2 and VP3 are encoded by the same ORF by the use of different initiation codons. They are identical at the carboxy-terminus, however, VP2 has an amino-terminal extension as compared to VP3. This amino-terminus of VP2 has been shown to be myristoylated (Schmidt et al., 1989). Recently, an additional small protein has been shown to be expressed by the use of another initiation codon within this ORF, which has functions in induction of cell lysis (Daniels et al., 2007).

Some of the polyomaviruses encode additional proteins. Late in the viral replication cycle the so-called agnoprotein is expressed by an ORF located upstream the VP2-encoding region of the primate polyomaviruses. This protein has functions in virus release, regulation of the cell cycle, and DNA repair (Khalili et al., 2005). At the same position of the genome, the so-called ORF-X encodes the VP4 in the case of polyomaviruses of birds. VP4 has been shown to interact with VP1 and the viral DNA, and to be incorporated into the viral capsid of APV (Johne and Müller, 2001). An additional function as a factor of pathogenesis has been suggested for this protein as it also efficiently induces apoptosis (Johne and Müller, 2007).

10.3 REPLICATION OF POLYOMAVIRUSES

Attachment of the polyomavirus particles to the cell is mediated by VP1. The different polyomaviruses use a variety of different receptor molecules (Dugan et al., 2006); however, most of them recognize sialic acids present at the cellular surface either as part of a ganglioside or a glycoprotein. While BKPyV, MPyV, and APV recognize $\alpha(2,3)$-linked sialic acid, JCPyV and LPyV use $\alpha(2,6)$-linked sialic acid residues. As additional receptors, MHC class I molecules and the $5HT_{2A}$ serotonin receptor have been identified for SV-40 and JCPyV, respectively.

Following binding to the cellular surface, two different major mechanisms have been described to be used by the different polyomaviruses for uptake into the cell (Dugan et al., 2006). SV-40 and BKPyV enter cells through caveolae-mediated endocytosis, whereas JCPyV enters the cell through clathrin-dependent endocytosis (Pho et al., 2000; Norkin et al., 2002; Dugan et al., 2006). Interestingly, although JCPyV enters the cells by clathrin-dependent endocytosis, it is then sorted from early endosomes to caveosomes (Querbes et al., 2006). The release of the virus particles from the vesicular compartments into the cytoplasm is thought to be mediated by VP2 involving its

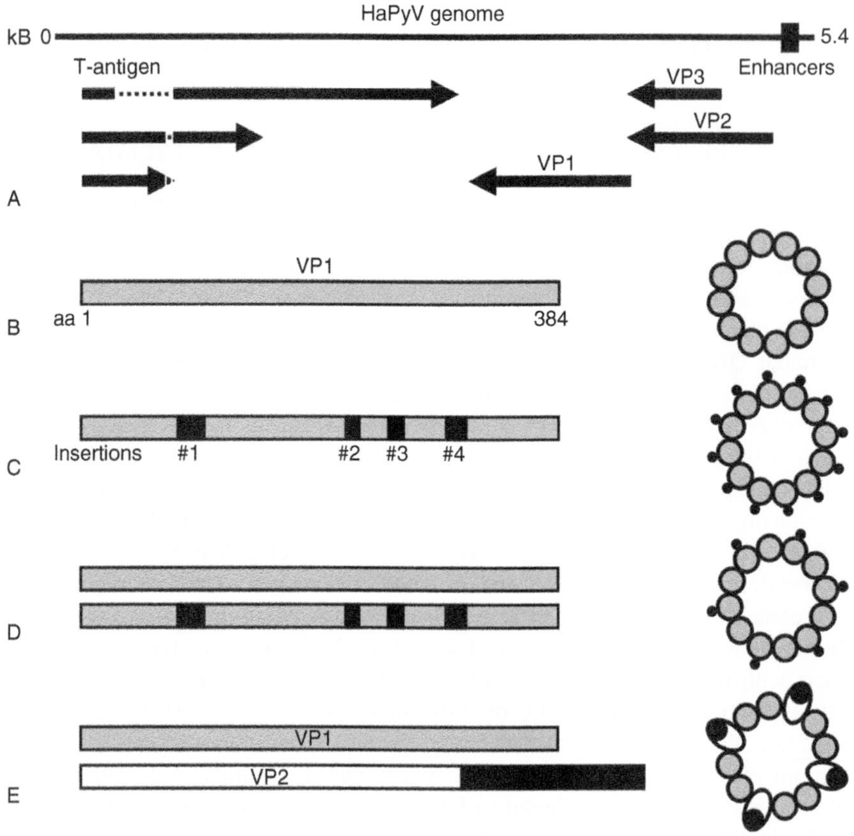

FIGURE 10.2 Genome organization of HaPyV and approaches to generate polyomavirus-derived VLPs. (A) The double-stranded circular DNA genome of HaPyV, 5366 bp in length, is encoding three structural proteins, the major capsid protein VP1, and the two minor capsid proteins VP2 and VP3. Like MPyV, the HaPyV genome also encodes three early proteins, small, middle, and large T antigens. (B) The authentic entire VP1 of HaPyV is 384 aa in length and forms autologous uniform single-shelled VLPs after expression in bacteria, yeast, and insect cells. (C) Insertion of foreign sequences into one of the insertion sites #1 (aa 80–89), #2 (aa 222–225), #3 (aa 243–247) or #4 (aa 288–295) can result in the formation of chimeric VLPs. (D) Co-expression of HaPyV-VP1 "helper" and a VP1 fusion protein can result in the formation of mosaic VLPs. (E) Co-expression of MPyV-VP1 "helper" and a carboxy-terminally extended MPyV-VP2 can result in the formation of pseudotype VLPs.

hydrophobic myristoylated N-terminus (Norkin et al., 2002). Thereafter, the virus traffics to the nuclear pore and the viral genome is released into the nucleus.

Within the nucleus, the cellular RNA polymerase II transcribes the early genes leading to expression of the T antigens. One of the major roles of these proteins is to induce the cell to enter the S phase thus expressing proteins required for viral genome replication (Sullivan and Pipas, 2002). By this, malignant transformation of the cell may occur if progression of the viral life cycle is blocked. Otherwise, the large T antigen thereafter binds to the origin of replication within the noncoding regulatory region of the viral genome and replicates the viral DNA by recruiting several cellular proteins.

In parallel to the onset of genome replication, the mode of transcription changes by means of large T antigen binding and expression of a viral microRNA to enable late gene expression. After synthesis of VP1, VP2, and VP3, the proteins are transported into the nucleus, most probably as

VP1 pentamers complexed with VP2 or VP3 molecules. All of the polyomavirus VP1 sequences with the exception of that of the bird polyomaviruses contain a nuclear localization signal at the amino-terminus. Also, the common carboxy-terminus of VP2 and VP3 of all polyomaviruses have a nuclear localization signal sequence, however, this sequence is more important for trafficking of incoming virus than for nuclear import of proteins during the late phase of infection (Nakanishi et al., 2007).

Within the nucleus, the viral particles assemble and the viral genome complexed with cellular histones is incorporated. Chaperons are involved in virus particle assembly (Chromy et al., 2003) and disulfide bonds (Li et al., 2005) as well as calcium ions (Li et al., 2003a) contribute to virion stability. The mechanism of virus particle release from the host cell is poorly understood until now, however, induction of cell lysis and apoptosis by accessory viral proteins has been shown to be involved in the case of SV-40 (Daniels et al., 2007) and APV (Johne et al., 2007), respectively.

10.4 GENERATION OF POLYOMAVIRUS-DERIVED VIRUS-LIKE PARTICLES

The simplest way of getting polyomavirus-derived virus-like particles (VLPs; in case of polyomaviruses synonymous to capsid-like particles, CLPs) is the heterologous expression of the major capsid protein VP1 which formed spontaneously *in vivo* or *in vitro* single-shelled VLPs. We designated this type of VLPs consisting of only target virus structural protein "autologous" uniform VLPs (Figure 10.2B). In general, these VLPs are very similar to native virus particles, namely in structure, size, and immunological properties. However, due to lacking of viral DNA these VLPs are noninfectious which may represent an important advantage for their use in vaccination and gene therapy approaches.

Initially, polyomavirus-derived VLPs were generated by the high-level expression of MPyV-VP1 in *Escherichia coli* which allowed the subsequent *in vitro* assembly of VLPs under certain ion conditions (Leavitt et al., 1985; Salunke et al., 1986). This was later on confirmed by *in vitro* assembly of APV-derived VLPs using bacterially expressed VP1 (Rodgers et al., 1994). Alternatively, baculovirus-driven expression of VP1 of rodent, primate, and human polyomaviruses as well as of GHPV result in the spontaneous formation of VLPs in insect cells (Montross et al., 1991; Kosukegawa et al., 1996; Pawlita et al., 1996; Chang et al., 1997; Touze et al., 2001; Tegerstedt et al., 2003; Zielonka et al., 2006). The nuclear localization of the VLPs in the recombinant insect cells suggested the necessity of a functional nuclear localization for the efficient *in vivo* formation of VLPs (Montross et al., 1991; Pawlita et al., 1996; Voronkova et al., 2007). Similarly, the addition of a SV-40 derived nuclear localization signal to the APV-VP1 synthesized in mammalian cells led to its transportation to the nucleus and resulted in VLP formation (Johne and Müller, 2004).

The generation of VLPs originating from HaPyV has strongly been motivated by the lack of an efficient cell culture propagation system for this virus. Synthesis of the authentic VP1 as well as an amino-terminally extended derivative in the yeast *Saccharomyces cerevisiae* resulted in the formation of VLPs which are similar in size and shape to those found in HaPyV-infected tumor cells in Syrian hamsters (Sasnauskas et al., 1999; Figure 10.3). Recently, plasmid-driven synthesis of HaPyV-VP1 in *E. coli* and insect cells also allowed the generation of VLPs (Voronkova et al., 2007).

The high efficiency of the *S. cerevisiae*-based expression system was confirmed by the production of VLPs based on the VP1 of different human (JCPyV, BKPyV), primate (SV-40), and rodent (MPyV) polyomaviruses (Sasnauskas et al., 2002). Recently, the yeast expression system was used for the generation of VLPs consisting of VP1 of novel polyomaviruses derived from chimpanzee, finch, and crow (A. Zielonka et al., unpublished data). As these viruses could not be isolated in cell culture (Johne et al., 2005; 2006), the use of heterologous expression systems is an important tool to provide antigens for diagnostic purposes and vaccines.

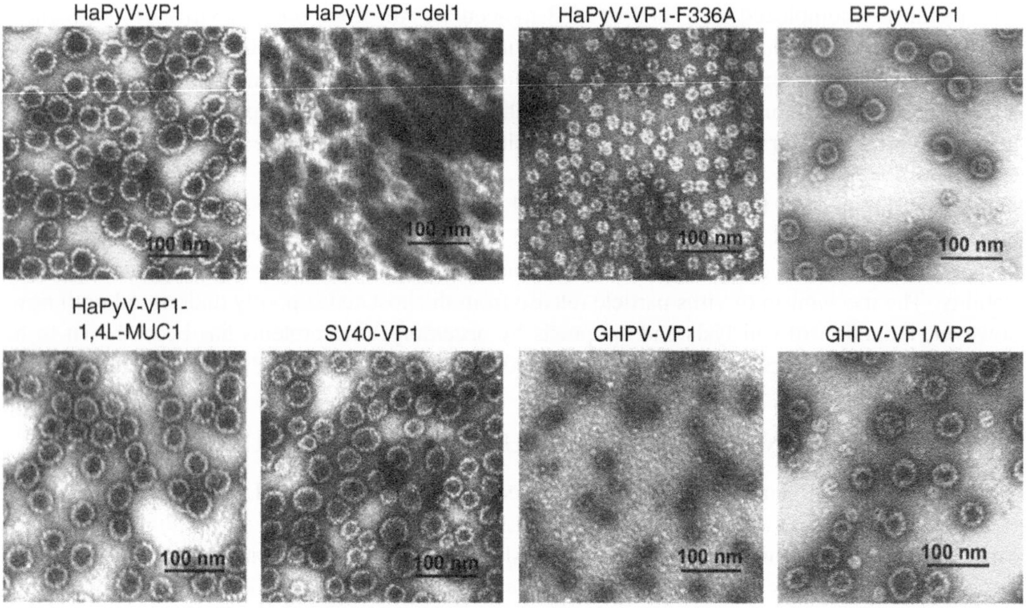

FIGURE 10.3 Negative stain electron microscopical images of yeast-expressed authentic, modified, and chimeric VLPs of different polyomaviruses. Entire HaPyV-VP1 and a derivative thereof harboring two mucin1-derived nonamer peptides (MUC1) flanked by GSSG linkers at sites #1 and #4 form VLPs of the structure and size typical for polyomaviruses. An amino acid exchange variant (F336A) and a deletion variant del1 (lacking amino acid residues 335–346) form structures of other size and shape. The authentic VP1 proteins of SV-40 and BFPyV also assemble to polyomavirus-typical VLPs, whereas for GHPV co-expression of VP1 and VP2 is required for formation of such VLPs.

Although in most cases the heterologous synthesis of the polyomavirus major capsid protein VP1 was demonstrated to be alone sufficient for VLP formation (Table 10.1), problems were observed for the efficient formation of VLPs of bird polyomaviruses. Thus, the synthesis of APV-VP1 in insect cells failed to form VLPs (An et al., 1999b); the yeast expression of the gene resulted in VLP formation, but with an about 10-fold lower efficiency as compared to the VP1 of other polyomaviruses (Sasnauskas et al., 2002). A missing nuclear localization signal in the APV-VP1 as well as the need for cell-type specific factors may be the reason for lacking or low efficient VLP formation (Johne and Müller, 2004).

Recent approaches to generate VLPs for GHPV were mainly driven by the lack of an efficient cell culture system and the necessity to generate diagnostic antigens. Baculovirus-driven expression of the VP1 gene resulted in a low efficient production of VLPs of the typical 45 nm diameter. By contrast, VP1 expression alone in the yeast expression system allowed the formation of 20 nm VLPs (Zielonka et al., 2006; Figure 10.3).

In addition to VLPs consisting of polyomavirus VP1 alone, autologous but complex VLPs were generated that contain also the minor capsid proteins VP2/VP3. Thus, baculovirus-driven co-expression of VP1, VP2, and VP3 of SV-40 resulted in the formation of VLPs (Sandalon et al., 1997). Interestingly a co-expression of VP1 and VP2 of GHPV in yeast resulted in the formation of typical 45 nm diameter VLPs (Zielonka et al., 2006; Figure 10.3). Co-expression of APV-VP1 with VP3 or VP3 and VP4 was demonstrated to recover the formation of VLPs in an influenzavirus-driven mammalian expression system (Johne and Müller, 2004).

10.5 GENERATION OF CHIMERIC, MOSAIC, AND PSEUDOTYPE POLYOMAVIRUS-DERIVED VLPs

For some reasons it might be desired to present nonassembly competent proteins and peptides of viral or other pathogenic origin as well as tumor antigen-derived epitopes or proteins on VLPs (see below). This can be achieved by two different ways, i.e., by chemical coupling or by genetic engineering techniques. Besides the classical chemical cross-linking, sophisticated procedures have been developed for a site-specific attachment of foreign proteins to the VLP carrier. Thus, antibody-binding domain protein Z was fused to MPyV-VP1 allowing the binding of antibodies for a cell-type specific targeting (Gleiter and Lilie, 2001). Alternatively, a WW domain was inserted into MPyV-VP1-VLPs allowing the decoration of the VLPs by proline-rich ligands. Using this approach a modified eGFP harboring a PPLP peptide was bound on the surface of these VLPs (Schmidt et al., 2001). Moreover, attachment of a fluorescence marker to MPyV-VP1 VLPs was accomplished by generation of a modified MPyV-VP1 where all cysteine residues of the wild-type protein were replaced by serines, and a new unique cysteine residue was introduced (Schmidt et al., 1999).

Strategies to generate chimeric, mosaic, and pseudotype VLPs are described in detail by Pumpens et al. (Chapter 9). Chimeric VLPs can be simply achieved by the generation of a fusion construct where the foreign sequence is inserted into the VLP carrier protein backbone (Figure 10.2C). Chimeric VLPs have been described for HaPyV, MPyV, LPyV, and SV-40 where the foreign sequence is added to VP1 or is substituting original VP1 residues. The size of the foreign sequences ranges from a few amino acid (aa) residues to entire proteins as enhanced green fluorescence proteins (eGFP) (fused to HaPyV-VP1) and dihydrofolate reductase (fused to MPyV-VP1). The high insertion capacity of HaPyV-VP1 was also confirmed by VLP formation of VP1 molecules harboring a simultaneous insertion of two or three peptide epitopes into the same VP1 molecule (Gedvilaite et al., 2000; Lawatscheck et al., 2007; Dorn et al., 2008; Aleksaite and Gedvilaite, 2006 and unpublished data; Table 10.2). Alternatively, foreign sequences can be presented on mosaic VLPs when the foreign sequences are inserted into the VP1 carrier and an autologous carrier VP1 protein is functioning as a "helper" mediating VLP formation of both protein species (Figure 10.2D). This novel approach was exploited for HaPyV- and MPyV-derived VLPs (Table 10.2). A similar approach is exploiting the assembly capacity of VP1 and its interaction with the minor capsid protein VP2 which is harboring a foreign sequence addition resulting in formation of pseudotype VLPs (Figure 10.2E). This approach seems to allow the presentation of very large-sized foreign sequences as demonstrated for the entire human tumor-associated antigen Her2 (Tegerstedt et al., 2005).

The selection of polyomavirus-derived VLPs as a carrier moiety follows some important reasons which are based on the biology of these viruses and discussed in detail in the corresponding chapters on vaccine development and gene therapy approaches. The generation of polyomavirus-derived chimeric, mosaic, and pseudotype VLPs is based either on empirical or rational-design-based approaches. The rational-design-based approaches exploited the known three-dimensional structure of SV-40 and MPyV capsids (Liddington et al., 1991; Stehle et al., 1994; 1996; Stehle and Harrison, 1996; Figure 10.4A–C). On the basis of these structural informations SV-40-derived chimeric VLPs were generated (A. Bulavaite and K. Sasnauskas, unpublished data; Table 10.2). On the basis of these resolved structures and the aa sequence homology, potential insertion sites for foreign sequences were predicted for the VP1 proteins of HaPyV and LPyV at different internal positions and at the carboxy-terminal part of the VP1 proteins (Gedvilaite et al., 2000; Langner et al., 2004). An aa sequence alignment of polyomavirus VP1 sequences demonstrated that the potential insertion sites are localized in or adjacent to highly variable regions (Figure 10.4D). These predictions were accompanied by B-cell epitope mapping studies in HaPyV-VP1 confirming an internal site as suitable for insertions and the carboxy-terminal region as immunodominant and highly cross-reactive (Siray et al., 2000).

Empirical studies revealed that the amino-terminus of HaPyV-VP1 tolerate the addition of a four aa-long peptide originating from an in-frame prolongation of VP1-ORF (Sasnauskas et al.,

TABLE 10.2
Summary of Polyomavirus-Derived Chimeric, Mosaic, and Pseudotype VLPs Harboring Foreign Inserts

Polyomavirus	Chimeric/Mosaic/Pseudotype	Insertion Sites	Foreign Insert	Formation of Chimeric VLPs	References
Hamster polyomavirus	Chimeric: VP1 with insert	#1 (between aa positions 80–89)	preS1 (5 aa)	+	Gedvilaite et al., 2000
			MUC1 (9 aa)	+	Dorn et al., submitted
			CEA (9 aa)	+	Lawatscheck et al., 2007
			PUUV-N (45 or 120 aa)	+	Gedvilaite et al., 2004
			hTERT (9 and 15 aa)	+/−	Aleksaite and A. Gedvilaite, unpublished data
		#2 (between aa positions 222–225)	preS1	+	Gedvilaite et al., 2000
			PUUV-N (45 or 120 aa)	−	Gedvilaite et al., 2004
		#3 (between aa positions 243–247)	preS1	+	Gedvilaite et al., 2000
			PUUV-N (45 or 120 aa)	−	Gedvilaite et al., 2004
		#4 (between aa positions 288–295)	preS1	+	Gedvilaite et al., 2000
			MUC1	+	Dorn et al., submitted
			CEA	+	Lawatscheck et al., 2007
			PUUV-N (45 or 120 aa)	+	Gedvilaite et al., 2004
			eGFP	+	Gedvilaite et al., 2006b
			TRP (9 aa) + MAGE (9 aa) + HTERT (9aa)	+/−	Aleksaite and Gedvilaite, 2006
			HTERT (9 and 15 aa)	+/−	Aleksaite and Gedvilaite, unpublished data
	VP1 with two inserts	#1 + #2	preS1	+	Gedvilaite et al., 2000
		#1 + #3	preS1	+	Gedvilaite et al., 2000
		#1 + #4	MUC1	+	Zvirbliene et al., 2006
			CEA	+	Lawatscheck et al., 2007
	VPI with three inserts	#1 + #3 + #4	TRP (9 aa) + MAGE (9 aa) + HTERT (9 aa)	+	Aleksaite and Gedvilaite, 2006
	VP1 with four inserts	#1 + #2 + #3 + #4	MUC1	−	Dorn et al., submitted
			CEA	−	Lawatscheck et al., 2007
	Mosaic: VP1 + VP1 with insert	#4 (between aa positions 288–295)	eGFP	+	Gedvilaite, unpublished data
Murine polyomavirus	Chimeric: VP1 with insert	aa positions 80 or 87	Human uPa (14 aa, or 60 aa or 135 aa)	nd	Shin and Folk, 2003

		150 aa position	Human uPa (14 aa, or 60 aa or 135 aa)	nd	Shin et al., 2003
		200 aa position	Human uPa (14 aa, or 60 aa or 135 aa)	nd	Shin et al., 2003
		293 aa position	Human uPa (14 aa, or 60 aa or 135 aa)	nd	Shin et al., 2003
			FLAG epitope (8 aa)	+	
		293–294 aa position	DHFR 18 kD	+	Gleiter et al., 1999
			Z protein 6.8 kD	+	Gleiter et al., 2001
	VP1 with two inserts	200 aa position and 293 aa position	Human uPa (60 aa) + FLAG epitope (8 aa)	+*	Shin et al., 2003
	Mosaic:				
	VP1 with two inserts +	VP1 with two inserts: 200 aa and 293 aa positions	Human uPa (60 aa) + FLAG epitope (8 aa)	+	Shin et al., 2003
	VP1 with one insert	VP1 with insert at 293 aa position	FLAG epitope (8 aa)		
	VP1+ VP2 with insert	Fusion with full-length VP2	Her2 (1-683 aa)	+	Tegerstedt et al., 2005
B-lymphotropic polyomavirus	Chimeric Substitution of 3 aa of the carrier by 3 foreign aa	aa 70–72, 72–74, 74–76 or 76–78 were altered	3 aa (RGB)	+	Langner et al., 2004
	Substitution of 3 aa of the carrier by 3 foreign aa	aa 133–135, 135–137 or aa 137–139 were altered	3 aa (RGB)	+	Langner et al., 2004
	Substitution of 3 aa of the carrier by 3 foreign aa	aa 272–274, 274–276, 276–278 or 278–280 were altered	3 aa (RGB)	+	Langner et al., 2004
SV-40	Chimeric	BC loop (between aa positions 73–76)	PreS1 (27 aa or 46 aa)	+	Bulavaite and Sasnauskas, unpublished data
		HI loop (between aa positions 273–276)	PreS1 (27 aa or 46 aa)	+	Bulavaite and Sasnauskas, unpublished data

* Only smaller particles approximately 20 nm in diameter.

Note: +/− mix of VLPs and pentamers; nd, not done.

Abbreviations: CEA, CTL-epitope of human tumor-associated carcinoembryonic antigen; DHFR, dihydrofolate reductase; eGFP, enhanced green fluorescent protein; FLAG peptide tag, 8-amino-acid leader peptide of the gene-10 product from bacteriophage T7; Her2, CTL epitope of human tumor-associated proto-oncogene Her2/*neu* which is a member of the epidermal growth factor receptor family with tyrosine kinase activity; HTERT, CTL-epitope of human telomerase reverse transcriptase; MAGE, CTL-epitope of human MAGE A family protein; MUC1, CTL-epitope of human tumor-associated mucin 1; PUUV-N; nucleocapsid protein of Puumala virus, strain Vranica-Hällnäs; preS1, B-cell epitope of the surface protein of the hepatitis B virus; TRP, CTL-epitope of human tyrosinase-related protein-2; uPA, human urokinase-type plasminogen activator; Z protein is an engineered antibody-binding domain derived from protein A of *Staphylococcus aureus*.

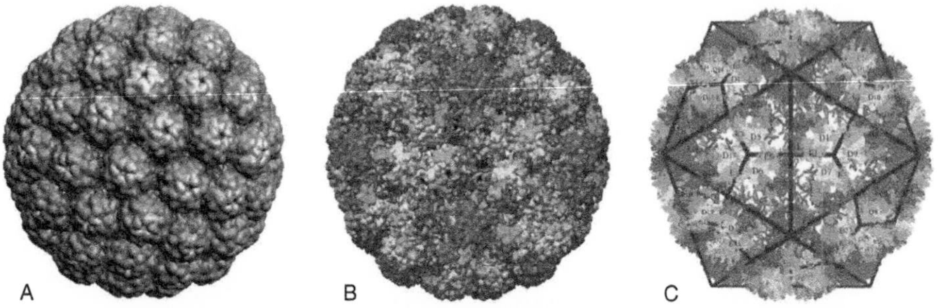

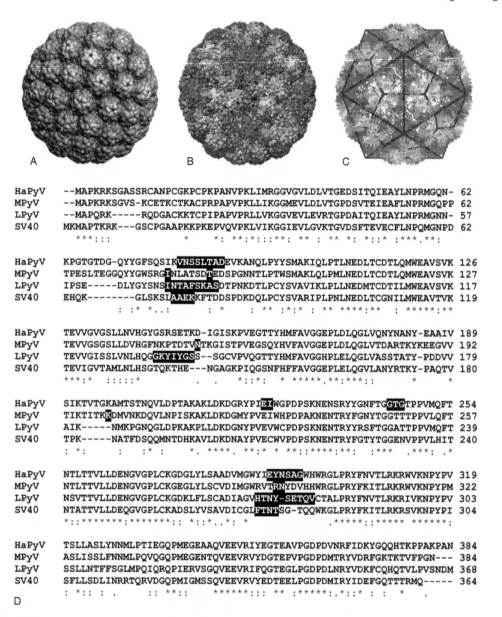

```
HaPyV     --MAPKRKSGASSRCANPCGKPCPKPANVPKLIMRGGVGVLDLVTGEDSITQIEAYLNPRMGQN- 62
MPyV      --MAPKRKSGVS-KCETKCTKACPRPAPVPKLLIKGGMEVLDLVTGPDSVTEIEAFLNPRMGQPP 62
LPyV      --MAPQRK-----RQDGACKKTCPIPAPVPRLLVKGGVEVLEVRTGPDAITQIEAYLNPRMGNN- 57
SV40      MKMAPTKRK---GSCPGAAPKKPKEPVQVPKLVIKGGIEVLGVKTGVDSFTEVECFLNPQMGNPD 62
            ***:::           *    *: **:*:::**: ** : ** *: *::* :***.**:

HaPyV     KPGTGTDG-QYYGFSQSIKVNSSLTADEVKANQLPYYSMAKIQLPTLNEDLTCDTLQMWEAVSVK 126
MPyV      TPESLTEGGQYYGWSRGINLATSDTEDSPGNNTLPTWSMAKLQLPMLNEDLTCDTLQMWEAVSVK 127
LPyV      IPSE-----DLYGYSNSINTAFSKASDTPNKDTLPCYSVAVIKLPLLNEDMTCDTILMWEAVSVK 117
SV40      EHQK--------GLSKSLAAEKKFTDDSPDKDQLPCYSVARIPLPNLNEDLTCGNILMWEAVTVK 119
                   : :* *..:       : *     : ** :*:* : ** ****:**  : *****:**

HaPyV     TEVVGVGSLLNVHGYGSRSETKD-IGISKPVEGTTYHMFAVGGEPLDLQGLVQNYNANY-EAAIV 189
MPyV      TEVVGSGSLLDVHGFNKPTDTVNTKGISTPVEGSQYHVFAVGGEPLDLQGLVTDARTKYKEEGVV 192
LPyV      TEVVGISSLVNLHQGGKYIYGSS--SGCVPVQGTTYHMFAVGGHPLELQGLVASSTATY-PDDVV 179
SV40      TEVIGVTAMLNLHSGTQKTHE---NGAGKPIQGSNFHFFAVGGEPLELQGVLANYRTKY-PAQTV 180
          ***:*  :::::*             .  .  *::*:  * *****.**:***::     *     *

HaPyV     SIKTVTGKAMTSTNQVLDPTAKAKLDKDGRYPIEIWGPDPSKNENSRYYGNFTGGTGTPPVMQFT 254
MPyV      TIKTITKKDMVNKDQVLNPISKAKLDKDGMYPVEIWHPDPAKNENTRYFGNYTGGTTTPPVLQFT 257
LPyV      AIK-----NMKPGNQGLDPKAKPLLDKDGNYPVEVWCPDPSKNENTRYYRSFTGGATTPPVMQFT 239
SV40      TPK-----NATFDSQQMNTDHKAVLDKDNAYPVECWVPDPSKNENTRYFGTYTGGENVPPVLHIT 240
          : *:     :   :* :     *. ****. **:* * **::****:**:  :***  .***: .*

HaPyV     NTLTTVLLDENGVGPLCKGDGLYLSAADVMGWYIEYNSAGWHWRGLPRYFNVTLRKRWVKNPYPV 319
MPyV      NTLTTVLLDENGVGPLCKGEGLYLSCVDIMGWRVTRNYDVHHWRGLPRYFKITLRKRWVKNPYPM 322
LPyV      NSVTTVLLDENGVGPLCKGDKLFLSCADIAGVHTNY-SETQVCTALPRYFNVTLRKRIVKNPYPV 303
SV40      NTATTVLLDEQGVGPLCKADSLYVSAVDICGLFTNTSG-TQQWKGLPRYFKITLRKRSVKNPYPI 304
          *::*******:*******:: *::*..*: *                .*****::***** ******:

HaPyV     TSLLASLYNNMLPTIEGQPMEGEAAQVEEVRIYEGTEAVPGDPDVNRFIDKYGQQHTKPPAKPAN 384
MPyV      ASLISSLFNNMLPQVQGQPMEGENTQVEEVRVYDGTEPVPGDPDMTRYVDRFGKTKTVFPGN--- 384
LPyV      SSLLNTFFSGLMPQIQRQPIERVSGQVEEVRIFQGTEGLPGDPDLNRYVDKFCQHQTVLPVSNDM 368
SV40      SFLLSDLINRRTQRVDGQPMIGMSSQVEEVRVYEDTEELPGDPDMIRYIDEFGQTTTRMQ----- 364
          : *:: : .      :: **::    ******::: ** :*****:.*::* : :  *   .
```

D

FIGURE 10.4 (See color insert following page 142.) Three-dimensional (x-ray) structure of MPyV and amino acid sequence alignment of VP1 proteins of HaPyV and other polyomaviruses demonstrating the sites used for foreign insertions. (A) Rendered surface, (B) subunit organization, and (C) $T = 7d$ lattice of the MPyV according to Stehle et al. 1996. The pictures were adapted from the VIPER site (Shepherd et al., 2006). (D) Alignment of the VP1 amino acid sequences of HaPyV (Acc. No. CAA06802), MPyV (Acc. No. P49302), LPyV (Acc. No. P04010), and SV-40 (Acc. No. AAK29047) generated by ClustalW (Thompson et al., 1994). Sites used for foreign insertions according to the data presented in Table 10.2 are labeled by black boxes.

1999; Voronkova et al., 2007). Although the carboxy-terminal region of MPyV-VP1 was demonstrated to be essential for VLP formation (Garcea et al., 1987), investigations on JCPyV- and HaPyV-VP1 demonstrated the potential of these carriers for substituting certain parts of the amino- and carboxy-terminal regions (Ou et al., 1999; Gedvilaite et al., 2006a). Because of these investigations, also the amino- and carboxy-terminal ends represent potential insertion sites for foreign sequences that need further characterization.

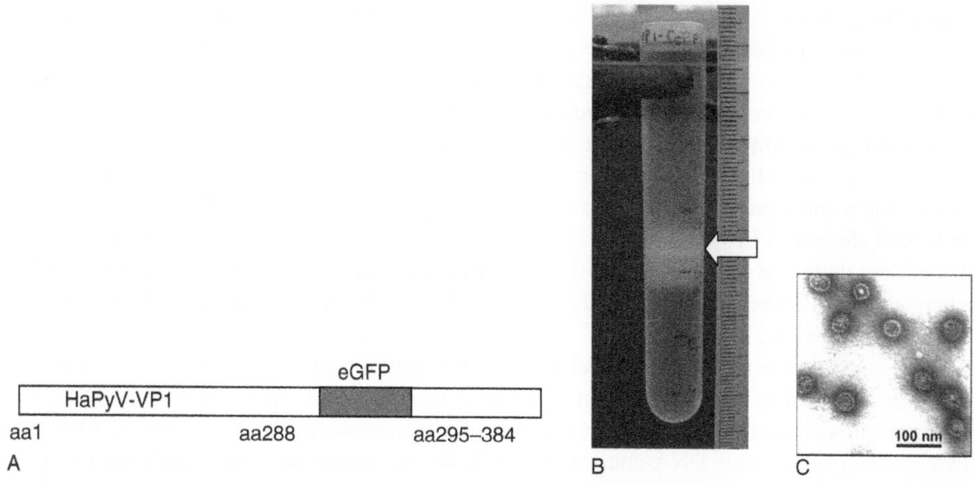

FIGURE 10.5 Generation, purification, and characterization of chimeric HaPyV-VP1 VLPs harboring green fluorescent protein (eGFP). (A) Schematic presentation of the structure of the HaPyV-VP1/eGFP fusion protein. (B) **(See color insert following page 142.)** Picture of the cesium chloride density gradient showing a fluorescent band representing VP1/eGFP VLPs. (C) Negative staining electron microscopic image of VP1/eGFP VLPs.

To prove the value of the predicted insertion sites model epitopes of different size and a RGD-motif were used to substitute original segments in the VP1 carrier proteins of HaPyV and LPyV, respectively (Table 10.2, Figure 10.5). Whereas in HaPyV-VP1 a five aa-long peptide originating from preS1 antigen of HBV was tolerated at four different sites, 45 and 120 aa-long sequences originating from a hantavirus nucleocapsid protein allowed only in positions #1 and #4 the efficient formation of chimeric VLPs (Gedvilaite et al., 2000; 2004). The large insertion capacity of site #4 was confirmed by the generation of chimeric VLPs with an eGFP insertion demonstrating a self-fluorescence in the cesium chloride gradient allowing an easy purification (Gedvilaite et al., 2006b; Figure 10.5). Recently, the potential influence of a flexible linker insertion on both sides of foreign nonamer peptides (two different CTL epitopes of human tumor-associated antigens mucin 1 and carcinoembryonic antigen) on the assembly of VLPs was studied (Lawatscheck et al., 2007; Dorn et al., submitted).

The insertion site in the carrier determines not only the assembly capacity but also the surface exposure, antigenicity, and immunogenicity of the inserted foreign epitope on VLPs (Gedvilaite et al., 2000; Lawatscheck et al., 2007). All problems related to these issues are discussed in the respective chapters on vaccine and gene therapy development.

10.6 BASIC RESEARCH APPLICATIONS OF POLYOMAVIRUS-DERIVED VLPs

Polyomavirus-derived VLPs were applied for a variety of basic research approaches. Thus, these VLPs have become an interesting tool for studying the receptor interaction and entry pathways, especially for viruses where an efficient cell culture system is still missing.

For VLP-based studies on cellular receptors, entry, and intracellular trafficking different procedures for labeling of polyomavirus-derived VLPs have been explored. On the one hand HaPyV-VP1-derived VLPs harboring an eGFP have been generated (Table 10.2). Alternatively, VLPs based on different polyomaviruses have been labeled by carboxyfluorescein diacetate succinimidyl ester (CFDA; Gedvilaite et al., 2006b; Dorn et al., submitted). Competition in cell entry experiments of nonmodified, eGFP-harboring and autologous, but CFDA-labeled HaPyV-VP1 VLPs demonstrated the strong structural similarities of the two modified VLPs to those nonmodified (Gedvilaite et al., 2006b).

Initially, polyomavirus-derived VLPs were used for studying the binding on mammalian cells and erythrocytes in particular. Competition of 3T6 cell binding of MPyV by MPyV-derived VP1/VP2-VLPs confirmed the structural similarities of VLPs to infectious virions highlighting their usefulness for receptor binding studies (An et al., 1999a).

First indications on differences of the receptor-mediated entry of some polyomaviruses were made using hemagglutination assays. In line with previous data on the corresponding infectious viruses (Bolen and Consigli, 1979; Knowles et al., 2003), yeast-expressed JCPyV-, BKPyV-, and MPyV-VP1-derived VLPs were found to hemagglutinate erythrocytes, and SV-40-derived VLPs did not (Gedvilaite et al., 2006b). Interestingly, yeast-expressed HaPyV-VP1-derived VLPs failed in the hemagglutination assay, as previously observed also for MPtV-derived VLPs (Tegerstedt et al., 2003; Gedvilaite et al., 2006b).

Major capsid protein VP1-derived VLPs from different human and nonhuman polyomaviruses have been applied for the identification of cellular receptors and entry pathways. The value of this approach was confirmed by electron and confocal microscopy investigations of MPyV and MPyV-derived empty and DNA-containing VLPs demonstrating no visible differences in their adsorption and internalization into mouse fibroblasts and epithelial cells (Richterova et al., 2001). Entry studies using BKPyV-VP1-derived VLPs demonstrated that cellular entry is dependent on cell surface sialic acid (Touze et al., 2001). Using JCPyV-derived VLPs oligosaccharides of glycoproteins and glycolipids were identified as putative JCPyV receptors (Komagome et al., 2002). The cellular entry of MPyV-VP1-derived VLPs was demonstrated to be mediated by glycolipids such as the sialic acid-containing gangliosides (Smith et al., 2003). In contrast, MPtV-VLP binding to cells was neuraminidase resistant but mostly trypsin and papain sensitive, indicating that the MPtV receptor lacks sialic acid components (Tegerstedt et al., 2003). The entry of yeast-expressed VP1-VLPs of different polyomaviruses to monocyte-derived human dendritic cells demonstrated in addition to an uptake by macropinocytosis also different receptor-mediated entry mechanisms (Dorn et al., submitted).

VLPs have also been used to study intracellular trafficking pathways. Thus JCPyV-VLPs have been found to bind to cellular importins via the nuclear localization site (NLS) of VP1 and transported into the nucleus through the nuclear pore complex (Qu et al., 2004). Studies on avian polyomavirus revealed that transport of VP1, or VP1–VP4 complex, into the nucleus is facilitated by the co-expression of VP3 and resulted in the formation of VLPs (Johne and Müller, 2004).

VLPs have been used not only for the identification of potential cellular receptors, but also for the identification of viral proteins and mapping of functional domains therein that are involved in receptor recognition and intracellular trafficking. This can be accomplished by insertion mutagenesis at surface-exposed loops as demonstrated for LPyV-VP1-derived VLPs (Langner et al., 2004). Entry studies using VLPs formed by SV-40 VP1 variants modified at the calcium-binding sites suggested that calcium coordination imparts not only stability but also structural flexibility to the virion, allowing the acquisition or loss of the ion at the two sites to control virion formation in the nucleus, as well as virion structural alterations at the cell surface and in the cytoplasm early during infection (Li et al., 2003a).

A second important basic research application of polyomavirus-derived VLPs is in investigations on the structure and assembly processes. Thus, the analysis of BKPyV-derived VLPs by electron cryomicroscopy and subsequent computer image reconstruction allow the generation of a more refined three-dimensional structure thereof (Li et al., 2003b). In contrast, crystallization of entire *E. coli*-expressed MPyV-VP1 failed due to its high assembly capacity; therefore, the high-resolution structure analysis was made using a truncated VP1 derivative (Stehle and Harrision, 1997).

In addition to structural investigations, VLPs were used to identify protein domains that are essential for VLP formation and responsible for shape and size determination. As mentioned above (Table 10.1), for the majority of polyomaviruses the major capsid protein VP1 is sufficient to form VLPs. To identify essential regions in VP1 for VLP formation, deletion mutagenesis approaches were followed for MPyV, JCPyV, and HaPyV. The deletion of 63 aa residues from the carboxy-terminal region of MPyV-VP1 (until residue P321 with a 6 aa-long additional linker resulting from cloning) resulted in a lack of VLP formation in the *E. coli* expression system (Garcea et al., 1987). Similarly, *E. coli*-expressed JCPyV-VP1 derivatives lacking 17 and 31 carboxy-terminal or 19 amino-terminal aa failed to form VLPs, whereas truncation of 16 carboxy-terminal or 12 amino-terminal aa still allowed VLP formation (Ou et al., 2001). Interestingly, yeast-expressed truncation variants of HaPyV-VP1 lacking 35, 45, and 56 carboxy-terminal aa residues also failed to form VLPs but those lacking 21, 69, and 79 aa residues at their carboxy-terminal region efficiently formed VLPs similar to those formed by the unmodified VP1 (diameter 40–45 nm). HaPyV-VP1 mutants with a single A336G aa exchange or internal deletions of aa 335–aa 346 and aa 335–aa 363 resulted in the formation of VLPs of a smaller size (diameter 20 nm) (Gedvilaite 2006a; Figure 10.3). Insertional mutagenesis using defined foreign insertions of different size and origin also allowed to confirm the structural flexibility of surface-exposed regions in VP1 (Table 10.2).

VLPs have also been used for the identification of intermediates of the assembly process, characterization of the interaction of the involved structural proteins, and the characterization of its environmental conditions. In addition to *in vivo* studies on the assembly process of different polyomaviruses, alternatively, in certain cases *in vitro* assembly was applied for characterization of the ion and pH conditions of the assembly process of polyomavirus VLPs, e.g., of SV-40 (Kosukegawa et al., 1996; Kanesashi et al., 2003). To study the functional roles of disulfide linkage and calcium ion-mediated interactions in SV-40 capsid formation, the assembly and disassembly of VP1 variants was studied *in vitro* (Ishizu et al., 2001). *In vitro* assembly studies on MPyV-VP1 demonstrated an equilibrium reaction followed by oxidation of intracapsomere disulfide bonds, which are not essential for the formation of VLPs but prevent their disassembly (Schmidt et al., 2000). These studies might also result in interesting insights for *in vitro* disassembly/reassembly for incorporation of nucleic acids or other compounds.

10.7 DIAGNOSTIC APPLICATIONS OF POLYOMAVIRUS-DERIVED VLPs

VLPs have been extremely useful for the development of diagnostic tools for serological detection of virus infections in human and animals. This is especially important for certain polyomaviruses, which very inefficiently replicate in cell culture or which could not be isolated so far. Moreover, VLPs usually provide a native protein folding exposing conformational epitopes. In the case of APV, it has been shown that most of the antibodies which exhibit neutralizing activity recognize conformation-dependent epitopes of VP1 (Fattaey et al., 1992). Using VLPs as antigens, those antibodies will be preferentially detected in ELISA tests in which the native antigen structure is preserved. In the case of hemagglutinating viruses, the particle structure of VLPs enables the development of hemagglutination inhibition (HI) tests, which show in most cases a stronger correlation to the neutralizing activity of the antibodies than ELISA tests. This is of special interest in the case of viruses for which the performance of neutralization tests is not practicable because of the lack of efficient tissue culture systems. Assessment of neutralizing antibodies is necessary for vaccination studies and for demonstration of protective immunity.

The use of polyomavirus-derived VLPs was so far limited to the corresponding autologous VP1-derived VLPs, although in natural infection not only antibodies against the major capsid protein are induced, but also antibodies directed against the minor capsid proteins and the T antigens can be detected, e.g., as demonstrated for HaPyV-infected hamsters (Ulrich et al., 1996). ELISA test formats based on autologous VP1-VLPs were developed for human polyomaviruses BKPyV strains, AS and SB, and JCPyV (Stolt et al., 2003). However, a problem for differentiation of infections by serological assays is the high level of cross-reactivity observed between the different polyomaviruses, especially that of the VP1 (Shah et al., 1977; Sasnauskas et al., 2002). For some polyomaviruses, e.g., JCPyV and BKPyV, but not SV-40, hemagglutination inhibition tests based on the corresponding VLPs might represent a tool to differentiate these infections. Interestingly, ELISA investigations of sera from HaPyV-infected hamsters demonstrated the highest endpoint titer for HaPyV-VP1 VLPs but lower titers for VLPs originating from JCPyV, BKPyV, and SV-40 (Foster et al., 2002). For harmonization of serological polyomavirus diagnostics and comparable seroepidemiological studies it might be important in the future to perform comparative ring test validation of the different test formats.

As HaPyV infections have recently been found in Syrian hamster colonies of different origin, ELISAs based on yeast- and *E. coli*-expressed HaPyV-VP1-VLPs were developed. VLP-based ELISA investigations of hamsters from these colonies indicated the presence of HaPyV-reactive antibodies which was found to be in line with the detection of viral antigen or HaPyV-DNA in these hamsters (Foster et al., 2002; Munoz, Lopez, Voronkova and Ulrich, unpublished data).

VLPs have been used for the development of serological tests for GHPV infection in geese (Zielonka et al., 2006), which represent the only available test systems for GHPV-specific antibodies at present. Purified VLPs consisting of VP1 expressed in the baculovirus system or in yeast as well as VLPs consisting of yeast-expressed VP1 and VP2 were compared in ELISA and HI tests. Using either type of antigen, the tests were able to identify geese flocks in which clinical hemorrhagic nephritis and enteritis of geese (HNEG) had been previously observed and showed absence of antibodies in a clinically healthy flock. However, the tests demonstrated GHPV-specific antibodies in two additional flocks without HNEG indicating for the first time that subclinical or unrecognized GHPV infections may occur. The results of the ELISA tests were comparable irrespective of the used antigen. In contrast, the VLPs consisting of VP1 and VP2 showed a significantly higher hemagglutinating activity as compared to the VP1-VLPs. It is not clear until now, whether the different size of the particles (VP1-VLPs have a diameter of only 20 nm, VP1/VP2-VLPs were regular 45 nm-sized) or the presence of VP2 within the particle is responsible for these differences. For APV, a blocking ELISA has been successfully developed for the detection of APV-specific antibodies in sera of different animal species on the basis of baculovirus-expressed VP1 (Khan et al., 2000). Despite the fact that this protein was not able to assemble into VLPs, the results of the ELISA correlated well with those of a neutralization test indicating that the structure of the expressed protein was sufficient for specific antibody detection. However, in contrast to yeast-expressed APV-VP1-VLPs, the baculovirus-expressed VP1 had no detectable hemagglutinating activity (R. Johne, unpublished observation).

As serum antibody responses to polyomavirus-derived VLPs are stable over time, VLP-based ELISAs have been applied, besides their direct use in serological diagnostics of acute virus infections, for different seroepidemiological studies (Konya and Dillner, 2001; Stolt et al., 2003). Investigations in humans using JCPyV-, BKPyV-, and SV40-derived VLPs demonstrated that some individuals have BKPyV and JCPyV-specific antibodies that cross-react with SV40, but do not provide support for SV-40 being a prevalent human pathogen (Carter et al., 2003). To investigate if polyomavirus infection during pregnancy is linked to development of neuroblastoma in the child, sera of a comprehensive cohort were studied for JCPyV- and BKPyV-specific antibodies using the corresponding VLPs. These serological investigations as well as polyomavirus DNA detection did not show any evidence for association between infections by these polyomaviruses and neuroblastoma development (Stolt et al., 2005). Moreover, VLP-based ELISAs for SV-40, BKPyV, and JCPyV are likely to contribute to a better understanding of the biology of these viruses (Viscidi et al., 2003).

10.8 GENE THERAPY APPLICATION OF POLYOMAVIRUS-DERIVED VLPs

Although significant progress has been made in the area of nonviral gene delivery, synthetic vectors remain less efficient by orders of magnitude than their viral counterparts (Roth and Sundaram, 2004). Noninfectious VLPs may represent an alternative, intermediate approach to viral and nonviral delivery systems. Especially VLPs derived from DNA viruses that replicate in the nucleus (polyomaviruses, papillomaviruses; Chapter 11) may represent promising vehicles for gene transfer approaches. The first investigations on the use of polyomavirus-derived VLPs for gene therapy approaches were made on SV-40 and MPyV, although currently VLP systems based on other polyomaviruses (HaPyV, JCPyV, BKPyV) have been evaluated (Forstová et al., 1995; Sandalon et al., 1997; Goldmann et al., 1999; Touze et al., 2001; Voronkova et al., 2007).

VLP-mediated DNA transfer *in vitro* depends on packaging of DNA within VLPs. The encapsidation of DNA to polyomavirus capsids is mediated by the nuclear localization signal mapped at the amino-terminus of VP1 (Wychowski et al., 1986; Moreland and Garcea, 1991). As this encapsidation signal is incorporating DNA in an unspecific way, an initial problem for use of polyomavirus-derived VLPs was the incorporation of host cell-derived DNA into VLPs expressed in baculovirus-driven insect cell systems (Pawlita et al., 1996; Gillock et al., 1997). Yeast-expressed VLPs were demonstrated to contain only traces of host-derived RNA (Sasnauskas et al., 1999). To overcome this problem, a baculovirus-expressed truncated variant of MPyV-VP1 lacking the amino-terminal 11 aa residues was generated resulting in destruction of DNA encapsidation, but also of the nuclear targeting which however could be recovered by coexpression of VP2/VP3 (Gillock et al., 1998). Alternatively, disassembly of VLPs followed by nuclease treatment or initial purification of pentamers and reassembly of VLPs was demonstrated to generate VLPs without unwanted incorporation of nucleic acid (Voronkova et al., 2007).

The incorporation of a "therapeutical" gene construct into VLPs could be achieved by different ways. The generation of VLP/DNA complexes for BKPyV-derived VLPs has been achieved by three different methods, disassembly/reassembly, osmotic shock, and direct interaction of VLPs and reporter plasmid DNA with the latter being the most efficient method for gene transfer. The gene transfer efficiency using BKPyV-derived VLPs was of the same order as that observed for HPV-16 L1 VLPs (Touze et al., 2001). Partial DNA encapsidation to MPyV-derived VLPs was simply achieved by mixing of plasmid DNA with VLPs *in vitro* (Stokrova et al., 1999). JCPyV-derived VLPs were loaded by a dissociation/reassociation process (Goldmann et al., 1999), whereas HaPyV-VP1-derived VLPs were prepared by initial purification of pentamers and subsequent loading of plasmid DNA during the reassembly process (Voronkova et al., 2007).

VLPs based on VP1 of different polyomaviruses have been demonstrated to transfer foreign DNA into a broad spectrum of cells. MPyV-VLP-mediated gene transfer to rat F111 cells was found to be 50–150 times more efficient than by the calcium phosphate precipitation method (Slilaty and Aposhin, 1983). In line, gene transfer by MPyV-VP1-VLPs to rat-2 cells was shown to have many advantages in terms of maintenance of DNA fidelity and increased efficiency of gene expression in comparisons with the calcium phosphate DNA coprecipitation procedure (or lipofectin route) (Forstová et al., 1995). This superior performance was confirmed by observing a significantly increased eGFP expression in mouse fibroblasts when plasmid DNA was delivered by MPyV-VP1-VLPs compared to control experiments with naked DNA (Henke et al., 2000). VP1-VLPs originating from BKPyV, JCPyV, and HaPyV were demonstrated to allow a functional transfer of exogenous DNA to COS-7 cells (Goldmann et al., 1999; Touze et al., 2001; Voronkova et al., 2007). MPyV-derived VLPs were also used for gene transfer to a human liver cell line (Forstová et al., 1995). SV-40 VLPs carrying the human multidrug-resistance gene MDR1 encoding P-glycoprotein (P-gp) or the green fluorescent protein (GFP) reporter gene were demonstrated to highly efficiently transduce different human, murine, and monkey cell lines. Interestingly, the percentage of expressing cells was proportional to the number of transducing VLPs, with close to 100% of cells transduced at optimal ratios of transducing VLPs to cells (Kimchi-Sarfaty et al., 2002). Recent *in vivo* investigations in mice indicated that SV-40 *in vitro* packaging is an effective system for cancer gene delivery in combination with chemotherapy (Kimchi-Sarfaty et al., 2006).

In addition to a "therapeutical" gene construct or reporter plasmid DNA, VLPs might be used for transfer of smaller DNA or RNA molecules, as encapsidation of oligonucleotides and dsDNA fragments of 1.5–1.8 kb were protected against degradation (Braun et al., 1999). Recently, SV-40 VLPs were demonstrated to deliver into human cells both principal types of RNAi effector molecules: plasmid-expressed short hairpin RNAs (shRNAs) and synthetic siRNAs (Kimchi-Sarfaty et al., 2005). The internalization of carboxyfluorescein diacetate succinimidyl ester-labeled VLPs of different polyomaviruses into human monocyte-derived dendritic cells and propidium iodide-loaded JCPyV-VLPs underlines that VLPs might represent promising vehicles for efficient delivery of various substances into cells (Goldmann et al., 2000; Gedvilaite et al., 2006b; Dorn et al., submitted). Moreover, studies using polyomavirus-derived VLPs harboring a GFP or methotrexate suggested their potential for intracellular delivery of various substances (Abbing et al., 2004).

A main problem for the repeated *in vivo* application of VLPs in gene transfer applications might represent the induced VLP-specific immune response and a preexisting immunity as the majority of the human population has a JCPyV/BKPyV immunity (Clark et al., 2001; Gedvilaite et al., 2004). Similarly, to the sequential application of VLPs derived from different HPV genotypes, a sequential application of polyomavirus-derived VLPs from different polyomaviruses may solve this problem (Tegerstedt et al., 2003).

A second drawback for the use of VLPs as gene transfer vehicles, as for other delivery systems as well, is its *per se* often lacking cell, tissue, and organ tropism. A potential solution for this problem might be the generation of chimeric VLPs harboring binding sites for cellular receptors as extensively studied for polyomavirus-derived VLPs. This have been achieved by generation of chimeric VLPs harboring an engineered antibody-binding domain from protein A of *Staphylococcus aureus* or a negatively charged peptide at a surface-exposed region of MPyV-VP1 and subsequent binding of a cell surface-specific antibody (Gleiter and Lilie, 2001; Schmidt et al., 2001; Stubenrauch et al., 2001). Similarly, cell tropisms might be achieved by binding of mosaic VLPs based on polyomavirus VP1 harboring a urokinase plasminogen activator (uPA) segment to uPA receptor of mammalian cells (Shin and Folk, 2003).

Although modified polyomavirus-derived VLPs allow specific targeting to antigen expressing cells, still only a low percentage of these cells showed a functional gene transfer, due to lysosomal degradation of particles representing a third current problem for VLP-based gene delivery systems (May et al., 2002).

10.9 CHIMERIC, MOSAIC, AND PSEUDOTYPE VLPs AS POTENTIAL VACCINES

The use of VLPs for vaccine development was stimulated by the successful introduction of the first human recombinant HBV vaccine based on yeast-expressed HBsAg-derived VLPs (Chapter 9). In general, due to the repetitive antigenic structure, VLPs are highly immunogenic making them promising vaccine candidates against different pathogens. This strong humoral immunogenicity may allow their successful application without additional adjuvants and in a single dose regimen. VLPs are not only able to induce potent B-cell responses, but also to generate strong T-cell responses suggesting their potential for prophylactic and therapeutical vaccine applications. A comparative study of MPyV-VP1-VLPs and a glutathione VP1 fusion protein in a MPyV challenge model in T-cell deficient mice demonstrated the superior performance of VLPs in terms of the induction of hemagglutination-inhibiting antibodies and protective immunity (Vlastos et al., 2003).

Because of the increasing importance of JCPyV-induced PML in AIDS patients and potential involvement of BKPyV reactivation in rejection of kidney transplants, vaccines based on the corresponding autologous VLPs have become an increasing issue for potential future immunization efforts. Thus, baculovirus-mediated expression of JCPyV-VP1 VLPs was demonstrated to raise a strong immune response in rabbits when applied with adjuvant. The induced antibodies were demonstrated to inhibit VLP binding to and JCPyV infection of SVG cells suggesting the possibility to develop prophylactic and therapeutic vaccines based on JCPyV-derived VLPs. Surprisingly, application of the VLPs without adjuvant failed to induce a significant antibody response (Goldmann et al., 1999). In addition, yeast-expressed JCPyV-derived VLPs have been used to generate monoclonal antibodies (mAbs) that can be used for detection of JCPyV-VP1 in brain tissues of PML patients (A. Zvirbliene, J.P. Teifke, R. Ulrich and F. van Landeghem, unpublished data). Similarly, VP1-specific mAbs generated against chimeric HaPyV-VP1-derived VLPs were applied for the detection of VP1 in tissue samples of a Syrian hamster colony in Spain (C. Munoz, D. Lopez, A. Zvirbliene and R. Ulrich, unpublished data). The potential application of VP1-derived autologous VLPs as vaccines may allow the differentiation of infected from vaccinated animals (DIVA) by the fact that a polyomavirus infection induced not only VP1-specific antibodies but also antibodies specific for VP2/VP3 and the T antigens as demonstrated for HaPyV infection in Syrian hamster (Ulrich et al., 1996; Foster et al., 2002).

As the polyomaviruses of birds are the causative agents of acute and chronic diseases, the development of efficient vaccines against these diseases is an important issue for veterinary medicine. Until now, however, no VLP-based vaccine has been tested for the bird polyomaviruses. The reasons behind that may be that most of these viruses have been detected very recently and that in most cases the generation of VLP vaccines is possible only now. Although APV is known for over 25 years, VLP production was difficult until the use of the yeast expression system in 2002. Taking into account the inability of most of the bird polyomaviruses to efficiently grow in cell culture and the novel ability to recombinantly express immunogenic VLPs with high efficiency, it is expected that VLP-based vaccines against bird polyomaviruses will be tested within the next few years (Zielonka et al., 2006; Johne and Müller, 2007).

A wide range of different VLPs have been used as carriers for foreign epitopes and protein segments (Chapter 9). Polyomavirus-derived VLPs offer a large panel of advantages making them a promising platform for vaccine development. Polyomavirus structural proteins demonstrate a strong flexibility at certain sites allowing the insertion of foreign sequences of different origin and size without disturbing the formation of VLPs. This includes also the possibility to use different approaches for presentation of foreign protein segments (chimeric, mosaic, and pseudotype VLPs) and the simultaneous insertion of the same or different foreign peptides into the same VP1 molecule. A limitation of the insertion capacity of HaPyV-VP1 was observed for VP1 fusion proteins harboring simultaneous foreign insertions at four different sites (Table 10.2).

A major reason for generation of chimeric or mosaic VLPs is to transfer the intrinsic strong immunogenicity of the VLP carrier to *per se* low immunogenic peptide sequences. This was evidenced for a mucin 1-derived nonamer peptide (MUC1) peptide presented on HaPyV-derived VLPs; whereas the MUC1 peptide complexed to BSA did not induce a specific antibody response, chimeric HaPyV-VP1 VLPs harboring two copies of the peptide induce MUC1-specific antibodies (Zvirbliene et al., 2006). Moreover, mAbs generated by this approach were demonstrated to react with human tumor cell lines expressing mucin 1 (Dorn et al., submitted).

Polyomavirus-derived VLPs have been used as carriers for a variety of different foreign peptides, protein segments, and entire proteins of different origin including virus- and cancer associated (Table 10.2). Comparative studies using defined model epitopes inserted into different positions of the same VLP carrier demonstrated that the insertion site in the carrier determines the surface exposure, antigenicity, and immunogenicity of the inserted foreign epitope on VLPs (Gedvilaite et al., 2000; 2004; Lawatscheck et al., 2007). These investigations demonstrated that insertion sites #1 and #4 of HaPyV-VP1 are superior in terms of insertion capacity and induction of a B-cell immunity. The insertion of a flexible GSSG linker seems to improve the immunogenicity of a HaPyV-VP1 harboring two insertions of a CEA epitope, but not of the corresponding VLPs harboring a single insert (Lawatscheck et al., 2007). Moreover, chimeric HaPyV-derived VLPs were demonstrated to be highly immunogenic, even when applied without adjuvant (Gedvilaite et al., 2004, Lawatscheck et al., 2007). In addition, chimeric HaPyV-VLPs harboring a CEA epitope at insertion site #1 induced a long-lasting B-cell immunity in mice when applied without adjuvant (Lawatscheck et al., 2007). A disadvantage of polyomavirus-derived VLPs compared to HBV core-derived VLPs (Milich and McLachlan, 1987; Fehr et al., 1998) might be the lacking ability to induce a T-cell independent (TI) antibody response as observed for MPyV-VP1-derived VLPs (Szomolanyi-Tsuda et al., 1998).

In addition to the humoral immunity, VLPs were demonstrated to induce also a strong T-cell immune response. Thus, mice immunized with HaPyV-VP1 derived VLPs harboring different-sized segments of a hantavirus nucleocapsid protein were shown to induce an antigen-specific T-cell help (Gedvilaite et al., 2004). Recently, yeast-expressed VP1-VLPs originating from rodent, primate and human polyomaviruses were tested for their interaction with human monocyte-derived dendritic cells and the induction of an *in vitro* T-cell response. HaPyV- and MPyV-derived VLPs were found to induce maturation of the dendritic cells as evidenced by increased levels of surface maturation markers and a reduced uptake of FITC dextran and Lucifer Yellow. Moreover, the dendritic cells stimulated with these VLPs produced interleukin-12 and stimulated CD8-positive T-cell responses *in vitro* (Gedvilaite et al., 2006b). In a similar study baculovirus-expressed MPyV-derived authentic VP1 and VP1/VP3-eGFP VLPs failed to induce an upregulation of maturation markers on human dendritic cells, but triggered an interleukin 12 secretion (Boura et al., 2005). Another previous study using baculovirus-expressed human polyomavirus JCPyV- and BKPyV-derived VP1 VLPs failed to induce maturation of murine dendritic cells, even at a concentration 100 times higher than that sufficient for phenotypic maturation by papillomavirus-derived VLPs (Lenz et al., 2001). The observed discrepancies between these investigations might be due to the different expression systems used as well as experimental details. Therefore additional studies would profit from the use of VLPs made in the same expression system and standardized protocols for generation and characterization of dendritic cells of murine and human origin.

In vivo data on the protective efficiency of polyomavirus-derived VLPs are rare so far. Immunization of mice with MPyV-VP1-derived VLPs abrogated the outgrowth of some MPyV tumors and delayed the outgrowth of non-MPyV tumors to some extent. Irradiation prior tumor challenge was demonstrated to avoid an unspecific immune response demonstrating exclusively MPyV-specific protection (Franzén et al., 2005). Pseudotype MPyV-VP1/VP2-derived VLPs harboring the extracellular and transmembrane domain of HER-2/neu protected mice from a lethal challenge with a mammary carcinoma transfected with human Her2 and against the outgrowth of autochthonous mammary carcinomas in mice transgenic for the activated rat Her2 oncogene (Tegerstedt et al., 2005).

In addition to direct use as vaccines (or for gene transfer) MPyV-VP1 VLPs have also been exploited as a carrier for a HIV-1 gag DNA vaccine demonstrating a 10fold increase of the antibody titer to the HIV-1 Gag proteins mediated by the VLP encapsidated DNA vaccine compared to the naked DNA vaccine (Rollman et al., 2003).

As mentioned above, VLPs are able to induce potent B- and T-cell responses requiring their uptake by antigen-presenting cells (APC), intracellular processing and presentation on MHC class I and II. Therefore, the delineation of the interaction of VLPs with professional APC is an important prerequisite for the development of efficient vaccines. Entry experiments using monocyte-derived human dendritic cells with VP1-VLPs from different mammalian polyomaviruses demonstrated macropinocytosis as a major entry pathway. Interestingly, in addition receptor-mediated entry pathways were also found to be operating for the VLPs of different polyomaviruses (Dorn et al., submitted).

The adjuvant effect of polyomavirus-derived VLPs might be mediated by recognition of pathogen-associated molecular patterns (PAMPs) which are recognized by pattern-recognizing receptors (e.g., Toll-like receptors—TLRs). Indeed, recent studies on MPyV and MPyV-derived VLPs suggested a role of TLR-2 and TLR-4 in the induction of innate immune responses and tumor-susceptibility/ resistance of different mice strains (Velupillai et al., 2006).

A major problem for the application of chimeric polyomavirus-derived VLPs as vaccines in humans (as for gene transfer as well, see Section 10.8) may represent a preexisting immunity directed against the VLP carrier as the majority of the human population worldwide is persistently infected by the human polyomaviruses JCPyV and BKPyV. Four different strategies may solve the problems associated with a carrier-specific preexisting immunity. The insertion of larger foreign protein segments into the VLP carrier were demonstrated to reduce the intrinsic antigenicity and immunogenicity of the carrier itself as observed for chimeric HaPyV-derived VLPs (Gedvilaite et al., 2004). Alternatively, the generation of hantavirus nucleocapsid protein-specific mAbs by alternating immunization of mice with HaPyV-VP1 VLPs harboring a 120 aa-long segment of the N protein at sites #1 or #4, may underline the possibility to overcome the potential inhibitory effect of a preexisting carrier-specific immunity on the antibody response against the foreign insertion (Zvirbliene et al., 2006). As documented for human papillomavirus-derived VLPs, an alternating use of carrier VP1 protein of different polyomaviruses may help to solve the potential inhibitory effect of VP1-specific antibodies. Moreover, recent deletion mutagenesis analysis on the carboxy-terminal region of HaPyV-VP1 suggested that partial deletion within or truncation of this immunodominant region also may pave a way to solve problems associated with the preexisting immunity.

10.10 SUMMARY AND OUTLOOK

Polyomavirus-derived VLPs are useful tools for different basic research applications, i.e., in structural and assembly studies, receptor identification, entry studies, and immunological investigations. Results of these investigations represent the basis for rational-based vaccine and gene therapy development. Application of the autologous VLP approach may result in efficient vaccines for human and nonhuman mammalian and avian polyomaviruses. Polyomavirus-derived VLPs harboring foreign peptides or proteins may represent promising vaccines not only for viral infections but also for cancer and autoimmune diseases. The efficiency of autologous and nonautologous VLP-based vaccines might be increased by complexation with novel adjuvants, e.g., immunostimulatory sequences such as CpG oligonucleotides, and by use of prime/boost vaccination strategies including DNA vaccines, VLPs, and live virus vaccines. Moreover, the acceptance of oral administration of VLP-based antiviral vaccines might be enhanced by the development of edible plant-derived vaccines. In addition to potential vaccine applications, polyomavirus-derived VLPs represent an intermediate alternative to nonviral and viral gene transfer vehicles. The major drawback for a broader *in vivo* application of VLPs in gene therapy is the still low efficiency of the transgene transfer. Additional efforts are required for improving the efficiency and targeting of the gene transfer *in vivo*. Moreover, VLPs may represent a vehicle for application of other antiviral or low-molecular weight substances. Additional investigations on the potential consequences of a preexisting polyomavirus-specific immunity on the *in vivo* efficiency of polyomavirus-VLP-derived vaccines and gene transfer vehicles are urgently required. Finally, VLPs represent an important tool for serological diagnostics, especially for polyomaviruses where efficient cell culture systems are lacking so far.

ACKNOWLEDGMENTS

The authors would like to thank Paul Pumpens for his help in preparing the manuscript and Aurelija Zvirbliene, Egle Aleksaite, Aiste Bulavaite (Vilnius), Dolores E. Lopez, Luis Munoz (Salamanca), Frank van Landeghem (Berlin), Jens Teifke (Greifswald–Insel Riems), Anja Zielonka (Leipzig), and David Dorn (New York) for provision of unpublished data. We wish to thank Yuri Khudyakov (Atlanta) for critical reading of the manuscript and Vijay V. Reddy (San Diego) for provision of the three-dimensional images used in Figure 10.4.

REFERENCES

Abbing, A., Blaschke, U.K., Grein, S., Kretschmar, M., Stark, C.M., Thies, M.J., Walter, J., Weigand, M., Woith, D.C., Hess, J., and Reiser, C.O., Efficient intracellular delivery of a protein and a low molecular weight substance via recombinant polyomavirus-like particles, *J. Biol. Chem.*, 279, 27410–27421, 2004.

Aleksaite, E. and Gedvilaite, A., Generation of chimeric hamster polyomavirus VP1 virus-like particles harboring three tumor-associated antigens, *Biologija* (Lithuania), 3, 83–87, 2006.

Allander, T., Andreasson, K., Gupta, S., Bjerkner, A., Bogdanovic, G., Petersson, M.A.A., Dalianis, T., Ramquist, T., and Andersson, B., Identification of a third human polyomavirus, *J. Virol.*, 81, 4130–4136, 2007.

An, K., Gillock, E.T., Sweat, J.A., Reeves, W.M., and Consigli, R.A., Use of the baculovirus system to assemble polyomavirus capsid-like particles with different polyomavirus structural proteins: Analysis of the recombinant assembled capsid-like particles, *J. Gen. Virol.*, 80, 1009–1016, 1999a.

An, K., Smiley, S.A., Gillock, E.T., Reeves, W.M., and Consigli, R.A., Avian polyomavirus major capsid protein VP1 interacts with the minor capsid proteins and is transported to the cell nucleus but does not assemble into capsid-like particles when expressed in the baculovirus system, *Virus Res.*, 64, 173–185, 1999b.

Asano, M., Iwakura, Y., and Kawade, Y., SV40 vector with early gene replacement efficient in transducing exogenous DNA into mammalian cells, *Nucleic Acids Res.*, 13, 8573–8586, 1985.

Bolen, J.B. and Consigli, R.A., Differential adsorption of polyoma virions and capsids to mouse kidney cells and guinea pig erythrocytes, *J. Virol.*, 32, 679–683, 1979.

Boura, E., Liebl, D., Spísek, R., Fric, J., Marek, M., Stokrová, J., Holán, V., and Forstová J. Polyomavirus EGFP-pseudocapsids: Analysis of model particles for introduction of proteins and peptides into mammalian cells, *FEBS Lett.*, 579, 6549–6558, 2005.

Braun, H., Boller, K., Lower, J., Bertling, W.M., and Zimmer, A., Oligonucleotide and plasmid DNA packaging into polyoma VP1 virus-like particles expressed in *Escherichia coli*, *Biotechnol. Appl. Biochem.*, 29, 31–43, 1999.

Carter, J.J., Madeleine, M.M., Wipf, G.C., Garcea, R.L., Pipkin, P.A., Minor, P.D., and Galloway, D.A., Lack of serologic evidence for prevalent simian virus 40 infection in humans, *J. Natl. Cancer Inst.*, 95, 1522–1530, 2003.

Chang, D., Fung, C.Y., Ou, W.C., Chao, P.C., Li, S.Y., Wang, M., Huang, Y.L., Tzeng, T.Y., and Tsai, R.T., Self-assembly of the JC virus major capsid protein, VP1, expressed in insect cells, *J. Gen. Virol.*, 78, 1435–1439, 1997.

Chen, P.L., Wang, M., Ou, W.C., Lii, C.K., Chen, L.S., and Chang, D., Disulfide bonds stabilize JC virus capsid-like structure by protecting calcium ions from chelation, *FEBS Lett.*, 500, 109–113, 2001.

Chromy, L.R., Pipas, J.M., and Barcea, R.L., Chaperone-mediated in vitro assembly of polyomavirus capsids, *Proc. Natl. Acad. Sci. U S A*, 100, 10477–10482, 2003.

Clark, B., Caparros-Wanderley, W., Musselwhite, G., Kotecha, M., and Griffin, B.E., Immunity against both polyomavirus VP1 and a transgene product induced following intranasal delivery of VP1 pseudocapsid-DNA complexes, *J. Gen. Virol.*, 82, 2791–2797, 2001.

Daniels, R., Sadowicz, D., and Hebert, D.N., A very late viral protein triggers the lytic release of SV40, *PLoS Pathogens*, 3, e98, 2007.

Dorn, D.C., Lawatscheck, R., Zvirbliene, A., Aleksaite, E., Pecher, G., Sasnauskas, K., Özel, M., Raftery, M., Schönrich, G., Ulrich, R.G., and Gedvilaite, A., Cellular and humoral immunogenicity of hamster polyomavirus-derived virus-like particles harboring a mucin1 cytotoxic T-cell epitope. *Viral Immunol.*, 21, 12–27, 2008.

Dugan, A.S., Eash, S., and Atwood, W.J., Update on BK virus entry and intracellular trafficking, *Transplant. Infect. Dis.*, 8, 62–67, 2006.

Fattaey, A., Lenz, L., and Consigli, R.A., Production, characterisation of monoclonal antibodies to budgerigar fledgling disease virus major capsid protein VP1, *Avian Dis.*, 36, 543–553, 1992.

Fehr, T., Skrastina, D., and Pumpens, P., Zinkernagel RM. T cell-independent type I antibody response against B cell epitopes expressed repetitively on recombinant virus particles, *Proc. Natl. Acad. Sci. U S A*, 95, 9477–9481, 1998.

Feng, H., Shuda, M., Chang, Y., and Moore, P.S., Clonal integration of a polyomavirus in human Merkel cell carcinoma. *Science*, 319, 1096–1100, 2008.

Forstová, J., Krauzewicz, N., Sandig, V., Elliott, J., Palková, Z., Strauss, M., and Griffin, B.E., Polyoma virus pseudocapsids as efficient carriers of heterologous DNA into mammalian cells, *Hum. Gene Ther.*, 6, 297–306, 1995.

Foster, A.P., Brown, P.J., Jandrig, B., Grosch, A., Voronkova, T., Scherneck, S., and Ulrich, R., Polyomavirus infection in hamsters and trichoepitheliomas/cutaneous adnexal tumours, *Vet. Rec.*, 151, 13–17, 2002.

Franzén, A.V., Tegerstedt, K., Holländerova, D., Forstová, J., Ramqvist, T., and Dalianis, T., Murine polyomavirus-VP1 virus-like particles immunize against some polyomavirus-induced tumours, *In Vivo*, 19, 323–326, 2005.

Garcea, R.L. and Imperiale, M.J., Simian virus 40 infection of humans, *J. Virol.*, 77, 5039–5045, 2003.

Garcea, R.L., Salunke, D.M., and Caspar, D.L., Site-directed mutation affecting polyomavirus capsid self-assembly in vitro, *Nature*, 329, 86–87, 1987.

Gaynor, A.M., Nissen, M.D., Whiley, D.N., Mackay, I.M., Lambert, S.B., Wu, G., Brennan, D.C., Storch, G.A., Sloots, T.P., and Wang, D., Identification of a novel polyomavirus from patients with acute respiratory tract infections, *PloS Pathogens*, 3, e64, 2007.

Gedvilaite, A., Aleksaite, E., Staniulis, J., Ulrich, R., and Sasnauskas, K., Size and position of truncations in the carboxy-terminal region of major capsid protein VP1 of hamster polyomavirus expressed in yeast determine its assembly capacity, *Arch. Virol.*, 151, 1811–1825, 2006a.

Gedvilaite, A., Dorn, D.C., Sasnauskas, K., Pecher, G., Bulavaite, A., Lawatscheck, R., Staniulis, J., Dalianis, T., Ramqvist, T., Schönrich, G., Raftery, M.J., and Ulrich, R., Virus-like particles derived from major capsid protein VP1 of different polyomaviruses differ in their ability to induce maturation in human dendritic cells, *Virology*, 354, 252–260, 2006b.

Gedvilaite, A., Frommel, C., Sasnauskas, K., Micheel, B., Ozel, M., Behrsing, O., Staniulis, J., Jandrig, B., Scherneck, S., and Ulrich, R., Formation of immunogenic virus-like particles by inserting epitopes into surface-exposed regions of hamster polyomavirus major capsid protein, *Virology*, 273, 21–35, 2000.

Gedvilaite, A., Zvirbliene, A., Staniulis, J., Sasnauskas, K., Krüger, D.H., and Ulrich, R., Segments of puumala hantavirus nucleocapsid protein inserted into chimeric polyomavirus-derived virus-like particles induce a strong immune response in mice, *Viral Immunol.*, 17, 51–68, 2004.

Gillock, E.T., An, K., and Consigli, R.A., Truncation of the nuclear localization signal of polyomavirus VP1 results in a loss of DNA packaging when expressed in the baculovirus system, *Virus Res.*, 58, 149–160, 1998.

Gillock, E.T., Rottinghaus, S., Chang, D., Cai, X., Smiley, S.A., An, K., and Consigli, R.A., Polyomavirus major capsid protein VP1 is capable of packaging cellular DNA when expressed in the baculovirus system, *J. Virol.*, 71, 2857–2865, 1997.

Gleiter, S. and Lilie, H., Coupling of antibodies via protein Z on modified polyoma virus-like particles, *Protein Sci.*, 10, 434–444, 2001.

Gleiter, S., Stubenrauch, K., and Lilie, H., Changing the surface of a virus shell fusion of an enzyme to polyoma VP1, *Protein Sci.*, 8, 2562–2569, 1999.

Goldmann, C., Petry, H., Frye, S., Ast, O., Ebitsch, S., Jentsch, K.D., Kaup, F.J., Weber, F., Trebst, C., Nisslein, T., Hunsmann, G., Weber, T., and Luke, W., Molecular cloning and expression of major structural protein VP1 of the human polyomavirus JC virus: Formation of virus-like particles useful for immunological and therapeutic studies, *J. Virol.*, 73, 4465–4469, 1999.

Goldmann, C., Stolte, N., Nisslein, T., Hunsmann, G., Lüke, W., and Petry, H., Packaging of small molecules into VP1-virus-like particles of the human polyomavirus JC virus, *J. Virol. Methods*, 90, 85–90, 2000.

Graffi, A., Schramm, T., Graffi, I., Bierwolf, D., and Bender, E., Virus-associated skin tumors of the Syrian hamster: Preliminary note, *J. Natl. Cancer Inst.*, 40, 867–873, 1968.

Greenlee, J.E., Pathogenesis of K virus infection in newborn mice, *Infect. Immun.*, 26, 705–713, 1979.

Guerin, J.L., Gelfi, J., Dubois, L., Vuillaume, A., Boucraut-Baralon, C., and Pingret, J.L., A novel polyomavirus (Goose Hemorrhagic Polyomavirus) is the agent of hemorrhagic nephritis of geese, *J. Virol.*, 74, 4523–4529, 2000.

Hale, A.D., Bartkeviciute, D., Dargeviviute, A., Jin, L., Knowles, W., Staniulis, J., Brown, D.W., and Sasnauskas, K., Expression and antigenic characterization of the major capsid proteins of human polyomaviruses BK and JC in *Saccharomyces cerevisiae*, *J. Virol. Methods*, 104, 93–98, 2002.

Henke, S., Rohmann, A., Bertling, W.M., Dingermann, T., and Zimmer, A., Enhanced in vitro oligonucleotide and plasmid DNA transport by VP1 virus-like particles, *Pharm. Res.*, 17, 1062–1070, 2000.

Hirsch, H.H., Polyomavirus BK nephropathy: A (re-)emerging complication in renal transplantation, *Am. J. Transplant.*, 2, 25–30, 2002.

Hou, J., Jens, P.J., Major, E.O., zur Hausen, H.J., Almeida, J., van der Noordaa, D., Walker, D., Lowy, D., Bernard, U., Butel, J.S., Cheng, D., Frisque, R.J., and Nagashima, K., Polyomaviridae, Fauquet, C.M., Mayo, M.A., Maniloff, J., Desselberger, U., and Ball, L.A. (Eds.), *Virus Taxonomy. Eighth Report of the ICTV*. Elsevier Academic Press, San Diego, California, 2005, pp. 231–238.

Ishizu, K.I., Watanabe, H., Han, S.I., Kanesashi, S.N., Hoque, M., Yajima, H., Kataoka, K., and Handa, H., Roles of disulfide linkage and calcium ion-mediated interactions in assembly and disassembly of virus-like particles composed of simian virus 40 VP1 capsid protein, *J. Virol.*, 75, 61–72, 2001.

Johne, R., Enderlein, D., Nieper, H., and Müller, H., Novel polyomavirus detected in the feces of a Chimpanzee by nested broad-spectrum PCR, *J. Virol.*, 79, 3883–3887, 2005.

Johne, R. and Müller, H., Avian polyomavirus in wild birds: Genome analysis of isolates from Falconiformes and Psittaciformes, *Arch. Virol.*, 143, 1501–1512, 1998.

Johne, R. and Müller, H., Avian polyomavirus agnoprotein 1a is incorporated into the virus particle as a fourth structural protein, VP4, *J. Gen. Virol.*, 82, 909–918, 2001.

Johne, R. and Müller, H., Nuclear localization of avian polyomavirus structural protein VP1 is a prerequisite for the formation of virus-like particles, *J. Virol.*, 78, 930–937, 2004.

Johne, R. and Müller, H., Polyomaviruses of birds: Etiologic agents of inflammatory diseases in a tumorvirus family, *J. Virol.*, 81, 11554–11559, 2007.

Johne, R., Paul, G., Enderlein, D., Stahl, T., Grund, C., and Müller, H., Avian polyomavirus mutants with deletions in the VP4-encoding region show deficiencies in capsid assembly, virus release and have reduced infectivity in chicken, *J. Gen. Virol.*, 88, 823–830, 2007.

Johne, R., Wittig, W., Fernández-de-Luco, D., Höfle, U., and Müller, H., Characterization of two novel polyomaviruses of birds by using multiply primed rolling-circle amplification of their genomes, *J. Virol.*, 80, 3523–3531, 2006.

Kanesashi, S.N., Ishizu, K., Kawano, M.A., Han, S.I., Tomita, S., Watanabe, H., Kataoka, K., and Handa, H., Simian virus 40 VP1 capsid protein forms polymorphic assemblies in vitro, *J. Gen. Virol.*, 84, 1899–1905, 2003.

Khalili, K., White, M.K., Sawa, H., Nagashima, K., and Safak, M., The agnoprotein of polyomaviruses: A multifunctional auxiliary protein, *J. Cell Physiol.*, 204, 1–7, 2005.

Khan, S.M.R., Johne, R., Beck, I., Kaleta, E.F., Pawlita, M., and Müller, H., Development of a blocking enzyme-linked immunosorbent assay for the detection of avian polyomavirus-specific antibodies, *J. Virol. Methods*, 89, 39–48, 2000.

Kimchi-Sarfaty, C., Ben-Nun-Shaul, O., Rund, D., Oppenheim, A., and Gottesman, M.M., In vitro-packaged SV40 pseudovirions as highly efficient vectors for gene transfer, *Hum. Gene Ther.*, 13, 299–310, 2002.

Kimchi-Sarfaty, C., Brittain, S., Garfield, S., Caplen, N.J., Tang, Q., and Gottesman, M.M., Efficient delivery of RNA interference effectors via in vitro-packaged SV40 pseudovirions, *Hum. Gene Ther.*, 16, 1110–1115, 2005.

Kimchi-Sarfaty, C., Vieira, W.D., Dodds, D., Sherman, A., Kreitman, R.J., Shinar, S., and Gottesman, M.M., SV40 Pseudovirion gene delivery of a toxin to treat human adenocarcinomas in mice, *Cancer Gene Ther.*, 13, 648–657, 2006.

Knowles, W.A. and Sasnauskas, K., Comparison of cell culture-grown JC virus (primary human fetal glial cells and the JCI cell line) and recombinant JCV VP1 as antigen for the detection of anti-JCV antibody by haemagglutination inhibition, *J. Virol. Methods*, 109, 47–54, 2003.

Komagome, R., Sawa, H., Suzuki, T., Suzuki, Y., Tanaka, S., Atwood, W.J., and Nagashima, K., Oligosaccharides as receptors for JC virus, *J. Virol.*, 76, 12992–13000, 2002.

Konya, J. and Dillner, J., Immunity to oncogenic human papillomaviruses, *Adv. Cancer Res.*, 82, 205–238, 2001.

Kosukegawa, A., Arisaka, F., Takayama, M., Yajima, H., Kaidow, A., and Handa, H., Purification and characterization of virus-like particles and pentamers produced by the expression of SV40 capsid proteins in insect cells, *Biochim. Biophys. Acta*, 1290, 37–45, 1996.

Langner, J., Neumann, B., Goodman, S.L., and Pawlita, M., RGD-mutants of B-lymphotropic polyomavirus capsids specifically bind to alphavbeta3 integrin, *Arch. Virol.*, 149, 1877–1896, 2004.

Lawatscheck, R., Aleksaite, E., Schenk, J.A., Micheel, B., Jandrig, B., Holland, G., Sasnauskas, K., Gedvilaite, A., and Ulrich, R.G., Chimeric polyomavirus-derived virus-like particles: The immunogenicity of an inserted peptide applied without adjuvant to mice depends on its insertion site and its flanking linker sequence, *Viral Immunol.*, 20, 453–460, 2007.

Leavitt, A.D., Roberts, T.M., and Garcea, R.L., Polyoma virus major capsid protein, VP1. Purification after high level expression in *Escherichia coli*, *J. Biol. Chem.*, 260, 12803–12809, 1985.

Lenz, P. et al., Papillomavirus-like particles induce acute activation of dendritic cells, *J. Immunol.*, 166, 5346, 2001.

Li, P.P., Nakanishi, A., Fontanes, V., and Kasamatsu, H., Pairs of VP1 cysteine residues essential for simian virus 40 infection, *J. Virol.*, 79, 3859–3864, 2005.

Li, P.P., Nakanishi, A., Trran, M.A., Ishizu, K., Kawano, M., Philips, M., Handa, H., Liddington, R.C., and Kasamatsu, H., Importance of VP1 calcium-binding residues in assembly, sell entry, and nuclear entry of simian virus 40, *J. Virol.*, 77, 7527–7538, 2003a.

Li, T.C., Takeda, N., Kato, K., Nilsson, J., Xing, L., Haag, L., Cheng, R.H., and Miyamura, T., Characterization of self-assembled virus-like particles of human polyomavirus BK generated by recombinant baculoviruses, *Virology*, 311, 115–124, 2003b.

Liddington, R.C., Yan, Y., Moulai, J., Sahli, R., Benjamin, T.L., and Harrison, S.C., Structure of simian virus 40 at 3.8-A resolution, *Nature*, 354, 278–284, 1991.

May, T., Gleiter, S., and Lilie, H., Assessment of cell type specific gene transfer of polyoma virus like particles presenting a tumor specific antibody Fv fragment, *J. Virol. Methods*, 105, 147–157, 2002.

Milich, D.R. and McLachlan, A., The nucleocapsid of hepatitis B virus is both a T-cell-independent and a T-cell-dependent antigen, *Science*, 234, 1398–1401, 1986.

Montross, L., Watkins, S., Moreland, R.B., Mamon, H., Caspar, D.L., and Garcea, R.L., Nuclear assembly of polyomavirus capsids in insect cells expressing the major capsid protein VP1, *J. Virol.*, 65, 4991–4998, 1991.

Moreland, R.B. and Garcea, R.L., Characterization of a nuclear localization sequence in the polyomavirus capsid protein VP1, *Virology*, 185, 513–518, 1991.

Nakanishi, A., Itoh, N., Li, P.P., Handa, H., Liddington, R.C., and Kasamatsu, H., Minor capsid proteins of simian virus 40 are dispensable for nucleocapsid assembly and cell entry but are required for nuclear entry of the viral genome, *J. Virol.*, 81, 3778–3785, 2007.

Norkin, L.C., Anderson, H.A., Wolfrom, S.A., and Oppenheim, A., Caveolar endocytosis of simian virus 40 is followed by brefeldin A-sensitive transport to the endoplasmatic reticulum, where the virus disassembles, *J. Virol.*, 76, 5156–5166, 2002.

Ou, W.C., Chen, L.H., Wang, M., Hseu, T.H., and Chang, D., Analysis of minimal sequences on JC virus VP1 required for capsid assembly, *J. Neurovirol.*, 7, 298–301, 2001.

Ou, W.C., Wang, M., Fung, C.Y., Tsai, R.T., Chao, P.C., Hseu, T.H., and Chang, D., The major capsid protein, VP1, of human JC virus expressed in *Escherichia coli* is able to self-assemble into a capsid-like particle and deliver exogenous DNA into human kidney cells, *J. Gen. Virol.*, 80, 39–46, 1999.

Padgett, B.L., Walker, D.L., ZuRhein, G.M., Eckroade, R.J., and Dessel, B.H., Cultivation of papova-like virus from human brain with progressive multifocal leucoencephalopathy, *Lancet*, 1, 1257–1260, 1971.

Pawlita, M., Müller, H., Oppenländer, M., Zentgraf, H., and Herrmann, M., DNA encapsidation by viruslike particles assembled in insect cells from the major capsid protein VP1 of B-lymphotropic papovavirus, *J. Virol.*, 70, 7517–7526, 1996.

Pho, M.T., Ashok, A., and Atwood, W.J., JC virus enters human glial cells by clathrin-dependent receptor-mediated endocytosis, *J. Virol.*, 74, 2288–2292, 2000.

Qu, Q., Sawa, H., Suzuki, T., Semba, S., Henmi, C., Okada, Y., Tsuda, M., Tanaka, S., Atwood, W.J., and Nagashima, K., Nuclear entry mechanism of the human polyomavirus JC virus-like particle: Role of importins and the nuclear pore complex, *J. Biol. Chem.*, 279, 27735–27742, 2004.

Querbes, W., O'Hara, B.A., Williams, G., and Atwood, W.J., Invasion of host cells by JC virus identifies a novel role for caveolae in endosomal sorting of noncaveolar ligands, *J. Virol.*, 80, 9402–9413, 2006.

Richterova, Z., Liebl, D., Horak, M., Palkova, Z., Stokrova, J., Hozak, P., Korb, J., and Forstova, J., Caveolae are involved in the trafficking of mouse polyomavirus virions and artificial VP1 pseudocapsids toward cell nuclei, *J. Virol.*, 75, 10880–10891, 2001.

Rodgers, R.E., Chang, D., Cai, X., and Consigli, R.A., Purification of recombinant budgerigar fledgling disease virus VP1 capsid protein and its ability for in vitro capsid assembly, *J. Virol.*, 68, 3386–3390, 1994.

Rollman, E., Ramqvist, T., Zuber, B., Tegerstedt, K., Kjerrström Zuber, A., Klingström, J., Eriksson, L., Ljungberg, K., Hinkula, J., Wahren, B., and Dalianis, T., Genetic immunization is augmented by murine polyomavirus VP1 pseudocapsids, *Vaccine*, 21, 2263–2267, 2003.

Roth, C.M. and Sundaram, S., Engineering synthetic vectors for improved DNA delivery: Insights from intracellular pathways, *Annu. Rev. Biomed. Eng.*, 6, 397–426, 2004.

Salunke, D.M., Caspar, D.L., and Garcea, R.L., Self-assembly of purified polyomavirus capsid protein VP1, *Cell*, 46, 895–904, 1986.

Sandalon, Z., Dalyot-Herman, N., Oppenheim, A.B., and Oppenheim, A., In vitro assembly of SV40 virions and pseudovirions: Vector development for gene therapy, *Hum. Gene Ther.*, 8, 843–849, 1997.

Sasnauskas, K., Bulavaite, A., Hale, A., Jin, L., Knowles, W.A., Gedvilaite, A., Dargeviciute, A., Barteviciute, D., Zvirbliene, A., Staniulis, J., Brown, D.W., and Ulrich, R., Generation of recombinant virus-like particles of human and non-human polyomaviruses in yeast Saccharomyces cerevisiae, *Intervirology*, 45, 308–317, 2002.

Sasnauskas, K., Buzaite, O., Vogel, F., Jandrig, B., Razanskas, R., Staniulis, J., Scherneck, S., Krüger, D.H., and Ulrich, R., Yeast cells allow high-level expression and formation of polyomavirus-like particles, *Biol. Chem.*, 380, 381–386, 1999.

Schmidt, M., Müller, H., Schmidt, M.F.G., and Rott, G., Myristoylation of budgerigar fledgling disease virus capsid protein VP2, *J. Virol.*, 63, 429–431, 1989.

Schmidt, U., Kenklies, J., Rudolph, R., and Böhm, G., Site-specific fluorescence labelling of recombinant polyomavirus-like particles, *Biol. Chem.*, 380, 397–401, 1999.

Schmidt, U., Rudolph, R., and Bohm, G., Mechanism of assembly of recombinant murine polyomavirus-like particles, *J. Virol.*, 74, 1658–1662, 2000.

Schmidt, U., Rudolph, R., and Bohm, G., Binding of external ligands onto an engineered virus capsid, *Protein Eng.*, 14, 769–774, 2001.

Shah, K.V., Daniel, R.W., and Kelly, T.J. Jr., Immunological relatedness of papovaviruses of the simian virus 40-polyoma subgroup, *Infect. Immun.*, 18, 558–560, 1977.

Shepherd, C.M., Borelli, I.A., Lander, G., Natarajan, P., Siddavanahalli, V., Bajaj, C., Johnson, J.E., Brooks, C.L. 3rd, and Reddy, V.S., VIPERdb: A relational database for structural virology, *Nucleic Acids Res.*, 34 (Database issue), D386–D389, 2006 Jan 1.

Shin, Y.C. and Folk, W.R., Formation of polyomavirus-like particles with different VP1 molecules that bind the urokinase plasminogen activator receptor, *J. Virol.*, 77, 11491–11498, 2003.

Siray, H., Frömmel, C., Voronkova, T., Hahn, S., Arnold, W., Schneider-Mergener, J., Scherneck, S., and Ulrich, R., An immunodominant, cross-reactive B-cell epitope region is located at the C-terminal part of the hamster polyomavirus major capsid protein VP1, *Viral Immunol.*, 13, 533–545, 2000.

Slilaty, S.N. and Aposhian, H.V., Gene transfer by polyoma-like particles assembled in a cell-free system, *Science*, 220, 725–727, 1983.

Smith, A.E., Lilie, H., and Helenius, A., Ganglioside-dependent cell attachment and endocytosis of murine polyomavirus-like particles, *FEBS Lett.*, 555, 199–203, 2003.

Soeda, E., Krauzewicz, N., Cox, C., Stokrová, J., Forstová, J., and Griffin, B.E., Enhancement by polylysine of transient, but not stable, expression of genes carried into cells by polyoma VP1 pseudocapsids, *Gene Ther.*, 5, 1410–1419, 1998.

Stehle, T., Gamblin, S.J., Yan, Y., and Harrison, S.C., The structure of simian virus 40 refined at 3.1 Å resolution, *Structure*, 4, 165–182, 1996.

Stehle, T. and Harrison, S.C., Crystal structures of murine polyomavirus in complex with straight-chain and branched-chain sialyloligosaccharide receptor fragments, *Structure*, 4, 183–194, 1996.

Stehle, T. and Harrison, S.C., High-resolution structure of a polyomavirus VP1-oligosaccharide complex: Implications for assembly and receptor binding, *EMBO J.*, 16, 5139–5148, 1997.

Stehle, T., Yan, Y., Benjamin, T.L., and Harrison, S.C., Structure of murine polyomavirus complexed with an oligosaccharide receptor fragment, *Nature*, 369, 160–163, 1994.

Stokrová, J., Palková, Z., Fischer, L., Richterová, Z., Korb, J., Griffin, B.E., and Forstová, J., Interactions of heterologous DNA with polyomavirus major structural protein, VP1, *FEBS Lett.*, 445, 119–125, 1999.

Stolt, A., Kjellin, M., Sasnauskas, K., Luostarinen, T., Koskela, P., Lehtinen, M., and Dillner, J., Maternal human polyomavirus infection and risk of neuroblastoma in the child, *Int. J. Cancer*, 113, 393–396, 2005.

Stolt, A., Sasnauskas, K., Koskela, P., Lehtinen, M., and Dillner, J., Seroepidemiology of the human polyomaviruses, *J. Gen. Virol.*, 84, 1499–1504, 2003.

Stubenrauch, K., Gleiter, S., Brinkmann, U., Rudolph, R., and Lilie, H., Conjugation of an antibody Fv fragment to a virus coat protein: Cell-specific targeting of recombinant polyoma-virus-like particles, *Biochem. J.*, 356, 867–873, 2001.

Sullivan, C.S. and Pipas, J.M., T antigens of simian virus 40: molecular chaperons for viral replication and tumorigenesis, *Microbiol. Mol. Biol. Rev.*, 66, 179–202, 2002.

Szomolanyi-Tsuda, E., Le, Q.P., Garcea, R.L., and Welsh, R.M., T-cell-independent immunoglobulin G responses in vivo are elicited by live-virus infection but not by immunization with viral proteins or virus-like particles, *J. Virol.*, 72, 6665–6670, 1998.

Tegerstedt, K., Andreasson, K., Vlastos, A., Hedlund, K.O., Dalianis, T., and Ramqvist, T., Murine pneumotropic virus VP1 virus-like particles (VLPs) bind to several cell types independent of sialic acid residues and do not serologically cross react with murine polyomavirus VP1 VLPs, *J. Gen. Virol.*, 84, 3442–3452, 2003.

Tegerstedt, K., Lindencrona, J.A., Curcio, C., Andreasson, K., Tullus, C., Forni, G., Dalianis, T., Kiessling, R., and Ramqvist, T., A single vaccination with polyomavirus VP1/VP2Her2 virus-like particles prevents outgrowth of HER-2/neu-expressing tumors, *Cancer Res.*, 65, 5953–5957, 2005.

Thompson, J.D., Higgins, D.G., and Gibson, T.J., CLUSTAL W: Improving the sensitivity of progressive multiple sequence alignment through sequence weighting, position-specific gap penalties and weight matrix choice, *Nucleic Acids Res.*, 22, 4673–4680, 1994.

Touze, A., Bousarghin, L., Ster, C., Combita, A.L., Roingeard, P., and Coursaget, P., Gene transfer using human polyomavirus BK virus-like particles expressed in insect cells, *J. Gen. Virol.*, 82, 3005–3009, 2001.

Ulrich, R., Sommerfeld, K., Schröder, A., Prokoph, H., Arnold, W., Krüger, D.H., and Scherneck, S., Hamster polyomavirus-encoded proteins: Gene cloning, heterologous expression and immunoreactivity, *Virus Genes*, 12, 265–274, 1996.

Velupillai, P., Garcea, R.L., and Benjamin, T.L., Polyoma virus-like particles elicit polarized cytokine responses in APCs from tumor-susceptible and -resistant mice, *J. Immunol.*, 176, 1148–1153, 2006.

Viscidi, R.P. and Clayman, B., Serological cross reactivity between polyomavirus capsids, *Adv. Exp. Med. Biol.*, 577, 73–84, 2006.

Viscidi, R.P., Rollison, D.E., Viscidi, E., Clayman, B., Rubalcaba, E., Daniel, R., Major, E.O., and Shah, K.V., Serological cross-reactivities between antibodies to simian virus 40, BK virus, and JC virus assessed by virus-like-particle-based enzyme immunoassays, *Clin. Diagn. Lab. Immunol.*, 10, 278–285, 2003.

Vlastos, A., Andreasson, K., Tegerstedt, K., Holländerová, D., Heidari, S., Forstová, J., Ramqvist, T., and Dalianis, T., VP1 pseudocapsids, but not a glutathione-S-transferase VP1 fusion protein, prevent polyomavirus infection in a T-cell immune deficient experimental mouse model, *J. Med. Virol.*, 70, 293–300, 2003.

Voronkova, T., Kazaks, A., Ose, V., Ozel, M., Scherneck, S., Pumpens, P., and Ulrich, R., Hamster polyomavirus-derived virus-like particles are able to transfer in vitro encapsidated plasmid DNA to mammalian cells, *Virus Genes*, 34, 303–314, 2007.

Wrobel, B., Yosef, Y., Oppenheim, A.B., and Oppenheim, A., Production and purification of SV40 major capsid protein (VP1) in *Esherischia coli* strains deficient for the GroELs chaperone machine. *J. Biotechnol.*, 84, 285–289, 2000.

Wychowski, C., Benichou, D., and Girard, M., A domain of SV40 capsid polypeptide VP1 that specifies migration into the cell nucleus, *EMBO J.*, 5, 2569–2576, 1986.

Zielonka, A., Gedvilaite, A., Ulrich, R., Lüschow, D., Sasnauskas, K., Müller, H., and Johne, R., Generation of virus-like particles consisting of the major capsid protein VP1 of goose hemorrhagic polyomavirus and their application in serological tests, *Virus Res.*, 120, 128–137, 2006.

zur Hausen, H. and Gissmann, L., Lymphotropic papovaviruses isolated from African green monkey and human cells, *Med. Microbiol. Immunol.*, 167, 137–153, 1979.

Zvirbliene, A., Samonskyte, L., Gedvilaite, A., Voronkova, T., Ulrich, R., and Sasnauskas, K., Generation of monoclonal antibodies of desired specificity using chimeric polyomavirus-derived virus-like particles, *J. Immunol. Methods*, 311, 57–70, 2006.

11 Papillomavirus-Derived Virus-Like Particles

Andris Kazaks and Tatyana Voronkova

CONTENTS

11.1 PAPILLOMAVIRUSES: INTRODUCTION

Papillomaviruses (PVs), members of the family Papillomaviridae, are icosahedral nonenveloped double-strand DNA viruses, which infect squamous or mucosal epithelia and produce a range of epithelial neoplasms, both benign and malignant, in most animals and humans (for details see Ref. [1]). PVs along with the polyomaviruses have belonged to former Papovaviridae family. On the basis of the later findings that the two virus groups have different genome sizes, completely different genome organizations, and no major nucleotide or amino acid (aa) sequence similarities, they are now recognized as two independent families. Until now, more than 130 genetically distinct PVs have been identified (for classification of PVs see Ref. [2]). Among them, over 100 different human papillomavirus (HPV) types have been characterized. Some HPV types cause benign skin warts, or papillomas, for which the virus family is named. A subset of HPVs cause virtually all cases of cervical cancer, the second most common cause of death from cancer after breast cancer, killing about 0.25 million women per year. In recent years, a lot of effort has been made for generation of both prophylactic and therapeutic HPV vaccines.

The viral genome is represented by 8,000 bases of double-stranded circular DNA packaged within the virion shell along with cellular histone proteins. In contrast to polyomaviruses, all the PV genes are encoded on one DNA strand (Figure 11.1A). The PV genome is divided into an early region (E), encoding various genes that are expressed immediately after initial infection of a host cell, and a late region (L) encoding the capsid genes L1 and L2. PVs infect the basal cells of epithelial surfaces, mainly through the microtrauma, and use the differentiation of the epithelium to regulate their replication. In basal cells and their progeny PVs replicate their DNA episomally, provided by the early proteins E1 and E2. The E4 protein is found to contribute to regulation of host

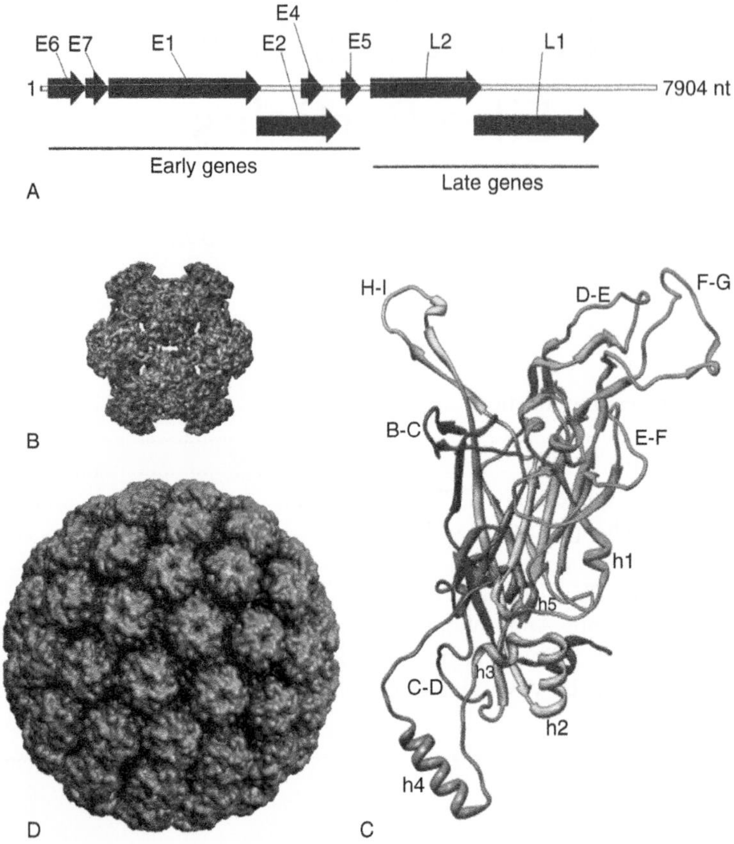

FIGURE 11.1 (See color insert following page 142.) The PV genome map (A) and the main structural features (B–D). (A) Schematic presentation of the HPV16 (GenBank accession No AF536179) genome organization. The respective genes are marked by arrows and scaled according to their relative sizes. (B) The structures of the HPV-16 small $T = 1$ VLPs and (C) L1 monomer fold are according to Ref. [7]. The hypervariable loops and α-helixes in (C) are marked according to Ref. [7], and individual domains are visualized using UCSF Chimera software package [157]. (D) Atomic model of the PV capsid according to Ref. [8]. The particle images in (B) and (D) are scaled according to their relative sizes, while the subunit (C) not. (Images B–D are adapted from the VIPER database from Reddy, V.S. et al., *J. Virol.*, 75, 11943, 2001.)

cell cycle control, while E5 protein aids in cellular transformation and viral replication. The early genes E6 and E7 of the high-risk HPV types encode the main transforming proteins. These genes are constitutively expressed in cervical cancer cells and are thought to play a role in both the initiation and maintenance of the oncogenic process. The synthesis of late viral structural proteins L1 and L2 and assembly of mature virions take place only in the most superficial, nondividing keratinocytes, which provides an explanation why suitable amounts of PVs cannot be obtained in tissue cultures.

Heterologous expression of the capsid genes L1 and L2 in vivo or in vitro may result in formation of noninfectious virus-like particles (VLPs), which exhibit structural and immunological properties similar to the native virions and resemble their tropism and cellular entry pathways. In recent years, VLPs have become a useful tools for a wide range of applications, both in basic research, e.g., for studies on the structure, assembly and encapsidation, as well as identification of cellular receptors, and applied research, e.g., for serodiagnostics, gene and pharmaceutical transfer, and vaccine development. In this chapter, we focus mainly on recent attempts for novel antiviral vaccine and gene therapy vector development based on recombinant PV VLP technologies.

11.2 STRUCTURAL DATA

Because of an inability to efficiently propagate the PVs in cultured cells, the x-ray crystallographic studies of native viruses have been limited. Instead, the structures of bovine papillomavirus (BPV) and HPV1 have been determined at 2.5 nm resolution by cryoelectron microscopy (cryoEM) and three-dimensional image reconstruction techniques [3]. These and previous data revealed that viral capsids, 50–60 nm in diameter, are comprised of 72 star-shaped capsomeres (pentamers) arranged on a $T=7$ icosahedral lattice. Each viral particle is composed from 360 subunits of 55 kDa L1 protein, which is a basic component of the PV capsid. The minor 70 kDa capsid protein L2 is present at 1/30 of L1, which means 12 copies of L2 per capsid [4]. Following cryoEM-based investigations led to determination of three-dimensional structure of BPV to 9 Å resolution, proposing location of the L2 minor capsid protein in the center of the pentavalent capsomeres [5].

Other structural research groups have tried to utilize self-assembly potential of the capsid proteins. Since investigations have confirmed that the structure of HPV1 VLPs produced by recombinant vaccinia virus appears identical to the capsid structure of native HPV1 [6], an alternative, VLP-based, way for structure determination of PVs has been explored. In following years, a number of VLPs representing various PVs have been generated in heterologous expression systems (Section 11.3); however, crystallization data are still unavailable, in part due to heterogeneity of particles obtained. Nevertheless, expression of N-terminally truncated HPV16 L1 in *Escherichia coli* led to a homogeneous preparation of "small," 12 pentamer, $T=1$ icosahedral VLPs, allowing the x-ray crystallographic analysis at 3.5 Å resolution [7] (Figure 11.1B). The L1 monomer resembled VP1 from polyomaviruses, the x-ray structure of which has been resolved previously, and contained 12 β-strands, 6 loops, and 5 α-helices (Figure 11.1C). These loops were found to be surface-exposed and highly variable among different papillomavirus types [7]. Combination of image reconstructions from cryoEM of BPV with coordinates from the crystal structure of the small $T=1$ HPV16 VLPs have led to generation of atomic model of the PV capsid [8] (Figure 11.1D). In this fitted $T=7$ model, the C-terminal arm exchange between the neighboring pentamers was proposed similar to the "invading arm" model of the polyomavirus VP1 $T=7$ virion structure. Very recently, it was demonstrated that α-helixes h2 and h3 are essential for L1 folding and pentamer formation, whereas h4 is indispensable for the assembly of not only $T=1$, but also of the $T=7$ VLPs [9]. Taken together, these data provide a useful tool for targeted gene and protein engineering manipulations within PV capsid.

11.3 GENERATION OF PV VLPs

Because of the persisting problem with propagation of native PVs in cell culture systems, generation of PV VLPs as viral surrogates has become the main strategy for investigation of PVs in both basic and applied research. This approach is based on ability of viral structural proteins to form "autologous" VLPs nearly indistinguishable from natural viruses in various expression systems (for general classification of VLPs see Chapter 9). Depending on the encapsidated nucleic acid, VLPs could be roughly divided as empty (free of DNA) or full (also called as pseudovirions; discussed more extensively in Section 11.5.3).

The first data about PV VLP formation in eukaryotes were obtained by recombinant vaccinia virus-directed coexpression of the HPV16 L1 and L2 genes in epithelial cells [10]. Following experiments revealed potential of the PV structural proteins from different genera to form VLPs in a broad spectrum of cells. Thus, the self-assembly competence of L1 protein alone was demonstrated in baculovirus-mediated insect cell expression system for both human, e.g., HPV16 [11], HPV11 [12], HPV6 [4], HPV33 [13], HPV45 [14], and animal PVs, e.g., BPV1 [11], cottontail rabbit papillomavirus (CRPV) [4], canine oral papillomavirus (COPV) [15], and horse papillomavirus (EcPV) [16] (Table 11.1). Only L1 and both L1/L2 genes together from HPV6 and HPV16 were successfully expressed in fission yeast *Schizosaccharomyces pombe*, though without integration of

TABLE 11.1
Overview on Formation of Autologous PV VLPs in Various Expression Systems

Genus/PVs	Formation of Autologous L1 and L1/L2 PV VLPs in Different Cells			
	Bacteria	**Yeast**	**Insect (*Spodoptera frugiperda*)**	**Mammalian/Plant**
α-PV				
HPV6	N/A	*S. pombe* [17]; L1 *S. cerevisiae* [18]; L1, L1/L2	L1 [4]	CV-1 kidney epithelial [155]; L1
HPV11	*E. coli* [23]; L1[a]	*S. cerevisiae* [133]; L1	L1 [12]	Transgenic plants [27]; L1 Epithelial and fibroblast [30]; L1
HPV16	*S. typhimurium* [22]; L1 *E. coli* [24]; L1[a] *L. casei* [139]; L1	*S. pombe* [17]; L1	L1 [11], L1/L2 [4]	CV-1 kidney epithelial [10]; L1/L2 BHK-21 [20]; L1/L2 Transgenic plants [26,28]; L1
HPV18	N/A	*S. cerevisiae* [156]; L1	L1 [130]	Human keratinocytes [116]; L1/L2[b]
HPV31	N/A	N/A	L1 [77]	N/A
HPV33	N/A	N/A	L1, L1/L2 [13]	COS-7 [21]; L1/L2[b]
HPV45	N/A	N/A	L1 [14]	N/A
HPV39, 58, 59	N/A	N/A	L1 [77]	N/A
δ-PV				
BPV1	N/A	N/A	L1 [11]	CV-1 kidney epithelial [114]; L1, L1/L2[b]
ζ-PV				
EcPV	N/A	N/A	L1 [16]	N/A
κ-PV				
CRPV	N/A	*S. cerevisiae* [19]; L1, L1/L2	L1 [4]	N/A
λ-PV				
COPV	N/A	N/A	L1 [15]	N/A
μ-PV				
HPV1	N/A	N/A	L1 [76]	Monkey kidney BSC-1 [32]; L1, L1/L2

Note: From HPVs, only those harboring most pronounced oncogenic potential were included in table. From other PVs, described are representatives of separate genera. N/A, data not available.

[a] Formation of VLPs after in vitro assembly.

[b] Formation of PV pseudovirions.

the L2 protein in VLPs [17]. In contrast, L1 and L2 of HPV6 and CRPV coexpressed in *Saccharomyces cerevisiae* showed clear complex formation and formed mixed autologous PV VLPs [18,19]. In line with these data, HPV16 L1 and L2 proteins were efficiently synthesized and coassembled in BHK-21 cells using Semliki Forest virus expression system [20]. In a more advanced approach, recombinant vaccinia virus-directed coexpression of the HPV33 L1/L2 genes led to formation of HPV pseudovirions capable of transferring the genetic marker located on the plasmid to COS-7 cells in a DNase-resistant way [21]. Nowadays, heterologously generated empty PV VLPs are recognized as main components of commercial prophylactic HPV vaccines or vaccine candidates (Section 11.5.4), while PV pseudovirions have been extensively used for gene transfer purposes and investigation of PV internalization pathways (Section 11.5.3).

Expression of the HPV16 L1 in *Salmonella typhimurium* was the first demonstration that papillomavirus VLPs can be formed also in prokaryotes [22]. In *E. coli*, the HPV11 L1 was synthesized and purified to near homogeneity as pentameric capsomeres capable of in vitro self-assembly into VLPs [23]. However, following attempts to obtain HPV16 L1 in *E. coli* found some difficulties since the protein appeared in a form of insoluble aggregates [24]. Solubility of the HPV16 L1 in *E. coli* was improved by fusion to glutathione-*S*-transferase, and nearly pure protein has been obtained in a form of pentamers, after thrombin cleavage from the fusion moiety [25].

More recently, HPV11 and HPV16 structural genes have been successfully expressed in transgenic tobacco and potato plants. Formation of VLPs was clearly demonstrated, which opens a way for generation of alternative edible HPV vaccines [26–28]. It should be noted, however, that in two latter cases only synthetic L1 genes with optimized codon composition have led to accumulation of target proteins in detectable amounts. Replacing the codons with those more frequently used in mammalian genes, without modifying the protein sequence (humanization of genes), has now become a general method to significantly increase formation of PV VLPs in eukaryotic cells [29,30]. In a promising novel approach, increased production of HPV16 VLPs was obtained in *Nicotiana benthamiana* via tobacco mosaic virus-derived vector, which is the first report of generation of PV VLPs using plant virus vector [31].

Taken together, the L1 from different PV types has demonstrated its capacity to self-assemble into VLPs in bacterial, yeast, insect, mammalian, and plant cells. Coexpression with the minor capsid protein L2, although not essential for particle assembly, was found to significantly increase the efficiency and quality of particle formation [4,13,17,32]. The role of the L2 in virus assembly, DNA encapsidation, cell entry, and nuclear transport is currently under intensive investigation (Section 11.4).

11.4 BASIC RESEARCH APPLICATIONS OF PV VLPs

Besides their significant role in structural investigations, VLPs have been extensively used to characterize the assembly processes of different viruses. At simplest, the studies were conducted for the identification of structural proteins sufficient for assembly of VLPs. Thus, single expression of the major capsid protein L1 of PVs was found to be sufficient for efficient generation of VLPs (Section 11.3). For characterization of the assembly process of PV VLPs, several in vitro attempts were applied [33,34]. Furthermore, deletion mutagenesis at the N- and C-termini of viral structural protein L1 of BPV1, COPV, HPV11, and HPV16 resulted in the identification of functional protein domains essential for VLP assembly [23,25,35,36]. Similar approach can be used for studying the interactions of L1 with the minor capsid protein L2 and the characterization of its environmental conditions. By analyzing the ability of truncated and mutated forms of L2 to form complexes, a proline-rich L1-binding domain near the carboxy terminus of L2 has been defined [34,37] (Figure 11.2).

All members of the Papillomaviridae multiply in the nucleus of the infected cell, suggesting presence of specific nuclear localization signals (NLS) directing the viral structural proteins from the cytoplasm to the nucleus. Indeed, series of deletion and substitution mutations of the HPV16 L1

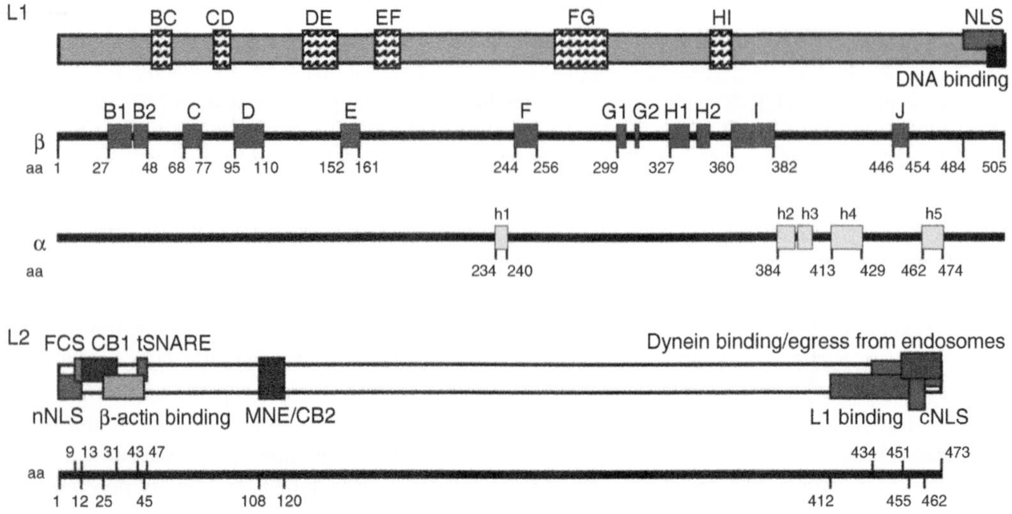

FIGURE 11.2 General portrait of the PV L1 and L2 proteins. Hypervariable regions and secondary structure elements in upper panel are marked according to Ref. [7]. Other important regions are depicted according to references in text (Section 11.4). Numbering is based on the HPV16 sequence (GenBank accession No AF536179). All the elements are marked according to their relative sizes. NLS, nuclear localization signal; FCS, furin cleavage site; CB1/2, cell-binding domains; tSNARE, eventual tSNARE syntaxin 18 binding domain; MNE, major neutralization epitope.

protein have led to finding that HPV16 L1 has two NLS sequences, each containing basic aa clusters located at the C-terminus of the L1 protein [38]. More recent data have demonstrated that L2 protein of various PVs also carries two NLS sequences located at the very N- and C-termini [39–41].

Heterologously expressed structural genes from different PVs have also been used for studying the interaction of viral nucleic acid and capsid proteins. These studies have located that the C-terminal part of L1 contains DNA-binding domain, which partially overlaps with NLS [23,42] (Figure 11.2). The DNA-binding activity of the minor capsid protein L2 has been discussed controversially. Whereas Zhou et al. [43] observed a DNA-binding activity at the N-terminus of the HPV16 L2, others have reported only weak or no DNA-binding capacity of L2 in Southwestern blot analysis [44]. In another approach, using DNA mobility shift assays, it was proposed that the N- and C- termini of the BPV L2 function separately in nuclear import and DNA binding, and that the C-terminal NLS is also the high-affinity DNA-binding site [41]. Latest data have supported this observation and strongly confirm that both the NLS of HPV11, HPV16, and HPV18 L2 proteins are also their DNA-binding sites, which ensure DNA binding by sequence-independent mechanism [45,46].

PV VLPs have become an interesting tool for studying the receptor interaction and entry pathways. Extensive studies were dedicated to the identification of putative receptors of HPV. An early approach using HPV6b L1 VLPs resulted in the identification of α6 integrin as a primary cell receptor for PVs [47]. In contrast, binding assays using HPV11, HPV16, and HPV33 L1 VLPs indicated that several heparan sulfate proteoglycans can serve as HPV receptors [48–50], and supported a putative role for syndecan-1, rather than α6 integrin, as a primary receptor in natural HPV infection in keratinocytes [51]. For studies of the interaction of VLPs with potential receptors, entry studies and intracellular trafficking, VLPs have been labeled by carboxyfluorescein diacetate succinimidyl ester [50,52]. However, recent genetic and biochemical studies have suggested that L2 facilitates PV infectivity by binding to a secondary cell surface receptor [53,54]. In fact, for HPV16 L2 its cell-binding capacity via two independent cell surface-binding domains was demonstrated [53,55]. Finally, PV VLPs have been extensively used for studies of their interaction with different antigen-presenting cells (APCs) [56–59]. These investigations are of special interest as modified PV VLPs are currently being explored in recent attempts to generate an efficient therapeutic vaccine against cervical cancer (Section 11.5.5).

PVs were found to infect cells via clathrin-dependent receptor-mediated endocytosis [60,61], therefore, they must escape from the endosomal compartment to the cytoplasm to initiate infection. Latest findings have demonstrated that the membrane-destabilizing 23 aa peptide at the C-terminus of L2 and the site-specific enzymatic cleavage at a consensus furin recognition site at the N-terminus of L2 might be required for egress of viral genomes from endosomes [62,63] (Figure 11.2). After then, L2 accompanies the viral DNA to the nucleus and subsequently to the subnuclear promyelocytic leukemia protein bodies [64]. Differences in the subcellular localization of HPV L1- and L1/L2-derived VLPs suggested an interaction of L2 with certain cellular components, namely, β-actin [65]. In addition, BPV1 L2 was found to interact with resident endoplasmic reticulum protein tSNARE syntaxin 18 [66], and with the microtubule network via the motor protein dynein [67]. Assembly studies have suggested a possible role of the chaperone Hsc70 for both assembly and disassembly of VLPs [68,69]. Taken together, L2 protein seems to fulfill several important functions during PV assembly and infection. These studies might be also explored for incorporation of nucleic acids or other compounds within PV VLPs following their transportation to target cells (Section 11.5.3).

11.5 APPLIED RESEARCH APPLICATIONS OF PV VLPs

11.5.1 Diagnostics

VLPs have been extremely useful for the development of diagnostic tools for serological detection of virus infections in humans and animals. Until development of the PV-VLP technique, serological studies on PV infection have been hampered by the lack of appropriate antigen targets. PV VLPs are highly immunogenic, present conformational epitopes, and induce high titers of type-specific neutralizing antibodies when injected into rabbits [70], dogs [15], and calves [71]. Numerous serologic studies mainly using HPV16 VLPs have demonstrated that infection with genital HPV is followed by a serologic immune response to viral capsid proteins. However, the titer of detectable serum antibodies to HPV VLPs is quite low. This immune response is largely HPV type specific and directed against conformational epitopes. Experimental data confirm the validity of HPV-VLP-based ELISA tests as markers of present or past HPV infection (see Ref. [72] and references herein).

Serum antibody responses to PV-VLPs are stable over time indicating that seropositivity is a valid marker of cumulative virus exposure. Therefore, VLP-based ELISAs have been applied, besides their direct use in serological diagnostics of acute virus infections, to different seroepidemiological studies (for a review see Ref. [73]. In HPV-infected individuals serum antibody response to VLPs is stable over time, also after HPV infection has been cleared, resulting in HPV serology being used as a marker of cumulative HPV exposure in spite of the fact that a significant proportion of HPV-exposed subjects fail to seroconvert. Despite the low sensitivity of HPV16 VLP-based ELISA, the assay appears to have adequate specificity and should be useful as an epidemiological marker of HPV16 infection and sexual behavior [73,74]. Interestingly, analysis of sera from women from 11 countries by IgG ELISAs using VLPs from three different HPV16 strains demonstrated a strong correlation between the reactivities of the three different VLP variants [75]. VLP-based serologic assays have been used for numerous approaches such as determination of the antibody status in prepubertal children [76], identification of increased risk groups for oncogenic HPV infections [72], and investigation of cross-reactivity between phylogenetically related HPV types, leading to suggestion that the antibodies detected by ELISA are not always type specific [77].

In addition to classical ELISA tests for detection of virus-specific antibodies, PV-VLPs were applied also for the development of more complex serological assays. Thus, an HPV-Luminex immunoassay using VLPs of different HPV types may represent a useful tool for simultaneously quantifying antibody responses to multiple HPV genotypes for natural history infection studies [78]. Recently developed bead-based multiplex serology method allows the simultaneous detection of antibodies against up to 100 in situ affinity-purified recombinant HPV proteins and has the potential to replace conventional VLP ELISA technology [79].

11.5.2 Chimeric PV VLPs

In recent years, VLPs of different origin carrying foreign epitopes or polypeptides in certain places (designed as chimeric VLPs; cVLPs) have been suggested as promising candidates for vaccine development (for reviews see Refs. [80,81]. VLP-based technologies were useful not only for antigen but also for gene transfer purposes (for reviews see Refs. [82,83]. It is likely, that development of novel genetic vaccines might require targeted combination of both strategies.

A number of experiments have confirmed that L1 protein alone or together with L2 is able to form autologous VLPs in the absence of other viral components (Table 11.1; Figure 11.3A). The first attempts to construct PV-based cVLPs were based on observation that, in contrast to N-terminus, the C-terminal region of L1 protein is not critical for $T=7$ capsid assembly, as the BPV L1 mutant, lacking the last 24 aa, formed VLPs even threefold more efficiently then wild-type (wt) BPV L1 [53]. This finding was supported by replacement of the last 34 aa of the HPV16 L1 with inserts of up to 60 aa, without disturbing VLP assembly [84], which opened a way for generation of PV-derived cVLPs. In following experiments, several strategies were developed for incorporation and exposition of peptide sequences within PV VLPs. Inserts could be, in principle, divided in two groups: (1) PV-derived sequences, and (2) "foreign" epitopes (presented separately in Table 11.2).

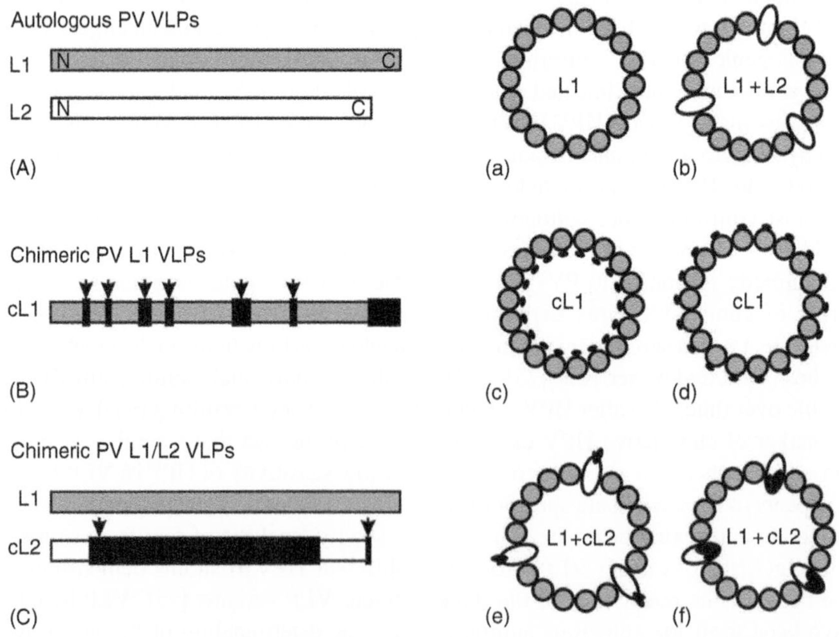

FIGURE 11.3 Schematic presentation of PV-derived VLPs. (A) autologous PV VLPs assembled either from L1 alone (a), or from both L1 and L2 proteins (b). (B) chimeric PV VLPs assembled from chimeric L1 alone. C-terminal inserts are likely to be situated inside the particle (c), while those inserted in hypervariable loop regions might be exposed on the surface of VLPs (d). (C) chimeric PV VLPs assembled by coexpression of wt L1 with chimeric L2 (e–f). Positions used for heterologous insertions are marked by arrows, while regions used for replacements are colored in black.

TABLE 11.2
Summary of PV-Derived Chimeric VLPs

PV L1	Generation of Chimeric VLPs L1 Derivatives	Description of the Insert PV-Derived	Reference
HPV16	Replacement of the C-terminus	HPV16 E7 protein (aa 1–55 or 1–60)	[84]^{T}
BPV1	Replacement of the C-terminus	HPV16 E7 CTL epitope (RAHYNIVTF)	[85]
HPV16	Replacement of the C-terminus	HPV16 E7 protein (aa 1–57)	[103]
HPV11	Replacement of hypervariable regions or complete functional domains	Hypervariable regions or complete functional domains of HPV16 L1 HPV16 L2 neutralization epitope aa 108–120 (LVEETSFIDAGAP)	[92]
HPV16	Replacement of selected internal regions	HPV16 L2 neutralization epitope aa 108–120 (LVEETSFIDAGAP)	[94]
BPV1	Insertion in loop region behind aa 133		[95]
		Heterologous	
BPV1	Replacement of the C-terminus	HIV-IIIB gp160 CTL epitope (RGPGRAFVTI)	[85]
BPV1	Replacement of selected internal regions	Mouse CCR5 epitope (HYAANEWVFGNIMCKV)	[88]
HPV16	Replacement of the C-terminus	Tumor-associated antigen P1A CTL epitope (LPYLGWLVF)	[86]
BPV1, HPV16	Insertions/replacements in loop regions	HIV1 gp41 B-cell epitope (LELDKWAS)	[89]
HPV16	Insertions in loop regions	HBc B-cell epitope (DPASRE)	[91]
BPV1	Replacement of aa 130–136	HIV1 gp41 B-cell epitope (ELDKWA)	[90]
HPV16*	Chemical coupling	Influenza type A M2 protein fragment (22 aa)	[96]
		Combined	
BPV1	Replacement of the C-terminus	Artificial "polytope" minigene, with HPV16 E7 B-cell epitope (QAEPD) following CTL epitopes from HPV16 E7 (RAHYNIVTF), HIV-IIIB gp120 (RGPGRAFVTI), Nef (PLTFGWCFKL), and RT (VIYQYMDDL) proteins	[87]
	L1 + L2 Derivatives	***PV-Derived***	
BPV1	Fusion to the C-terminus of BPV1 L2	Full-length HPV16 E7 protein (98 aa)	[97]^{T}
HPV16	Fusion to the C-terminus of HPV16 L2	Full-length HPV16 E7 protein (98 aa)	
HPV16	Fusion to the C-terminus of HPV16 L2	Full-length CRPV E2 (391 aa)	
HPV16	Fusion to the C-terminus of HPV16 L2	HPV16 E7 protein (aa 1–57)	[103]
CRPV	Fusion to the C-terminus of CRPV L2	Full-length HPV16 E7 protein (98 aa)	[98]
HPV16*	Replacement of the internal part of HPV16 L2	Full-length HPV16 E1, E2, and E7 fusion (649 + 368 + 98 aa)	[102]^{T}
HPV16	Fusion to the C-terminus of HPV16 L2	Full-length HPV16 E7 and E2 fusion (98 + 368 aa)	[100]^{T}
HPV6b	Fusion to the C-terminus of HPV6b L2	Artificial shuffled HPV16 E7 gene (158 aa)	[99]
		Heterologous	
BPV1	Insertion/replacement in N-terminal region of L2	Green fluorescent protein	[101]

Note: Described are only those attempts resulting in particle formation. All the VLPs were obtained in insect cells using recombinant baculoviruses except those marked by (*) produced in yeast *S. cerevisiae*. (^{T}), VLPs designed to construct therapeutic HPV vaccine.

Replacement of the C-terminus of the L1 protein has become the first method for generation of PV cVLPs (Figure 11.3B). HPV16 L1-based cVLPs harboring part of E7 protein were still able to induce a neutralizing antibody response and followed a similar intracellular pathway as wt VLPs; therefore, they were regarded to deliver peptides into mammalian cells in vitro and in vivo [84]. To evaluate this antigen delivery system, a single HPV16 E7 CTL epitope was fused to the C-terminus of BPV1 L1 sequence. Mice immunized with cVLPs elicited high titer-specific antibodies against BPV1 L1 VLPs and strong CTL responses able to protect mice against challenge with E7-transformed tumor cells [85]. Similarly, immunization with HPV16 L1 cVLPs containing peptide from tumor-associated antigen P1A induced a protective immune response in mice against a lethal tumor challenge with a progressor tumor cell line. What is important, the authors have demonstrated that these cVLPs could be used also therapeutically to suppress the growth of established tumors [86]. To investigate capacity of the PV VLP delivery system for inducing multiple immune responses, the C-terminus of the BPV1 L1 was replaced with an artificial polytope minigene, containing multiple HPV16 and human immunodeficiency virus (HIV) B- and T-cell epitopes. Serum antibodies obtained against cVLPs in mice reacted with both cVLPs and wt BPV1 L1 VLPs. In addition, CTL epitope-specific precursors were also detected in the spleen of immunized mice [87]. Taken together, these data demonstrated that cVLPs as self-adjuvanting immunogen delivery system might be used to target antigenic epitopes to both the MHC class I and II pathways, with the potential to elicit therapeutic virus-specific and antitumor immune responses.

However, because the highly basic C-terminus of the PV L1 probably interacts with DNA in the virion, it is likely that the C-terminally fused inserts will be packed preferably into the interior of the capsid [7]. Following investigations revealed that L1 protein tolerates insertions also in other sites along the molecule. The first attempt to incorporate short peptide representing an extracellular loop of the mouse chemokine receptor CCR5 into internal regions of BPV1 L1 has met limited success since only one out of three chimeras assembled in small 28 nm VLPs [88]. With resolving of the L1 monomer structure [7], a special interest was given to insertions and replacements in hypervariable loop regions since they are likely to be exposed on the surface of VLPs (see Figures 11.1C, 11.2, and 11.3B). To characterize the regions where a foreign epitope could be introduced, several studies were conducted to generate chimeric BPV- and HPV16-derived VLPs. Thus, the B-cell epitope from the HIV1 gp41 envelope was introduced into three distinct surface loops (DE, FG, and HI) of the L1 monomer and the cVLPs obtained were regarded to prevent and to treat HIV1 infection [89,90]. In similar approach, all the L1 protein loops were scanned by insertion of short HBc-derived B-cell epitope, and all the respective chimeras were capable of self-assembly into VLPs [91]. Generally, the PV-specific antigenicity and immunogenicity of this type of cVLPs were reduced compared to the wt VLPs confirming the involvement of loop residues (especially FG and HI) in L1 neutralizing epitopes, while placing a foreign peptide in the context of an assembled structure (except the one with insertion in the BC loop) markedly enhanced its immunogenicity.

To characterize type-specific conformational neutralizing epitopes and generate antibody responses to antigenically unrelated HPV types, several hypervariable regions in HPV11 L1 backbone were replaced by complementary regions from HPV16 L1 [92]. Furthermore, a well-characterized 13 aa-long HPV16 L2 epitope, which is a common-neutralizing epitope for HPV types 6 and 16 [93] was inserted in various regions of the L1 structure of HPV16 [94] or BPV1 [95]. Construction of such multivalent cVLPs was suggested as an alternative strategy to generate broad-spectrum vaccines to protect against multiple HPV types.

A novel method for generation of cVLPs is chemical coupling of antigenic peptides via a linker to an antigen carrier (see Chapter 6). For PVs, there was only one report dealing with this strategy, where extracellular peptide fragment of influenza type A M2 protein was coupled to the HPV VLPs of approximately 4000 copies per VLP. The conjugate vaccine was found to be highly immunogenic and conferred good protection against lethal challenge of influenza virus in mice [96].

However, due to the structural and technical reasons, the size of epitopes or polypeptides incorporated in chimeric L1 VLPs is limited, and large proteins cannot be delivered through this way. This led to the development of an alternative strategy by exploiting potential of the minor capsid protein L2. Coexpression of the wt L1 with the chimeric L2 derivatives might result in appearance of VLPs, composed of L1, but containing also few copies of L2 chimeras (Figure 11.3C). Fusion to the C-terminus of the L2 allowed incorporation of such long sequences as full-length 98 aa E7 [97,98], artificial 158 aa-long E7 [99], 43 kDa E2 [97], or even E7 and E2 fusion proteins [100] into VLPs. It was also found that 240 aa-long green fluorescent protein can be inserted into the N-terminal region of BPV1 L2, without disruption of the VLP structure [101]. Such fluorescent VLPs might be a useful tool for study of virus–cell interactions. Moreover, it was showed that replacement of the internal area of HPV16 L2 by approximately 130 kDa HPV16-derived E1–E2–E7 fusion protein did not disrupt VLP assembly and that the cVLPs had good immunogenicity [102]. Despite the relatively low insert copy number in comparison with L1-derived cVLPs (approximately 1/30), L1/L2-derived cVLPs have demonstrated only slightly reduced tumor prevention capacity. At the same time, however, L1-derived cVLPs are limited by the size of the antigen that can be incorporated and in the amount of particles that can be obtained from cultures when compared to L1/L2 cVLPs [103]. Therefore, the latter strategy seems to be more promising for construction of the novel therapeutic HPV vaccine (discussed in Section 11.5.5).

11.5.3 Gene Therapy Applications

Despite significant progress in the area of nonviral gene delivery, synthetic vectors remain less efficient by orders of magnitude than their viral counterparts. Nonviral gene delivery is currently a subject of increasing attention because of its relative safety and simplicity of use; however, its use is still far from being ideal because of its comparatively low efficiency (for reviews see Refs. [104,105]). Noninfectious VLPs may represent an alternative, intermediate approach to viral and nonviral delivery systems. Especially, VLPs derived from DNA viruses that replicate in the nucleus may represent promising vehicles for gene transfer approaches. Therefore, the most experiments in this direction were performed using papilloma- and polyomavirus-derived VLPs (see Chapter 10), not only for incorporation of DNA but also of other substances (for reviews see Refs. [82,106]). PV pseudovirions (PSVs) composed of L1 alone or containing also L2 have been applied for gene transfer studies. Importantly, the foreign DNA can be packaged into VLPs either by in vitro dis- and reassembly of VLPs, or with the help of packaging cell lines.

In a first dis- and reassembly approach, HPV16-derived VLPs composed of L1 protein alone were able to package unrelated DNA in vitro and then deliver this foreign DNA to eukaryotic cells with subsequent expression of the encoded gene. The gene transfer efficiency has been found higher than for naked DNA or liposome-associated DNA [107]. Later, authors have demonstrated that gene transfer efficiency might be improved by direct interaction between a reporter plasmid and the VLPs [108]. Because native PVs naturally infect mucosa and skin, PV PSVs have been suggested as novel oral vaccine candidates. It was found that oral but not subcutaneous immunization with the HPV16 L1-derived PSVs carrying HPV16 E7 gene-induced mucosal and systemic CTL responses and protected mice against mucosal challenge [109]. Extending their experiments, authors have generated PSVs encoding lymphocytic choriomeningitis virus gp33–41 CTL epitope; however, oral immunization induced low protection in aged mice compared with young adult mice. In contrast, when coadministrated with PSVs encoding human interleukin-2, effective immune responses were induced providing a basis for clinical trials using PV PSVs for vaccination of the elderly [110]. It was also reported that encapsidation of an interleukin-2-expressing plasmid could significantly enhance the induction of mucosal and systemic antibody responses to HPV16 PSVs even after intramuscular immunization into mice [111]. As the mucosal surfaces are the primary portals for HIV transmission, PV PSVs encoding HIV1 Gag were regarded as mucosal HIV vaccine candidate. Data have demonstrated that oral immunization of mice with Gag-PSVs induced Gag-specific memory CTLs and protected mice against a rectal mucosal challenge with a recombinant vaccinia virus expressing HIV1 Gag [112].

Although PV PSVs can be successfully generated from L1 protein alone by dis- and reassembly of VLPs, the efficiency of DNA packaging and infectivity might be enhanced in the presence of minor capsid protein L2, as monitored by PSV-transferred β-galactosidase activity [113]. This is supported by data obtained from packaging cell lines. Thus, the first generation of infectious PV PSVs has been described as early as in 1993 in epithelial cells infected with double vaccinia virus recombinant for BPV1 L1 and L2 but not with the recombinant for BPV1 L1 alone [114]. Similarly, L2 was required for DNA encapsidation in BPV1 or HPV16 PSVs using Semliki Forest virus-mediated expression system [115]. Consistent with this, the HPV18-derived L1 protein alone was unable to encapsidate DNA, while the L2 was found to be attached to plasmid mini-chromosomes, suggesting a role for L2 in encapsidation [116]. In contrast, Unckell et al. have demonstrated that the L2 is not required for DNA encapsidation but is essential for efficient pseudoinfection of HPV33-derived PSVs [21]. The fact that L2 contributes for PV infection was supported also by others, because PSVs containing the L1 protein only displayed a strongly reduced infectivity [113,117]. Since the L2 apparently does not play a role in the initial binding of PVs to the cell surface [118,119] and internalization [120], it seems to be important for later steps in the infection process, which is supported by multifunctionality of the L2 (see Figure 11.2). Generally, it was confirmed that DNA encapsidation in PV PSVs is largely sequence independent. Because PV PSV infections could be inhibited by type-specific antibodies, PSVs containing such reporter genes as green fluorescent protein, β-galactosidase, or secreted alkaline phosphatase have been extensively used to investigate PV internalization pathways, to detect the level of neutralizing antibodies, and for population-based HPV screening [21,107,113,115–117,121].

Recently, a highly efficient method for producing BPV L1/L2 PSVs in transiently transfected mammalian cells has been described, however, intracellular packaging of target plasmids into VLPs by this approach was limited to approximately 6–7 kb in size, which is much less than the natural viral genome size of approximately 8 kb [117]. This method was further improved by using codon-optimized synthetic genes in 293TT human embryonic kidney cells, to achieve the efficient packaging of the full-length HPV genomes into HPV16 capsids. Obtained virus particles were highly infectious in epithelial cells, providing a novel useful tool for studies of HPV replication and vaccine and drug development [122]. Furthermore, it was demonstrated that such infectious PV particles, named as "quasivirions," can cause papillomatous growths in the natural host animal. Moreover, species-matched PV genomes were successfully delivered in vivo by a heterologous, species-mismatched PV capsid [123]. Preclinical animal models cannot be used to test for protection against HPV infections due to species restriction, therefore authors have developed a model using particles composed from HPV capsid with encapsidated CRPV genome to permit the direct testing of HPV VLP vaccines in rabbits [124].

11.5.4 Generation of Prophylactic HPV Vaccine

In recent years, a lot of attention has been paid to both prophylactic and therapeutic HPV vaccine development (see recent reviews in Refs. [125–128]. Prophylactic viral vaccines that are currently licensed for use in humans work through the induction of virus-neutralizing antibodies and, as a rule, immunologically mimic the infections they prevent. Papillomavirus L1 VLPs resemble the natural virus structurally and immunologically. The results obtained with animal models have demonstrated that prophylactic immunization with VLPs can be very effective in vivo. Moreover, passive transfer of immune sera conferred protection to naive animals, indicating that an antibody-mediated response plays a major role in preventing PV infection [15,70]. However, studies with infectious HPV virions and VLPs of different HPV types have revealed the immune response as predominantly type specific [129,130]. An explanation was provided by definition of the crystal structure of the L1 molecule, demonstrating that variability between PV genotypes is concentrated in the solvent-facing apices of the L1 protein loops [7], harboring the sites to which neutralizing antibodies are directed (see Figure 11.1C). Therefore, an ideal anti-HPV vaccine should include VLPs able to elicit an immune response directed against all the serotypes of interest. Since approximately 70% of cervical cancer is caused by infection with HPV types 16 and 18 while 90% of genital warts are caused by HPV types 6 and 11, most attempts for vaccine construction were focused on these particular subtypes.

Several strategies have been developed for construction of polyvalent prophylactic HPV vaccine. One of them focused on construction of hybrid VLPs harboring immunologic determinants from distinct PV subtypes. Thus, a set of cVLPs were prepared using complementary regions of the major capsid L1 proteins of HPV types 11 and 16. Notably, one of the cVLPs containing hypervariable FG and HI loops of HPV16 L1 replaced into an HPV11 L1 background provoked neutralizing activity against both HPV11 and HPV16 [92]. Others have suggested exploiting the cross-neutralization potential of the L2 protein. To induce a strong anti-L2 Ab response, HPV16 L2 neutralization epitope comprising L2 aa residues 108–120 was inserted within the L1 immunodominant surface loops either of HPV16, or BPV1 [94,95]. HPV16-derived cVLPs reacted with the L2 serum and elicited high-titer anti-L1 immune responses in mice [94], while BPV-derived cVLPs-induced L2-specific antisera in rabbits, capable for partial neutralization of HPV16 PSVs [95]. In another study, a diploid yeast strain has been generated for coexpression of the L1/L2 capsid proteins from HPV types 6 and 16. The data indicated that all the four proteins coassembled into mixed HPV6/16 cVLPs, which induced conformational antibodies against both L1 protein types in mice [131].

Yeast *S. cerevisiae* as a host system being attributed to GRAS (generally regarded as safe) microorganisms and used successfully to produce the first licensed VLP vaccine against hepatitis B [132] was an attractive candidate for development of the HPV vaccine. A recombinant yeast expression system was engineered to express single HPV capsid proteins [18,133] and establish a rapid purification procedure of assembled VLPs [134]. The first randomized placebo-controlled trial of a monovalent yeast-derived HPV16 VLP vaccine showed 100% efficacy to prevent HPV16 infection in young, previously uninfected women [135]. Investigations in this direction, conducted by Merck, were extended to generate a quadrivalent (HPV6, 11, 16, and 18) VLP vaccine, Gardasil, which was the first-ever cancer vaccine approved by U.S. Food and Drug Administration in June, 2006. To obtain the final formulation of Gardasil, the four respective HPV genes were separately expressed in *S. cerevisiae*, individually purified and finally adsorbed on aluminium hydroxyphophate sulphate adjuvant (reviewed in Ref. [136]. In parallel, GlaxoSmithKline has developed bivalent (HPV16 and 18) insect cell-derived VLP vaccine, Cervarix, which has already received approval in Australia and Europe and is expected to be validated by Food and Drug Administration soon. Each type of virus-like particle was produced on *Spodoptera frugiperda* Sf-9 and *Trichoplusia ni* Hi-5 cell substrate with AS04 adjuvant containing aluminium hydroxide and 3-deacylated monophosphoryl lipid A [137]. Both vaccines are given as three intramuscular injections over a 6 month period and demonstrated 100% efficiency in protecting women against incident and persistent HPV infections.

To avoid parenteral injection, alternative HPV vaccine candidates have been proposed by other research groups. Since attenuated strains of *Salmonella* are effective vehicles for delivering heterologous antigens to the mucosal and systemic immune systems, formation of HPV VLPs in *Salmonella* raised interest as eventual live HPV vaccine candidate [22]. Extending their investigations, authors optimized the HPV16 L1 gene to fit with the most frequently used codons in *Salmonella*. A new generation of recombinant *Salmonella* organisms assembled HPV16 VLPs and induced high titers of neutralizing antibodies in mice after a single nasal or oral immunization with live bacteria [138]. Recently, an attempt to develop live mucosal prophylactic HPV vaccine was carried by expression of the HPV16 L1 in *Lactobacillus casei*. The protein self-assembled into VLPs intracellularly and sera from mice that were subcutaneously immunized with *L. casei* expressing L1 efficiently reacted with insect cell-derived HPV16 VLPs, confirming their immunological identity [139].

Transgenic plants might be another system to be used as edible vaccines to induce an L1-specific prophylactic immune response. Thus, transgenic tobacco and potato plants were adapted for expression of the HPV16 L1 by changing the L1 gene codon usage and increasing its transcript stability. The plant-derived L1 protein displayed conformation-specific epitopes and assembled into VLPs. However, feeding of tubers from transgenic potatoes to mice induced an anti-L1 antibody response only in minority of animals [28]. In a parallel study, rabbits immunized with plant-produced VLPs also elicited a weak anti-HPV16 L1 immune response [26], suggesting that this particular system needs a significant improvement. Such an alternative might be usage of plant virus vectors as it allows to produce L1 protein at levels about 10 times higher than previously obtained by transgenic expression of the same gene in tobacco, and produced L1-specific antibodies in rabbits [31]. As a general problem for plant-mediated expression of PV structural genes, only part of L1 was found assembled in particles, which probably explains the low immunogenicity observed.

11.5.5 Therapeutic HPV Vaccine Candidates

A large number of attempts have been undertaken to construct therapeutic HPV vaccines including live-vector vaccines, peptide or protein vaccines, and nucleic acid vaccines, each of them exhibiting defined advantages and disadvantages (for a review see Ref. [128]). Therapeutic vaccines are based on the induction of cellular immunity directed against cells presenting viral antigens. The goal of effective therapeutic HPV vaccine therefore should be the clearance of infected cells and the rejection of HPV-associated lesions and tumors, however, there is no such a vaccine currently available. Despite a success of PV VLPs as prophylactic HPV vaccines, they are unlikely to have therapeutic effects because the virion capsid proteins are not detected in the proliferating cells of the infected epithelia or in cervical carcinomas. In fact, vaccination of women positive for HPV DNA with HPV16/18 vaccine, Cervarix, did not accelerate clearance of the virus demonstrating this approach as not useful to treat prevalent HPV infections [140]. In contrast, the early viral E6 and E7 oncoproteins are constitutively expressed in cervical cancer cells and are required for maintenance of the transformed phenotype in cell lines [141]; therefore, they apparently represent the most promising target molecules for therapeutic HPV vaccine construction.

However, purified antigens per se do not control immunity, especially T-cell immunity. Dendritic cells (DCs), a family of most professional APCs, are important intermediaries between antigens and better control of the immune system. An eventual HPV vaccine administered into human skin will interact with a high concentration of DCs and Langerhans cells (LCs). LCs form a three-dimensional network in the epidermis of skin, whereas DCs reside below the basal layer of the epidermis in the dermis (for a review see Ref. [142]). Interactions between VLPs and various APCs have been extensively studied in recent years. PV VLPs are highly immunogenic and can induce strong Th-cell responses. It was demonstrated that fully assembled autologous L1- or L1/L2-derived PV VLPs but not capsomeres can activate DCs in the absence of adjuvants [143]. There is also evidence that immunization with HPV6b L1 VLPs without adjuvant may accelerate regression of warts in infected humans [144].

Incorporation of HPV early proteins into VLPs is now considered as one of the possibilities to promote the induction or enhancement of E-protein-specific cellular immune responses. The attempts to evaluate cVLPs as a therapeutic vaccination strategy against cervical cancer have started with generation of the first HPV cVLPs [84] (see also Table 11.2 and Figure 11.3B). These L1-derived cVLPs carrying N-terminal part of E7 protein induced an E7-specific CTL response in mice [145] and E7-specific CTLs in humans by in vitro vaccination [146], indicating that cVLPs might be useful for tumor prevention as well as for cancer therapy. However, the data from a clinical trial in women suffering from high-grade cervical intraepithelial neoplasia suggested that improvement of the immunogenicity is still required [147]. One of the possible solutions might be the use of appropriate adjuvants, as the lipopolysaccharides, unmethylated CpG motifs, and sorbitol increased the immunogenicity of cVLPs and activated bone marrow-derived DCs [148].

Importantly, genetic vaccination with a corresponding DNA revealed that only short C-terminal deletions of the L1 were tolerated for eliciting a humoral immune response against viral capsids. DNA vaccination with prolonged L1-E7 gene encoding L1 aa 1–498 induced not only L1-specific antibodies but also L1- and E7-specific CTL responses in mice [149].

Another therapeutic vaccine strategy utilizes the L2 protein (Figure 11.3C). HPV16-derived cVLPs comprised of the L1 and a fusion of entire E7 to the L2 capsid protein have been shown to initiate a potent E7-specific CTL response in vaccinated mice in the absence of adjuvant and induced protection from tumor challenge [97]. These cVLPs efficiently activated human peripheral blood-derived DCs and induced a significant upregulation of the CD80 and CD83 molecules as well as secretion of interleukin-12 [150]. LCs were able to bind and internalize cVLPs in a manner quantitatively equivalent to DCs, however, failed to activate, suggesting possible immune escape mechanism used by HPVs [151]. In contrast, analogous BPV- or CRPV-derived cVLPs activated both DCs and LCs and initiated an immune response against cVLP-derived antigens [58], indicating that animal PV-derived cVLPs might be more advantageous as therapeutic HPV vaccine candidates. Importantly, it was reported that preexisting neutralization antibodies prevented the induction of E7-specific cell-mediated immunogenicity by subsequent immunization with cVLPs [152], thus these therapeutic vaccines may be inefficient in patients with preexisting HPV-specific humoral immunity. Increase in the efficacy of cVLP-based vaccines may be achieved through the combined use of different cVLP types for prime/boost regimens [98] and generation of cVLP immune complexes with anti-L1 antibodies, as this might result in the antigen being taken up by APCs via Fcγ receptors [58].

To improve CTL responses, others have suggested including E2 protein within cVLPs since the E2 is expressed in early dysplasia and neoplasia lesions (for a review see Ref. [153]), where E7 is expressed in more advance lesions. In an attempt to develop combined prophylactic and therapeutic vaccine, authors have generated HPV16-based cVLPs incorporated with both the E2 and E7 proteins and evaluated the strategies to both improve primary CTL responses to the cVLPs and effectively boost these responses in mice. Data have shown that such cVLPs coinjected with immune modulators has a therapeutic potential for the treatment of HPV-associated cervical intraepithelial neoplasia and cancer, without diminishing its potential as a prophylactic vaccine [100]. Even more complex HPV16 cVLPs were constructed by replacement of the internal part of L2 with approximately 130 kDa E1–E2–E7 fusion protein and coexpression with L1 in yeast *S. cerevisiae*. cVLPs induced strongly neutralizing antibodies and strong L1-specific Th2 response in immunized rhesus macaques, however, in contrast to murine experiments failed to induce a $CD8^+$ T-cell-mediated IFN-γ response, indicating that T-cell response in mice may be greater than in primates [102].

Recently, it was shown that HPV-derived VLPs are able to mediate delivery and expression of plasmid DNA both in vitro and in vivo in the presence of L2 in VLPs. Moreover, expression of transgene was observed in migrating APCs recovered from mice inoculated with plasmid-VLP complexes, but not with the plasmid alone, while coadministration of VLPs with a plasmid encoding HPV16 E6 oncoprotein was associated with significant enhancement of E6-specific cellular immune responses [154]. These data indicate that a particular combination of gene and protein delivery systems provides another opportunity for improving CTL responses that still remain the major problem for generating an effective therapeutic VLP-based vaccine against cervical cancer. Recent advances in this area, however, indicate that significant progress towards generation of the therapeutic HPV vaccine will be achieved in the nearest future.

ACKNOWLEDGMENTS

The work of A.K. was supported by European Social Fund. The authors wish to thank Paul Pumpens (Riga) and Yury Khudyakov (Atlanta) for critical reading of the manuscript and helpful discussions. We apologize to the authors for the important work not cited in this chapter.

REFERENCES

1. Howley, P.M. and Lowy, D.R., Papillomaviruses, in *Fields Virology* 5th edn., Knipe, D.M., (Ed.), Lippincott Williams & Wilkins, Philadelphia, 2007, Chapter 62.
2. De Villiers, E.M. et al., Classification of papillomaviruses, *Virology*, 324, 17, 2004.
3. Baker, T.S. et al., Structures of bovine and human papillomaviruses, analysis by cryoelectron microscopy and three-dimensional reconstruction, *Biophys. J.*, 60, 1445, 1991.
4. Kirnbauer, R. et al., Efficient self-assembly of human papillomavirus type 16 L1 and L1-L2 into virus-like particles, *J. Virol.*, 67, 6929, 1993.
5. Trus, B.L. et al., Novel structural features of bovine papillomavirus capsid revealed by a three-dimensional reconstruction to 9 A resolution, *Nat. Struct. Biol.*, 4, 413, 1997.
6. Hagensee, M.E. et al., Three-dimensional structure of vaccinia virus-produced human papillomavirus type 1 capsids, *J. Virol.*, 68, 4503, 1994.
7. Chen, X.S. et al., Structure of small virus-like particles assembled from the L1 protein of human papillomavirus 16, *Mol. Cell*, 5, 557, 2000.
8. Modis, Y., Trus, B.L., and Harrison, S.C., Atomic model of the papillomavirus capsid, *EMBO J.*, 21, 4754, 2002.
9. Bishop, B., Dasgupta, J., and Chen, X.S., Structure-based engineering of papillomavirus major capsid L1: Controlling particle assembly, *Virol. J.*, 4, 3, 2007.
10. Zhou, J. et al., Expression of vaccinia recombinant HPV 16 L1 and L2 ORF proteins in epithelial cells is sufficient for assembly of HPV virion-like particles, *Virology*, 185, 251, 1991.
11. Kirnbauer, R.F. et al., Papillomavirus L1 major capsid protein self-assembles into virus-like particles that are highly immunogenic, *Proc. Natl. Acad. Sci. U S A*, 89, 12180, 1992.
12. Rose, R.C. et al., Expression of human papillomavirus type 11 L1 protein in insect cells: in vivo and in vitro assembly of viruslike particles, *J. Virol.*, 67, 1936, 1993.
13. Volpers, C. et al., Assembly of the major and the minor capsid protein of human papillomavirus type 33 into virus-like particles and tubular structures in insect cells, *Virology*, 200, 504, 1994.
14. Touzé, A. et al., Production of human papillomavirus type 45 virus-like particles in insect cells using a recombinant baculovirus, *FEMS Microbiol. Lett.*, 141, 111, 1996.
15. Suzich, J.A. et al., Systemic immunization with papillomavirus L1 protein completely prevents the development of viral mucosal papillomas, *Proc. Natl. Acad. Sci. U S A*, 92, 11553, 1995.
16. Ghim, S.J. et al., Equine papillomavirus type 1: Complete nucleotide sequence and characterization of recombinant virus-like particles composed of the EcPV-1 L1 major capsid protein, *Biochem. Biophys. Res. Commun.*, 324, 1108, 2004.
17. Sasagawa, T. et al., Synthesis and assembly of virus-like particles of human papillomaviruses type 6 and type 16 in fission yeast *Schizosaccharomyces pombe*, *Virology*, 206, 126, 1995.
18. Hofmann, K.J. et al., Sequence determination of human papillomavirus type 6a and assembly of virus-like particles in Saccharomyces cerevisiae, *Virology*, 209, 506, 1995.
19. Jansen, K.U. et al., Vaccination with yeast-expressed cottontail rabbit papillomavirus (CRPV) virus-like particles protects rabbits from CRPV-induced papilloma formation, *Vaccine*, 13, 1509, 1995.
20. Heino, P., Dillner, J., and Schwartz, S., Human papillomavirus type 16 capsid proteins produced from recombinant Semliki Forest virus assemble into virus-like particles, *Virology*, 214, 349, 1995.
21. Unckell, F., Streeck, R.E., and Sapp, M., Generation and neutralization of pseudovirions of human papillomavirus type 33, *J. Virol.*, 71, 2934, 1997.
22. Nardelli-Haefliger, D. et al., Human papillomavirus type 16 virus-like particles expressed in attenuated *Salmonella typhimurium* elicit mucosal and systemic neutralizing antibodies in mice, *Infect. Immun.*, 65, 3328, 1997.
23. Li, M. et al., Expression of the human papillomavirus type 11 L1 capsid protein in Escherichia coli: Characterization of protein domains involved in DNA binding and capsid assembly, *J. Virol.*, 71, 2988, 1997.
24. Zhang, W. et al., Expression of human papillomavirus type 16 L1 protein in *Escherichia coli:* Denaturation, renaturation, and self-assembly of virus-like particles in vitro, *Virology*, 243, 423, 1998.
25. Chen, X.S. et al., Papillomavirus capsid protein expression in *Escherichia coli*: Purification and assembly of HPV11 and HPV16 L1, *J. Mol. Biol.*, 307, 173, 2001.

26. Varsani, A. et al., Expression of human papillomavirus type 16 major capsid protein in transgenic *Nicotiana tabacum* cv. Xanthi, *Arch. Virol.*, 148, 1771, 2003.
27. Warzecha, H. et al., Oral immunogenicity of human papillomavirus-like particles expressed in potato, *J. Virol.*, 77, 8702, 2003.
28. Biemelt, S. et al., Production of human papillomavirus type 16 virus-like particles in transgenic plants, *J. Virol.*, 77, 9211, 2003.
29. Leder, C. et al., Enhancement of capsid gene expression: Preparing the human papillomavirus type 16 major structural gene L1 for DNA vaccination purposes, *J. Virol.*, 75, 9201, 2001.
30. Mossadegh, N. et al., Codon optimization of the human papillomavirus 11 (HPV 11) L1 gene leads to increased gene expression and formation of virus-like particles in mammalian epithelial cells, *Virology*, 326, 57, 2004.
31. Varsani, A. et al., Transient expression of human papillomavirus type 16 L1 protein in Nicotiana benthamiana using an infectious tobamovirus vector, *Virus Res.*, 120, 91, 2006.
32. Hagensee, M.E., Yaegashi, N., and Galloway, D.A., Self-assembly of human papillomavirus type 1 capsids by expression of the L1 protein alone or by coexpression of the L1 and L2 capsid proteins, *J. Virol.*, 67, 315, 1993.
33. Iyengar, S. et al., Self-assembly of in vitro-translated human papillomavirus type 16 L1 capsid protein into virus-like particles and antigenic reactivity of the protein, *Clin. Diagn. Lab. Immunol.*, 3, 733, 1996.
34. Finnen, R.L. et al., Interactions between papillomavirus L1 and L2 capsid proteins, *J. Virol.*, 77, 4818, 2003.
35. Paintsil, J. et al., Carboxyl terminus of bovine papillomavirus type-1 L1 protein is not required for capsid formation, *Virology*, 223, 238, 1996.
36. Chen, Y. et al., Mutant canine oral papillomavirus L1 capsid proteins which form virus-like particles but lack native conformational epitopes, *J. Gen. Virol.*, 79, 2137, 1998.
37. Okun, M.M. et al., L1 interaction domains of papillomavirus L2 necessary for viral genome encapsidation, *J. Virol.*, 75, 4332, 2001.
38. Zhou, J. et al., Identification of the nuclear localization signal of human papillomavirus type 16 L1 protein, *Virology*, 185, 625, 1991.
39. Sun, X.Y. et al., Sequences required for the nuclear targeting and accumulation of human papillomavirus type 6B L2 protein, *Virology*, 213, 321, 1995.
40. Darshan, M.S. et al., The L2 minor capsid protein of human papillomavirus type 16 interacts with a network of nuclear import receptors, *J. Virol.*, 78, 12179, 2004.
41. Fay, A. et al., The positively charged termini of L2 minor capsid protein required for bovine papillomavirus infection function separately in nuclear import and DNA binding, *J. Virol.*, 78, 13447, 2004.
42. Touze, A. et al., The nine C-terminal amino acids of the major capsid protein of the human papillomavirus type 16 are essential for DNA binding and gene transfer capacity, *FEMS Microbiol. Lett.*, 189, 121, 2000.
43. Zhou, J. et al., Interaction of human papillomavirus (HPV) type 16 capsid proteins with HPV DNA requires an intact L2 N-terminal sequence, *J. Virol.*, 68, 619, 1994.
44. Schafer, F., Florin, L., and Sapp, M., DNA binding of L1 is required for human papillomavirus morphogenesis in vivo, *Virology*, 295, 172, 2002.
45. Bordeaux, J. et al., The L2 minor capsid protein of low-risk human papillomavirus type 11 interacts with host nuclear import receptors and viral DNA, *J. Virol.*, 80, 8259, 2006.
46. Klucevsek, K. et al., Nuclear import strategies of high-risk HPV18 L2 minor capsid protein, *Virology*, 352, 200, 2006.
47. Evander, M. et al., Identification of the alpha6 integrin as a candidate receptor for papillomaviruses, *J. Virol.*, 71, 2449, 1997.
48. Joyce, J.G. et al., The L1 major capsid protein of human papillomavirus type 11 recombinant virus-like particles interacts with heparin and cell-surface glycosaminoglycans on human keratinocytes, *J. Biol. Chem.*, 274, 5810, 1999.
49. Giroglou, T. et al., Human papillomavirus infection requires cell surface heparan sulfate, *J. Virol.*, 75, 1565, 2001.
50. Drobni, P. et al., Carboxy-fluorescein diacetate, succinimidyl ester labeled papillomavirus virus-like particles fluoresce after internalization and interact with heparan sulfate for binding and entry, *Virology*, 310, 163, 2003.

51. Shafti-Keramat, S. et al., Different heparan sulfate proteoglycans serve as cellular receptors for human papillomaviruses, *J. Virol.*, 77, 13125, 2003.
52. Bergsdorf, C. et al., Highly efficient transport of carboxyfluorescein diacetate succinimidyl ester into COS7 cells using human papillomavirus-like particles, *FEBS Lett.*, 536, 120, 2003.
53. Kawana, Y. et al., Human papillomavirus type 16 minor capsid protein l2 N-terminal region containing a common neutralization epitope binds to the cell surface and enters the cytoplasm, *J. Virol.*, 75, 2331, 2001.
54. Roden, R.B. et al., Positively charged termini of the L2 minor capsid protein are necessary for papillomavirus infection, *J. Virol.*, 75, 10493, 2001.
55. Yang, R. et al., Cell surface-binding motifs of L2 that facilitate papillomavirus infection, *J. Virol.*, 77, 3531, 2003.
56. Da Silva, D.M. et al., Physical interaction of human papillomavirus virus-like particles with immune cells, *Int. Immunol.*, 13, 633, 2001.
57. Fausch, S.C., Da Silva, D.M., and Kast, W.M., Differential uptake and cross-presentation of human papillomavirus virus-like particles by dendritic cells and Langerhans cells, *Cancer Res.*, 63, 3478, 2003.
58. Fausch, S.C., Da Silva, D.M., and Kast, W.M., Heterologous papillomavirus virus-like particles and human papillomavirus virus-like particle immune complexes activate human Langerhans cells, *Vaccine*, 23, 1720, 2005.
59. Da Silva, D.M. et al., Uptake of human papillomavirus virus-like particles by dendritic cells is mediated by Fcgamma receptors and contributes to acquisition of T cell immunity, *J. Immunol.*, 178, 7587, 2007.
60. Selinka, H.C., Giroglou, T., and Sapp, M., Analysis of the infectious entry pathway of human papillomavirus type 33 pseudovirions, *Virology*, 299, 279, 2002.
61. Day, P.M., Lowy, D.R., and Schiller, J.T., Papillomaviruses infect cells via a clathrin-dependent pathway, *Virology*, 307, 1, 2003.
62. Kämper, N. et al., A membrane-destabilizing peptide in capsid protein L2 is required for egress of papillomavirus genomes from endosomes, *J. Virol.*, 80, 759, 2006.
63. Richards, R.M. et al., Cleavage of the papillomavirus minor capsid protein, L2, at a furin consensus site is necessary for infection, *Proc. Natl. Acad. Sci. U S A*, 103, 1522, 2006.
64. Day, P.M. et al., Establishment of papillomavirus infection is enhanced by promyelocytic leukemia protein (PML) expression, *Proc. Natl. Acad. Sci. U S A*, 101, 14252, 2004.
65. Yang, R. et al., Interaction of L2 with beta-actin directs intracellular transport of papillomavirus and infection, *J. Biol. Chem.*, 278, 12546, 2003.
66. Bossis, I. et al., Interaction of tSNARE syntaxin 18 with the papillomavirus minor capsid protein mediates infection, *J. Virol.*, 79, 6723, 2005.
67. Florin, L. et al., Identification of a dynein interacting domain in the papillomavirus minor capsid protein L2, *J. Virol.*, 80, 6691, 2006.
68. Florin, L. et al., Nuclear translocation of papillomavirus minor capsid protein L2 requires Hsc70, *J. Virol.*, 78, 5546, 2004.
69. Chromy, L.R. et al., Chaperone-mediated in vitro disassembly of polyoma- and papillomaviruses, *J. Virol.*, 80, 5086, 2006.
70. Breitburd, F. et al., Immunization with viruslike particles from cottontail rabbit papillomavirus (CRPV) can protect against experimental CRPV infection, *J. Virol.*, 69, 3959, 1995.
71. Kirnbauer, R. et al., Virus-like particles of bovine papillomavirus type 4 in prophylactic and therapeutic immunization, *Virology*, 219, 37, 1996.
72. Touzé, A. et al., Prevalence of anti-human papillomavirus type 16, 18, 31, and 58 virus-like particles in women in the general population and in prostitutes, *J. Clin. Microbiol.*, 39, 4344, 2001.
73. Konya, J. and Dillner, J., Immunity to oncogenic human papillomaviruses, *Adv. Cancer Res.*, 82, 205, 2001.
74. Viscidi, R.P. et al., Prevalence of antibodies to human papillomavirus (HPV) type 16 virus-like particles in relation to cervical HPV infection among college women, *Clin. Diagn. Lab. Immunol.*, 4, 122, 1997.
75. Touze, A. et al., The L1 major capsid protein of human papillomavirus type 16 variants affects yield of virus-like particles produced in an insect cell expression system, *J. Clin. Microbiol.*, 36, 2046, 1998.
76. Cubie, H.A. et al., Presence of antibodies to human papillomavirus virus-like particles (VLPs) in 11–13-year-old schoolgirls, *J. Med. Virol.*, 56, 210, 1998.

77. Combita, A.L. et al., Serologic response to human oncogenic papillomavirus types 16, 18, 31, 33, 39, 58 and 59 virus-like particles in colombian women with invasive cervical cancer, *Int. J. Cancer*, 97, 796, 2002.
78. Opalka, D. et al., Simultaneous quantitation of antibodies to neutralizing epitopes on virus-like particles for human papillomavirus types 6, 11, 16, and 18 by a multiplexed luminex assay, *Clin. Diagn. Lab. Immunol.*, 10, 108, 2003.
79. Waterboer, T. et al., Multiplex human papillomavirus serology based on in situ-purified glutathione *S*-transferase fusion proteins, *Clin. Chem.*, 51, 1845, 2005.
80. Pumpens, P. and Grens, E., HBV core particles as a carrier for B cell/ T cell epitopes, *Intervirology*, 44, 98, 2001.
81. Pumpens, P. and Grens, E., Artificial genes for chimeric virus-like particles, In: *Artificial DNA: Methods and Applications*, Khudyakov, Y.E. and Fields, H.A., (Eds.), CRC Press LLC, Boca Raton, Florida, 2002, Chapter 8.
82. Petry, H. et al., The use of virus-like particles for gene transfer, *Curr. Opin. Mol. Ther.*, 5, 524, 2003.
83. Xu, Y.F. et al., Papillomavirus virus-like particles as vehicles for the delivery of epitopes or genes, *Arch. Virol.*, 151, 2133, 2006.
84. Müller, M. et al., Chimeric papillomavirus-like particles, *Virology*, 234, 93, 1997.
85. Peng, S. et al., Papillomavirus virus-like particles can deliver defined CTL epitopes to the MHC class I pathway, *Virology*, 240, 147, 1998.
86. Nieland, J.D. et al., Chimeric papillomavirus virus-like particles induce a murine self-antigen-specific protective and therapeutic antitumor immune response, *J. Cell. Biochem.*, 73, 145, 1999.
87. Liu, W.J. et al., Papillomavirus virus-like particles for the delivery of multiple cytotoxic T cell epitopes, *Virology*, 273, 374, 2000.
88. Chackerian, B., Lowy, D.R., and Schiller, J.T., Induction of autoantibodies to mouse CCR5 with recombinant papillomavirus particles, *Proc. Natl. Acad. Sci. U S A*, 96, 2373, 1999.
89. Slupetzky, K. et al., Chimeric papillomavirus-like particles expressing a foreign epitope on capsid surface loops, *J. Gen. Virol.*, 82, 2799, 2001.
90. Zhang, H. et al., Induction of mucosal and systemic neutralizing antibodies against human immunodeficiency virus type 1 (HIV-1) by oral immunization with bovine Papillomavirus-HIV-1 gp41 chimeric virus-like particles, *J. Virol.*, 78, 8342, 2004.
91. Sadeyen, J.R. et al., Insertion of a foreign sequence on capsid surface loops of human papillomavirus type 16 virus-like particles reduces their capacity to induce neutralizing antibodies and delineates a conformational neutralizing epitope, *Virology*, 309, 32, 2003.
92. Christensen, N.D. et al., Hybrid papillomavirus L1 molecules assemble into virus-like particles that reconstitute conformational epitopes and induce neutralizing antibodies to distinct HPV types, *Virology*, 291, 324, 2001.
93. Kawana, K. et al., Common neutralization epitope in minor capsid protein L2 of human papillomavirus types 16 and 6, *J. Virol.*, 73, 6188, 1999.
94. Varsani, A. et al., Chimeric human papillomavirus type 16 (HPV-16) L1 particles presenting the common neutralizing epitope for the L2 minor capsid protein of HPV-6 and HPV-16, *J. Virol.*, 77, 8386, 2003.
95. Slupetzky, K. et al., A papillomavirus-like particle (VLP) vaccine displaying HPV16 L2 epitopes induces cross-neutralizing antibodies to HPV11, *Vaccine*, 25, 2001, 2007.
96. Ionescu, R.M. et al., Pharmaceutical and immunological evaluation of human papillomavirus viruslike particle as an antigen carrier, *J. Pharm. Sci.*, 95, 70, 2006.
97. Greenstone, H.L. et al., Chimeric papillomavirus virus-like particles elicit antitumor immunity against the E7 oncoprotein in an HPV16 tumor model, *Proc. Natl. Acad. Sci. U S A*, 95, 1800, 1998.
98. Da Silva, D.M., Schiller, J.T., and Kast, W.M., Heterologous boosting increases immunogenicity of chimeric papillomavirus virus-like particle vaccines, *Vaccine*, 21, 3219, 2003.
99. Xu, Y.F. et al., Encapsidating artificial human papillomavirus-16 mE7 protein in human papillomavirus-6b L1/L2 virus like particles, *Chin. Med. J. (Engl)*, 120, 503, 2007.
100. Qian, J. et al., Combined prophylactic and therapeutic cancer vaccine: Enhancing CTL responses to HPV16 E2 using a chimeric VLP in HLA-A2 mice, *Int. J. Cancer*, 118, 3022, 2006.
101. Peng, S., Zhou, J., and Frazer, I.H., Construction and production of fluorescent papillomavirus-like particles, *J. Tongji Med. Univ.*, 19, 170, 1999.

102. Tobery, T.W. et al., Effect of vaccine delivery system on the induction of HPV16L1-specific humoral and cell-mediated immune responses in immunized rhesus macaques, *Vaccine*, 21, 1539, 2003.
103. Wakabayashi, M.T. et al., Comparison of human papillomavirus type 16 L1 chimeric virus-like particles versus L1/L2 chimeric virus-like particles in tumor prevention, *Intervirology*, 45, 300, 2002.
104. Roth, C.M. and Sundaram, S., Engineering synthetic vectors for improved DNA delivery: Insights from intracellular pathways, *Annu. Rev. Biomed. Eng.*, 6, 397, 2004.
105. Khalil, I.A. et al., Uptake pathways and subsequent intracellular trafficking in nonviral gene delivery, *Pharmacol. Rev.*, 58, 32, 2006.
106. Georgens, C., Weyermann, J., and Zimmer, A., Recombinant virus like particles as drug delivery system, *Curr. Pharm. Biotechnol.*, 6, 49, 2005.
107. Touze, A. and Coursaget, P., In vitro gene transfer using human papillomavirus-like particles, *Nucleic Acids Res.*, 26, 1317, 1998.
108. Bousarghin, L. et al., Detection of neutralizing antibodies against human papillomaviruses (HPV) by inhibition of gene transfer mediated by HPV pseudovirions, *J. Clin. Microbiol.*, 40, 926, 2002.
109. Shi, W. et al., Papillomavirus pseudovirus: A novel vaccine to induce mucosal and systemic cytotoxic T-lymphocyte responses, *J. Virol.*, 75, 10139, 2001.
110. Fayad, R. et al., Oral administration with papillomavirus pseudovirus encoding IL-2 fully restores mucosal and systemic immune responses to vaccinations in aged mice, *J. Immunol.*, 173, 2692, 2004.
111. Oh, Y.K. et al., Enhanced mucosal and systemic immunogenicity of human papillomavirus-like particles encapsidating interleukin-2 gene adjuvant, *Virology*, 328, 266, 2004.
112. Zhang, H. et al., Human immunodeficiency virus type 1 gag-specific mucosal immunity after oral immunization with papillomavirus pseudoviruses encoding gag, *J. Virol.*, 78, 10249, 2004.
113. Kawana, K. et al., In vitro construction of pseudovirions of human papillomavirus type 16: Incorporation of plasmid DNA into reassembled L1/L2 capsids, *J. Virol.*, 72, 10298, 1998.
114. Zhou, J. et al., Synthesis and assembly of infectious bovine papillomavirus particles in vitro, *J. Gen. Virol.*, 74, 763, 1993.
115. Roden, R.B. et al., In vitro generation and type-specific neutralization of a human papillomavirus type 16 virion pseudotype, *J. Virol.*, 70, 5875, 1996.
116. Stauffer, Y. et al., Infectious human papillomavirus type 18 pseudovirions, *J. Mol. Biol.*, 283, 529, 1998.
117. Buck, C.B. et al., Efficient intracellular assembly of papillomaviral vectors, *J. Virol.*, 78, 751, 2004.
118. Roden, R.B. et al., Interaction of papillomaviruses with the cell surface, *J. Virol.*, 68, 7260, 1994.
119. Müller, M. et al., Papillomavirus capsid binding and uptake by cells from different tissues and species, *J. Virol.*, 69, 948, 1995.
120. Selinka, H.C. et al., Further evidence that papillomavirus capsids exist in two distinct conformations, *J. Virol.*, 77, 12961, 2003.
121. Rossi, J.L. et al., Assembly of human papillomavirus type 16 pseudovirions in Saccharomyces cerevisiae, *Hum. Gene Ther.*, 11, 1165, 2000.
122. Pyeon, D., Lambert, P.F., and Ahlquist, P., Production of infectious human papillomavirus independently of viral replication and epithelial cell differentiation, *Proc. Natl. Acad. Sci. U S A*, 102, 9311, 2005.
123. Culp, T.D. et al., Papillomavirus particles assembled in 293TT cells are infectious in vivo, *J. Virol.*, 80, 11381, 2006.
124. Mejia, A.F. et al., Preclinical model to test human papillomavirus virus (HPV) capsid vaccines in vivo using infectious HPV/cottontail rabbit papillomavirus chimeric papillomavirus particles, *J. Virol.*, 80, 12393, 2006.
125. Frazer, I.H., Prevention of cervical cancer through papillomavirus vaccination, *Nat. Rev. Immunol.*, 4, 46, 2004.
126. Mahdavi, A. and Monk, B.J., Vaccines against human papillomavirus and cervical cancer: Promises and challenges, *Oncologist*, 10, 528, 2005.
127. Lowy, D.R. and Schiller, J.T., Prophylactic human papillomavirus vaccines, *J. Clin. Invest.*, 116, 1167, 2006.
128. Lin, Y.Y. et al., Vaccines against human papillomavirus, *Front. Biosci.*, 12, 246, 2007.
129. Christensen, N.D. et al., Human papillomavirus types 6 and 11 have antigenically distinct strongly immunogenic conformationally dependent neutralizing epitopes, *Virology*, 205, 329, 1994.
130. Rose, R.C. et al., Serological differentiation of human papillomavirus types 11, 16 and 18 using recombinant virus-like particles, *J. Gen. Virol.*, 75, 2445, 1994.

131. Buonamassa, D.T. et al, Yeast coexpression of human papillomavirus types 6 and 16 capsid proteins, *Virology*, 293, 335, 2002.
132. McAleer, W.J. et al., Human hepatitis B vaccine from recombinant yeast, *Nature*, 307, 178, 1984.
133. Neeper, M.P., Hofmann, K.J., and Jansen, K.U., Expression of the major capsid protein of human papillomavirus type 11 in *Saccharomyces cerevisae*, *Gene*, 180, 1, 1996.
134. Cook, J.C. et al., Purification of virus-like particles of recombinant human papillomavirus type 11 major capsid protein L1 from *Saccharomyces cerevisiae*, *Protein Expr. Purif.*, 17, 477, 1999.
135. Koutsky, L.A. et al., A controlled trial of a human papillomavirus type 16 vaccine, *N. Engl. J. Med.*, 347, 1645, 2002.
136. Bryan, J.T., Developing an HPV vaccine to prevent cervical cancer and genital warts, *Vaccine*, 25, 3001, 2007.
137. Harper, D.M. et al., Sustained efficacy up to 4.5 years of a bivalent L1 virus-like particle vaccine against human papillomavirus types 16 and 18: Follow-up from a randomised control trial, *Lancet*, 367, 1247, 2006.
138. Baud, D. et al., Improved efficiency of a Salmonella-based vaccine against human papillomavirus type 16 virus-like particles achieved by using a codon-optimized version of L1, *J. Virol.*, 78, 12901, 2004.
139. Aires, K.A. et al., Production of human papillomavirus type 16 L1 virus-like particles by recombinant *Lactobacillus casei* cells, *Appl. Environ. Microbiol.*, 72, 745, 2006.
140. Hildesheim, A. et al., Effect of human papillomavirus 16/18 L1 viruslike particle vaccine among young women with preexisting infection: A randomized trial, *JAMA*, 298, 743, 2007.
141. DeFilippis, R.A. et al., Endogenous human papillomavirus E6 and E7 proteins differentially regulate proliferation, senescence, and apoptosis in HeLa cervical carcinoma cells, *J. Virol.*, 77, 1551, 2003.
142. Banchereau, J. et al., Immunobiology of dendritic cells, *Annu. Rev. Immunol.*, 18, 767, 2000.
143. Lenz, P. et al., Papillomavirus-like particles induce acute activation of dendritic cells, *J. Immunol.*, 166, 5346, 2001.
144. Zhang, L.F. et al., HPV6b virus like particles are potent immunogens without adjuvant in man, *Vaccine*, 18, 1051, 2000.
145. Schafer, K. et al., Immune response to human papillomavirus 16 L1E7 chimeric virus-like particles: Induction of cytotoxic T cells and specific tumor protection, *Int. J. Cancer*, 81, 881, 1999.
146. Kaufmann, A.M. et al., HPV16 L1E7 chimeric virus-like particles induce specific HLA-restricted T cells in humans after in vitro vaccination, *Int. J. Cancer*, 92, 285, 2001.
147. Kaufmann, A.M. et al., Vaccination trial with HPV16 L1E7 chimeric virus-like particles in women suffering from high grade cervical intraepithelial neoplasia (CIN 2/3), *Int. J. Cancer*, 121, 2794, 2007.
148. Freyschmidt, E.J. et al., Activation of dendritic cells and induction of T cell responses by HPV 16 L1/E7 chimeric virus-like particles are enhanced by CpG ODN or sorbitol, *Antivir. Ther.*, 9, 479, 2004.
149. Kuck, D. et al., Efficiency of HPV 16 L1/E7 DNA immunization: Influence of cellular localization and capsid assembly, *Vaccine*, 24, 2952, 2006.
150. Rudolf, M.P. et al., Human dendritic cells are activated by chimeric human papillomavirus type-16 virus-like particles and induce epitope-specific human T cell responses in vitro, *J. Immunol.*, 166, 5917, 2001.
151. Fausch, S.C. et al., Human papillomavirus virus-like particles do not activate Langerhans cells: A possible immune escape mechanism used by human papillomaviruses, *J. Immunol.*, 169, 3242, 2002.
152. Da Silva, D.M. et al., Effect of preexisting neutralizing antibodies on the anti-tumor immune response induced by chimeric human papillomavirus virus-like particle vaccines, *Virology*, 290, 350, 2001.
153. Hegde, R.S., The papillomavirus E2 proteins: Structure, function, and biology, *Annu. Rev. Biophys. Biomol. Struct.*, 31, 343, 2002.
154. Malboeuf, C.M. et al., Human papillomavirus-like particles mediate functional delivery of plasmid DNA to antigen presenting cells in vivo, *Vaccine*, 25, 3270, 2007.
155. Fang, N.X. et al., Post translational modifications of recombinant human papillomavirus type 6b major capsid protein, *Virus Res.*, 60, 113, 1999.
156. Hofmann, K.J. et al., Sequence conservation within the major capsid protein of human papillomavirus (HPV) type 18 and formation of HPV-18 virus-like particles in Saccharomyces cerevisiae, *J. Gen. Virol.*, 77, 465, 1996.
157. Pettersen, E.F. et al., UCSF chimera—A visualization system for exploratory research and analysis, *J. Comput. Chem.*, 25, 1605, 2004.
158. Shepherd, C.M. et al., VIPERdb: A relational database for structural virology, *Nucleic Acids Res.*, 34, D386, 2006.

12 Recombinant Vaccines against Hepatitis Viruses

Jonny Yokosawa, Néstor O. Perez, and Lívia M.G. Rossi

CONTENTS

12.1 INTRODUCTION

Hepatitis is an inflammation of the liver, commonly caused by a virus, and is an important cause of mortality and morbidity worldwide. Hepatitis A and E viruses (HAV and HEV, respectively) usually cause self-limited infection. The patients completely recover from hepatitis A and E after a few weeks of the onset of the disease symptoms [1,2]. The HAV and HEV infections are especially prevalent in developing countries, where poor sanitation promotes transmission of these viruses. On the other hand, hepatitis B and C viruses (HBV and HCV, respectively), mainly transmitted via contaminated body fluids, may cause chronic disease and lead to more severe complications such as liver cirrhosis and hepatocellular carcinoma [3,4]. In any case, viral hepatitis has an important impact on public health and, although measures to control virus transmission have been used, vaccination is the most effective practice. Therefore, enormous efforts have been made to develop vaccines to prevent viral hepatitis. Currently, vaccines are available only against HAV and HBV. The first hepatitis vaccine available, against hepatitis B, was formulated using empty particles containing hepatitis B surface antigen (HBsAg) purified from plasma of patients. The use of this plasma derived vaccine raised concerns about its safety. With the advance of recombinant DNA methodology, current HBV vaccine was developed by expressing HBsAg in yeast cells and, until 2007, when a vaccine against human papillomavirus infection was approved, it was the only recombinant vaccine available. Although this vaccine is highly immunogenic, offering protection to approximately 95% of vaccinees, some individuals do not respond or respond poorly, leading researchers to improve the current vaccine by adding other epitopes to the HBsAg and increase its immunogenicity.

Due to the difficulties in producing current HAV vaccine, which is made with inactivated or attenuated virus, development of a recombinant HAV vaccine seems in order. However, HAV neutralizing epitopes are conformational dependent. For this reason, development of an expression system for the production of HAV proteins that assemble into virus-like particles (VLP) for the correct presentation of these epitopes is important, although it is a challenging task.

Similarly, although a vaccine is not yet commercially available to prevent hepatitis E, there are significant advances in the development of such vaccine using truncated HEV capsid proteins self-assembled into particles or aggregates. These vaccine candidates elicit a strong immune response in vaccinated nonhuman primates; however, after challenge, some of the animals acquire infection, but did not present disease symptoms. Although not as efficient against the disease this vaccine may be important to reduce viral transmission.

One of the difficulties in developing an effective vaccine against HCV is related to the high variability of this virus. One of the most characterized neutralizing antigenic epitopes is located in the HCV hypervariable region 1 (HVR1). Researchers are still looking for more conserved neutralizing determinants in HCV proteins. It is conceivable that several yet unknown epitopes with ability to elicit neutralizing immune response may exist. Some of these epitopes may offer protection against a number of different HCV variants and, therefore, may be used for engineering HCV vaccine. However, current knowledge of the HCV neutralizing responses does not allow for a clear prediction of the way to the robust HCV vaccine.

The development of recombinant hepatitis vaccines was mainly focused on addition or removal of amino acid (aa) regions from prospective immunogens. This chapter reviews some of the developments in this field.

12.2 RECOMBINANT HEPATITIS A VACCINE

Hepatitis A virus (HAV), the etiological agent of hepatitis A, was first visualized by immune electron microscopy more than 30 years ago [5]. By early 1990s, a safe and robust inactivated HAV vaccine became available and is still currently used [6,7]. This vaccine led to a declining interest in the study of the hepatitis A viral pathogenesis and molecular virology [1]. However, due to the fact that the HAV vaccine based on the cell culture derived inactivated [6,7] or attenuated [8] virus has a list of inherent problems, mainly associated with safety, cost, and manufacture, the idea of engineering and development of the HAV recombinant vaccine still attracts although a small but enthusiastic group of researchers who face a unique and interesting challenge.

HAV produced in cell culture is used for either vaccine or diagnostic purposes. The virus does not cause cytopathic effects, except for some strains, and has a slow growth and relatively low yield [9–11]. The cumbersome and expensive way of HAV production in conjunction with increased knowledge of the HAV biology continues to stimulate the efforts toward development of a vaccine based on recombinant viral structural proteins (VP0, VP1, and VP3). One of the main advantages of a recombinant vaccine is a reduced cost for its manufacturing and, as a consequence, the increased availability of HAV vaccine to the resource limited parts of the world for the implementation of broad immunization programs.

Indeed, HAV is still a serious public health problem around the world with close to 1.5 million [1] cases reported annually. The clinical manifestation of HAV infection in humans varies from asymptomatic infection to fulminant hepatic failure [12–14].

HAV has been classified within the family Picornaviridae, in the genus *Hepatovirus*. HAV genome is a positive-strand RNA of approximately 7.5 kb and contains a 5′ noncoding region (NCR) with an identified internal ribosome entry site (IRES), a single open-reading frame encoding a large polyprotein (P0) and a 3′ NCR connected to a poly(A) sequence. The polyprotein P0 of 2,227 aa is divided into P1, P2, and P3 regions. P1 region encodes the structural (capsid) proteins—1A (also known as VP4), 1B (VP2), 1C (VP3), and 1D (VP1). VP4–VP2 is also known as VP0, a structural precursor for VP2 and VP4. VP4 is a small protein of 21–23 aa long and is cleaved from VP2 in the final steps of viral morphogenesis [14]. P2 and P3 regions encode nonstructural proteins required for HAV RNA replication: 2B, 2C, 3A, 3B (a small protein, also known as VPg, located at the 5′ end of the genomic RNA and probably serves as the primer for RNA synthesis), 3C (a protease responsible for the cleavage within the polyprotein), and 3D (RNA-dependent RNA polymerase). In this chapter, we pay a special attention to P2A, because unlike other picornaviruses, whose P2A is a protease, in HAV it is a structural protein which has an essential role in the viral capsid assembly [1,14]. Similar to other picornaviruses, HAV RNA is packaged within a nonenveloped icosahedral capsid composed of 60 copies of each of 3 major structural proteins, VP1, VP2, and VP3 (P1 region).

In the natural course of HAV infection, nonstructural proteins can induce antibody response as an indicator of viral replication, but these antibodies are short-lived and non-neutralizing [15–17].

HAV neutralizing epitopes have been mapped in VP1 and VP3 by using HAV mutants that escaped neutralization by monoclonal antibodies [18–20].

HAV structural proteins have been expressed in *Escherichia coli* [21–25]. These proteins induced production of specific antibodies; however, they were not able to elicit high-titer neutralizing antibodies. Although individual structural proteins are immunogenic, only the HAV particle can induce robust neutralizing antibodies. This observation was interpreted that the critical epitopes within the immunodominant neutralization site of HAV are conformation-dependent and require assembly of the structural proteins into subviral particles or virus-like particles (SVP and VLP, respectively) [26,27].

The HAV specific antigenic material produced by growing HAV in cell culture predominantly sediments at 156S. It was shown that this material represents the HAV infectious virions. The 70S antigenic material was identified as empty viral particles [28]. HAV morphogenesis involves several important steps. After 3C protease-mediated primary cleavage, the VP0, VP3, and VP1-2A proteins are associated into a unit containing one copy of each protein. Five of these units assemble into 14S pentamers. These 14S particles are capable of eliciting a neutralizing immunoresponse, which indicates that at this stage of assembly the structural proteins form some conformational neutralizing antigenic epitopes [27]. Twelve pentamers aggregate to assemble the procapsid 70S, which exposes additional epitopes. In the presence of newly synthesized viral RNA a pre-provirion is formed [27,29,30]. At this moment, the precursor VP0 in the provirion 130S is cleaved into VP2 and VP4 to produce the 156S infectious virion [29–32]. The VP4/VP2 cleavage occurs only after the encapsidation of HAV full-length genomic RNA, a crucial step required for the HAV maturation [33].

The identification of a single immunodominant neutralization antigenic site that is strictly conformation dependent stirred researchers from simple protein expression experiments to modeling HAV morphogenesis in different heterologous expression systems to obtain HAV-like particles (HAV-LP) preserving a fully functional neutralizing epitopes suitable for the development of recombinant HAV vaccines.

Recombinant 14S pentamers and 70S empty capsids were obtained by expressing the complete HAV open reading frame in *E. coli* [34–36]. Both structures were immunogenic in mice [36]. Inclusion bodies that contained only 14S pentamers did not elicit a strong immune response. Sera from mice immunized with 14S particles demonstrated only 66% HAV neutralization at dilution 1:4, while the response to the mixture 14S/70S showed 73%–90% neutralization at dilutions 1:500 to 1:2000. The major setback for this expression system was, though, a low yield of HAV-LP, which, according to the authors, was most probably caused by incorrect folding of the HAV capsid. The other factor that could contribute into a low level of HAV-LP production and incorrect protein folding is dissimilarity in codon usage between HAV genome and *E. coli* [36].

Insect cell-based protein expression system was found to be more efficient than *E. coli* for the HAV-LP production. This system has several important advantages. Besides being easily amenable to scale-up, this eukaryotic system is capable of producing large amounts of recombinant proteins in high-density cell culture conditions, resulting in high recovery of correctly folded antigens. Also, compared to mammalian cell expression systems, baculovirus expression system is considered to be safer because the insect cells can be cultured without mammalian cell-derived supplements. Despite these advantages, proteins expressed using this system have not yet been approved for the human use.

The HAV proteins expressed in insect cells were used to study HAV morphogenesis and evaluate their immunological properties [37–39]. These proteins expressed from the full-length HAV ORF were found to be correctly processed and formed multiprotein aggregates or particles with buoyant density similar to the HAV empty capsids. It was shown that these particles are fully immunogenic. Mice immunized with these HAV particles produced antibodies that showed neutralizing activity in a plaque reduction assay [40].

Expression of HAV proteins has been done also in HeLa cells using the vaccinia virus system [41,42]. After expression of the entire HAV polyprotein in these cells the authors found that the HAV structural proteins were properly processed and organized into VLP's that could be detected using immune electron microscopy. Unfortunately, the authors did not test immunogenic properties of these particles. Using this expression system, Probst and colleagues [43] made a very important observation that brings a new light on the important role of P2A in morphogenesis of mature HAV virions and opens a door for engineering HAV proteins that can be organized into VLP's maintaining the neutralizing antigenic epitopes. The authors elucidated the role of 3C protease in the processing of P1-2A and viral morphogenesis [43]. As mentioned above, P2A plays an important role at the initial steps of the HAV particle assembly. However, P2A is not present in mature virions excreted by infected individuals, which suggests that this protein is removed by unknown mechanism from HAV particles before completion of the virion maturation. It seems that this final process in the virion maturation can be modeled, albeit with variable efficiently, in different heterologous expression systems. To get control over this important step, Rachow and colleagues [44] engineered the factor Xa cleavage site at the VP1-2A junction and demonstrated an efficient assembly of HAV particles after removal of P2A using the protease Xa cleavage. However, there is no data on comparison of immunogenic and neutralizing properties of HAV-LP with or without 2A. Although theoretically interesting, the practical significance of such expression system with the artificial protease cleavage remains to be evaluated.

12.3 RECOMBINANT HEPATITIS B VACCINE

Approximately 2 billion people have serological evidence of past or present hepatitis B virus (HBV) infection and 350 million people are chronically infected (reviewed in Ref. [4]). HBV infection is a major cause of cirrhosis and hepatocellular carcinoma (HCC), leading to over 1 million deaths each year. HBV is transmitted by percutaneous and mucous membrane exposures to contaminated blood and body fluids that contain blood. First generation HBV vaccine, developed in the late 1970s, contained hepatitis B virus surface antigen (HBsAg) particles obtained from plasma of infected individuals (reviewed in Ref. [45]). One early vaccine, produced in France [46], contained particles with all three forms of the surface antigens (HBsAg) [47]: Small-size antigen (SHBs) is 226 aa long. It is the major protein present in the particle; middle-size antigen (MHBs) shares the 226 aa sequence with SHBs and contains an additional 55 aa sequence, known as the preS2 domain, at the N-terminus; and large-size antigen (LHBs) contains, in addition to SHBs and preS2 aa sequences, the N-terminal preS1 domain of 108 or 119 aa, depending on the HBV subtype. Another plasma derived HBV vaccine produced in the United States contained only SHBs [48]. Significant concerns about safety of plasma-derived HBV antigens, which could be contaminated with infectious hepatitis B particles or other viral pathogens, as well as advances in recombinant DNA techniques led to the development of a recombinant HBV vaccine. Although expression of a recombinant HBsAg was first reported in a bacterial system, this expression system proved to be highly inefficient due to a low level of expression of a poorly immunoreactive antigen [49]. Only recently, *E. coli* was successfully used for the expression of HBsAg, however, only after the adjustment of codons to the preferred *E. coli* codons and fusion of HBsAg to anthrax lethal factor [50]. This hybrid protein was used as a potential candidate for a therapeutic vaccine eliciting cell-mediated immune responses.

SHBs expressed in *Saccharomyces cerevisiae* forms particles that resemble HBsAg particles obtained from HBV infected patients [51]. Most probably, the assembly into VLP is the major feature of this antigen that makes it highly immunogenic [52]. It seems that yeast cells as many other eukaryotic expression systems are completely suitable for correct folding of HBsAg and many other eukaryotic proteins. The yeast-expressed HBV-LP has been used to formulate a highly efficient vaccine. In early experiments, it was shown that immunization with the recombinant yeast-derived HBsAg prevents chimpanzees from HBV infection [53]. Properties of yeast-derived SHBs and its application for HBV vaccine production were thoroughly reviewed elsewhere [54]. Many molecular parameters of the recombinant SHBs such as aa composition, N- and C-terminal sequences and pattern of migration in SDS-PAGE were found identical to the native antigen. The recombinant SHBs was extensively tested using a panel of monoclonal antibodies (MAbs) to show the presence of all epitopes [55]. This antigen was also proven to be highly immunogenic in a variety of animals and in humans [56]. In 1986, the yeast recombinant hepatitis B vaccine was licensed for use in humans. It was also the first vaccine developed by recombinant DNA technology [54].

Vaccination with the yeast-derived vaccine produces seroprotection in up to 95% of immunized individuals [57] and elicits neutralizing antibodies that protect against all HBV subtypes. The neutralizing antibodies are directed against the main HBsAg hydrophilic region representing a common determinant "a" (reviewed in Ref. [58]).

Despite the high immunogenicity of the yeast recombinant vaccine, it was found to be less immunogenic than the plasma-derived vaccine [59]. Approximately 5% of vaccinated individuals, including subjects such as elderly people, smokers, overweight people, and even some healthy individuals are non-responders to the available recombinant vaccine and are at risk of being infected with HBV [45,60,61]. The risk is especially high for the health care workers who are more often exposed to contaminated blood.

This issue of non-responders to the yeast-derived SHBs vaccine was addressed by engineering novel proteins containing additional neutralizing antigenic epitopes [62–68]. PreS1 domain contains T- and B-cell epitopes and elicits antibodies earlier than SHBs in experimentally infected chimpanzees [62,64]. It also contains a sequence involved in the attachment of the virus to hepatocytes and elicits neutralizing antibodies [66,67]. PreS2 domain contains a binding site for human serum albumin [69,70], which in turn binds to hepatocytes. This domain also contains a highly immunogenic epitopes. Incubation of HBV with antibodies from rabbits immunized with a synthetic peptide based on this region protected chimpanzees from infection [66]. Also, chimpanzees immunized with a preS2 peptide were protected upon challenge [63].

Tests carried out with the third-generation vaccines containing preS2 and/or preS1 antigenic regions have shown advantageous immunogenic properties. Experiments with mouse strains that are non-responders to the SHBs showed that immunization with particles containing SHBs and MHBs antigens elicited antibodies to preS2 and SHBs in most of them [71]. Also, antibodies to preS2 domain were detected earlier and at higher titer in comparison to anti-SHBs. Similarly, particles containing S, M, and LHBs showed an increased seroconversion rate to S domain and higher anti-HBs titer [72–75]. Results with human subjects have also shown the increased immunogenicity of S/MHBs and S/M/L/HBs vaccines [68,76–78]. The increased immunogenicity of these antigens, however, seems attributable to their expression in mammalian cell systems, since yeast derived HBsAg is less immunogenic. Yeast cells have a different pattern of posttranslational modifications, including protein folding and glycosylation. Additionally, HBV particles obtained by mammalian cell expression system are secreted while those produced by yeast cells are expressed intracellularly. The yeast cells need to be disrupted to release particles. Despite these disadvantages, though, yeast-derived MHBs was found to produce a faster seroconversion rate in non-responders in clinical trials [79–81]. LHBs expressed using a yeast secretion system also showed a rapid seroconversion rate and higher titer of anti-SHBs antibodies [82].

The issue of improvement of the recombinant HBsAg immunogenicity was also recently addressed using a novel molecular technique of DNA shuffling or molecular breeding [83]. This technology models the evolutionary process in vitro that involves an iterative application of DNA recombination and selection for recombinant genes with improved selectable properties. The authors used a large set of the HBV S gene sequence variants to select for a recombinant gene with improved immunogenicity. Although not yet thoroughly validated this technology has a significant potential to generate an HBV vaccine with a high rate of immunological response among vaccinated individuals.

A serious concern related to the vaccine containing only the small HBs is the appearance of the so-called vaccine escape mutants. These mutants have been first detected in children born from HBV-carrier mothers and, despite vaccination and immunoglobulin therapy, some of them developed hepatitis [84–87]. With the increasing efforts to establish global immunization programs to prevent hepatitis B, the selection pressure favors HBV mutants that can escape vaccine-induced immunity. Thus, vaccine may become ineffective against such mutants and the number of hepatitis B cases may rebound. Additionally, such mutants may also escape detection by immunodiagnostic assays. Indeed, yeast-expressed HBs antigens carrying the most common "escape" mutations showed different reactivity patterns when tested against a panel of monoclonal antibodies directed to the determinant "a" [88,89]. Furthermore, some commercially available kits do not detect G145R (glycine-to-arginine substitution at position 145 of the small HBsAg), the most common mutant HBs antigen [90]. One study demonstrated that a currently available commercial vaccine can protect against infection with this mutant [91]. However, many researchers take this information with a very careful optimism. It is still conceivable that there are some HBV mutants that are less sensitive to neutralization with available vaccines. Further surveillance needs to be conducted to investigate the protectiveness of these vaccines against other HBs mutants. Therefore, since it remains unknown whether or not conventional vaccines, i.e., SHBs-containing vaccines, would be effective against these other mutants, third generation vaccines may offer a higher level of protection. Even though that there are concerns about presence of proto-oncogene in the antigen preparations, these vaccines have proven safe in adults, children, and neonates [68,76–78]. However, due to its high cost, it has been suggested to recommend these vaccines only for those people who do not respond to the conventional vaccines.

Finally, due to the ability to form VLPs with enhanced immunogenicity [92,93], HBs antigen has been explored as a carrier for epitopes of other viruses (see Chapter 9), such as herpes simplex virus [94], poliovirus [95–97], hepatitis C virus (HCV) [98,99], and HIV [100]. Such chimeric antigens form particles that upon immunization elicit antibodies to both HBs and foreign epitope. However, the degree of HBV specific immunoresponse depends on the site of insertion of epitopes in the HBsAg sequence. When foreign sequences are inserted at the N-terminus of HBsAg, both HBsAg and foreign epitope are immunogenic [100]. However, immunogenicity to the HBsAg part is lost or lowered when sequences are inserted within or near the determinant "a" [95,98]. These observations offer an opportunity to formulate combined vaccines against HBV and other pathogen, whose neutralizing epitope is inserted at the N-terminus of HBsAg. For those individuals that have already been vaccinated against HBV, a vaccine can be formulated with the epitope inserted within the determinant "a" of HBsAg [101].

12.4 RECOMBINANT HEPATITIS C VACCINE

Hepatitis C virus is an important cause of morbidity and mortality; around 170 million persons are chronically infected worldwide and there are 3 to 4 million new cases per year [3]. HCV has been associated with chronic liver disease, cirrhosis, and liver cancer. Approximately 70%–80% of hepatitis C cases progresses to chronicity, about 20% of patients with chronic hepatitis C develop liver cirrhosis, and 1%–4% of these patients develop hepatocellular carcinoma (reviewed in Ref. [102]). The virus is transmitted via the parenteral route, primarily by direct contact with human blood [103].

Vaccination of risk groups exposed to HCV infection, including health care workers, clinical laboratory technicians, paramedics, sex workers, and intravenous drug users, is of utmost importance. The first antiviral treatment based on use of interferon-α (IFN-α) has only limited success. With the development of pegylated (polyethylene glycol) interferon in combination with ribavirin, the sustained response rate increases to 54%–56% [104]. However, the combination treatment is expensive and is associated with side effects. HCV is an enveloped RNA virus [105] that belongs to the family Flaviviridae. The HCV genome is a linear positive sense RNA of about 9.5 kb that encodes a polyprotein of about 3000 aa [106]. The polyprotein is processed into 10 mature proteins [107]. The structural proteins, capsid or core, envelope glycoproteins E1 and E2, are cleaved from the polyprotein by host proteases [108]. HCV genome is highly heterogeneous. HCV has been classified into six genotypes that differ considerably from one another and show different geographic distribution [109]. The primary structure of the structural glycoproteins is most variable between HCV genotypes [110]. The N-terminus of the HCV E2 protein contains the HVR1. This region contains the HCV neutralizing epitopes [111]. It was hypothesized that sequence variation within HVR1 is related to HCV immune escape [112]. Only 20%–30% of infected persons resolve the infection. Viral clearance in acutely infected persons is associated with a strong $CD4^+$ and $CD8^+$ helper T-cell response (reviewed in Refs. [113–115]). All these findings clearly demonstrate that HCV infection can be in some cases controlled by antiviral immune response. However, harnessing these immune responses to achieve protection against HCV infections is a difficult task. HCV seems to explore several potential mechanisms to escape host immune responses [116]. A continuous generation of escape variants [117] make the development of an effective vaccine a great challenge for the scientific community.

Core protein is the most conserved protein of HCV. It is a non-glycosylated 191 aa-long protein [118]. Since core is highly immunogenic, it was considered as a potential candidate for vaccine research. This protein was expressed in *E. coli* as a GST-fusion protein [119] or by itself [120,121] as well as in yeast [122]. The expression of core protein under the T7 promoter in *E. coli* [120] or AOX promoter in *Pichia pastoris* [122] leads to formation of VLPs that can be visualized using electron microscopy and characterized using buoyant density gradient centrifugation. The expression of just the N-terminal 79 aa of core protein in *P. pastoris* results in forming nucleocapsid-like particles [123]. The *E. coli*-produced core protein adsorbed onto ISCOMATRIX (a structure composed of saponins, cholesterol, and phospholipids) induces strong $CD4^+$ and $CD8^+$ T-cell responses as well as a high antibody titer, suggesting its use as adjuvant for the HCV envelope proteins E1E2 [121]. However, no experiments to prove its neutralizing properties have been pursued.

Recombinant HCV glycoproteins E1 and E2 were the first HCV antigens explored as vaccine candidates in experiments on chimpanzees [124]. Both proteins were expressed in HeLa cells using a vaccinia virus expression system. The animals that had high antibody titers to E1 and E2 at the time of challenge were fully protected against the homologous HCV strain. In contrast, two animals with the lowest antibody titers became infected. However, in one of them HCV detection after challenge was delayed, the virus load was low and the animal showed a minor ALT elevation. Both animals, however, were HCV RNA negative at the end of the study, i.e., 35 weeks after the challenge. Similarly, a chimpanzee immunized with the recombinant E1/E2 produced using Sindbis virus expression system showed a shorter and milder viremia following challenge with homologous strain [125].

A therapeutic vaccine, based on E1 glycoprotein only, was tested in chronically infected patients [126]. After a second round of vaccination humoral responses against E1 were detected together with a relative decrease in hepatic fibrosis and ALT levels. However, HCV-RNA levels in blood remained unchanged.

The HCV E2 protein has been also expressed alone in mammalian cell system [127] and used for immunization [128]. The C-terminal transmembrane region was removed to facilitate secretion of this protein. Nevertheless, some fraction of the E2 protein remained intracellular. Both fractions were highly immunogenic in guinea pigs, but only the intracellular form elicited antibodies that prevented binding of recombinant E2 to CD81 onto human cells, probably due to a less complex glycosylated structure [128,129]. CD81, a member of the tetraspannin family of proteins, is a component of the HCV receptor [130].

Both E1 and E2 glycoproteins were also expressed in bacterial cells [131]. The transmembrane regions were removed from both proteins. These proteins were expressed either separately or as a polyprotein in tandem. The chimeric E1E2 protein was found to be more immunogenic in immunized rabbits than E1 and E2 proteins expressed separately. Interestingly, antibodies elicited against the chimeric E1E2 protein inhibited binding of E2 protein to the recombinant CD81.

Several truncated HCV polyprotein constructions were expressed in yeast cells [132]. A set of truncated E2 proteins fused to the sucrose invertase 2 leader sequence to facilitate secretion into the culture medium was successfully expressed and all proteins were tested positive with HCV specific monoclonal and polyclonal antibodies. However, none of the constructs was evaluated for immunogenicity or neutralizing activity [132].

The use of VLP is a very attractive approach to vaccine development because VLP's closely resemble the structure of virions. Therefore, significant efforts were made to generate and study immunological properties of HCV-LP. It was found that expression of core, E1, and E2 proteins in insect cells results in forming HCV-LP. Strong humoral and cellular immunoresponses were detected in mice immunized with these HCV-LP's [133]. After challenge with a recombinant vaccinia expressing HCV structural proteins the HCV-LP immunized mice demonstrated a noticeable reduction in vaccinia titer [134].

Since the discovery of the HCV neutralizing epitopes located in the HVR1, this region of the HCV E2 protein became a target for vaccine designs. Several lines of research prove that antibody to HVR1 are important for the viral clearance. Early appearance of HVR1 antibodies was associated with self-limiting hepatitis C [135–137]. HVR1 specific antibodies have been shown to prevent infection with homologous HCV strains in chimpanzees [111,138] and block HCV binding in cell culture [139]. However, protection was not observed against HCV with heterologous HVR1 variants present in the specimen used for challenge [111,140]. On the other hand, despite high sequence variability, some HVR1 variants have shown various degree of cross-immunoreactivity [141,142], suggesting that one or several broadly cross-reactive HVR1 sequences could be used as vaccines.

As was described above (see also Chapter 9), HBsAg is a very strong immunogen when assembled into VLP [143,144]. These advantageous properties have been explored by using HBsAg as a carrier for expression of HCV HVR1 sequences. Lee and colleagues [98] inserted individual HVR1 sequences and other E2 protein domains at aa position 113 of HBsAg expressed using vaccinia virus system. The results showed that all HCV domains were expressed on the outer surface of the HBsAg particles. This is a very interesting and encouraging study. Unfortunately, the authors did not perform any immunogenicity experiments. In another study, Netter and colleagues [99] inserted HVR1 sequences of different length at position 127 within the "a" determinant of HBsAg. In this case, HVR1 inserts showed immunoreactivity with human anti-HCV positive sera and were found immunogenic in mice [99,145]. Antibodies against these constructs were used to block HCV pseudoparticles (HCVpp) from binding to Huh7 cells. It was found that these antibodies block HCVpp containing either homologous or heterologous HVR1 sequences [145].

To overcome sequence heterogeneity of HVR1, several attempts were made to generate HVR1 mimotopes [141,146,147]. In one study, an HVR1 consensus sequence selected using the phage display technology was shown capable of inducing cross-neutralizing antibodies against different HCV variants [141]. This consensus HVR1 was fused to the B subunit of cholera toxin (CTB) and expressed in plants [148]. The HVR1 nucleotide coding sequence was designed using plant optimized codons and was inserted at the 3′-end of the CTB gene. After immunization of mice the HVR1/CTB chimeric protein elicited antibodies that showed reactivity against insect cell-derived HCV-LP (the HCV-LP binding assay described in Refs. [149,150]). However, neutralization properties of this protein remain to be tested.

12.5 RECOMBINANT HEPATITIS E VACCINE

Hepatitis E is an enterically transmitted disease caused by HEV and affects mainly developing countries, causing a number of large epidemics [151–153]. Transmission in these outbreaks probably occurs via consumption of contaminated water. In countries with low prevalence, in cases that are not linked to travel history to endemic areas, HEV is considered to be zoonotic, especially among farm workers who handle pigs and might be more exposed to swine HEV. Indeed, human and swine HEV are genetically closely related and some human HEV strains were capable of infecting pigs [154,155]. HEV is a nonenveloped virus that was recently reclassified as a species of the new *Hepevirus* genus within the Hepeviridae family [156] (http://www.ncbi.nlm.nih.gov/8th report ICTV/). It contains a positive strand RNA genome of approximately 7.2 kb and has three open-reading frames (ORFs) [2,157]. ORF1 is about 5 kb long and codes for non-structural proteins that contain motifs of methyltransferase, protease, helicase, and RNA-dependent RNA polymerase. ORF2 is about 2 kb long and codes for a protein of 660 aa that forms the virus capsid. ORF3 is 369 bases long, overlaps partially with ORF1 and ORF2 and codes for a small protein (123 aa) of unknown function. So far, there is no efficient cell culture system for replication of HEV and, thus, it has not been possible to develop an attenuated or killed virus vaccine.

Recombinant ORF2 and ORF3 proteins (pORF2 and pORF3, respectively) have been used as reagents for immunodiagnostic assays [158]. On the other hand, because of a low neutralizing activity of pORF3 [159], pORF2 has been the main target for vaccine development. Antigenic properties of pORF2 were thoroughly characterized [160–162]. Full-length and truncated ORF2 proteins have been successfully expressed in both *E. coli* and baculovirus expression systems. In both systems, pORF2 assembles into VLPs, which is an important characteristic for maintaining the structural and immunogenic properties of an antigen [52,163].

The first experiments demonstrating a neutralization potential of the pORF2 protein expressed in bacteria were conducted almost right after the HEV discovery [164]. This antigen contained the C-terminal two-thirds of pORF2 attached to tryptophan synthetase as a carrier. Cynomolgus macaques immunized with this protein were completely protected against challenge with the homologous (Burmese strain, genotype 1) HEV strain. However, the animal challenged with a heterologous (Mexican strain, genotype 2) strain exhibited HEV RNA in its stool and HEV antigen in its liver. The neutralization potential of this antigen was additionally investigated in a polymerase chain reaction (PCR)-based cell culture neutralization assay [165]. Serum samples of guinea pigs immunized with the same antigen could efficiently block binding of HEV strains from Burma, Pakistan (both of genotype 1) and Mexico (genotype 2), but failed to neutralize the HEV Morocco strain (also genotype 1) [166].

Using the PCR-based cell culture neutralization assay and a large set of truncated recombinant ORF2 proteins of various sizes, Meng and collaborators [162] mapped the HEV neutralizing epitope within the region of 166 aa at position 452–617 near the C-terminus of pORF2. This protein, named pB166, was expressed as a fusion with glutathione *S*-transferase (GST). The GST part can be removed using a specific protease before immunization. A similar protein, HEVp179, containing additional 13 aa at the N-terminus (aa residues 439–617) of pB166 elicited neutralizing antibodies against HEV in immunized mice [163]. It was shown that HEVp179 forms VLP upon expression in bacteria, which most probably explains its high immunogenicity.

Although there are many examples of expression of pORF2 in the immunogenic form in bacterial cells, the early observations of assembly of pORF2 into VLP after expression in insect cells digressed the field of HEV vaccine development toward the use of the baculovirus expression system. Tsarev and collaborators [167] expressed the complete ORF2 of the HEV Pakistani strain in insect cells and observed that the expressed protein assembled into VLP's, which were secreted into the cell culture medium. The full-length pORF2 is about 72 kDa. However, this protein is rapidly processed in insect cells into a 53 kDa protein (aa residues 112–578) that forms the secreted VLP [168]. Another product of baculovirus expression, a 56 kDa protein, accumulates in the insect cell cytoplasm. This product was used to immunize cynomolgus macaques [169]. All animals in this experiment developed HEV specific antibodies and did not show any signs of hepatitis following challenge. However, while two animals that received two 50 μg antigen doses did not show any signs of HEV infection, three of the four animals that received only one dose excreted virus in feces. The intracellular product of the ORF2 expression in insect cells is approximately 3 kDa bigger than the secreted protein. It was suggested that this difference is due to glycosylation [168,170]. Indeed, three potential N-linked glycosylation sites have been identified [171], but their importance for the VLP structure and immunogenicity is not known.

The 56 kDa antigen was further evaluated as vaccine in a preclinical trial involving 44 rhesus macaques [172]. One or 10 μg in one or two doses were administered intramuscularly. For challenge, three HEV variants were used: Pakistani (genotype 1, homologous to the vaccine), Mexican (genotype 2), and United States (genotype 3) strains. All animals vaccinated with the 56 kDa antigen developed anti-HEV antibodies. Those that received two doses of the vaccine, regardless of the amount of antigen (1 or 10 μg/dose), showed higher anti-HEV titers and did not show signs of disease. However, HEV RNA was detected in serum samples of some of these animals. The infection occurred more often in animals challenged with the homologous and Mexican strains than with U.S. strain. On the other hand, those monkeys that received only one dose of 10 μg of immunogen showed lower antibody titers and following the challenge were more prone to infection. Interestingly, two 1 μg dose regimen was more effective in preventing infection than with two 10 μg dose regimen (83% and 64% protection, respectively). Both regimens were 100% effective in preventing hepatitis, while one-dose regimen was 78% effective in preventing the disease and only 25% in preventing the infection. This vaccine was tested on volunteers in the United States and proved to be safe and immunogenic [173]. Its efficacy and safety was further evaluated in a phase 2 trial on 2000 healthy adults in Nepal, a high-risk area for HEV infection [174]. The subjects received three doses of either the recombinant HEV (rHEV) vaccine containing 20 μg antigen per dose or placebo. For a period of over 2 years after the immunization 66 subjects (out of 896 subjects) in the placebo group and only 3 subjects (out of 898) in the vaccinated group developed hepatitis E. The vaccine efficacy was estimated to be 95.5% (95% confidence interval, 85.6–98.6).

Another group of researchers also expressed a truncated pORF2 (112–660 aa) in insect cells [175] and used it for immunization of cynomolgus macaques [176]. Two animals were immunized by oral administration of five doses containing 10 mg of antigen without adjuvant. Anti-HEV IgM and IgG in the serum and IgA in the stool were detected indicating that the HEV-LP was immunogenic when delivered orally. Following challenge, the animals did not show any signs of hepatitis. However, HEV RNA was detected in stool of the monkey with a lower IgG titer at the time of challenge.

Although the HEV vaccine development field is dominated by antigens derived from insect cells, the baculovirus expression system is not as widely accepted for the development of vaccine for human use as yeast expression system. Recently, pB166 derived from the HEV Moroccan strain was expressed in yeast cells using *Hansenula polymorpha* as a host (Yokosawa, personal observation). This protein was fused to the yeast mating-type α factor to facilitate secretion of pB166 in the media. The most interesting observation is that the secreted pB166 is assembled into VLP, which may additionally aggregate into VLP clusters. The preparations of this protein are highly immunogenic in mice and elicit antibodies immunoreacting with recombinant pORF2 of different HEV genotypes.

Li and collaborators [177] reported on the use of HBsAg as a carrier for the HEV neutralizing antigenic epitope. The authors inserted a short fragment of the HEV pORF2 at position 551–607 aa into the C-terminus of HBsAg. This region contains several HEV antigenic epitopes. One epitope was mapped at position 578–607 aa using monoclonal antibodies that neutralized HEV in experimentally infected rhesus macaques [178]. In addition, this region contains an immunodominant epitope identified using synthetic peptides immunoreactive with human serum specimens obtained from HEV infected patients [179]. The expressed chimeric antigen formed VLP's, which immunoreacted with both HBsAg- and HEV-specific antibodies. These chimeric particles can be viewed as a step toward a bivalent vaccine against HBV and HEV. However, an extensive further evaluation of immunogenic and neutralizing properties of this hybrid antigen should be pursued before considering it as a candidate for such bivalent vaccine.

REFERENCES

1. Martin, A. and S.M. Lemon, Hepatitis A virus: From discovery to vaccines. *Hepatology*, 2006, 43(2 Suppl 1): S164–S172.
2. Panda, S.K., D. Thakral, and S. Rehman, Hepatitis E virus. *Rev Med Virol*, 2007, 17(3): 151–180.
3. Global surveillance and control of hepatitis C. Report of a WHO Consultation organized in collaboration with the Viral Hepatitis Prevention Board, Antwerp, Belgium. *J Viral Hepat*, 1999. 6(1): 35–47.
4. Alter, M.J., Epidemiology and prevention of hepatitis B. *Semin Liver Dis*, 2003, 23(1): 39–46.
5. Feinstone, S.M., A.Z. Kapikian, and R.H. Purceli, Hepatitis A: Detection by immune electron microscopy of a viruslike antigen associated with acute illness. *Science*, 1973, 182(116): 1026–1028.
6. Innis, B.L., et al., Protection against hepatitis A by an inactivated vaccine. *JAMA*, 1994, 271(17): 1328–1334.
7. Werzberger, A., et al., A controlled trial of a formalin-inactivated hepatitis A vaccine in healthy children. *N Engl J Med*, 1992, 327(7): 453–457.
8. Zhao, Y.L., et al., H2 strain attenuated live hepatitis A vaccines: Protective efficacy in a hepatitis A outbreak. *World J Gastroenterol*, 2000, 6(6): 829–832.
9. Ashida, M., et al., Propagation of hepatitis A virus in hybrid cell lines derived from marmoset liver and Vero cells. *J Gen Virol*, 1989, 70(Pt 9): 2487–2494.
10. Cromeans, T., M.D. Sobsey, and H.A. Fields, Development of a plaque assay for a cytopathic, rapidly replicating isolate of hepatitis A virus. *J Med Virol*, 1987, 22(1): 45–56.
11. Daemer, R.J., et al., Propagation of human hepatitis A virus in African green monkey kidney cell culture: Primary isolation and serial passage. *Infect Immun*, 1981, 32(1): 388–393.
12. Rezende, G., et al., Viral and clinical factors associated with the fulminant course of hepatitis A infection. *Hepatology*, 2003, 38(3): 613–618.
13. Taylor, R.M., et al., Fulminant hepatitis A virus infection in the United States: Incidence, prognosis, and outcomes. *Hepatology*, 2006, 44(6): 1589–1597.
14. Totsuka, A. and Y. Moritsugu, Hepatitis A virus proteins. *Intervirology*, 1999, 42(2–3): 63–68.
15. Kabrane-Lazizi, Y., et al., Detection of antibodies to HAV 3C proteinase in experimentally infected chimpanzees and in naturally infected children. *Vaccine*, 2001, 19(20–22): 2878–2883.
16. Robertson, B.H., et al., Antibody response to nonstructural proteins of hepatitis A virus following infection. *J Med Virol*, 1993, 40(1): 76–82.
17. Stewart, D.R., et al., Detection of antibodies to the nonstructural 3C proteinase of hepatitis A virus. *J Infect Dis*, 1997, 176(3): 593–601.
18. Nainan, O.V., M.A. Brinton, and H.S. Margolis, Identification of amino acids located in the antibody binding sites of human hepatitis A virus. *Virology*, 1992, 191(2): 984–987.
19. Ping, L.H., et al., Identification of an immunodominant antigenic site involving the capsid protein VP3 of hepatitis A virus. *Proc Natl Acad Sci U S A*, 1988, 85(21): 8281–8285.
20. Ping, L.H. and S.M. Lemon, Antigenic structure of human hepatitis A virus defined by analysis of escape mutants selected against murine monoclonal antibodies. *J Virol*, 1992, 66(4): 2208–2216.
21. Gauss-Muller, V., G.G. Frosner, and F. Deinhardt, Propagation of hepatitis A virus in human embryo fibroblasts. *J Med Virol*, 1981, 7(3): 233–239.
22. Gauss-Muller, V., et al., Recombinant proteins VP1 and VP3 of hepatitis A virus prime for neutralizing response. *J Med Virol*, 1990, 31(4): 277–283.
23. Haro, I., et al., Anti-hepatitis A virus antibody response elicited in mice by different forms of a synthetic VP1 peptide. *Microbiol Immunol*, 1995, 39(7): 485–490.
24. Johnston, J.M., et al., Antigenic and immunogenic properties of a hepatitis A virus capsid protein expressed in *Escherichia coli*. *J Infect Dis*, 1988, 157(6): 1203–1211.
25. Melnick, J.L., Properties and classification of hepatitis A virus. *Vaccine*, 1992, 10(Suppl 1): S24–S26.
26. Stapleton, J.T. and S.M. Lemon, Neutralization escape mutants define a dominant immunogenic neutralization site on hepatitis A virus. *J Virol*, 1987, 61(2): 491–498.
27. Stapleton, J.T., et al., Antigenic and immunogenic properties of recombinant hepatitis A virus 14S and 70S subviral particles. *J Virol*, 1993, 67(2): 1080–1085.
28. LaBrecque, F.D., et al., Recombinant hepatitis A virus antigen: Improved production and utility in diagnostic immunoassays. *J Clin Microbiol*, 1998, 36(7): 2014–2018.
29. Anderson, D.A. and B.C. Ross, Morphogenesis of hepatitis A virus: Isolation and characterization of subviral particles. *J Virol*, 1990, 64(11): 5284–5289.

30. Probst, C., M. Jecht, and V. Gauss-Muller, Intrinsic signals for the assembly of hepatitis A virus particles. Role of structural proteins VP4 and 2A. *J Biol Chem*, 1999, 274(8): 4527–4531.
31. Borovec, S.V. and D.A. Anderson, Synthesis and assembly of hepatitis A virus-specific proteins in BS-C-1 cells. *J Virol*, 1993, 67(6): 3095–3102.
32. Ruchti, F., G. Siegl, and M. Weitz, Identification and characterization of incomplete hepatitis A virus particles. *J Gen Virol*, 1991, 72(Pt 9): 2159–2166.
33. Bishop, N.E. and D.A. Anderson, RNA-dependent cleavage of VP0 capsid protein in provirions of hepatitis A virus. *Virology*, 1993, 197(2): 616–623.
34. Bosch, A., et al., Recombinant hepatitis A virus polyprotein expressed in *E. coli* assembles in subviral structures. In *Viral Hepatitis and Liver Disease*. Edited by M. Rizzetto, R.H. Purcell, J.L. Gerin, and G. Verme. Turin: Edizioni Minerva Medica, 1997: 27–31.
35. Pinto, R.M., et al., Hepatitis A virus polyprotein processing by *Escherichia coli* proteases. *J Gen Virol*, 2002, 83(Pt 2): 359–368.
36. Sanchez, G., et al., Antigenic hepatitis A virus structures may be produced in *Escherichia coli*. *Appl Environ Microbiol*, 2003, 69(3): 1840–1843.
37. Harmon, S.A., et al., Expression of hepatitis A virus capsid sequences in insect cells. *Virus Res*, 1988, 10(2–3): 273–280.
38. Rosen, E., J.T. Stapleton, and J. McLinden, Synthesis of immunogenic hepatitis A virus particles by recombinant baculoviruses. *Vaccine*, 1993, 11(7): 706–712.
39. Stapleton, J.T., J. Frederick, and B. Meyer, Hepatitis A virus attachment to cultured cell lines. *J Infect Dis*, 1991, 164(6): 1098–1103.
40. Lemon, S.M., L.N. Binn, and R.H. Marchwicki, Radioimmunofocus assay for quantitation of hepatitis A virus in cell cultures. *J Clin Microbiol*, 1983, 17(5): 834–839.
41. Probst, C., M. Jecht, and V. Gauss-Muller, Proteinase 3C-mediated processing of VP1–2A of two hepatitis A virus strains: In vivo evidence for cleavage at amino acid position 273/274 of VP1. *J Virol*, 1997, 71(4): 3288–3292.
42. Winokur, P.L., J.H. McLinden, and J.T. Stapleton, The hepatitis A virus polyprotein expressed by a recombinant vaccinia virus undergoes proteolytic processing and assembly into viruslike particles. *J Virol*, 1991, 65(9): 5029–5036.
43. Probst, C., M. Jecht, and V. Gauss-Muller, Processing of proteinase precursors and their effect on hepatitis A virus particle formation. *J Virol*, 1998, 72(10): 8013–8020.
44. Rachow, A., V. Gauss-Muller, and C. Probst, Homogeneous hepatitis A virus particles. Proteolytic release of the assembly signal 2A from procapsids by factor Xa. *J Biol Chem*, 2003, 278(32): 29744–29751.
45. Shouval, D., Hepatitis B vaccines. *J Hepatol*, 2003, 39(Suppl 1): S70–S76.
46. Maupas, P., et al., Active immunization against hepatitis B in an area of high endemicity. Part I: Field design. *Prog Med Virol*, 1981, 27: 168–184.
47. Tiollais, P., C. Pourcel, and A. Dejean, The hepatitis B virus. *Nature*, 1985, 317(6037): 489–495.
48. Szmuness, W., et al., Hepatitis B vaccine: Demonstration of efficacy in a controlled clinical trial in a high-risk population in the United States. *N Engl J Med*, 1980, 303(15): 833–841.
49. Burrell, C.J., et al., Expression in *Escherichia coli* of hepatitis B virus DNA sequences cloned in plasmid pBR322. *Nature*, 1979, 279(5708): 43–47.
50. Shu, L., et al., Recombinant hepatitis B large surface antigen, successfully produced in Escherichia coli, stimulates T-cell response in mice. *Vaccine*, 2006, 24(20): 4409–4416.
51. Valenzuela, P., et al., Synthesis and assembly of hepatitis B virus surface antigen particles in yeast. *Nature*, 1982, 298(5872): 347–350.
52. Kruger, D.H., R. Ulrich, and W.H. Gerlich, Chimeric virus-like particles as vaccines. *Biol Chem*, 1999, 380(3): 275–276.
53. McAleer, W.J., et al., Human hepatitis B vaccine from recombinant yeast. *Nature*, 1984, 307(5947): 178–180.
54. Stephenne, J., Development and production aspects of a recombinant yeast-derived hepatitis B vaccine. *Vaccine*, 1990, 8 Suppl: S69–S73; discussion S79–S80.
55. Iwarson, S., et al., Neutralization of hepatitis B virus infectivity by a murine monoclonal antibody: an experimental study in the chimpanzee. *J Med Virol*, 1985, 16(1): 89–96.
56. Hauser, P., et al., Immunological properties of recombinant HBsAg produced in yeast. *Postgrad Med J*, 1987, 63(Suppl 2): 83–91.

57. Coates, T., et al., Hepatitis B vaccines: Assessment of the seroprotective efficacy of two recombinant DNA vaccines. *Clin Ther*, 2001, 23(3): 392–403.
58. Torresi, J., The virological and clinical significance of mutations in the overlapping envelope and polymerase genes of hepatitis B virus. *J Clin Virol*, 2002, 25(2): 97–106.
59. Heijtink, R.A., et al., Immune response after vaccination with recombinant hepatitis B vaccine as compared to that after plasma-derived vaccine. *Antiviral Res*, 1985, Suppl 1: 273–279.
60. From the Centers for Disease Control. Inadequate immune response among public safety workers receiving intradermal vaccination against hepatitis B—United States, 1990–1991. *JAMA*, 1991, 266(10): 1338–1339.
61. Shaw, F.E., Jr., et al., Effect of anatomic injection site, age and smoking on the immune response to hepatitis B vaccination. *Vaccine*, 1989, 7(5): 425–430.
62. Gerlich, W.H., et al., Protective potential of hepatitis B virus antigens other than the S gene protein. *Vaccine*, 1990, 8 Suppl: S63–S68; discussion S79–S80.
63. Itoh, Y., et al., A synthetic peptide vaccine involving the product of the pre-S(2) region of hepatitis B virus DNA: protective efficacy in chimpanzees. *Proc Natl Acad Sci U S A*, 1986, 83(23): 9174–9178.
64. Klinkert, M.Q., et al., Pre-S1 antigens and antibodies early in the course of acute hepatitis B virus infection. *J Virol*, 1986, 58(2): 522–525.
65. Milich, D.R., et al., Immune response to the pre-S(1) region of the hepatitis B surface antigen (HBsAg): a pre-S(1)-specific T cell response can bypass nonresponsiveness to the pre-S(2) and S regions of HBsAg. *J Immunol*, 1986, 137(1): 315–322.
66. Neurath, A.R., et al., Antibodies to a synthetic peptide from the preS 120–145 region of the hepatitis B virus envelope are virus neutralizing. *Vaccine*, 1986, 4(1): 35–37.
67. Neurath, A.R., et al., Identification and chemical synthesis of a host cell receptor binding site on hepatitis B virus. *Cell*, 1986, 46(3): 429–436.
68. Soulie, J.C., et al., Immunogenicity and safety in newborns of a new recombinant hepatitis B vaccine containing the S and pre-S2 antigens. *Vaccine*, 1991, 9(8): 545–548.
69. Krone, B., et al., Interaction between hepatitis B surface proteins and monomeric human serum albumin. *Hepatology*, 1990, 11(6): 1050–1056.
70. Machida, A., et al., A polypeptide containing 55 amino acid residues coded by the pre-S region of hepatitis B virus deoxyribonucleic acid bears the receptor for polymerized human as well as chimpanzee albumins. *Gastroenterology*, 1984, 86(5 Pt 1): 910–918.
71. Milich, D.R., et al., Enhanced immunogenicity of the pre-S region of hepatitis B surface antigen. *Science*, 1985, 228(4704): 1195–1199.
72. Jones, C.D., et al., T-cell and antibody response characterisation of a new recombinant pre-S1, pre-S2 and SHBs antigen-containing hepatitis B vaccine; demonstration of superior anti-SHBs antibody induction in responder mice. *Vaccine*, 1999, 17(20–21): 2528–2537.
73. Leroux-Roels, G., et al., Hepatitis B vaccine containing surface antigen and selected preS1 and preS2 sequences. 2. Immunogenicity in poor responders to hepatitis B vaccines. *Vaccine*, 1997, 15(16): 1732–1736.
74. Leroux-Roels, G., et al., Hepatitis B vaccine containing surface antigen and selected preS1 and preS2 sequences. 1. Safety and immunogenicity in young, healthy adults. *Vaccine*, 1997, 15(16): 1724–1731.
75. Shouval, D., et al., Improved immunogenicity in mice of a mammalian cell-derived recombinant hepatitis B vaccine containing pre-S1 and pre-S2 antigens as compared with conventional yeast-derived vaccines. *Vaccine*, 1994, 12(15): 1453–1459.
76. Eyigun, C.P., et al., A comparative trial of two surface subunit recombinant hepatitis B vaccines vs a surface and PreS subunit vaccine for immunization of healthy adults. *J Viral Hepat*, 1998, 5(4): 265–269.
77. Shapira, M.Y., et al., Rapid seroprotection against hepatitis B following the first dose of a Pre-S1/Pre-S2/S vaccine. *J Hepatol*, 2001, 34(1): 123–127.
78. Zuckerman, J.N., et al., Evaluation of a new hepatitis B triple-antigen vaccine in inadequate responders to current vaccines. *Hepatology*, 2001, 34(4 Pt 1): 798–802.
79. Fujisawa, Y., et al., Protective efficacy of a novel hepatitis B vaccine consisting of M (pre-S2 + S) protein particles (a third generation vaccine). *Vaccine*, 1990, 8(3): 192–198.
80. Suzuki, H., et al., Safety and efficacy of a recombinant yeast-derived pre-S2 + S-containing hepatitis B vaccine (TGP-943): Phase 1, 2 and 3 clinical testing. *Vaccine*, 1994, 12(12): 1090–1096.
81. Yamada, K., et al., Efficacy of hepatitis B vaccines with and without preS2-region product in Sumo wrestlers in Japan. *Hepatol Res*, 1998, 12: 3–11.

82. Han, X., et al., Expression, purification and characterization of the Hepatitis B virus entire envelope large protein in *Pichia pastoris*. *Protein Expr Purif*, 2006, 49(2): 168–175.
83. Locher, C.P., et al., DNA shuffling and screening strategies for improving vaccine efficacy. *DNA Cell Biol*, 2005, 24(4): 256–263.
84. Carman, W.F., et al., Vaccine-induced escape mutant of hepatitis B virus. *Lancet*, 1990, 336(8711): 325–329.
85. Hsu, H.Y., et al., Changes of hepatitis B surface antigen variants in carrier children before and after universal vaccination in Taiwan. *Hepatology*, 1999, 30(5): 1312–1317.
86. Nainan, O.V., et al., Genetic variation of hepatitis B surface antigen coding region among infants with chronic hepatitis B virus infection. *J Med Virol*, 2002, 68(3): 319–327.
87. Zanetti, A.R., et al., Hepatitis B variant in Europe. *Lancet*, 1988, 2(8620): 1132–1133.
88. Waters, J.A., et al., Loss of the common "A" determinant of hepatitis B surface antigen by a vaccine-induced escape mutant. *J Clin Invest*, 1992, 90(6): 2543–2547.
89. Yokosawa, J., et al., Antigenic properties of wild type and mutant hepatitis B virus surface antigens. *Proceedings of the 10th International Symposium on Viral Hepatitis and Liver Disease*, 2002, 155–159.
90. Coleman, P.F., Y.C. Chen, and I.K. Mushahwar, Immunoassay detection of hepatitis B surface antigen mutants. *J Med Virol*, 1999, 59(1): 19–24.
91. Ogata, N., et al., Licensed recombinant hepatitis B vaccines protect chimpanzees against infection with the prototype surface gene mutant of hepatitis B virus. *Hepatology*, 1999, 30(3): 779–786.
92. Jagadish, M.N., et al., Chimeric potyvirus-like particles as vaccine carriers. *Intervirology*, 1996, 39(1–2): 85–92.
93. Muster, T., et al., Cross-neutralizing activity against divergent human immunodeficiency virus type 1 isolates induced by the gp41 sequence ELDKWAS. *J Virol*, 1994, 68(6): 4031–4034.
94. Valenzuela, P., et al., Antigen engineering in yeast: Synthesis and assembly of hybrid hepatitis B surface antigen-herpes simplex 1 gD particles. *Nat Biotechnol*, 1985, 3: 323–326.
95. Delpeyroux, F., et al., A poliovirus neutralization epitope expressed on hybrid hepatitis B surface antigen particles. *Science*, 1986, 233(4762): 472–475.
96. Delpeyroux, F., et al., Presentation and immunogenicity of the hepatitis B surface antigen and a poliovirus neutralization antigen on mixed empty envelope particles. *J Virol*, 1988, 62(5): 1836–1839.
97. Delpeyroux, F., et al., Structural factors modulate the activity of antigenic poliovirus sequences expressed on hybrid hepatitis B surface antigen particles. *J Virol*, 1990, 64(12): 6090–6100.
98. Lee, I.H., C.H. Kim, and W.S. Ryu, Presentation of the hydrophilic domains of hepatitis C viral E2 envelope glycoprotein on hepatitis B surface antigen particles. *J Med Virol*, 1996, 50(2): 145–151.
99. Netter, H.J., et al., Antigenicity and immunogenicity of novel chimeric hepatitis B surface antigen particles with exposed hepatitis C virus epitopes. *J Virol*, 2001, 75(5): 2130–2141.
100. Michel, M.L., et al., Induction of anti-human immunodeficiency virus (HIV) neutralizing antibodies in rabbits immunized with recombinant HIV—hepatitis B surface antigen particles. *Proc Natl Acad Sci U S A*, 1988, 85(21): 7957–7961.
101. Netter, H.J., et al., Immunogenicity of recombinant HBsAg/HCV particles in mice pre-immunised with hepatitis B virus-specific vaccine. *Vaccine*, 2003, 21(21–22): 2692–2697.
102. Moradpour, D., et al., Hepatitis C: An update. *Swiss Med Wkly*, 2001, 131(21–22): 291–298.
103. Zanetti, A.R., L. Romano, and S. Bianchi, Primary prevention of hepatitis C virus infection. *Vaccine*, 2003, 21(7–8): 692–695.
104. Feld, J.J. and J.H. Hoofnagle, Mechanism of action of interferon and ribavirin in treatment of hepatitis C. *Nature*, 2005, 436(7053): 967–972.
105. Choo, Q.L., et al., Isolation of a cDNA clone derived from a blood-borne non-A, non-B viral hepatitis genome. *Science*, 1989, 244(4902): 359–362.
106. Choo, Q.L., et al., Genetic organization and diversity of the hepatitis C virus. *Proc Natl Acad Sci U S A*, 1991, 88(6): 2451–2455.
107. Drazan, K.E., Molecular biology of hepatitis C infection. *Liver Transpl*, 2000, 6(4): 396–406.
108. Hijikata, M., et al., Gene mapping of the putative structural region of the hepatitis C virus genome by in vitro processing analysis. *Proc Natl Acad Sci U S A*, 1991, 88(13): 5547–5551.
109. Simmonds, P., Variability of hepatitis C virus. *Hepatology*, 1995, 21(2): 570–583.
110. Salemi, M. and A.M. Vandamme, Hepatitis C virus evolutionary patterns studied through analysis of full-genome sequences. *J Mol Evol*, 2002, 54(1): 62–70.

111. Farci, P., et al., Prevention of hepatitis C virus infection in chimpanzees by hyperimmune serum against the hypervariable region 1 of the envelope 2 protein. *Proc Natl Acad Sci U S A*, 1996, 93(26): 15394–15399.
112. Korenaga, M., et al., A possible role of hypervariable region 1 quasispecies in escape of hepatitis C virus particles from neutralization. *J Viral Hepat*, 2001, 8(5): 331–340.
113. Bertoletti, A. and C. Ferrari, Kinetics of the immune response during HBV and HCV infection. *Hepatology*, 2003, 38(1): 4–13.
114. Cerny, A. and F.V. Chisari, Pathogenesis of chronic hepatitis C: Immunological features of hepatic injury and viral persistence. *Hepatology*, 1999, 30(3): 595–601.
115. Gremion, C. and A. Cerny, Hepatitis C virus and the immune system: A concise review. *Rev Med Virol*, 2005, 15(4): 235–268.
116. Dreux, M. and F.L. Cosset, The scavenger receptor BI and its ligand, HDL: Partners in crime against HCV neutralizing antibodies. *J Viral Hepat*, 2007, 14(Suppl 1): 68–76.
117. von Hahn, T., et al., Hepatitis C virus continuously escapes from neutralizing antibody and T-cell responses during chronic infection in vivo. *Gastroenterology*, 2007, 132(2): 667–678.
118. Santolini, E., G. Migliaccio, and N. La Monica, Biosynthesis and biochemical properties of the hepatitis C virus core protein. *J Virol*, 1994, 68(6): 3631–3641.
119. Lo, S.Y., et al., Differential subcellular localization of hepatitis C virus core gene products. *Virology*, 1995, 213(2): 455–461.
120. Lorenzo, L.J., et al., Assembly of truncated HCV core antigen into virus-like particles in Escherichia coli. *Biochem Biophys Res Commun*, 2001, 281(4): 962–965.
121. Polakos, N.K., et al., Characterization of hepatitis C virus core-specific immune responses primed in rhesus macaques by a nonclassical ISCOM vaccine. *J Immunol*, 2001, 166(5): 3589–3598.
122. Acosta-Rivero, N., et al., Characterization of the HCV core virus-like particles produced in the methylotrophic yeast *Pichia pastoris*. *Biochem Biophys Res Commun*, 2001, 287(1): 122–125.
123. Majeau, N., et al., The N-terminal half of the core protein of hepatitis C virus is sufficient for nucleocapsid formation. *J Gen Virol*, 2004, 85(Pt 4): 971–981.
124. Choo, Q.L., et al., Vaccination of chimpanzees against infection by the hepatitis C virus. *Proc Natl Acad Sci U S A*, 1994, 91(4): 1294–1298.
125. Puig, M., et al., Immunization of chimpanzees with an envelope protein-based vaccine enhances specific humoral and cellular immune responses that delay hepatitis C virus infection. *Vaccine*, 2004, 22(8): 991–1000.
126. Nevens, F., et al., A pilot study of therapeutic vaccination with envelope protein E1 in 35 patients with chronic hepatitis C. *Hepatology*, 2003, 38(5): 1289–1296.
127. Spaete, R.R., et al., Characterization of the hepatitis C virus E2/NS1 gene product expressed in mammalian cells. *Virology*, 1992, 188(2): 819–830.
128. Heile, J.M., et al., Evaluation of hepatitis C virus glycoprotein E2 for vaccine design: An endoplasmic reticulum-retained recombinant protein is superior to secreted recombinant protein and DNA-based vaccine candidates. *J Virol*, 2000, 74(15): 6885–6892.
129. Rosa, D., et al., A quantitative test to estimate neutralizing antibodies to the hepatitis C virus: Cytofluorimetric assessment of envelope glycoprotein 2 binding to target cells. *Proc Natl Acad Sci U S A*, 1996, 93(5): 1759–1763.
130. Pileri, P., et al., Binding of hepatitis C virus to CD81. *Science*, 1998, 282(5390): 938–941.
131. Xiang, Z.H., et al., Purification and application of bacterially expressed chimeric protein E1E2 of hepatitis C virus. *Protein Expr Purif*, 2006, 49(1): 95–101.
132. Martinez-Donato, G., et al., Expression and processing of hepatitis C virus structural proteins in Pichia pastoris yeast. *Biochem Biophys Res Commun*, 2006, 342(2): 625–631.
133. Lechmann, M., et al., Hepatitis C virus-like particles induce virus-specific humoral and cellular immune responses in mice. *Hepatology*, 2001, 34(2): 417–423.
134. Murata, K., et al., Immunization with hepatitis C virus-like particles protects mice from recombinant hepatitis C virus-vaccinia infection. *Proc Natl Acad Sci U S A*, 2003, 100(11): 6753–6758.
135. Zibert, A., et al., Characterization of antibody response to hepatitis C virus protein E2 and significance of hypervariable region 1-specific antibodies in viral neutralization. *Arch Virol*, 1997, 142(3): 523–534.
136. Zibert, A., et al., Epitope mapping of antibodies directed against hypervariable region 1 in acute self-limiting and chronic infections due to hepatitis C virus. *J Virol*, 1997, 71(5): 4123–4127.

137. Zibert, A., et al., Early antibody response against hypervariable region 1 is associated with acute self-limiting infections of hepatitis C virus. *Hepatology*, 1997, 25(5): 1245–1249.
138. Goto, J., et al., Prevention of hepatitis C virus infection in a chimpanzee by vaccination and epitope mapping of antiserum directed against hypervariable region 1. *Hepatol Res*, 2001, 19(3): 270–283.
139. Zibert, A., E. Schreier, and M. Roggendorf, Antibodies in human sera specific to hypervariable region 1 of hepatitis C virus can block viral attachment. *Virology*, 1995, 208(2): 653–661.
140. Esumi, M., et al., In vivo and in vitro evidence that cross-reactive antibodies to C-terminus of hypervariable region 1 do not neutralize heterologous hepatitis C virus. *Vaccine*, 2002, 20(25–26): 3095–3103.
141. Puntoriero, G., et al., Towards a solution for hepatitis C virus hypervariability: Mimotopes of the hypervariable region 1 can induce antibodies cross-reacting with a large number of viral variants. *Embo J*, 1998, 17(13): 3521–3533.
142. Scarselli, E., et al., Occurrence of antibodies reactive with more than one variant of the putative envelope glycoprotein (gp70) hypervariable region 1 in viremic hepatitis C virus-infected patients. *J Virol*, 1995, 69(7): 4407–4412.
143. Gedvilaite, A., et al., Formation of immunogenic virus-like particles by inserting epitopes into surface-exposed regions of hamster polyomavirus major capsid protein. *Virology*, 2000, 273(1): 21–35.
144. Schirmbeck, R., W. Bohm, and J. Reimann, Virus-like particles induce MHC class I-restricted T-cell responses. Lessons learned from the hepatitis B small surface antigen. *Intervirology*, 1996, 39(1–2): 111–119.
145. Vietheer, P.T., et al., Immunizations with chimeric hepatitis B virus-like particles to induce potential anti-hepatitis C virus neutralizing antibodies. *Antivir Ther*, 2007, 12(4): 477–487.
146. Shang, D., W. Zhai, and J.P. Allain, Broadly cross-reactive, high-affinity antibody to hypervariable region 1 of the hepatitis C virus in rabbits. *Virology*, 1999, 258(2): 396–405.
147. Zhou, Y.H., M. Moriyama, and M. Esumi, Multiple sequence-reactive antibodies induced by a single peptide immunization with hypervariable region 1 of hepatitis C virus. *Virology*, 1999, 256(2): 360–370.
148. Nemchinov, L.G., et al., Development of a plant-derived subunit vaccine candidate against hepatitis C virus. *Arch Virol*, 2000, 145(12): 2557–2573.
149. Baumert, T.F., et al., Hepatitis C virus structural proteins assemble into viruslike particles in insect cells. *J Virol*, 1998, 72(5): 3827–3836.
150. Baumert, T.F., et al., Hepatitis C virus-like particles synthesized in insect cells as a potential vaccine candidate. *Gastroenterology*, 1999, 117(6): 1397–1407.
151. Balayan, M.S., et al., Evidence for a virus in non-A, non-B hepatitis transmitted via the fecal-oral route. *Intervirology*, 1983, 20(1): 23–31.
152. Krawczynski, K., S. Kamili, and R. Aggarwal, Global epidemiology and medical aspects of hepatitis E. *Forum (Genova)*, 2001, 11(2): 166–179.
153. Worm, H.C., W.H. van der Poel, and G. Brandstatter, Hepatitis E: An overview. *Microbes Infect*, 2002, 4(6): 657–666.
154. Meng, X.J., et al., Experimental infection of pigs with the newly identified swine hepatitis E virus (swine HEV), but not with human strains of HEV. *Arch Virol*, 1998, 143(7): 1405–1415.
155. Meng, X.J., et al., Genetic and experimental evidence for cross-species infection by swine hepatitis E virus. *J Virol*, 1998, 72(12): 9714–9721.
156. Mayo, M., Changes to virus taxonomy 2004. *Arch Virol*, 2005, 150: 189–198.
157. Tam, A.W., et al., Hepatitis E virus (HEV): Molecular cloning and sequencing of the full-length viral genome. *Virology*, 1991, 185(1): 120–131.
158. Mast, E.E., et al., Evaluation of assays for antibody to hepatitis E virus by a serum panel. Hepatitis E Virus Antibody Serum Panel Evaluation Group. *Hepatology*, 1998, 27(3): 857–861.
159. Tam, A.W., et al., In vitro infection and replication of hepatitis E virus in primary cynomolgus macaque hepatocytes. *Virology*, 1997, 238(1): 94–102.
160. Khudyakov, Y.E., et al., Epitope mapping in proteins of hepatitis E virus. *Virology*, 1993, 194(1): 89–96.
161. Khudyakov, Y.E., et al., Antigenic domains of the open reading frame 2-encoded protein of hepatitis E virus. *J Clin Microbiol*, 1999, 37(9): 2863–2871.
162. Meng, J., et al., Identification and characterization of the neutralization epitope(s) of the hepatitis E virus. *Virology*, 2001, 288(2): 203–211.
163. Dong, C., X. Dai, and J.H. Meng, The first experimental study on a candidate combined vaccine against hepatitis A and hepatitis E. *Vaccine*, 2007, 25(9): 1662–1668.

164. Purdy, M.A., et al., Expression of a hepatitis E virus (HEV)-trpE fusion protein containing epitopes recognized by antibodies in sera from human cases and experimentally infected primates. *Arch Virol*, 1992, 123(3–4): 335–349.
165. Meng, J., P. Dubreuil, and J. Pillot, A new PCR-based seroneutralization assay in cell culture for diagnosis of hepatitis E. *J Clin Microbiol*, 1997, 35(6): 1373–1377.
166. Meng, J., et al., Neutralization of different geographic strains of the hepatitis E virus with anti-hepatitis E virus-positive serum samples obtained from different sources. *Virology*, 1998, 249(2): 316–324.
167. Tsarev, S.A., et al., ELISA for antibody to hepatitis E virus (HEV) based on complete open-reading frame-2 protein expressed in insect cells: Identification of HEV infection in primates. *J Infect Dis*, 1993, 168(2): 369–378.
168. Robinson, R.A., et al., Structural characterization of recombinant hepatitis E virus ORF2 proteins in baculovirus-infected insect cells. *Protein Expr Purif*, 1998, 12(1): 75–84.
169. Tsarev, S.A., et al., Successful passive and active immunization of cynomolgus monkeys against hepatitis E. *Proc Natl Acad Sci U S A*, 1994, 91(21): 10198–10202.
170. Emerson, S.U. and R.H. Purcell, Hepatitis E virus. *Rev Med Virol*, 2003, 13(3): 145–154.
171. Zafrullah, M., et al., Mutational analysis of glycosylation, membrane translocation, and cell surface expression of the hepatitis E virus ORF2 protein. *J Virol*, 1999, 73(5): 4074–4082.
172. Purcell, R.H., et al., Pre-clinical immunogenicity and efficacy trial of a recombinant hepatitis E vaccine. *Vaccine*, 2003, 21(19–20): 2607–2615.
173. Safary, A., Perspectives of vaccination against hepatitis E. *Intervirology*, 2001, 44(2–3): 162–166.
174. Shrestha, M.P., et al., Safety and efficacy of a recombinant hepatitis E vaccine. *N Engl J Med*, 2007, 356(9): 895–903.
175. Li, T.C., et al., Expression and self-assembly of empty virus-like particles of hepatitis E virus. *J Virol*, 1997, 71(10): 7207–7213.
176. Li, T.C., et al., Protection of cynomolgus monkeys against HEV infection by oral administration of recombinant hepatitis E virus-like particles. *Vaccine*, 2004, 22(3–4): 370–377.
177. Li, H.Z., et al., Production in *Pichia pastoris* and characterization of genetic engineered chimeric HBV/HEV virus-like particles. *Chin Med Sci J*, 2004, 19(2): 78–83.
178. Schofield, D.J., et al., Identification by phage display and characterization of two neutralizing chimpanzee monoclonal antibodies to the hepatitis E virus capsid protein. *J Virol*, 2000, 74(12): 5548–5555.
179. Khudyakov Yu, E., et al., Immunodominant antigenic regions in a structural protein of the hepatitis E virus. *Virology*, 1994, 198(1): 390–393.

Part III

Protein Engineering Targets: Diagnostics and Therapy

13 Designer Diagnostic Reagents and Allergens

Anna Obriadina

CONTENTS

13.1 INTRODUCTION

Today, diagnostics is a key tool for maintenance of public health and one of the most dynamic research areas. Tremendous progress has been made in genetic engineering of diagnostic reagents. Numerous new proteins with desired properties have been devised and developed for the detection of specific immunoresponses to viral and bacterial infections and identification of other biologically important targets such as hormones and allergens. The trend is set—artificial designer proteins are on the way to replacing traditional antigens and antibodies in the laboratory assays and immunotherapy.

This chapter reviews several new developments in the application of designer diagnostic proteins in in vitro diagnostics and immunotherapy.

13.2 DESIGNER PROTEINS AS DIAGNOSTIC REAGENTS

13.2.1 Designer Diagnostic Antigens

Rapid strides over the last two decades in the field of chemistry of nucleic acids and genetic engineering made possible generating long polynucleotides of any sequence. These developments opened the door for engineering synthetic genes encoding proteins with any desired properties. Since the very beginning of protein engineering, the design and construction of synthetic genes encoding immunoreactive proteins specifically tailored for the applications in diagnostics, immunotherapy, and vaccine development became the area of a very intense research. Currently, the list of synthetic genes constructed for the expression of artificial immunoreactive proteins is rather impressive. However, insufficient knowledge on the relationship between protein structure and function drastically reduces the rate of successful applications of synthetic DNA methodology to the design and construction of artificial antigens. Many experimental works in this area are forced to rely on the trial-and-error approach. The major issues in immunodiagnostics are related to sensitivity and specificity of antigen binding to antibodies. The so-called natural antigens, or proteins and their fragments expressed in a heterologous expression system with the goal of being used as diagnostic targets without any manipulation of the primary structure, may pose a problem to their diagnostic applications due to diagnostically irrelevant or "nonspecific" immunological reactivity and strain-specific immunoreactivity. Forces of natural selection do not mold pathogen proteins into efficient diagnostic targets. Therefore, although the antigens are efficiently recognized by the host immuno-system, they may not be suitable for diagnostic development due to the presence of epitopes that can be recognized by antibodies elicited against a different antigen or may have properties hindering antibody binding. Viruses such as human immunodeficiency virus and hepatitis C virus contain significantly heterogeneous genomes. This heterogeneity is one of the major mechanisms for viral escape from the host immunosurveillance system. Pathogen sequence variations cause a significant heterogeneity in immunoresponses against different strains and, as a result, reduction in sensitivity of diagnostic assays based on antigens from a single strain.

13.2.1.1 Cross-Reactivity and Strain-Specific Variation

The cross-reactivity can be found in many viral and bacterial infections and frequently renders false-positive or undetermined results. Immunodominant epitopes of viral proteins may share some amino acid sequences with phylogenetically related relatives or with orthologs in other organism. Molecular mimicry was found between HCV antigen and human nitrogen-oxide synthase, as well as Tyrosine kinase Lck and hepatic growth factor activator [1]. The HIV-1 gag precursor shares antigenic sites with the major capsid protein of human cytomegalovirus [2]. Such cross-immuno-reactive epitopes found among different organisms and pathogens affect the diagnostic relevance of corresponding antigens and require some engineering intervention to reduce its effect on specificity of antibody binding. The most straightforward approach is to excise the nonspecific epitopes from diagnostic antigens. Although it is a rather elementary task for the modern genetic engineering, it cannot be frequently implemented because of the difficulties in exact mapping of such nonspecific epitopes. The alternative way of antigen engineering from selected short antigenic epitopes relevant to immunodiagnostics seems more appropriate and was developed into a novel promising platform for assay development.

The strategy of excluding nonspecific epitopes from artificial proteins was first used in the development of the HEV mosaic protein as a novel designer diagnostic antigen [3,4]. The polypeptide comprises a collection of three diagnostically relevant regions from the protein encoded by open reading frame 2 (ORF2), one region from the protein encoded by ORF3 of the Burmese HEV strain and one region from the protein encoded by ORF3 of the Mexican strain. The mosaic protein was expressed in *Escherichia coli* as a chimera with glutathione-*S*-transferase and used for the development of a sensitive and specific EIA for the detection of IgG anti-HEV activity in human sera [5].

The following examples illustrate a general trend in engineering of artificial antigens from diagnostically relevant protein regions with reduced nonspecific immunoreactivity. The Lyme disease is the most common vector-born disease in North America and Europe, caused by spirochete *Borrelia burgdorferi*. None of the clinical manifestations of Lyme disease are pathognomonic and laboratory diagnosis of the infection is primarily dependent on the detection of specific antibodies. The use of whole cells of *Borrelia* spp. as the source of antigen has posed problems in both the sensitivity and specificity. At the same time, a test based on a single antigenic protein proved to be not sufficiently sensitive [6–9]. In 2000, Gomes–Solecki and coworkers described the cloning, expression, and serologic characterization of a recombinant chimera containing key sequences from a number of *B. burgdorferi* proteins, including OspA, OspB, OspC, flagellin, and p93 [6]. Both the sensitivity and specificity of the assay were increased after removing cross-reactive epitopes of these proteins and presentation of only the regions that induce a strong specific immune response to *B. burgdorferi*. The assay based on this recombinant chimera demonstrated a superior sensitivity in comparison to the most sensitive of whole-cell Borrelia assays [6].

Constructing synthetic genes from short DNA fragments is a rather daunting task. To improve on the engineering of complex antigens from diagnostically relevant epitopes, a new approach designated as restriction enzyme-assisted ligation (REAL) was developed and applied for the first time to the assembly of an artificial antigen composed of 17 small antigenic regions derived from the NS4-protein of hepatitis C virus (HCV) genotypes 1 through 5. This artificial antigen was found to specifically detect anti-NS4 activity earlier in two of four seroconversion panels than did the antigen used in a commercially available supplemental assay. Furthermore, the artificial NS4 antigen demonstrated an equivalent anti-NS4 immunoreactivity with serum specimens obtained from patients infected with different HCV genotypes, whereas the NS4 recombinant protein derived from genotype 1, used in the commercial supplemental test, was less immunoreactive with serum specimens containing HCV genotypes 2, 3, and 4 [10].

The second generation HEV mosaic protein (MP-II) was designed to contain a set of antigenic regions located at position 31–66 aa, 85–114 aa, 95–119 aa, 398–427 aa, 614–638 aa, 626–655 aa, 631–660 aa, 635–660 aa of the ORF2 protein and two sequence variants of an epitope from the ORF3 protein at position 91–123 aa, one from the Burmese strain and one from the Mexican strain. This new composite protein was used for developing EIA for specific anti-HEV detection in a combination with a new HEV recombinant antigen that contains the neutralizing antigenic epitope (s) of the HEV ORF2-encoded protein [11,12]. The immunoassay based on these two antigens detected anti-HEV IgG in 81 out of 81 (100%) serum specimens obtained from patient acutely infected with HEV and in serum specimens obtained from experimentally infected chimpanzees after more than 2.5 years following HEV inoculation with specificity 97.6%. This assay specifically detected the HEV antibodies with equal efficiency in patients infected with different HEV strains and residing in different endemic areas [5].

Chaga's disease is a parasitic illness affecting 16–18 million people worldwide. It is routinely diagnosed by detecting specific antibodies. A cross-reactivity with antibodies elicited during infections caused by *Leishmania spp.* or *Tripanosoma rangeli* and autoimmune diseases results frequently in misdiagnosis [13,14]. A synthetic gene encoding a chimeric protein containing the C-terminal region of *T. cruzi* calflagin (C29) and the N-terminal of TcP2β protein was constructed by Aguirre and coworkers [15]. Both components of the mosaic protein were optimized in terms of specificity by removing low specificity protein fragments. In case of *T. cruzi* calflagin, it was a fragment responsible for cross-reactivity with sera from patients infected with *Leishmania spp.* For the TcP2β protein, it was the C-terminal fragment sharing a 12 aa-region with a homologs human protein. This homology was responsible for a high cross-reactivity with sera from patients with autoimmune or related parasitic diseases [16]. A diagnostic assay constructed on the basis of this chimera protein was tested against a panel of 104 specific and 112 nonspecific sera for Chagas' disease including 65 sera from patients with autoimmune diseases or infected with *Leishmania brasiliensis, Brucella abortus, Streptococcus pyogenes*, and *Toxoplasma gondii*. The assay developed by Aguirre et al. [15] demonstrated a 100% sensitivity and specificity in diagnosing Chagas' disease.

A fusion of DNA sequences encoding selected specific antigenic epitopes derived from different proteins or from different strains of pathogens was successfully used to develop composite proteins for the use in diagnostic detection of visceral leishmaniasis [17], toxoplasmosis [18], HCV [10,19,20], HEV [3–5], insulin-dependent diabetes mellitus [21], and HTLV-2 infection [22]. Currently, at the beginning of the twenty-first century the concept of artificial designer diagnostic antigens is a focus of a very intensive research that promises a significant improvement in diagnostic specificity and sensitivity.

The protein antigens discussed above were designed for the use as diagnostic targets. Another notable trend in the application of protein engineering to diagnostics is development of designer antibodies.

13.2.2 Designer Diagnostic Antibodies

Bispecific, chimeric, and bifunctional antibodies produced by conventional chemical or somatic methods are successfully used in the immunodiagnostics. Today, through genetic engineering, it is possible to create designer antibody molecules with two or more desired functions and there are numerous examples in the extant literature describing the usage of such proteins as calibrators, detection tools, and other diagnostic reagents.

13.2.2.1 Bispecific Antibodies

Bispecific antibody or diabody is antibody with ability to recognize two different epitopes of different [23,24] or same target molecules [25]. Antibodies usually used in the development of immunoassays have specificity for a reporter molecule such as enzyme or for a target antigen [26,27]. A very interesting application of recombinant bispecific diabodies was recently described by Chen and coworkers [23]. The authors successfully used a bispecific diabody against an epitope on human red blood cells and hepatitis B virus surface antigen (HBsAg) for the development of a rapid agglutination assay in blood specimens.

13.2.2.2 Chimeric Antibodies as Calibrators

Chimeric antibodies are genetically engineered by fusion of parts of mouse and human antibodies. Generally, chimeric antibodies are approximately 33% mouse protein and 67% human protein. Engineered chimeric antibodies may be used in clinical laboratory at least in three areas: As reference proteins to document the specificity of clinical assay reagents [28,29], calibration proteins [28–33], and interference proteins to study the effects of naturally occurring autoantibody on the accuracy of immunoassays [28]. Calibrators in diagnostic kits for the detection of antibodies to infectious agents are usually prepared from a positive plasma or serum. However, this source has several drawbacks including difficulty in obtaining a large volume plasma or serum with a high antibody titer, lot-to-lot variability and dangerous nature of infectious samples. Mouse–human chimeric antibodies have previously been shown to be useful for quantification of specific antibody in reference standards mostly by heterologous interpolation [28,30]. The first demonstration of the utility of chimeric antibody calibrators for homologous interpolation of human antibodies specific to *Toxoplasma gondii* was made by Hackett and colleagues [31]. Murine antibodies specific for P30 and P66 were selected for conversion to mouse–human chimera. Functional immunoglobulin V region genes were cloned and transferred into expression vectors containing human constant region genes to generate recombinant chimeric antibodies. The calibrators were compared to anti-*T. gondii* Ig-positive human plasma matched to World Health Organization International Standard. Immunoreactivity of the anti-P30 IgG1 was comparable to that of human plasma-derived calibrators. The signal generated using the anti-P66 IgG1 was to plateau below the maximal reactivity level. Nevertheless, the dilution profile of a combined calibrator based on both anti-P30 and anti-P66 chimeric antibodies was similar to that of the human plasma-derived Toxo IgG calibrator. When evaluated against a panel of patient samples, the correlation to results obtained using human calibrators was $\geq$0.985. In those cases when a closer match between immunoreactivity of chimeric antibodies and human samples is desired, a careful selection of the epitope specificity, heavy chain isotype and affinity of chimeric antibodies is required.

13.2.2.3 Bifunctional Antibodies

Analyte detection in immunoassays is carried out mostly with primary or secondary antibodies that are chemically coupled with sensitive reporter molecules, like enzymes. An attractive alternative to the chemical coupling of these proteins is the construction of bifunctional antibodies—genetically engineered fusion proteins consisting of an enzyme and analyte-specific antibody. An example of a simple form of bifunctional antibody is a single-chain variable fragment (scFv) fused to short oligopeptides, such as streptavidin tag (biotin-specific peptide) [34]. Biotin–streptavidin-based assay can be used for detection and antigen binding study. The streptavidin tag permits the attachment to other functional molecules via streptavidin–biotin bridge, thus, providing a convenient way to generate immunologic reagents with dual functions. Bifunctional enzyme-antibodies have been shown to be highly efficient detection reagents in enzyme immunoassays, dot-blot hybridization, and other diagnostic formats [35–38].

One of the most successful examples of the application of bifunctional antibodies to diagnostics is the development of highly sensitive and specific hemagglutination-based assay for the detection of antibodies to HIV in whole blood [39]. This bifunctional antibody was engineered from a recombinant monovalent Fab fragment specific for human red blood cells and two protein fragments derived from HIV-1 gp41 (590–620 aa) and HIV-2 gp36 (581–611 aa). A major problem faced by the researchers in constructing this antibody was the presence of disulfide bonds in both env1 and env2 proteins, which might affect the correct inter- and intrachain disulfide formation. In order to avoid such a complication, undesirable cysteines could be replaced for serine residues. Since it was not known whether this replacement may cause a functional inactivation of the antibody, the authors designed a large number of fusion proteins with different combination of cysteine–serine replacements. The Fabenv12CS containing sequence of the native HIV-1 gp41 followed by a linker and sequence of the HIV-2 gp36 having serine residues in place of cysteine residues was shown to retain a highly specific immunoreactivity [39]. The technology described in this chapter is applicable for development of chimeric molecules for detection of various infections in whole blood samples.

One of the most promising substitutes for secondary antibodies in immunoassays is bacterial Fc receptors like protein A and protein G. It was shown that a protein A–enzyme chimera has the same advantages as bifunctional antibodies. It is also thermostable. When attached to a modified alkaline phosphatase (K328 changed to A328), which has a higher catalytic turnover [40], this chimeric protein was proven to be a very powerful diagnostic reagent. A recombinant artificial protein combining the immunoglobulin binding sites of proteins A and G was constructed and used as a universal detection reagent in a serological survey of human sera and samples from laboratory, domestic and wild animals [41–45]. An added advantage was a universal cutoff for all tests using this protein A–protein G chimera, which significantly simplifies diagnostic interpretation of the data [43].

13.2.2.3.1 New Immunoassay Formats

Fusion proteins containing both binding and reporter domains have not only been used extensively in the conventional immunoassay formats, but were explored for the development of novel immunoassay approaches. The most widely used sandwich immunoassay cannot quantify low-molecular weight antigens, such as hormones or toxic chemicals, because they are too small to be recognized simultaneously by two antibodies. Recently, a novel "open sandwich" immunoassay was developed to overcome this limitation [46,47]. The open sandwich (OS) immunoassay exploits the phenomenon of stabilization of an antibody variable region by the bound antigen. It was shown that the V_L and V_H components of the antibody variable region are weakly associated ($K_a < 10^5$/M) with each other in the absence of the antigen. However, this association is markedly strengthened by the bound antigen ($K_a \sim 10^9$/M) [48]. The applicability of the OS immunoassay to the quantification of low-molecular weight (monovalent) haptens was first reported by Suzuki et al. [47]. The V_H and V_L chains of the variable region from the antibody recognizing 4-hydroxy-3-nitrophenylacetyl (NP) were separately fused with *E. coli* alkaline phosphatase (PhoA) and *Streptococcal* protein G (proG), producing hybrid proteins V_H–PhoA and V_L–proG. The proG domain was attached to V_L to facilitate its immobilization on an IgG-coated plate. When NP and V_H–PhoA were incubated with V_L–proG attached to the microtiter plates, a significant increase was observed in the signal generated by the reporter PhoA. More recently, an extremely sensitive homogeneous OS assay was reported for the detection of small haptens [49]. The method has been shown to have superior sensitivity. It seems that both heterogeneous and homogeneous types of the OS assay will find a wide application in the area of clinical diagnostics or in the detection of trace chemicals in the environment.

Innovative DNA technologies have enhanced the diagnostic efficiency of murine monoclonal antibodies (mAbs). Nowadays, recombinant mAbs have been dissected into minimal binding fragment, rebuilt into multivalent high avidity reagents and fused with a range of molecules limited only by the imagination. Numerous fusion proteins containing immunologically relevant binding domains such as protein A or protein G, as well as portions of immunoglobulin molecules such as single-chain Fv and Fab, and reporter domain such as alkaline phosphatase, luciferase [50–52], and green fluorescent protein (GFP) [53,54] have been successfully applied in diagnostic assay development. Artificial antibodies seem on their way to join the rank of mAbs as powerful diagnostic agents and as the basis for the development of novel robust assay formats such as OS, microarray, and biosensors.

However, some problems still remain. For example, recombinant bispecific and bifunctional antibodies contain some foreign peptide sequences (linker, heterodimerization domain, etc.) that may be responsible for some false-positive result [55]. It has been found that some artificial composite antibodies have a 5–10-fold lower antigen binding affinity compared to that of the parent monoclonal antibodies [56–58]. There are many factors that may affect affinity and specificity of artificial antibody binding such as a possible conformational flexibility that was found to be responsible for a slight shift in antigen specificity observed by Huston et al. [56]. This finding as many other similar observations point towards the future directions in the rational design of artificial antibodies (see Chapter 15 for more extensive review of recombinant antibodies).

13.2.3 Designer Allergens

Allergic diseases continue to increase in prevalence and, currently, approximately 25% of the population in industrialized countries is affected [59]. Type I allergy is caused by the formation of immunoglobulin E (IgE) antibodies against allergens, which are highly soluble proteins of 10–100 kD [60,61]. Allergic sensitization occurs early in childhood and leads to the development of IgE-producing B cells and plasma cells [59]. The produced allergen-specific IgE antibodies are bound to the surface of effector cells (mast cells and basophils) via the high affinity receptor for IgE (FcεRI). Allergen exposure leads to cross-linking of IgE antibodies on the surface of effector cells, which then triggers the release of histamine and other mediators, causing an inflammatory tissue reaction [59].

Allergen-specific immunotherapy (SIT) is one of few efficient treatments and the only curative approach for allergies to some known allergens [61]. The main disadvantages for the application of native allergens in SIT are associated with the risk of anaphylactic and other IgE-mediated side effects. The recombinant molecular technology has made possible to develop safer forms of allergen vaccines (hypoallergens). Hypoallergens are substances with decreased IgE reactivity and allergenicity but conserved immunogenicity (T-cell reactivity) [62,63]. Current molecular approaches attempt to engineer allergens, aiming at a reduction of their IgE-binding epitopes while maintaining the T-cell activating capacity (T-cell epitopes) and the structural motives necessary for IgG antibody (blocking antibodies) induction [61].

13.2.3.1 Genetically Engineered Hypoallergens

Studies on the structure and IgE-binding sites of common allergens revealed that the presence of conformational epitopes on intact and folded allergens is of crucial importance for the IgE recognition [64,65]. By contrast, T-cell epitopes and most of IgG-binding sites can be of the continuous type and remain preserved, even if the conformation of protein is destroyed. Various strategies were used to construct hypoallergens with altered protein folds and IgE-binding capacity (reviewed in Refs. [61,64,66,67]). Many different approaches such as amino acid replacement

[68–77], and modification of disulfide bonds [78–80] or calcium-binding domain [81–83] were successfully applied to construct hypoallergens.

Bet v 1 is a single major allergenic protein of birch pollen. More than 95% of birch pollen-allergic individuals display IgE against Bet v 1 [77]. Recently, it has been shown that even one single amino acid substitution in Bet v 1 (Glu to Ser in position 45) inhibits IgE binding for 2%–50% [73]. More recently, two genetically engineered forms of the Bet v 1 with four and nine point mutations have been characterized [74]. All modified amino acid positions were located on the conserved surface of the allergen. As was shown earlier, majority of birch pollen allergic patients have serum IgG cross-reacting with major allergens of Fagales including birch, alder, hazel, and hornbeam. The selected amino acid residues were substituted to maximize changes in the surface properties (acidic or basic amino acid residues were substituted to neutral or vice versa). Both mutants were cloned and expressed in *E coli.* One was designed with four amino acid substitutions in positions 28, 32, 45, and 108, another Bet v 1 mutant was designed with eight point mutation at positions 5, 42, 45, 78, 103, 123, 134, 156, and one C-terminal extension. Both mutants demonstrate reduced capacity to bind human-specific serum IgE in inhibition assay with pool of sera from seven birch pollen allergic patients. Immunization of mice with Bet v1 mutants induced strong Bet v 1-specific IgG responses. Both mutants induce a significantly less histamine release compared to unmodified Bet v 1. The authors did not observe any significant difference in activity between these Bet v 1 mutants.

A different strategy for selecting positions for modification was employed by Ferreira et al. [77] to construct hypoallergenic variants of Bet v 1. Using an algorithm developed by Casari et al. [84] the authors predicted functional residues responsible for IgE recognition in 14 tree pollen isoallergens by analyzing IgE-binding activity of these isoallergens tested with serum specimens obtained from 9 birch pollen–allergic patients. This method uses a transformation of sequences into vectors in a generalized "sequence space." Projection of these vectors onto a lower-dimensional space reveals groups of residues specific for particular subfamilies that are predicted to be involved in protein function (in this case, IgE-binding activity).

The amino acid residues at positions 30, 57, 112, 113, and 125 of a high IgE-binding isoform of Bet v 1 were predicted to influence IgE binding and were selected for substitution with amino acids found at same positions in low IgE-binding isoforms. It was observed that recognition of Bet v 1 by IgE antibodies from allergic patients depended on at least six crucial amino acid residues/positions. The Bet v 1 six-point mutant exhibited an extremely low IgE binding activity in vitro. In vivo (skin prick) tests showed significantly decreased potency of the mutated form of Bet v 1 to induce typical urticarial type I reactions. Proliferation assays of allergen-specific T-cell clones demonstrated that these amino acid substitutions did not affect T-cell recognition.

Potential disadvantages of these strategies might be the solubility of hypoallergens because denatured or unfolded proteins can potentially form aggregates.

Oligomerization is another approach to engineer a nonimmunogenic form of recombinant allergens. Dimers and trimers of Bet v 1, the major allergen from birch pollen, were produced and analyzed in terms of structure, allergenic activity, and T- and B-cell activating potential by Vrtala and colleagues [85]. The recombinant Bet v 1 trimer consisting of three copies of the major birch pollen allergen had greatly reduced allergenic activity in skin tests and the experiment involving basophil histamine release [86,87]. The trimer represents a hypoallergenic form of Bet v 1 with profoundly reduced allergenic activity but contains the same B- and T-cell epitopes and similar secondary structure as the wild-type allergens. The trimer was able to induce Th1-cytocine release [85]. This is most likely caused by the decreased IgE-binding affinity because of a reorientation of IgE epitopes on the trimeric molecule or may be a result of microaggregation, steric or charge interactions hiding IgE epitopes. A similar effect has been shown for Derf 2, the major house dust mite allergen [88]. A mutant protein of Derf2 C8/119S containing serine in place of cysteine at positions 8 and 119 to abrogate an intramolecular disulfide bond shows a much reduced IgE binding with increased immunogenicity. Structural analyses have revealed that the degenerated secondary structure of C8/119S leads to stable molecular polymerization (around 125 molecules) in physiological solution that may be responsible for the loss of IgE epitopes and markedly augmented immunogenicity of mutant protein.

As pointed out above, disruption of the three-dimensional allergen structure destroys conformational IgE epitopes for the most allergens. If fragments of allergen sequences are reassembled in the order different from that in the folded wild-type allergen, an artificial antigen with altered or lacking fold may be obtained which will fail to react with IgE and will not induce degranulation of mast cells [89]. The development of artificial allergens constructed out of two or more epitopes of a given allergen or antigens/epitopes from other molecules or domain have been reported [90–96].

It has been shown that artificially constructed allergens consisting of two or four of the most important grass pollen allergens demonstrate a strong induction of T-cell response and induce high levels of allergen-specific IgG [91,93]. A fusion protein of the major cat allergen Fel d 1 with a truncated human IgGFcγ1 was constructed with the idea of inhibition of allergen-induced basophil and mast cell degradation by cross-linking of FcεRI and Fcγ receptors. The artificial protein shows reduced allergenic activity and ability to inhibit histamine release [95]. Artificial composite hypoallergens have been produced for several antigen sources. Most of these hypoallergens exhibit reduced allergenic activity in comparison to the corresponding wild-type allergen and retain T-cell reactivity.

13.2.3.2 Recombinant Allergens in Diagnostics

Currently, diagnosis of type I allergy is mostly performed using crude allergen extracts which allow the identification of the allergen-containing source, but not the disease-eliciting molecules [97]. The introduction of a large panel of recombinant allergens for in vitro tests allows for establishing a patient's individual reactivity profile, a.k.a. component resolved diagnosis (CRD) [98–100]. Some of these studies were performed in specific geographical population, revealing that reactivity profiles are determined by locally distributed allergens [101–103]. The increasing number of recombinant allergens has allowed the development of a powerful chip-based technology for simultaneous detection up to 5000 different allergens and epitopes [104–108]. Also, recombinant allergens have been used for follow-up and for specific IgG-antibody (blocking antibodies) titer determination [109–115].

In difference to the area of immunotherapy where designed artificial composite allergens with desired property are already successfully applied, the use of such artificial allergens in diagnostics is only at the very first discussion steps [89]. The advent of recombinant DNA technology has led to the development of individual recombinant allergens from the most important allergen sources. Recombinant allergens can be used to determine the individual sensitization profile of allergic patients and have allowed for developing novel therapeutic tools.

A number of different strategies have been applied to develop artificial allergens with reduced propensity to induce IgE-mediated side effects for immunotherapy. Of these, artificial composite hypoallergens are considered as the most promising agents. Recently, a clinical immunotherapy trial was carried out with a genetically engineered birch pollen allergen. Treatment with such modified allergen significantly improved the symptoms of birch pollen allergy, and induced a normal allergen-specific IgG response [116,117].

In conclusion, the development of artificial allergens (hypoallergens) is an attractive strategy for the improvement of both safety and efficiency of immunotherapy.

13.3 CONCLUSION

In the beginning of the twenty-first century the concept of artificial designer diagnostic antigens was already widely used in serologic assay development. Among other notable trends in protein engineering for diagnostics is the development of designer calibrators and detection reagents.

Novel artificial proteins not only provide a reliable source for highly efficient diagnostic reagents but generate unexpected opportunities for the development of new highly efficient diagnostic formats such as OS, microarray or biosensors, and new approaches to immunotherapy of allergies, all of which would be inconceivable or inefficient and impractical without the magic of protein engineering.

REFERENCES

1. Vasiljevic, N. and Markovic, L., Gene similarity between hepatitis C virus and human proteins—a blood transfusion problem, *Med. Pregl.*, 58, 582, 2005.
2. Gibson, W. et al., Evidence that HIV-1 gag precursor shares antigenic sites with the major capsid protein of human cytomegalovirus, *Virology*, 175, 595, 1990.
3. Fields, H.A. et al., Artificial mosaic proteins as new immunodiagnostic reagents: The hepatitis E virus experience, *Clin. Diagn. Virol.*, 5, 167, 1996.
4. Khudyakov, Y.E. et al., Artificial mosaic protein containing antigenic epitopes of hepatitis E virus, *J. Virol.*, 68, 7067, 1994.
5. Obriadina, A. et al., A new enzyme immunoassay for the detection of antibody to hepatitis E virus, *J. Gastroentorol. Hepatol.*, 17, 360, 2002.
6. Gomes-Solecki, M.J.C. et al., Recombinant chimeric borrelia proteins for diagnosis of Lyme disease, *J. Clin. Microbiol.*, 38, 2530, 2000.
7. Gerber, M.A. et al., Recombinant outer surface protein C ELISA for the diagnosis of early Lyme disease, *J. Infect. Dis.*, 171, 724, 1995.
8. Hansen, K., Hindersson, P., and Pedersen, N.S., Measurement of antibodies to the *Borrelia burgdorferi* flagellum improves serodiagnosis in Lyme disease, *J. Clin. Microbiol.*, 26, 338, 1988.
9. Mathiesen, M.J. et al., Peptide-based OspC enzyme-linked immunosorbent assay for serodiagnosis of Lyme borreliosis, *J. Clin. Microbiol.*, 36, 3474, 1998.
10. Chang, J.C. et al., Artificial NS4 mosaic antigen of hepatitis C virus, *J. Med. Virol.*, 59, 437, 1999.
11. Meng, J. et al., Identification and characterization of the neutralization epitope(s) of the hepatitis E virus, *Virology*, 288, 203, 2001.
12. Obriadina, A. et al., Antigenic properties of recombinant proteins of hepatitis E virus. In *Proceedings of the 10th International Symposium on Viral Hepatitis and Liver Disease*, Margolis, H.S., Alter, M.J., Conlon, R., Dienstag, J., and Liang, J. (Eds.), International Medical Press, London, 2002, 117.
13. Araujo, F.G., Analysis of trypanosoma cruzi antigens bound by specific antibodies and by antibodies to related trypanosomatids, *Infect. Immun.*, 53, 179, 1986.
14. Camargo, M.E. et al., Three years of collaboration on the standardization of Chagas' disease serodiagnosis in the Americas: An appraisal, *Bull. Pan. Am. Health. Organ.*, 20, 233, 1986.
15. Aguirre, S. et al., Design, construction, and evaluation of a specific chimeric antigen to diagnose chagasic infection, *J. Clin. Microbiol.*, 44, 3768, 2006.
16. Levitus, G. et al., Humoral autoimmune response to ribosomal P proteins in chronic Chagas heart disease, *Clin. Exp. Immunol.*, 85, 413, 1991.
17. Boarino, A. et al., Development of recombinant chimeric antigen expressing immunodominant B epitopes of *Leishmania infantum* for serodiagnosis of visceral leishmaniasis, *Clin. Diagn. Lab. Immunol.*, 12, 647, 2005.
18. Beghetto, E. et al., Chimeric antigens of *Toxoplasma gondii*: Toward standardization of toxoplasmosis serodiagnosis using recombinant products, *J. Clin. Microbiol.*, 44, 2133, 2006.
19. Kalamvoki, M. et al., Expression of immunoreactive forms of the hepatitis C NS5A protein in E. coli and their use for diagnostic assays, *Arch. Virol.*, 147, 1733, 2002.
20. Pereboeva, L.A. et al., Hepatitis C epitopes from phage-displayed cDNA libraries and improved diagnosis with a chimeric antigen, *J. Med. Virol.*, 60, 144, 2000.
21. Rickert, M. et al., Fusion proteins for combined analysis of autoantibodies to the 65-kDa isoform of glutamic acid decarboxylase and islet antigen-2 in insulin-dependent diabetes mellitus, *Clin. Chem.*, 47, 926, 2001.
22. Hernandez, M.M. et al., Use of a chimeric synthetic peptide from the core p19 protein and the envelope gp46 glycoprotein in the immunodiagnosis of HTLV-II virus infection, *Prep. Biochem. Biotechnol.*, 33, 29, 2003.

23. Chen, Y.P. et al., Rapid detection of hepatitis B virus surface antigen by an agglutination assay mediated by a bispecific diabody against both human erythrocytes and hepatitis B virus surface antigen, *Clin. Vaccine Immunol.,* 14, 720, 2007.
24. Kricka, L.J., Selected strategies for improving sensitivity and reliability of immunoassays, *Clin. Chem.,* 40, 347, 1994.
25. Robert, B. et al., Tumor targeting with newly designed biparatopic antibodies directed against two different epitopes of the carcinoembryonic antigen (CEA), *Int. J. Cancer.,* 81, 285, 1999.
26. Kricka, L.J., Chemiluminescent and bioluminescent techniques, *Clin. Chem.,* 37, 1472, 1991.
27. Behrsing, O. et al., Bispecific IgA/IgM antibodies and their use in enzyme immunoassay, *J. Immunol. Methods,* 156, 69, 1992.
28. Hamilton, R.G., Molecular engineering: Applications to the clinical laboratory, *Clin. Chem,* 39, 1988, 1993.
29. Hamilton, R.G., Engineered human antibodies as immunologic quality control reagents, *Ann. Biol. Clin. (Paris),* 48, 473, 1990.
30. Hamilton, R.G., Application of engineered chimeric antibodies to the calibration of human antibody standards, *Ann. Biol. Clin. (Paris),* 49, 242, 1991.
31. Hackett, J.J. et al., Recombinant mouse-human chimeric antibodies as calibrators in immunoassays that measure antibodies to *Toxoplasma gondii, J. Clin. Microbiol.,* 36, 1277, 1998.
32. Naess, L.M. et al., Quantitation of IgG subclass antibody responses after immunization with a group B meningococcal outer membrane vesicle vaccine, using monoclonal mouse-human chimeric antibodies as standards, *J. Immunol. Methods,* 196, 41, 1996.
33. Ichikawa, K. et al., A chimeric antibody with the human gamma1 constant region as a putative standard for assays to detect IgG beta2-glycoprotein I-dependent anticardiolipin and anti-beta2-glycoprotein I antibodies, *Arthritis Rheum.,* 42, 2461, 1999.
34. Skerra, A. and Schmidt, T.G., Applications of a peptide ligand for streptavidin: The Strep-tag, *Biomol. Eng.,* 16, 79, 1999.
35. Ducancel, F. et al., Recombinant colorimetric antibodies: Construction and characterization of a bifunctional F(ab)2/alkaline phosphatase conjugate produced in *Escherichia coli, Bio/Technology,* 11, 601, 1993.
36. Muller, B.H. et al., Recombinant single-chain Fv antibody fragment-alkaline phosphatase conjugate for one-step immunodetection in molecular hybridization, *J. Immunol. Methods,* 227, 177, 1999.
37. Rau, D., Kramer, K., and Hock, B., Single-chain Fv antibody-alkaline phosphatase fusion proteins produced by one-step cloning as rapid detection tools for ELISA, *J. Immunoassay and Immunochemistry,* 23, 129, 2002.
38. Kim, J.Y. et al., Genetically engineered elastin-protein A fusion as a universal platform for homogeneous, phase-separation immunoassay, *Anal. Chem.,* 77, 2318, 2005.
39. Gupta, A. and Chaudhary, V.K., Bifunctional recombinant fusion proteins for rapid detection of antibodies to both HIV-1 and HIV-2 in whole blood, *BMC Biotechnol.,* 6, 39, 2006.
40. Chowdhury, P.S. et al., An expression system for secretion and purification of a genetically engineered thermostable chimera of protein A and alkaline phosphatase, *Protein Expr. Purif.,* 5, 89, 1994.
41. Schuurman, J. et al., Production of a mouse/human chimeric IgE monoclonal antibody to the house dust mite allergen Der p 2 and its use for the absolute quantification of allergen-specific IgE, *J. Allergy Clin. Immunol.,* 99, 545, 1997.
42. Eliasson, M. et al., Chimeric IgG-binding receptors engineered from staphylococcal protein A and streptococcal protein G, *J. Biol. Chem.,* 263, 4323, 1988.
43. Nielsen, K. et al., Enzyme immunoassays for the diagnosis of brucellosis: Chimeric Protein A-Protein G as a common enzyme labeled detection reagent for sera for different animal species, *Vet. Microbiol.,* 101, 123, 2004.
44. Inoshima, Y. et al., Use of protein AG in an enzyme-linked immunosorbent assay for screening for antibodies against parapoxvirus in wild animals in Japan, *Clin. Diagn. Lab. Immunol.,* 6, 388, 1999.
45. Kelly, P.J. et al., Reactions of sera from laboratory, domestic and wild animals in Africa with protein A and a recombinant chimeric protein AG, *Comp. Immun. Microbiol. Infect. Dis.,* 16, 299, 1993.
46. Suzuki, C. et al., Open sandwich ELISA with V(H)-/V(L)-alkaline phosphatase fusion proteins, *J. Immunol. Methods,* 224, 171, 1999.
47. Suzuki, C. et al., Open sandwich enzyme-linked immunosorbent assay for the quantitation of small haptens, *Anal. Chem.,* 286, 238, 2000.

48. Ueda, H. et al., Open sandwich ELISA: A novel immunoassay based on the interchain interaction of antibody variable region, *Nat. Biotechnol.*, 14, 1714, 1996.
49. Yokozeki, T. et al., A homogeneous noncompetitive immunoassay for the detection of small haptens, *Anal. Chem.*, 74, 2500, 2002.
50. Zhang, X.M. et al., Genetically fused protein A-luciferase for immunological blotting analyses, *Anal. Biochem.*, 282, 65, 2000.
51. Maeda, Y. et al., Expression of a bifunctional chimeric protein A-Vargula hilgendorfii luciferase in mammalian cells, *Biotechniques*, 20, 116, 1996.
52. Maeda, Y. et al., Engineering of functional chimeric protein G-Vargula luciferase, *Anal. Biochem.*, 249, 147, 1997.
53. Aoki, T. et al., Construction of a fusion protein between protein A and green fluorescent protein and its application to western blotting, *FEBS Lett.*, 384, 193, 1996.
54. Ueda, H., Open sandwich immunoassay: A novel immunoassay approach based on the interchain interaction of an antibody variable region, *J. Biosci. Bioeng.*, 94, 614, 2002.
55. Kriangkum, J. et al., Bispecific and bifunctional single chain recombinant antibodies, *Biomol. Eng*, 18, 31, 2001.
56. Huston, J. et al., Protein engineering of antibody binding sites: Recovery of specific activity in an anti-digoxin single chain fv analogue produced in *Escherichia coli*, *Proc. Nat. Acad. Sci. U S A*, 85, 5879, 1988.
57. Reiter, Y. et al., Engineering antibody Fv fragments for cancer detection and therapy: Disulfide stabilized Fv fragments, *Nat. Biotechnol.*, 14, 1239, 1996.
58. Wels, W. et al., Construction, bacterial expression and characterization of a bifunctional single-chain antibody-phosphatase fusion protein targeted to the human erbB-2 receptor, *Biotechnology*, 10, 1128, 1992.
59. Niederberger, V. and Valenta, R., Molecular approaches for new vaccines against allergy, *Vaccines*, 5, 103, 2006.
60. Valenta, R. and Kraft, D., Recombinant allergen molecules: Tools to study effector cell activation, *Immunol. Rev.*, 179, 119, 2001.
61. Mutschlechner, S. and Ferreira, F., Engineering allergens, *Euro. Annals of Allergy Clin. Immunol.*, 38, 226, 2006.
62. Durham, S.R. and Till, S.J., Immunologic changes associated with allergen immunotherapy, *J. Allergy Clin. Immunol.*, 102, 157, 1998.
63. Frew, A.J., Immunotherapy of allergic disease, *J. Allergy Clin. Immunol.*, 111, 712, 2003.
64. Niederberger, V. and Valenta, R., Genetically modified allergens, *Immunol. Allergy Clin. N. Am.*, 24, 727, 2004.
65. Valenta, R. and Kraft, D., From allergen structure to new forms of allergen-specific immunotherapy, *Curr. Opin. Immunol.*, 14, 718, 2002.
66. Linhart, B. and Valenta, R., Molecular design of allergy vaccines, *Curr. Opin. Immunol*, 17, 646, 2005.
67. Bhalla, P.L. and Singh, M.B., Engineered allergens for immunotherapy, *Curr. Opin. Allergy Clin. Immunol.*, 4, 569, 2004.
68. Swoboda, I. et al., Mutants of the major ryegrass pollen allergen, Lol p 5, with reduced IgE-binding capacity: Candidates for grass pollen-specific immunotherapy, *Eur. J. Immunol.*, 32, 270, 2002.
69. Ferreira, F. et al., Dissection of immunoglobulin E and T lymphocyte reactivity of isoforms of the major birch pollen allergen Bet v 1: Potential use of hypoallergenic isoforms for immunotherapy, *J. Exp. Med.*, 183, 599, 1996.
70. Ferreira, F. et al., Isoforms of atopic allergens with reduced allergenicity but conserved T cell antigenicity: Possible use for specific immunotherapy, *Int. Arch. Allergy Immunol.*, 113, 125, 1997.
71. Takai, T. et al., Effects of proline mutations in the major house dust mite allergen Der f 2 on IgE-binding and histamine-releasing activity, *Eur. J. Biochem.*, 267, 6650, 2000.
72. Son, D.Y. et al., Pollen-related food allergy: Cloning and immunological analysis of isoforms and mutants of Mal d 1, the major apple allergen, and Bet y 1, the major birch pollen allergen, *Eur. J. Nutr.*, 38, 201, 1999.
73. Spangfort, M.D. et al., Dominating IgE-binding epitope of Bet v 1, the major allergen of birch pollen, characterized by x-ray crystallography and site-directed mutagenesis, *J. Immunol.*, 171, 3084, 2003.
74. Holm, J. et al., Allergy vaccine engineering: epitope modulation of recombinant bet v 1 reduces IgE binding but retains protein folding pattern for induction of protective blocking-antibody responses, *J. Immunol.*, 173, 5258, 2004.

75. Chan, S.L. et al., Nuclear magnetic resonance structure-based epitope mapping and modulation of dust mite group 13 allergen as a hypoallergen, *J. Immunol.*, 176, 4852, 2006.
76. Karisola, P. et al., Construction of hevein (hev b 6.02) with reduced allergenicity for immunotherapy of latex allergy by comutation of six amino acid residues on the conformational IgE Epitopes, *J. Immunol.*, 172, 2621, 2004.
77. Ferreira, F. et al., Modulation of IgE reactivity of allergens by site-directed mutagenesis: potential use of hypoallergenic variants for immunotherapy, *FASEB J.*, 12, 231, 1998.
78. Bonura, A. et al., Hypoallergenic variants of the *Parietaria judaica* major allergen Par j 1: A member of the non-specific lipid transfer protein plant family, *Int. Arch. Allergy Immunol.*, 126, 32, 2001.
79. Smith, A.M. and Chapman, M.D., Reduction in IgE binding to allergen variants generated by site-directed mutagenesis: Contribution of disulfide bonds to the antigenic structure of the major house dust mite allergen Der p 2, *Mol. Immunol.*, 33, 399, 1996.
80. Orlandi, A. et al., The recombinant major allergen of *Parietaria judaica* and its hypoallergenic variant: In vivo evaluation in a murine model of allergic sensitization, *Clin. Exp. Allergy*, 34, 470, 2004.
81. Engel, E. et al., Immunological and biological properties of bet v 4, a novel birch pollen allergen with two EF-hand calcium-binding domains, *J. Biol. Chem.*, 272, 28630, 1997.
82. Okada, T. et al., Engineering of hypoallergenic mutants of the Brassica pollen allergen, Bra r 1, for immunotherapy, *FEBS Lett.*, 434, 255, 1998.
83. Westritschnig, K. et al., Generation of an allergy vaccine by disruption of the three-dimensional structure of the cross-reactive calcium-binding allergen, Phl p 7, *J. Immunol.*, 172, 5684, 2004.
84. Casari, G., Sander, C., and Valencia, A., A method to predict functional residues in proteins, *Nat. Struct. Biol.*, 2, 171, 1995.
85. Vrtala, S. et al., Genetic engineering of a hypoallergenic trimer of the major birch pollen allergen, Bet v 1, *FASEB J.*, 15, 2045, 2001.
86. Pauli, G. et al., Comparison of genetically engineered hypoallergenic rBet v 1 derivatives with rBet v 1 wild-type by skin prick and intradermal testing: Results obtained in a French population, *Clin. Exp. Allergy*, 30, 1076, 2000.
87. Hage-Hamsten, M. et al., Skin test evaluation of genetically engineered hypoallergenic derivatives of the major birch pollen allergen, Bet v 1: Results obtained with a mix of two recombinant Bet v 1 trimerin a Swedish population before the birch pollen season, *J. Allergy Clin. Immunol.*, 104, 969, 1999.
88. Korematsu, S. et al., Mutation of major mite allergen derf-2 leads to degenerate secondary structure and molecular polymerization and induces potent and exclusive th1 cell differentiation, *J. Immunol.*, 165, 2895, 2000.
89. Linhart, B. and Valenta, R., Vaccine engineering improved by hybrid technology, *Int Arch. Allergy Immunol.*, 134, 324, 2004.
90. King, P. et al., Recombinant allergens with reduced allergenicity but retaining immunogenicity of the natural allergens: Hybrids of yellow jacket and paper wasp venom allergen antigen 5s, *J. Immunol.*, 166, 6057, 2001.
91. Linhart, B. et al., Combination vaccines for the treatment of grass pollen allergy consisting of genetically engineered hybrid molecules with increased immunogenicity, *FASEB J.*, 16, 1301, 2002.
92. Kussebi, F. et al., A major allergen gene-fusion protein for potential usage in allergen-specific immunotherapy, *J. Allergy Clin. Immunol.*, 115, 323, 2005.
93. Linhart, B. et al., A hybrid molecule resembling the epitope spectrum of grass pollen for allergy vaccination, *J. Allergy Clin. Immunol.*, 115, 1010, 2005.
94. Bonura, A. et al., A hybrid expressing genetically engineered major allergens of the Parietaria pollen as a tool for specific allergy vaccination, *Int. Arch. Allergy Immunol.*, 142, 274, 2007.
95. Zhu, D. et al., A chimeric human-cat fusion protein blocks cat-induced allergy, *Nat. Med.*, 11, 381, 2005.
96. Bohle, B. et al., A novel approach to specific allergy treatment: The recombinant fusion protein of a bacterial cell surface (S-Layer) protein and the major birch pollen allergen bet v 1 (rSbsC-Bet v 1) combines reduced allergenicity with immunomodulating capacity, *J. Immunol.*, 172, 6642, 2004.
97. Mothes, N., Valenta, R., and Spitzauer, S., Allergy testing: The role of recombinant allergens, *Clin. Chem. Lab. Med.*, 44, 125, 2006.

98. Mari, A., Skin test with a timothy grass (Phleum pretense) pollen extract vs. IgE to a timothy extract vs. IgE to rPhl p 1, rPhl p 2, nPhl p 4, rPhl p 5, rPhl p 6, rPhl p 7, rPhl p 11, and rPhl p 12: Epidemiological and diagnostic data, *Clin. Exp. Allergy,* 33, 43, 2003.
99. Satinover, S.M. et al., Specific IgE and IgG antibody-binding patterns to recombinant cockroach allergens, *J. Allergy Clin. Immunol.*, 115, 803, 2005.
100. Valenta, R. et al., The recombinant allergen-based concept of component-resolved diagnostics and immunotherapy (CRD and CRIT), *Clin. Exp. Allergy*, 29, 896, 1999.
101. Westritschnig, K. et al., Analysis of the sensitization profile towards allergens in central Africa, *Clin. Exp. Allergy*, 33, 22, 2003.
102. Ghunaim, N. et al., Antibody profiles and self-reported symptoms to pollen-related food allergens in grass pollen-allergic patients from northern Europe, *Allergy*, 60, 185, 2005.
103. Movérare, R. et al., IgE reactivity pattern to timothy and birch pollen allergens in Finnish and Russian Karelia, *Int. Arch. Allergy Immunol.,* 136, 33, 2005.
104. Hiller, R. et al., Microarrayed allergen molecules: Diagnostic gatekeepers for allergy treatment, *FASEB J.*, 16, 414, 2002.
105. Fall, B.I. et al., Microarrays for the screening of allergen-specific IgE in human serum, *Anal. Chem.*, 75, 556, 2003.
106. Jahn-Schmid, B. et al., Allergen microarray: Comparison of microarray using recombinant allergens with conventional diagnostic methods to detect allergen-specific serum immunoglobulin E, *Clin. Exp. Allergy*, 33, 1443, 2003.
107. Deinhofer, K. et al., Microarrayed allergens for IgE profiling, *Methods*, 32, 249, 2004.
108. Suck, R. et al., Rapid method for arrayed investigation of IgE-reactivity profiles using natural and recombinant allergens, *Allergy*, 57, 821, 2002.
109. Ball, T. et al., Induction of antibody responses to new B cell epitopes indicates vaccination character of allergen immunotherapy, *Eur. J. Immunol.*, 29, 2026, 1999.
110. Mothes, N. et al., Allergen-specific immunotherapy with a monophosphoryl lipid A-adjuvanted vaccine: Reduced seasonally boosted immunoglobulin E production and inhibition of basophil histamine release by therapy-induced blocking antibodies, *Clin. Exp. Allergy*, 33, 1198, 2003.
111. Neerven, R.J. et al., Blocking antibodies induced by specific allergy vaccination prevent the activation of CD4 + T cells by inhibiting serum-IgE-facilitated allergen presentation, *J. Immunol.*, 163, 2944, 1999.
112. Flicker, S. and Valenta, R., Renaissance of the blocking antibody concept in type I allergy, *Int. Arch. Allergy Immunol.*, 132, 13, 2003.
113. Kayhan, T. et al., Grass pollen immunotherapy induces mucosal and peripheral IL-10 responses and blocking IgG activity, *J. Immunol.*, 172, 3252, 2004.
114. Wachhols, P.A. et al., Inhibition of allergen-IgE binding to B cells by IgG antibodies after grass pollen immunotherapy, *J. Allergy Clin. Immunol.*, 113, 1225, 2004.
115. Wachhols, P.A. and Durham, S.R., Mechanisms of immunotherapy: IgG revisited, *Curr. Opin. Allergy Clin. Immunol.*, 4, 313, 2004.
116. Niederberger, V. et al., Vaccination with genetically engineered allergens prevents progression of allergic disease, *PNAS*, 101, 14677, 2004.
117. Reisinger, J. et al., Allergen-specific nasal IgG antibodies induced by vaccination with genetically modified allergens are associated with reduced nasal allergen sensitivity, *J. Allergy Clin. Immunol.*, 116, 347, 2005.

14 Peptide-Based Diagnostic Assays

María José Gómara and Isabel Haro

CONTENTS

14.1 INTRODUCTION

Health care largely depends on good diagnosis. There is increasing interest in the development of good preventive medicine, which has prompted research into new diagnostic markers allowing early detection of physiological disorders, months or even years before clinical symptoms appear and enabling doctors to treat diseases with a greater chance of success or even preventing progression to pathological manifestations.

In particular, there is an urgent need to improve the diagnosis of cancer, autoimmune, and neurological diseases. Early diagnosis of these primarily adult disorders facilitates the administration of shorter and more effective treatments with fewer side effects, improving the patient's quality of life. The development of new, highly specific, and sensitive diagnosis systems with prognostic significance would therefore enable patients requiring more aggressive therapies to be identified from the time of diagnosis.

For example, to prevent joint destruction in patients of rheumatoid arthritis (RA), early diagnosis and treatment are required. Although in some patients the diagnosis of early RA can be made during the first consultation, in many of them, it would be very useful for clinicians to have a serological test that can distinguish RA from other types of rheumatic diseases at an early stage of the disease. In this sense, new peptides bearing domains of fibrinogen and filaggrin proteins have been designed and synthesized [1]. These peptides can be used as a vehicle for the design of useful robust diagnostic strategies in RA (Figure 14.1). On the other hand, the investigation of the clinical significance of antibody responses against multiple tumor-associated antigens and their correlation with the clinical course of malignant diseases might provide important information regarding cancer prognosis and progression. As an example of a peptide-based immunoassay useful in cancer diagnosis, it has been reported a synthetic peptide from a dominant B-cell epitope of a shared tumor antigen NY-ESO-1 which detects specific antibodies present in the sera of patients with melanoma,

RA diagnosis → Citrullinated peptides

(α) Fibrin peptide – Cyclic filaggrin peptide

HSTKRGHAKSRPVCitG – HQCHQESTCitGRSRGRCGRSGS

HSTKRGHAKSCitPVCitG – HQCHQESTCitGRSRGRCGRSGS

HSTKCitGHAKSRPVCitG – HQCHQESTCitGRSRGRCGRSGS

(α) Fibrin peptide – (β) Fibrin peptide

ARGHCitPLDKKREEAPSLCitPA – GGG – HSTKCitGHAKSRPVCitG

FIGURE 14.1 Peptides bearing domains of fibrinogen and filaggrin proteins for the design of new diagnostic tests of rheumatoid arthritis.

prostate cancer, nonsmall cell lung cancer, esophageal cancer, gastric cancer, and hepatocellular carcinoma [2].

The need for diagnostic systems specific to a single analyte of interest and with a high degree of sensitivity requires tests to be optimized, and this involves extensive multidisciplinary research. Synthetic peptides are being used successfully here in a large number of biosensor applications ranging from conventional immunoenzymatic ELISA tests through to the development of miniaturized equipment for parallel detection of numerous analytes (high-throughput analysis) in a highly sensitive and specific way using complex biological samples such as peptide microarrays. Currently, the development of new detection systems based on the application of emergent technologies that fall within the field of nanotechnology is one of the most influential areas of biomedical research. These methods of detection that do not require the use of markers for biomolecules are characterized by having much greater resolution, resulting in an increase in sensitivity and specificity. These methods include mass spectrometry (MS), surface plasmon resonance (SPR), quartz crystal microbalance (QCM) analysis, atomic force microscopy (AFM) techniques, microelectromechanical systems (MEMS) or cantilever sensors, nanotube and nanowire sensors, and nanoparticles or quantum dots (diagnostic imaging). Many of these technologies are still at a preliminary research phase, but nevertheless there are already signs of the enormous potential for their application in the field of clinical diagnosis.

14.2 PEPTIDE-BASED IMMUNOASSAYS

The use of synthetic peptides as biosensors that are useful for the development of new diagnosis systems is a valid alternative to minimize the nonspecific reactions that arise in traditional immunoassays, and to remove the difficult reproducibility and the high variability sometimes observed in these assays. Another advantageous property of synthetic peptides is the possibility of modifying their chemical structure by (1) the covalent attachment of a variety of chemical moieties such as biotin, cysteine residue, fatty acids, and carrier proteins, (2) the incorporation of posttranslational modifications such as phosphorylation or citrullination, (3) the incorporation of non-natural building blocks such as D-amino acids, organic residues, and (4) the cyclization or (5) the multimerization of the antigenic epitope sequence.

In the last 20 years, several peptide sequences have been used to effectively increase the sensitivity and specificity of immunological assay systems that use recombinant proteins or native proteins as substrates [3–9]. Thus, a great number of peptide-based diagnostic systems are at present in development and some have already been commercialized. However, the use of relatively short linear peptides as antigenic substrates in immunoassay techniques is often not exempt from certain problems. One of the main drawbacks is the difficulty of passive adsorption on the polystyrene titration plates (standard ELISA procedure), which can potentially be overcome by covalent coupling to protein carriers [10]. Cysteinyl-glycyl-glycine (CGG) moiety derivatization [11], the use of derivatized peptides with hydrophobic tails of fatty acids that orientate and encourage the adsorption of the peptidic sequences on plates [12] or the covalent bonding of the peptide to the solid surface [13] are other alternatives to improve the adsorption of small peptides on the plates.

In recent years, it has been reported that there was an improvement in sensitivity and specificity when sera were tested not only with monomeric but also with multimeric peptides which contain more than one putative epitope [14–16]. There is a tendency toward using chimeric peptides that contain several epitopes from different proteins within the same chemical molecule to improve the sensitivity and specificity of the assays [17]. Published results demonstrate that chimeric peptides are more antigenic than the monomeric peptides and these can be used to detect antibodies to more than one epitope simultaneously [18–21]. In Figure 14.2, two chimeric peptides (Qm1, Qm2) with two different epitope orientations and a triglycyl spacer between the epitopes obtained in our group are shown [18].

Another issue of short linear peptides is that they do not usually adopt any stabilized conformation in solution, which is clearly detrimental to the affinity these have for antibodies, although the synthetic region constitutes an antigenic site [22]. Literature reports thus describe the restriction of mobility by the formation of disulfide bridges or intramolecular amide bonds, as an alternative that improves the peptide reactivity against antibodies [23–24]. Since epitope sequences are frequently located in accessible regions within the protein (i.e., β-turn or loop regions), cyclic peptides can mimic better the native secondary structure than the linear ones, thus being more appropriate candidates for the development of synthetic antigens. Reported results demonstrated significant statistical differences in the antigenicity observed for linear and cyclic constructs suggesting that the constrained peptides adopt a certain degree of structural organization that could be related with a better resemblance between the peptide epitope and the corresponding region within the whole viral protein [25–26].

Another attempt to increase the antigenicity of peptide sequences is the covalent attachment of epitopes to synthetic linear or branched polymeric scaffold of controlled chemical composition. Generations of branched poly-α-amino acid chains with a poly(L-lysine) backbone have been introduced for the rational design of polymeric polypeptides for construction of synthetic antigens [27–29]. Similarly, improvements in the antigenic capacity of peptides in homogeneous or heterogeneous multimeric form (multiple antigenic peptides, MAPs) have also been described [30–33]. The MAP system first described by Tam is based on a small immunogenically inert core matrix of lysine residues bearing radially branching synthetic peptides [34]. The methodology of MAP synthesis as well as the diagnostic uses of these macromolecular structures has been extensively reviewed [35–38]. The chemical synthesis of MAPs has changed from stepwise solid-phase peptide synthesis to ligation strategies, which are carried out with free soluble reunified fragments in solution. Therefore, the purification and characterization of the ligated products is much easier than previously reported. Extensive efforts have been made to develop new methodologies of

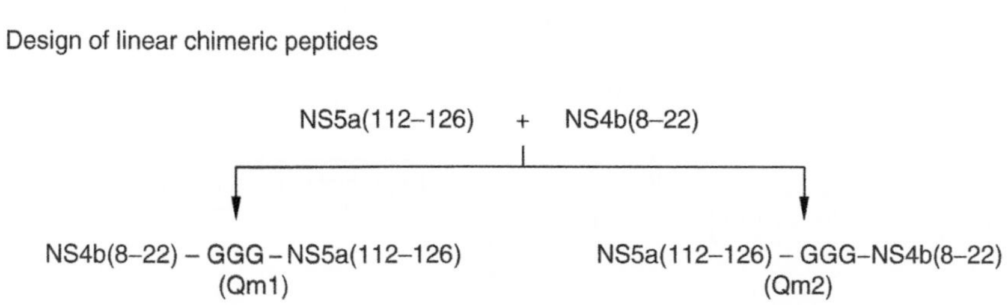

FIGURE 14.2 Colinear linking of NS4b(8–22) and NS5a(112–126) GBV-C/HGV antigens. The two chimeric peptides (Qm1, Qm2) contain different epitope orientations and a triglycyl spacer between the epitopes.

ligation chemistry that involve the fewest steps possible [39–41]. These chemistries are based on chemoselectivity and orthogonality of unprotected intermediates for assembling peptide dendrimers. The chemoselective ligation approaches can be subdivided into two main categories; those that generate a nonpeptide bond at the ligation site which include thioether, disulfide, thioester, hydrazone, oxime, and thiazolidine linkages and, the native chemical ligation that forms a peptide bond between unprotected peptide segments. The generation of nonpeptide bond by the unprotected segment ligation approaches has the advantage of a greater efficiency of the ligation reaction due to high and specific reactivity between the two functional groups. Nevertheless, the drawback of these ligation strategies relies on some of these linkages that are not stable as an amide bond or basic pH. Native chemical ligation, introduced by Kent [42], involves the chemoselective coupling of two protein fragments, one containing a C-terminal thioester and the other containing an N-terminal cysteine residue (Figure 14.3). This highly chemoselective reaction occurs in an aqueous solution at physiological pH in two steps. At the first step a transthioesterification reaction occurs between the thioester in one peptide and the sulfhydryl group from the N-terminal cysteine residue in the other. The intramolecular acylation rate, which is first order and often spontaneous, minimizes side reactions associated with enthalpic activation methods. This chemical strategy has been widely used for chemical synthesis of a multitude of medium-sized proteins.

Synthetic peptides of different complexity have been used for detection of specific antibodies of human diseases and have been recently reviewed [43]. As a result of the many advances in the synthesis of solid-phase peptide, there is currently great interest in working on the design of peptidic sequences that are three-dimensionally similar to the most significant protein epitopes and meanwhile ensure total homogeneity of the antigen used in immunoassays.

Regarding the most significant bibliography recently published, several reports have appeared comparing the diagnostic accuracy of peptide-based immunoassays for the diagnosis of celiac

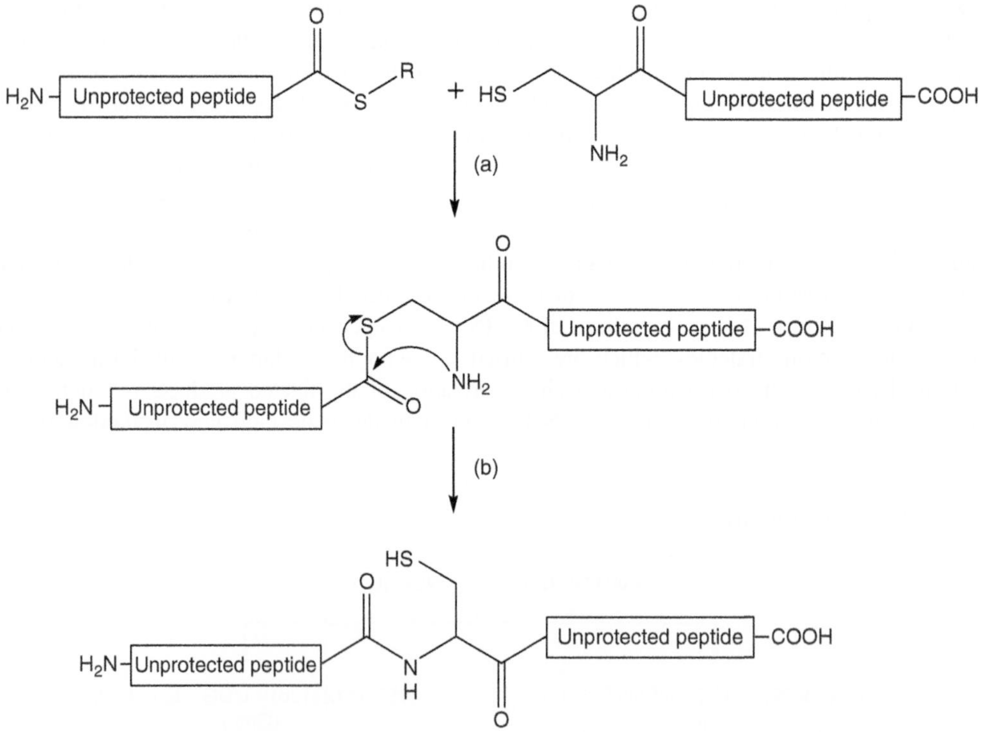

FIGURE 14.3 Synthesis of peptides by native chemical ligation of unprotected peptide fragments. (a) Transthioesterification. (b) Intramolecular acylation (S→N acyl shift).

disease [44], human herpesvirus 8 [45], and rheumatoid arthritis [46–47]. Extensive bibliography has also been published in the last year regarding the diagnostic value of autoantibodies to a cyclic citrullinated peptide in patients with rheumatoid arthritis [48–53]. Moreover, it has been described novel citrullinated synthetic peptides as autoantigens with diagnostic value in rheumatoid arthritis, such as the citrullinated telopeptide of type II collagen [54–55] or a citrullinated fibrin-filaggrin chimeric peptide [1].

Also, it should be pointed out the studies performed for optimization of the antigenic probes in the development of a simple diagnostic test based on a glycopeptide for detection of specific autoantibodies in multiple sclerosis [56–58].

Regarding human viral infections, two chimeric multiple antigenic peptides have been described for the simultaneous detection of specific antibodies to HIV-1 groups M, N, O, and HIV-2 [59] as well as a synthetic gag p24 epitope chemically coupled to BSA through a decaalanine peptide for serodiagnosis of HIV-1 [60]. Also, a synthetic peptide from the E1 glycoprotein of rubella virus has been used to study the humoral immune response to primary rubella virus infection [61]. In Table 14.1 the most recent peptide-based ELISA assays for clinical diagnosis of human diseases are summarized.

14.3 PEPTIDE-BASED MICROARRAYS

In recent years, microarrays have become an invaluable tool in biomedical research and specifically their use in the diagnosis sector is one of the most promising emerging fields in medicine [62]. Microarrays consist of biomolecules immobilized on different types of substrates in a spatially directed way. These substrates can be flat surfaces such as, for example, glass microscope slide, the wells of a microplate or arrays of beads. Immobilized molecules include mainly oligonucleotides, PCR products, proteins, peptides, and carbohydrates. Ideally, once these biomolecules are immobilized on the surface they should maintain their activity, be stable, and should not be eliminated in experimental steps. Many of these arrays are difficult to prepare due to the complexity and heterogeneity of some biomolecules, for example, proteins, and the need to preserve their conformational folding after binding to the surface. These microarrays enable the analysis of highly heterogeneous samples such as cellular extracts, serums, PCR products, or other types of samples allowing molecular recognition. One of the most important criteria in establishing quality in these type of experiments, therefore, is that the nonspecific binding of biomolecules should be minimal.

Despite the fact that this field has traditionally been dominated by DNA microarray technology, in recent years a wide variety of protein microarray applications has appeared and has recently been revised [63–66]. These revisions have mainly focused on the description of the technical approaches undertaken both for the development of new surfaces and in the use of new detection systems. Moreover, the most recent applications have been selected to illustrate future prospects in this field.

The production of protein microarrays generally requires cloning, overexpression, isolation, and purification of the proteins of interest. Despite the development of standardized protocols for the production and purification of proteins, the preparation of high-quality functionally active proteins is a major problem for the industrial production of protein microarrays at reasonable prices, so it is useful to emphasize that peptide arrays are an attractive alternative to the use of protein arrays since they are easier and cheaper to prepare in view of the less complex nature of the immobilized molecules. The development of chemical methodologies and surfaces enabling the immobilization and presentation of the peptides in a controlled way, the prevention of nonspecific interactions on the surface, and the development of analytical methods to determine the activity of the chip have in recent years driven the implementation of peptide arrays with applications in biomedicine. Peptide arrays have therefore been used in the identification of enzymatic inhibitors and substrates; the identification of potential protein ligands in research into new drugs (e.g., active ligands in cellular adhesion), in molecular immunology applications, for example, the identification of linear epitopes, and in the diagnosis of antibodies [67–70]. Specifically, in the field of clinical diagnosis, the limitations involved in conventional diagnostic techniques based mainly on immunoenzymatic

TABLE 14.1
Recent Peptide-Based Immunosorbent Assays for Clinical Diagnosis of Human Diseases

Human Diseases	Synthetic Antigen	Epitope Sequences	References
Viral infections			
Rubella virus	Linear peptide	E1(208–239): MNYTGNQQSRWGLGSPNCHGPDWASPVCQRHS	[61]
HIV-1, HIV-2	Chimeric MAP5	HIV-1 group M: DQQLLGIWGCSKLICTTA	[59]
	Chimeric MAP8	HIV-1 group N: DQQILSLWGCSGKTICYTT	
		HIV-1 group O: NQQLLNLWGCKGRLICYTS	
		HIV-2: DQAQLNSWGCAFRQVCHTT	
HIV-1	p24-Decapeptide-BSA conjugate	HIV-1 p24: IRQGPKEPFRDYVDRFFKTLRAEQA	[60]
Human herpesvirus 8	MAP4 (PK8.1)	orf K8.1: RSHLGFWQEGWSG	[45]
	Chimeric peptide (PK8.1-orf65)	orf 65: QVYQDWLGRMNCSYETAAPAVADARKPPSGKKK	
Autoimmune diseases			
Multiple sclerosis	Glycopeptide bearing a glucosyl moiety N-linked to an Asn.	CSF114(Glc): TPRVERN(Glc)GHSVFLAPYGWMVK	[57]
Rheumatoid arthritis	Citrullinated linear peptides	α1 chain type I and II collagens: EKAHDGGRYYX[a]A EKGPDPLQYMX[a]A	[54]
	Citrullinated cyclic chimeric peptide	Fibrin: HSTKRGHAKSRPVX[a]G	[1]
		Filaggrin: HQCHQESTX[a]GRSRGRCGRSGS	

[a] X: Citrulline.

tests (ELISA) as regards the requirements both for reagents and biological samples and the lack of capacity to analyze multiple samples have driven the development of miniaturized devices for the high-throughput, highly sensitive, and specific detection of antibodies from complex biological samples. In this respect, peptide microarrays have been described for the parallel detection of different specific antibodies from samples of serum with application in the diagnosis of infectious diseases [71–74], autoimmune diseases [75–76], and cancer [66,77].

In view of the restricted capacity of peptides to imitate the three-dimensional structure of the antigenic determinants of proteins, correct orientation of the peptides on the surface is required in order to promote their interaction with the target molecule. The immobilization of short peptides therefore generally requires covalent linking of the compounds to solid surfaces. In this respect, various methods for the preparation of peptide arrays have been described. These methods can be classified by the binding of the peptides on the surface either by in situ synthesis of the peptides or by the immobilization of functionalized peptides on chips.

The in situ synthesis of peptides on surface avoids the conventional preparation of hundreds of peptide sequences separately which involves a cost in terms of reagents and time, particularly in the purification processes. The main advantage of this approach is in the use of small quantities of reagents avoiding the purification and immobilization processes for each of the peptides making up an array. A general method for the synthesis of peptides in situ consists of the photolithographic synthesis described by Fodor et al. [78]. This is based on the solid-phase synthesis methods for peptides applied directly on an array surface using photolabile protective groups, so that synthesis only takes place in the illuminated areas of the surface. Later, Pellois et al. [79] described a strategy for the in situ synthesis of peptides using conventional units of amino acids based on the labile protective group for the tert-butoxycarbonyl acids. This strategy focuses on the use of a photoacid giving chemical deprotection of the peptide in illuminated areas, but the need for light sources and optical systems has limited the use of this type of technique.

A different approach in the development of peptide arrays in situ is the SPOT synthesis developed by Frank and based on the parallel synthesis of peptides directly on cellulose membranes [80]. In this strategy, the peptides are synthesized by sequentially spotting small volumes (generally in the range of microliters) of activated amino acids onto a porous membrane. The advantage of this method is that it is based on the same chemistry used in conventional solid phase synthesis. In addition, the arrays can be prepared by means of manual or automated distribution of the reagents.

An alternative method to the in situ synthesis of peptides is the immobilization of peptides that have previously been synthesized using chemical or recombinant methods. The advantage of this approach lies in the capacity to prepare a large number of identical peptide arrays because it enables numerous spots of the same peptide to be prepared from the same peptide solution, synthesizing the compound just once.

There are various strategies available for immobilizing peptides in arrays. It is important that the method does not require tedious protocols for chemically modifying the peptides with tags that allow reaction with the surface. The chemoselective binding of the N-terminal end of peptides to solid surfaces using glycoxylic acid and peptides containing a residue of N-terminal Cys [81], the use of the Diels–Alder reaction [82], and the native chemical ligation and avidine–biotin interaction have been described [71,83].

However, the properties of the surface can be critical in the development of the peptide array. The surfaces that avoid or minimize nonspecific adsorption can reduce the level of false positive results and ensure that a larger fraction of immobilized peptide is available for interaction with the target compound. The most common strategy for preventing nonspecific interactions consists of treating the array with a block protein such as BSA which is adsorbed on the surface to prevent later unwanted interactions. However, BSA can also block the interaction of the immobilized peptides; so it is preferable to use a nonprotein blocking agent such as triethanolamine [68].

Recently, the application of new substrates as well as derivatized surfaces was reported for the preparation of miniaturized devices for the high-throughput analysis of biomolecular interactions.

Particularly, a novel method was described for the chemical micropatterning based on printing of functionalized silica nanoparticles. The micropatterns are used for site-specific immobilization of peptides by Schiff-base chemistry. The immobilized peptides were shown to efficiently bind antibodies, thus demonstrating their accessibility for biomolecular interaction. The nanoparticle layer was characterized using a wide-field optical imaging technique called Sarfus allowing for visualizing nanometer films. The semicarbazide groups present on the surface of the nanoparticles were used for the site-specific semicarbazone ligation of unprotected peptides derivatized by an α-oxoaldehyde group [84].

In addition, a surface chip has been developed with a 1,1-carbonyldiimidazole-activated Tween 20-functionalized (CDI-Tween 20) high-yield single-walled carbon nanotube (SWNT) film: CDI-Tween20/SWCN [85]. These films are efficient substrates for microarray chip applications. A simple one-step treatment of carbon nanotubes (SWNT) films with CDI-Tween 20 allows for both the immobilization of probe proteins or peptides and efficient suppression of nonspecific binding without the need for BSA blocking. Furthermore, SWNT substrates help preserving the protein conformation. Because of its high aspect ratio and pseudo-3D-network structure, SWNT film can significantly reduce the contact area for protein immobilization, leading to minimal shape deformation for protein molecules.

Another important point in regard to the generation of peptide arrays is the uniform density of peptides at each spot that is necessary to enable a direct comparison of the measured activities between spots in the array. The capacity of surface to bind analytes depends on the quantity of immobilized molecules. The less functionalized the surface with the molecule, the lower the interaction with the target compound will be. Thus, a nonhomogeneous spot gives a lower signal than a homogeneous spot. Recently, the Brock's group studied the multivalency and heterogeneity of spots in microarrays based on the measurements of the binding constants [86]. They demonstrated that high-resolution visualization combined with the segmentation of images and separate analysis of heterogeneous spots gives valuable information on the method of binding the target molecules to the molecules of the array. The authors of the study suggested that the resolution obtained using visualization techniques (imaging resolution) can be a key factor in regard of the use of miniaturized binding tests. These authors highlighted the importance of using new techniques for obtaining high-resolution images together with segmentation of images, for example, visualization by surface plasmon resonance imaging (SPRi).

Conventionally, detection strategies used in protein microarrays are based on the use of labeled molecules, either fluorescent, chemiluminiscent or radioactive markers derived from the standardized protocols for immunotests or radioimmunotests. Currently, the development of new detection systems based on the application of nanotechnology is one of the most influential areas of biomedical research. These methods of detection that do not require the use of markers for biomolecules have much greater resolution, resulting in an increase in sensitivity and specificity. These methods include MS, SPR, QCMA, AFM techniques, microelectromechanical systems or cantilever sensors, nanotube, and nanowire sensors. Many of these technologies are still at a preliminary research phase. Nevertheless, there are already signs of the enormous potential for their application in the field of clinical diagnosis. In this chapter we shall only look at applications that have been described to date for these new technologies based on the use of peptides for diagnosis or detection with clinical implications.

14.4 PEPTIDE-BASED NANOTECHNOLOGIES

14.4.1 Surface Plasmon Resonance

The technology leading the field of detection without markers for molecular interactions is SPR. This technique consists of an optical biosensor that measures molecular interactions on the surface of a metal by detecting changes in the local refractive index. In this technique, a population of the molecules to be analyzed (ligand) is bound to a sensor surface whilst the other population of molecules (analyte) is injected in solution onto this surface. During the association phase, the analyte is exposed to the surface of the sensor chip. The optical detection system measures the change in the refractive index of the buffer next to the surface of the sensor which occurs when the mass of the analyte accumulates on the surface. If the reaction is prolonged sufficiently, equilibrium is achieved. The dissociation phase in which the surface is washed with buffer and the complexes dissociate over time, takes place later. The resonance signal during the association and dissociation phase is monitored against time, producing an interaction profile that we know as a sensorgram.

The SPR technique offers a number of advantages over conventional techniques such as fluorescence or the ELISA tests. Because of the fact that the interaction measurements between the biomolecules are based on the changes in the refractive index on the surface of the sensor, the detection of an analyte takes place directly and without the presence of markers. The analyte does not therefore require special characteristics (scattering bands) or markers (radioactive or fluorescent) and can be detected directly without the need to use detection protocols in which multiple steps are used. In addition, unlike the type of tests that require wash stages prior to the quantification of the interaction, SPR technology measures the formed complexes in the presence of free material without disturbing the equilibrium reaction of the complex and, thus, allows for monitoring weak or transient interactions. This advantage translates into a great versatility of the technique that enables analyzing molecules with a wide range of molecular weights and binding affinities. Furthermore, the interaction measures can be carried out in real time, enabling the user to obtain both kinetic and thermodynamic data. All these advantages have made the SPR technique a very useful tool in studying interactions between biomolecules.

SPR technology has been used to analyze antibody–antigen interactions, specific disease markers, protein–DNA interactions, DNA hybridization, and cellular ligation. A detailed review of the SPR applications is published every year by Myzska [87]. Particularly, SRP-based BIA technology was shown to have a high analytical performance in the diagnostics of several infectious diseases. Peptide-based-biosensor assays have been developed for the detection of specific antibodies against hepatitis A virus [12], hepatitis G virus [18,88], herpes simplex virus [89], and, recently, Epstein–Barr virus [90] and type I diabetes [91] in crude plasma samples. The ability of the SPR biosensor technology to directly detect specific antibodies at clinically relevant levels illustrates the potential of this technology for medical diagnostics. Figure 14.4 shows that the SPR biosensor can be used for monitoring rheumatoid arthritis autoantibodies by measuring the immunochemical reaction in real time. As shown, higher resonance unit values were measured when the assayed sera circulate through the citrulline-containing peptide channel, thus indicating an enhanced binding of [$Cit^{312,314}$] Filaggrin (306–324) to the autoantibodies in rheumatoid arthritis. The main disadvantage of this technology, however, is that up to now the commercially available SPR equipment offered a very limited number of independent sensor surfaces or channels. These systems are very useful for conducting studies with a small number of ligands but are not practical for high-throughput applications.

A recent development of SPR imaging combines the advantages of the traditional SPR technique with the capacity to analyze multiple samples in parallel (high-throughput analysis), which means that thousands of biomolecular interactions can be monitored simultaneously.

Typically, an SPR imaging apparatus consists of a coherent p-polarized light source expanded with a beam expander and consequently reflected from an SPR active medium to a detector. A CCD (charge-coupled device) camera collects the reflected light intensity in an image. SPR imaging measurements are performed at a fixed angle of incidence that falls within a linear region of the SPR dip, such that changes in light intensity are proportional to the changes in refractive index caused by binding of biomolecules to the surface. As a result, gray-level intensity correlates with the amount of material bound to the sensing region [92]. Numerous studies have been conducted to develop the SPR imaging technology. Today, there are various commercially available SPR instruments capable of analyzing multiple compounds (e.g., FLEXCHIP (BIAcore); MultiSPRinter (TOYOBO)).

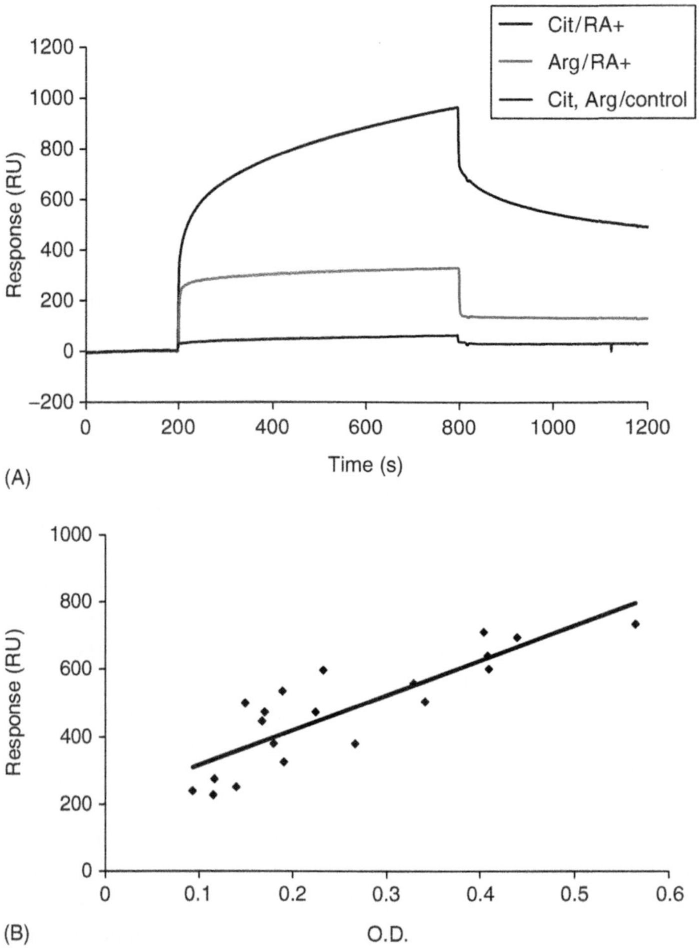

FIGURE 14.4 (A) Typical sensorgrams obtained when control and RA positive (RA+) patients' sera were injected over peptides immobilized on the dextran matrix. Arg: filaggrin(306–324); Cit: [$Cit^{312,314}$]filaggrin (306–324). (From Perez, T. et al., *Lett. Peptide Sci.*, 9, 291, 2002.) (B) Correlation between biospecific interaction analysis (BIA) and enzyme-linked immunosorbent assay (ELISA) using as immunoreagent [$Cit^{312,314}$]filaggrin(306–324) (R^2: 0.72).

Rich and Myszka have updated the recent technology on higher throughput systems [93]. Although the described instruments employ different optical systems, they have focussed their attention mainly on the developments in sample delivery and processing. These authors also notice that currently the sensitivity of the two-dimensional array-based systems is not sufficiently high to allow direct detection of small molecules bound to immobilized targets. Enzymatic-, radioligand-, and fluorescent-based assays can be 1000 times more sensitive than optical biosensors when analyzing targets that are not especially pure or highly active. Nevertheless, Cherif et al. [94] have developed a new method based on polypyrrole-peptide chips and SPRi that allows parallel analysis of multiple analytes in biological fluids. This technology is suitable for screening biological samples and for large-scale studies. This group has generated a chip bearing a large panel of peptide probes by successive electro-directed copolymerization of pyrrole–peptide conjugates on a gold surface. These peptide chips are easy to generate and can be regenerated several times without loss of efficiency. The usefulness of this system has been validated in a clinical setting by detecting anti-hepatitis C antibodies in patient-derived sera. The results have provided evidence that the signal produced by antibody binding is highly specific and reflects the antibody concentration of the tested solution in a dose-dependent manner. Moreover, when the SPRi method was compared to a classic enzyme-immunoassay, it was found that the results obtained by both assays are equivalent. The high specificity of this technique makes it applicable to the study and screening of complex biological samples such as undiluted sera, for which currently used methods are not suitable.

Inamori et al. have proposed a novel detection system for on-chip phosphorylation of peptides by SPRi using a newly synthesized phosphate capture molecule, particularly a biotinylated zinc (II) complex [95]. This complex is suitable for accessing phosphate anions as a bringing ligand on the two zinc (II) ions. The compound was exposed on the peptide array and detected with streptavidin (SA) via biotin–SA interaction by SPRi. This method can be applied to high-throughput analysis of kinase activity assays and kinase expression profiling. This detection method is independent of the amino acid residues and allows the detection of phosphorylation of all peptides on the array using a single-probe complex. As another advantage, this system is quantitative for phosphorylation efficiency, and the kinetics of on-chip phosphorylation can be studied using a peptide array.

14.4.2 Nanowire and Nanotube Sensors

This sensitive technology is based on a change in electrical conductance as proteins bind to a functionalized nanowire that bridges between two electrodes. Nanowires have a mean diameter of 30–100 nm, lengths of 5–10 μm, and sensor spacing from 50 nm to 2 μm, with arrays containing 2400 fabricated nanowire sensors. A miniature size and high density of sensors offer such advantages for the application of this technology as the use of small sample volume for large arrays. It has been demonstrated that these nanobiosensors provide ultrahigh sensitivity, fast response, and high-degree multiplex detection, yet with minimum sample requirements [96]. Antibody-functionalized silicon-nanowire sensors have been used for label-free, real-time multiplexed detection of protein cancer markers with high selectivity and femtomolar sensitivity [97]. Another described application of this nanotechnology is protein recognition via surface molecularly imprinted polymer nanowires, which selectively recognize various proteins such as albumin, hemoglobin, and cytochrome *c* [98]. It is based on the forming of a versatile recognition interface using peptides for electrically conducting carbon nanotubes, which enable sensing of soluble biomolecular targets. Recently, in order to develop nanoelectronic sensors for biological targets new self-assembling site-specific peptide linkers have been synthesized and evaluated for generation of stable and regenerable carbon nanotube biosensors interfaces [99].

An alternative to the carbon nanotubes in sensing applications are the peptide nanotubes. As a reported example, diphenylalanine peptides can self-assemble into discrete and well-ordered tubular structures since geometrically restricted aromatic interactions contribute order and directionality [100]. Peptide nanotubes were used in several biosensors applications. A peptide nanotube-based assay has been recently developed for attomolar detection of viruses [101]. Another peptide-based nanotube amperometric biosensor was developed for the sensitive determination of glucose and ethanol by the electrocatalytic oxidation of enzymatically liberated hydrogen peroxide and NADH, respectively [102]. The development of peptide nanotube electrodes for the sensitive measurement of enzymatically generated hydrogen peroxide could have clinical applications in detection of oxidoreductase substrates such as glucose, lactate, choline, and cholesterol.

14.4.3 Cantilevers

Cantilevers are silicon strips attached at one end to the capture molecule (antibody or protein) bound to the surface. The analyte binding onto the cantilever causes its bending due to the imposed weight. This bending can be measured by detecting the deflection of an optical beam, electrical resistance in a piezoelectric thin film on the cantilever, or mechanical resonant frequency (frequency shift due to uploaded mass). The application of the cantilever-based sensors offers unprecedented sensitivity and dynamic range for sensing biomolecular interactions. Recently, the nanomechanical biodetection for whole *B. subtilis* spores was developed using peptide-functionalized silicon cantilever arrays in two modes: stress/deflection measurements as well as frequency measurements [103], suggesting the application of peptide-funtionalized microcantilevers arrays for the real-time detection of multiple pathogenic organisms.

Additionally, nanomechanical cantilever sensors based on the electrical measurements of the resonant frequency change generated by the molecular interaction between antigen and antibody or inhibitor and receptor have been used for analytical and clinical applications. These sensors were used for the label-free detection of prostate-specific antigen (PSA) at the level of only 10 pg/mL [104]. Recently, the same research group developed a peptide-based cantilever sensor assay for detecting activated cyclic adenosine monophosphate-dependent protein kinase using the nanomechanical cantilever previously applied for electrical measurements of antibody–antigen interactions [105]. These authors suggested that quantitative analysis on the cantilevers has improved dynamic response and sensitivity over conventional assays, which would allow their use in the future for the investigation of activated protein kinase (PKA) in real samples without amplification or labeling of the samples.

14.5 PEPTIDE-BASED TOOLS FOR DIAGNOSTIC IMAGING

Diagnostic-imaging techniques make it possible to noninvasively evaluate certain areas of the human body. The positron emission tomography (PET) and single photon emission computed tomography (SPECT) techniques enable three-dimensional images to be obtained after detection of signals generated by a radioactively labeled molecular probe injected into the organism. PET is used mainly in clinical oncology, cardiology, and neurology, whilst SPECT is used in the detection of tumors and infections, for obtaining thyroid images and in the studies of certain bone problems. As has been widely described, the PET technique is in general superior to SPECT in terms of sensitivity, resolution, and quantification [106].

In recent years, a large number of radiopharmaceuticals have been developed for diagnostic imaging of tumors and other diseases. Small size peptides (3–50 amino acids) are the agents of choice in nuclear medicine for diagnostic imaging of dysfunctions in various organs as they are usually easy to synthesize, penetrate the tissue effectively, and are rapidly eliminated from the circulation. They have been used for diagnosing cancer, in cardiology, neurology, for detecting inflammation/infection, and for diagnosing arteriosclerosis and thrombosis. The choice of label is probably the most important factor in successful application of a radiopeptide. Second to that is the way the label is attached to the ligand, and how this affects and changes the biological properties of the molecule. Finally, the in vivo stability and the distribution pattern of the label are of utmost importance. The specific PET nuclide ^{11}C can be used to study the fate of biogenic peptides in vivo, which mainly is of basic research interest. The more long-lived ^{18}F has a high potential for clinical use but the present laborious time-consuming preparation techniques prevent a routine use of this radionuclide.

One of the major applications of peptides is the diagnosis of tumors. Depending on their origin, tumors can overexpress different types of receptors [107]. The design of radioactively labeled peptides with the ability to bind to those receptors can be used to visualize these lesions. It has, for example, been seen that various tumors express receptors of somatostatin, vasoactive intestinal peptide (VIP), bombesin/gastrin-releasing peptide (BN/GRP), α-melanocyte stimulating hormone (α-MSH), neurotensin (NT), α-M2 (EPPT), and cholecystokinin, and that these can be efficiently visualized by using labeled analogs of these peptides [108]. For example, tissue positive to the somatostatin receptor can be visualized in vivo by means of scintigraphy with analogs labeled with ^{111}indium [109], and the VIP peptide labeled with ^{123}iodine has been used to detect colorectal cancer and primary adenocarcinomas [110–111]. Recently, the synthesis of ^{111}In-labeled RGD-dendrimers consisting of multiple branched monomers to obtain better tumor targeting properties and better in vivo imaging properties than its dimeric and monomeric congeners has been described [112].

The VIP receptors are expressed in various tumors in a much greater density than the somatostatin receptors. In recent years, the production of various radioactively labeled analogs of VIP (e.g., ^{123}I-VIP and ^{18}F-(Arg^{15}, Arg^{21})-VIP to name a few) and their use in diagnosing tumors that overexpress receptors of this peptide have been described, but the rapid proteolytic degradation has severely limited their clinical application. Bhargava and collaborators have performed a complete substitutional analysis of VIP for better tumor-imaging properties [113].

More recently, a new analog, [R^{8,15,21}, L^{17}]-VIP, in which the substitution of the amino acids Asp^{8}, Lys^{15}, and Lys^{21} by Arg and Met^{17} by Leu and labeled with ^{18}F has given the VIP a greater capacity to bind with its receptor and a greater proteolytic stability than the native sequence [114].

The main disadvantage of this approach, however, is that radioactive labeling requires the use of special tools and highly qualified personnel who on occasion are not easily accessible. As a result, imaging techniques based on fluorescence are frequently used instead to study certain diseases at molecular level, requiring the use of epifluorescence and confocal microscopes in combination with fluorescent probes. Fluorescent labeling is simple and does not require any special equipment. The conjugation of peptides with fluorescent probes has a great many advantages as it can be done using well-established chemical reactions [115].

Frequently, peptides are conjugated to a cyanine derivative, such as ICG, Cy5, Cy5, 5, Cypate or NIR820 [116]. Becker et al. describe the in vivo diagnostic use of a peptide–dye conjugate consisting of a cyanine dye and the somatostatin analog octreotate as a contrast agent for optical tumor imaging. They claimed that this imaging approach, combining the specificity of ligand/receptor interaction with near-infrared fluorescence detection, may be applied in various other fields of cancer diagnosis [117].

A more recent approach and one of the main areas of investment, both public and private, is the use of nanotechnology in the field of diagnostic imaging. One of the most promising lines of investigation currently is quantum dots (qdots). The qdots used in bioapplications are nanoscale crystalline structures synthesized from various types of semiconductor materials (CdSe, CdS, CdTe, InP, InAs, etc.) that have composition- and size-dependent absorption and emission. The unique optical properties of qdots make them appealing as in vivo and in vitro fluorophores in biological investigations, in which current organic fluorophores fall short of providing long-term stability and simultaneous detection of multiple signals. Most dye conjugates are synthesized by attaching one or more fluorophores to a single biomolecule; however, the large surface area afforded by the nanocrystal fluorophore allows simultaneous conjugation of many biomolecules to a single qdot nanocrystal. Advantages conferred by this approach include increased avidity for targets, the potential for cooperative binding in some cases, and the use of efficient signal amplification methodologies. Qdots particles enable powerful new approaches to genetic analysis, drug discovery, and disease diagnostics. Cellular labeling using qdots has made the most progress and attracted the greatest interest. Within the last years, numerous reports describe the ability of one or more "colour/size" of biofunctionalized qdots to label cells.

A number of approaches enabling coating nanocrystals with water-soluble materials are developed to facilitate the use of qdots in imaging techniques in biological environment [118]. In this sense, synthetic peptides are used both to improve the solubility of the qdots in aqueous solutions and to direct them towards biological targets [119]. For example, Pinaud et al. [120] coated qdots with peptides composed of two different domains: A hydrophobic domain derived from phytochelatin rich in cysteines to carry out thiolation on the qdots and hydrophilic domain composed of glycine, serine, and glutamate to give the qdots the necessary solubility in water. Various peptide-coating approaches are used to modulate the nanocrystal surface properties and bioactivate the nanoparticles. CdSe/ZnS nanocrystals coated with biotinylated peptides efficiently bind to streptavidin and are specifically targeted to GPI-anchored avidin-CD14 chimeric proteins expressed on the membranes of live HeLa cells. This peptide-coating surface chemistry provides a novel approach for the production of biocompatible photoluminescent nanocrystal probes. Delehanty et al. [121] describe a bifunctional oligoarginine cell penetrating peptide (based on the HIV-1 Tat protein motif) bearing a terminal polyhistidine tract that facilitates the transmembrane delivery of the qdots bioconjugates. The polyhistidine sequence allows the peptide to self-assemble onto the qdots surface via metal–affinity interactions while the oligoarginine sequence allows specific qdots delivery across the cellular membrane and intracellular labeling as compared to nonconjugated qdots. In 2005, Chang et al. [122] designed qdots coated with a polyethylene glycol polymer to increase the solubility of the nanocrystals that were functionalized with carboxylic acids later used to detect collagenase activity (metalloprotease of therapeutic interest).

Medintz et al. [123] later developed probes based on qdots for analysis of different proteases and even more recently Shi et al. [124] designed qdots as radiometric sensors based on FRET (fluorescence resonance energy transfer) to measure protease activity in vivo. Qdots labeled peptides are also emerging as promising candidates for determination of protease activities. The authors have synthesized rhodamine-labeled peptide-coated CdSe/ZnS qdots and use them as FRET probes to monitor the proteolytic activity of matrix metalloproteinases (MMPs) in normal and cancerous cell cultures and were able to discriminate between a normal and cancerous tissue in less than 15 min. The method can be extended to other applications involving overexpression of proteolytic activity. Changing the peptide sequence would enable measuring the activity of specific proteolytic enzymes. It could also enable high-throughput screening of protease inhibitors and activators in an array format.

A recent example of the use of qdots for detecting tumor tissue both in vitro and in vivo is described by Cai et al. [125] who used for the first time a cyclic peptide containing the integrin binding motif (RGD tripeptide) conjugated to CdTe qdots. Chan et al. [126], exploiting the dependence of fluorescent emission on the dimension of nanocrystals, used different sizes of qdots for the cellular labeling of three different biological targets.

Finally, it is important to realize that one of the main questions to consider is the biocompatibility of qdots since Cd and Se, which are known to be toxic, occur in the most common composition of these nanocrystals. However, the large number of already conducted studies have shown that qdots are not cytotoxic because coating using ZnS and other biocompatible organic materials protect the oxidation and release of Cd and Se during the course of the test [127]. In fact, the few studies that signal a moderate cytotoxicity in the cells are those using qdots coated with bifunctional ligands that are unstable and undergo constant absorption/desorption processes [128,129]. In addition to the noncytotoxicity, it is also very important that the qdots do not interfere with other physiological or cellular processes, whether due to nonspecific bonds, ester impediment, or to changes in the functionality or diffusion of the biomolecules. Although nonspecific binding can be minimized by using hydrophilic polymers such as polyethylene glycol (PEG), a great many studies, both basic and applied in nanoengineering, still need to be carried out before intravenous injection of these nanocrystals and, therefore, their capacity as cellular markers for tissues and diseased organs in humans becomes reality. However, it is a confirmed fact that qdots as biological probes have met the expectations that had initially been suggested. Although they cannot replace the well-established technologies that use fluorophors or fluorescent proteins, they do in fact complement them in applications requiring greater photostability. In conclusion, it is fair to say that qdots are powerful tools for biological detection and imaging.

14.6 CONCLUSION

The immunodiagnostic methods have expanded over time and now are one of the most powerful tools for individual diagnosis as well as for epidemiological studies. Immunoenzymatic assays have become the most widely used diagnostic techniques. These assays are used either for detection of antibodies or circulating antigens. They are simple, sensitive, and cost effective. Although immunoenzymatic assays enjoy high popularity as diagnostic tools, the application of these assays has significant limitations including very specific requirements for the quality and preparation of diagnostic reagents and biological samples and the lack of capacity to efficiently analyze multiple samples. All these limitations spurred the development of miniaturized devices for the high-throughput, highly sensitive, and specific detection of biomarkers from complex biological samples. The advent of nanotechnology is changing the scale and methodology of diagnostics. The range of potential diagnostic applications is unlimited. This chapter, however, describes only the most important current applications for nanoscale devices using specially arranged synthetic peptides in diagnosis of human diseases such as cancer, infectious diseases, and autoimmune disorders.

ACKNOWLEDGMENTS

This research was supported by grants BQU2000-0793-CO2-02, BQU2003-05070-CO2-02, CTQ2006-15396-CO2 from the Ministerio de Ciencia y Tecnología (Spain) and by the Foundation La Marató de TV3 (Catalonia) (project 030331).

We greatly appreciate the valuable contribution to the projects of all colleagues in our research group at the Department of Peptides & Protein Chemistry, IIQAB-CSIC.

REFERENCES

1. Pérez, M.L. et al., Antibodies to citrullinated human fibrinogen synthetic peptides in diagnosing rheumatoid arthritis, *J. Med. Chem.*, 50, 3573, 2007.
2. Zeng, G. et al., Dominant B cell epitope from NY-ESO-1 recognized by sera from a wide spectrum of cancer patients: Implications as a potential biomarker, *Int. J. Cancer*, 114, 268, 2005.
3. Mahler, M., Bluthner, M., and Pollard, K.M., Advances in B-cell epitope analysis of autoantigens in connective tissue diseases, *Clin. Immunol.*, 107, 65, 2003.
4. Gnann, J.W. et al., Synthetic peptide immunoassay distinguishes HIV type 1 and HIV type 2 infections, *Science*, 237, 1346, 1987.
5. Favorov, M.O. et al., Enzyme immunoassay for the detection of antibody to hepatitis E virus based on synthetic peptides, *J. Virol. Methods*, 46, 237, 1994.
6. Gevorkian, G. et al., Serologic reactivity of a synthetic peptide from human immunodeficiency virus type 1 gp41 with sera from a Mexican population, *Clin. Diagn. Lab. Immunol.*, 3, 651, 1996.
7. Shin, S.Y. et al., The use of multiple antigenic peptide (MAP) in the immunodiagnosis of human immunodeficiency virus infection, *Biochem. Mol. Biol. Int.*, 43, 713, 1997.
8. Hernández, M. et al., Cysticercosis: Towards the design of a diagnostic kit based on synthetic peptides, *Immunol. Lett.*, 71, 13, 2000.
9. El Awady, M.K. et al., Synthetic peptide-based immunoassay as a supplemental test for HCV infection, *Clin. Chim. Acta*, 325, 39, 2002.
10. Hudecz, F., Manipulation of epitope function by modification of peptide structure: A minireview, *Biologicals*, 29, 197, 2001.
11. Manocha, M. et al., Comparing modified and plain peptide linked enzyme immunosorbent assay (ELISA) for detection of human immunodeficiency virus type-1 (HIV-1) and type-2 (HIV-2) antibodies, *Immunol. Lett.*, 85, 275, 2003.
12. Gómara, M.J. et al., Use of linear and multiple antigenic peptides in the immunodiagnosis of acute hepatitis A virus infection, *J. Immunol. Methods*, 234, 23, 2000.
13. Schellekens, G.A. et al., Citrulline is an essential constituent of antigenic determinants recognized by rheumatoid arthritis-specific autoantibodies, *J. Clin. Invest.*, 101, 273, 1998.
14. Marin, M.H. et al., Antigenic activity of three chimeric synthetic peptides of the transmembrane (Gp41) and the envelope (Gp120) glycoproteins of HIV-1 virus, *Prep. Biochem. Biotech.*, 34, 227, 2004.
15. Hernandez, M. et al., Antigenicity of chimeric synthetic peptides based on HTLV-1 antigens and the impact of epitope orientation, *Biochem. Biophys. Res. Commun.*, 276, 1085, 2000.
16. Drakopoulou, E. et al., Synthesis and antibody recognition of mucin 1 (MUC1)-alpha-conotoxin chimera, *J. Pept. Sci.*, 6, 175, 2000.
17. Hernandez, M. et al., Chimeric synthetic peptide as antigen for immunodiagnosis of HIV-1 infection, *Biochem. Biophys. Res. Commun.*, 272, 259, 2000.

18. Perez, T. et al., Antigenicity of chimeric and cyclic synthetic peptides based on nonstructural proteins of GBV-C/HGV, *J. Pept. Sci.*, 12, 267, 2006.
19. Hernandez, M. et al., Chimeric synthetic peptides as antigens for detection of antibodies to Trypanosoma cruzi, *Biochem. Biophys. Res. Commun.*, 339, 89, 2006.
20. Hernandez, M. et al., Chimeric synthetic peptides containing two immunodominant epitopes from the envelope gp46 and the transmembrane gp21 glycoproteins of HTLV-I virus, *Biochem. Biophys. Res. Commun.*, 289, 1, 2001.
21. Hernandez, M. et al., Chimeric synthetic peptides from the envelope (gp46) and the transmembrane (gp21) glycoproteins for the detection of antibodies to human T-cell leukemia virus type II, *Biochem. Biophys. Res. Commun.*, 289, 7, 2001.
22. Dyson, H.J. and Wright, P.E., Antigenic peptides, *FASEB J.,* 9, 37, 1995.
23. Leonetti, M. et al., Immunization with a peptide having both T-cell and conformationally restricted B-cell epitopes elicits neutralizing antisera against a snake neurotoxin, *J. Immunol.,* 145, 4214, 1990.
24. Conley, A.J. et al., Immunogenicity of synthetic HIV-1 gp120 V3-loop peptide-conjugate immunogens, *Vaccine,* 12, 445, 1994.
25. Schellekens, G.A. et al., The diagnostic properties of rheumatoid arthritis antibodies recognizing a cyclic citrullinated peptide, *Arth. Rheum.*, 43, 155, 2000.
26. Mezo, G. et al., Synthesis and comparison of antibody recognition of conjugates containing herpes simplex virus type 1 glycoprotein D epitope VII, *Bioconjug. Chem.,* 14, 1260, 2003.
27. Hudecz, F., Alteration of immunogenicity and antibody recognition of B-cell epitopes by synthetic branched chain polypeptide carriers with poly[L-lysine] backbone, *Biomed. Pept. Prot. Nucleic Acids,* 1, 213, 1995.
28. García, M. et al., Analysis of the interaction with biomembrane models of the HAV-VP3(101–121) sequence conjugated to synthetic branched chain polypeptide carriers with a poly[L-lysine] backbone, *Langmuir*, 14, 1861, 1998.
29. Sospedra, P. et al., Physicochemical behavior of polylysine-[HAV-VPS peptide] constructs at the air-water interface, *Langmuir*, 15, 5111, 1999.
30. Tam, J.P., Recent advances in multiple antigen peptides, *J. Immunol. Methods*, 196, 17, 1996.
31. Nardin, E.H. et al., The use of multiple antigen peptides in the analysis and induction of protective immune responses against infectious diseases, *Adv. Immunol.*, 60, 105, 1995.
32. Pinto, R.M. et al., Enhancement of the immunogenicity of a synthetic peptide bearing a VP3 epitope of hepatitis A virus, *FEBS Lett.*, 438, 106, 1998.
33. Firsova, T. et al., Synthesis of a diepitope multiple antigen peptide containing sequences from VP1 and VP3 proteins of hepatitis A virus and its use in hepatitis A diagnosis, in *Peptides 1996*, Ramage, R. and Epton, R. (Eds.), Mayflower Scientific Ltd., Kingswinford, England, 1998, 383.
34. Tam, J.P., Synthetic peptide vaccine design-synthesis and properties of a high-density multiple antigenic peptide system, *Proc. Natl. Acad. Sci. U S A,* 85, 5409, 1988.
35. Sadler, K. and Tam, J.P., Peptide dendrimers: Applications and synthesis, *Rev. Mol. Biotechnol.*, 90, 195, 2002.
36. Tam, J.P., Xu, J., and Eorn, K.D., Methods and strategies of peptide ligation, *Biopolymers,* 60, 194, 2001.
37. Niederhafner, P., Sebestik, J., and Jezek, J., Peptide dendrimers, *J. Pept. Science*, 11, 757, 2005.
38. Papas, S., Strongylis, C., and Tsikaris, V., Synthetic approaches for total chemical synthesis of proteins and protein-like macromolecules of branched architecture, *Curr. Org. Chem.*, 10, 1727, 2006.
39. Tam, J.P., Yu, Q., and Miao, Z., Orthogonal ligation strategies for peptide and protein, *Biopolymers*, 51, 311, 1999.
40. Kimmerlin, T. and Seebach, D., '100 years of peptide synthesis': Ligation methods for peptide and protein synthesis with applications to beta-peptide assemblies, *J. Peptide Res.*, 65, 229, 2005.
41. Bode, J.W., Emerging methods in amide- and peptide-bond formation, *Curr. Opin. Drug Disc.*, 9, 765, 2006.
42. Dawson, P.E. et al., Synthesis of proteins by native chemical ligation, *Science*, 266, 776, 1994.
43. Gómara, M.J. and Haro, I., Synthetic peptides for the immunodiagnosis of human diseases, *Curr. Med. Chem.*, 14, 531, 2007.
44. Villalta, D. et al., Testing for IgG class antibodies in celiac disease patients with selective IgA deficiency. A comparison of the diagnostic accuracy of 9 IgG anti-tissue transglutaminase, 1 IgG anti-gliadin and 1 IgG anti-deaminated gliadin peptide antibody assays, *Clin. Chim. Acta*, 382, 95, 2007.

45. Pérez, C. et al., Correlations between synthetic peptide-based enzyme immunoassays and immunofluorescence assay for detection of human herpesvirus 8 antibodies in different Argentine populations, *J. Med. Virol.*, 78, 806, 2006.
46. Coenen, D. et al., Technical and diagnostic performance of 6 assays for the measurement of citrullinated protein/peptide antibodies in the diagnosis of rheumatoid arthritis, *Clin. Chem.*, 53, 498, 2007.
47. Cruyssen, B.V. et al., Diagnostic value of anti-human citrullinated fibrinogen ELISA and comparison with four other anti-citrullinated protein assays, *Arthritis Res. Ther.*, 8, R122, 2006.
48. Van Venrooij, W.J., Zendman, A.J.W., and Pruijn, G.J.M., Autoantibodies to citrullinated antigens in (early) rheumatoid arthritis, *Autoimmun. Rev.*, 6, 37, 2006.
49. Ates, A., Karaaslan, Y., and Aksaray, S., Predictive value of antibodies to cyclic citrullinated peptide in patients with early arthritis, *Clin. Rheumatol.*, 26, 499, 2007.
50. Raptopoulou, A. et al., Anti-citrulline antibodies in the diagnosis and prognosis of rheumatoid arthritis: Evolving concepts, *Crit. Rev. Cl. Lab. Sci.*, 44, 339, 2007.
51. Niewold, T.B., Harrison, M.J., and Paget, S.A., Anti-CCP antibody testing as a diagnostic and prognostic tool in rheumatoid arthritis, *QJM- Int. J. Med.*, 100, 193, 2007.
52. Vannini, A. et al., Anti-cyclic citrullinated peptide positivity in non-rheumatoid arthritis disease samples: Citrulline-dependent or not? *Ann. Rheum. Dis.*, 66, 511, 2007.
53. Ingegnoli, F. et al., Use of antibodies recognizing cyclic citrullinated peptide in the differential diagnosis of joint involvement in systemic sclerosis, *Clin. Rheumatol.*, 26, 510, 2007.
54. Koivula, M.K. et al., Autoantibodies binding to citrullinated telopeptide of type II collagen and to cyclic citrullinated peptides predict synergistically the development of seropositive rheumatoid arthritis, *Ann. Rheum. Dis.*, 66, 1450, 2007.
55. Koivula, M.K. et al., Inhibitory characteristics of citrullinated telopeptides of type I and II collagens for autoantibody binding in patients with rheumatoid arthritis, *J. Rheumatology*, 45, 1364, 2006.
56. Carotenuto, A. et al., Conformation-activity relationship of designed glycopeptides as synthetic probes for the detection of autoantibodies, biomarkers of multiple sclerosis, *J. Med. Chem.*, 49, 5072, 2006.
57. Papini, A.M., Simple test for multiple sclerosis, *Nat. Med.*, 11, 13, 2005.
58. Lolli, F. et al., An N-glucosylated peptide detecting disease-specific autoantibodies, biomarkers of multiple sclerosis, *Proc. Nat. Acad. Sci. U S A*, 102, 10273, 2005.
59. Pau, C.P., Luo, W., and McDougal, J.S., Chimeric multiple antigenic peptides for simultaneous detection of specific antibodies to HIV-1 groups M, N, O, and HIV-2, *J. Immunol. Methods*, 318, 59, 2007.
60. Singh, S.K., Shah, N.K., and Bisen, P.S., A synthetic gag p24 epitope chemically coupled to BSA through a decaalanine peptide enhances HIV type 1 serodiagnostic ability by several folds, *AIDS Res. Hum. Retrov.*, 23, 153, 2007.
61. Wilson, K.M. et al., Humoral immune response to primary rubella virus infection, *Clin. Vaccine Immunol.*, 13, 380, 2006.
62. Ling, M.M., Ricks, C., and Lea, P., Multiplexing molecular diagnostics and immunoassays using emerging microarray technologies, *Expert. Rev. Mol. Diagn.*, 7, 87, 2007.
63. Cretich, M. et al., Protein and peptide arrays: Recent trends and new directions, *Biomol. Eng.*, 23, 77, 2006.
64. Kricka, L.J. et al., Current perspectives in protein array technology, *Ann. Clin. Biochem.*, 43, 457, 2006.
65. Tomozaki, K., Usui, K., and Mihara, H., Protein-detecting microarrays: Current accomplishments and requirements, *Chembiochem*, 6, 782, 2005.
66. Ahmed, F.E., Expression microarray proteomics and the search for cancer biomarkers, *Curr. Genomics*, 7, 399, 2006.
67. Panicker, R.C., Huang, X., and Yao, S.Q., Recent advances in peptide-based microarray technologies, *Comb. Chem. High T. Scr.*, 7, 547, 2004.
68. Min, D.H. and Mrksich, M., Peptide arrays: Towards routine implementation, *Curr. Opin. Chem. Biol.*, 8, 554, 2004.
69. Frank, R., High-density synthetic peptide microarrays: Emerging tools for functional genomics and proteomics, *Comb. Chem. High T. Scr.*, 5, 429, 2002.
70. Gannot, G. et al., Layered peptide arrays—A diverse technique for antibody screening of clinical samples, *Ann. N Y Acad. Sci.*, 451, 1098, 2007.
71. Andresen, H. et al., Peptide microarrays with site-specifically immobilized synthetic peptides for antibody diagnostics, *Sensor. Actuat. B*, 113, 655, 2006.

72. Andresen, H. et al., Functional peptide microarrays for specific and sensitive antibody diagnostics, *Proteomics,* 6, 1376, 2006.
73. Duburcq, X. et al., Peptide-protein microarrays for the simultaneous detection of pathogen infections, *Bioconjugate Chem.*, 15, 307, 2004.
74. Melnyl, O. et al., Peptide arrays for highly sensitive and specific antibody-binding fluorescence assays, *Bioconjugate Chem.*, 13, 713, 2002.
75. Robinson, W.H. et al., Autoantigen microarrays for multiplex characterization of autoantibody responses, *Nat. Med.*, 8, 295, 2002.
76. Hueber, W. et al., Antigen microarray profiling of autoantibodies in rheumatoid arthritis, *Arth. Rheum.*, 52, 2645, 2005.
77. Chen, G.A. et al., Autoantibody profiles reveal ubiquilin 1 as a humoral immune response target in lung adenocarcinoma, *Cancer Res.,* 67, 3461, 2007.
78. Fodor, S.P. et al., Light-directed, spatially addressable parallel chemical synthesis, *Science,* 251, 767, 1991.
79. Pellois, J.P. et al., Individually addressable parallel peptide synthesis on microchips, *Nat. Biotechnol.*, 20, 922, 2002.
80. Frank, R., The SPOT-synthesis technique synthetic peptide arrays on membrane supports—principles and applications, *J. Immunol. Methods*, 267, 13, 2002.
81. Falsey, J.R. et al., Peptide and small molecule microarray for high throughput cell adhesion and functional assays, *Bioconjug. Chem.*, 12, 346, 2001.
82. Houseman, B.T. et al., Peptide chips for the quantitative evaluation of protein kinase activity, *Nat. Biotechnol.*, 20, 270, 2002.
83. Lesaicherre, M.L. et al., Developing site-specific immobilization strategies of peptides in a microarray, *Bioorg. Med. Chem. Lett.*, 12, 2079, 2002.
84. Carion, C. et al., Chemical micropatterning of polycarbonate for site-specific peptide immobilization and biomolecular interactions, *Chembiochem*, 8, 315, 2007.
85. Byon, H.R. et al., Pseudo 3D single-walled carbon nanotube film for BSA-free protein chips, *Chembiochem*, 6, 1331, 2005.
86. Elbs, M. et al., Multivalence and spot heterogeneity in microarray-based measurement of binding constants, *Anal. Bioanal. Chem.*, 387, 2017, 2007.
87. Rich, R.L. and Myszka, D.G., Survey of the year 2005 commercial optical biosensor literature, *J. Mol. Recognit.*, 19, 478, 2006.
88. Rojo, N., Ercilla, G., and Haro, I., GB virus C (GBV-C) hepatitis G virus (HGV): Towards the design of synthetic peptides-based biosensors for immunodiagnosis of GBV-C/HGV infection, *Curr. Protein Pept. Sci.*, 4, 291, 2003.
89. Wittekindt, C. et al., Detection of human serum antibodies against type-specifically reactive peptides from the N-terminus of glycoprotein B of herpes simplex virus type 1 and type 2 by surface plasmon resonance, *J. Virol. Methods,* 87, 133, 2000.
90. Vaisocherova, H. et al., Surface plasmon resonance biosensor for direct detection of antibody against Epstein-Baff virus, *Biosens. Bioelectron.*, 22, 1020, 2007.
91. Ayela, C. et al., Antibody-antigenic peptide interactions monitored by SPR and QCM-D—A model for SPR detection of IA-2 autoantibodies in human serum, *Biosens. Bioelectron.*, 22, 3113, 2007.
92. Boozer, C. et al., Looking towards label-free biomolecular interaction analysis in a high-throughput format: A review of new surface plasmon resonance technologies, *Curr. Opin. Biotech.*, 17, 400, 2006.
93. Rich, R.L. and Myszka, D.G., Higher-throughput, label-free, real-time molecular interaction analysis, *Anal. Biochem.,* 361, 1, 2007.
94. Cherif, B. et al., Clinically related protein-peptide interactions monitored in real time on novel peptide chips by surface plasmon resonance imaging, *Clin. Chem.,* 52, 255, 2006.
95. Inamori, K. et al., Detection and quantification of on-chip phosphorylated peptides by surface plasmon resonance imaging techniques using a phosphate capture molecule, *Anal. Chem.*, 77, 2979, 2005.
96. Li, J., Ng, H.T., and Chen, H. Carbon nanotubes and nanowires for biological sensing, in *Methods in Molecular Biology,* Tuan Vo-Dinh, Ed., Clifton, New Jersey, 300, 191, 2005.
97. Zheng, G. et al., Multiplexed electrical detection of cancer markers with nanowire sensor arrays, *Nat. Biotechnol.*, 23, 1294, 2005.

98. Li, Y. et al., Protein recognition via surface molecularly imprinted polymer nanowires, *Anal. Chem.*, 78, 317, 2006.
99. Contarino, M.R. et al., Modular, self-assembling peptide linkers for stable and regenerable carbon nanotube biosensor interfaces, *J. Mol. Recognit.*, 19, 363, 2006.
100. Yemini, M. et al., Novel electrochemical biosensing platform using self-assembled peptide nanotubes, *Nano Lett.*, 5, 183, 2005.
101. MacCuspie, R.I. et al., Multiplexed peptide nanotube pathogen assays, *Abstract paper Am. Chem. Soc.*, 231, 5-IEC, 2006.
102. Yemini, M. et al., Peptide nanotube-modified electrodes for enzyme-biosensor applications, *Anal. Chem.*, 77, 5155, 2005.
103. Dhayal, B. et al., Detection of *Bacillus subtilis* spores using peptide-functionalized cantilever arrays, *J. Am. Chem. Soc.*, 128, 3716, 2006.
104. Lee, J.H. et al., Immunoassay of prostate-specific antigen (PSA) using resonant frequency shift of piezoelectric nanomechanical microcantilever, *Biosens. Bioelectron.*, 20, 2157, 2005.
105. Kwon, H.S. et al., Development of a peptide inhibitor-based cantilever sensor assay for cyclic adenosine monophosphate-dependent protein kinase, *Anal. Chim. Acta*, 585, 344, 2007.
106. Lundqvist, H. and Tolmachev, V., Targeting peptides and positron emission tomography, *Biopolymers*, 66, 381, 2002.
107. Reubi, J.C., Peptide receptors as molecular targets for cancer diagnosis and therapy, *Endocr. Rev.*, 24, 389, 2003.
108. Okarvi, S.M., Peptide-based radiopharmaceuticals: Future tools for diagnostic imaging of cancers and other diseases, *Med. Res. Rev.*, 24, 357, 2004.
109. Hofland, L.J. et al., Internalisation of isotope-coupled somatostatin analogues, *Digestion*, 57, 2–6 Suppl 1, 1996.
110. Signore, A. et al., Peptide radiopharmaceuticals for diagnosis and therapy, *Eur. J. Nucl. Med.*, 28, 1555, 2001.
111. Jong, M. et al., Radiolabelled peptides for tumour therapy: Current status and future directions, *Eur. J. Nucl. Med.*, 30, 463, 2003.
112. Dijkgraaf, I. et al., Synthesis of DOTA-conjugated multivalent cyclic-RGD peptide dendrimers via 1,3-dipolar cycloaddition and their biological evaluation: Implications for tumor targeting and tumor imaging purposes, *Org. Biomol. Chem.*, 5, 935, 2007.
113. Bhargava, S. et al., A complete substitutional analysis of VIP for better tumor imaging properties, *J. Mol. Recognit.*, 15, 145, 2002.
114. Cheng, D.F. et al., Radiolabeling and in vitro and in vivo characterization of $[^{18}F]FB\text{-}[R^{8,15,21}, L^{17}]$-VIP as a PET imaging agent for tumor overexpressed VIP receptors, *Chem. Biol. Drug. Des.*, 68, 319, 2006.
115. Brinkley, M., A brief survey of methods for preparing protein conjugates with dyes, haptens, and cross-linking reagents, *Bioconj. Chem.*, 3, 2, 1992.
116. Tung, C.H., Fluorescent peptide probes for in vivo diagnostic imaging, *Biopolymers*, 76, 391, 2004.
117. Becker, A. et al., Receptor-targeted optical imaging of tumors with near-infrared fluorescent ligands, *Nat. Biotechnol.*, 19, 327, 2001.
118. Medintz, I.L. et al., Quantum dot bioconjugates for imaging, labelling and sensing, *Nat. Mater.*, 4, 435, 2005.
119. Zhou, M. and Ghosh, I., Quantum dots and peptides: A bright future together, *Biopolymers*, 88, 325, 2007.
120. Pinaud, F. et al., Bioactivation and cell targeting of semiconductor CdSe/ZnS nanocrystals with phyto-chelatin-related peptides, *J. Am. Chem. Soc.*, 126, 6115, 2004.
121. Delehanty, J.B. et al., Self-assembled quantum dot-peptide bioconjugates for selective intracellular delivery, *Bioconjugate Chem.*, 17, 920, 2006.
122. Chang, E. et al., Protease-activated quantum dot probes, *Biochem. Biophys. Res. Commun.*, 334, 1317, 2005.
123. Medintz, I.L. et al., Proteolytic activity monitored by fluorescence resonance energy transfer through quantum-dot-peptide conjugates, *Nat. Mater.*, 5, 581, 2006.
124. Shi, L.F. et al., Synthesis and application of quantum dots FRET-based protease sensors, *J. Am. Chem. Soc.*, 128, 10378, 2006.

125. Cai, W.B. et al., Peptide-labeled near-infrared quantum dots for imaging tumor vasculature in living subjects, *Nano Lett.*, 6, 669, 2006.
126. Chan, P.M. et al., Method for multiplex cellular detection of mRNAs using quantum dot fluorescent in situ hybridization, *Nucleic Acids Res.*, 33, e161, 2005.
127. Derfus, A.M., Chan, W.C.W., and Bhatia, S.N., Probing the cytotoxicity of semiconductor quantum dots, *Nano Lett.*, 4, 11, 2004.
128. Hoshino, A. et al., Physicochemical properties and cellular toxicity of nanocrystal quantum dots depend on their surface modification, *Nano Lett.*, 4, 2163, 2004.
129. Shiohara, A. et al., On the cyto-toxicity caused by quantum dots, *Microbiol. Immunol.*, 48, 669, 2004.

15 Recombinant Antibodies

Alla Likhacheva and Tatyana Ulanova

CONTENTS

15.1 ANTIBODY STRUCTURE

Antibodies are specific proteins, termed immunoglobulins, which are produced by B cells under stimulation by antigens. The main function of the specific immune response is recognition of foreign antigens. Two types of molecules participate in the recognition: immunoglobulins and T-cell receptors. B cells specifically produce immunoglobulins. These proteins may be expressed in the form of integral membrane proteins on the surface of B lymphocytes. Alternatively, they may be secreted into the bloodstream by plasma cells. This soluble form of immunoglobulins is what the immunologists call the antibodies. Each antibody carries two functions. The main and primary function of antibody is binding to antigen. This interaction with antigen may lead directly to such effects as neutralizing the bacterial toxin or preventing the virus invasion into the cells. The interaction of antibody with antigen frequently results in effects mediated by the antibody secondary or effector functions. Effector functions of antibodies include complement system activation, selective interaction with different cell surface receptors, etc. One part of the antibody molecule is responsible for the interaction with antigen, while another part realizes the effector functions [1,2].

It is known that antigens with different physicochemical properties such as proteins, nucleic acids, polysaccharides, etc., may cause production of antibodies. Potentially, the immune system is capable to produce antibodies against any macromolecule of the living world. How is it possible to produce a necessary huge number of antibodies required to interact specifically with all the potential antigens? To answer this question let us review the structure of antibody.

The typical antibody is a symmetric molecule formed by two identical, light (L) and heavy (H) chains (Figure 15.1). The disulfide bonds covalently link these chains. Each H–L pair contains a combined site for antigen recognition. Light chains are approximately 23 kD. There are two different forms of the light chains: kappa (κ) and lambda (λ). Heavy chains are 50–77 kD. They differ structurally between the classes and subclasses of immunoglobulins. Thus, five classes of immunoglobulins in higher vertebrates IgG, IgM, IgA, IgD, and IgE contain γ (γ_1, γ_2, γ_3, or γ_4), μ, α (α_1 or α_2), δ, and ε heavy chains, respectively. Classes of antibodies differ from each other by the

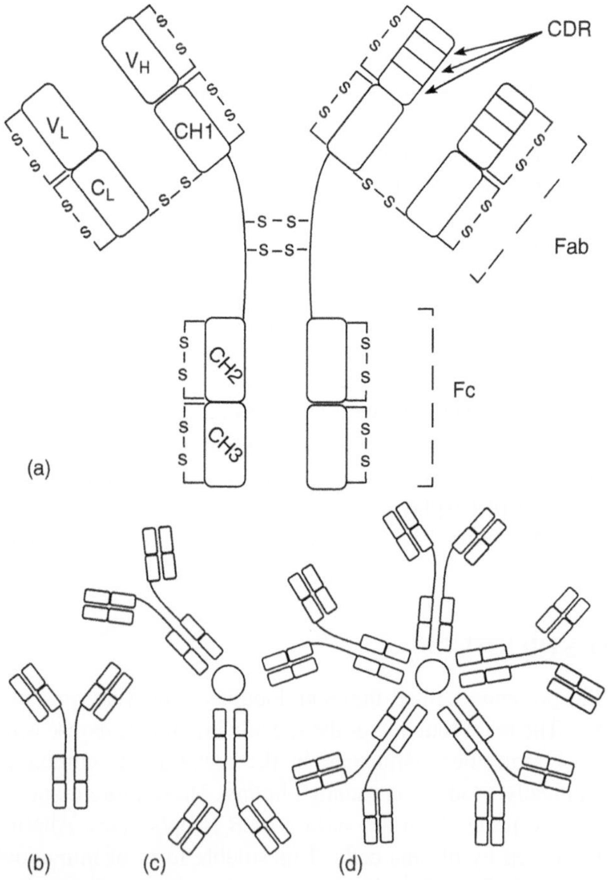

FIGURE 15.1 Antibody structure: (a) Schematic model of the "typical" antibody—IgG; (b) monomeric IgG; (c) dimeric form of IgE; and (d) pentameric form of IgM.

type of the heavy chain and the degree of oligomerization. For example, IgA is mostly a dimer. IgM forms a pentamer in the serum (Figure 15.1). Other classes of antibodies are monomers [1–3].

Each chain (both L and H) consists of variable (V) and constant (C) regions. Variable regions have a great number of amino acid sequence variations. They are responsible for the binding with antigens. The treatment of antibody with the protease papain leads to digestion of the molecule into three fragments. Two of them (named as Fab) are identical. They contain the V regions and are capable of binding the antigen. The third fragment (Fc) includes the heavy chain constant regions. It does not recognize the antigen, but it fulfills some effector functions. For example, it fixes the first component of the complement cascade. Thus, cleavage with papain separates structural regions of antibodies responsible for the antigen binding (Fab fragments) and effector functions (Fc fragment) [1,2,4].

The antibody molecule has a globular domain structure (Figure 15.1). There are two domains in the light chains. One of them is variable and another is constant. Heavy chains have 4–5 domains, depending on the class of the heavy chain. Similar to the light chains, one of these domains is variable (V_H), whereas other domains are constant. Each domain has a typical structure known as the "immunoglobulin fold." The domain consists of two layers of β sheets. Three or four antiparallel segments of the chain form the layer. The entire globule has a hydrophobic internal core. The disulfide bridge covalently links layers to each other approximately in the middle of the domain [3,5,6].

The variable domains from the light (V_L) and heavy chains (V_H) are located at the amino-terminus of the chains. The antigen-binding site is assembled from both of them. Some short subregions within the V regions have an extraordinary diversity of the amino acid sequence. These regions of hypervariability could be identified around the positions of 30, 50, and 100. They are also called complementarity-determining regions (CDRs), due to their direct involvement into antigen recognition. Relatively invariant segments of V regions, not included in CDRs, are named framework regions (FRs). There are three CDRs (CDR1–CDR3) and four FRs (FR1–FR4) in the variable domains of both light and heavy chains [1,4,6]. An extraordinary hypervariability of CDRs could explain a huge repertoire of antibodies. There are several sources for amino acid sequence diversity of antibodies discussed in the extant literature: a large number of gamete genes for V regions, random recombination of the V-gene elements, gene conversions, somatic mutagenesis, and insertions of the additional nucleotides [3].

15.2 ANTIBODY–ANTIGEN INTERACTION

Chemical interaction between antibody and antigen is similar to other protein–ligand interactions. Different types of noncovalent bonds participate in the antibody–antigen contact such as hydrogen, electrostatic, van der Waals, and hydrophobic bonds. Salt bridges are also sometimes involved in antigen binding.

Binding of an antigenic determinant (epitope) is determined by the structure of the antigen-binding site (paratope). There is an extremely close steric complementarity between the antibody and antigen surfaces during their contact. Sometimes gaps between the antibody and the antigen surfaces are filled with water molecules [1,2].

Folding of six CDRs composes the antigen-binding site. Three CDRs are derived from the light chain (L1, L2, L3), and the remaining three from the heavy chain (H1, H2, H3). CDRs form loops, which adjoin to each other, and are located at the end of V domains. Each CDR has a fixed orientation in the framework depending on its length and sequence characteristic of the domain. CDRs form the majority of intermolecular contacts. Framework regions are responsible for supporting the structural conformations of V domains. However, sometimes they can also form contacts [1,2,7]. CDRs carry a great variety of the structural repertoire. Despite this variability, a structural similarity can be found among many CDRs. Thus, they can be classified into small families on the base of "canonical structure" [7]. The classification is based on the presence of the key residues at defined positions, determining the CDR conformation. Among the CDRs, the heavy chain CDR H3 is known to have the largest variety in sequence, length, and structure [8,9]. It usually plays the most important role in antigen binding. It occupies the central position in the antigen-binding site and affects critically its topography.

The main characteristics of antibody binding to the antigen are specificity and affinity. The antigen–antibody reaction is highly specific. The antibody recognizes its epitope from the astronomical number of other epitopes. Sometimes antigen 1 and antigen 2 have common epitopes. In that case the antibody specific to antigen 1 also reacts with antigen 2. This phenomenon is called cross-reactivity. Affinity is a binding strength of an antigen-binding site with the individual antigenic epitope. The total force of the interaction between the antibody and antigen is the antibody's avidity (determined by the number of antigen-binding sites and their capacity to bind the numerous epitopes of the antigen).

The structure of the antigen-binding site determines the antibody specificity and affinity. Structural modifications caused, for example, by point mutations provide the mechanism for the generation of the antigen binding specificity repertoire and the antibody affinity maturation.

15.3 ANTIBODY APPLICATIONS: ADVANTAGES OF RECOMBINANT ANTIBODIES

Antibodies have found a wide range of applications (Table 15.1). Among well-known antibody-based laboratory assays are immunoelectrophoresis, enzyme-linked immunosorbent assay (ELISA),

TABLE 15.1
Examples of Antibody Applications

Antigen identification	Individual antigens in complex mixtures and solutions
	Antigen purity
	Production, expression, localization, and activity of antigens in tissues and cells
	Relationship between the structure and function of antigen
	Identification of cell populations, including tumor cells
	Cross-immunoreactivity between different molecules
	Titration of lymphocytes and erythrocytes
	Monitoring expression of antigens
	Antigen diagnostics of different diseases
Quantifying the antigens	Diagnostically relevant components of blood plasma
	Hormones (diseases, menstrual cycle, pregnancy)
	Specific antibodies (infections, immune status, autoimmune diseases)
	Mediators of cell activity
	Drugs and toxins
Purification of antigens and cells	Immunoaffinity purification of antigens
	Precipitation of RNA/antigen complexes from cytoplasm
	Labeling of cells for flow cytometry
	Binding to cell membranes for isolation
Therapy	Immunosuppression
	Intrabodies/gene therapy
	Targeting/neutralizing
	Signaling/cross-linking
	Cancer therapy

Source: From Catty, D., *Antibodies: A Practical Approach*, IRL Press, Oxford-Washington DC, 1989, Vol. 1, Chap. 1; Roque, A.C.A., Lowe, C.R., and Taipa, M.A., *Biotechnol. Prog.*, 20, 639, 2004.

different techniques of immunohistochemistry, and chromatographic immunoassays [10,11]. Recently, flow cytometry, proteomics, and protein microarrays have also started exploiting antibodies [12,13].

Antibodies offer a great possibility for their use in different fields of biotechnology and biomedicine, both in research and clinic. Examples of such applications include diagnostics of a variety of diseases, in vivo and in vitro therapy agents, neutralization of toxins, immunosuppression, passive immunization, drug and radionuclide delivery for the treatment and imaging, and so on. Food and environmental industries also exploit antibodies by utilizing them as biosensors for the detection and removal of contaminants and organic pollutants [14–19].

The origin and nature of the antibody define its application. Antibodies coming from the immunized animals (such as goat or rabbit) are described as polyclonal. Polyclonal antiserum contains antibodies with different specificity against various epitopes of polydeterminant immunogens. Therefore, polyclonal antibodies are heterogeneous by their specificity. This is the main limitation for the application of polyclonal antibodies in many research, industrial, and clinical settings. Monoclonal antibodies are targeted against a single antigenic site and, therefore, are monospecific. Antibody monospecificity has revolutionized the area of the antibody applications.

Monoclonal antibody production was achieved by introducing the hybridoma technology in 1975 [20]. The method is based on the fusion of the antibody-producing spleen cells from an immunized animal with the myeloma cells to obtain hybridoma. Hybridoma cells grow and divide indefinitely, because they are immortal. A particular hybridoma expresses and mass-produces the antibody of only one specificity. Hybridoma methodology provides an opportunity for generating a large amount of monoclonal antibodies with defined specificities. Monoclonal antibodies can be generated against any antigen of interest. The hybridoma is one of the main technologies in biotechnology today. Unfortunately, this technology is not free of some shortcomings. First of all, the hybridoma-derived monoclonal antibodies are limited in their clinical applications. Majority of the hybridoma-produced antibodies are murine antibodies and, therefore, are immunogenic in humans. Application of these antibodies as therapeutic agents may induce a human antimouse antibody [HAMA] response, which quickly reduces the effectiveness of therapy by clearing the murine antibody from the bloodstream [19,21]. Another problem is that this technology requires a large number of B lymphocytes due to the inefficiency of the transformation and immortalization. Additionally, an individual immune response always has some variability, thus limiting a possibility for standardization of antibody preparations. On top of these, in some cases immunization of experimental animals fails to produce antibodies of desired specificity. The process of immortalization cannot be easily manipulated in experiments and frequently results in producing cells with multiple karyotypes [15]. The shortcomings of the hybridoma technology can, however, be largely overcome through the use of the genetic engineering techniques and production of the recombinant antibodies. Recombinant antibody molecules have several advantages over natural monoclonal antibodies [4,14–17,22]:

- Recombinant antibodies may have reduced immunogenicity. Genetic manipulations provide the mechanism for the construction of different chimeric and humanized molecules. For example, it is possible to replace much of the rodent-derived sequence of an antibody with human immunoglobulin-derived sequences without loss of the function. This procedure leads to the decrease of immunogenic properties in humans.
- Antibody specificity and affinity may be enhanced.
- Antibodies may be engineered into small molecules (fragments and even single variable domains are produced). The smaller molecules have some advantages in tissue distribution during therapy.
- Recombinant antibodies can be easily modified to generate bispecific antibodies or antibodies conjugated to drugs, toxins, isotopes, or any other molecule. These recombinants have already found application in immunotherapy, immunohistochemistry, and other areas.

15.4 DESIGN OF THE RECOMBINANT ANTIBODIES

15.4.1 Antibody Fragments

Whole antibody molecules and their fragments are widely used in numerous applications in different fields of biotechnology and biomedicine. For some therapeutic applications intact antibodies are preferred because of their extended half-life in bloodstream and ability to trigger humoral and cellular effector mechanisms. However, the use of antibody fragments is more suitable when only the antibody antigen-binding site is required and the effector functions of the Fc part are dispensable. Fragments may be successfully used, for example, in diagnostic and therapeutic applications. Many advantages of antibody fragments in different applications spin off from their small size and low immunogenicity. Because of their small size, antibody fragments have a better penetration and greater homogeneous distribution within tissues (which is of great importance, for example, in the case of tumor diagnostics and therapy). Small fragments are also cleared rapidly from serum and other tissues through kidneys. This property makes small antibody fragments very suitable as toxin removers and targeting agents [15–17,22,23].

There are a few different methods used to produce antibody fragments. Some fragments may be prepared by the proteolytic hydrolysis of the antibody (Figure 15.2). For instance, treatment with pepsin leads to the generation of bivalent $(Fab')_2$ fragment and digestion with papain gives a monovalent Fab fragment. Proteolytic hydrolysis remained the only method for obtaining the antibody fragments until the advent of new tools of the recombinant DNA technology. This technology provided an opportunity for constructing different types of recombinant antibodies. The other advantages of this technology are standardization of reagents with the potential for large-scale production and simple methods for purification of the created products.

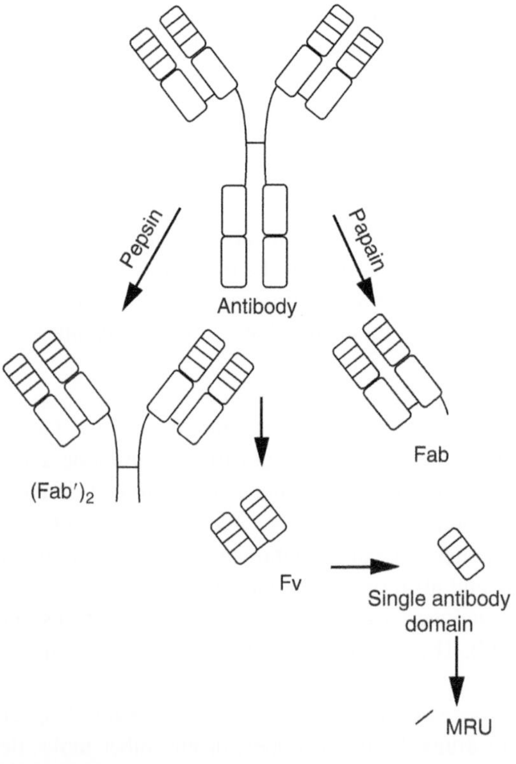

FIGURE 15.2 Antigen-binding fragments of antibody.

15.4.1.1 Fab Fragment

Fab (the antigen-binding fragment) is a heterodimer of V_H–C_{H1} and V_L–C_L domains linked together through a disulfide bond (Figures 15.1 and 15.2). Nowadays, the recombinant DNA technology is preferred in Fab production over the isolation of these fragments by proteolysis of antibodies. The antigen-binding properties of the recombinant Fab fragments have been reported to be comparable to the proteolytically produced corresponding Fab fragments [24]. Compared to the intact antibody molecules, Fab fragments, which lack Fc portion, may offer potential advantages in therapy due to their lower immunogenicity and better pharmacokinetic properties [18,23]. In addition, the use of Fab fragments rather than complete antibodies in diagnostic immunoassays may decrease unwanted interferences, which are sometimes observed [25]. Recombinant Fab fragments are mainly produced in the bacterial and mammalian cell expression systems. Bacterial expression provides a convenient source for large amounts of relatively homogenous protein preparation. There are two successful strategies for the expression of Fab fragments in bacterial cells. One strategy takes advantage of the co-translational periplasmic secretion when two peptides (V_H–C_{H1} and V_L–C_L) are simultaneously expressed in *E. coli* in the equivalent amounts and secreted into the periplasm to form functional Fab molecules [26,27]. The other method of phage combinatorial libraries was initially described for the construction and rapid analysis of Fab [28]. In this case, a large number of Fab fragments are expressed on the surface of the filamentous phage particles as gIII fusion proteins. This library can be screened to select Fab fragments with desirable specificity and affinity. A combination of Fab expression and phage display technology is widely used today. Many different vector systems have been developed for the expression and selection of Fab using phage display. Corisdeo and Wang [29] compared three commonly used phagemid vector systems for Fab expression based on:

- Two separate expression cassettes for V_L–C_L and V_H–C_{H1} genes.
- Bi-cistronic expression cassette with the V_H–C_{H1} gene being the first cistron and V_L–C_L the second.
- Bi-cistronic expression cassette with the V_L–C_L gene as the first cistron and V_H–C_{H1} as the second cistron.

The authors found that the last variant was the most efficient for the expression of active Fab fragments in *E. coli*. Over the years, many different alternatives to the bacterial Fab expression have been developed. Variety of recombinant Fab fragments has been successfully expressed in yeasts [30–32], transgenic plants [33,34], and insect cells [35].

15.4.1.2 Fv Fragment

Fv is the smallest antibody fragment that carries the whole antibody-binding site. It consists of only V_H and V_L domains (Figure 15.2). The reduced size of Fv has some advantages over the Fab fragment. For example, Fv may be applied in the cases where the presence of the constant domains is not required or may be detrimental. Fv is also less immunogenic and has a better penetration into the dense tissues. These features have provided a unique use for Fv, especially in the diagnostics, treatment of cancer, structural studies, and in vivo imaging [15,16,18].

Active Fv fragments can be obtained by proteolytic digestion of antibodies [36]. However, the proteolysis often does not result in a homogeneous population of Fv fragments. It rather produces a number of partially digested protein molecules. Instead, recombinant antibody engineering provides a more efficient and generally applicable method for Fv fragments production.

Despite many advantages, recombinant Fv is not as readily exploitable as recombinant Fab due to its low stability. To stabilize Fv fragments, a number of approaches have been developed (Figure 15.3) such as linkage of V_H and V_L domains with short peptide regions, "knob-into-hole" mutations, and chemical cross-linking by glutaraldehyde or disulfide bridges [16,22]. Two Fv domains can be fused genetically by a flexible linker peptide [37–50]. The resulting protein is

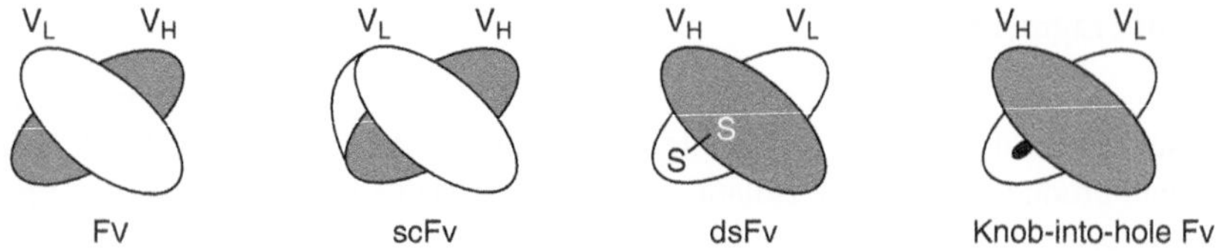

FIGURE 15.3 Variants of Fv fragment stabilization.

termed a single-chain Fv (scFv) fragment. Both orientations, V_H–linker–V_L and V_L–linker–V_H can be used. This approach to the Fv stabilization is frequently applied. The other efficient approach is based on intermolecular disulfide bond linkage that produces a disulfide-stabilized (ds) Fv fragment. dsFv is more stable than scFv or chemically cross-linked Fv. Two sites, V_H 44–V_L 100 and V_H 105–V_L 43, are the most suitable for introducing the disulfide bridges.

As mentioned above, construction of scFv is the most popular way for obtaining Fv. Many different genetic systems have been developed for the scFv expression and production. Various scFv fragments have been successfully expressed in yeast [38,39], plants [40,41], mammalian, and insect cells [39,42–44]. One of the most suitable and commonly used approaches is bacterial scFv expression. However, the expressed scFv is frequently insoluble and inactive when this expression system is used. Several approaches have been developed to overcome this problem. In order to obtain functional scFvs, these proteins are expressed in the format exportable to the periplasmic space, where the correct folding may occur. Also, several strategies for the functional cytoplasmic expression have been developed, including renaturation/refolding from the inclusion bodies [45,46], using mutant strains with more oxidized cytoplasm [47,48], construction of the different type fusion proteins (MBP, thioredoxin, etc.) [43,47,49]. Recently, in addition to *E. coli*, *Bacillus megaterium* has been used for the functional scFv expression [50]. As for Fab fragments, phage-display technology is also commonly used for the selection of highly affinitive and specific scFv fragments. Phage display was also used for the selection of scFv with improved stability [51].

One of the disadvantages of scFv as well as Fab fragments is their monovalency. The scFv usually shows monovalent-binding affinity similar to the Fab of the parent antibody. Native antibodies gain significant avidity due to the presence of two antigen-binding sites. A dramatic increase in the functional affinity may be achieved by oligomerization, as in the case of pentameric IgM. Similarly, the avidity of scFv and Fab may be increased when these molecules are complexed into aggregates [52]. In order to increase functional affinity, several approaches have been applied to produce dimeric and multimeric antibody fragments (Figure 15.4). These approaches include chemical cross-linking, coupling by adhesive protein domains, or an additional linker peptide. In one work, a bivalent $(scFv')_2$ dimer was generated by formation of a disulfide bridge using additional C-terminal cysteine residues [53]. The other popular approach is joining scFv into tandem by an additional linker peptide [15,16,37,54]. One of the simplest and commonly used methods for scFv multimerization is manipulation of the linker length between V_H and V_L domains [Figures 15.4 and 15.5]. The linker is generally designed with glycine and serine residues, which provide flexibility and protease resistance. scFv antibody fragments are predominantly monomeric when the V_H and V_L domains are joined by the linker of more than 12 amino acids in length. Usually, a linker of 15 residues with the structure of $((Ser)_4Gly)_3$ is used. In this case, the V_H and V_L domains are in the natural Fv orientation. If the linker is shortened to 3–12 residues, the V_H domain is unable to bind to its attached V_L domain in the natural Fv orientation. Monomeric configuration of scFv is prevented and intermolecular V_H/V_L pairing takes place with formation of a noncovalent scFv dimer, termed as a diabody. Reduction of the linker length below 3 or its complete removal from V_H–V_L orientation leads to the formation of trimers (termed triabodies) or tetramers (tetrabodies). scFv multimers show a significant gain in the functional affinity. Gall et al. [55] obtained 1.5- and 2.5-fold increase in affinities by diabody and tetrabody specific to CD 19 B-cell antigen, respectively,

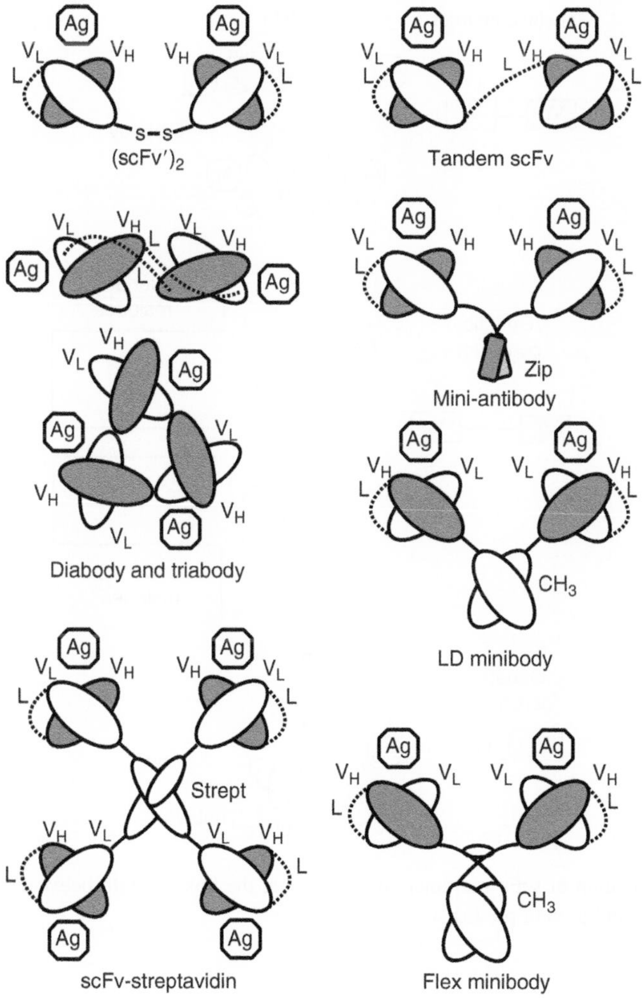

FIGURE 15.4 Approaches for increasing the avidity of scFv.

in comparison to scFv monomer. The gain in functional affinity for scFv dia/triabodies is seen primarily in reduced off-rates, which results from multiple binding to two or more target antigens. Multiple binding to the surface-bound antigens is dependent on the correct alignment and orientation of the Fv modules in diabodies and triabodies. Antigen orientation is also important for simultaneous binding of diabodies and triabodies to multiple antigens [37]. Therefore, orientation of the binding sites of multimeric scFv is of a significant importance. This orientation depends on the configuration of the target antigen. It has been shown, that the order of the variable domains (V_H–linker–V_L or V_L–linker–V_H) greatly affects the orientation of the Fv binding sites. Crystal studies of diabodies constructed in the order of V_H–linker–V_L revealed that these dimers assemble via back-to-back interactions. It results in the coplanar Fv units. The antigen-binding sites are oriented at 170°–180° angles. By contrast, crystal structure of an anti-CEA diabody constructed in the V_L–linker–V_H order showed a strikingly different orientation of the Fv unit. The antibody combining regions are oriented at about 105° angle to each other, and V_L–V_H diabody adopts a compact, stacked configuration [56]. The order of the variable domains, and the length and sequence of the linker is important for the functional scFv assembly. It was shown that the specificity of the N- and C-terminal residues of the V domains should be also taken into consideration when

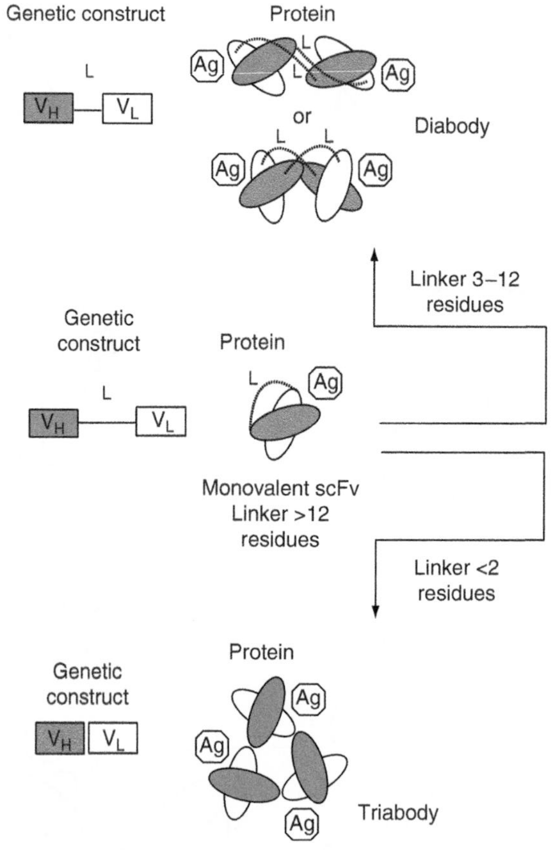

FIGURE 15.5 Generation of scFv multimers by shortening the linker length between V_L and V_H domains. ScFv gene constructs and protein products.

designing the linker length. Hudson and Kortt [37] were guided by the x-ray diffraction analysis of the anti-neuraminidase (NC10) Fab-NA complexes when they designed multimeric NC10 scFv by shortening the linker length between the variable domains. These authors selected $SerH^{112}$ as the V_H C-terminus for the V_H–V_L orientation scFv, because this residue was not in any direct hydrogen bonding contact with any other V_H domain residues. For V_L domain, the N-terminal residue Asp^{L1} was chosen. In the reverse V_L–V_H orientation of NC10 scFv, Arg^{L107} was taken as the C-terminal residue of the V_L domain and Gln^{H1} as the V_H domain N-terminal residue (Table 15.2).

Construction of larger multivalent scFv fragments can also be achieved by the genetic fusion with protein domains normally involved in the protein association (Figure 15.4). Dimerizing motifs such as amphipathic helices are used to produce "mini-antibody" molecules. For example, mini-antibodies were obtained with 4-helix bundles or coiled coils as dimerization domains. Various versions of coiled coil sequences such as "leucine zippers" were fused to scFv to produce the dimeric mini-antibodies [57]. scFv can also be fused to immunoglobulin C_{H3} domain, resulting in a bivalent protein, called a "minibody" [39,58]. Minibodies are reported to recognize target molecules with high affinity and specificity. ScFv may be joined to C_{H3} domain either by a two amino acid spacer (LD, resulting in a noncovalent dimer), or by the incorporation of the human IgG1 hinge region (Flex, resulting in a disulfide-linked dimer). Other proteins are also exploited for assembling scFv fragments, including immunoglobulin Cκ and C_{H1}/Cκ domains or the entire hinge-Fc [39]. Interesting data were obtained by fusion of scFv with streptavidin. Tetravalent scFv-streptavidin

TABLE 15.2
Sequences of Linkers and Regions of V-Domains Flanking the Linker Used by Hudson and Kortt for Anti-Neuraminidase scFv Multimers

Order of V Domains	Linker Length Between V Domains	Sequences of the Linker (Underlined) and Regions of V Domains Flanking the Linker
V_H–V_L	0 residue	TVTVS - - - - - DIELT
	1 residue	TVTVSG - - - - DIELT
	2 residues	TVTVSGG - - - DIELT
	3 residues	TVTVSGGG - - DIELT
	4 residues	TVTVSGGGG – DIELT
	5 residues	TVTVSGGGGSDIELT
	10 residues	TVTVSGGGGSGGGGSDIELT
	15 residues	TVTVSGGGGSGGGGSGGGGSDIELT
V_L–V_H	0 residue	KLEIR - - - - - QVQLQ
	1 residue	KLEIRS - - - - QVQLQ
	2 residues	KLEIRSG - - - QVQLQ
	3 residues	KLEIRSGG - - QVQLQ
	4 residues	KLEIRSGGG - QVQLQ
	5 residues	KLEIRSGGGGQVQLQ
	15 residues	KLEIRSGGGGSGGGGSGGGGQVQLQ

Source: Adapted from Hudson, P.J. and Kortt, A.A., *J. Immunol. Methods*, 231, 177, 1999.

demonstrated both antigen- and biotin-binding activities. It also was stable over a wide range of pH, and did not dissociate at high temperatures [59].

Increased avidity is only one of scFv multimers advantages. Another benefit is their large size. ScFv multimers are significantly larger than scFv monomers. For example, diabodies are approximately 60 kDa, triabodies are ~90 kDa, and tetrabodies are ~120 kDa, while monomeric scFv are ~30 kDa. There is a correlation between the size of the molecule and its blood clearance rate through the kidneys. Monomeric scFv have a rapid clearance from the circulation while scFv multimers have a better retention rate. A reduced renal clearance and higher avidity have been reported for diabodies in comparison to monomeric scFv [60].

15.4.1.3 Single Antibody Domains

The antigen-binding site comprises six CDRs; however, all six may not be necessary for antigen binding. In some cases, the single antibody V domains alone are able to bind antigen [61,62]. There are only few reports of functional single V domains. For example, functional antigen-binding V_H domains were isolated from the lymphocytes of immunized mice. Unfortunately, isolated single V domains have reduced affinity and expression/solubility [15]. Thus, additional protein engineering is required to solve this problem. There are natural examples of single V-domain scaffolds known. Camelids display natural repertories of single heavy-chain domains attached to Fc domains. The light chain is totally absent. These antibodies interact with antigen through only V_{HH} domain. These antibodies exhibit, however, a broad antigen-binding repertoire by enlarging their hypervariable regions [62]. The nurse shark displays antigen receptor which contains single monomeric V domains attached to five C domains. The extracellular domain of the human cytotoxic T lymphocyte-associated protein-4 (CTLA-4) contains a single V domain, which functions as an effective scaffold [62,63]. These natural scaffolds justify and encourage the development and engineering of the recombinant single antibody domains. Potential advantages of the recombinant single antibody domains are simplicity of expression and production, enhanced stability, potential for targeting certain antigen types, and rapid engineering into multimeric or multivalent reagents [16]. The technology for engineering of simple-fold proteins has been recently developed. Several single-fold proteins were engineered, including a three-helix bundle domain derived from the IgG-binding staphylococcal protein A, lipocalins, the α-amylase inhibitor tendamistat, and the tenth fibronectin type III domain [16]. Also, recombinant single-domain V_{HH} constructs have been engineered against the conserved carbohydrate epitope of the trypanosomes surface glycoprotein [62]. Recently, Irving et al. [64] have offered a cell-free technology of ribosomal display for the efficient selection of functional single V domain proteins.

15.4.1.4 Molecular Recognition Units

Molecular recognition units (MRU) are peptides represented by just one CDR. It was reported that MRU may mimic the specificity of the parent molecule. Several peptides with sequences that correspond to CDR2 or CDR3 were produced. These peptides retained the antibody specificity and ability to mimic the antibody binding [15,65]. MRU may be included in the class of imaging and targeting molecules. Recently, a 17 amino acid 1.8 kDa MRU fragment was approved for clinical use to image thrombi in cardiovascular disorders [15]. Owing to their small size, MRU can be also used for structure–functional analysis. Heap et al. [65] investigated a microantibody by alanine scanning, mass spectroscopy, and surface plasmon resonance. The microantibody was represented by a 17-residue cyclized peptide derived from CDR-H3 of IgG1 specific for the gp120 envelope glycoprotein of human immunodeficiency virus type 1. This study demonstrated how the minimal recognition unit of an antibody could make a useful contribution to the understanding of antigen–antibody interactions.

15.4.2 Whole Antibody Molecules

15.4.2.1 Phage-Display-Derived Antibodies

In some cases, whole antibody molecules are more suitable due to their extended stability and serum half-life. More importantly, they may be used when the effector function of the Fc part is required. The hybridoma technology is the method of choice for production of whole monoclonal antibodies with defined specificity. However, HAMA response [19,21] reduces the range of applications for monoclonal antibodies. New recombinant DNA techniques have been developed to overcome this limitation. The integration of hybridoma technology with recombinant DNA and phage-display techniques has enabled designing and constructing human-like antibodies of desirable affinity and specificity combined with low immunogenicity [22] (Figure 15.6).

Phage-display technique mimics the strategy used by the humoral immune system to produce whole human antibodies in vivo. Using this technique, V region genes from immunoglobulin light and heavy chains are displayed in the functional form on the phage surface. After selection for a strong and specific binding to the antigen, the V domains can be fused to the human antibody constant region to reconstruct a whole monoclonal antibody. Genes encoding such recombinant antibodies can be expressed using myeloma cell lines producing secreted monoclonal antibodies. The phage-display method is described in more detail in Section 15.6.1.

15.4.2.2 Chimeric Antibodies

The first generation of the recombinant monoclonal antibodies contained a rodent or murine variable region fused to a human constant region. The resulting molecule is termed the chimeric antibody (Figure 15.7) [15,16,22,66]. It is known that the most immunogenic regions of antibodies are located in the conserved C domains. The chimeric antibodies represented the first attempt to reduce the immunogenicity of heterologous monoclonal antibodies. Another advantage of chimeric antibodies is that while retaining their binding specificity and affinity for an antigen they maintain the

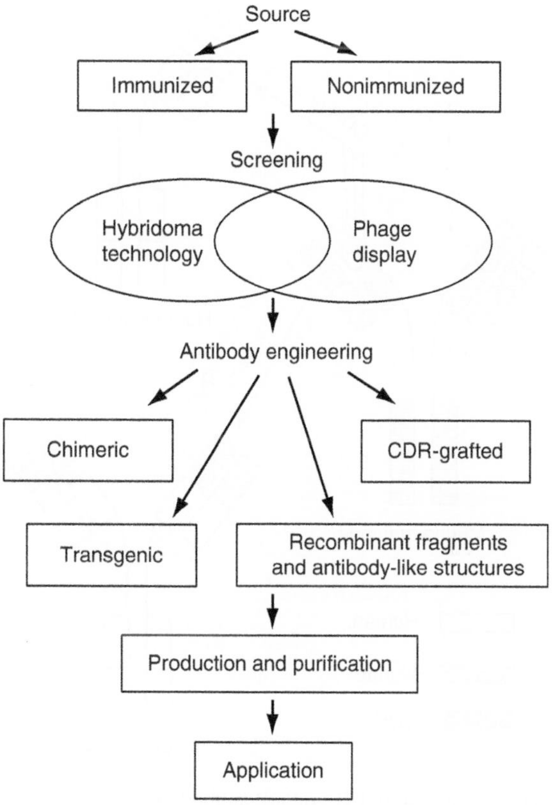

FIGURE 15.6 Integration of hybridoma and phage-display technologies for production of recombinant antibodies.

functions of the Fc region. Thus, the specificity of the variable region can be combined with effector functions in recombinant antibodies. In almost all cases, the chimeric antibodies possess the antigen specificity derived from mouse, and have the effector function determined by human constant regions.

Two functional mouse/human chimeric antibodies (IgMκ and IgG1κ isotypes) have been recently constructed by inserting genomic DNA fragments encoding V_H and Vκ variable regions of the murine monoclonal antibody IgMK-83D4 (specific for Tn determinant) into mammalian expression vector. The vector contained human μ, γ1, and κ constant exons. The construction was transfected into the nonsecreting mouse myeloma X-63 cell line [66]. These chimeric antibodies retained the binding affinities against synthetic Tn glycopeptides. On the other hand, the replacement of mouse C regions with human C regions conferred both chimeric antibodies the ability to activate human complement. Another example of a successful application of this technique is assembly of a new chimeric monoclonal antibody (ch806) with specificity for an epitope of the epidermal growth factor receptor (EGFR) [67]. Ch806 retained the antigen-binding specificity and affinity of the murine parental antibody and displayed enhanced antibody-dependent cellular cytotoxicity against target cells expressing the 806 antigen in the presence of human effector cells.

Chimeric antibodies may be used as calibrators or controls in immunoassays. Data obtained by Hackett et al. [68] demonstrate that chimeric mouse/human antibodies are the adequate alternative to high-titer positive human plasma for the manufacture of calibrators and controls for diagnostic assays (Chapter 13).

The choice of isotype of constant region for chimeric antibodies depends on the intended application for the antibody. For example, the use of an inert isotype, such as IgG2, is appropriate

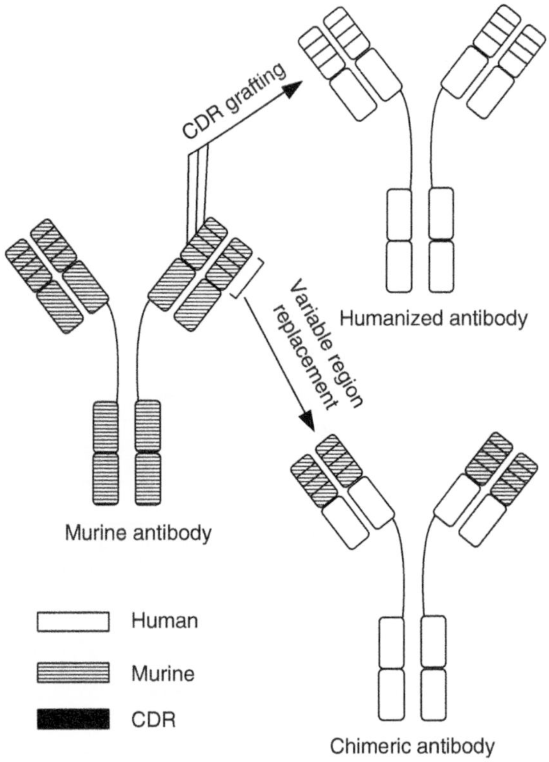

FIGURE 15.7 Chimeric and humanized (CDR grafted) antibodies.

for receptor blocking or drug delivery. In the case when the desired effect is a cellular damage, the isotypes with the effective antibody-depended cell-mediated cytotoxicity properties, such as IgG1 or IgG3, would be more effective [15]. In general IgG is the preferred immunoglobulin class for therapeutic use, because of its complement fixation properties, serum half-life, Fc binding capacity, stability, and ease of purification [16].

15.4.2.3 CDR-Grafted Antibodies

As mentioned above, chimeric antibodies can diminish HAMA response and in general provide favorable effector functions. Nevertheless, chimeric antibodies may still be immunogenic because of their nonself variable region. To further reduce immunogenicity of monoclonal antibodies, a new method of CDR grafting or antibody reshaping has been developed. In this approach, CDRs from the variable region of a murine monoclonal antibody are grafted into the human framework and C regions, resulting in a humanized antibody (Figure 15.7). A humanized antibody is 90%–95% of human origin, but it retains the affinity and specificity similar to the original murine antibody. This approach results in the significant decrease in the immunogenicity. The first attempt to generate a CDR-grafted antibody has been achieved by replacing the CDRs from human myeloma protein NEWM by the heavy chain variable region CDRs from the murine monoclonal B1–8 specific to hapten NP-cap (4-hydroxy-3-nitrophenacetyl caproic acid). The resulting antibody contained only a small fraction of murine-derived sequences, but showed the affinity similar to the parent murine antibody [69]. Currently, CDR grafting is a well-established and commonly used technique for the antibody humanization. A large number of humanized antibodies were obtained by this method. For example, to extend the therapeutic efficacy, humanization by CDR grafting was employed for the chimeric T84.66 monoclonal antibody specific to the tumor-associated carcinoembryonic antigen [70]. CDR grafting was used for construction of humanized antiepidermal growth factor receptor/anti-CD3 bispecific diabody [71].

It has been found, however, that direct CDR grafting may not result in complete retention of the antigen-binding properties. In general, murine CDRs cannot be simply transplanted into the human framework without alteration of their specificity or affinity [16]. To overcome this problem, different strategies can be used for effective humanization. Some of these strategies were recently discussed in [15]:

- Use of human framework regions that bear the greatest homology to the murine counterpart
- The "veneering" murine antibody CDR surface residues to reassemble a human profile
- Use of minimal strategic changes in framework regions to complement CDR grafting

15.4.2.4 Production of the Whole Antibodies

A functional whole antibody can be produced only in expression systems that support a proper protein folding and glycosylation [22]. The point is that the presence of an accessible carbohydrate structure within the Fc region is essential for the efficient effector function. Antibody glycosylation does not influence antigen binding, but the lack of glycosylation does affect some effector functions. Bacterial system cannot glycosylate antibody molecules. Such unglycosylated molecules are expected to bind the antigen, but they may have folding or stability deficiencies in the Fc part and might not be able to fulfill some effector functions. Nevertheless, in some cases *E. coli* derived unglycosylated whole antibodies can be used for biochemical investigations [72].

A whole glycosylated antibody can be successfully expressed in yeast. However, these antibodies may have defects in the effector functions due to the different glycosylation patterns [16]. Regardless, the yeast expression is most commonly used for the whole antibody generation. Some commercially available proprietary yeast expression systems have been developed to make fully functional whole antibodies, such as a system from Alder Biopharmaceuticals (http://www.alderbio.com). Plants can also produce full-length antibodies. Recently, transgenic tobacco plants have been used for antibody generation [33,73,74]. However, whole antibodies expressed in plants (plantabodies) usually have an unnatural carbohydrate composition and structure. One of the strategies to overcome this problem is engineering host cells capable of producing glycosyl antibodies or cells with modified carbohydrate synthesis pathways [22]. The mammalian expression system is currently the most widely used system for fully functional whole antibody production. It possesses the molecular mechanisms required for the correct immunoglobulin assembly, posttranslational modification, and secretion of antibodies. The effector functions and serum half-life of the mammalian-derived antibodies are equivalent to those observed for naturally occurring antibodies. The major disadvantage of mammalian expression systems is a high cost of the manufacturing process [16,22]. Irrespectively, both plant and mammalian expression systems are widely used for the large-scale full-length antibody generation. There is currently an intensive interest from biotechnological companies in scaling up the production to commercial levels (Figure 15.8).

Another strategy for the production of whole human antibodies involves the use of transgenic animals. In this case, antibodies can be obtained by genetically engineering the immune system of mouse [22]. The basic idea is to generate an immune response in engineered "knock-out" (producing human antibodies) mice and to fuse spleen cells from these mice or lymph nodes with myeloma cells (hybridoma technology). There are several approaches to obtain knock-out mice. The first approach is to inactivate endogenous mouse Ig genes and to introduce subsequently human Ig segments, which will enable the mouse to produce entirely human whole antibodies upon immunization with a specific target antigen. The second approach is to genetically engineer mice that carry human mini-chromosomes in every cell. The third approach is to introduce human lymphocytes from immune donors or cancer patients to SCID (severe combined immunodeficient) mice or

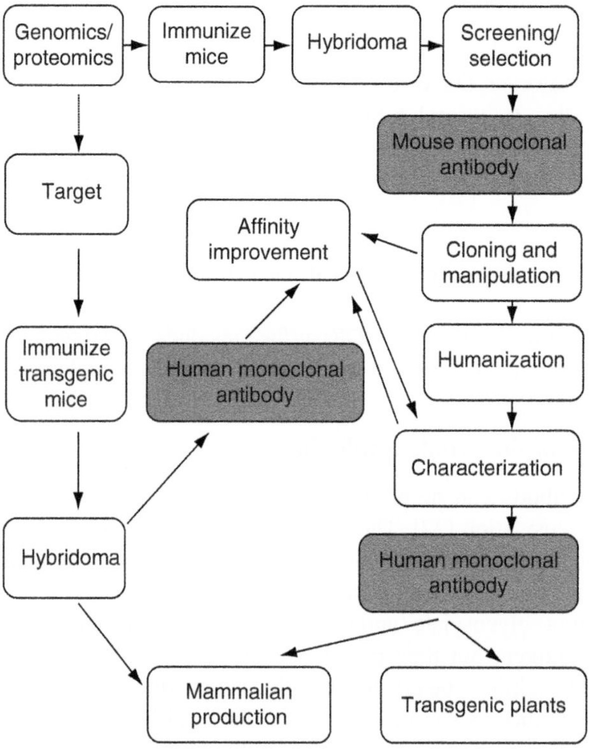

FIGURE 15.8 A strategy used by the Sinol biotechnology group for the large-scale recombinant antibody production.

"Trimera" mice (mice lethally irradiated and radio-protected by transplantation of bone-marrow cells from SCID mice). Other transgenic animals can also be used as bioreactors to produce therapeutic antibodies. For example, a transgenic chicken is suggested as a promising system. Several animal species can be used for producing the recombinant antibodies by secretion into the milk, such as a transgenic dairy or goat [16,22].

15.4.3 Bispecific and Bifunctional Antibodies

15.4.3.1 Bispecific Antibodies

The bispecific antibody is a man-made antibody that binds two different epitopes. It incorporates the binding specificities of two different antibodies into a single molecule [75]. Bispecific antibodies are attractive immunotherapy agents for cancer treatment. They can be designed to bind both the tumor and immune cells in order to utilize cellular immune defense mechanisms and to destroy the tumor target cells [75,76]. Bispecific antibodies can also be used in different immunoassays for enhancing the binding affinity. They may be designed to bind two nonoverlapping epitopes on the same target molecule. It gains a significant increase in avidity [77]. These resulting antibodies may be used with an obvious advantage not only for immunoassays, but also for therapy (for example, for improving tumor targeting and retention). In addition, the bispecific antibodies are employed in various immunochemical and immunohistochemical assays. One arm of the antibody can be designed to bind the target antigen, while another arm to bind the indicator molecule, such as marker enzyme. This approach greatly enhances the performance of the assay.

Bispecific antibodies may be produced using different methods. An early approach for their generation exploited conventional chemical conjugation. But it was accompanied with some instability and batch-to-batch variations. The other strategy used for production of bispecific antibodies represented the somatic fusion of two hybridomas (quadroma). This method provided more stable bispecific antibodies. However, quadroma had some limitations. The major limitation is the production of inactive antibodies due to the random L–H and H–H association, which greatly decreases the yield of the desired production (only 15%). Another limitation of the quadroma bispecific antibodies obtained from the rodent cell lines is their immunogenicity [16,75]. Nevertheless, the quadroma technology is still in a wide use. Thus, Das and Suresh [78] have recently described a rapid method of somatic fusion for hybrid hybridoma generation. Bispecific antibodies, obtained by this method, may be used in the area of ultrasensitive immunodetection and immunodiagnostics. Limitations of quadroma methodology can be successfully overcome by the recombinant antibody engineering.

Currently, it is a preferred technique for bispecific antibodies generation. Current investigations focus on the use of the antibody fragments, instead of the whole antibodies, to create recombinant bispecific antibodies. Two main approaches for engineering the scFv-based bispecific antibodies have been described [75] (Figure 15.9). The first one is the noncollinear approach, where two or

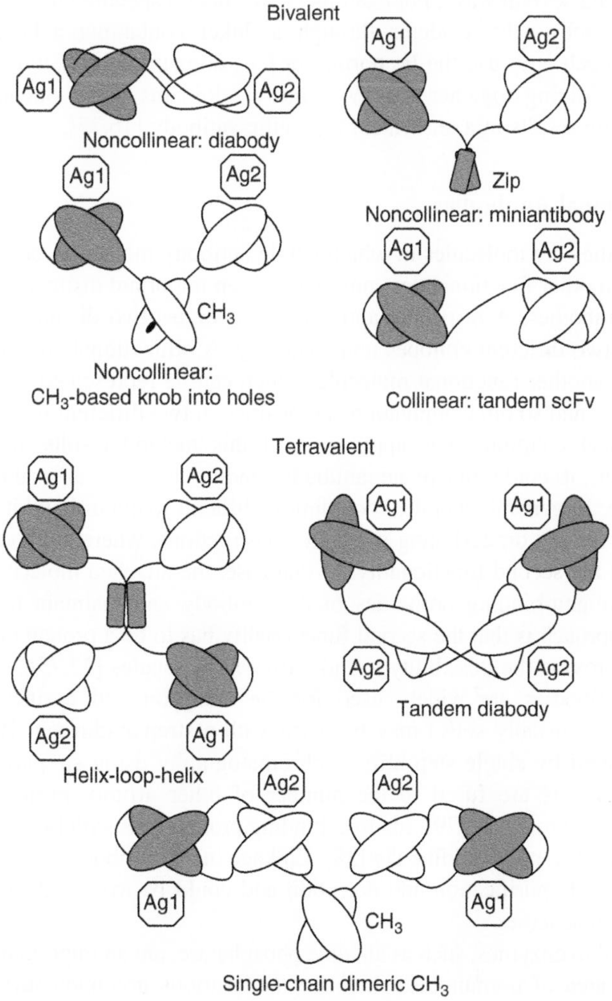

FIGURE 15.9 Strategies for engineering the scFv-based bispecific antibodies.

more chains capable of heterodimerization create bivalent or multivalent bispecific scFvs. The second approach is the collinear approach, where two different scFv are fused together as a single polypeptide chain. Two general strategies are used to generate a noncollinear bispecific scFv. The first method is based on the diabody assembly by shortening the length of the linker between V_H and V_L domains. The association of two different scFv molecules, each comprising V_H and V_L domains derived from different parent Ig, forms a fully functional bispecific diabody [37]. The second method exploits the natural ability of certain protein domains to associate as heterodimers. It includes mini-antibody and minibody formation using leucine zipper (Fos/Jun) and C_L/C_{H1} or C_{H3} domains. Noncollinear bispecific antibodies can also be designed by the use of C_{H3} domain-based "knob into holes" strategy [75]. The collinear method of bispecific scFv antibody formation is based on the tandem joining different scFvs by a linker. These antibodies can be secreted into periplasm of *E. coli* as inclusion bodies and obtained as active antigen-binding proteins after refolding. The length and sequence of the linker may determine the flexibility and correct folding of the molecule. Several linker sequences are used to join the scFvs, such as gly–ser linker (gly4ser or (gly4ser) 3), helical peptide linker, and others [53,75].

Conventionally produced bispecific antibodies have only one binding domain for each specificity. However, a bivalent binding is an important means of enhancing avidity. The increase in valency may be achieved using several ways. For example, tetravalent bispecific molecules may be obtained by dimerization of scFv–scFv tandems through a linker containing a helix-loop-helix motif. An alternative approach is to use the Fc portion or C_{H3} domain of IgG1 to dimerize single-chain bispecific diabody. Joining together four V_H and V_L domains in an orientation that prevents intramolecular pairing results in a tetravalent bispecific antibody [16,75].

15.4.3.2 Bifunctional Antibodies

Bifunctional antibodies are molecules in which whole antibody molecules or their antigen-binding domains are fused to other functional domains. There is an important distinction between bispecific and bifunctional antibodies. A bispecific antibody incorporates two distinct paratopes, which are capable of binding two different epitopes noncovalently. A bifunctional antibody has one or more paratopes linked to another functional molecule, which can be represented by a different peptide, protein, radionuclide, and so on. Conjugation can be done in two different ways. The first approach is based on chemical coupling. The application of this method results in increased antibody efficiency. However, it could inactivate antibody-binding sites or alter the properties of the functional molecule. In general, chemical coupling is difficult to control and it is not site-directed. Another approach is based on designing antibody constructions, where antibody or its fragment is genetically fused to the second functionality. In that case, the attached molecule must be designed not to block the antigen-binding properties of the antibody and maintain its own activity. The limitation of this approach is that the second functionality has to be a protein or a peptide, whereas chemical methods provide the possibility to link different molecules [22,75].

Bifunctional antibodies are widely used for the detection and purification purposes. For example, an antibody (usually scFv) may be fused with a streptavidin tag. Highly purified scFv fragments are prepared by single step affinity chromatography using streptavidin–agarose. Similarly, antibody fragments are fused to the number of other affinity purification tags, such as *Staphylococcus aureus* protein A [79], maltose-binding protein of *E. coli* [43,49], cellulose-binding domain, thioredoxin [47], polyhistidine tag [38]. Linkage of the antibody to the affinity purification tag greatly simplifies its purification and detection and could be exploited to develop large-scale industrial purification schemes.

Antibodies fused to enzymes, such as alkaline phosphatase, are an important source for different applications in the area of immunohistochemistry and various immunoassays. Bifunctional antibodies are also used in a wide range of biomedical applications including diagnostics of a variety of diseases, drug and radionuclides delivery for the treatment, imaging, immunosuppression, and

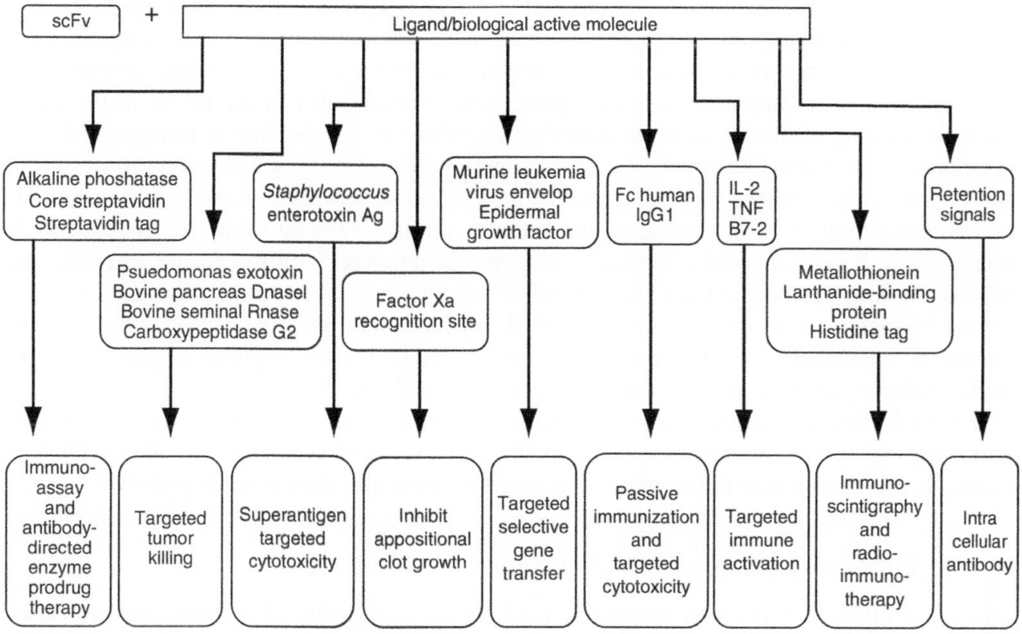

FIGURE 15.10 Examples of bifunctional scFv antibodies applications.

passive immunization. Some examples of the bifunctional antibody applications are shown in Figure 15.10. One of the most interesting uses is targeted cytotoxicity and immune activation. A recombinant antibody can be fused with a protein toxin or its effector fragment. The resulting immunotoxin is capable of killing the target cell (for example, the tumor cell) efficiently, while causing little or no harm to other tissues. For instance, scFv can be fused to *Pseudomonas* exotoxin A, which inhibits protein synthesis and causes the death of the target cell [80]. Recently, a specific targeted immunotoxin application has been improved by the development of antibody-directed enzyme prodrug therapy. In this method an enzyme linked to a tumor-specific antibody is first administered, followed by administration of a nontoxic prodrug. The prodrug is then converted at the tumor site into a cytotoxic drug by the prelocated enzyme [81].

15.5 CLONING, EXPRESSION, AND PURIFICATION OF RECOMBINANT ANTIBODIES

15.5.1 Cloning of Antibody Genes

The expression of antibody genes may be carried out by cloning the antibody encoding genomic DNA with its own promoter and regulatory sequences followed by transfection of plasma cells. However, this method has several disadvantages [4]. Alternatively, it is possible to express antibodies from cDNA clones. Nowadays, the genetic information for antibody V regions is generally retrieved from the total cDNA preparations using a polymerase chain reaction. Hybridoma cells are frequently employed as a source of mRNA. Nevertheless, human peripheral blood or spleen lymphocytes and even a single human B cell may also be used. The last approach provides human antibody fragments with original V_H/V_L pairing and avoids the cumbersome hybridoma technology [16]. Initially, the PCR primers for specific amplification of V domains were generated using known DNA sequences. However, it is possible today to clone the antibody genes without complete knowledge of their sequences. Since antibody-combining sites are based on protein frameworks that are highly conserved, sets of oligonucleotide primers can be designed that will bind essentially to all natural murine (human, etc.) variable regions. Because there are a relatively few different constant regions, designing the primers for 3′-end of the V domains is straightforward. However, the 5′-end primer design is usually more difficult. Two approaches have been developed [4]. These approaches are based on designing primers that anneal either in the FR or in the leader sequence. Priming in the FR gives an ability to use relatively few degenerate primers to clone a majority of the V-domain repertoire, due to the conserved nature of these sequences. However, the strategy has a disadvantage. The problem is that its application may result in amino acid changes in the framework region. It may influence the antibody affinity. Priming in the leader sequence is usually more successful because the leader sequence is removed from the mature antibody, and variation in this sequence will not affect antibody affinity. Primers are usually designed with the internal restriction sites and the obtained PCR products may be then cloned, selected, and sequenced. A number of existing antibody library techniques, such as phage display, allow for fast screening of antibody variants with a suitable specificity and affinity.

For antibodies with already known sequences the method of overlap PCR may be used for assembling the gene segments encoding variable regions, constant regions, or other protein of interest. It also can be used for introducing convenient restriction sites or linker peptides.

15.5.2 Expression of Recombinant Antibodies

A range of different expression systems is used for the production of recombinant antibodies: bacteria, yeast, plants, insect, and mammalian cells. Each system has advantages and disadvantages over other systems depending on the antibody or its fragment to be expressed and potential application of the obtained product. For example, a bacterial system is convenient to use for the production of antigen-binding fragments, however, it cannot provide proper glycosylation of whole antibodies. The whole antibodies may be obtained from mammalian, yeast, and plant expression systems. Despite the advantages of the mammalian system, which supports a proper assembly, posttranslational modifications and secretion of antibodies, its main limitation is a high cost of manufacturing. Plant system restrictions are related to differences in carbohydrate structures. Expression in yeast may result in different glycosylation patterns.

15.5.2.1 Mammalian and Yeast Expression Systems

The mammalian cell expression system is currently a mainstay for engineered antibody expression, including antibody fragments, intact antibodies and multi-chain proteins in general [22,42–44]. Folding, assembly, glycosylation, and secretion of recombinant antibodies are accomplished in a variety of mammalian cells. However, recombinant antibodies produced by cultured mammalian cells may be heterogeneously glycosylated, which can greatly affect antibody properties. There are, though, approaches to overcome this problem [16]. Antibodies can be expressed in mammalian cells either transiently or stably. It depends on the amount of the required protein. Chinese hamster ovary (CHO) cells, mouse myeloma cell lines (SP2/0, NSO, etc.), and human embryonic kidney cells (HEK-293) are widely used to obtain stably transfected cell lines. Different improvements in vector construction, the choice of selectable markers and high-throughput screening strategies have resulted in generation of the cell lines with high productivity of antibodies (20–60 pg/cell/d). However, due to the slow growth of mammalian cells, mammalian expression requires a prolonged time frame and high costs. Another approach for the rapid production of recombinant antibodies has been developed on the basis of transient expression system. This strategy allows for harvesting recombinant proteins from the culture supernatant within a short period of time. The productivity of antibodies obtained by transient expression system has been reported as high as 20 mg/L [16].

The yeast expression system can produce correctly folded, glycosylated complex eukaryotic proteins. It is commonly used both for synthesis of intact antibodies and their antigen-binding fragments. The expression of active antibody fragments (scFv and Fab) has been described in *Pichia pastoris* and *Saccoharomyces cerevisiae* [30–32,38,39]. Yeast expression of antibodies requires different vectors and culture conditions. Miller et al. [38] have compared expression of a human scFv in *S. cerevisiae*, *P. pastoris*, and *E. coli*. For yeast expression, the scFv gene was subcloned into an expression vector, which encodes a fusion protein with the N-terminal α-factor signal peptide, directing recombinant proteins to the media. This system produced soluble scFv antibodies with high yield. Shusta et al. [82] have reported that co-expression of chaperonin and enzyme disulfide isomerase increases the capacity of *S. cerevisiae* to secrete scFv.

One of the advantages of the yeast expression system is the possibility to use the yeast two-hybrid technology for the selection and isolation of antibodies with appropriate specificity and affinity. The two-hybrid system is a genetic tool to study protein–protein interactions. If two proteins interact, then a reporter gene (e.g., gal1-lacZ, the beta-galactosidase gene) is transcriptionally activated. Visintin et al. [83] have developed the intracellular antibody capture protocol. It involves a combination of the yeast two-hybrid and phage-display technologies for the selection of antibodies. Another advantage of the yeast system is that the homologous recombination mechanism may be used for production of chimeric genes and antibody libraries, screened further by the yeast surface display [84]. The yeast system is often used not only for the expression of antibodies, but also for the construction of antibody libraries, expressed on the yeast surface. The yeast surface display as well as other display systems represent a powerful technology for the antibody affinity maturation [31,84]. A mammalian cell display system has also been recently developed [44].

15.5.2.2 Expression in Plants

Transgenic plants expressing recombinant antibodies have arisen as a convenient technology for the large-scale production of antibodies. Plants have been successfully used for the expression of both the whole antibodies and smaller antibody fragments. Thus, anti-HBsAg (anti-hepatitis B virus surface antigen) mouse IgG1 monoclonal antibody was expressed in transgenic tobacco plants. For this purpose, F1 plants were generated using the sweet potato sporamin signal peptide, and a heavy- and light-chain gene tandem construction. The expression of the antibody product in these plants reached up to 0.5% of the total soluble protein [74]. Also, full-size IgGs and Fab fragments were expressed in transgenic potato tubers. Genetic constructions were engineered for accumulating antibodies in the plant cell apoplast or endoplasmic reticulum. Accumulation levels in tubers of up to 0.5% of total soluble protein were found for antibodies targeted to the endoplasmic reticulum, whereas fivefold lower accumulation levels were found for antibodies targeted for secretion [33]. Single-chain Fv antibodies were successfully expressed in the *Arabidopsis* plants. They had an ability to bind the target antigen with high affinity, comparable to the bacterially produced scFv antibodies and the parent monoclonal antibodies [41].

The plant system has several advantages for the antibody production, since plants can be grown easily and inexpensively in large quantities, can be harvested, stored, and processed by existing infrastructures. Plants are also safe and free of human pathogens [22]. These features have spurred an interest from the biotechnological companies in scaling up production to commercial levels (Figure 15.11). Antibodies can be harvested from transgenic potato tubers and even from seeds, where they can be stored safely for long periods. Tubers could be stored for up to 6 months without a significant loss of antibody amount or activity [33].

The integration of the gene and all regulatory elements into the plant genome is required for the stable expression over the life of the plant and in the progeny. Two approaches are commonly used for transforming plants. The first approach uses the Ti plasmid of *Agrobacterium tumefacient* as an introducing agent for the recombinant DNA into the plant cell nucleus. The second method is based on the bombardment with DNA-coated gold particles, which allows the simultaneous introduction of multiple constructions [16].

The plant antibody expression can be used not only for the antibody production but also for protection of plants themselves from the pathogens. The idea is based on obtaining resistance by

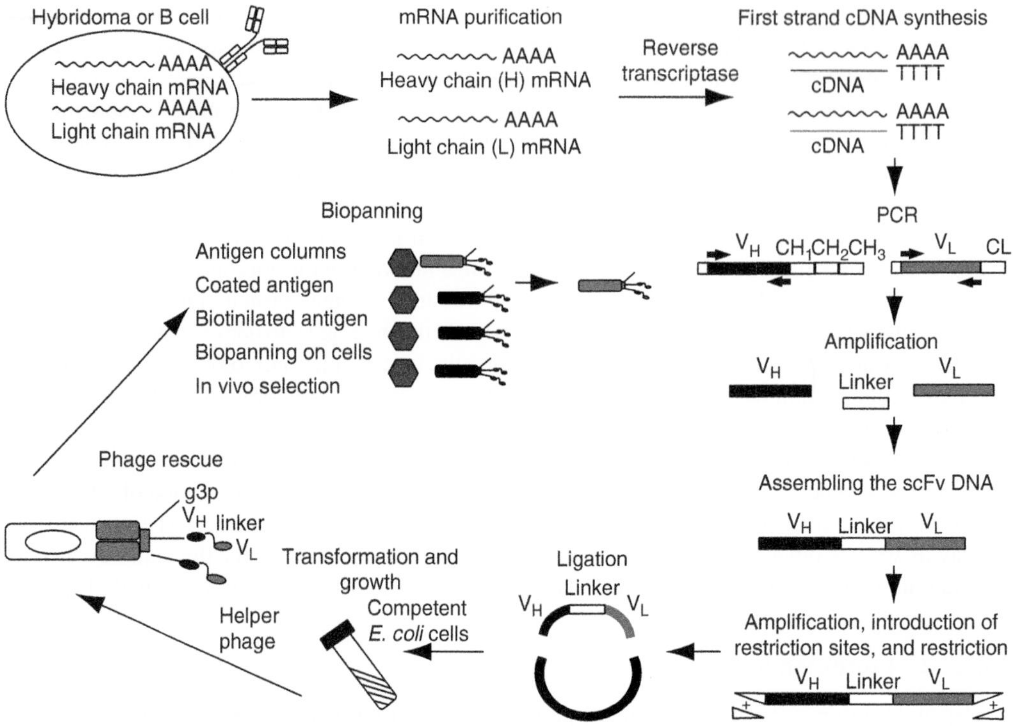

FIGURE 15.11 A scheme for scFv phage-display library generation and screening.

transfecting plants with genes, which encode monoclonal antibodies against pathogen-specific proteins. These monoclonal antibodies inhibit the biological activity of molecules that are essential for the pathogenesis. Several attempts in this area have been carried out with tobacco and *Arabidopsis* plants [40,41].

Despite many advantages, the plant expression system has several drawbacks. The first disadvantage is the length of time required from the initial transformation event to the expression and characterization of the recombinant protein (may take 5–10 mo). The second limitation of the expression of therapeutic antibodies in food crops is a potential gene flow (via pollen) to the surrounding crops. These problems may be overcome by the alternative plant expression system, where proteins may be expressed in the soluble form in the chloroplasts [34]. The use of plants for antibody production is also limited due to the difference in protein glycosylation between mammalian and plant cells. For more detailed review of plant expression systems see Chapters 18 and 19 in this book.

15.5.2.3 Expression in Bacteria

The bacterial expression system is commonly used for the production of antibody antigen-binding fragments. Whole antibodies can also be produced in bacteria when glycosylation is not an issue. Bacteria-derived antibody products are only used in research [72]. *E. coli* is the most popular host due to its fast growth rate, ease of genetic manipulation, and low cost of manufacturing. However, the antibodies expressed in *E. coli* are often insoluble and inactive. It happens because the reduced cytoplasm environment prevents the proper folding of complex antibody molecules, mainly, due to the inability to form disulfide bonds [43,45,46]. There are several approaches to obtain functional antibody fragments from *E. coli*. The most popular strategy is to secrete these proteins into the periplasmic space [26,27,85,86]. The periplasmic space is known to contain proteins such as chaperons and disulfide isomerases, which facilitate the correct folding of recombinant proteins [87]. The scFv and Fab fragments are usually well expressed in the periplasm; they contain correctly folded disulfide bonds and are soluble. However, the efficiency of folding appears to depend on the physicochemical properties of a protein, and in some cases antibodies are observed as nonsoluble and nonfunctional products even when expressed in periplasm. A high-level expression often also results in the accumulation of insoluble antibodies in the periplasmic space [15,16,88]. To overcome this problem, a co-expression with different proteins such as bacterial periplasmic protein Skp/OmpH or PPIase FkpA was suggested [89]. A substitution of the hydrophobic residues by hydrophilic ones can also increase the functional periplasmic expression [90]. Another strategy is based on manipulation of bacterial growth rate and induction conditions by lowering the temperature [91]. Long incubation time can result in releasing a soluble protein into the cultural medium. Secretion into the medium is greatly enhanced in special mutant strains. For instance, Chames et al. [92] obtained almost completely insoluble scFv in a wild-type strain of *E. coli*. However, when scFv was produced in a tol R leaky strain, it was secreted into the culture medium as an active soluble protein.

Another approach for the antibody production in *E. coli* is expression of antibodies in the form of inclusion bodies in cytoplasm. The inclusion bodies contain proteins in a nonnative and nonactive conformation. Thus, these proteins must be dissolved and folded into the right shape. Different improved protocols are currently used for the in vivo and in vitro antibody refolding [45,46]. Several attempts to obtain soluble and active antibodies in the *E. coli* cytoplasm have also been carried out. Expression in *E. coli* mutant strains carrying mutations in the major intracellular disulfide bond reduction systems (i.e., thioredoxin and glutathione/glutaredoxin pathways) facilitates oxidation and, therefore, a proper folding of antibody fragments in *E. coli* cytoplasm [47,48,93]. Co-expression of disulfide bond chaperons can also be used for enhancing antibody solubility and functional antibody production [47]. Another strategy is a genetic fusion of antibody with a protein that has enhanced solubility and oxidation properties [43,47,49].

15.5.3 Purification of Recombinant Antibodies

Developing the strategies for antibody purification is of great importance today. The purification process is highly dependent on the quality of starting material that contains the protein of interest. In addition to the purity of the obtained product, its activity and proper structure must be controlled. Different strategies based on classical chromatography have been described for antibody purification. Nonchromatographic techniques are also used separately or in combination with chromatography. The examples of strategies to antibody purification include filtration methods based on thiophilic and affinity interactions, two-phase aqueous systems, preparative electrophoresis and isoelectric focusing [94].

Affinity separation is one of the most suitable methods for the recovery of antibodies. Affinity chromatography is commonly used for antibody purification. However, it has some limitations when applied to the large-scale purification. These limitations are related to column operation and clogging of the packed bed of adsorbents with complex feeds. To overcome these problems different modes of affinity interactions are used such as expanded-bed affinity chromatography, affinity extraction, and affinity precipitation [22,94].

The purification of antibodies by affinity techniques includes two general strategies. One strategy is based on the specificity of antigen binding. Another approach targets the constant domain (Fc fragment). The second approach is based on the use of the natural immunoglobulin-binding ligands such as protein G and protein A. Both bind immunoglobulins primarily through the Fc region. Thus, they can be used mainly for the whole antibody or its Fc part purification. Isolation of antigen-binding antibody fragments such as Fab or scFv is difficult using this methodology.

However, in some cases low-affinity binding sites in the Fab region can be suitable for efficient purification of some Fab fragments using protein G and protein A [22].

A great advantage of recombinant antibodies is that their purification can be improved by incorporation of genetically engineered peptide tags. Thus, in addition to standard antibody purification methods, it is possible to purify antibody by targeting the nonantibody region. Popular tags include a c-myc peptide or FLAG tag (either of which can be recognized by monoclonal antibodies), and polyhistidine tag, which can be utilized for immobilized metal affinity chromatography. Some other peptide tags such as *S. aureus* protein A, maltoso-binding protein of *E. coli* and cellulose-binding domain are also used [38,43,49,56,79].

15.6 ANTIBODY LIBRARIES SCREENING AND AFFINITY MATURATION

15.6.1 Phage Display

Nowadays, antibody engineering relies on the ability to generate and screen large combinatorial libraries of antibody genes. To isolate a highly specific antibody appropriate for most biomedical applications, the library of 10^5–10^{11} clones has to be screened. Several attempts to develop high throughput screening techniques were made during last years [95,96]. However, the great progress was achieved only after development of various display systems. Display system allows for establishing a direct physical link between a gene, the protein encoded by this gene, and the molecule recognized by this protein. This approach greatly simplifies screening of large recombinant protein libraries. Different in vivo and in vitro display systems are in a wide use today. Among them, phage display is currently the most exploitable method for screening peptide and protein libraries in general, and antibody libraries in particular. Phage display is a powerful molecular technology that allows for high throughput screening of peptide or/and protein libraries displayed on the surface of a filamentous phage. DNA sequences of interest are fused in the frame to phage genes encoding surface-exposed proteins, most commonly gene III (g3). The engineered phage infects bacteria, and the resulting phage particles express active products on their surface. The phage display library expresses many different variants of protein of interest. This library can easily be used to select and purify phage particles containing peptides/proteins with desired specificities [14,97–99].

Since introduction by Smith in 1985, phage display has been proven to be a very effective way for isolating proteins that perform specific functions. Moreover, it is also used today to improve or modify the affinity of proteins for their binding partners and to study protein–ligand interactions, receptor and antibody-binding sites [98]. As far as antibody phage display concerned, two essential steps were significant for its development. The first one is the discovery that foreign DNA fused to g3 is expressed on the surface of the phage particle. The second is improvement in the antibody fragment engineering and successful production in the functional and soluble form in the *E. coli* periplasm [98].

Phage display technology has greatly improved the antibody engineering process. The following examples of the most successful application of phage display to antibody engineering were reviewed in [14,98]:

- De novo isolation of high affinity human antibodies from libraries
- Isolation of antigen-specific antibodies from immunized animals and hybridomas
- Generation of human monoclonal antibodies or humanization of mouse antibodies useful for cancer immunotherapy
- Isolation of human antibodies from patients exposed to certain viral pathogens for a better understanding of immune response during infection and protective antibodies
- Isolation of antibodies with increased resistance to chemical or thermal denaturation
- Elucidation of specificity of autoimmune antibodies

- Affinity maturation of antibodies
- Construction of catalytic antibodies
- Isolation of antibodies specific to cell surface markers or receptors by direct panning against whole cells and tissue samples

Both Fab and scFv fragments may be obtained by phage display [23,27–29,51,100]. After selecting the appropriate variants, the primary structures of the variable domain may be determined. These domains may be fused with the Fc part to reconstruct the whole antibody. For scFv phage library, V_H and V_L chains are assembled into a single gene by a peptide linker. This gene is fused via its carboxyl terminus to g3. In the case of Fab fragment display, either light or heavy chain is fused to g3, whereas the partner chain is unfused. In the periplasmic space of *E. coli* the association of two chains occurs to form an intact Fab.

The use of phage-display library for antibody screening can be reviewed on the following example (Figure 15.11). First, the antibody variable light and heavy chain genes are prepared by reverse transcription of mRNA, which is isolated from hybridoma or B-cells. mRNA is affinity purified using oligo(dT)-cellulose. Reverse transcription of mRNA into cDNA is initiated by a random hexamer primer. The V_H and V_L genes are amplified from cDNA by PCR with a special set of primers. The V_H and V_L encoding PCR fragments are connected into a single gene. The assembled scFv fragments are amplified by PCR with introduction of restriction sites. Finally, scFv fragments are inserted into the display vector. Both phage and phagemid vectors have been used for antibody phage display. However, phagemid vectors are the most popular and in general they superseded phage vectors. Phagemids, as hybrids of phage and plasmid vectors, may be grown as plasmids or may be packaged as the phage particles using helper phage [99]. The phagemid library containing scFv variants is transformed into the *E. coli* cells. The transformed bacterial cells are infected with the helper phage (M13VCS or KO7). The recombinant phage particles displaying scFv fragments on its surface are propagated (the process, known as phage rescue). The resulting phage-displayed library is subjected to screening and selection for antibody-specific clones by the method known as biopanning. For this purpose, the phage library is incubated with an immobilized target antigen. Unbound phages are removed by washing steps. Phage particles that bind specifically to antigen are eluted. Usually several rounds of panning are required for the selection of phage displayed antigen-specific antibody. Several strategies of biopanning are in widespread use today [98]:

- Selection using column with immobilized antigens
- Selection using antigen adsorbed on the plastic surface
- Selection using biotinylated antigens in solution and avidin-coated magnetic beads
- Selection using antigen expressed on the surface of cells
- In vivo selection using injection of phage libraries into animals with following recovery of tissues for collection of phage particles bound to the tissue-specific antigens

To improve the effectivity of phage display methodology and achieve the desired results, a number of key points must be considered when constructing the antibody phage library. The size of the library and its quality are essential for performance. The method of electroporation is preferred for introduction of phagemid library into the *E. coli* cells over chemical transformation. The source of antibody genes is of great importance. Two fundamentally different sources of antibody genes may be used [98,99]. The first source is the natural V-gene repertoires derived from human or animal donors. The second source is based on the in vitro construction of synthetic V-gene repertoires. The source of natural repertoires may be subdivided into the groups of naïve and immune repertoires. In naïve repertoire library, V genes are isolated from the IgM mRNA of B cells from unimmunized human donors or animal sources. These libraries have a series of advantages. First, antibodies can be isolated to self, nonimmunogenic and toxic antigens. Second, a single library can be used for all antigens. Third, less than two weeks is needed for antibody generation. Finally, donor immunization is not required. It is especially important in the case of human antibody generation. The major disadvantage of naïve library is that a larger library size is required for isolation of high affinity antibodies in comparison with the relatively small size of antibody library made from the immunized donors, where isolation of antibodies with similar affinities is possible. Also, content and quality of the library are influenced by unequal expression of the V-genes repertoire, and exact nature of V-genes repertoire is largely unknown and uncontrollable.

V-genes for the immune library construction are derived from the IgG mRNA of B cells from immunized animals or, in some cases, humans. These libraries are enriched in antibodies specific to immunogen. Therefore, a relatively small library can be used for selection. Additionally, some of these antibodies may be already affinity maturated by the immune system. However, immune libraries also have several drawbacks. Thus, it is difficult or impossible to isolate antibodies to self and toxic antigens. Long time is required for animal immunization. There is a little control over the immune response. In addition, for each antigen a new library has to be constructed.

The advantage of synthetic repertoires is that they have the potential to control and define the contents and diversity of synthetic V regions. Synthetic library can be made from germline V genes. To increase library size, diversity and so possibility to isolate high affinity antibodies, randomization of the VH-CDR3 region using oligonucleotide-directed mutagenesis or PCR is used. Other approaches were also described [98].

Several trends in the development of antibody-phage technology can be found described in the extant literature. The most promising ones include the following [14]:

- Systems with only one copy of antibody per phage particle displayed
- Methods for guaranteed release of bound phage from a solid support
- Methods for enrichment of phage binding
- Selective infectivity with phage particles bound to ligand

15.6.2 Other Types of Display

Antibodies can also be displayed on the surface of the cell. Different types of the cell surface display have been reported in literature, including the display on the surface of microbial cells such as *E. coli* [101], yeast cells [31,84] and even on the surface of mammalian cells [44]. For screening purposes, a library of cells, with each cell displaying multiple copies of a different antibody variant, is incubated with a fluorescently tagged ligand in buffer. The cells become fluorescently labeled and are isolated by fluorescence-activated cell sorting. However, the limitation of the technology is the limited size of the library in comparison with phage display libraries.

Ribosomal display is a method of cell-free selection of antibodies from antibody libraries [64]. It relies on the formation of a ternary complex between ribosomes, mRNA, and the polypeptide. In this system large libraries with virtually unlimited diversity can be constructed, and the particular proteins are translated, displayed and selected completely in vitro on the ribosomes. Different other types of the antibody display are also in use, such as baculovirus display, ProFusion, CDT (covalent display technology) and polysome display [22].

15.6.3 Affinity Maturation

As a general rule, antibodies generated either by animal immunization or by repertoire library screening exhibit antigen affinities in the range from 10^{-6} to 10^{-9} M [14,102]. Obtaining higher-affinity antibodies is important for efficient binding to the antigenic target. To improve antibody affinity, various strategies are currently in use [14]:

- Site-specific mutagenesis
- Chain shuffling

- Combinatorial mutagenesis
- Extension of the heavy chain CDR3 loop
- Random mutagenesis over the entire V-gene
- Recombination of beneficial mutations isolated from all the variants listed above

Site-specific mutagenesis is based on the structural information. Therefore, it requires a crystal structure or a computer model of antibody/antigen complex [103]. In chain shuffling methodology, gene for one V domain (e.g., V_H) of the Fv module is cloned into the repertoire for another domain (V_L). The resulting library contains scFvs with random V_L chains and V_H chains specific for target antigen. After panning and isolation of clones with good binding properties, the new V_L chain is cloned into the repertoire of V_H genes [84,104]. Combinatorial mutagenesis is applied to one or more CDR loops. Randomizing a select set of CDR amino acid residues to all possible 20 amino acids creates the libraries. That randomization can be performed by degenerate oligonucleotides. Extension of the heavy chain CDR3 loop is used to artificially extend the region, responsible for contact of antibody with antigen. The method is especially attractive for antihapten antibodies [14]. Instead of focusing on mutagenesis of specific regions, random mutagenesis of the entire gene can be carried out. Randomly mutagenized libraries [105] can be constructed by in vitro DNA amplification under conditions, which increase the possibility of mis-incorporation of nucleotides (error-prone PCR) [106]. Homologous gene re-arrangement [84] is also often used for this purpose, as well as propagation of V genes in *E. coli* mutator cells [64]. It is now obvious that the most successful methods of affinity maturation rely on several cycles of mutation, display, selection, and gene amplification (Figure 15.12). These cycles can be carried out using either in vitro or in vivo strategies [64].

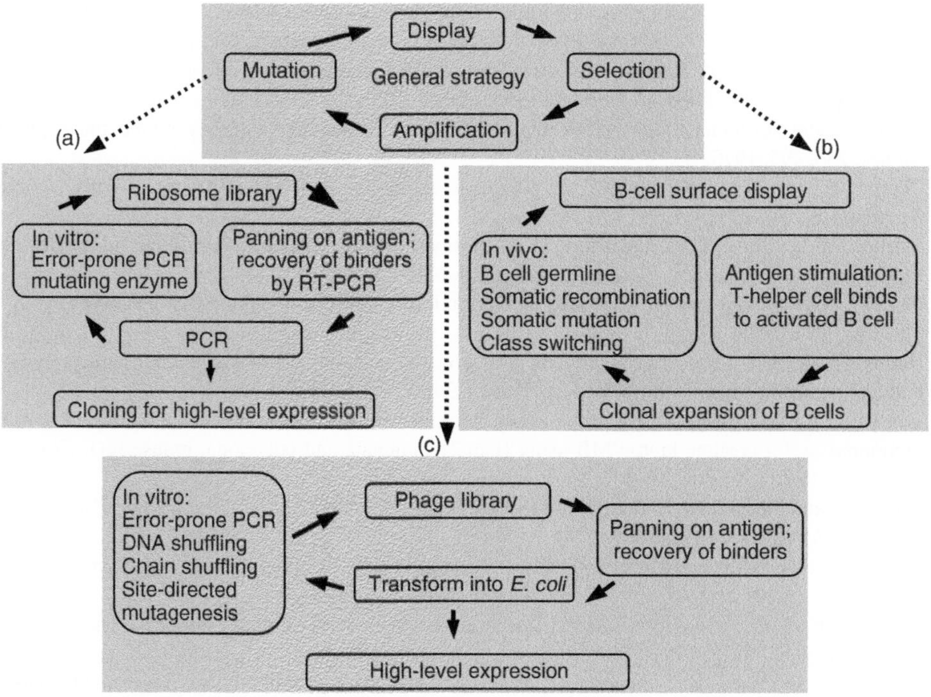

FIGURE 15.12 Cycles of antibody affinity maturation. (a) In vitro cycle of mutation and selection by ribosome display of the engineered antibody library; (b) in vivo antibody maturation using mammalian cells; and (c) in vitro cycle of mutation and selection by phage display of the engineered antibody library.

REFERENCES

1. Greighton, T.E., *Encyclopedia of Molecular Biology*, John Wiley & Sons, Inc., New York/Chichester/Weinheim/Brisbane/Singapore/Toronto, 1999, vol. 1–4, p. 284.
2. Wang, W. et al., Antibody structure, instability, and formulation, *J. Pharm. Sci.*, 96, 1, 2007.
3. Roitt, I., Brostoff, J., and Male D., *Immunology*, 5th edn., Mosby Int., London/Philadelphia/St Louis/Sydney/Tokyo, 1998, Chap. 6.
4. Sandhu, J.S., Protein engineering of antibodies, *Crit. Rev. Biotechnol.*, 12, 1992.
5. Catty, D., *Antibodies: A Practical Approach*, IRL press, Oxford-Washington DC, 1989, Vol. 1, Chap. 1.
6. Rees, A.R. et al., Antibody design: Beyond the natural limits, *TIBTECH*, 12, 199, 1994.
7. Al-Lazikani, B., Chothia, C., and Lesk, A.M., Standard conformations for the canonical structures of immunoglobulins, *J. Mol. Biol.*, 273, 927, 1997.
8. Furukawa, K. et al., A role of the third complementarity-determining region in the affinity maturation of an antibody, *J. Biol. Chem.*, 276, 27622, 2001.
9. Xu, J.L. and Davis, M.M., Diversity in the CDR3 region of VH is sufficient for most antibody specificities, *Immunity*, 13, 37, 2000.
10. Hage, D.S. and Nelson, M.A., Chromatographic immunoassays, *Anal. Chem.*, 73, 198, 2001.
11. Blank, K. et al., Self-immobilizing recombinant antibody fragments for immunoaffinity chromatography: Generic, parallel, and scalable protein purification, *Protein Expr. Purif.*, 24, 313, 2002.
12. Wingren, C. and Borrebaeck, C., Antibody microarrays: Current status and key technological advances, *J. Integr. Biol.*, 10, 411, 2006.
13. Ohara, R. et al., Antibodies for proteomic research: Comparison of traditional immunization with recombinant antibody technology, *Proteomics*, 6, 2638, 2006.
14. Maynard, J. and Georgiou, G., Antibody engineering, *Annu. Rev. Biomed. Eng.*, 2, 339, 2000.
15. Rapley, R., The biotechnology and applications of antibody engineering, *Mol. Biotechnol.*, 3, 139, 1995.
16. Kipriyanov, S.M. and LeGall, F., Generation and production of engineered antibodies, *Mol. Biotechnol.*, 26, 39, 2004.
17. Filpula, D., Antibody engineering and modification technologies, *Biomol. Eng.*, 24, 201, 2007.
18. Dübel, S., Recombinant therapeutic antibodies, *Appl. Microbiol. Biotechnol.*, 74, 723, 2007.
19. Sa'Adu, A. and Zumla, A., Human monoclonal antibodies: Production, use, problems, in *Monoclonal Antibodies—Production, Engineering and Clinical Applications*, Ritter, M. and Ladyman, H. (Eds.), Cambridge University Press, Cambridge, 1995, pp. 85–106.
20. Kohler, C. and Milstein, C., Continuous cultures of fused cells secreting antibody of predefined specificity, *Nature*, 256, 495, 1975.
21. Thorpe, S.J. et al., Clonal analysis of a human antimouse antibody (HAMA) response, *Scand. J. Immunol.*, 57, 85, 2003.
22. Roque, A.C.A., Lowe, C.R., and Taipa, M.A., Antibodies and genetically engineered related molecules: Production and purification, *Biotechnol. Prog.*, 20, 639, 2004.
23. Colcher, D. et al., Pharmacokinetics and biodistribution of genetically-engineered antibodies, *Q. J. Nucl. Med.*, 43, 225, 1998.
24. Hemminki, A. et al., Specificity improvement of a recombinant anti-testosterone Fab fragment by DCRIII mutagenesis and phage display selection, *Protein Eng.*, 11, 311, 1998.
25. Vaidya, H.C. and Beatty, B.G., Eliminating interference from heterophilic antibodies in a two-site immunoassay for creatine kinase MB using (Fab′2) conjugate and polyclonal mouse IgG, *Clin. Chem.*, 38, 1742, 1992.
26. Skerra, A. and Plückthun, A., Assembly of a functional immunoglobulin Fv fragment in *Escherichia coli* surface, *Science*, 240, 1038, 1988.
27. Saad, A. et al., Cloning and expression in *E. coli* of a functional Fab fragment obtained from single human lymphocyte against *anthrax toxin*, *Molec. Immunol.*, 8, 2101, 2007.
28. Barbas, C.F. et al., Assembly of combinatorial antibody libraries on phage surfaces: The gene III site, *Biochemistry*, 88, 7978, 1991.
29. Corisdeo, S. and Wang, B., Functional expression and display of an antibody Fab fragment in *Escherichia coli*: Study of vector designs and culture conditions, *Protein Expr. Purif.*, 34, 270, 2004.
30. Ning, D. et al., Expression, purification, and characterization of humanized Anti-HBs Fab fragment, *J. Biochem.*, 134, 813, 2003.

31. Blaise, L. et al., Construction and diversification of yeast cell surface displayed libraries by yeast mating: Application to the affinity maturation of Fab antibody fragments, *Gene*, 342, 211, 2004.
32. Kozyr, A.V. et al., Production of DNA-hydrolyzing antibody BV04-01 Fab fragment in methylotrophic yeast *Pichia pastoris*, *Molec. Biol.*, 38, 914, 2004.
33. Wilde, C.D. et al., Expression of antibodies and Fab fragments in transgenic potato plants: A case study for bulk production in crop plants, *Molec. Breed.*, 9, 271, 2002.
34. Nguyen, V., Olave, M., and Cohen, A., Expression of the Anti-cocaine Antibody Fab Fragments in *Chlamydomonas reinhardtii*, Report 1011, Botany and Plant Biology Joint Congress, Chicago, 2007.
35. Guttieri, M.C. et al., Cassette vectors for conversion of Fab fragments into full-length human IgG1 monoclonal antibodies by expression in stably transformed insect cells, *Hybrid Hybridomics*, 22, 135, 2003.
36. Hochman, J., Inbar, D., and Givol, D., An active antibody fragment (Fv) composed of the variable portions of the heavy and light chains, *Biochemistry*, 12, 1130, 1973.
37. Hudson, P.J. and Kortt, A.A., High avidity scFv multimers; diabodies and triabodies, *J. Immunol. Methods*, 231, 177, 1999.
38. Miller, K.D. et al., Production, purification, and characterization of human scFv antibodies expressed in *Saccharomyces cerevisiae*, *Pichia pastoris*, and *Escherichia coli*, *Protein Expr. Purif.*, 42, 255, 2005.
39. Braren, I. et al., Comparative expression of different antibody formats in mammalian cells and *Pichia pastoris*, *Biotechnol. Appl. Biochem.*, 47, 205, 2007.
40. LeGall, F., Bové, J.-M., and Garnier, M., Engineering of a single-chain variable-fragment (scFv) antibody specific for the *Stolbur Phytoplasma* (Mollicute) and its expression in *Escherichia coli* and tobacco plants, *Appl. Environ. Microbiol.*, 64, 4566, 1998.
41. Yuan, Q. et al., Expression of a functional antizearalenone single-chain Fv antibody in transgenic *Arabidopsis* plants, *Appl. Environm. Microbiol.*, 66, 3499, 2000.
42. Reavy, B. et al., Expression of functional recombinant antibody molecules in insect cell expression systems, *Protein Expr. Purif.*, 18, 221, 2000.
43. Shaki-Loewenstein, S. et al., A universal strategy for stable intracellular antibodies, *J. Immunol. Methods*, 303, 19, 2005.
44. Ho, M., Nagata, S., and Pastan, I., Isolation of anti-CD22 Fv with high affinity by Fv display on human cells, *Proc. Natl. Acad. Sci. U S A*, 103, 9637, 2006.
45. Kouhei, T. et al., Immobilized oxidoreductase as an additive for refolding inclusion bodies: Application to antibody fragments, *Protein Eng.*, 16, 535, 2003.
46. Das, D. et al., Development of a biotin mimic tagged ScFv antibody against western equine encephalitis virus: Bacterial expression and refolding, *J. Virol. Methods*, 117, 169, 2004.
47. Jurado, P., DeLorenzo, V., and Fernánde, L.A., Thioredoxin fusions increase folding of single chain Fv antibodies in the cytoplasm of *Escherichia coli*: Evidence that chaperone activity is the prime effect of thioredoxin, *J. Mol. Biol.*, 357, 49, 2006.
48. Xiong, S. et al., Solubility of disulfide-bonded proteins in the cytoplasm of *Escherichia coli* and its "oxidizing" mutant, *World J. Gastroenterol.*, 11, 1077, 2005.
49. Hayhurst, A., Improved expression characteristics of single-chain Fv fragments when fused downstream of the *Escherichia coli* maltose-binding protein or upstream of a single immunoglobulin-constant domain, *Protein Expr. Purif.*, 18, 1, 2000.
50. Jordan, E. et al., Production of recombinant antibody fragments in *Bacillus megaterium*, *Microb. Cell Factor.*, 6, 1475, 2007.
51. Jung, S., Honegger, A., and Plückthun, A., Selection for improved protein stability by phage display, *J. Mol. Biol.*, 294, 163, 1999.
52. Plückthun, A. and Pack, P., New protein engineering approaches to multivalent and bispecific antibody fragments, *Immunotechnology*, 3, 83, 1997.
53. Albrecht, H., DeNardo, G.L., and DeNardo, S.J., Monospecific bivalent scFv-SH: Effect of linker length and location of an engineered cysteine on production, antigen binding activity and free SH accessibility, *J. Immunol. Methods*, 310, 100, 2006.
54. Kumada, Y. et al., Polypeptide linkers suitable for the efficient production of dimeric scFv in *Escherichia coli*, *Biochem. Engineer. J.*, 35, 158, 2007.
55. LeGall, F. et al., Di-, tri- and tetrameric single chain Fv antibody fragments against human CD19: Effect of valency on cell binding, *FEBS Lett.*, 453, 164, 1999.

56. Wu, P.J. and Yazaki, P.J., Designer genes: Recombinant antibody fragments for biological imaging, *Q. J. Nucl. Med.*, 43, 268, 1999.
57. DeKruif, J. and Logtenberg, T., Leucine zipper dimerized bivalent and bispecific scFv antibodies from a phage display library, *Immunotechnology*, 2, 298, 1996.
58. Olafsen, T. et al., Characterization of engineered anti-p185$^{HER\text{-}2}$ (scFv-$C_H3)_2$ antibody fragments (minibodies) for tumor targeting, *PEDS*, 17, 315, 2004.
59. Kipriyanov, S.M. et al., Affinity enhancement of a recombinant antibody: Formation of complexes with multiply valency by a single-chain Fv fragment-core streptavidin fusion, *Protein Eng.*, 9, 203, 1996.
60. Gregory, P.A. et al., Avidity-mediated enhancement of in vivo tumor targeting by single-chain Fv dimers, *Clin. Cancer Res.*, 12, 1599, 2006.
61. Mireille, D. et al., Single-domain antibody fragments with high conformational stability, *Protein Sci.*, 11, 500, 2002.
62. Stijlemans, B. et al., Efficient targeting of conserved cryptic epitopes of infectious agents by single domain antibodies: African trypanosomes as paradigm, *J. Biol. Chem.*, 279, 1256, 2004.
63. Purohit, S. et al., Lack of correlation between the levels of soluble cytotoxic T-lymphocyte associated antigen-4 (CTLA-4) and the CT-60 genotypes, *J. Autoimm. Dis.*, 2, 8, 2005.
64. Irving, R.A., Kortt, A.A., and Hudson, P.J., Affinity maturation of recombinant antibodies using *E. coli* mutator cells, *Immunotechnology*, 2, 127, 1996.
65. Heap, K.J. et al., Analysis of a 17-amino acid residue, virus-neutralizing microantibody, *J. Gen. Virol.*, 86, 1791, 2005.
66. Oppezzo, P. et al., Production and functional characterization of two mouse/human chimeric antibodies with specificity for the tumor-associated Tn-antigen, *Hibridoma*, 19, 229, 2000.
67. Panousis, C. et al., Engineering and characterization of chimeric monoclonal antibody 806 (ch806) for targeted immunotherapy of tumours expressing de2–7 EGFR or amplified EGFR, *Br. J. Cancer*, 92, 1069, 2005.
68. Hackett, J. et al., Recombinant mouse-human chimeric antibodies as calibrators in immunoassays that measure antibodies to Toxoplasma gondii, *J. Clin. Microbiol.*, 36, 1277, 1998.
69. Jones, P.T. et al., Replacing the complementary determining region in a human antibody with those from a mouse, *Nature*, 321, 522, 1886.
70. Yazaki, P.J. et al., Humanization of the anti-CEA T84.66 antibody based on crystal structure data, *Protein Sci.*, 17, 481, 2004.
71. Asano, R. et al., Humanization of the bispecific epidermal growth factor receptor x CD3 diabody and its efficacy as a potential clinical reagent, *Clin. Cancer Res.*, 12, 4036, 2006.
72. Plückthun, A., Antibody engineering: Advances from the use of *Escherichia* expression systems, *Bio/Technology*, 9, 545, 1991.
73. Bakker, H. et al., Galactose-extended glycans of antibodies produced by transgenic plants, *Proc. Natl. Acad. Sci. U S A*, 98, 2899, 2001.
74. Ramírez, N. et al., Expression and characterization of an anti-(hepatitis B surface antigen) glycosylated mouse antibody in transgenic tobacco (*Nicotiana tabacum*) plants and its use in the immunopurification of its target antigen, *Biotechnol. Appl. Biochem.*, 38, 223, 2003.
75. Kriangkum, J. et al., Bispecific and bifunctional single chain recombinant antibodies, *Biomol. Eng.*, 18, 31, 2001.
76. Xie, Z. et al., A new format of bispecific antibody: Highly efficient heterodimerization, expression and tumor cell lysis, *J. Immunol. Methods*, 296, 95, 2004.
77. Robert, B. et al., Tumor targeting with newly designed biparatopic antibodies directed against epitopes of the carcinoembryonic antigen (CEA), *Int. J. Cancer*, 81, 285, 1999.
78. Das, D. and Suresh, M.R., Producing bispecific and bifunctional antibodies, *Methods Mol. Biol.*, 109, 329, 2005.
79. Gandecha, A.R. et al., Production and secretion of a bifunctional staphylococcal protein A::antiphytochrome single chain Fv fusion protein in *Escherichia coli*, *Gene*, 122, 361, 1992.
80. Kreitman, R.J., Immunotoxins in cancer therapy, *Curr. Opin. Immunol.*, 11, 570, 1999.
81. Bhatia, J. et al., Catalytic activity of an in vivo tumor targeted anti-CEA scFv::carboxypeptidase G2 fusion protein, *Int. J. Cancer*, 85, 571, 2000.
82. Shusta, E.V. et al., Increasing the secretory capacity of *Saccharomyces cerevisiae* for production of single-chain antibody fragments, *Nat. Biotechnol.*, 16, 773, 1998.

83. Visintin, M., Quondam, M., and Cattaneo, A., The intracellular antibody capture technology: Towards the high-throughput selection of functional intracellular antibodies for target validation, *Methods*, 2, 2000, 2004.
84. Swers, J.S., Kellogg, B.A. and Wittrup, K.D., Shuffled antibody libraries created by in vivo homologous recombination and yeast surface display, *Nucleic Acids Res.*, 32, 1, 2004.
85. Humphreys, D.P. et al., Engineering of *Escherichia coli* to improve the purification of periplasmic Fab' fragments: Changing the pl of the chromosomally encoded Phos/PstS protein, *Protein Expr. Purif.*, 37, 109, 2004.
86. Muramatsu, H. et al., Production and characterization of an active single-chain variable fragment antibody recognizing CD25, *Cancer Lett.*, 28, 225, 2005.
87. Baneyx, F., Recombinant protein expression in *Escherichia coli*, *Curr. Opin. Biotechnol.*, 10, 411, 1999.
88. Smallshaw, J.E. et al., Synthesis, cloning and expression of the single-chain Fv gene of the HPr-specific monoclonal antibody, Jel42. Determination of the binding constants with wild-type and mutant HPrs, *Protein Eng.*, 12, 623, 1999.
89. Bothmann, H. and Plückthun, A., The periplasmic *Escherichia coli* peptidylprolyl cis,trans-isomerase FkpA. Increased functional expression of antibody fragments with or without cis-prolines, *J. Biol. Chem.*, 275, 17100, 2000.
90. Nieba, L. et al., Disrupting the hydrophobic paths at the antibody variable/constant domain interface: Improved in vivo folding and physical characterization of an engineered scFv fragment, *Protein Eng.*, 10, 435, 1997.
91. Wülfing, C. and Plückthun, A., Correctly folded T-cell receptor fragments in the periplasm of *Escherichia coli*. Influence of folding catalysts, *J. Mol. Biol.*, 242, 655, 1994.
92. Chames, P., Fieschi, J., and Baty, D., Production of a soluble and active MBP-scFv fusion: Favorable effect of the leaky tolR strain, *FEBS Lett.*, 405, 224, 1997.
93. Venturi, M., Seifert, C., and Hunte, C., High level production of functional antibody Fab fragments in an oxidizing bacterial cytoplasm, *J. Mol. Biol*, 315, 1, 2002.
94. Josi, D. and Lim, Y.-P., Analytical and preparative methods for purification of antibodies, *Food Technol. Biotechnol.*, 39, 215, 2001.
95. Shusta, E.V., VanAntwerp, J., and Wittrup, K.D., Biosynthetic polypeptide libraries, *Curr. Opin. Biotechnol.*, 10, 117, 1999.
96. Huse, W.D. et al., Generation of a large combinatorial library of the immunoglobulin repertoire in phage lambda, *Science*, 246, 1275, 1989.
97. Smith, G.P. and Petrenko, V.A., Phage Display, *Chem. Rev.*, 97, 391, 1997.
98. Azzazy, H.M.E. and Highsmith, Jr. W.E., Phage display technology: Clinical applications and recent innovations, *Clin. Biochem.*, 35, 425, 2002.
99. Griffiths, A.D. and Duncan, A.R., Strategies for selection of antibodies by phage display, *Curr. Opin. Biotechnol.*, 9, 102, 1998.
100. Zhao, X. et al., Selection and characterization of an internalizing epidermal-growth-factor-receptor antibody, *Biotechnol. Appl. Biochem.*, 46, 27, 2007.
101. Dhillon, J.K., Drew, P.D., and Porter, A.J.R., Bacterial surface display of an anti-pollutant antibody fragment, *Lett. Appl. Microbiol.*, 28, 350, 1999.
102. Hudson, P.J. and Souriau, C., Engineered antibodies, *Nat. Med.*, 9, 129, 2003.
103. Rumbley, C.A. et al., Construction, characterization, and selected site-specific mutagenesis of an anti-single stranded DNA single-chain autoantibody, *J. Biol. Chem.*, 268, 13667, 1993.
104. Park, S.-G., et al., Affinity maturation of natural antibody using a chain shuffling technique and the expression of recombinant antibodies in *Escherichia coli*, *Biochem. Biophys. Res. Commun.*, 275, 553, 2000.
105. Hong, W.W.L., Ya-Huang Yen, Y.-H., and Wu, S.C., Enhanced antibody affinity to Japanese encephalitis virus E protein by phage display, *Biochem. Biophys. Res. Commun.*, 356, 124, 2007.
106. Daugherty, P.S. et al., Quantitative analysis of the effect of the mutation frequency on the affinity maturation of single chain Fv antibodies, *PNAS*, 97, 2029, 2000.

16 Protein Engineering for HIV Therapeutics

James M. Smith, Dennis Ellenberger, and Salvatore Butera

CONTENTS

16.1 INTRODUCTION

It has been more than 25 years since the first cases of AIDS were described, and even though the etiological agent responsible, HIV, was identified early in 1983, we are still in the midst of a global pandemic [1–3]. The development of specific inhibitors aimed at essential steps for viral infectivity and replication is a rapidly expanding field. The first therapeutic attempts at controlling replication and limiting transmission were focused primarily on the unique enzymes encoded by HIV: reverse transcriptase (RT), protease (PR), and more recently integrase (IN) [4]. The focus has broadened to include strategies that target structural proteins of HIV, and specifically the mechanisms used by HIV to infect a cell and replicate.

The replication cycle of HIV has been described in detail in numerous reviews [5–9]. A brief description of each step would be (1) viral attachment and binding to CD4, the primary receptor;

(2) binding to a coreceptor (CCR5 or CXCR4); (3) insertion of the fusion peptide of gp41 into the cell membrane; (4) fusion with the host cell membrane, entry and uncoating; (5) reverse transcription of the RNA genome and formation of the pre-integration complex (PIC); (6) translocation of the PIC to the nucleus; (7) integration, transcription, and mRNA transport; (8) translation of new viral proteins; (9) assembly and budding; (10) release and maturation of new virus. As depicted in Figure 16.1, newer classes of antivirals are being developed that will expand the pool of viral targets. Inhibitors of viral attachment and entry into the host cell, assembly of new virions, and maturation of mature particles are all being actively investigated [5,10,11].

Advances in protein engineering coupled with a more thorough understanding of the intricate protein–protein interactions between HIV and the host cell has resulted in the development of numerous inhibitors of HIV entry and fusion. This chapter focuses on therapeutic peptide and protein-based strategies designed to inhibit attachment and entry of HIV into the cell (steps 1–4 of Figure 16.1) and briefly discuss inhibition strategies aimed at the latter points in the life cycle.

16.2 BINDING AND FUSION OF HIV TO THE HOST CELL

16.2.1 Viral Attachment to Host Target Cell

The detailed understanding of the processes involved in receptor-specific HIV attachment and fusion to the host cell has allowed for a systematic development of peptide-based antiviral therapies. The descriptions of CD4 as the cellular receptor used by HIV for attachment to the host cell

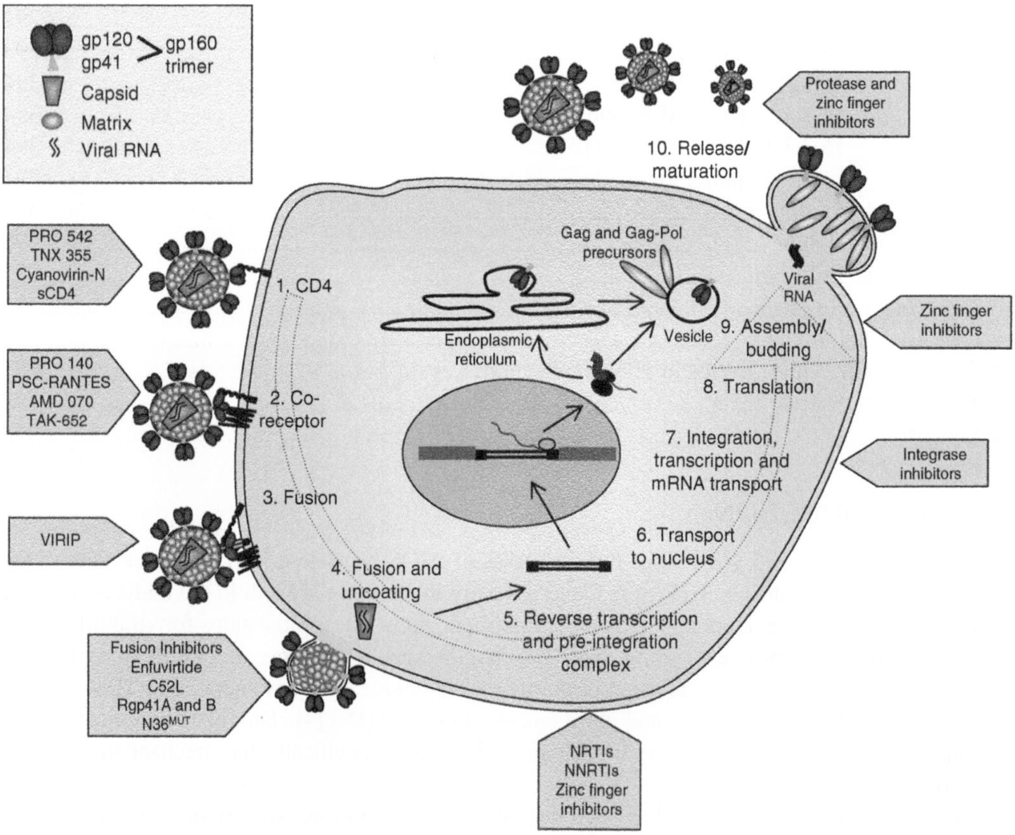

FIGURE 16.1 Schematic of the HIV lifecycle. Simplified schematic of events in the replication of HIV highlighting key points of inhibition and examples of inhibitors.

represented the first major target for strategies to stop infection independent of an immune response [12–14]. The discovery of the need for coreceptors for efficient binding of HIV 10 years after the first description of CD4 as the primary receptor launched a new discipline focusing on inhibition strategies to block this second event in virion binding to the host cell [15,16].

16.2.1.1 CD4-gp120 Binding

The envelope complex on the virion consists of trimers of heterodimers of an exposed glycoprotein, gp120, and a transmembrane protein, gp41, that are non-covalently linked. The gp120 molecule contains a negatively charged binding pocket, called the Phe-43 cavity, which is attracted to the positively charged end of CD4. Although monoclonal antibodies targeting this site have been insightful, the Phe-43 cavity is normally hidden within the gp120 molecule precluding potentially neutralizing antibodies from binding to the free virion.

16.2.1.2 Coreceptor Binding

HIV also uses a second receptor for efficient binding and fusion to its target cell. This two stage binding event has been postulated to be an evolutionary adaptation of HIV to further escape the humoral immune response [15,16]. The two coreceptors most commonly used by HIV are the chemokine receptors CCR5 (R5) and CXCR4 (X4). Following the initial CD4 binding event, the envelope gp120 undergoes conformational changes that expose gp120 domains hidden in the free virion to allow binding of the coreceptor (Figure 16.1). Most primary isolates of HIV use CCR5 (R5 tropic) as the coreceptor and R5 viruses are the dominant species for transmission, although X4 and R5/X4 isolates have also been described emerging late in disease. Further evidence of the important role played by CCR5 in infection is the clear genetic advantage for decreased susceptibility to HIV infection among individuals with a homozygous 32 base pair deletion in the CCR5 gene [17–19]. Both R5- and X4-tropic viruses bind their respective coreceptor via the V3 loop of gp120 with the V3 amino acid charge determining which coreceptor is utilized. Binding of CD4 and either R5 or X4 by the virion envelope promotes a second conformational change in gp120 setting the stage for fusion with the plasma membrane.

16.2.2 Fusion of the Virion with the Host Cell

The conformational changes in the gp120 trimer induced by binding of the virion to CD4 and the coreceptor relax the non-covalent interaction between gp120 and gp41. This change allows the amino terminus fusion peptide (FP) of gp41 to "open" via a hinge mechanism and engage the host cell membrane, essentially anchoring the virion to it. Once the fusion peptide is anchored, two heptad repeat regions (HR1 and HR2) each form trimeric coiled-coil structures leading to the extended pre-hairpin structure. This is the critical point in the process targeted by the current classes of fusion inhibitors. The pre-hairpin intermediate of gp41 briefly exposes the HR1 and HR2 trimers that interact to form a hairpin structure of a six-helix bundle which transforms the trimer of gp41 into a fusogenic state (Figure 16.2). The free energy released by this process most likely provides the energy needed to fuse the viral and cell membranes.

16.2.3 Summary

The binding of HIV to the host cell and the fusion of the viral and cellular membrane represent the first targets for inhibition of infection. The seminal publications describing CD4 as the major receptor and the chemokine receptors CCR5 and CXCR4 as the coreceptors laid the groundwork for the development of multiple strategies to interfere with the protein–protein interactions necessary for HIV to begin its entry into the cell. The molecular details involved in the fusion process are now understood to such a degree that the fusion process itself has now been targeted. Section 16.3

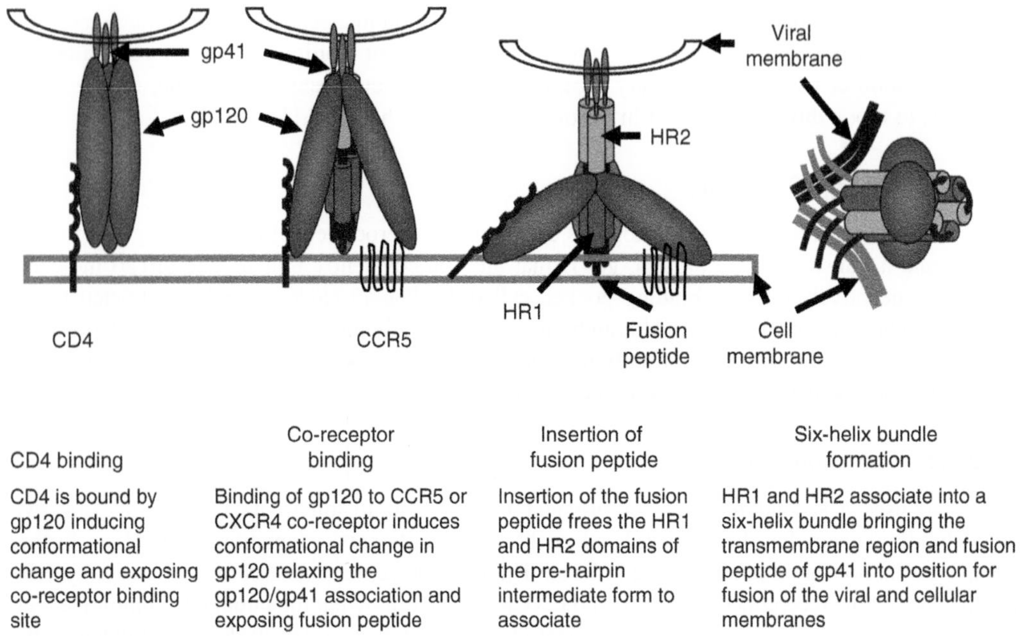
gp41
gp120
Viral membrane
HR2
HR1
Fusion peptide
Cell membrane
CD4
CCR5
CD4 binding
Co-receptor binding
Insertion of fusion peptide
Six-helix bundle formation
CD4 is bound by gp120 inducing conformational change and exposing co-receptor binding site
Binding of gp120 to CCR5 or CXCR4 co-receptor induces conformational change in gp120 relaxing the gp120/gp41 association and exposing fusion peptide
Insertion of the fusion peptide frees the HR1 and HR2 domains of the pre-hairpin intermediate form to associate
HR1 and HR2 associate into a six-helix bundle bringing the transmembrane region and fusion peptide of gp41 into position for fusion of the viral and cellular membranes

FIGURE 16.2 Schematic of attachment and fusion of HIV with the cell membrane.

describes both past and current avenues of protein engineering for inhibitors of attachment and fusion of HIV to the host cell.

16.3 INHIBITION OF BINDING AND FUSION

16.3.1 Blocking CD4-gp120 Binding

There are several different strategies utilizing peptides and proteins, both naturally occurring and synthetic, engineered to efficiently block the initial CD4–gp120 binding event (Table 16.1). The strategies described below are either based on the recognition of CD4 by gp120, or antibody recognition of gp120 coupled with CD4 or an exotoxin. Historically, the first attempts used soluble CD4 (sCD4), a truncated version of CD4, as a decoy to bind gp120 before it could attach to the cell [20]. However, several studies reported problems with this approach [21–23]. At low concentrations sCD4 was found not only to have poor neutralizing activity towards primary isolates in vitro, but it was also shown to enhance viral infection at suboptimal concentrations. In addition, the in vivo half-life of the truncated protein was very short. Therefore, investigators quickly engineered modifications of sCD4 in an attempt to overcome these problems.

Recently, bivalent and trivalent forms of a CD4-mimetic miniprotein, CD4M9, were designed to bind gp120 in its natural trimeric form on the virion. This multivalent approach dramatically increased potency (up to 140-fold) over the monovalent CD4M9. Furthermore, the trivalent form is superior to both monovalent CD4M9 and sCD4 by having a much longer half-life and being effective against both X4- and R5-tropic isolates [24,25]. Clinical trials of the trivalent CD4M9 product are currently in the planning stages.

Several alternative strategies have further employed protein engineering to therapeutically target the CD4–gp120 interaction. A novel strategy was developed by coupling the gp120-binding capacity of CD4 with an exotoxin to kill HIV infected cells. A chimeric protein, CD4(178)-PE40, links *Pseudomonas* exotoxin to the gp120-binding domain of CD4 [26–28]. The pseudomonas exotoxin enters the infected cell and effectively inhibits protein synthesis. This combination was

TABLE 16.1
Entry and Fusion Inhibitors of HIV-1

Target	Inhibitor	Description	Mechanism	References
CD4/gp120	Soluble CD4 (sCD4)	Recombinant protein	Inhibits binding to CD4	[20]
	PRO-542	Tetravalent CD4/Ig fusion	Inhibits binding to CD4	[30,31,33,34]
	sCD4-17b	CD4/recombinant Ab fusion	Inhibits binding to CD4	[29]
	CD4-exotoxin	CD4-Pseudomonas exotoxin (PE) hybrid protein	Selectively inhibits protein synthesis in gp120 expressing cells	[26–28,38]
	3B3(Fv)-PE38	Fv fragment of 3B3 Ab fused to a truncated version of PE	Specifically kills gp120-expressing cells	[37–39]
	CD4M33	27-amino acid CD4 mimic	Inhibits binding to CD4	[24]
	CD4M9	CD4-mimetic mini-proteins	Inhibits binding to CD4	[24,25]
	TNX-355	mAb to CD4	Inhibits binding to CD4	[113,114]
	HGP44	Peptide derived from *P. gingivalis*	Inhibits binding to CD4	[115]
	12p1, HNG105	Peptide	Inhibits binding to CD4	[116–119]
	D1D2-Igαtp	Dodecameric CD4–Ig fusion protein	Binds gp120 molecules, inhibits binding to CD4	[22]
	Cyanovirin-N (CV-N)	Lectin from cyanobacterium-101 amino acids	Carbohydrate-binding agent (CBA)—binds gp120	[40–42, 46–48,117]
	SVN	Lectin from cyanobacterium-95 amino acids	CBA—binds gp120, 160, 41	[120–122]
	GRFT	Lectin from red alga	CBA—binds gp120	[44,45]
	MVL	13kDa protein from cyanobacterium	CBA—binds gp120	[43]
	BMS-806	Piperazine compound	Blocks conformational changes in CD4 after gp120 binding	[11,123]
Coreceptor	−2 RANTES	Truncated fragment of RANTES	CCR5 ligand	[52,53]
	PRO-140	mAb to CCR5	Binds to CCR5	[57]
	PSC-RANTES	Modified truncated RANTES	CCR5 ligand	[54,55]
	SCH-D (Vicrivoc)	Oxime-piperidine	CCR5 antagonist	[56]
	Maraviroc	Cyclohexanecarboxamide	CCR5 antagonist	[56]
	Tak-220, TAK-652, TAK-779	Small molecule		[56]
	AMD070	Bicyclam	CXCR4 antagonist	[56]
	KRH-2731	Small molecule	CXCR4 antagonist	[56,58]
Fusion	T20(Enfuvirtide), C34, SC34, T649, T-1249	Peptides	Mimic HR2 domain of gp41—Inhibits fusion	[64,66,73,92]
	C52L	52-mer peptide	Mimic HR2 domain of gp41—Inhibits fusion	[75–77]
	D-peptides	Peptides containing D-amino acids	Prevents six-helix bundle formation of gp41	[88]
	C_{52}-HlyA_{218}	C52 peptide expressed from *E. coli*	Mimic HR2 domain of gp41—Inhibits fusion	[78]
	DP106, N36, IQN23, $N36^{mut}$	Peptides	Mimic HR1 domain of gp41—Inhibits fusion	[63]
	Retrocyclins	Peptides	Prevents six-helix bundle formation of gp41	[124, 125]
	Rgp41A, Rgp41B	Trimeric recombinant gp41	Prevents six-helix bundle formation of gp41	[82]
	VIRIP (virus-inhibitory peptide)	20-mer from α1-antitrypsin	Binds fusion peptide of gp41	[83]

Note: Shaded areas are not peptide or protein based.

very effective at eliminating HIV-infected cells displaying gp120 on their surface and limited the spread of new virus in vitro but its effectiveness in vivo has yet to be determined.

Another strategy employs chimeric proteins consisting of sections of select immunoglobulin proteins combined with the gp120-binding region of CD4 [22,29–32]. The most advanced of these is Pro 542, a CD4-IgG fusion [31,33–35]. Domains 1 and 2 (D1 and D2) of CD4 were replaced by the Fv portions of the heavy and light chains of human IgG in a tetravalent motif. A dodecameric CD4–Ig fusion complex has also been developed capable of binding multiple gp120 molecules, effectively eliminating the problem of enhancement of infection seen with monomeric sCD4 [22]. Pro 542 has an excellent safety record in human trials and was effective at reducing HIV-1 RNA in infected subjects during phase II clinical trials [30,33,34]. Although primarily developed and used as a salvage therapy agent for HIV positive individuals, the success of the phase II trials coupled with evidence of synergy with other entry inhibitors may allow their therapeutic use for accidental occupational exposures as well [36].

Dey and colleagues have taken the CD4–IgG fusion proteins one step further with sCD4–17b [29]. This chimeric protein is engineered with a flexible polypeptide linker to join D1 and D2 of soluble CD4 to a variable chain fragment from the 17b antibody. The 17b antibody recognizes an epitope that overlaps the coreceptor binding domain on gp120 that is only accessible when a conformational change is induced in gp120 by binding to CD4. The neutralizing capability of the chimeric protein is substantially more potent than when sCD4 and 17b are used as separate components. However, some primary isolates were found to be insensitive to neutralization thereby limiting the overall therapeutic potential of this product.

A unique combination of *Pseudomonas* exotoxin joined with the single chain Fv fragment of 3B3, another anti-gp120 antibody that can neutralize several primary isolates, is also being investigated for its therapeutic potential [37–39]. This product has enhanced neutralizing capabilities against primary isolates over the sCD4–17b product and is well tolerated in macaques [39].

Another class of compounds that have been recruited into the battle to inhibit CD4/gp120 binding are lectins, a class of nonspecific carbohydrate binding agents. Cyanovirin-N (CV-N) is a naturally occurring lectin of 101 amino acids isolated from the cyanobacterium *Nostoc ellipsosporum* which binds with high selectivity to high mannose carbohydrates on gp120 [40–42]. Other lectin candidates from cyanobacteria and red alga have also shown antiviral activities against HIV in vitro and are currently being evaluated [43–45]. A unique advantage of CV-N is its ability to inhibit cell-to-cell fusion of HIV infected cells and uninfected cells in vitro by blocking gp120 binding to the coreceptors of the new target cell [46]. CV-N efficiently blocks infection of both T-cell and macrophage-tropic primary isolates of HIV [42]. The results from animal models of infection are impressive. When used as a topical microbicide in a gel formulation, 15 of 18 macaques in the treatment arm were protected from infection when a single high virus dose was used in a vaginal challenge [47].

To overcome the limitations of vaginal gel application and the high costs associated with production of a molecularly engineered CV-N, Liu et al. recently reported stable in vivo expression in mice of CV-N from *Lactobacillus jensenii*, a common microbial colonist of the human vaginal mucosa [48]. The *L. jensenii* recombinant was well tolerated and the level of CV-N expression at the mucosal surface was sufficient to potentially block heterosexual transmission of HIV to women. Future studies in a non-human primate model for vaginal transmission will provide valuable insight into this novel approach.

Non-peptide-based compounds and small molecules are also being investigated to block the protein–protein associations involved in viral attachment [11,49]. BMS-806 [Piperazine, 4-benzoyl-1-(2-(4-methoxy-1H-pyrrolo(2,3-b)pyridin-3-yl)-1,2-dioxoethyl)-2-methyl-, (2R)] was initially thought to inhibit binding of gp120 to CD4. However, it was later revealed BMS-806 blocks infection by disallowing the proper gp120 conformational changes to take place after CD4 binding that expose the coreceptor binding domain [50,51]. The limitations of BMS-806 to inhibit a diverse spectrum of

HIV isolates and concerns about resistance have stopped further development, but additional compounds based on this strategy to therapeutically block the gp120–CD4 association are worth pursuing.

16.3.2 Blocking Coreceptor Binding

Most attempts at blocking coreceptor binding by gp120 have focused on CCR5, the major coreceptor for the R5 viruses which are the dominant species transmitted mucosally. The three main inflammatory chemokines that are the natural ligands for CCR5 (Regulated on Activation, Normal T Expressed and Secreted (RANTES), macrophage inflammatory protein (MIP)-1α, and MIP-1β) clearly have HIV-suppressive activity. These ligands compete with HIV for binding to CCR5 and therefore have been examined as potential therapeutic agents to block HIV entry. The prime candidate for development of a new class of antivirals has been RANTES, and two modifications of RANTES in particular have shown the most promise in the macaque transmission model.

RANTES naturally binds to several different chemokine receptors, but a truncated form missing two N-terminal residues, −2 RANTES, is a dedicated CCR5 ligand. In addition, −2 RANTES is less effective than native RANTES in its ability to signal through chemokine receptors that lead to an inflammatory response. Preclinical toxicity studies done in mice and rabbits using two different vaginal delivery systems, a nonphospholipid liposome (Novasomes 7474) and a methylcellulose gel, showed that −2 RANTES is safe and well tolerated [52]. The product was then advanced to the nonhuman primate vaginal challenge model using the Novasomes 7474 delivery [53]. Although four of six animals in the treatment group were protected from viral infection after challenge, the same number of animals were protected in the Novasomes 7474 placebo group making it impossible to ascribe efficacy to this product. When −2 RANTES was applied in PBS without the Novasomes, two out of five animals were protected, while all animals in the naïve group became infected suggesting some efficacy with the drug alone.

Kawamura et al., screened a series of synthetic RANTES analogues for similar or better antiviral activity than native RANTES in immature human Langerhans cells, a prime CCR5$^+$ target during mucosal transmission [54]. Of the three compounds tested, aminooxypentane (AOP)-RANTES, N-nanoyl (NYY)-RANTES, and PSC-RANTES (substitutes the first three N-terminal amino acids with a nanoyl, praline, and cyclohexylglycine), the PSC-RANTES exhibited the highest potency in blocking infection by R5-tropic viruses [54]. In a subsequent study of the topical use of this compound, rhesus macaques received PSC-RANTES in PBS intravaginally prior to intravaginal challenge with SHIV162 [55]. Nine out of twelve animals that received either 100 μM or 300 μM doses were protected from infection. A lower PSC-RANTES dose (≤33 μM) resulted in limited efficacy. Although the PSC-RANTES appears to perform better in a nonhuman primate model, there were significant differences in the experimental conditions and only a side-by-side comparison under the same experimental conditions can address this issue.

A viable concern in using RANTES derivatives is the possibility of inflammatory side effects due to signaling through CCR5. This has led to the successful development of several CCR5- and CXCR4-specific small molecule antagonists that function by physically occupying the receptor without activation of the signaling pathway and subsequent receptor down-modulation [11,56]. Takeda Chemicals developed the first CCR5 antagonist, TAK-779, but it was subsequently replaced by second-generation products (TAK-220 and TAK-652) with increased bioavailability and potency [56]. Similarly, SCH-D (Vicriviroc, Schering-Plough) is a second-generation oxime-piperidine CCR5 antagonist that replaced SCH-C due to a better safety profile and increased potency. Testing of Vicriviroc has successfully completed phase I and II trials and is currently enrolling for a phase III trial. Recently approved for therapeutic use, Maraviroc (Pfizer) is another small molecule CCR5 antagonist that has a similar potency as Vicriviroc towards R5 isolates. Direct antibody targeting to CCR5 has been explored using PRO 140 (Progenics Pharmaceuticals, Inc.), a murine monoclonal antibody with high affinity to CCR5 [57]. When bound to its epitope, PRO 140 does not downregulate CCR5 or induce aberrant signaling, therefore its mechanism of action is thought to be physical blocking. Because of anticipated limitation for repeated use, PRO140 is being proposed as post-exposure prophylaxis and is currently in phase I clinical trials.

Although the majority of transmission events of HIV occur with R5 isolates, the existence of X4 and R5/X4 isolates, especially in the end stages of disease, has prompted the development of CXCR4 inhibitors [11,56,58]. The most promising of the candidates from in vitro studies, AMD070 (AnorMED), has been advanced into a phase Ib/IIa clinical trial (Table 16.1) [56].

16.3.3 Blocking Fusion

Because of the close proximity of the fusogenic structure to the cell membrane and brief exposure to antibodies, the fusion event largely escapes the immune response prompting the investigation of alternative strategies to block fusion [59]. There are two viable targets in the fusion process that are under close scrutiny for inhibitor design: inhibition of the formation of the gp41 six-helix bundle required for fusion of the virion with the cell membrane and direct inhibition of membrane insertion of the fusion peptide of gp41 (Figure 16.2). Fusion inhibitors targeting the formation of the six-helix bundle were first described in 1992 and represent one of the most advanced areas of investigation in HIV therapy [60]. Peptide-based fusion inhibitors allow for interaction with the viral target region outside of the cell, with no requirement of the inhibitor to be taken up by the target cell. However, an inhibitory molecule must be small enough to access the fusion complex and bind with high affinity to inhibit the fusion process.

The fusion process following binding of the virion to CD4 and the coreceptor is described in Figure 16.2. The HIV gp41 protein can be divided into distinct domains starting with the fusion peptide at the N-terminus, followed by two heptad repeat (HR) regions, HR1 and HR2, separated by a short linker section, and ending at the C-terminus with the transmembrane domain (Figure 16.3A).

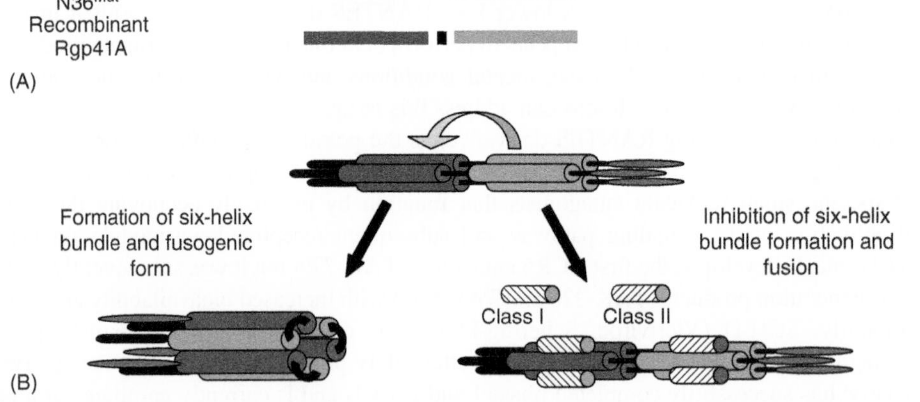

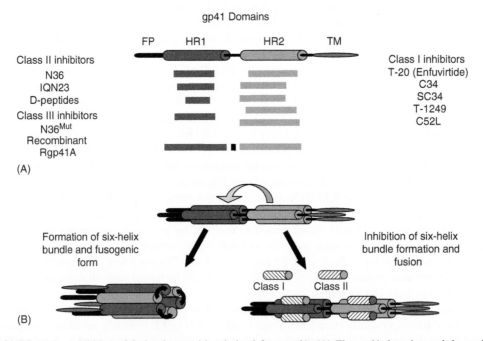

FIGURE 16.3 Inhibition of fusion by peptides derived from gp41. (A) The gp41 domains and the regions duplicated by fusion inhibitors. FP, fusion peptide; HR1, heptad repeat 1; HR2, heptad repeat 2; TM, transmembrane. (B) The trimers of HR1 and HR2 of the pre-hairpin intermediate form of gp41 associate to form the six-helix bundle fusogenic form of gp41. Binding by fusion inhibitors deter formation of the six-helix bundle.

Small peptide fusion inhibitors directly derived from either HR1 or HR2 have been separated into three distinct classes: class I (C-peptides) are derived from HR2 and bind to HR1; class II (N-peptides) are derived from HR1 and target HR2; and although $N36^{mut}$ targets HR1, because it differs from the class I inhibitors in its mechanism of action, it is the lone representative of the class III peptide inhibitor of fusion [61,62]. These small peptide inhibitors tightly bind to either the HR1 or HR2 domain of gp41 preventing the association of HR1 with HR2 necessary for formation of the six-helix bundle [63,64]. This essentially locks the pre-hairpin structure in place, stopping the fusion process (Figure 16.3B). The following sections describe in greater detail the class I, II, and III peptide inhibitors. In addition, a direct inhibitor of the fusion peptide and the utility of larger recombinant polypeptide inhibitors of fusion are also discussed.

16.3.3.1 Class I Peptides

The most advanced of the Class I peptides is enfuvirtide (also known as DP178, T20, Fuzeon) [63,65–68]. Enfuvirtide is a synthetic peptide of 36 amino acids that corresponds to amino acids 127–162 of the HR2 of gp41. As previously stated, T20 functions by binding to HR1 at the pre-hairpin intermediate stage, thereby preventing the formation of the fusogenic six-helix bundle (Figure 16.3). It is effective in blocking cell-to-cell transmission of primary isolates of HIV with an IC_{90} of 1.5 ng/mL, and is active against all HIV subtypes within group M [69]. It was developed in 1993 and was the first HIV-1 fusion inhibitor approved by the FDA in March of 2003 [70]. Enfuvirtide has an excellent safety and efficacy profile in clinical trials and has been added to several ART therapies currently in clinical trials (discussed in Section 16.4). A modified version of enfuvirtide, T-1249 or tifuvirtide, differs from enfuvirtide by only a few amino acids at the N-terminus and is more potent in its inhibitory activity than enfuvirtide [35,63]. Another positive attribute of tifuvirtide is that it is effective against enfuvirtide-resistant HIV. Although limited clinical trials with tifuvirtide provided promising results, the trials were halted because of difficulties in the manufacturing process that were not seen with enfuvirtide [71].

Several other Class I inhibitors are also being investigated for potential therapeutic use (Table 16.1). C34 is another Class I peptide that differs from T-20 by incorporating more of the N-terminal amino acids of HR2. In an effort to reduce the emergence of escape mutants, the C34 design encompasses a more conserved region of HR2, although mutants have arisen in vitro [72]. In addition to standard measurements of efficacy and potency in vitro, C34 also blocks infection of Langerhans cells in an explant model and transmission of HIV from Langerhan's cells to T-cells [73]. A modified version of C34, SC34, has been engineered by introducing a few key amino acid substitutions that not only make the peptide more soluble, but also more potent in its inhibitory properties [74]. C52L is a Class I inhibitor that encompasses amino acids 114–162 of gp41 which contains all of HR2 and 3 amino acids of the loop region between HR1 and HR2 including the tryptophan-rich region [75]. It is effective against multiple subtypes of HIV irrespective of coreceptor usage, and is also effective against T-20-resistant mutants [76,77]. C52L has also been tested in the macaque model of vaginal transmission and alone it protected three of five macaques from a vaginal viral challenge with a simian-human immunodeficiency hybrid virus [77]. In this same study, the efficacy increased to 100% when C52L was used in conjunction with BMS-378806, an inhibitor targeting gp120 binding to CD4 and CCR5.

C52L may have a distinct advantage over other fusion inhibitors in that it has been engineered to be expressed from *E. coli* making production much easier and less costly. Rao et al. reported a novel delivery method in which the C52 peptide can be expressed from *E. coli* Nissle 1917, a well-characterized commensal strain that grows well in mice and humans and is not pathogenic [78]. The *E. coli* Hly system was used to promote secretion of a recombinant C52-HlyA protein and this fusion protein retains its inhibitory effect on HIV. In a mouse model, the bacteria was able to colonize several tissues in the gastrointestinal tract as well as the vagina and successfully secrete C52-HlyA [78].

16.3.3.2 Class II Peptides

Class II peptides are derived from HR1 and competitively inhibit the binding of HR1 to HR2 of gp41 (Figure 16.3). Due to poor solubility and higher inhibitory concentrations (micromolar rather than the nanomolar range of Class I peptides), Class II peptides were initially very poor at inhibiting HIV fusion and entry and modifications were required to overcome these issues [61,79]. IQN23, a modified HR1 mimic, has 23 amino acids from the N-terminal domain of HR1 combined with 28 amino acids from the core region of HR1 [63,79]. This peptide is more soluble than smaller peptides comprised of the wild-type HR1 region and inhibits fusion of HIV in the nanomolar range [79,80]. Further development of class II peptides will be needed before they are viable candidates for clinical trials.

16.3.3.3 Class III Peptide

One of the original class II candidates, N36, was modified to enhance its inhibitory activity via a series of amino acid substitutions to block formation of the coiled-coil intermediate [61]. However, the modified $N36^{Mut}$ peptide did not bind HR2 as expected for a typical class I inhibitor, but instead inhibited fusion at nanomolar levels by forming heterotrimers with HR1. Because of this difference in mechanism, the $N36^{Mut}$ peptide has been given the distinction of being the sole class III inhibitor [62]. The natural interaction of HR1 and HR2 domains in gp41 results in the formation of the six-helix bundle. Class I and II inhibitors derived from HR1 and HR2 can also associate with each other and are therefore antagonistic when used in combination. Due to the different inhibitory mechanism of $N36^{Mut}$, the class III inhibitor can work synergistically with the class II inhibitors allowing for combination therapy [62].

16.3.3.4 Helix Inhibitors

An alternative approach in inhibiting HIV fusion by targeting HR2 has been developed by Root et al. [81]. A recombinant polypeptide, called five-helix, was designed by stringing three HR1 domains and two HR2 domains together with short linkers. Instead of forming the natural fusion intermediate of a trimer of hairpins and subsequently the six-helix bundle, five-helix leaves a vacancy for an HR2 domain which binds with high affinity, thereby halting the formation of the six-helix bundle and inhibiting fusion. A potential additional advantage of this approach may be the induction of neutralizing antibodies to the pre-hairpin intermediate form of gp41 generated in response to the five-helix as a foreign immunogen.

Delcroix-Genete et al. described another attempt to use a form of the six-helix bundle to inhibit fusion [82]. The Rgp41 A protein contains the HR1 and HR2 domains, but 13 amino acids of the loop between HR1 and HR2 have been replaced with a 7 amino acid hydrophilic linker. The recombinant protein spontaneously folds into a pair of trimers that mimic the six-helix bundle. The recombinant is produced in *E. coli*, thereby reducing production costs, and was able to inhibit entry of both R5 and X4 isolates of HIV in vitro.

16.3.3.5 Inhibition of the Fusion Peptide

An effective peptide inhibitor that binds the fusion peptide (FP) of gp41 was isolated from a human hemofiltrate peptide library and contains the C-proximal region of α1-antitrypsin [83]. The FP primary sequence is highly conserved in HIV and the peptide, termed virus-inhibitory peptide (VIRIP), has inhibitory effects in the low micromolar range in vitro against primary HIV-1 isolates. VIRIP does not alter the binding of gp120 with CD4 or the coreceptor, or formation of the six-helix bundle of gp41. Uniquely, VIRIP directly binds FP with high affinity and inhibits insertion of FP into the cell membrane. Therefore, VIRIP may be very useful in combination with other fusion inhibitors, or when viral resistance to other fusion inhibitors occurs with other therapy regimens. Although in vivo efficacy studies have not yet been completed, preliminary studies in animal models have shown a lack of apparent toxicity even at very high concentrations [83].

16.3.4 Summary

Design of inhibitors by protein engineering against HIV binding and fusion to the host cell has closely followed the discoveries elucidating the detailed mechanisms of HIV attachment and entry. Lessons learned from the initial attempts at blocking CD4 binding by using sCD4 as a decoy have led to the development of inhibitors with more antiviral potency and greater stability in vivo. Other approaches based on the gp120–CD4 interaction have utilized recombinant technology to couple the gp120-binding domains of CD4 to antibody fragments or exotoxins. Recombinant protein engineering will likely play a pivotal role in the development of new combinations with even greater potency and stability.

Following the discovery of the coreceptors for HIV binding, investigators turned to the natural ligands of CCR5. Modifications of RANTES, in particular PSC-RANTES and -2 RANTES, have shown great promise in vitro and in vivo in animal models. However, some concern has been expressed as to whether using a CCR5 inhibitor will result in a rapid evolution in vivo from R5-tropic to more pathogenic X4-tropic HIV in infected people treated with RANTES products. Further development of X4-based inhibitors and analogues of other CCR5 ligands warrants further investigation.

Development of inhibitors of the fusion event has been rapid from laboratory to clinic with enfuvirtide leading the way in clinical trials. Efforts to increase stability and better formulations that will allow oral dosing are needed for both class I and II fusion inhibitors. For example, polyethylene glycol modification (pegylation) of biologically active molecules and peptides is a well-established technique to enhance stability while retaining function [84]. Modifications to the pegylation process and coupling this technology with nanoparticle delivery mechanisms designed to control the release of peptides and recombinant proteins may further improve the clinical potential of entry and fusion inhibitors [85,86]. Improvements in half-life, thermal stability, and pharmacokinetics have already been accomplished for class I peptides by increasing the helical stability while retaining or improving potency [87].

Another intriguing modification of small peptide inhibitors of fusion is the use of peptides made of D-amino acids (D-peptides). D-peptides are nonimmunogenic and resistant to proteolytic activity. Therefore, they are more amenable to oral formulations. The combination of D-peptides with pegylation has yielded some very promising results with inhibitory concentrations in the picomolar range being reported [88].

16.4 CLINICAL TRIALS

16.4.1 Enfuvirtide

There are currently several clinical trials testing peptide- or protein-based entry inhibitors for treating HIV infection that are ongoing or recruiting patients (Table 16.2 and clinical trials.gov/ct). Currently, the most clinically advanced peptide-based entry inhibitor is enfuvirtide (T-20, Fuzeon). In phase I/II trials using enfuvirtide alone, the average reduction in HIV viral load was 1.9 $\log_{10}$ copies/mL at the 100 mg twice daily dose [66]. The subsequent phase III trial of T-20 versus Optimized Regimen Only (TORO 1 and TORO 2) evaluated the efficacy of enfuvirtide as an additional drug in a background of 3–5 traditional reverse transcriptase (RT) and protease inhibitors

TABLE 16.2
Clinical Trials of Peptide or Protein Based Entry Inhibitors

Inhibitor	Type of Compound	Administration	Clinical Trials
PRO 542	Tetravalent CD4/Ig fusion	Intravenous infusion	2 completed
PRO 140	mAb to CCR5	Intravenous infusion	1 completed
TNX 355	mAb to CD4	Injection	1 completed
T-1249	Class I peptide fusion inhibitor	Injection: Intravenous, subcutaneous, Biojector	1 suspended
T-20 (Enfuvirtide)	Class I peptide fusion inhibitor	Injection: Intravenous, subcutaneous, Biojector	11 completed; 10 ongoing; 7 recruiting; 3 planned; 1 suspended

Source: From U.S. Department of Health and Human Services http//aidsinfo.nih.gov.

(PI) [89,90]. Significant clinical improvements were observed in the groups that received a T-20-boosted regimen. Enfuvirtide is now being added to many therapeutic regimens and the synergistic effects of enfuvirtide with other inhibitors specific for RT and protease are being examined (http//aidsinfo.nih.gov). In addition, enfuvirtide is also being tested as salvage therapy to treat patients that have failed ART.

The extensive clinical data with enfuvirtide have emphasized the advantages and disadvantages in using peptides to inhibit fusion, and this knowledge will be of great benefit for future therapies. For example, most of the current peptide inhibitors of fusion are restricted as to how they can be administered. They are subject to degradation by peptidases in the gastrointestinal tract and therefore are not amenable to oral formulations, complicating the translation of enfuvirtide from clinical trials to the public domain [35]. The mean half-life reported for enfuvirtide is 3.8 h requiring a twice daily dosage by injection. Adherence to this regimen has proven difficult for patients, but enfuvirtide can be administered by subcutaneous injection with only minor side effects being reported. To alleviate one of the most common adverse reactions, irritation at the injection site, clinical trials have also been completed using a needle-free injection device for subcutaneous injections.

16.4.2 Resistance

Emergence of antiviral resistance has always been a problem with ART, and the fusion inhibitors also subject HIV-1 to selective pressures which can lead to emergence of resistant virus populations [49,64]. This has been observed in enfuvirtide monotherapy trials in as little as 6 weeks [66,68]. The resistance to enfuvirtide has been mapped to the Gly–Ile–Val motif in HR1 (amino acids 36–38 of gp41). A change from Gly–Ile–Val to Ser–Ile–Val gives the virus some resistance, and enhanced resistance occurs with a second change to Ser–Ile–Met, but these changes also result in a decrease in viral fitness [91]. It is therefore important to develop multiple compounds to the same target without cross-resistance, like the improved version of T-20, T-1249 (tifuvirtide). However, due to production and technical problems, T-1249 clinical trials were halted [71,92,93].

Viruses resistant to T-20 are also susceptible to C52L, and different mutations are required to confer resistance to C52L [75]. Combination therapy with peptides and non-peptide-based entry and fusion inhibitors are currently in clinical trials and monitoring for resistant virus will give insight into the fitness cost of multiple mutations.

16.4.3 Summary

Several of the other inhibitors mentioned in this chapter are currently being tested in animal models and will be following closely behind enfuvirtide in clinical trials (Table 16.2). The cost efficiency of the administration and formulation of entry and fusion inhibitors may unfortunately limit their utility. Subcutaneous injections or infusions can be tolerated by those in salvage therapy, but for most patients this is not a very appealing option. The half-life of peptides and recombinant proteins precludes oral dosing, but there are other options being explored [87]. The expression of peptide inhibitors by recombinant commensal bacteria, especially at mucosal sites, is an intriguing concept and warrants further study [48,78]. The development of gel formulations that stabilize the inhibitor and can be easily applied is another viable option. This has been a successful strategy for other antivirals and for a few peptides as well [94].

16.5 TARGETING OTHER EVENTS IN HIV REPLICATION

The list of antivirals that are approved for use by the FDA for therapeutic treatment of HIV is expanding, but the viral targets of these approved compounds are limited. Therefore, the need to identify additional targets for inhibition is apparent and has become a very active field. Advances in several disciplines, including x-ray crystallography, combinatorial chemistry, high throughput screening of compounds, and nuclear magnetic resonance, has allowed a more detailed examination of the interaction between viral and cellular proteins. Protein engineering can now utilize these methods as the foundation for the design of new peptides and drugs to inhibit HIV at other points in its replication cycle (Figure 16.1 and Table 16.3) [4–6,11,24,95–105]. For example, the nucleocapsid (NC) p7 protein of HIV has several essential functions in the replication cycle making it a very attractive target for inhibition. NCp7 binds to viral RNA and incorporates it into the mature virion, aids in reverse transcription, in transcription of viral mRNA, and in dimerization of gag–pol precursor protein allowing release of the active protease. These functions are facilitated by the zinc finger motif of NCp7. Several compounds have been developed to inhibit binding or ejection of zinc from NCp7 (Table 16.3).

Another very active area of investigation is the indirect inhibition of HIV by targeting essential interactions of viral proteins with host proteins. Tripartite motif (TRIM)-5α has been identified as a key block in replication of HIV in macaque cells by sequestering viral cores before uncoating [106,107]. Although human TRIM-5α does not block HIV replication, a thorough understanding of the mechanisms involved in macaque cells may lead to a new class of inhibitors [108].

TABLE 16.3
New and Future Targets of Inhibition

Target	Inhibitor	Mechanism	References
Integrase	Raltegravir, GS-9137	Blocks the putative acceptor DNA-binding site of the integrase-DNA complex	[4,97,126]
Nucleocapsid (NC)p7	NOBA, DIBA, PATES, ADA	Inhibits or ejects zinc binding to zinc finger motif of NCp7 [127]	[5,100–104]
RNase-H	β-thujaplicinol, manicol	Inhibits RNase H activity. Specific for the C-terminal RNase H domain	[128]
gp160	UK-201844	Interferes with processing of gp160	[96]
Tat	Benzyldiazepine	Inhibits initiation of transcription	[129–131]
Rev	Cyclic peptides	Competitively inhibits Rev binding	[129,132]
vpr	Fumagillin	Inhibits vpr-dependent G2 arrest and gene expression	[133]

The interaction of HIV viral infectivity factor (Vif) with cellular apolipoprotein B messenger RNA-editing enzyme catalytic polypeptide-like (APOBEC) 3G leading to degradation of APOBEC3G in infected cells is also being pursued [109–111]. APOBEC3G acts on viral reverse transcripts by converting cytosine to uracil leading to G to A hypermutation and introduction of potentially lethal mutations into the viral genome [112]. By protecting APOBEC3G from degradation, another escape strategy of HIV could be thwarted, and this approach may prove useful as a new tool for therapeutic treatment of HIV-infected patients.

These are but a few examples of the potential targets available to investigation. Advances in protein engineering, recombinant protein production and purification, and basic research to identify new targets and further refine present strategies, are all very important in the development of new inhibitors.

16.6 CONCLUSION

Although current therapeutic strategies have been quite successful in slowing progression to AIDS and lowering transmission rates, especially with the utilization of different combinations of drugs for ART, the limitations of this approach have become apparent. Some of these limitations can be overcome with the development of newer therapies that have greater potency, specificity, bioavailability, and lower toxicity, as well as increased funding from sources aimed at making the drugs more accessible. Protein engineering will play a pivotal role in the research and development of new products, and advancing existing products by refining them through various modifications to be more amenable to clinical applications.

Entry and fusion inhibitors are a promising new class of antiviral agents that is being added to our armamentarium against HIV. By acting on the virus before entry can occur, they have the potential to produce the same outcome as a vaccine eliciting sterilizing immunity. This attribute also gives entry and fusion inhibitors an inherent temporal advantage over RT, IN, and protease inhibitors. Fusion inhibitors are currently being evaluated in conjunction with current ART and the results have been very promising. Patients that fail with established ART due to resistant virus or tolerance issues related to the RT and protease inhibitors will benefit the most from the salvage therapy being offered with the fusion inhibitors.

The synergistic effects of multiple inhibitors have become readily apparent with ART, and expanding the pool of viable targets is an encouraging positive sign for the future of HIV therapeutics. As the mechanisms behind the interaction of viral and host proteins are better understood through basic science techniques, more inhibitors, both viral and cellular, will be developed. Protein–protein interactions are key to several steps in HIV replication and the details of these mechanisms will guide us in development of new strategies. The development of new strategies employing peptide and protein engineering coupled with novel delivery methods must keep pace as new targets are identified.

The findings and conclusions of this chapter are those of the authors and do not necessarily represent the views of the Center for Disease Control and Prevention. The use of trade names does not signify endorsement of the product and are used solely for identification.

REFERENCES

1. Barre-Sinoussi, F., Chermann, J.C., Rey, F., Nugeyre, M.T., Chamaret, S., Gruest, J., Dauguet, C., Axler-Blin, C., Vezinet-Brun, F., Rouzioux, C., Rozenbaum, W., and Montagnier, L., Isolation of a T-lymphotropic retrovirus from a patient at risk for acquired immune deficiency syndrome (AIDS). *Science* 1983, 220(4599), 868–871.

2. Chermann, J.C., Barre-Sinoussi, F., Dauguet, C., Brun-Vezinet, F., Rouzioux, C., Rozenbaum, W., and Montagnier, L., Isolation of a new retrovirus in a patient at risk for acquired immunodeficiency syndrome. *Antibiot Chemother* 1983, 32, 48–53.
3. Beyrer, C., HIV epidemiology update and transmission factors: Risks and risk contexts—16th International AIDS Conference epidemiology plenary. *Clin Infect Dis* 2007, 44(7), 981–987.
4. Savarino, A., A historical sketch of the discovery and development of HIV-1 integrase inhibitors. *Expert Opin Investig Drugs* 2006, 15(12), 1507–1522.
5. Turpin, J.A., The next generation of HIV/AIDS drugs: Novel and developmental antiHIV drugs and targets. *Expert Rev Anti Infect Ther* 2003, 1(1), 97–128.
6. Finzi, D., Dieffenbach, C.W., and Basavappa, R., Defining and solving the essential protein-protein interactions in HIV infection. *J Struct Biol* 2007, 158(2), 148–155.
7. Freed, E.O. and Mouland, A.J., The cell biology of HIV-1 and other retroviruses. *Retrovirology* 2006, 3, 77.
8. Gomez, C. and Hope, T.J., The ins and outs of HIV replication. *Cell Microbiol* 2005, 7(5), 621–626.
9. Greene, W.C. and Peterlin, B.M., Charting HIV's remarkable voyage through the cell: Basic science as a passport to future therapy. *Nat Med* 2002, 8(7), 673–680.
10. Cohen, M.S., Gay, C., Kashuba, A.D., Blower, S., and Paxton, L., Narrative review: Antiretroviral therapy to prevent the sexual transmission of HIV-1. *Ann Intern Med* 2007, 146(8), 591–601.
11. Reeves, J.D. and Piefer, A.J., Emerging drug targets for antiretroviral therapy. *Drugs* 2005, 65(13), 1747–1766.
12. Maddon, P.J., Dalgleish, A.G., McDougal, J.S., Clapham, P.R., Weiss, R.A., and Axel, R., The T4 gene encodes the AIDS virus receptor and is expressed in the immune system and the brain. *Cell* 1986, 47(3), 333–348.
13. Sattentau, Q.J., Dalgleish, A.G., Weiss, R.A., and Beverley, P.C., Epitopes of the CD4 antigen and HIV infection. *Science* 1986, 234(4780), 1120–1123.
14. Sattentau, Q.J. and Weiss, R.A., The CD4 antigen: Physiological ligand and HIV receptor. *Cell* 1988, 52(5), 631–633.
15. Alkhatib, G., Combadiere, C., Broder, C.C., Feng, Y., Kennedy, P.E., Murphy, P.M., and Berger, E.A., CC CKR5: A RANTES, MIP-1alpha, MIP-1beta receptor as a fusion cofactor for macrophage-tropic HIV-1. *Science* 1996, 272(5270), 1955–1958.
16. Feng, Y., Broder, C.C., Kennedy, P.E., and Berger, E.A., HIV-1 entry cofactor: Functional cDNA cloning of a seven-transmembrane, G protein-coupled receptor. *Science* 1996, 272(5263), 872–877.
17. Liu, R., Paxton, W.A., Choe, S., Ceradini, D., Martin, S.R., Horuk, R., MacDonald, M.E., Stuhlmann, H., Koup, R.A., and Landau, N.R., Homozygous defect in HIV-1 coreceptor accounts for resistance of some multiply-exposed individuals to HIV-1 infection. *Cell* 1996, 86(3), 367–377.
18. Dean, M., Carrington, M., Winkler, C., Huttley, G.A., Smith, M.W., Allikmets, R., Goedert, J.J., Buchbinder, S.P., Vittinghoff, E., Gomperts, E., Donfield, S., Vlahov, D., Kaslow, R., Saah, A., Rinaldo, C., Detels, R., and O'Brien, S.J., Genetic restriction of HIV-1 infection and progression to AIDS by a deletion allele of the CKR5 structural gene. Hemophilia Growth and Development Study, Multicenter AIDS Cohort Study, Multicenter Hemophilia Cohort Study, San Francisco City Cohort, ALIVE Study. *Science* 1996, 273(5283), 1856–1862.
19. Huang, Y., Paxton, W.A., Wolinsky, S.M., Neumann, A.U., Zhang, L., He, T., Kang, S., Ceradini, D., Jin, Z., Yazdanbakhsh, K., Kunstman, K., Erickson, D., Dragon, E., Landau, N.R., Phair, J., Ho, D.D., and Koup, R. A., The role of a mutant CCR5 allele in HIV-1 transmission and disease progression. *Nat Med* 1996, 2(11), 1240–1243.
20. Berger, E.A., Fuerst, T.R., and Moss, B., A soluble recombinant polypeptide comprising the amino-terminal half of the extracellular region of the CD4 molecule contains an active binding site for human immunodeficiency virus. *Proc Natl Acad Sci U S A* 1988, 85(7), 2357–2361.
21. Sullivan, N., Sun, Y., Sattentau, Q., Thali, M., Wu, D., Denisova, G., Gershoni, J., Robinson, J., Moore, J., and Sodroski, J., CD4-Induced conformational changes in the human immunodeficiency virus type 1 gp120 glycoprotein: Consequences for virus entry and neutralization. *J Virol* 1998, 72(6), 4694–4703.
22. Arthos, J., Cicala, C., Steenbeke, T.D., Chun, T.W., Dela Cruz, C., Hanback, D.B., Khazanie, P., Nam, D., Schuck, P., Selig, S.M., Van Ryk, D., Chaikin, M.A., and Fauci, A.S., Biochemical and biological characterization of a dodecameric CD4-Ig fusion protein: Implications for therapeutic and vaccine strategies. *J Biol Chem* 2002, 277(13), 11456–11464.

23. Salzwedel, K., Smith, E.D., Dey, B., and Berger, E.A., Sequential CD4-coreceptor interactions in human immunodeficiency virus type 1 Env function: Soluble CD4 activates Env for coreceptor-dependent fusion and reveals blocking activities of antibodies against cryptic conserved epitopes on gp120. *J Virol* 2000, 74(1), 326–333.
24. Li, H., Guan, Y., Szczepanska, A., Moreno-Vargas, A.J., Carmona, A.T., Robina, I., Lewis, G.K., and Wang, L.X., Synthesis and anti-HIV activity of trivalent CD4-mimetic miniproteins. *Bioorg Med Chem* 2007, 15(12), 4220–4228.
25. Li, H., Song, H., Heredia, A., Le, N., Redfield, R., Lewis, G.K., and Wang, L.X., Synthetic bivalent CD4-mimetic miniproteins show enhanced anti-HIV activity over the monovalent miniprotein. *Bioconjug Chem* 2004, 15(4), 783–789.
26. Berger, E.A., Clouse, K.A., Chaudhary, V.K., Chakrabarti, S., FitzGerald, D.J., Pastan, I., and Moss, B., CD4-Pseudomonas exotoxin hybrid protein blocks the spread of human immunodeficiency virus infection in vitro and is active against cells expressing the envelope glycoproteins from diverse primate immunodeficiency retroviruses. *Proc Natl Acad Sci U S A* 1989, 86(23), 9539–9543.
27. Chaudhary, V.K., Mizukami, T., Fuerst, T.R., FitzGerald, D.J., Moss, B., Pastan, I., and Berger, E.A., Selective killing of HIV-infected cells by recombinant human CD4-Pseudomonas exotoxin hybrid protein. *Nature* 1988, 335(6188), 369–372.
28. Kennedy, P.E., Moss, B., and Berger, E.A., Primary HIV-1 isolates refractory to neutralization by soluble CD4 are potently inhibited by CD4-pseudomonas exotoxin. *Virology* 1993, 192(1), 375–379.
29. Dey, B., Del Castillo, C.S., and Berger, E.A., Neutralization of human immunodeficiency virus type 1 by sCD4-17b, a single-chain chimeric protein, based on sequential interaction of gp120 with CD4 and coreceptor. *J Virol* 2003, 77(5), 2859–2865.
30. Fletcher, C.V., Deville, J.G., Samson, P.M., Moye, J.H., Jr., Church, J.A., Spiegel, H.M., Palumbo, P., Fenton, T., Smith, M.E., Graham, B., Kraimer, J.M., and Shearer, W.T., Nonlinear pharmacokinetics of high-dose recombinant fusion protein CD4-IgG(2) (PRO 542) observed in HIV-1-infected children. *J Allergy Clin Immunol* 2007, 119(3), 747–750.
31. Zhu, P., Olson, W.C., and Roux, K.H., Structural flexibility and functional valence of CD4-IgG2 (PRO 542): Potential for cross-linking human immunodeficiency virus type 1 envelope spikes. *J Virol* 2001, 75(14), 6682–6686.
32. Allaway, G.P., Davis-Bruno, K.L., Beaudry, G.A., Garcia, E.B., Wong, E.L., Ryder, A.M., Hasel, K.W., Gauduin, M.C., Koup, R.A., McDougal, J.S., et al., Expression and characterization of CD4-IgG2, a novel heterotetramer that neutralizes primary HIV type 1 isolates. *AIDS Res Hum Retroviruses* 1995, 11(5), 533–539.
33. Jacobson, J.M., Israel, R.J., Lowy, I., Ostrow, N.A., Vassilatos, L.S., Barish, M., Tran, D.N., Sullivan, B. M., Ketas, T.J., O'Neill, T.J., Nagashima, K.A., Huang, W., Petropoulos, C.J., Moore, J.P., Maddon, P.J., and Olson, W.C., Treatment of advanced human immunodeficiency virus type 1 disease with the viral entry inhibitor PRO 542. *Antimicrob Agents Chemother* 2004, 48(2), 423–429.
34. Jacobson, J.M., Lowy, I., Fletcher, C.V., O'Neill, T.J., Tran, D.N., Ketas, T.J., Trkola, A., Klotman, M.E., Maddon, P.J., Olson, W.C., and Israel, R.J., Single-dose safety, pharmacology, and antiviral activity of the human immunodeficiency virus (HIV) type 1 entry inhibitor PRO 542 in HIV-infected adults. *J Infect Dis* 2000, 182(1), 326–329.
35. Castagna, A., Biswas, P., Beretta, A., and Lazzarin, A., The appealing story of HIV entry inhibitors: From discovery of biological mechanisms to drug development. *Drugs* 2005, 65(7), 879–904.
36. Nagashima, K.A., Thompson, D.A., Rosenfield, S.I., Maddon, P.J., Dragic, T., and Olson, W.C., Human immunodeficiency virus type 1 entry inhibitors PRO 542 and T-20 are potently synergistic in blocking virus-cell and cell-cell fusion. *J Infect Dis* 2001, 183(7), 1121–1125.
37. Bera, T.K., Kennedy, P.E., Berger, E.A., and Barbas, C.F., 3rd; Pastan, I., Specific killing of HIV-infected lymphocytes by a recombinant immunotoxin directed against the HIV-1 envelope glycoprotein. *Mol Med* 1998, 4(6), 384–391.
38. Goldstein, H., Pettoello-Mantovani, M., Bera, T.K., Pastan, I.H., and Berger, E.A., Chimeric toxins targeted to the human immunodeficiency virus type 1 envelope glycoprotein augment the in vivo activity of combination antiretroviral therapy in thy/liv-SCID-Hu mice. *J Infect Dis* 2000, 181(3), 921–926.
39. Kennedy, P.E., Bera, T.K., Wang, Q.C., Gallo, M., Wagner, W., Lewis, M.G., Berger, E.A., and Pastan, I., Anti-HIV-1 immunotoxin 3B3(Fv)-PE38: Enhanced potency against clinical isolates in human PBMCs and macrophages, and negligible hepatotoxicity in macaques. *J Leukoc Biol* 2006, 80(5), 1175–1182.

40. Botos, I., O'Keefe, B.R., Shenoy, S.R., Cartner, L.K., Ratner, D.M., Seeberger, P.H., Boyd, M.R., and Wlodawer, A., Structures of the complexes of a potent anti-HIV protein cyanovirin-N and high mannose oligosaccharides. *J Biol Chem* 2002, 277(37), 34336–34342.
41. Botos, I., Wlodawer, A., and Cyanovirin-N: A sugar-binding antiviral protein with a new twist. *Cell Mol Life Sci* 2003, 60(2), 277–287.
42. Boyd, M.R., Gustafson, K.R., McMahon, J.B., Shoemaker, R.H., O'Keefe, B.R., Mori, T., Gulakowski, R. J., Wu, L., Rivera, M.I., Laurencot, C.M., Currens, M.J., and Cardellina, J.H., 2nd, Buckheit, R.W., Jr., Nara, P.L., Pannell, L.K., Sowder, R.C., 2nd, Henderson, L.E., Discovery of cyanovirin-N, a novel human immunodeficiency virus-inactivating protein that binds viral surface envelope glycoprotein gp120: Potential applications to microbicide development. *Antimicrob Agents Chemother* 1997, 41(7), 1521–1530.
43. Bewley, C.A., Cai, M., Ray, S., Ghirlando, R., Yamaguchi, M., and Muramoto, K., New carbohydrate specificity and HIV-1 fusion blocking activity of the cyanobacterial protein MVL: NMR, ITC and sedimentation equilibrium studies. *J Mol Biol* 2004, 339(4), 901–914.
44. Mori, T., O'Keefe, B.R., Sowder, R.C., 2nd, Bringans, S., Gardella, R., Berg, S., Cochran, P., Turpin, J.A., Buckheit, R.W., Jr., McMahon, J.B., and Boyd, M.R., Isolation and characterization of Griffithsin, a novel HIV-inactivating protein, from the red alga Griffithsia sp. *J Biol Chem* 2005, 280(10), 9345–9353.
45. Giomarelli, B., Schumacher, K.M., Taylor, T.E., Sowder, R.C., 2nd, Hartley, J.L., McMahon, J.B., and Mori, T., Recombinant production of anti-HIV protein, griffithsin, by auto-induction in a fermentor culture. *Protein Expr Purif* 2006, 47(1), 194–202.
46. Dey, B., Lerner, D.L., Lusso, P., Boyd, M.R., Elder, J.H., and Berger, E.A., Multiple antiviral activities of cyanovirin-N: Blocking of human immunodeficiency virus type 1 gp120 interaction with CD4 and coreceptor and inhibition of diverse enveloped viruses. *J Virol* 2000, 74(10), 4562–4569.
47. Tsai, C.C., Emau, P., Jiang, Y., Agy, M.B., Shattock, R.J., Schmidt, A., Morton, W.R., Gustafson, K.R., and Boyd, M.R., Cyanovirin-N inhibits AIDS virus infections in vaginal transmission models. *AIDS Res Hum Retroviruses* 2004, 20(1), 11–18.
48. Liu, X., Lagenaur, L.A., Simpson, D.A., Essenmacher, K.P., Frazier-Parker, C.L., Liu, Y., Tsai, D., Rao, S.S., Hamer, D.H., Parks, T.P., Lee, P.P., and Xu, Q., Engineered vaginal lactobacillus strain for mucosal delivery of the human immunodeficiency virus inhibitor cyanovirin-N. *Antimicrob Agents Chemother* 2006, 50(10), 3250–3259.
49. Briz, V., Poveda, E., and Soriano, V., HIV entry inhibitors: Mechanisms of action and resistance pathways. *J Antimicrob Chemother* 2006, 57(4), 619–627.
50. Lin, P.F., Blair, W., Wang, T., Spicer, T., Guo, Q., Zhou, N., Gong, Y.F., Wang, H.G., Rose, R., Yamanaka, G., Robinson, B., Li, C.B., Fridell, R., Deminie, C., Demers, G., Yang, Z., Zadjura, L., Meanwell, N., and Colonno, R., A small molecule HIV-1 inhibitor that targets the HIV-1 envelope and inhibits CD4 receptor binding. *Proc Natl Acad Sci U S A* 2003, 100(19), 11013–11018.
51. Si, Z., Phan, N., Kiprilov, E., and Sodroski, J., Effects of HIV type 1 envelope glycoprotein proteolytic processing on antigenicity. *AIDS Res Hum Retroviruses* 2003, 19(3), 217–226.
52. Kish-Catalone, T.M., Lu, W., Gallo, R.C., and DeVico, A.L., Preclinical evaluation of synthetic -2 RANTES as a candidate vaginal microbicide to target CCR5. *Antimicrob Agents Chemother* 2006, 50(4), 1497–1509.
53. Kish-Catalone, T., Pal, R., Parrish, J., Rose, N., Hocker, L., Hudacik, L., Reitz, M., Gallo, R., and Devico, A., Evaluation of -2 RANTES vaginal microbicide formulations in a nonhuman primate simian/human immunodeficiency virus (SHIV) challenge model. *AIDS Res Hum Retroviruses* 2007, 23(1), 33–42.
54. Kawamura, T., Bruse, S.E., Abraha, A., Sugaya, M., Hartley, O., Offord, R.E., Arts, E.J., Zimmerman, P.A., and Blauvelt, A., PSC-RANTES blocks R5 human immunodeficiency virus infection of Langerhans cells isolated from individuals with a variety of CCR5 diplotypes. *J Virol* 2004, 78(14), 7602–7609.
55. Lederman, M.M., Veazey, R.S., Offord, R., Mosier, D.E., Dufour, J., Mefford, M., Piatak, M., Jr., Lifson, J.D., Salkowitz, J.R., Rodriguez, B., Blauvelt, A., and Hartley, O., Prevention of vaginal SHIV transmission in rhesus macaques through inhibition of CCR5. *Science* 2004, 306(5695), 485–487.
56. Schols, D., HIV coreceptor inhibitors as novel class of anti-HIV drugs. *Antiviral Res* 2006, 71(2–3), 216–226.
57. Trkola, A., Ketas, T.J., Nagashima, K.A., Zhao, L., Cilliers, T., Morris, L., Moore, J.P., Maddon, P.J., and Olson, W.C., Potent, broad-spectrum inhibition of human immunodeficiency virus type 1 by the CCR5 monoclonal antibody PRO 140. *J Virol* 2001, 75(2), 579–588.

58. Ichiyama, K., Yokoyama-Kumakura, S., Tanaka, Y., Tanaka, R., Hirose, K., Bannai, K., Edamatsu, T., Yanaka, M., Niitani, Y., Miyano-Kurosaki, N., Takaku, H., Koyanagi, Y., and Yamamoto, N., A duodenally absorbable CXC chemokine receptor 4 antagonist, KRH-1636, exhibits a potent and selective anti-HIV-1 activity. *Proc Natl Acad Sci U S A* 2003, 100(7), 4185–4190.
59. Hamburger, A.E., Kim, S., Welch, B.D., and Kay, M.S., Steric accessibility of the HIV-1 gp41 N-trimer region. *J Biol Chem* 2005, 280(13), 12567–12572.
60. Wild, C., Oas, T., McDanal, C., Bolognesi, D., and Matthews, T., A synthetic peptide inhibitor of human immunodeficiency virus replication: Correlation between solution structure and viral inhibition. *Proc Natl Acad Sci U S A* 1992, 89(21), 10537–10541.
61. Bewley, C.A., Louis, J.M., Ghirlando, R., and Clore, G.M., Design of a novel peptide inhibitor of HIV fusion that disrupts the internal trimeric coiled-coil of gp41. *J Biol Chem* 2002, 277(16), 14238–14245.
62. Gustchina, E., Louis, J.M., Bewley, C.A., and Clore, G.M., Synergistic inhibition of HIV-1 envelope-mediated membrane fusion by inhibitors targeting the N and C-terminal heptad repeats of gp41. *J Mol Biol* 2006, 364(3), 283–289.
63. Baldwin, C.E., Sanders, R.W., and Berkhout, B., Inhibiting HIV-1 entry with fusion inhibitors. *Curr Med Chem* 2003, 10(17), 1633–1642.
64. Chinnadurai, R., Rajan, D., Munch, J., and Kirchhoff, F., Human immunodeficiency virus type 1 variants resistant to first- and second-version fusion inhibitors and cytopathic in ex vivo human lymphoid tissue. *J Virol* 2007, 81(12), 6563–6572.
65. Chen, R.Y., Kilby, J.M., and Saag, M.S., Enfuvirtide. *Expert Opin Investig Drugs* 2002, 11(12), 1837–1843.
66. Kilby, J.M., Hopkins, S., Venetta, T.M., DiMassimo, B., Cloud, G.A., Lee, J.Y., Alldredge, L., Hunter, E., Lambert, D., Bolognesi, D., Matthews, T., Johnson, M.R., Nowak, M.A., Shaw, G.M., and Saag, M.S., Potent suppression of HIV-1 replication in humans by T-20, a peptide inhibitor of gp41-mediated virus entry. *Nat Med* 1998, 4(11), 1302–1307.
67. Wild, C., Greenwell, T., Shugars, D., Rimsky-Clarke, L., and Matthews, T., The inhibitory activity of an HIV type 1 peptide correlates with its ability to interact with a leucine zipper structure. *AIDS Res Hum Retroviruses* 1995, 11(3), 323–325.
68. Kilby, J.M., Lalezari, J.P., Eron, J.J., Carlson, M., Cohen, C., Arduino, R.C., Goodgame, J.C., Gallant, J. E., Volberding, P., Murphy, R.L., Valentine, F., Saag, M.S., Nelson, E.L., Sista, P.R., and Dusek, A., The safety, plasma pharmacokinetics, and antiviral activity of subcutaneous enfuvirtide (T-20), a peptide inhibitor of gp41-mediated virus fusion, in HIV-infected adults. *AIDS Res Hum Retroviruses* 2002, 18 (10), 685–693.
69. Wild, C.T., Shugars, D.C., Greenwell, T.K., McDanal, C.B., and Matthews, T.J., Peptides corresponding to a predictive alpha-helical domain of human immunodeficiency virus type 1 gp41 are potent inhibitors of virus infection. *Proc Natl Acad Sci U S A* 1994, 91(21), 9770–9774.
70. Wild, C., Greenwell, T., and Matthews, T., A synthetic peptide from HIV-1 gp41 is a potent inhibitor of virus-mediated cell-cell fusion. *AIDS Res Hum Retroviruses* 1993, 9(11), 1051–1053.
71. Martin-Carbonero, L., Discontinuation of the clinical development of fusion inhibitor T-1249. *AIDS Rev* 2004, 6(1), 61.
72. Armand-Ugon, M., Gutierrez, A., Clotet, B., and Este, J.A., HIV-1 resistance to the gp41-dependent fusion inhibitor C-34. *Antiviral Res* 2003, 59(2), 137–142.
73. Sugaya, M., Hartley, O., Root, M.J., and Blauvelt, A., C34, a membrane fusion inhibitor, blocks HIV infection of Langerhans cells and viral transmission to T cells. *J Invest Dermatol* 2007, 127(6), 1436–1443.
74. Otaka, A., Nakamura, M., Nameki, D., Kodama, E., Uchiyama, S., Nakamura, S., Nakano, H., Tamamura, H., Kobayashi, Y., Matsuoka, M., and Fujii, N., Remodeling of gp41-C34 peptide leads to highly effective inhibitors of the fusion of HIV-1 with target cells. *Angew Chem Int Ed Engl* 2002, 41(16), 2937–2940.
75. Deng, Y., Zheng, Q., Ketas, T.J., Moore, J.P., and Lu, M., Protein design of a bacterially expressed HIV-1 gp41 fusion inhibitor. *Biochemistry* 2007, 46(14), 4360–4369.
76. Ketas, T.J., Schader, S.M., Zurita, J., Teo, E., Polonis, V., Lu, M., Klasse, P.J., and Moore, J.P., Entry inhibitor-based microbicides are active in vitro against HIV-1 isolates from multiple genetic subtypes. *Virology* 2007, 364(2), 431–440.

77. Veazey, R.S., Klasse, P.J., Schader, S.M., Hu, Q., Ketas, T.J., Lu, M., Marx, P.A., Dufour, J., Colonno, R. J., Shattock, R.J., Springer, M.S., and Moore, J.P., Protection of macaques from vaginal SHIV challenge by vaginally delivered inhibitors of virus-cell fusion. *Nature* 2005, 438(7064), 99–102.
78. Rao, S., Hu, S., McHugh, L., Lueders, K., Henry, K., Zhao, Q., Fekete, R.A., Kar, S., Adhya, S., and Hamer, D.H., Toward a live microbial microbicide for HIV: Commensal bacteria secreting an HIV fusion inhibitor peptide. *Proc Natl Acad Sci U S A* 2005, 102(34), 11993–11998.
79. Eckert, D.M., Malashkevich, V.N., Hong, L.H., Carr, P.A., and Kim, P.S., Inhibiting HIV-1 entry: Discovery of D-peptide inhibitors that target the gp41 coiled-coil pocket. *Cell* 1999, 99(1), 103–115.
80. Eckert, D.M. and Kim, P.S., Design of potent inhibitors of HIV-1 entry from the gp41 N-peptide region. *Proc Natl Acad Sci U S A* 2001, 98(20), 11187–11192.
81. Root, M.J., Kay, M.S., and Kim, P.S., Protein design of an HIV-1 entry inhibitor. *Science* 2001, 291 (5505), 884–888.
82. Delcroix-Genete, D., Quan, P.L., Roger, M.G., Hazan, U., Nisole, S., and Rousseau, C., Antiviral properties of two trimeric recombinant gp41 proteins. *Retrovirology* 2006, 3, 16.
83. Munch, J., Standker, L., Adermann, K., Schulz, A., Schindler, M., Chinnadurai, R., Pohlmann, S., Chaipan, C., Biet, T., Peters, T., Meyer, B., Wilhelm, D., Lu, H., Jing, W., Jiang, S., Forssmann, W. G., and Kirchhoff, F., Discovery and optimization of a natural HIV-1 entry inhibitor targeting the gp41 fusion peptide. *Cell* 2007, 129(2), 263–275.
84. Francis, G.E., Fisher, D., Delgado, C., Malik, F., Gardiner, A., and Neale, D., PEGylation of cytokines and other therapeutic proteins and peptides: The importance of biological optimisation of coupling techniques. *Int J Hematol* 1998, 68(1), 1–18.
85. Basu, A., Yang, K., Wang, M., Liu, S., Chintala, R., Palm, T., Zhao, H., Peng, P., Wu, D., Zhang, Z., Hua, J., Hsieh, M.C., Zhou, J., Petti, G., Li, X., Janjua, A., Mendez, M., Liu, J., Longley, C., Zhang, Z., Mehlig, M., Borowski, V., Viswanathan, M., and Filpula, D., Structure-function engineering of interferon-beta-1b for improving stability, solubility, potency, immunogenicity, and pharmacokinetic properties by site-selective mono-PEGylation. *Bioconjug Chem* 2006, 17(3), 618–630.
86. Schmidt, P.G., Campbell, K.M., Hinds, K.D., and Cook, G.P., PEGylated bioactive molecules in biodegradable polymer microparticles. *Expert Opin Biol Ther* 2007, 7(9), 1427–1436.
87. Dwyer, J.J., Wilson, K.L., Davison, D.K., Freel, S.A., Seedorff, J.E., Wring, S.A., Tvermoes, N.A., Matthews, T.J., Greenberg, M.L., and Delmedico, M.K., Design of helical, oligomeric HIV-1 fusion inhibitor peptides with potent activity against enfuvirtide-resistant virus. *Proc Natl Acad Sci U S A* 2007, 104(31), 12772–12777.
88. Welch, B.D., Vandemark, A.P., Heroux, A., Hill, C.P., and Kay, M.S., Potent D-peptide inhibitors of HIV-1 entry. *Proc Natl Acad Sci U S A* 2007, 104(43), 16828–16833.
89. Lalezari, J.P., Henry, K., O'Hearn, M., Montaner, J.S., Piliero, P.J., Trottier, B., Walmsley, S., Cohen, C., Kuritzkes, D.R., Eron, J.J., Jr., Chung, J., DeMasi, R., Donatacci, L., Drobnes, C., Delehanty, J., and Salgo, M., Enfuvirtide, an HIV-1 fusion inhibitor, for drug-resistant HIV infection in North and South America. *N Engl J Med* 2003, 348(22), 2175–2185.
90. Lazzarin, A., Clotet, B., Cooper, D., Reynes, J., Arasteh, K., Nelson, M., Katlama, C., Stellbrink, H.J., Delfraissy, J.F., Lange, J., Huson, L., DeMasi, R., Wat, C., Delehanty, J., Drobnes, C., and Salgo, M., Efficacy of enfuvirtide in patients infected with drug-resistant HIV-1 in Europe and Australia. *N Engl J Med* 2003, 348(22), 2186–2195.
91. Reeves, J.D., Lee, F.H., Miamidian, J.L., Jabara, C.B., Juntilla, M.M., and Doms, R.W., Enfuvirtide resistance mutations: Impact on human immunodeficiency virus envelope function, entry inhibitor sensitivity, and virus neutralization. *J Virol* 2005, 79(8), 4991–4999.
92. Lalezari, J.P., Bellos, N.C., Sathasivam, K., Richmond, G.J., Cohen, C.J., Myers, R.A., Jr., Henry, D.H., Raskino, C., Melby, T., Murchison, H., Zhang, Y., Spence, R., Greenberg, M.L., Demasi, R.A., and Miralles, G.D., T-1249 retains potent antiretroviral activity in patients who had experienced virological failure while on an enfuvirtide-containing treatment regimen. *J Infect Dis* 2005, 191(7), 1155–1163.
93. Eron, J.J., Gulick, R.M., Bartlett, J.A., Merigan, T., Arduino, R., Kilby, J.M., Yangco, B., Diers, A., Drobnes, C., DeMasi, R., Greenberg, M., Melby, T., Raskino, C., Rusnak, P., Zhang, Y., Spence, R., and Miralles, G.D., Short-term safety and antiretroviral activity of T-1249, a second-generation fusion inhibitor of HIV. *J Infect Dis* 2004, 189(6), 1075–1083.
94. Hart, C.E. and Evans-Strickfaden, T., HIV-1 entry inhibitors as microbicides. *Entry Inhibitors in HIV Therapy*, Reeves, J.D. and Derdeyn, C.A., (Eds.), Birrkhauser Verlag AG, Basel, Boston, Berlin 2007, 99–117.

95. Harrich, D., McMillan, N., Munoz, L., Apolloni, A., and Meredith, L., Will diverse Tat interactions lead to novel antiretroviral drug targets? *Curr Drug Targets* 2006, 7(12), 1595–1606.
96. Blair, W.S., Cao, J., Jackson, L., Jimenez, J., Peng, Q., Wu, H., Isaacson, J., Butler, S., Chu, A., Graham, J., Malfait, A.M., Tortorella, M., and Patick, A.K., Identification and characterization of UK-201844, a novel inhibitor that interferes with HIV-1 gp160 processing. *Antimicrob Agents Chemother* 2007, 51(10), 3554–3561.
97. DeJesus, E., Berger, D., Markowitz, M., Cohen, C., Hawkins, T., Ruane, P., Elion, R., Farthing, C., Zhong, L., Cheng, A.K., McColl, D., and Kearney, B.P., Antiviral activity, pharmacokinetics, and dose response of the HIV-1 integrase inhibitor GS-9137 (JTK-303) in treatment-naive and treatment-experienced patients. *J Acquir Immune Defic Syndr* 2006, 43(1), 1–5.
98. Carr, J.M., Davis, A.J., Feng, F., Burrell, C.J., and Li, P., Cellular interactions of virion infectivity factor (Vif) as potential therapeutic targets: APOBEC3G and more? *Curr Drug Targets* 2006, 7(12), 1583–1593.
99. Li, F. and Wild, C., HIV-1 assembly and budding as targets for drug discovery. *Curr Opin Investig Drugs* 2005, 6(2), 148–154.
100. Schito, M.L., Soloff, A.C., Slovitz, D., Trichel, A., Inman, J.K., Appella, E., Turpin, J.A., and Barratt-Boyes, S.M., Preclinical evaluation of a zinc finger inhibitor targeting lentivirus nucleocapsid protein in SIV-infected monkeys. *Curr HIV Res* 2006, 4(3), 379–386.
101. Barbaro, G., Scozzafava, A., Mastrolorenzo, A., and Supuran, C.T., Highly active antiretroviral therapy: Current state of the art, new agents and their pharmacological interactions useful for improving therapeutic outcome. *Curr Pharm Des* 2005, 11(14), 1805–1843.
102. Turpin, J.A., Song, Y., Inman, J.K., Huang, M., Wallqvist, A., Maynard, A., Covell, D.G., Rice, W.G., and Appella, E., Synthesis and biological properties of novel pyridinioalkanoyl thiolesters (PATE) as anti-HIV-1 agents that target the viral nucleocapsid protein zinc fingers. *J Med Chem* 1999, 42(1), 67–86.
103. Basrur, V., Song, Y., Mazur, S.J., Higashimoto, Y., Turpin, J.A., Rice, W.G., Inman, J.K., and Appella, E., Inactivation of HIV-1 nucleocapsid protein P7 by pyridinioalkanoyl thioesters. Characterization of reaction products and proposed mechanism of action. *J Biol Chem* 2000, 275(20), 14890–14897.
104. Schito, M.L., Goel, A., Song, Y., Inman, J.K., Fattah, R.J., Rice, W.G., Turpin, J.A., Sher, A., and Appella, E., In vivo antiviral activity of novel human immunodeficiency virus type 1 nucleocapsid p7 zinc finger inhibitors in a transgenic murine model. *AIDS Res Hum Retroviruses* 2003, 19(2), 91–101.
105. Mugnaini, C., Petricci, E., Corelli, F., and Botta, M., Combinatorial chemistry as a tool for targeting different stages of the replicative HIV-1 cycle. *Comb Chem High Throughput Screen* 2005, 8(5), 387–401.
106. Anderson, J.L., Campbell, E.M., Wu, X., Vandegraaff, N., Engelman, A., and Hope, T.J., Proteasome inhibition reveals that a functional preintegration complex intermediate can be generated during restriction by diverse TRIM5 proteins. *J Virol* 2006, 80(19), 9754–9760.
107. Javanbakht, H., Diaz-Griffero, F., Stremlau, M., Si, Z., and Sodroski, J., The contribution of RING and B-box 2 domains to retroviral restriction mediated by monkey TRIM5alpha. *J Biol Chem* 2005, 280(29), 26933–26940.
108. Javanbakht, H., Yuan, W., Yeung, D.F., Song, B., Diaz-Griffero, F., Li, Y., Li, X., Stremlau, M., and Sodroski, J., Characterization of TRIM5alpha trimerization and its contribution to human immunodeficiency virus capsid binding. *Virology* 2006, 353(1), 234–246.
109. Bishop, K.N., Holmes, R.K., Sheehy, A.M., and Malim, M.H., APOBEC-mediated editing of viral RNA. *Science* 2004, 305(5684), 645.
110. Malim, M.H., Natural resistance to HIV infection: The Vif-APOBEC interaction. *C R Biol* 2006, 329(11), 871–875.
111. Chiu, Y.L. and Greene, W.C., APOBEC3 cytidine deaminases: Distinct antiviral actions along the retroviral life cycle. *J Biol Chem* 2006, 281(13), 8309–8312.
112. Mangeat, B., Turelli, P., Caron, G., Friedli, M., Perrin, L., and Trono, D., Broad antiretroviral defence by human APOBEC3G through lethal editing of nascent reverse transcripts. *Nature* 2003, 424(6944), 99–103.

113. Kuritzkes, D.R., Jacobson, J., Powderly, W.G., Godofsky, E., DeJesus, E., Haas, F., Reimann, K.A., Larson, J.L., Yarbough, P.O., Curt, V., and Shanahan, W.R., Jr., Antiretroviral activity of the anti-CD4 monoclonal antibody TNX-355 in patients infected with HIV type 1. *J Infect Dis* 2004, 189(2), 286–291.
114. Zhang, X.Q., Sorensen, M., Fung, M., and Schooley, R.T., Synergistic in vitro antiretroviral activity of a humanized monoclonal anti-CD4 antibody (TNX-355) and enfuvirtide (T-20). *Antimicrob Agents Chemother* 2006, 50(6), 2231–2233.
115. Xie, H., Belogortseva, N.I., Wu, J., Lai, W.H., and Chen, C.H., Inhibition of human immunodeficiency virus type 1 entry by a binding domain of Porphyromonas gingivalis gingipain. *Antimicrob Agents Chemother* 2006, 50(9), 3070–3074.
116. Ferrer, M. and Harrison, S.C., Peptide ligands to human immunodeficiency virus type 1 gp120 identified from phage display libraries. *J Virol* 1999, 73(7), 5795–5802.
117. McFadden, K., Cocklin, S., Gopi, H., Baxter, S., Ajith, S., Mahmood, N., Shattock, R., and Chaiken, I., A recombinant allosteric lectin antagonist of HIV-1 envelope gp120 interactions. *Proteins* 2007, 67(3), 617–629.
118. Biorn, A.C., Cocklin, S., Madani, N., Si, Z., Ivanovic, T., Samanen, J., Van Ryk, D.I., Pantophlet, R., Burton, D.R., Freire, E., Sodroski, J., and Chaiken, I.M., Mode of action for linear peptide inhibitors of HIV-1 gp120 interactions. *Biochemistry* 2004, 43(7), 1928–1938.
119. Cocklin, S., Gopi, H., Querido, B., Nimmagadda, M., Kuriakose, S., Cicala, C., Ajith, S., Baxter, S., Arthos, J., Martin-Garcia, J., and Chaiken, I.M., Broad-spectrum anti-human immunodeficiency virus (HIV) potential of a peptide HIV type 1 entry inhibitor. *J Virol* 2007, 81(7), 3645–3648.
120. Xiong, C., O'Keefe, B.R., Byrd, R.A., and McMahon, J.B., Potent anti-HIV activity of scytovirin domain 1 peptide. *Peptides* 2006, 27(7), 1668–1675.
121. Xiong, C., O'Keefe, B.R., Botos, I., Wlodawer, A., and McMahon, J.B., Overexpression and purification of Scytovirin, a potent, novel anti-HIV protein from the cultured cyanobacterium Scytonema varium. *Protein Expr Purif* 2006, 46(2), 233–239.
122. Bokesch, H.R., O'Keefe, B.R., McKee, T.C., Pannell, L.K., Patterson, G.M., Gardella, R.S., Sowder, R. C., 2nd; Turpin, J., Watson, K., Buckheit, R.W., Jr., and Boyd, M.R., A potent novel anti-HIV protein from the cultured cyanobacterium Scytonema varium. *Biochemistry* 2003, 42(9), 2578–2584.
123. Madani, N., Perdigoto, A.L., Srinivasan, K., Cox, J.M., Chruma, J.J., LaLonde, J., Head, M., and Smith, A.B., 3rd; Sodroski, J.G., Localized changes in the gp120 envelope glycoprotein confer resistance to human immunodeficiency virus entry inhibitors BMS-806 and #155. *J Virol* 2004, 78(7), 3742–3752.
124. Wang, W., Owen, S.M., Rudolph, D.L., Cole, A.M., Hong, T., Waring, A.J., Lal, R.B., and Lehrer, R.I., Activity of alpha- and theta-defensins against primary isolates of HIV-1. *J Immunol* 2004, 173(1), 515–520.
125. Cole, A.L., Herasimtschuk, A., Gupta, P., Waring, A.J., Lehrer, R.I., and Cole, A.M., The retrocyclin analogue RC-101 prevents human immunodeficiency virus type 1 infection of a model human cervicovaginal tissue construct. *Immunology* 2007, 121(1), 140–145.
126. Grinsztejn, B., Nguyen, B.Y., Katlama, C., Gatell, J.M., Lazzarin, A., Vittecoq, D., Gonzalez, C.J., Chen, J., Harvey, C.M., and Isaacs, R.D., Safety and efficacy of the HIV-1 integrase inhibitor raltegravir (MK-0518) in treatment-experienced patients with multidrug-resistant virus: A phase II randomised controlled trial. *Lancet* 2007, 369(9569), 1261–1269.
127. Jenkins, L.M., Byrd, J.C., Hara, T., Srivastava, P., Mazur, S.J., Stahl, S.J., Inman, J.K., Appella, E., Omichinski, J.G., and Legault, P., Studies on the mechanism of inactivation of the HIV-1 nucleocapsid protein NCp7 with 2-mercaptobenzamide thioesters. *J Med Chem* 2005, 48(8), 2847–2858.
128. Budihas, S.R., Gorshkova, I., Gaidamakov, S., Wamiru, A., Bona, M.K., Parniak, M.A., Crouch, R. J., McMahon, J.B., Beutler, J.A., and Le Grice, S.F., Selective inhibition of HIV-1 reverse transcriptase-associated ribonuclease H activity by hydroxylated tropolones. *Nucleic Acids Res* 2005, 33(4), 1249–1256.
129. Ptak, R.G., HIV-1 regulatory proteins: Targets for novel drug development. *Expert Opin Investig Drugs* 2002, 11(8), 1099–1115.
130. Cupelli, L.A. and Hsu, M.C., The human immunodeficiency virus type 1 Tat antagonist, Ro 5–3335, predominantly inhibits transcription initiation from the viral promoter. *J Virol* 1995, 69(4), 2640–2643.

131. Hsu, M.C., Dhingra, U., Earley, J.V., Holly, M., Keith, D., Nalin, C.M., Richou, A.R., Schutt, A.D., Tam, S.Y., Potash, M.J., et al., Inhibition of type 1 human immunodeficiency virus replication by a tat antagonist to which the virus remains sensitive after prolonged exposure in vitro. *Proc Natl Acad Sci U S A* 1993, 90(14), 6395–6399.
132. Hariton-Gazal, E., Rosenbluh, J., Zakai, N., Fridkin, G., Brack-Werner, R., Wolff, H., Devaux, C., Gilon, C., and Loyter, A., Functional analysis of backbone cyclic peptides bearing the arm domain of the HIV-1 Rev protein: Characterization of the karyophilic properties and inhibition of Rev-induced gene expression. *Biochemistry* 2005, 44(34), 11555–11566.
133. Watanabe, N., Nishihara, Y., Yamaguchi, T., Koito, A., Miyoshi, H., Kakeya, H., and Osada, H., Fumagillin suppresses HIV-1 infection of macrophages through the inhibition of Vpr activity. *FEBS Lett* 2006, 580(11), 2598–2602.

17 Recombinant Enzymes with New and Improved Properties

Lilia Milkova Ganova-Raeva

CONTENTS

17.1 INTRODUCTION: WHY CHOOSE A RECOMBINANT ENZYME?

Enzymes have two outstanding features that distinguish them from all other types of potential drugs. Enzymes can bind and act on their targets with great affinity and specificity and can catalytically convert multiple target molecules into a desired product. These characteristics have allowed the development of many enzymes to be used to treat a wide range of disorders [1,2].

Recombinant technologies have contributed immensely to the development of novel protein based pharmaceuticals, especially with the development of large array of supporting technologies (e.g., hybridoma lines, directed evolution, posttranslational modifications, *in silico* combinatorial analysis, etc.). The products of recombinant technologies have much to offer in terms of preparation consistency, purity of the product, and affordability. Recombinant technologies have an invaluable advantage—the possibility of engineering the natural structure in ways that can make them easier to deliver, to control their release in the organism, and improve their efficacy. The new technologies have allowed the development of medicinal proteins in several directions: generation of natural or modified proteins from human, bacterial, fungal, or plant descent; production of monoclonal antibodies (MAbs); high throughput screening for inhibitors or new active compounds; and secondary chemical modification of active protein recombinant compounds. The first attempts to treat cancer by proteolytic pancreatic enzymes were proposed by John Beard in 1902. Insulin was the first introduced and approved recombinant protein in 1982, followed by human growth hormone in 1985, blood clotting factor in 1987 and the first recombinant enzyme drug—Activasel (an alteplase-recombinant human tissue plasminogen activator). By the end of 2003, there were already 148 biopharmaceuticals approved for medicinal use in Europe and the United States [3]. A number of important therapeutic enzymes like L-glutaminase, glucosidase and L-asparaginase with antitumor activity, lipase and laccase for lipid digestion and detoxification have been derived from bacteria, yeast, fungi, or plants [4–9]. Recombinant technologies are now making possible the idea of actively pursuing personalized medicine for the treatment of rare genetic diseases [10].

This chapter summarizes current trends in recombinant protein engineering, discusses the different underlying recombinant technologies, and highlights the various applications of the products that the technologies generate.

17.2 RECOMBINANT ENZYME TECHNOLOGIES

Recombinant DNA technologies have several different aspects that deal with the cloning, expression, purification, and modification of the protein target of interest. Numerous relevant examples could be found in the literature [11–17]. In their classical version recombinant proteins are created from cloned DNA sequences that encode an enzyme or protein with known function. Several specialized vectors (bacterial, bacculoviral, yeast, etc.) have been designed and are used for production of the proteins from specific DNA sequences cloned into them. The designs can include features that help control replication and expression, and tags to facilitate the post-expression purification of the expressed protein. The cloned DNA itself may differ from the wild type and include modifications than enhance the sought after properties of the protein product. Highlighted below are different technologies based on this classical strategy that have been refined for the generation of proteins with properties superior to those encoded by the wild-type genes.

17.2.1 Site-Directed Mutagenesis

Mutagenesis has developed into a major tool in biology/genetics since many discoveries were made when functions that deviate from the expected "norm" were found and ascribed to particular mutations. The idea and method for generating specific nucleotide changes in a predetermined site of interest was introduced by Michael Smith, who in 1993 was awarded the Nobel Prize for his work, together with the inventor of PCR, Kary Mullis. The efficiency of the method has undergone multiple improvements and now, coupled with high throughput cloning, expression and screening techniques, has become important for generation of mutants for the study, re-design, and utilization of protein properties. Site-directed mutagenesis is usually done using a mismatched oligo; however, other techniques that use PCR or restriction enzymes have also emerged and all those are commercially available products.

One of the early examples of systematically replacing amino acids in an enzyme, in order to elucidate their molecular roles in substrate binding or catalysis and eventually facilitate the engineering of new enzymatic activities, was the work on tyrosyl tRNA synthetase [18]. It was shown that cysteine (Cys) 35 [19] at the enzyme's active site converted to serine leads to a reduction in enzymatic activity due to a lowered K_m for adenosine triphosphate (ATP) [18]. Another single point mutation engineered to affect amino acid position 51(Thr 51 Ala/Pro) in the same enzyme improved its K_m by a factor of 100 [20]. Site-directed mutagenesis studies are essential for the study of the functional sites of a protein as highlighted in the following examples. The recombinant metalloproteinase domain of the human asthma susceptibility gene (ADAM33) has been purified and tested for its substrate-cleavage specificity using peptides derived from β-amyloid precursor protein. A single substitution in the precursor yields a 20-fold more efficient substrate cleavage. Terminal truncation studies have identified a minimal nine-residue core (P5-P4′) important for ADAM33 recognition and cleavage at this site [21]. Mutations in the gene encoding copper-zinc superoxide dismutase 1 (SOD1) cause approximately 20% cases of familial amyotrophic lateral sclerosis (FALS), characterized by selective loss of motor neurons. Mutant SOD1 forms inclusions in tissues from FALS patients. A modification of SOD1 at residue 75 (named SUMO-1) that causes change of the wild-type lysine was found to participate in regulating the protein stability and the aggregation process involved in FALS pathogenesis [22]. Nucleoside phosphorylases (NP) have been targeted for engineering for the purposes of treating adenosine deaminase deficiencies. Recombinant purine NP with two active site substitutions where Asn243Asp; Lys244Gln can accept 6 amino acid-substituted purine nucleosides [23]. Adenosine phosphorylase conjugate to branched PEG has dramatically increased plasma half-life while retaining the native catalytic activity. The human topoisomerase I-mediated DNA relaxation reaction was studied following modification of the enzyme at the active site tyrosine 723. The relaxation activities of the modified topoisomerase were observed after incorporation of series of unnatural tyrosine analogues into the active site of the enzyme. Modifications involving replacing of the nucleophilic tyrosine OH group with NH2, SH, or I eliminated DNA relaxation activity [24].

Studies of natural glycoproteins and their recombinants have shown that the oligosaccharide structures and site-occupancy of any glycosylated polypeptides are predetermined by the protein conformation. Some human tissue-specific terminal carbohydrate motifs are not synthesized by other mammalian host cells, so transfection of these hosts with genes encoding terminal human glycosyltransferases allows obtaining products with human tissue-specific glycosylation. Using site-directed mutagenesis, unglycosylated polypeptides have been successfully converted to N- and/or O-glycoproteins by glycosylation domains transfer from donor glycoproteins to different regions of the acceptor proteins. The genetic engineering by a rational design of glycoproteins and of host cell lines has thus provided a versatile tool to obtain therapeutic glyco-products with novel/improved properties [25].

The cytokine erythropoietin controls the production of red blood cells as an apoptosis inhibitor. Maintaining high levels of erythropoietin (Epo) activity is crucial when treating anemic patients. The modification of human Epo with mutations His32Gly, Cys33Pro, Trp88Cys, and Pro90Ala leads to the rearrangement of the disulfide bonding pattern. The resulting Fc-Epo (NDS) fusion protein containing those mutations has significantly improved properties. It is secreted almost exclusively as a dimer, is relatively stable, has improved pharmacokinetic and enhances red blood cell production [26]. Such results indicated that rearrangement of the disulfide bonding pattern in a therapeutic protein can have a significant effect on pharmacokinetics and, potentially, the dosing schedule of a protein drug.

Recombinant technologies to engineer ordinary monoclonal antibodies (MAbs) may cause loss of antibody affinity, increased tendency to aggregate, increased temperature sensitivity, and low yield of active protein. The well-characterized MAb H7 raised against placental alkaline phosphatase (PLAP) has been engineered by site-directed mutagenesis of the VL and the VH fragment [27]. It was found that a linker length of 25/30 amino acids leads to improved solubility but additional disulfide bonds does not affect stability substantially. The yield of the antibody in *E. coli* was improved approximately 10-fold by such modifications. Genetically engineered disulfide bonds in B-domain-deleted factor (F) VIII variants (C662-C1828 FVIII and C664-C1826 FVIII) were found to improve FVIIIa stability by blocking A2 domain dissociation because the new disulfide bond links covalently the A2 and A3 domains in FVIIIa [27,28]. The mutants reportedly had physiologically relevant, superior clot-forming properties in a whole blood environment, most likely due to their increased half-life.

17.2.2 DNA Shuffling and Directed Evolution

Proteins have an enormous potential to evolve in such a manner as to perform novel molecular functions. By recombination of their structural elements, new properties emerge that are optimized by molecular evolution under the pressure of natural selection. The redesign of proteins for desired functions is a promising emerging area in biotechnology. Combinatorial protein chemistry can mimic natural evolution through recombination of genetic materials followed by selection of gene products, but can also add new dimensions by chemical semi-synthesis, including nonbiological building blocks.

In vitro recombination has been used to generate novel sequences in a process known as DNA shuffling technique that allows accelerated and directed protein evolution for desired properties *in vitro*, by recombining and evolving genes to rapidly obtain molecules with improved biological activity and fitness. It is achieved by DNaseI treatment and PCR, and the procedure has been significantly developed to the point that today it is a simplified low cost protocol [29]. In contrast to point mutation techniques, like the site-directed mutagenesis described above, DNA shuffling exchanges large functional domains of sequences to search for the best candidate molecule, thus mimicking and accelerating the process of sexual recombination. DNA shuffling has been used to increase the antiviral and Th1-inducing activity, while simultaneously decreasing the antiproliferative activity of interferons (IFNs). The obtained hybrids of IFN-α generated an activity profile that may result in an improved therapeutic index and, thus, better clinical efficacy for the treatment of chronic viral diseases such as those caused by hepatitis B and C viruses, human papilloma virus, and HIV [30].

The fragmentation-based method for DNA shuffling is labor intensive. Another approach of recombination-dependent PCR (RD-PCR) is an easy to perform fragmentation-free shuffling protocol. A test system to compare and reveal biases of both protocols has been developed and showed that both approaches performed similarly with slight advantages for RD-PCR. When applied to homologous genes of varying DNA identities, however, the RD-PCR has a less pronounced bias of the crossovers in regions with high sequence identity. Template variations, including engineered terminal truncations, exert significant influence on the position of the crossovers in the RD-PCR. DNA shuffling can produce higher crossover numbers, while the RD-PCR frequently results in one crossover. Both methods produce counterproductive by-products (associated with parental sequences and overrepresented chimeras) and RD-PCR alone yields chimeras even when the templates have low homology [31].

The directed evolution [32,33] technology resembles natural evolution as it relies on the natural ability of nucleic acid polymerases to proofread reliably, i.e., maintain the correctness of the genetic information. On the other hand, some enzymes, e.g., T7 RNA polymerase, have relaxed fidelity or substrate tolerance and may be used successfully to tailor an enzyme with given function to a new version that meets specific requirements. The principle of the directed evolution approach does not require prior knowledge of the enzymes' structure or mechanism of action. The process of directed evolution involves multiple iterations of mutation, amplification, and selection. A polymerase with 20-fold higher error rate has been designed for the purpose [34]. The selection step usually involves screening by a phage-display system.

Directed evolution has contributed to the substantial advances in the engineering of glycosidases and glycosyltransferases for the synthesis and degradation of glycan structures. Improvement of the thermostability of xylanase, creation of the first glycosynthase derived from an inverting glycosidase, and the emergence of a new class of modified glycosidases capable of efficiently synthesizing thioglycosidic linkages have been achieved by comprehensive random mutagenesis and directed evolution [35]. In the past decade, the methods of directed molecular evolution have proven revolutionary and constitute a powerful new strategy for improving the characteristics of enzymes in a targeted manner. By coupling various protocols for generating large variant libraries of genes, together with high-throughput screens that select for specific properties of an enzyme, such as thermostability, catalytic activity, and substrate specificity, it is possible to optimize biocatalysts for specific applications. Further work is now directed to broadening the range of usable screening that can be utilized in terms of reaction types (e.g., hydroxylation and carbon–carbon bond formation) and functional characteristics (e.g., enantio-selectivity and regio-selectivity) [36].

Directed evolution is being used increasingly in both industrial and academic laboratories to improve commercially important enzymes as the catalytic properties of many naturally occurring enzymes often need to be further tailored to meet the specific requirements of a given application. Directed evolution has been applied to alter various enzyme properties such as activity, selectivity, substrate specificity, stability, and solubility. Directed evolution is now considered key to enzyme engineering and biocatalysis [37,38]. More research in this evolving field is also focusing on the quality and comprehensiveness of library construction and analysis, and studies into evolutionary mechanisms, limitations, and consequences of the various methodologies are ongoing [39]. A critical review of methods for creating new functional proteins with altered specificity profiles and some practical case studies have recently been presented by Antikinen et al. [40].

17.2.3 Combinatorial Biosynthesis and Chimeric Enzymes

Combinatorial biosynthesis utilizes the enzymes from known natural biosynthetic pathways to create novel chemical structures that would otherwise be difficult to obtain. In both theory and practice, the number of combinations possible from different types of natural product pathways ranges widely. Enzymes that have been the most amenable to this technology synthesize polyketides, nonribosomal peptides, and hybrids of the two [4]. The modular polyketide synthases (PKSs) has been manipulated to produce several erythromycin analogs. Tools for manipulating and studying multifunctional enzymes have resulted in more rational and faster methods of engineering new chemotherapeutic agents from natural products [41].

An alternative way to construct discriminating substructures is the reassembly of common medicinal chemistry building blocks. Cross et al. have proposed an algorithm that can be designed to meet different objectives, e.g., to build features that discriminate for biological activity in a local structural neighborhood, to build scaffolds for R-group analysis, to make cluster signatures that determine a cluster membership, and to identify substructures that characterize major classes in a set of heterogeneous compounds. This algorithm was applied successfully to a dataset of 118 compounds with in vitro inhibition data against recombinant human protein tyrosine phosphatase 1B [42].

Combinatorial peptide libraries have been primarily employed in the screening of combinatorial antibody libraries. As a result of the relative ease of screening and identifications of products with desirable features [43], the application of phage-display libraries has extended further into the science of protein–receptor or protein–ligand interactions, granting the opportunity to design and create many novel proteins.

Current protein engineering methods are somewhat limited by lack of appropriate high throughput physical or computational tests that can accurately predict protein activity under conditions relevant to its final application. To address this issue, Liao et al. [44] developed a new approach to protein engineering that combines high throughput gene synthesis with machine learning-based design algorithms. The work was done on proteinase K, in which 24 amino acid substitutions were selected from alignments of homologous sequences. Fifty-nine specific variants containing different combinations of the selected substitutions have been designed to use as training set, synthesized and tested for their ability to hydrolyze a tetrapeptide substrate under restrictive conditions. The activity data coupled with the corresponding sequences were analyzed using machine-learning algorithms.

The results of the analysis were used to design and synthesize a new set of variants predicted to have increased activity over the training set. After two cycles of machine-learning analysis and variant design, 20-fold improved proteinase K variants were obtained.

Molecular engineering approaches for the generation and production of new antibiotics have been in development for about 25 years [45,46]. Advances in cloning and analysis of antibiotic gene clusters, engineering biosynthetic pathways in *E. coli*, transfer of engineered pathways from *E. coli* into *Streptomyces* expression hosts, and stable maintenance and expression of cloned genes have streamlined the process in recent years. Advances in understanding the mechanisms and substrate specificities during assembly by polyketide synthases, nonribosomal peptide synthetases, glycosyltransferases, and other enzymes have made molecular engineering design and outcomes more predictable. Complex molecular scaffolds not amenable to synthesis by medicinal chemistry (for example, vancomycin [Vancocin], daptomycin [Cubicin], and erythromycin) are now tractable by molecular engineering. Medicinal chemistry can further embellish the properties of engineered antibiotics, making the two disciplines complementary.

One of the potential approaches for the improvement of the existing therapeutic thrombolytic agents is the construction of chimeric plasminogen activators and of conjugates of plasminogen activators with MAbs. Tissue-type plasminogen activators' (tPA) mutants obtained by deletion or substitution of functional domains or of single amino acids have been found to result in markedly reduced clearances *in vivo*, but also reduced specific thrombolytic potencies. Mutants of single-chain urokinase-type plasminogen activators (scuPA) with improved thrombolytic potencies have not been reported. Chimeric molecules containing functional domains of both tPA and scuPA, however, have been demonstrated to retain intact enzymatic properties of scuPA, some of the fibrin affinity of tPA and unaltered or reduced thrombolytic potencies.

A chimeric enzyme (GST121) of the human α-glutathione *S*-transferases GST1-1 and GST2-2, which has improved catalytic efficiency and thermostability over its wild-type parent proteins, has been crystallized in a space group that is isomorphous with that reported for crystals of GST1-1 [47,48].

Chimeric, predominantly human/murine monoclonal high-affinity antibody (MAb 2E2) was created that inhibits the distribution of cocaine to the murine brain. Mice infused with 2E2 also produce a dramatic dose-dependent increase in plasma cocaine concentrations and a concomitant decrease in the brain cocaine concentrations introduced by an intravenous injection of cocaine HCl (0.56 mg kg^{-1}). These data support the concept of immunotherapy for drug abuse [49,50]. The class I hyaluronan synthase (HAS) is a unique enzyme that uses one catalytic domain to elongate two different monosaccharides. This enzyme has great biotechnology potential for the generation of functional carbohydrates for medicinal purposes, which has prompted the creation of a recombinant human/*E. coli* chimera containing the human catalytic domain expressed in the bacterium. The engineered enzyme retains polymerization properties [51]. Another very interesting example of developed chimera is that related to polyhydroxylkanoate synthase [52]. The chimera was constructed from the phaA gene of *Pseudomonas oleovoranse*. A large internal fragment was removed and replaced with random PCR fragments generated with primers designed to flank the missing region and environmental DNA as template. The recombinant clones were selected based on increased in vivo enzyme activity and up to threefold enhancement was achieved.

In contrast to this randomized approach, an example of deliberate chimera construct is the human endostatin engineering work. On the basis of the knowledge that there is higher aminopeptidase expression in blood vessel tumors and the asparagine–glycine–arginine (NGR) sequences home to the enzyme, human endostatin was modified by introduction of the NGR motif and expressed in yeast system. This modification improved tumor binding and decreased angiogenesis [53]. Impressive work has been done with transgenic animals in which the human lysozyme gene was introduced in dairy goats. In this proof-of-concept study, the chimeras expressed and secreted the enzyme in milk with its antibacterial activity and effect on the micro-intestinal flora preserved [54].

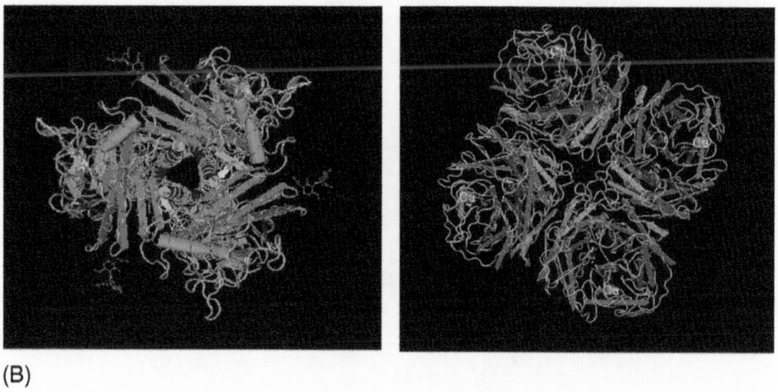

FIGURE 8.2 (B) Ribbon representations of three-dimensional structures of the HA trimer (left) and NA tetramer (right). HA (H5 serotype) shown with the ligands as seen on Cn3D image of PDB Id 2IBX at NCBI (From Yamada, S., et al., *Nature*, 444, 378, 2006.). NA (N1 serotype) monomers contain canonical six-bladed β-propeller structure; also shown are locations of bound ligands and Ca^{++} as seen on Cn3D image of PDB Id 2HTY at NCBI (From Russell, R.J., et al., *Nature*, 443, 45, 2006.).

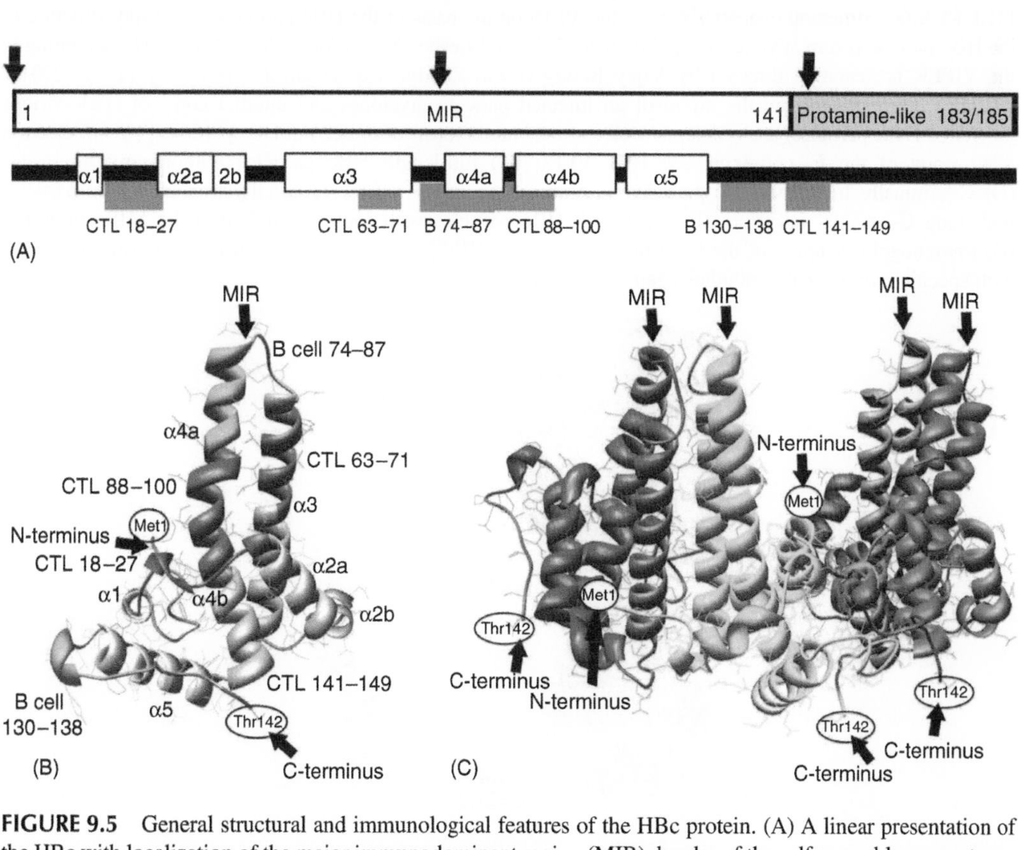

FIGURE 9.5 General structural and immunological features of the HBc protein. (A) A linear presentation of the HBc with localization of the major immunodominant region (MIR), border of the self-assembly competence before the protamine-like C-terminal region. Location of the α-helices and major immunodominant epitopes (B cell, blue; CTL, orange) is presented. For details of the immunological epitopes, see reviews of Refs. [174,175]. (B) A three-dimensional presentation of the HBc chain A according to the x-ray structure [13] with localization of α-helices, immunodominant epitopes (the same colors as in A). (C) Tetrameric asymmetric unit of the HBc particles. Chains are colored as follows: A, orange red; B, gold; C, green; D, blue. Major targets for foreign insertions are depicted by blue arrows on the linear and three-dimensional presentations of the HBc protein. When located on the forefont of the map, the N- and C-terminal amino acid residues are deciphered.

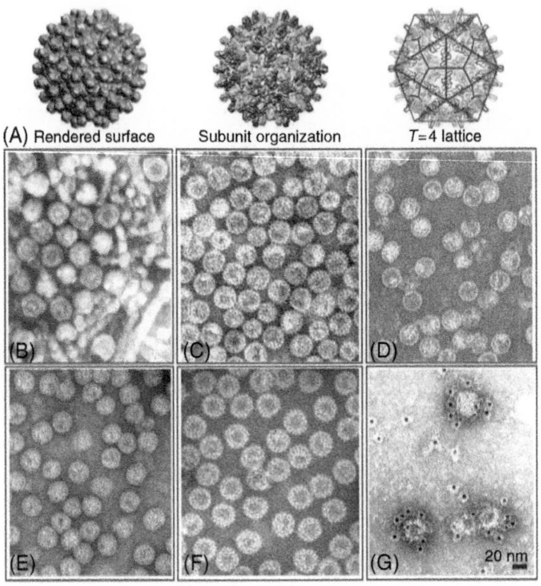

FIGURE 9.6 Structural characteristics of the VLPs on the basis of the HBc particles. (A) X-ray structure of the HBc particle (From Wynne, S.A., Crowther, R.A., and Leslie, A.G., *Mol. Cell*, 3, 771, 1999.) according to the VIPER presentation directed by Vijay Reddy (From Reddy, V.S. et al., *J. Virol.*, 75, 11943, 2001.). (B) HBV preparation from the blood of an infected patient, envelopes and internal cores of HBV virions (so-called Dane particles) are visible, as well as filamentous forms of the 22 nm HBs particles, (C–G), electron microscopy of the *E. coli*-expressed HBc VLPs: (C) full-length HBc particles without any insertions; (D) C-terminally truncated HBc particles without any insertions; (E) C-terminally truncated HBc particles with long C-terminal insertion; (F) full-length HBc particles with long internal (into the MIR) insertion; (G) immunogold mapping of the C-terminally truncated HBc particles with internal insertion (into the MIR) with specific monoclonal antibodies against inserted epitope.

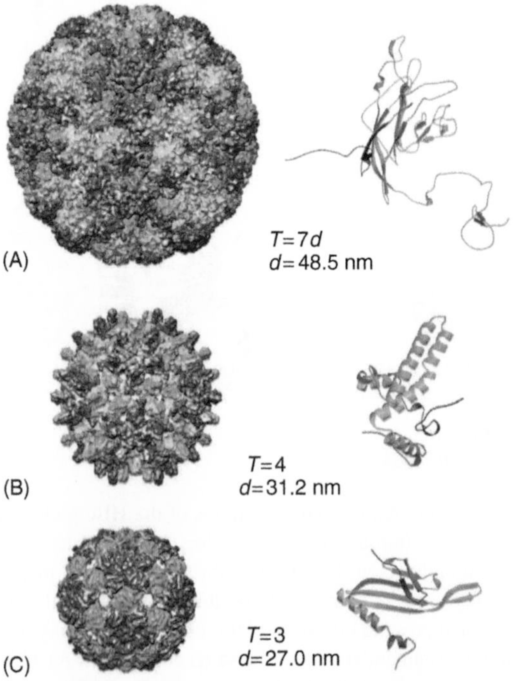

FIGURE 9.7 Comparative size and structure of the most popular VLP carriers: (A) Mouse polyomavirus; (B) HBc; and (C) RNA phage Qβ. Subunit organization (left) and subunit folds (right) taken from the VIPER program are presented (From Reddy, V.S. et al., *J. Virol.*, 75, 11943, 2001.). The VLP representations are scaled according to their relative sizes, while the subunits are not. The triangulation (*T*) numbers and average diameters of the VLPs are given in the middle.

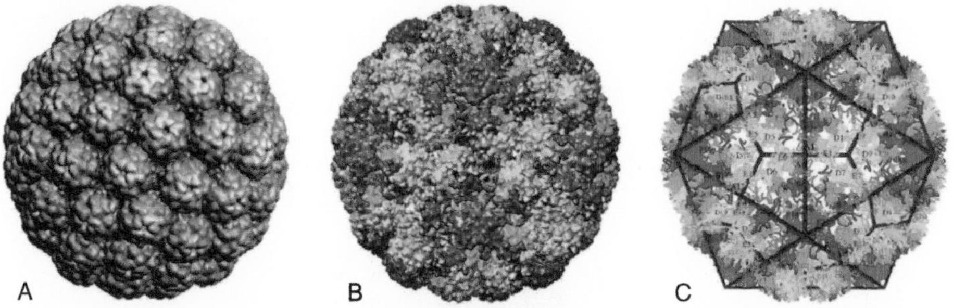

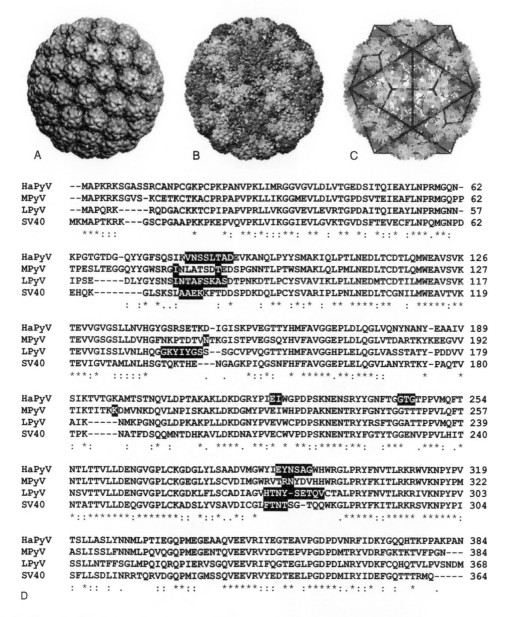

```
HaPyV   --MAPKRKSGASSRCANPCGKPCPKPANVPKLIMRGGVGVLDLVTGEDSITQIEAYLNPRMGQN-  62
MPyV    --MAPKRKSGVS-KCETKCTKACPRPAPVPKLLIKGGMEVLDLVTGPDSVTEIEAFLNPRMGQPP  62
LPyV    --MAPQRK-----RQDGACKKTCPIPAPVPRLLVKGGVEVLEVRTGPDAITQIEAYLNPRMGNN-  57
SV40    MKMAPTKRK---GSCPGAAPKKPKEPVQVPKLVIKGGIEVLGVKTGVDSFTEVECFLNPQMGNPD  62
          ***:::              *    *: **:*:::**: **  : **  *: *::* :***.**:

HaPyV   KPGTGTDG-QYYGFSQSIKVNSSLTADEVKANQLPYYSMAKIQLPTLNEDLTCDTLQMWEAVSVK 126
MPyV    TPESLTEGGQYYGWSRGINLATSDTEDSPGNNTLPTWSMAKLQLPMLNEDLTCDTLQMWEAVSVK 127
LPyV    IPSE-----DLYGYSNSINTAFSKASDTPNKDTLPCYSVAVIKLPLLNEDMTCDTILMWEAVSVK 117
SV40    EHQK--------GLSKSLAAEKKFTDDSPDKDQLPCYSVARIPLPNLNEDLTCGNILMWEAVTVK 119
                 : :* *..:       : *    : ** :*:* : ** ****:**  : *****:**

HaPyV   TEVVGVGSLLNVHGYGSRSETKD-IGISKPVEGTTYHMFAVGGEPLDLQGLVQNYNANY-EAAIV 189
MPyV    TEVVGSGSLLDVHGFNKPTDTVNTKGISTPVEGSQYHVFAVGGEPLDLQGLVTDARTKYKEEGVV 192
LPyV    TEVVGISSLVNLHQGGKYIYGSS--SGCVPVQGTTYHMFAVGGHPLELQGLVASSTATY-PDDVV 179
SV40    TEVIGVTAMLNLHSGTQKTHE---NGAGKPIQGSNFHFFAVGGEPLELQGVLANYRTKY-PAQTV 180
        ***:*  :::::*          .  .     *::*:  * *****.**:***::     *    *

HaPyV   SIKTVTGKAMTSTNQVLDPTAKAKLDKDGRYPIEIWGPDPSKNENSRYYGNFTGGTGTPPVMQFT 254
MPyV    TIKTITKKDMVNKDQVLNPISKAKLDKDGMYPVEIWHPDPAKNENTRYFGNYTGGTTTPPVLQFT 257
LPyV    AIK-----NMKPGNQGLDPKAKPLLDKDGNYPVEVWCPDPSKNENTRYYRSFTGGATTPPVMQFT 239
SV40    TPK-----NATFDSQQMNTDHKAVLDKDNAYPVECWVPDPSKNENTRYFGTYTGGENVPPVLHIT 240
        : *:    :   :* :     *. ****. **:* * **::****:**:  :***  .***: .*

HaPyV   NTLTTVLLDENGVGPLCKGDGLYLSAADVMGWYIEYNSAGWHWRGLPRYFNVTLRKRWVKNPYPV 319
MPyV    NTLTTVLLDENGVGPLCKGEGLYLSCVDIMGWRVTRNYDVHHWRGLPRYFKITLRKRWVKNPYPM 322
LPyV    NSVTTVLLDENGVGPLCKGDKLFLSCADIAGVHTNY-SETQVCTALPRYFNVTLRKRIVKNPYPV 303
SV40    NTATTVLLDEQGVGPLCKADSLYVSAVDICGLFTNTSG-TQQWKGLPRYFKITLRKRSVKNPYPI 304
        *::*******:*******:: *::*..*: *           .*****::***** ******:

HaPyV   TSLLASLYNNMLPTIEGQPMEGEAAQVEEVRIYEGTEAVPGDPDVNRFIDKYGQQHTKPPAKPAN 384
MPyV    ASLISSLFNNMLPQVQGQPMEGENTQVEEVRVYDGTEPVPGDPDMTRYVDRFGKTKTVFPGN--- 384
LPyV    SSLLNTFFSGLMPQIQRQPIERVSGQVEEVRIFQGTEGLPGDPDLNRYVDKFCQHQTVLPVSNDM 368
SV40    SFLLSDLINRRTQRVDGQPMIGMSSQVEEVRVYEDTEELPGDPDMIRYIDEFGQTTTRMQ----- 364
        : *:: : .     :: **::    ******::: ** :*****:.*::* : :  *    .
D
```

FIGURE 10.4 Three-dimensional (x-ray) structure of MPyV and amino acid sequence alignment of VP1 proteins of HaPyV and other polyomaviruses demonstrating the sites used for foreign insertions. (A) Rendered surface, (B) subunit organization, and (C) $T = 7d$ lattice of the MPyV according to Stehle et al., 1996. The pictures were adapted from the VIPER site (Shepherd et al., 2006). (D) Alignment of the VP1 amino acid sequences of HaPyV (Acc. No. CAA06802), MPyV (Acc. No. P49302), LPyV (Acc. No. P04010), and SV-40 (Acc. No. AAK29047) generated by ClustalW (Thompson et al., 1994). Sites used for foreign insertions according to the data presented in Table 10.2 are labeled by black boxes.

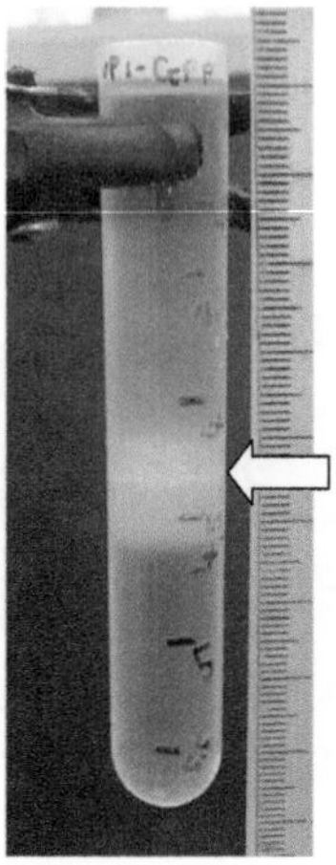

B

FIGURE 10.5 (B) Picture of the cesium chloride density gradient showing a fluorescent band representing VP1/eGFP VLPs.

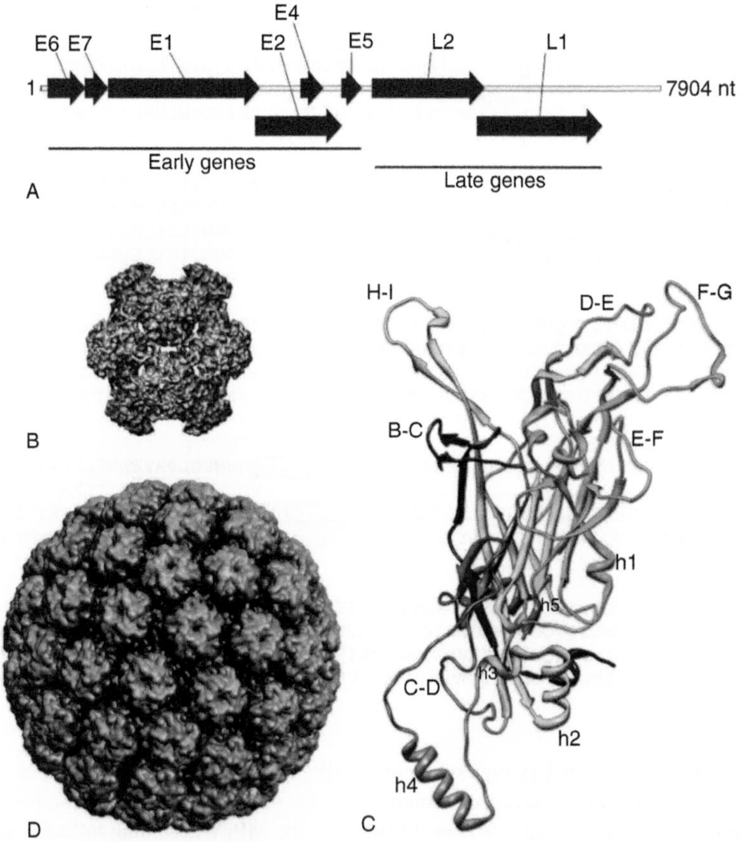

FIGURE 11.1 The PV genome map (A) and the main structural features (B–D). (A) Schematic presentation of the HPV16 (GenBank accession No AF536179) genome organization. The respective genes are marked by arrows and scaled according to their relative sizes. (B) The structures of the HPV-16 small $T = 1$ VLPs and (C) L1 monomer fold are according to Ref. [7]. The hypervariable loops and α-helixes in (C) are marked according to Ref. [7], and individual domains are visualized using UCSF Chimera software package [157]. (D) Atomic model of the PV capsid according to Ref. [8]. The particle images in (B) and (D) are scaled according to their relative sizes, while the subunit (C) not. Images B–D are adapted from the VIPER database. (From Reddy, V.S. et al., *J. Virol.*, 75, 11943, 2001.)

The reading frame selection of recombinant nucleic acids is another feature with important implications for protein engineering. The solubility of the gene product can limit the selection of the gene of interest. The pInSALect vector provides strict reading frame selection without concomitant selection for protein solubility or folding. This plasmid has been inserted to the *Saccharomyces* cis-splicing VMA intein sequence to facilitate the posttranslational self-excision of the protein of interest. Two libraries of chimeric glycinamide ribonucleotide formyltransferases have been developed and they confirm the superior performance of the vector over other reading frame selection systems [55].

Engineered protein-based sensors of ligand binding have emerged as attractive tools for the discovery of therapeutic compounds through simpler screening systems. Chimeric enzymes comprised of nuclear hormone receptors ligand-binding domains and a sensitive thymidylate synthase reporter act as simple sensors that can screen for the presence of hormone-like compounds through changes in bacterial growth. This mechanism of ligand dependence is not specific to nuclear hormone receptors; thus this protein engineering strategy could be applicable to the construction of simple sensors for different classes of (therapeutic) binding proteins [56].

Another interesting example of chimeric protein application is the construction of allergy multivaccine by random in vitro recombination of T- and B-cell epitopes of a family of related allergens (14 genes of the Bet v 1 family order Fagales) [57]. Two chimeric proteins selected from the library of shuffled clones displayed low allergenicity and high immunogenicity, and thus generated molecules suitable for specific immunotherapy not only against birch pollen allergy but also against allergies caused by other cross-reactive tree pollens.

17.2.4 Monoclonal Antibodies

Antibodies can be generated against almost any substance and they have the unique ability to bind with high specificity to this substance. Monoclonal antibodies (MAbs) are homogeneous identical antibody molecules produced by a single parental B-cell clone presented with one single antigen. The technology to generate MAbs, developed in the mid 1970s [58,59] has greatly advanced the science fields of both cancer treatment and molecular diagnostics. Herceptin, a humanized MAb for breast cancer treatment, was designed to target HER2 + cancer cells. Herceptin attaches to a HER2 + cancer cell, signals the body's defense system to target the cancer cell and may also interfere with the development of HER2 + cancer cell itself by stopping its proliferation. Herceptin became the first drug designed by a bimolecular engineering approach and was approved by the FDA in 1995.

Mesothelin is a cell surface protein highly expressed in some malignant tumors and is suspected to have a role in the metastases of ovarian cancer, and hence is a promising target for immunotherapy [60]. Existing mesothelin MAb had shown certain limitations for the study of expression of native mesothelin because of their low affinity or reactivity only with denatured mesothelin protein. Novel MAbs to mesothelin have been induced by immunizing mesothelin-deficient mice with plasmid cDNA encoding mesothelin, and boosted with a mesothelin-rabbit IgG Fc fusion protein prior to cell fusion. Hybridomas producing anti-mesothelin antibodies were then established by ELISA screening and shown to react with two epitopes on mesothelin. One of the epitopes was the same originally used to identify mesothelin and the other was a new epitope. The newly established and improved MAbs can be used for the detection of mesothelin by immunohistochemistry, fluorescence-activated cell sorting, ELISA, and Western blotting.

Recombinant technologies that engineer MAbs to single-chain fragment variable (scFv) may cause loss of antibody affinity, increased tendency to aggregate, increased temperature sensitivity, and low yield of active protein. MAb H7 raised against placental alkaline phosphatase (PLAP) is a well-described MAb model that has been engineered to improve solubility and stability of scFv with retained high affinity. This single-chain antibody was obtained by specific site-directed mutations and introducing a linker of 30 amino acids. The yield of the antibody was improved approximately

10-fold by these modifications and it could be efficiently expressed in a bacterial system using the PET-32a TrxA vector [27].

Exiting novel area in the antibody research is the generation of supercatalytic antibodies [61]. They have the unique ability to decompose the target molecule rather than just sequestering it. By immunizing ground-state proteins or peptides, supercatalytic antibodies possessing serine protease-like characteristics were produced that destroy the HIV-1 envelope protein gp41, chemokine receptor CCR5 peptide, and *Helicobacter pylori* urease, etc. Some of them can degrade antigens at high catalytic reaction rates and the catalytic function resides in its light chain. Germline analysis shows that supercatalytic antibodies are generated from some discrete germlines in which at least one catalytic triad composed of three amino acid residues, namely, Asp1, Ser27a, and His93, is encoded. Such antibody light chains can be referred to as antigenase (antigen-decomposing enzyme) and may have arisen during the evolution of antibodies to acquire a higher ability than that of enzymes for developing a better self-defense system for survival [62,63].

Current treatments for cocaine addiction are not effective. The development of a catalytic MAb has provided a strategy not only for binding, but also degrading cocaine. MAbs that hydrolyze the benzoate ester of cocaine were obtained using a well-designed hapten linker essential for the correct and specific catalysis programming [49]. Recently, and for first time, the production of MAb against beta human chorionic gonadotropin (hCG) with proteolytic activity was reported. MAb 7D9 was raised in Balb/C mice using purified hCG. The epitope for this antibody is in the C-terminal of β chain as suggested by the absence of cross-reactivity with other glycoprotein hormones. This MAb, however, has autodegradation characteristics and a high level of hydrolytic activity (casein and gelatin) that consequently explains its instability [64]. An extensive list of therapeutic MAbs is found in Table 17.1.

TABLE 17.1
Enzymes and Other Recombinant Proteins under Investigation or Approved for Therapy

Proteins	Product	Application	Source	Company
Hormones				
Insulin	Humulin, Humalog (lispro), NovoLog (aspart)	Diabetes	*E. coli*	Eli Lilly
	Novolin, Levemir (rh insulin detemir)		*S. cerevisiae*	Novo Nordisk
	Lantus (-glargin), Apidra (insulin glulisine)		*E. coli*	Aventis
hGH[a]	Protropin (somatrem), Nutropin (somatropin)	hGH deficiency	*E. coli*	Genentech
	Humatrope (somatropin)	hGH deficiency	*E. coli*	Eli Lilly
	Genotropin (somatropin)	hGH deficiency	*E. coli*	Pfizer/Pharmacia
	Norditropin (somatropin)	hGH deficiency	*E. coli*	Novo Nordisk
	Bio-tropin (somatropin)	hGH deficiency	*E. coli*	Biotechnology General
	Serostim (somatropin)	AIDS wasting	Mouse C127	Serono
Growth hormone analogue	Somavert (pegvisomant)	Acromegaly	*E. coli*	Pfizer/Pharmacia
Follicle-stimulating hormone	Follistim (follitropin-β)	Infertility	CHO[b]	Organon
	Gonal-F (follitropin-α)	Infertility	CHO	Serono
rh chorionic gonadotropin	Ovidrel (choriogonadotropin α)	Infertility	CHO	Serono
Glucagon	GlucaGen (glucagon)	Hypoglycemia	*S. cerevisiae*	Novo Nordisk
	Glucagon	Hypoglycemia	*E. coli*	Eli Lilly
rh[c] B-type Natriuretic Peptide	Natrecor (nesiritide)	Heart failure	*E. coli*	Scios
Parathyroid hormone	Forteo (teriparatide)	Osteoporosis	*E. coli*	Eli Lilly

TABLE 17.1 (continued)
Enzymes and Other Recombinant Proteins under Investigation or Approved for Therapy

Proteins	Product	Application	Source	Company
Enzymes				
Human tPA[d]	Activase (alteplase), TNKase (tenecteplase)	Thrombolysis	CHO	Genentech
	Retavase (reteplase)	Thrombolysis	*E. coli*	Centocor/J&J
Deoxyribonuclease	Pulmozyme (dornase α)	Cystic fibrosis	CHO	Genentech
β-Glucocerebrosidase	Ceredase (aglucerase)	Gaucher's disease type I	CHO	Genzyme
	Cerezyme (imiglucerase)	Gaucher's disease type I, II, and III	CHO	Genzyme
	Lysodase (PEG[e]-glucoserebrosidase)	Gaucher's disease type I, II, and III		National Institute of Mental Health
Urate oxidase	Elitek (rasburicase)	Ureate management	*S. cerevisiae*	Sanofi-Synthelabo
	Fasturtec	Chemo-induced hyperuricemia		Sanofi-Synthelabo
PEG-uricase	Puricase	Hyperuricemia/Gout		Bio Technology General
α-L-Iduronidase	Aldurazyme (laronidase)	Mucopolysaccharoidosis I	CHO	Genzyme/BioMarin
Sarcosidase	Sucraid	Sucrase-isomaltase deficiency		Orphan Medical
Human-α-galactosidase A	Fabrazyme Agalsidase β	Fabry disease	CHO	Genzyme
	Replagal (α-galactosidase A)	Fabry disease		Transkaryotic Therapies
	Fabrase (α-galactosidase A)	Fabry disease		Mount Sinai School of Medicine
	Human α-galactosidase A	Fabry disease	Plant	Large Scale Biology
rh α-glucosidase precursor	Pompase	Pompe's disease, GSD II[f]		Pharming/Genzyme
	Myozyme	Pompe's disease, GSD II		Genzyme
Rh highly phosphorylated acid α-glucosidase	TBD	Pompe's disease, GSD II		Novazyme Pharmaceuticals
Pegademase bovine	Adagen	ADA in SCID patients		Enzon
Pegaspargase	Oncaspar	Acute lymphocytic leukemia		Enzon
L-asparaginase		Leukemia	Erwinia	Lyphomed
	Erwinase		Erwinia	Porton International
Blood Factors				
Factor VIII	Recombinate	Hemophilia	CHO	Baxter Healthcare
	Kogenate	Hemophilia	BHK[g]	Bayer
	ReFacto (moroctocog-α)	Hemophilia	CHO	Genetics Institute/ Wyeth
	Advate	Hemophilia	CHO	Baxter Healthcare
Factor VIIa	NovoSeven	Hemophilia	BHK	Nov Nordisk
Factor IX	BeneFIX	Hemophilia	CHO	Genetics Institute/ Wyeth
Human-activated protein C	Xigris (drotrecogin α)	Sepsis	Mammalian	Eli Lilly

(continued)

TABLE 17.1 (continued)
Enzymes and Other Recombinant Proteins under Investigation or Approved for Therapy

Proteins	Product	Application	Source	Company
Cytokines/Antagonists/Growth Factors				
IFN-α-2b	Intron A,	Cancer, hepatitis	*E. coli*	Schering-Plough
IFN-α-2a	Roferon A	Cancer, hepatitis	*E. coli*	Hoffman-La Roche
IFN consensus	Infergen (consensus IFN)	Hepatitis C	*E. coli*	InterMune
	PEG-Intron	Hepatitis C	*E. coli*	Schering-Plough
PEG-IFN a-2a	Pegasys	Hepatitis C	*E. coli*	Hoffman-La Roche
IFN g-1b	Actimmune	Chronic granulomatosis, osteoporosis	*E. coli*	InterMune
IFN b-1b	Betaseron	Multiple sclerosis	*E. coli*	Bertex/Chiron
IFN b-1a	Avonex	Multiple sclerosis	CHO	Biogen Idec
	Rebif	Multiple sclerosis	CHO	Serono
Epoetin a (EPO)	Epogen	Anemia	CHO	Amgen
	Procrit	Anemia	CHO	Ortho Biotech/J&J
G-CSF[d]	Neupogen (filgastim)	Neutropenia	*E. coli*	Amgen
	Neulasta (pegfilgrastin), PEG-filgrastin	Neutropenia	*E. coli*	Amgen
G/M-CSF[e]	Leukine (sargramostin)	Cancer/bone marrow transplantation	*S. cerevisiae*	Berlex
Interleukin 2-IL2	Proleukin (aldesleukin)	Renal carcinoma	*E. coli*	Chiron
	Ontak (IL2-difteria toxin fusion protein)	T-cell lymphoma	*E. coli*	Ligand/Seragen
Interleukin 11-IL11	Neumega (oprelvekin)	Thrombocytopenia (chemotherapy)	*E. coli*	Genetics Institute/Wyeth
IL 1 receptor antagonist	Kineret (anakinra)	Rheumatoid arthritis	*E. coli*	Amgen
Platelet growth factor	Regranex (becaplermin)	Diabetic ulcers	*S. cerevisiae*	Ortho-McNeil/J&J
EPO analog	Aranesp (Darbepoetin-alfa)	Anemia	CHO	Amgen
Bone morphogenic protein 7	Osteogenic protein 1	Bone repair	CHO	Stryker Biotech
Bone morphogenic protein 2	InFuse (Dibotermin alfa)		CHO	Medronic Sofamor Danek
MAbs/ Fusion Proteins				
CD3[h]	Orthoclone OKT3 (Muromonab)	Transplant rejection	HCL[i]	Ortho Bioteck/J&J
IL2 receptor[h]	Zenapax (daclizumab)	Transplant rejection	NS0[j]	Hoffman La Roche
	Simulect (basiliximab)	Transplant rejection	NS0	Novartis
CD20[h]	Rituxan (rituximab)	Non-Hodgkin's lymphoma	CHO	Genentech/Biogen Idec
	Zevalin (conjugate with Yttrium90)	Non-Hodgkin's lymphoma	CHO	Biogen Idec
	Bexxar (conjugate with Iodine 131)	Non-Hodgkin's lymphoma		GSK/Corixa
CD33[h]	Mylotarg (calicheamicin cytotoxin conj.)	Acute myeloid leukemia	NS0	Wyeth Pharmaceuticals
CD52[h]	Campath (alemtuzumab)	Chronic lymphocytic leukemia	CHO	Berlex/Ilex Oncology
CD11a[h]	Raptiva (efalizumab)	Psoriasis	CHO	Genentech/Xoma

TABLE 17.1 (continued)
Enzymes and Other Recombinant Proteins under Investigation or Approved for Therapy

Proteins	Product	Application	Source	Company
CD15[h]	NeutroSpec (fanolesomab)	Appendicitis imaging	HCL	Pallatin Technologies/ Mallinckrodt
Platelet GPIIb/ IIIa receptor[h]	ReoPro (abciximab)	Cardiovascular disease	Sp2/0[k]	Eli Lilly/Centocor/J&J
Respiratory Syncytial Virus[h]	Synagis (palivizumab)	RSV infection	NS0	Medimmune
TNF[g,h]	Remicade (infliximab)	Chrons disease/ rheumatoid arthritis	Sp2/0	Centocor/J&J
TNF-receptor/Ig G1 Fc fusion[h]	Enbrel (etanercept)	Rheumatoid arthritis	CHO	Amgen/Wyeth
TNF-alfa[h]	Humira (adalimumab)	Rheumatoid arthritis	CHO	Abbot Laboratories
HER2 receptor[h]	Herceptin (trastuzumab)	Breast cancer	CHO	Genentech
LF[l]Ag-3/IgG fusion[h]	Amevive (alefacept)	Psoriasis	CHO	Genentech/Xoma
IgE[h]	Xolair (omalizumab)	Asthma	CHO	Genentech/Novartis/ Tanox
EG[m] factor receptor[h]	Erbitux (cetuximab)	Colorectal cancer	MMCL[n]	ImClone/Bristol-Myers Squibb
VEG[o] factor[h]	Avastin (bavacizumab)	Colorectal cancer	CHO	Genentech
Leukocyte integrins[h]	Tysabri (natalizumab)	Multiple sclerosis relapse	MMCL	Biogen Idec/Elan

Source: From Vellard, M., *Curr. Opin. Biotechnol.*, 14, 444, 2003; Birch, J.R. and Onakunle, Y., in *Therapeutic Proteins: Methods and Protocols*, Humana press, Totowa, NJ, 2005, 1–16; Walsh, G., *Trends Biotechnol.*, 23(11), 553, 2005.

[a] hGH, human growth hormone.
[b] CHO, Chinese hamster ovary cell line.
[c] hr, human recombinant.
[d] tPA, tissue-type plasminogen activator.
[e] PEG, polyethilenglycol.
[f] GSDII,Glycogen storage disease type II.
[g] BHK, hamster kidney cell line.
[h] – antibody target.
[i] NS0, BALB/c myeloma murine cell line.
[j] HCL, human cancer cell line.
[k] SP2/0, BALB/c myeloma hybrid cell line.
[l] LF, leukocyte function associated.
[m] EG, epidermal growth factor.
[n] MMCL, melanoma cell line.
[o] VEG, vascular endothelial growth factor.

17.2.5 Expression and Delivery

The most common expression system for recombinant proteins is *E. coli* [65]. Examples abound in the literature and are highlighted in Table 17.1. Other expression systems include yeasts (*Saccharomyces* and *Pichia*), baculovirus [66,67], infection of insect cell lines [68], and retroviral expression in animal cell lines [69].

The ability to stably deliver recombinant proteins to the systemic circulation would facilitate the treatment of a variety of acquired and inherited diseases. Many delivery vehicles have been explored including engineered myoblasts [70,71], erythrocytes [72], subviral particles [73], liposomes [74–76], microencapsulation [77–79], etc. Some of their applications are illustrated below.

To explore the feasibility of the use of genetically engineered myoblasts as a recombinant protein delivery system, stable transfectants of the murine C2C12 myoblast cell line were produced that synthesize and secrete high levels of human growth hormone (hGH) in vitro. They were injected with hGH-transfected myoblasts in mice and remained stable for at least 3 weeks. Histological examination of the injected muscles showed that many of the injected cells had fused to form multinucleated myotubes. Such experimental data shows that genetically engineered myoblasts have the potential to be used for the stable delivery of recombinant proteins into the circulation [70].

Subviral dense bodies (DB) of human cytomegalovirus (a β-herpesvirus) can transport viral proteins directly into target cells by membrane fusion [80]. These non-infectious particles provide a candidate delivery system for the prophylactic and therapeutic application of proteins. DB can be modified genetically. The proof-of-principle design involved a 55 kDa fusion protein consisting of the green fluorescent protein and the neomycin phosphotransferase. This fusion could be packed in and delivered into cells by recombinant DB in a functional fashion. Efficient transfer of protein into fibroblasts and dendritic cells has been reported [73], indicating that direct protein delivery is an another emerging technology in vaccine development and gene therapy.

A very interesting work was the encapsulation of recombinant human erythropoietin (rHuEpo) in human and mouse red blood cells (RBCs) that was achieved by a hypotonic dialysis-isotonic resealing procedure. The hypoosmotic resistance of carrier erythrocytes was studied by osmotic fragility measurements. Encapsulation yielded more human than mouse RBCs with cell recovery of around 70%. Carrier-RBCs exhibited a tendency to spherocytic morphology, and showed the typical higher hypoosmotic resistance than normal RBCs. The presence of rHuEpo inside carrier RBCs was identified by radioimmunoassay. This work suggests that carrier RBC-preparations may serve as an alternative sustained cell delivery system for the *in vivo* administration of rHuEp.

Similar strategy has been investigated for nucleoside kinases from several species to serve as suicide genes for treatment of malignant tumors by combined gene- and chemo-therapy. For their delivery a new protein–lipid complexes strategy has been developed. It does not entail delivery of the genetic material but the nucleoside kinase proteins are directly provided to the cells. A mix of a trifluoroacetylated lipopolyamine and dioleoyl phosphatidylethanolamine is used to form the protein–lipid complexes containing either recombinant herpes simplex virus type-1 thymidine kinase or *Drosophila melanogaster* multisubstrate deoxyribonucleoside kinase. The nucleoside kinase containing protein–lipid complexes is imported into human osteosarcoma and Chinese hamster ovary cell lines by endocytosis and by that the enzymes are delivered to the cytosol and the nucleus. The enzymatic activity is retained, and the treated cells acquire increased sensitivity to nucleoside analogues. Such results demonstrate that direct delivery of suicide gene proteins to cells may be an alternative approach to conventional suicide gene therapy strategies [74]. Research of non-parenteral protein delivery has been done on several routes: transdermal, nasal, buccal, and oral, but the best results have been demonstrated via pulmonary delivery [81,82]. Exubera is a powder-formulated insulin administered via pulmonary route that had completed phase III clinical trials [83].The selective delivery of human recombinant IFN-γ to isolated rat hepatocytes was also studied with liposomes (asialofetuin (AF)-labeled). Experimental delivery resulted in virtually no IFN-γ internalized into hepatocytes, whereas AF-liposomes containing IFN-γ were taken up to a significant degree. This liposome preparation method provided useful means for the encapsulation of unstable macromolecules into AF-liposomes and they were found to carry effectively IFN-γ into hepatocytes *in vitro* [76]. Phospholipid nanosomes are smaller, uniform liposomes manufactured utilizing supercritical fluid technologies. They can encapsulate hydrophilic molecules such as proteins and nucleic acids. Hydrophobic therapeutics are co-solvated with phospholipid raw materials in supercritical fluids that, when decompressed, form phospholipid nanosomes encapsulating these drugs in their lipid bilayers. Phospholipid nanosomes are projected to find utility in the enhanced delivery of hydrophilic drugs, such as recombinant proteins and nucleic acid, as well as hydrophobic anticancer and anti-HIV drugs [75].

The sleeping beauty (SB) transposon is a natural nonviral gene transfer system that can mediate long-term transgene expression. A recombinant transposon carrying human hIDO gene was delivered with a construct containing overactive transposase [81,84]. The transfection, aided by the cationic polymer polyethylamine, resulted in uniform transgene expression in rat model without noticeable toxicity. The enzyme (indoleamine-2,3-dioxygeanse), that has both T cell–suppressive and antioxidant properties, sustained therapeutic response in the rat system, and, inhibited fibroblast proliferation *in vitro*.

17.2.6 Screening, Metagenomes, and Posttranslational Modifications

Are new strategies for modification and improvement needed, or it will be better to invest in improving screening approaches and take advantage of the diversity that nature already offers? The economy with which nature uses genes presents an important scientific paradox. Research shows that in all studied higher animals far fewer genes than expected were found. The functional outputs are, however, far more complex than would be expected from the genomic blueprint and one of the reasons is that most proteins are modulated by posttranslational modifications.

A popular, older approach to protein modifications and improvement has been the generation and screening of mutant libraries from which protein variants with valuable properties are isolated. The strategy can involve either rational mutation of targeted sites with or without stochastic recombinations of evolving "quasi-species." Peptide aptamers are powerful chemical genetic tools for the dissection of biological networks, but their application to *in vivo* systems has been limited by targeting peptides to a specific site on a single target protein. A new approach to this challenge is targeting combinatorial peptide aptamers to a protein-binding site using a small-molecule binding-partner or "Trojan horse." The validity of the approach has been tested for peptide-based inhibitors for *Plasmodium falciparum* myristoyl-CoA-*N*-myristoyl transferase (PfNMT). The inhibitor was selected from a one-bead, one-compound library using a high-throughput on-bead screening methodology, targeted to the active site of NMT with a myristate (C14:0 fatty acid) substrate analogue. The screening approach has been further refined by developing a scintillation-proximity assay that allows easy automation and high throughput screening [85].

The phage-display system of combinatorial peptide libraries is being exploited extensively to design and create many novel proteins, owing to its relative ease of screening and identifying desirable proteins. Its application has extended from screening of combinatorial antibody libraries to the investigation of protein–receptor and protein–ligand interactions. There is no one efficient library screening method that can be recommended because the success of the process is tightly linked to the nature of the properties of interest and the availability of robust reporter system for those specific properties. More modern approaches to screening involve computational and mathematical analyses. The initial molecular procedures involve recursive isolation of improved mutants followed by mutagenesis (molecular breeding) to tailor proteins to a particular function. The sought-after properties of the emerging proteins are then studied and clusters of mutants with optimal evolutionary potential are identified in multi-dimensional factor space by means of multivariate analysis [86]. Most recent approaches to screening involve *in silico* "prescreening" by various methods like genetic algorithms [87,88], machine learning [44], multivariate modeling [89], and protein library design [90] based on the results of the upstream steps.

Biocatalysts are widely accepted as useful alternative tools to classic organic synthetic techniques for regio- and enantio-selective synthesis. The development of techniques for the rational or evolutionary design of new or modified enzyme exemplifies the need for fast and reliable methods for catalyst identification of powerful catalysts. Various optical, separation, NMR, and mass spectroscopy methods exist for the screening of biocatalysts [91]. Quantitative matrix-assisted laser desorption/ionization-time-of-flight mass spectrometry (MALDI-ToF MS) can be used as a tool for screening of sugar-converting enzymes and has been applied successfully on pyranose oxidase variants [91]. Directed evolution is an extremely useful method for biocatalyst improvement on the molecular level in combination with high-throughput screening systems. Methods for accessing "nonculturable" biodiversity using metagenome approaches have brought on great progress to sequence-based biocatalyst discovery. New carriers and tools for immobilization of enzymes have also been developed [92].

The metagenomes of complex microbial communities are rich sources of novel biocatalysts and enzymes [7,81,93]. The exploitation of the metagenome of a mixed microbial population has led, in one study alone, to isolation of more than 15 different genes encoding novel biocatalysts by using a combined cultivation and direct cloning strategy [9]. A microbial consortium (by 16S rRNA analysis) comprised of bacteria closely related to the genera *Pseudomonas*, *Agrobacterium*, *Xanthomonas*, *Microbulbifer*, and *Janthinobacterium* was used for the construction of DNA libraries from total genomic DNA and those were functionally searched for novel enzymes of biotechnological value [94]. The search identified novel agarase genes (mostly organized in gene clusters), stereoselective amidase genes (amiA), cellulases (gnuB and uvs080), an α-amylase (amyA), a 1,4-α-glucan branching enzyme (amyB), and pectate lyases (pelA and uvs119) and lipase genes. The development of novel cultivation-dependent and molecular cultivation-independent approaches has been very gratifying in the area of product recovery from soil microorganisms. These strategies have resulted in the isolation of novel biocatalysts and bioactive molecules like agarase, amidases, DNases, lipases, glycerol dehydratase, indirubin, polyketide synthase, terragine, violacein, etc. [95]. Esterase EstA3 derived from a drinking water metagenome, and esterase EstCE1 derived from a soil metagenome show similarity to β-lactamases (family VIII of the lipases/esterases). The enzymes display remarkable characteristics like high activity, enantio-selectiveness, activity against a wide range of substrates and stability over a wide temperature and pH range that makes them highly useful for biotechnological applications [96]. A novel variant of direct expression cloning was explored with metagenomic DNA from compost. DNA isolated by a modified direct lysis method was purified by size-exclusion chromatography and cloned directly using a dual-orientation expression vector that allows bidirectional transcription. The functional diversity of the constructed metagenomic expression library was screened by different approaches based on functional heterologous expression leading to the identification of large number of active clones, including lipolytic enzymes, amylases, phosphatases, and dioxygenases. The newly found lipolytic enzymes exhibited only limited similarity to known enzymes. This new approach provides an efficient way for the rapid generation of large libraries of hitherto unknown enzyme candidates which could be screened for different specific target reactions [97].

The majority of protein-based biopharmaceuticals approved or currently undergoing clinical trials bear some form of posttranslational modification (PTM). As the genomic blueprint is not unlimited, posttranslational modifications create a next level of protein structural and functional variety. PTM can profoundly affect protein properties relevant to their therapeutic application. The relationship between structure and function is understood for many PTMs but remains incomplete for others, especially more complex ones such as glycosylation. Glycosylation is the most common and most diverse modification [98], but carboxylation, hydroxylation, sulfation, and amidation are involved in many products. A better understanding of structural–functional relationships can facilitate the development of second-generation products displaying a PTM profile engineered to optimize therapeutic usefulness [99]. Natural protein production methods typically produce mixtures of PTM products whose functions are difficult to characterize and control. A chemical tagging approach that enables the attachment of multiple modifications to bacterially expressed (bare) protein scaffolds has been proposed to allow reconstitution of functionally effective mimics of higher organism PTMs. A LacZ reporter enzyme scaffold, with appropriate modifications at suitable distances, has been used to create sensitive protein probes for detection of mammalian brain inflammation and disease. Combining targeted chemical control of PTM with readily available protein scaffolds provides a systematic platform for creating probes of protein–PTM interactions [100].

Chemical "mutation" of the serine protease subtilisin was accomplished in the 1960s by chemical conversion of the catalytic triad serine residue of subtilisin to a cysteine [101]. Interest in the chemical modification of enzymes to enhance and modify functionality reemerged in the

1980s with studies like the covalent linking of flavin to papain that resulted in the conversion of a protease into an oxido-reductase [102]. Covalent chemical modification for tailoring proteins and enzymes is a powerful complementary approach to mutagenesis. Glutaraldehyde crosslinking of enzyme crystals generates easily recoverable insoluble biocatalysts and polyethylene glycol. Modification of enzyme surface amino groups (PEGylation) increases solubility in organic solvents. Such chemical modification has been exploited for the incorporation of cofactors onto protein templates and for atom replacement to generate new functionality, such as the conversion of a hydrolase into a peroxidase. Despite the breadth of applicability of chemically modified enzymes, a major difficulty that has previously impeded their implementation is the lack of chemical- or regional-specificity of the modifications. This lack of specificity creates heterogeneous and irreproducible product mixtures with inconsistent performance where the modification of choice is underrepresented. This challenge has been addressed by the introduction of a unique position for modification method of site-directed mutagenesis [103]. The basic idea of this method is to create proteins by site-directed mutagenesis that carry specific amino acid residue (e.g., cysteine) at the site of interest. This residue is then thioalkylated by methanethiosulfonate that reacts very specifically with sulfhydryls. This residue can subsequently be modified to introduce a side chain in a highly specific manner.

17.3 APPLICATIONS

17.3.1 Enzyme Replacement Therapy for Treatment of Inheritable Diseases

Lysosomal storage disorders (LSDs) are monogenic inborn errors of metabolism. Various disease groups have been delineated according to the affected pathway and the accumulated substrate, and new entities continue to be identified. They cause severe disorders with a heterogeneous clinical spectrum and high morbidity and mortality. Most of the genes encoding the lysosomal enzymes have been cloned, and animal models have been obtained for almost each disease. Gaucher, Fabri, and Pompe diseases are the most predominant inheritable conditions involving defects of the lysosomal storage and the available recombinant products for their treatment are listed in Table 17.1.

17.3.1.1 Glucocerebrosidase/Gaucher Disease

Gaucher disease is a result of inborn error of glycosphingolipid metabolism and is the most frequent in the lysosomal storage diseases group [104–106]. Various point mutations, deletions, and insertions within the glucocerebrosidase (acid β-glucosidase, EC 3.2.1.45) locus at chromosome 1q21 result in a deficiency of this lysosomal enzyme [107,108]. The accumulation of its substrate, glucocerebroside (glucosylceramide), in cells of monocyte–macrophage lineage leads to the visceral manifestations of anemia, thrombocytopenia, hepatosplenomegaly, and skeletal disease. Gaucher disease was the first disorder to be treated by enzyme replacement therapy (ERT). The successful development of the enzyme replacement approach was possible because of substantial achievements in number of fundamental areas: phenotype definition, genetic cause, biology of the disease, enzyme structure and purification, and how to effectively deliver the enzyme by specific drug-targeting mechanism to tissue macrophages [109,110]. The decisive step in the development of enzyme replacement therapy was the discovery of receptors for glycoproteins and the purification of glucocerebrosidase. Lysosomal endocytic activity is the basis for the original hypothesis that lysosomes could take up exogenous enzyme and thereby correct the disease. Thereafter, multiple approaches to ERT were employed [111]. Mannose binds to the mannose receptor on Kupffer cells and on hepatocytes, hence glucocerebrosidase can gain entry into both hepatocytes and Kupffer cells, but the delivery to Kupffer cells is much better for mannose-modified glucocerebrosidase. As it turned out, the mannose moieties in the oligosaccharides were the key to successful enzyme replacement therapy. Ceredase (enzyme purified from placenta–alglucerase) is a commercial form of glucocerebrosidase that has been modified for targeting mannose receptor sites on macrophages and other cells. A theoretical limitation to its use is the remote possibility of infective contaminants in the preparation from human placenta. A practical limitation has been the finite availability of acceptable placentae. Heterologous expression enzyme production of human complementary DNA (cDNA) for glucocerebrosidase in eukaryotic cells eliminates both limitations [109–112]. Gaucher disease has become a prototype genetic disease for the development of prenatal diagnosis [113], genotype–phenotype correlations [114,115], and effective therapy. On the basis of the clear efficacy of targeted enzyme therapy [116] studies have established that regular infusions lead to regression of the clinical manifestations. The efficacy of recombinant glucocerebrosidase was tested in a randomized, double-blind, parallel trial with Ceredase and the human enzyme produced in Chinese hamster ovary (CHO) cells [106].

17.3.1.2 α-Galactosidase A/Fabri Disease

Fabry disease is an X-linked inborn error of glycosphingolipid metabolism resulting from mutations in the α-galactosidase A (α-Gal A) gene [117]. The long-term safety and efficacy of ERT have been well documented [118]. Two differently produced recombinant α-galactosidase A (α-gal A) preparations have been used independently. Agalsidase α is obtained from human fibroblasts that have been modified by gene activation and agalsidase β is obtained from CHO cells transduced with human α-gal A cDNA. It has been suspected that α-gal A mRNA undergoes editing, which may result in coproduction of an edited protein, that might have a relevant physiological function, but no indications of that have been found at the protein or RNA level. Both recombinant enzymes used in therapy are unedited and are capable of functionally correcting cultured fibroblasts from patients in their excessive globotriaosylceramide accumulation [119]. The emergence of antibodies with in vivo neutralizing capacities is frequently encountered in treated Fabry disease patients. They are completely cross-reactive so it is unlikely that switching from one to the other recombinant protein prevents the immune response and related effects [120]. The two products have been tested in pivotal trials [121] for several years. ERT with agalsidase for 30–36 months has been demonstrated to decrease plasma GL-3 levels, to promote sustained endothelial GL-3 clearance, stable kidney function, and have a favorable safety profile [122]. Potential issues of this ERT are that not all patients are suitable for treatment and some organs and tissues are corrected more readily than others. Problems with gauging efficacy are also present as these disorders are highly variable and the therapies can be cost prohibitive [10].

17.3.1.3 Acid-α-Glucosidase/Pompe Disease

Pompe disease, also known as glycogen storage disease type II (GSD-II), is a lethal, autosomal recessive metabolic myopathy caused by a lack of acid-α-glucosidase (GAA) activity in the cardiac and skeletal muscles which results in massive amounts of glycogen accumulation in muscles. It has been shown in a mouse model of GSD-II that infection of the liver with a modified adenovirus (Ad) vector encoding human GAA (hGAA) results in long-term persistence (for 6 months) of the vector in liver tissues. The persistence correlates with long-term correction of pathologic intramuscular glycogen accumulations in all muscle groups tested. Such gene therapy strategies may have the potential to significantly improve the clinical course for GSD-II patients [123]. Recombinant human (rh) GAA has been purified from genetically engineered CHO cells overproducing GAA. The safety and efficacy of the rhGAA enzyme therapy has been tested as a phase I/II, single-dose rhGAA infusions intravenously twice weekly in small number of infants with infantile GSD-II. The results of more than 250 infusions showed that rhGAA is generally well tolerated. This first study showed that rhGAA is capable of improving cardiac and skeletal muscle functions in infantile GSD-II patients [124]. Further reports describe clinical studies on the safety and efficacy of the ERT. Other infants with rhGAA have received recombinant human α-glucosidase from transgenic rabbit milk. This product was also well tolerated, and was reported to reach the primary target tissues [125,126]. Following these promising studies and to enhance the delivery of rhGAA to affected muscles, the carbohydrate moieties on the enzyme have been remodeled to exhibit a high-affinity ligand for the CI-MPR (cation-independent mannose 6-phosphate receptor). This has been achieved by chemically conjugating six synthetic oligosaccharide ligands to each rhGAA enzyme. The resulting modified enzyme had near-normal specific activity in mouse models and significantly increased affinity for the CI-MPR with unaltered binding to the mannose receptor. Uptake studies showed that the modified enzyme was internalized 20-fold more efficiently. The reductions in tissue glycogen levels in mice were realized using eightfold lower dose in the heart and diaphragm, and fourfold lower dose in the skeletal muscles. Such results show that enzyme remodeling to improve its affinity is a feasible approach to enhance enzyme efficacy [71].

17.3.1.4 Cystic Fibrosis

The cause of cystic fibrosis (CF) has been traced to gene mutations that generate a defective cystic fibrosis transmembrane conductance regulator (CFTR) protein. This defect does not allow epithelial cells to regulate the way chloride passes across cell membranes, thereby disrupting the balance needed to maintain a normal mucosal coating inside the lungs, pancreas, and passage ways in other organs. The prognosis of CF has dramatically improved owing to optimization of symptomatic treatment with antibiotics, nutritional support, bronchial drainage by physiotherapy and aerosols.

Viscoelastic secretions in cystic fibrosis impair mucus clearance and increase bacterial persistence within the lung. In addition to actin, the presence of high molecular weight DNA originating from inflammatory cells contributes to the thickness of airway secretions. Actin can bind to DNA-rich fibers and potently inhibit the enzymatic activity of rhDNase. The ability of recombinant human DNase I (DNase I) to degrade DNA to lower molecular weight fragments is the basis for its therapeutic use in CF patients and its potential use as a treatment for systemic lupus erythematosus (SLE). A recombinant human DNase (rhDNAse) has become available recently for clinical use (Table 17.1). When aerosolized in the patients, it facilitates bronchial drainage and clearing of the airways [127]. RhDNase reduces the viscosity of respiratory mucus, improves pulmonary function, and therefore reduces the risk of respiratory tract infectious exacerbations [128]. A phase II double blind placebo controlled study showed that rhDNase improves pulmonary function in CF patients by reduction in the proportion of high-molecular-weight DNA [129].

To increase the potency of human DNase I, three classes of mutants have been generated and characterized: hyperactive variants, with one to six additional positively charged residues that digest DNA much more efficiently than wild type; G-actin-resistant variants; and combination variants that are both hyperactive and G-actin-resistant. Combinations of these three types of mutants yield 20- to 30,000-fold increased activity over wild type for large DNA size and concentration and also for digesting lower concentrations of DNA complexed to anti-DNA antibodies in human serum and very low concentration of endogenous DNA. It has been observed that the actin resistance property of the combination mutants further enhances the degree of potency in human serum. These engineered variants have greater therapeutic potential for treatment of both CF and SLE [130] and similar in vitro studies supported the conclusion that the actin-resistance variant improves the DNase activity [131].

To investigate DNases with improved properties and efficient production of the recombinant, a molecular fusion of human DNase I with the hinge and Fc region of human IgG1 heavy chain has been constructed. This fusion protein was expressed and secreted by Sf9 insect cells infected with the recombinant baculovirus. The fusion protein can be purified from the culture medium by affinity chromatography and gel filtration and was shown to contain DNase I, and an intact antibody Fc region that retained their specific activity [132].

Nebulized rhDNAseI has been successfully used for the treatment of infants with respiratory insufficiency caused by respiratory syncytial virus (RSV) and bronchiolitis that also causes abundance of thickened mucus due to the presence of highly polymerized DNA. Such studies have been done by two independent groups [133,134] and one of them was done on a large scale in randomized, double-blind, placebo-controlled investigation of 75 patients with RSV bronchiolitis. Both studies recorded significant improvement of outcomes.

An innovative delivery method has been tested in a double-blind, dose-increase gene transfer in subjects with CF. A vector consisting of single molecules of plasmid DNA carrying the CFTR gene is compacted into DNA nanoparticles, using polyethylene glycol-substituted 30-mer lysine peptides. Placebo or the compacted DNA was superfused onto the inferior turbinate of the nostrils. Corrections persisting for as long as 6 days could be observed after the gene transfer [135,136].

CF also causes exocrine pancreatic insufficiency, leading to malabsorption. Supplemental pancreatic enzyme therapy can alleviate malnutrition experienced by CF patients. To investigate the safety and efficacy of supplemental pancreatic enzyme therapy in CF, two different drug concentrations (Ultrase MT12 and Ultrase MT20) were compared to placebo in two separate safety and efficacy studies. Excellent fat and protein absorption was achieved with minimal adverse events and safe doses, further supporting the idea that the use of enzymes to treat pancreatic insufficiency is beneficial in CF [135,137].

17.3.1.5 Phenylalanine Conversion/Phenylketonuria

Phenylketonuria (PKU) is an autosomal recessive genetic disease caused by the defects in the phenylalanine hydroxylase (PAH) gene. Defective PAH alleles result in increased levels of plasma Phe leading to increased risk of neuronal damage [13]. It has become clear that blood Phe concentration needs to be controlled for the life of the patient, therefore alternative approaches to therapy, other than dietary, are being actively pursued [138].

Characteristic aspects of PAH enzymatic activity are determined by a single gene and can be constituted in the absence of any specific accessory functions of the cell [139]. A full-length human PAH cDNA has been recombined with a prokaryotic expression vector and introduced into *E. coli*. The transformed bacteria expressed phenylalanine hydroxylase immunoreactive protein that catalyzed pterin-dependent conversion of Phe to Tyr. This rhPAH has been partially purified, and biochemical studies of the activity and kinetics of the recombinant enzyme have demonstrated that the rhPAH is an authentic and complete phenylalanine hydroxylase enzyme. Much progress has been made in ERT for PKU. Other than PAH, phenylalanine ammonia-lyase (PAL) can also be used for ERT. PAL converts Phe to nontoxic t-cinnamic acid. Despite difficulties in developing stable and functional forms of both enzymes, there has been marked success in producing their PEG-modified forms [138]. Recombinant PAL originating from parsley has been chemically conjugated with activated PEG2. It has shown greatly enhanced stability in the circulation and has been effective in reducing the plasma Phe concentration in mice [13].

17.3.1.6 Insulin/Diabetes

Insulin, produced by the pancreatic Langerhans' Islets, is a 51 amino acid hormone that is responsible for the uptake of glucose in the liver and the muscle and its storage as glycogen, and for the fat cells to convert blood lipids into triglycerides. Insulin deficiency causes diabetes which if left unchecked can lead to severe complications (renal failure, cardiovascular failure, blindness, gangrene, etc.) in the affected person.

Insulin lispro, a recombinant insulin analog, is identical to human insulin except for the transposition of Pro and Lys at positions 28 and 29, located at the C-terminus of the B chain. This change causes reduced capacity for self-association in solution; hence the analog absorbs more rapidly than regular human insulin from subcutaneous sites. Insulin lispro has a quicker onset and a shorter duration of glucose-lowering activity. These pharmacological properties are the basis for comparative clinical trials of subcutaneous insulin lispro and subcutaneous human regular insulin in patients with type 1 (insulin-dependent) and type 2 (noninsulin-dependent) diabetes mellitus. Clinical experience supports findings from early trials that had showed that insulin lispro offers important advantages over regular insulin [140]. Lantus is insulin analog whose molecular structure has been modified to result in a 24 h long therapeutic effect. It has been approved for use in children with diabetes [141]. Other insulin analogs, such as Humalog, are engineered to provide a very rapid effect. A list of available insulin recombinants is found in Table 17.1.

17.3.1.7 Recombinant Enzymes for Other Diseases

Tay-Sachs disease is a fatal autosomal lipid storage disorder in which harmful quantities of ganglioside GM2 build up in the brain. The condition is caused by insufficient activity of the β-hexosaminidase A (HexA) that catalyzes the biodegradation of acidic fatty materials. HexA is a heterodimeric glycoprotein composed of α- and β-subunits active in lysosomes. There is no current treatment available for the condition. To prepare a large amount of HexA for ERT, recombinant HexA has been produced in the methylotrophic yeast *Ogataea minuta* instead of mammalian cells [142]. The problem of antigenicity due to differences in N-glycan structures between mammalian and yeast glycoproteins has been resolved by using och1δ yeast (mannose outer-chain-elongation deficient). Genes encoding the α- and β-subunits of HexA were integrated into the yeast cell, and the heterodimer was expressed together with its isozymes. The enzyme preparation was incorporated dose dependently into fibroblasts from patients with GM2 gangliosidosis via the M6P receptors on the cell surface, and degradation of accumulated GM2 ganglioside has been observed.

Autism is a developmental disease usually manifesting within the first 3 years of life. To date, the cause is unknown, and so options for available treatment options are limited. Enzyme therapy for the digestion of neuroactive peptides known as exorphins, suspected to underlie its pathogenesis, combined with dietary enzyme therapy has been considered. A novel enzyme formula, ENZYMAID, has been found to be beneficial and safe [143].

Horseradish peroxidase (HP) is an important heme-containing enzyme with a known three-dimensional structure and catalytic intermediates, mechanisms of catalysis and function of the specific amino acid residues. The most widely studied peroxidase is isoform C from horseradish roots (*Armoracia rusticana*) mainly due to its many diverse uses in biotechnology. Site-directed mutagenesis has been used extensively with HP to investigate its structure and function and to develop engineered enzymes for practical applications. A combination of horseradish peroxidase and indole-3-acetic acid or its derivatives is currently being evaluated as an agent for use in targeted cancer therapies [144]. A tryptophanless mutant of HP (W117F) has been constructed and expressed in *E. coli*. The mutation affects the enzyme's folding and stability. The mutant was found to be more stable in acid media, during the reaction course and toward irradiation. The effect of hydrogen peroxide pretreatment on radiation-induced inactivation of the wild-type and mutant enzyme has indirectly indicated participation of Trp-117 in electron transfer pathways. These observations are in agreement with the enzyme's steady-state kinetic data [145]. An engineered horseradish peroxidase isozyme C (HRP C) gene has been constructed by the addition of a 6xArg fusion tail to 6xHis–HRP C by PCR strategy. The results have provided evidence that the poly-Arg tag is more effective than a standard poly-His tag for peroxidase purification from the baculovirus expression system [146]. Peroxidases are prone to suicide inactivation by the H_2O_2 substrate. This becomes particularly problematic in high-value applications such as diagnostics, biosensors, and in wastewater treatment; hence improvement of HRP stability would further increase its potential application range. The only attempt to date of genetic manipulation of HRP to improve its peroxide stability by using directed evolution was reported by Fagain et al. Multiple single- and double-mutants and one pentuple-mutant of solvent exposed, proximal lysine and glutamic acid residues have been analyzed for enhanced H_2O_2 stability. Mutants (K232N, K241F and T110V) showed significantly increased H_2O_2 tolerances of 25- (T110V), 18- (K232N), and 12-fold (K241F). This improved stability was speculated to be due to an altered enzyme-H_2O_2 catalysis pathway or to removal of potentially oxidizable residues [147].

Lysosomal acid lipase (LAL) is essential for the hydrolysis of triglycerides (TG) and cholesteryl esters (CE) in lysosomes. A lal −/− mouse model has been used to demonstrate that massive storage of TG and CE is observed in the liver, adrenal glands, and small intestine. Lal−/− mice lack white adipose tissue and brown adipose tissue is progressively lost. Their plasma free fatty acid levels are significantly changed and so is their energy intake [148,149]. In humans LAL deficiency produces two phenotypes, Wolman disease (WD) and cholesteryl ester storage disease (CESD). The potential for enzyme therapy has been tested in lal−/− mice. Mannose terminated human LAL expressed in *Pichia pastoris* has been purified and administered by tail vein injections. The treated mice had demonstrated nearly complete resolution of the discoloration and reduction of hepatic weight and the histology had shown reductions in macrophage lipid storage. TG and cholesterol levels have been decreased by approximately 50% in liver, 69% in spleen, and 50% in small intestine. These studies suggest that LAL enzyme therapy in human may also be feasible [148].

Clinical problems associated with nucleotide metabolism in humans are predominantly the result of abnormal catabolism of the purines. The clinical consequences range from mild to severe and even fatal disorders. Relevant diseases are gout (hypoxanthine–guanine phosphoribosyltransferase deficiency), Lesh–Nyhan syndrome (lack of hypoxanthine–guanine phosphoribosyltransferase), severe combined immunodeficiency disease (SCID) (lack of adenosine deaminase), immunodeficiency (absence of purine nucleotide phosphorylase), and von Gierke's disease (glucose-6-phosphatase deficiency). Clinical manifestations of abnormal purine catabolism arise from the insolubility of the degradation byproduct, uric acid. Deficiency in the enzyme adenosine deaminase, a purine catabolic enzyme that manages levels of the biologically active purines in tissues and cells (ADA), in humans manifests as severe lymphopenia and immunodeficiency, resulting in death by 6 months of age, if untreated. A mouse model in which levels of ADA can be biochemically and genetically manipulated has provided possibilities for uncovering the mechanisms by which this purine catabolic enzyme affects lymphopoiesis. ADA-deficient mice develop a combined immunodeficiency as well as severe pulmonary insufficiency. ADA enzyme therapy has been used in two protocols. The "low-dose" protocol prevents the pulmonary phenotype but does not improve the immune status. The "high-dose" protocol results in decreased metabolic disturbances in the thymus and spleen and improves the immune status. The findings in this model suggest that the pulmonary and immune phenotypes are separable and are related to the severity of metabolic disturbances in the corresponding tissues and that it can be used to study the efficacy of ADA enzyme therapy [150,151]. Adenosine phosphorylase (AP) has been shown to protect human CEM cells in culture from the toxic effects of 2′-deoxyadenosine. This purine nucleoside phosphorylase has been engineered to accept 6 amino-substituted purine nucleosides by two active site substitutions, Asn243Asp; Lys244Gln. A recombinant AP conjugated to branched polyethylene glycol (PEG) polymers has been reported to retain greater than 90% of the native catalytic activity and a 67-fold increased plasma half-life. PEG-AP therapy in PNP-deficient mice resulted in a 2.7-fold increase in urine urate. Such studies provide evidence for consideration of PEG-AP as an alternative enzyme therapy for the inherited deficiency of adenosine deaminase [23]. Abbreviated purine nucleoside phosphorylase (PNP) genes were engineered to determine the effect of their introns on human PNP gene expression. This work found that first two introns or all five PNP introns resulted in substantial human PNP isozyme expression after transient transfection of murine NIH 3T3 cells and construct of PNP minigene devoid of introns resulted in undetectable human PNP activity [152]. The research was further developed with the construction of retroviral vectors for the delivery of PNP genes [153]. Purine nucleoside phosphorylase has been engineered to accept 6 amino-substituted purine nucleosides by two active site substitutions, Asn243Asp; Lys244Gln [154]. Adenosine phosphorylase (AP) conjugated to branched polyethylene glycol (PEG) polymers has been demonstrated to retain greater than 90% of the native catalytic activity. Administration of the enzyme to mice demonstrated that it was principally confined to the plasma and it had increased plasma half-life, minimal activity detected in tissues, and no activity in urine. PEG-AP was also shown to protect human CEM cells in culture from the toxic effects of 2′-deoxyadenosine. These studies provide evidence for consideration of PEG-AP as an alternative enzyme therapy for the inherited deficiency of adenosine deaminase [23].

Glucose oxidase (GOX) is a hydrogen peroxide-generating flavoprotein catalyzing the oxidation of β-D-glucose to D-glucono-1,5-lactone. GOX is used primarily in the food industry but also in biosensors for the quantitative determination of D-glucose in samples such as blood, food, and fermentation products. The enzyme has been purified from both *Aspergillus niger* and *Penicillium* spp. A problem with utilizing GOX from its native sources is the presence of impurities such as catalase, cellulase, and amylase, which may impair some of its applications. To overcome these difficulties and simplify purification *A. niger* GOX has been cloned and expressed in *Saccharomyces cerevisiae* as a highly glycosylated form [155].

The D-amino acid oxidase holoenzyme form of *Rhodotorula gracilis*, reacts only to a limited extent with general thiol reagents but the monomeric apoprotein is completely inactivated and denatured by them. Cys208 has been identified as the only cysteine to react with the flavin analogue [156]. The replacement of histidine 307 with leucine in pig kidney D-amino acid oxidase perturbs its active site conformation accompanied by dramatic losses in protein–flavin interactions and enzymatic activity. However, the negative effect of this mutation on the holoenzyme structure was shown to be essentially eliminated in the presence of glycerol, resulting in up to 50% activity recovery and greater than 16-fold increase in the flavin affinity. Further studies had shown that glycerol assists in the rearrangement of the protein toward its holoenzyme-like conformation together with reduction in the solvent-accessible protein hydrophobic area [157]. Artificial flavinylation was used by the same research group to replace a native noncovalent flavin prosthetic group with a covalently attached flavin analogue in recombinant human D-amino acid oxidase. The residue 281(Gly) was replaced with Cys by site-directed mutagenesis, followed by reaction between the mutated apoenzyme and the thiol-reactive flavin analogue, 8-methylsulfonyl FAD. The stoichiometric process of flavin attachment was reported to be accompanied by gain in enzymatic activity, reaching up to 26% activity of the recombinant native enzyme [158].

The enzyme arylamine *N*-acetyltransferase (NAT2) catalyses the N-acetylation of primary arylamine in hydrazine drugs and chemicals. N-Acetylation is subject to polymorphism and many genetic variants at the polymorphic NAT2 locus have been described. Humans can be categorized as either fast or slow acetylators according to their ability to N-acetylate certain arylamine substrates *in vivo*. Five of the most common human NAT2 variants have been expressed in *E. coli* as a convenient source and their apparent K_m values for different substrates determined. The data show that the K_m of the slow variant NAT2 7B for the arylamine sulfamethazine is 10-fold lower than all the other allotypes. The apparent K_m for a structurally related sulfone antibiotic, dapsone, was fivefold lower for the slow variant NAT2 7B when compared with the wild-type NAT2 4. These results indicate that the NAT2 7B specific amino acid substitution, Gly286-Glu, is important in promoting the binding of sulfamethazine and dapsone to the active site [159].

17.3.2 Repair Therapy

17.3.2.1 Thrombolytic Agents and Plasminogen Activators

The therapeutic efficacy of thrombolytic therapy for the treatment of acute myocardial infarction was first demonstrated convincingly in 1985 by GISSI investigators. The high incidence of thromboembolic diseases has long sustained the search for new agents which are able to stimulate the natural fibrinolytic system. The optimal thrombolytic drug is sought to deliver rapid, complete, and sustained recovery of the infarct artery, be safe (low incidence of severe bleedings), easy to administer (e.g., bolus application), and cost effective. Attempts to improve thrombolytic treatment include the search for better fibrinolytic agents and more effective adjunctive therapies. Until two decades ago, only two plasminogen activators were available for clinical use: a bacterial streptokinase and human urokinase were the first-generation antithrombotic agents. They were characterized by limited thrombolytic potencies and major side effects including systemic fibrinogen breakdown, bleeds, and stroke. These molecules lacked specificity for the fibrin clot, so recombinant

DNA technology has been used to produce agents with higher fibrin clot selectivity such as t-PA (tissue-type plasminogen activator) and scu-PA (single) [66,67].

In addition to the use of adjunctive agents such as inhibitors of platelet aggregation and thrombin, new genetically engineered "second-generation" thrombolytic agents have been developed that offer the promise of improved clinical outcomes [160]. Potential approaches to improve thrombolytic agents include the construction of mutants and variants of t-PA, of scu-PA, of chimeric plasminogen activators, and of conjugates of plasminogen activators with MAbs. Mutants of tPA with altered pharmacokinetic or functional properties have been constructed, including ones that have enhanced binding to and improved stimulation by fibrin and also possess resistance to plasmin and to protease inhibitors. The t-PA mutants obtained by deletion/substitution of functional domains or of single amino acids, have shown markedly reduced clearances but also reduced specific thrombolytic potencies. A chimera consisting of amino acids 87–274 of t-PA and amino acids 138–411 of scu-PA, with negligible fibrin affinity, has a 10-fold higher thrombolytic potency than scu-PA in animal models of venous thrombosis, as a result of a delayed in vivo clearance and a relatively maintained specific thrombolytic activity. Plasminogen activators conjugated with antifibrin or antiplatelet monoclonal antibodies, either chemically or by recombinant DNA technology, when targeted to blood clots, result in a 5- to 10-fold increased thrombolytic potency. Thus, it is possible to develop plasminogen activators with improved thrombolytic potency. An acylated, recombinant plasminogen/t-PA hybrid has sufficiently slow clearance to be administered by bolus and is more potent and fibrin selective than t-PA in vivo [48,161,162]. Further improved plasminogen activator with enhanced thrombolytic properties is the bifunctional protein, HSL-2, which exhibits both a plasminogen-activating and an anticoagulative activity. The chimeric protein is comprised of four elements: a derivative of thrombin inhibitor hirudin, a 6 amino acid spacer, the sequence of plasminogen-activator staphylokinase (Sak), and a 13 amino acid expression tag at the C-terminus [163].

Early intervention with thrombolytic agents unequivocally reduces mortality after acute myocardial infarction. As highlighted above the agents used initially had disadvantages such as short half-life, immunogenicity, hypotension, and bleeding complications. To address this, a hybrid plasminogen/t-PA cDNA has been constructed and expressed in CHO cells. The chimera protein, comprising the fibrin-binding domains of plasminogen covalently linked to the catalytic domain of t-PA, has been purified and evaluated in vitro and in vivo. The hybrid was inhibited rapidly by α 2-antiplasmin. Acylation with inverse acylating agent, 4′-amidinophenyl-4-chloroanthranilic acid (AP-CLAN) generates a CLAN-plasminogen/t-PA hybrid. Upon evaluation of such protected acylated chimera in guinea pig model the plasminogen was shown to have sufficiently slow clearance to be administered by bolus and is more potent and fibrin selective than t-PA *in vivo* [162].

Over 10 years ago a lot of promise was seen in the recombinant production of naturally occurring plasminogen activators as well as the development of "designer drugs," created by altering the natural t-PA and scu-PA. Recombinant production of naturally occurring plasminogen activators seems at least as promising as the production of designer drugs. Recombinant hirudin has been found to reduce re-occlusions and re-infarctions after t-PA thrombolysis in dose-finding studies. t-PA itself contains four domains with different functional properties and from a variety of mutants and variants of t-PA with altered fibrin-affinity, half-life, or fibrin specificity one recombinant plasminogen activator (r-PA) had undergone clinical investigation. Chimeric plasminogen activators (consisting of parts of t-PA and the single-chain urokinase-type plasminogen activator (scu-PA)) do not significantly improve at the same time fibrin specificity and thrombolytic potency of the natural occurring molecules. Complexes of plasminogen activators and monoclonal antibodies against platelets or fibrin improved the specificity and thrombolytic activity of the plasminogen activators, however, these molecules are potentially antigenic and costly [48].

Several lines of research towards improvement of thrombolytic agents have been explored including the construction of mutants and variants of plasminogen activators (PAs), chimeric PAs, conjugates of PAs with monoclonal antibodies, and PAs from animal or bacterial origin. Some of those have shown promise in animal models. Good candidates have been numerous of t-PA mutants with prolonged *in vivo* half-life or resistance to protease inhibitors, and chimeric PAs consisting of different regions of t-PA and of urokinase-type PA (u-PA). Vampire bat PA is 85% homologous to human t-PA but lacks kringle 2 and the plasmin-sensitive cleavage site. Several molecular forms of the thrombolytic substance in the saliva of the vampire bat have been characterized and cloned. A thrombolytic enzyme, fibrolase, found in the venom of a southern copperhead snake is now produced by recombinant technology. Fibrolase does not activate plasminogen or protein C, but directly degrades the α and β chains of fibrin and fibrinogen. Recombinant staphylokinase is not an enzyme, but it forms a 1:1 stoichiometric complex with plasminogen, which becomes active after conversion of plasminogen to plasmin. The complex has been demonstrated to be a potent and highly fibrin specific thrombolytic agent [164,165]. Alteplase (recombinant tissue-type plasminogen activator) (Table 17.1) stimulates the fibrinolysis of blood clots by converting plasminogen to plasmin and it seems the effect can be improved when administered in combination with heparin [166].

Anistreplase was one of the first new plasminogen activators of the second generation. It is a chemically modified version of the streptokinase-plasminogen-activator-complex and tissue-type plasminogen-activator produced by recombinant technology. Both new substances have fueled the development in modern thrombolytic treatment. While the clinical progress with t-PA was confirmed in large clinical trials, no real superiority of anistreplase over the traditional plasminogen activators urokinase and streptokinase has been substantiated [167].

Very interesting recombinant designs of plasminogen activators with improved thrombolytic properties are found in several bifunctional proteins with both plasminogen-activating and anti-coagulative activity. They were constructed by fusing a thrombin-inhibitory moiety comprised of four elements: linker 1, thrombin's active site motif, linker 2 and a hirudin fragment that binds to the fibrinogen-recognition site of thrombin. To improve the anticoagulative activity of this construct further, linker 2 was substituted by a linker (FLLRNP) from the human thrombin receptor. This modification conferred about a 10-fold increase in anticoagulative activity compared with the parent molecule M23 carrying an aliphatic linker. The increased thrombin-inhibitory activity of M37 was explained by the presence of an arginine in the linker from the thrombin receptor which may interact with one of two glutamic acid residues located at the exit of the thrombin substrate binding pocket [168]. The recombinant form of tissue-type plasminogen activator (rt-PA) has been shown superior over first-generation fibrinolytic agents. Biochemically modified mutants of wild-type t-PA, e.g., r-PA (reteplase), n-PA (lanoteplase), and TNK-t-PA have altered biochemical and pharmacokinetic characteristics which make them more easy to use (as double or single bolus injection) and have an efficacy benefit [169]. Other chimeric plasminogen activators with improved thrombolytic properties combine both a plasminogen-activating and an anticoagulative activity.

Reteplase is the most recent thrombolytic agent derived from human tissue plasminogen activator and approved by FDA for use in the management of acute myocardial infarction in adults. Its mechanism of action is similar to that of alteplase, but due to some structural changes it differs in pharmacokinetic and pharmacodynamic properties, i.e., has a prolonged half-life. Reteplase has fibrin specificity similar to that of alteplase, but with a lower binding affinity for fibrin which enables it to bind to the thrombus repeatedly thus increasing its fibrinolytic potential. In clinical trials, reteplase has demonstrated more rapid and complete coronary patency compared with alteplase, without a significant increase in clinical adverse events [170].

The effect of recombinant human microplasmin was studied in mice ischemic stroke models and in a rabbit loop thrombosis model. This human microplasminogen (fPlg), which lacks the five "kringle" domains of plasminogen was expressed with high yield in *Pichia pastoris* [171].

Factor VIII is another essential blood clotting factor that lack of which causes hemophilia in affected people. The thrombin generation capabilities of wild-type FVIII and FVIII variants have been determined. Genetically engineered disulfide bonds in B-domain-deleted factor (F) VIII variants (C662-C1828 FVIII and C664-C1826 FVIII) were found to improve FVIIIa stability by blocking A2 domain dissociation, because the new disulfide covalently links the A2 and A3 domains in FVIIIa. The two disulfide bond-stabilized FVIII variants and WT FVIII had comparable clotting times at all studied concentrations. However, at low concentrations, the FVIII variants required only 10% as much FVIII to achieve comparable clot-formation rates, clot-formation times, and clot firmness values and the disulfide bond-stabilized FVIIIa variants had approximately fivefold increased half-lives relative to wild type [28].

17.3.2.2 Atherosclerosis

Atherosclerosis is the name of the process in which deposits of fatty substances, cholesterol, cellular waste products, calcium, and other substances build up in the inner lining of an artery to create plaque and can harden over time. Atherosclerosis is the leading cause of death in the developed countries. In addition to lipid-lowering drugs (statins), dietary control, and exercise, new approaches are needed for the treatment and prevention of atherosclerosis. Lysosomal acid lipase has been tested in a mouse model as an enzyme therapy to reduce atherosclerotic lesions and to examine the molecular basis supporting this novel strategy and its mechanism of effect. Administration of human lysosomal acid lipase via tail vein into mice with atherosclerosis eliminates early aortic and coronary ostial lesions, and reduces lesional size in advanced disease. These studies have indicated that the enzyme lipase affects the atherogenesis by at least two mechanisms, either direct targeting of lesional macrophages with resultant decreases in cholesteryl esters and triglyceride in the lysosomes of macrophages in the lesions, or systemic effects that mediate the liver to reduce the hepatic cholesteryl ester and triglyceride release [172]. α-Mannosyl-terminated human lysosomal acid lipase (phLAL) produced in *Pichia pastoris*, purified, and administered intravenously in a mouse model of atherosclerosis, reduces plasmal, hepatic, and splenic cholesteryl esters and triglycerides. phLAL has been detected in hepatic Kupffer cells and in atheromatous foam cells. Repeated enzyme injections have been tolerated well, with no obvious adverse effects. The coronary and aortic lesions are either eliminated in their early stages or reduced in their advanced stages. Results like these have supported the utility of lysosomal acid lipase supplementation for the treatment of atherosclerosis [173].

Alternative approach for the treatment of atherosclerosis has been suggested by studies of the human paraoxonase (PON) gene family. The family consists of three members, PON1, PON2, and PON3, residing on chromosome 7. Its most-studied member is the serum PON1, a high-density lipoprotein-associated esterase/lactonase. Its capability to hydrolyze toxic organophosphates gave its name as a derivative from one of its most commonly used in vitro substrates, paraoxon. Extensive studies have demonstrated PON1's ability to protect against atherosclerosis by hydrolyzing specific derivatives of oxidized cholesterol and phospholipids in oxidized low-density lipoprotein and in atherosclerotic lesions. More recently PON2 and PON3 have also been shown to have antioxidant properties [174].

17.3.2.3 Alzheimer's Disease

Alzheimer disease (AD) is a progressive dementia. One of the most consistent abnormalities in AD brain is a severe loss of basal forebrain cholinergic neurons and cortical cholinergic innervations, together with other pathological features (amyloid plaques and neurofibrillary tangles). Cognitive dysfunction in AD patients is correlated with decline in choline acetyltransferase (ChAT) activity and loss of cholinergic neurons. Therapies based on acetylcholine esterase (AChE) inhibition have produced small, but well-attested improvements in AD patients [175]. Phosphorylation events appear critical to the understanding of the pathogenesis and treatment of AD. Proteolytic processing and phosphorylation of amyloid precursor protein (APP), and hyperphosphorylation of tau protein, have been shown to be increased in AD, leading to increased production of β-amyloid (Abeta) peptides and neurofibrillary tangles. Pin-1 is peptidyl-prolyl isomerase that catalyzes the isomerization of the peptide bond between pSer/Thr-Pro in proteins, thereby regulating their assembly, folding, intracellular transport, signaling, transcription, cell cycle progression and apoptosis. Pin1 is co-localized with phosphorylated tau in AD brain, and has an inverse relationship to the expression of tau. Pin1 is oxidatively modified and has reduced activity in the hippocampus in mild cognitive impairment and AD. Because of the diverse functions of Pin1, and the discovery that this protein is one of the oxidized proteins common to both mild cognitive impairment (MCI) and AD brain, the question arises whether Pin1 is one of the reasons for the initiation or progression of AD pathogenesis making it a good target for therapeutic applications [176].

17.3.2.4 Superoxide Dismutase

The antithrombotic activities of superoxide dismutase and catalase are determined by their effects on reactive oxygen species. It has been postulated that oxygen-derived free radicals are produced in significant quantities upon in the ischemic brain and that they play critical role in triggering the ischemic neuronal damage causing delayed neuronal death. To examine the effects of human recombinant superoxide dismutase (hrSD) on the delayed neuronal death and on the expression of mRNA for endogenous copper-zinc superoxide dismutase, hrSD and apo-SD have been tested in gerbil model. Apo-SD treatment caused almost complete destruction of CA1 neurons [177]. Gerbils treated with hr superoxide dismutase showed only mild lesions. The protective effects of human recombinant superoxide dismutase against ischemic neuronal damage support the hypothesis that free radicals induce delayed neuronal death.

Modification of SD and catalase with chondroitin sulphate (CS) enhances their effect on reactive oxygen species due to accumulation of the derivatives on the surface of the vascular wall cells. The effects of covalently modified biocatalysts were observed to exceed those of native enzymes, free CS and their mixtures. The SD-CS conjugate reduce the thrombus mass significantly and catalase-CS conjugate preserves blood flow in a rat arterial thrombosis model. A combination of SD-CS and catalase-CS covalent conjugates had a significantly lower antithrombotic effect compared with that of the SD-CD bienzymatic covalent conjugate, explained by different surface distribution of the conjugates in the circulation [178]. Such approach promises a new antioxidant therapeutic strategy of simple and effective protection of the vascular wall against injuries with the use of the covalent conjugate.

17.3.2.5 Organophosphate Poisoning

Organophosphate compounds are esters, amides, or simple derivatives of phosphoric and thiophosphoric acids and include some of the most toxic insecticides used in agriculture. Organophosphate toxicity is due to the ability of these compounds to inhibit acetylcholinesterase at cholinergic junctions of the nervous system. Acetylcholinesterase is the enzyme that degrades acetylcholine following stimulation of a nerve, its inhibition allows acetylcholine to accumulate and result in initial excessive stimulation followed by depression. Standard treatment employs atropine and 2-pyridine aldoxime methiodide. Enzyme therapy for the prevention and treatment of organophosphate poisoning depends on the availability of large amounts of cholinesterases. Transgenic plants have been evaluated for their efficiency and cost-effectiveness as a system for the bioproduction of therapeutically valuable proteins. Mor et al. [179] reported production of a recombinant isoform of human acetylcholinesterase in transgenic tomato plants. The human enzyme accumulated in tomato plants remained active and stable and retained its kinetic characteristics. High levels of specific activity were registered in both the leaves and the tomato fruits.

17.3.2.6 Universal Blood Supply

ABO-incompatible red blood cells (RBCs) transfusion is a leading cause of fatal reactions. The idea of converting blood group A and B antigens to H, i.e., to create a universal blood supply, using specific exo-glycosidases to remove the immunodominant sugar residues, was pioneered by Goldstein and colleagues at the New York Blood Center in the early 1980s [180–182]. Conversion of group B to O was initially attempted with coffee bean α-galactosidase and the converted RBCs appeared to survive normally in all recipients. Multiple-unit and second transfusions of red cells enzymatically converted from group B to group O have passed phase I trials [183]. A successful phase II clinical trial utilizing recombinant enzyme was reported by Kruskall et al. [184], however enzymatic conversion of group A-RBCs has lagged behind due to lack of appropriate glycosidases and the more complex nature of A antigens. The field has been advanced by the identification of some bacterial glycosidases with improved kinetic properties and specificities for the A and B antigens [185].

17.3.3 Infectious Diseases and Cancer

Interferons (IFNs) are a family of naturally occurring proteins that are produced by cells of the immune system. Three classes of IFNs with overlapping activities have been identified: α, β, and γ. Together, the IFNs direct the immune system's attack on viruses, bacteria, tumors and other foreign substances that may invade the body. IFN, its action and its recombinant derivatives have been studied extensively. Polyethylene glycol (PEG) causes the IFN to remain in the body longer and thus prolongs and enhances the effects of the IFN. IFN is used for the treatment of hepatitis C [186] and B [187], virus infections, human papilloma virus [188], melanoma [189], hepatocellular carcinoma [190] multiple sclerosis [191] and leukemia [192].

Type I IFNs are broadly pleiotropic cytokines that bind to a single heterodimeric receptor and have potent antiviral, antiproliferative, and immune modulators activities. Studies have shown that various members of the natural IFN-α family and engineered variants, such as IFN-con1, vary in the ratios between various IFN-mediated cellular activities [30]. Type I IFNs are produced by fibroblasts in response to stimulation by live or inactivated virus, by double-stranded RNA or micro-organisms. At least 23 different variants of IFN-α are known. The individual proteins have molecular masses between 19–26 kDa and consist of proteins with lengths of 156–166 and 172 amino acids. All IFN-α subtypes possess a common conserved sequence region between amino acid positions 115–151 while the amino-terminal ends are variable. Many IFN-α subtypes differ in their sequences at only one or two positions. Naturally occurring variants include proteins truncated by 10 amino acids at the carboxy-terminus. Recombinant human IFN α-1b obtained from human leukocytes has been produced in *E.coli* as a single, non-glycosylated, 166 amino acids polypeptide chain [30].

IFN-β is also induced by some cytokines such as TNF and IL-1. The human fibroblasts INF-β gene has been cloned and altered to substitute Serine for the cysteine residue at position 17. This recombinant human IFN-β 1b has been produced in *E. coli* and is a single, non-glycosylated variant form of the human IFN-1b 165 amino acids polypeptide chain.

Therapeutic genes, antibodies and their derivatives, as well as small molecular weight compounds, have been used for to target directly the genetic and biochemical causes of malignant cancer transformation. Small peptides might be able to complement these agents because of their ability to recognize specific protein domains and thus to interfere with enzymatic functions or protein–protein interactions. A variation of the yeast-two-hybrid procedure allows the selection of specifically binding peptides (aptamers) from a high complexity peptide library. This selection procedure can be adapted to any protein or protein fragment used as bait construct and selects for the intracellular interaction between the bait of choice and the peptide aptamer prey. Aptamers selected in that manner can be cloned, provided with a protein transduction domain, expressed in bacteria and introduced into cancer cells. Example of selective inhibition of receptor signaling is an aptamer that causes the growth arrest of EGF receptor-dependent tumor cells. Such aptamer constructs can be additionally supplemented with more functional domains to enhance their inhibitory effects [6,193].

The signal transducer and activator of transcription (STAT) proteins are a family of latent cytoplasmic transcription factors, which form dimers when activated by cytokine receptors, tyrosine kinase growth factor receptors as well as non-receptor tyrosine kinases. Dimeric STATs translocate to the nucleus, where by binding to specific DNA-response promoters elements they induce unique gene expression. Certain activated STATs have been implicated in human carcinogenesis and there is enough evidence of their molecular interplays and complex regulation to suggest that some STATS are appealing targets for lung cancer treatment with new developing strategies [194].

Bacterial L-asparaginases catalyze the conversion of L-asparagine into L-aspartate and ammonia, and are widely used for the treatment of acute lymphoblastic leukemia. Using protein chemistry and engineering approach a trypsin-resistant PEGylated variant was created based on *Erwinia carotovora* L-asparaginase by site directed mutagenesis of Arg206 to His and covalent coupling of methoxyPEG K [8,195]. Colaspase is also L-asparaginase, or L-asparaginase amidohydrolase. It is an enzyme produced from cultures of *E. coli* HAP. Colaspase is a monomer thought to consist of four subunits. Colaspase hydrolyses the amino acid L-asparagine to L-aspartic acid and ammonia, and thus interferes with the growth of certain tumor cells, which, unlike healthy cells, are unable to synthesize L-asparagine for their metabolism [196]. *Aeromonas* L-asparaginase had been found to be antileukaemic. Levels of protein bound hexose, fucose and sialic acid which are known to increase during leukemia, attain normal levels when treated with L-asparaginase in murine model. Effects of L-asparaginase are comparable with "Leunase", a commercially available drug used in the treatment of leukemia. L-asparaginases (l-ASNases) have also been cloned from *Erwinia chrysanthemi* 3937 (Erl-ASNase) in *E. coli* BL21(DE3)pLysS. The purified enzyme was studied and the kinetic parameters (K_m, k_{cat}) for a number of substrates have been determined. Those studies led to the observation that the enzyme can be efficiently immobilized on epoxy-activated Sepharose CL-6B and retain most of its activity and stability at 4°C. This approach offers the possibility of designing an Erl-ASNase bioreactor that can be operated over a long period of time with high efficiency, which can be used in leukemia therapy [195].

17.4 CONCLUSION

The advances in high throughput screening technology for discovery of target molecules and the accumulation of functional genomics and proteomics data at accelerating rates enable us to design and discover novel biomolecules and proteins on a rational basis in diverse areas of pharmaceutical, agricultural, industrial, and environmental applications. The bioinformatics for genome and proteome analyses are indispensable for accelerating progress in the pharmaceutical industry. Biomolecular engineering is a very important scientific discipline, because it enables systematic and comprehensive analyses of gene expression patterns in both normal and diseased cells, as well as the discovery of novel molecular targets and agents. The immense advances in massive parallel sequencing and complementary technologies allow the exploration of life's diversity without culturing and had brought on many enlightening discoveries from various extreme environments like the polar cap, sulfur and hyperthermal wells, the rumen, and depths of the ocean. When functional genomics databases, EST techniques, microarrays, and proteome analysis in combination with mass spectrometry and sensitive nucleic acid detection and amplification technologies intersect with advanced computational approaches, biomolecular engineering research can yield new drug discoveries, improved therapies, and significantly improved bioprocess technology.

REFERENCES

1. Vellard, M. The enzyme as drug: Application of enzymes as pharmaceuticals. *Curr. Opin. Biotechnol.*, 2003, 14(4), 444–450.
2. Walsh, G. Biopharmaceuticals: Recent approvals and likely directions. *Trends Biotechnol.*, 2005, 23(11), 553–558.

3. Birch, J.R. and Onakunle, Y. Biopharmaceutical proteins: Opportunities and challenges. In *Therapeutic Proteins: Methods and Protocols*, 1st edn., Smales, C.M. and James, D.C., (Eds.), Humana Press, Totowa, NJ, 2005, pp. 1–16.
4. Reeves, C.D. The enzymology of combinatorial biosynthesis. *Crit. Rev. Biotechnol.*, 2003, 23(2), 95–147.
5. Seco, E.M., Cuesta, T., Fotso, S., Laatsch, H., and Malpartida, F. Two polyene amides produced by genetically modified Streptomyces diastaticus var. 108. *Chem. Biol.*, 2005, 12(5), 535–543.
6. Buerger, C. and Groner, B. Bifunctional recombinant proteins in cancer therapy: Cell penetrating peptide aptamers as inhibitors of growth factor signaling. *J. Cancer Res. Clin. Oncol.*, 2003, 129(12), 669–675.
7. Lorenz, P., Liebeton, K., Niehaus, F., and Eck, J. Screening for novel enzymes for biocatalytic processes: Accessing the metagenome as a resource of novel functional sequence space. *Curr. Opin. Biotechnol.*, 2002, 13(6), 572–577.
8. Kotzia, G.A. and Labrou, N.E. L-Asparaginase from Erwinia Chrysanthemi 3937: Cloning, expression and characterization. *J. Biotechnol.*, 2007, 127(4), 657–669.
9. Voget, S., Leggewie, C., Uesbeck, A., Raasch, C., Jaeger, K.E., and Streit, W.R. Prospecting for novel biocatalysts in a soil metagenome. *Appl. Environ. Microbiol.*, 2003, 69(10), 6235–6242.
10. Wraith, J.E. Limitations of enzyme replacement therapy: Current and future. *J. Inherit. Metab. Dis.*, 2006, 29(2–3), 442–447.
11. Akram, K. and Pearlman, B.L. Congestive heart failure-related anemia and a role for erythropoietin. *Int. J. Cardiol.*, 2007, 117(3), 296–305.
12. Dodhia, V.R., Fantuzzi, A., and Gilardi, G. Engineering human cytochrome P450 enzymes into catalytically self-sufficient chimeras using molecular Lego. *J. Biol. Inorg. Chem.*, 2006, 11(7), 903–916.
13. Ikeda, K., Schiltz, E., Fujii, T., Takahashi, M., Mitsui, K., Kodera, Y., Matsushima, A., Inada, Y., Schulz, G.E., and Nishimura, H. Phenylalanine ammonia-lyase modified with polyethylene glycol: Potential therapeutic agent for phenylketonuria. *Amino Acids*, 2005, 29(3), 283–287.
14. Keppler, B.R. and Jarstfer, M.B. Inhibition of telomerase activity by preventing proper assemblage. *Biochemistry*, 2004, 43(2), 334–343.
15. Lonnerdal, B. Recombinant human milk proteins. *Nestle Nutr. Workshop Ser. Pediatr. Program*, 2006, 58, 207–215.
16. Makino, T., Matsumoto, M., Suzuki, Y., Kitajima, Y., Yamamoto, K., Kuramoto, M., Minamitake, Y., Kangawa, K., and Yabuta, M. Semisynthesis of human ghrelin: Condensation of a Boc-protected recombinant peptide with a synthetic O-acylated fragment. *Biopolymers*, 2005, 79(5), 238–247.
17. Moll, S., Kenyon, P., Bertoli, L., De Maio, J., Homesley, H., and Deitcher, S.R. Phase II trial of alfimeprase, a novel-acting fibrin degradation agent, for occluded central venous access devices. *J. Clin. Oncol.*, 2006, 24(19), 3056–3060.
18. Winter, G., Fersht, A.R., Wilkinson, A.J., Zoller, M., and Smith, M. Redesigning enzyme structure by site-directed mutagenesis: Tyrosyl tRNA synthetase and ATP binding. *Nature*, 1982, 299(5885), 756–758.
19. Wilkinson, A.J., Fersht, A.R., Blow, D.M., and Winter, G. Site-directed mutagenesis as a probe of enzyme structure and catalysis: Tyrosyl-tRNA synthetase cysteine-35 to glycine-35 mutation. *Biochemistry*, 1983, 22(15), 3581–3586.
20. Wilkinson, A.J., Fersht, A.R., Blow, D.M., Carter, P., and Winter, G. A large increase in enzyme-substrate affinity by protein engineering. *Nature*, 1984, 307(5947), 187–188.
21. Zou, J., Zhang, R., Zhu, F., Liu, J., Madison, V., and Umland, S.P. ADAM33 enzyme properties and substrate specificity. *Biochemistry*, 2005, 44(11), 4247–4256.
22. Fei, E., Jia, N., Yan, M., Ying, Z., Sun, Q., Wang, H., Zhang, T., Ma, X., Ding, H., Yao, X., Shi, Y., and Wang, G. SUMO-1 modification increases human SOD1 stability and aggregation. *Biochem. Biophys. Res. Commun.*, 2006, 347(2), 406–412.
23. Brewerton, L.J., Fung, E., and Snyder, F.F. Polyethylene glycol-conjugated adenosine phosphorylase: Development of alternative enzyme therapy for adenosine deaminase deficiency. *Biochim. Biophys. Acta*, 2003, 1637(2), 171–177.
24. Gao, R., Zhang, Y., Dedkova, L., Choudhury, A.K., Rahier, N.J., and Hecht, S.M. Effects of modification of the active site tyrosine of human DNA topoisomerase I. *Biochemistry*, 2006, 45(27), 8402–8410.
25. Grabenhorst, E., Schlenke, P., Pohl, S., Nimtz, M., and Conradt, H.S. Genetic engineering of recombinant glycoproteins and the glycosylation pathway in mammalian host cells. *Glycoconj. J.*, 1999, 16(2), 81–97.

26. Way, J.C., Lauder, S., Brunkhorst, B., Kong, S.M., Qi, A., Webster, G., Campbell, I., McKenzie, S., Lan, Y., Marelli, B., Nguyen, L.A., Degon, S., Lo, K.M., and Gillies, S.D. Improvement of Fc-erythropoietin structure and pharmacokinetics by modification at a disulfide bond. *Protein Eng. Des. Sel.*, 2005, 18(3), 111–118.
27. Sheikholvaezin, A., Sandstrom, P., Eriksson, D., Norgren, N., Riklund, K., and Stigbrand, T. Optimizing the generation of recombinant single-chain antibodies against placental alkaline phosphatase. *Hybridoma (Larchmt.)*, 2006, 25(4), 181–192.
28. Radtke, K.P., Griffin, J.H., Riceberg, J., and Gale, A.J. Disulfide bond-stabilized factor VIII has prolonged factor VIIIa activity and improved potency in whole blood clotting assays. *J. Thromb. Haemost.*, 2007, 5(1), 102–108.
29. Pokhrel, S., Lamsal, M., and Wang, Z.X. Easy approach to DNA shuffling: Its potential implication in health sciences. *Nepal Med. Coll. J.*, 2006, 8(2), 136–139.
30. Brideau-Andersen, A.D., Huang, X., Sun, S.C., Chen, T.T., Stark, D., Sas, I.J., Zadik, L., Dawes, G.N., Guptill, D.R., McCord, R., Govindarajan, S., Roy, A., Yang, S., Gao, J., Chen, Y.H., Skartved, N.J., Pedersen, A.K., Lin, D., Locher, C.P., Rebbapragada, I., Jensen, A.D., Bass, S.H., Nissen, T.L., Viswanathan, S., Foster, G.R., Symons, J.A., and Patten, P.A. Directed evolution of gene-shuffled IFN-alpha molecules with activity profiles tailored for treatment of chronic viral diseases. *Proc. Natl. Acad. Sci. U S A*, 2007, 104(20), 8269–8274.
31. Chaparro-Riggers, J.F., Loo, B.L., Polizzi, K.M., Gibbs, P.R., Tang, X.S., Nelson, M.J., and Bommarius, A.S. Revealing biases inherent in recombination protocols. *BMC Biotechnol.*, 2007, 7(1), 77.
32. Clarke, P.H. Experiment in microbial evolution. In *The Bacteria*, Vol. 6, Gunsalos, T.C., Ornston, L.N., and Sokatch, J.R., (Eds.), Academic Press: New York, 1978, pp. 137–218.
33. Kricker, M. and Hall, B.G. Directed evolution of cellobiose utilization in *Escherichia coli* K12. *Mol. Biol. Evol.*, 1984, 1(2), 171–182.
34. Patel, P.H., Kawate, H., Adman, E., Ashbach, M., and Loeb, L.A. A single highly mutable catalytic site amino acid is critical for DNA polymerase fidelity. *J. Biol. Chem.*, 2001, 276(7), 5044–5051.
35. Hancock, S.M., Vaughan, M.D., and Withers, S.G. Engineering of glycosidases and glycosyltransferases. *Curr. Opin. Chem. Biol.*, 2006, 10(5), 509–519.
36. Turner, N.J. Directed evolution of enzymes for applied biocatalysis. *Trends Biotechnol.*, 2003, 21(11), 474–478.
37. Kaur, J. and Sharma, R. Directed evolution: An approach to engineer enzymes. *Crit. Rev. Biotechnol.* 2006, 26(3), 165–199.
38. Rubin-Pitel, S.B. and Zhao, H. Recent advances in biocatalysis by directed enzyme evolution. *Comb. Chem. High Throughput Screen*, 2006, 9(4), 247–257.
39. Lutz, S. and Patrick, W.M. Novel methods for directed evolution of enzymes: Quality, not quantity. *Curr. Opin. Biotechnol.*, 2004, 15(4), 291–297.
40. Antikainen, N.M. and Martin, S.F. Altering protein specificity: Techniques and applications. *Bioorg. Med. Chem.*, 2005, 13(8), 2701–2716.
41. Rodriguez, E. and McDaniel, R. Combinatorial biosynthesis of antimicrobials and other natural products. *Curr. Opin. Microbiol.*, 2001, 4(5), 526–534.
42. Cross, K.P., Myatt, G., Yang, C., Fligner, M.A., Verducci, J.S., and Blower, P.E., Jr. Finding discriminating structural features by reassembling common building blocks. *J. Med. Chem.*, 2003, 46(22), 4770–4775.
43. Ryu, D.D. and Nam, D.H. Recent progress in biomolecular engineering. *Biotechnol. Prog.*, 2000, 16(1), 2–16.
44. Liao, J., Warmuth, M.K., Govindarajan, S., Ness, J.E., Wang, R.P., Gustafsson, C., and Minshull, J. Engineering proteinase K using machine learning and synthetic genes. *BMC Biotechnol.*, 2007, 7, 16.
45. Baltz, R.H. Molecular engineering approaches to peptide, polyketide and other antibiotics. *Nat. Biotechnol.*, 2006, 24(12), 1533–1540.
46. Baltz, R.H., Brian, P., Miao, V., and Wrigley, S.K. Combinatorial biosynthesis of lipopeptide antibiotics in Streptomyces roseosporus. *J. Ind. Microbiol. Biotechnol.*, 2006, 33(2), 66–74.
47. Zeng, K., Rose, J.P., Chen, H.C., Strickland, C.L., Tu, C.P., and Wang, B.C. A surface mutant (G82R) of a human alpha-glutathione S-transferase shows decreased thermal stability and a new mode of molecular association in the crystal. *Proteins*, 1994, 20(3), 259–263.
48. Zeymer, U. and Neuhaus, K.L. Development of new thrombolytic substances. *Herz*, 1994, 19(6), 314–325.

49. Matsushita, M., Hoffman, T.Z., Ashley, J.A., Zhou, B., Wirsching, P., and Janda, K.D. Cocaine catalytic antibodies: The primary importance of linker effects. *Bioorg. Med. Chem. Lett.*, 2001, 11(2), 87–90.
50. Nachon, F., Nicolet, Y., Viguie, N., Masson, P., Fontecilla-Camps, J.C., and Lockridge, O. Engineering of a monomeric and low-glycosylated form of human butyrylcholinesterase: Expression, purification, characterization and crystallization. *Eur. J. Biochem.*, 2002, 269(2), 630–637.
51. Hoshi, H., Nakagawa, H., Nishiguchi, S., Iwata, K., Niikura, K., Monde, K., and Nishimura, S. An engineered hyaluronan synthase: Characterization for recombinant human hyaluronan synthase 2 *Escherichia coli. J. Biol. Chem.*, 2004, 279(4), 2341–2349.
52. Niamsiri, N., Delamarre, S.C., Kim, Y.R., and Batt, C.A. Engineering of chimeric class II polyhydroxyalkanoate synthases. *Appl. Environ. Microbiol.*, 2004, 70(11), 6789–6799.
53. Yokoyama, Y. and Ramakrishnan, S. Addition of an aminopeptidase N-binding sequence to human endostatin improves inhibition of ovarian carcinoma growth. *Cancer*, 2005, 104(2), 321–331.
54. Maga, E.A., Walker, R.L., Anderson, G.B., and Murray, J.D. Consumption of milk from transgenic goats expressing human lysozyme in the mammary gland results in the modulation of intestinal microflora. *Transgenic Res.*, 2006, 15(4), 515–519.
55. Gerth, M.L., Patrick, W.M., and Lutz, S. A second-generation system for unbiased reading frame selection. *Protein Eng. Des. Sel.*, 2004, 17(7), 595–602.
56. Skretas, G., Meligova, A.K., Villalonga-Barber, C., Mitsiou, D.J., Alexis, M.N., Micha-Screttas, M., Steele, B.R., Screttas, C.G., and Wood, D.W. Engineered chimeric enzymes as tools for drug discovery: Generating reliable bacterial screens for the detection, discovery, and assessment of estrogen receptor modulators. *J. Am. Chem. Soc.*, 2007, 129(27), 8443–8457.
57. Wallner, M., Stocklinger, A., Thalhamer, T., Bohle, B., Vogel, L., Briza, P., Breiteneder, H., Vieths, S., Hartl, A., Mari, A., Ebner, C., Lackner, P., Hammerl, P., Thalhamer, J., and Ferreira, F. Allergy multivaccines created by DNA shuffling of tree pollen allergens. *J. Allergy Clin. Immunol.*, 2007, 120(2), 374–380.
58. Kohler, G. and Milstein, C. Continuous cultures of fused cells secreting antibody of predefined specificity. *Nature*, 1975, 256(5517), 495–497.
59. Schwaber, J. and Cohen, E.P. Human x mouse somatic cell hybrid clone secreting immunoglobulins of both parental types. *Nature*, 1973, 244(5416), 444–447.
60. Onda, M., Willingham, M., Nagata, S., Bera, T.K., Beers, R., Ho, M., Hassan, R., Kreitman, R.J. and Pastan, I. New monoclonal antibodies to mesothelin useful for immunohistochemistry, fluorescence-activated cell sorting, Western blotting, and ELISA. *Clin. Cancer Res.*, 2005, 11(16), 5840–5846.
61. Hifumi, E., Okamoto, Y., and Uda, T. Super catalytic antibody [I]: Decomposition of targeted protein by its antibody light chain. *J. Biosci. Bioeng.*, 1999, 88(3), 323–327.
62. Uda, T. and Hifumi, E. Super catalytic antibody and antigenase. *J. Biosci. Bioeng.*, 2004, 97(3), 143–152.
63. Ohara, K., Munakata, H., Hifumi, E., Uda, T., and Matsuura, K. Improvement of catalytic antibody activity by protease processing. *Biochem. Biophys. Res. Commun.*, 2004, 315(3), 612–616.
64. Mirshahi, M., Shamsipour, F., Mirshahi, T., Khajeh, K., and Naderi-Manesh, H. A novel monoclonal antibody with catalytic activity against beta human chorionic gonadotropin. *Immunol. Lett.*, 2006, 106(1), 57–62.
65. Hidari, K.I., Horie, N., Murata, T., Miyamoto, D., Suzuki, T., Usui, T., and Suzuki, Y. Purification and characterization of a soluble recombinant human ST6Gal I functionally expressed in *Escherichia coli. Glycoconj. J.*, 2005, 22(1–2), 1–11.
66. Beyer, B.M. and Dunn, B.M. Self-activation of recombinant human lysosomal procathepsin D at a newly engineered cleavage junction, "short" pseudocathepsin D. *J. Biol. Chem.*, 1996, 271(26), 15590–15596.
67. Sandoval, C., Curtis, H., and Congote, L.F. Enhanced proliferative effects of a baculovirus-produced fusion protein of insulin-like growth factor and alpha(1)-proteinase inhibitor and improved anti-elastase activity of the inhibitor with glutamate at position 351. *Protein Eng.*, 2002, 15(5), 413–418.
68. Tomita, S., Kawai, Y., Woo, S.D., Kamimura, M., Iwabuichi, K., and Imanishi, A.S. Ecdysone-inducible foreign gene expression in stably-transformed lepidopteran insect cells. *In Vitro Cell Dev. Biol. Anim.*, 2001, 37(9), 564–571.
69. Recillas-Targa, F. Multiple strategies for gene transfer, expression, knockdown, and chromatin influence in mammalian cell lines and transgenic animals. *Mol. Biotechnol.*, 2006, 34(3), 337–354.
70. Barr, E. and Leiden, J.M. Systemic delivery of recombinant proteins by genetically modified myoblasts. *Science*, 1991, 254(5037), 1507–1509.

71. Zhu, Y., Li, X., Vie-Wylie, A., Jiang, C., Thurberg, B.L., Raben, N., Mattaliano, R.J., and Cheng, S.H. Carbohydrate-remodelled acid alpha-glucosidase with higher affinity for the cation-independent mannose 6-phosphate receptor demonstrates improved delivery to muscles of Pompe mice. *Biochem. J.*, 2005, 389 (Pt 3), 619–628.
72. Garin, M.I., Lopez, R.M., Sanz, S., Pinilla, M., and Luque, J. Erythrocytes as carriers for recombinant human erythropoietin. *Pharm. Res.*, 1996, 13(6), 869–874.
73. Pepperl-Klindworth, S., Frankenberg, N., Riegler, S., and Plachter, B. Protein delivery by subviral particles of human cytomegalovirus. *Gene Ther.*, 2003, 10(3), 278–284.
74. Zheng, X., Lundberg, M., Karlsson, A., and Johansson, M. Lipid-mediated protein delivery of suicide nucleoside kinases. *Cancer Res.*, 2003, 63(20), 6909–6913.
75. Castor, T.P. Phospholipid nanosomes. *Curr. Drug Deliv.*, 2005, 2(4), 329–340.
76. Ishihara, H., Hara, T., Aramaki, Y., Tsuchiya, S., and Hosoi, K. Preparation of asialofetuin-labeled liposomes with encapsulated human interferon-gamma and their uptake by isolated rat hepatocytes. *Pharm. Res.*, 1990, 7(5), 542–546.
77. Barsoum, S.C., Milgram, W., Mackay, W., Coblentz, C., Delaney, K.H., Kwiecien, J.M., Kruth, S.A., and Chang, P.L. Delivery of recombinant gene product to canine brain with the use of microencapsulation. *J. Lab. Clin. Med.*, 2003, 142(6), 399–413.
78. Tai, I.T. and Sun, A.M. Microencapsulation of recombinant cells: A new delivery system for gene therapy. *FASEB J.*, 1993, 7(11), 1061–1069.
79. Zhang, Y., Wang, W., Xie, Y., Yu, W., Teng, H., Liu, X., Zhang, X., Guo, X., Fei, J., and Ma, X. In vivo culture of encapsulated endostatin-secreting Chinese hamster ovary cells for systemic tumor inhibition. *Hum. Gene Ther.*, 2007, 18(5), 474–481.
80. Pepperl, S., Munster, J., Mach, M., Harris, J.R., and Plachter, B. Dense bodies of human cytomegalovirus induce both humoral and cellular immune responses in the absence of viral gene expression. *J. Virol.*, 2000, 74(13), 6132–6146.
81. Owens, D.R., Zinman, B., and Bolli, G. Alternative routes of insulin delivery. *Diabet. Med.*, 2003, 20(11), 886–898.
82. Walsh, G.M. Novel therapies for asthma—advances and problems. *Curr. Pharm. Des.*, 2005, 11(23), 3027–3038.
83. Patton, J.S., Bukar, J.G., and Eldon, M.A. Clinical pharmacokinetics and pharmacodynamics of inhaled insulin. *Clin. Pharmacokinet.*, 2004, 43(12), 781–801.
84. Liu, H., Liu, L., Fletcher, B.S., and Visner, G.A. Sleeping Beauty-based gene therapy with indoleamine 2,3-dioxygenase inhibits lung allograft fibrosis. *FASEB J.*, 2006, 20(13), 2384–2386.
85. Bowyer, P.W., Gunaratne, R.S., Grainger, M., Withers-Martinez, C., Wickramsinghe, S.R., Tate, E.W., Leatherbarrow, R.J., Brown, K.A., Holder, A.A., and Smith, D.F. Molecules incorporating a benzothiazole core scaffold inhibit the N-myristoyltransferase of *Plasmodium falciparum*. *Biochem. J.*, 2007, 408(2), 173–180.
86. Kurtovic, S., Runarsdottir, A., Emren, L.O., Larsson, A.K., and Mannervik, B. Multivariate-activity mining for molecular quasi-species in a glutathione transferase mutant library. *Protein Eng. Des. Sel.*, 2007, 20(5), 243–256.
87. Ho, S.Y., Hsieh, C.H., Yu, F.C., and Huang, H.L. An intelligent two-stage evolutionary algorithm for dynamic pathway identification from gene expression profiles. *IEEE/ACM. Trans. Comput. Biol. Bioinform.*, 2007, 4(4), 648–660.
88. Tse, T.H., Chan, B.P., Chan, C.M., and Lam, J. Mathematical modeling of guided neurite extension in an engineered conduit with multiple concentration gradients of nerve growth factor (NGF). *Ann. Biomed. Eng.*, 2007, 35(9), 1561–1572.
89. Kriegl, J.M., Eriksson, L., Arnhold, T., Beck, B., Johansson, E., and Fox, T. Multivariate modeling of cytochrome P450 3A4 inhibition. *Eur. J. Pharm. Sci.*, 2005, 24(5), 451–463.
90. Chaparro-Riggers, J.F., Polizzi, K.M., and Bommarius, A.S. Better library design: Data-driven protein engineering. *Biotechnol. J.*, 2007, 2(2), 180–191.
91. Tholey, A. and Heinzle, E. Methods for biocatalyst screening. *Adv. Biochem. Eng. Biotechnol.*, 2002, 74, 1–19.
92. Bornscheuer, U.T. Trends and challenges in enzyme technology. *Adv. Biochem. Eng. Biotechnol.*, 2005, 100, 181–203.

93. Yun, J. and Ryu, S. Screening for novel enzymes from metagenome and SIGEX, as a way to improve it. *Microb. Cell Fact*, 2005, 4(1), 8.
94. Janssen, D.B., Dinkla, I.J., Poelarends, G.J., and Terpstra, P. Bacterial degradation of xenobiotic compounds: Evolution and distribution of novel enzyme activities. *Environ. Microbiol.*, 2005, 7(12), 1868–1882.
95. Daniel, R. The soil metagenome—a rich resource for the discovery of novel natural products. *Curr. Opin. Biotechnol.*, 2004, 15(3), 199–204.
96. Elend, C., Schmeisser, C., Leggewie, C., Babiak, P., Carballeira, J.D., Steele, H.L., Reymond, J.L., Jaeger, K.E., and Streit, W.R. Isolation and biochemical characterization of two novel metagenome-derived esterases. *Appl. Environ. Microbiol.*, 2006, 72(5), 3637–3645.
97. Lammle, K., Zipper, H., Breuer, M., Hauer, B., Buta, C., Brunner, H., and Rupp, S. Identification of novel enzymes with different hydrolytic activities by metagenome expression cloning. *J. Biotechnol.*, 2007, 127(4), 575–592.
98. van Kasteren, S.I., Kramer, H.B., Gamblin, D.P., and Davis, B.G. Site-selective glycosylation of proteins: Creating synthetic glycoproteins. *Nat. Protoc.*, 2007, 2(12), 3185–3194.
99. Walsh, G. and Jefferis, R. Post-translational modifications in the context of therapeutic proteins. *Nat. Biotechnol.*, 2006, 24(10), 1241–1252.
100. van Kasteren, S.I., Kramer, H.B., Jensen, H.H., Campbell, S.J., Kirkpatrick, J., Oldham, N.J., Anthony, D.C., and Davis, B.G. Expanding the diversity of chemical protein modification allows post-translational mimicry. *Nature*, 2007, 446(7139), 1105–1109.
101. Neet, K.E. and Koshland, D.E., Jr. The conversion of serine at the active site of subtilisin to cysteine: A "chemical mutation". *Proc. Natl. Acad. Sci. U S A*, 1966, 56(5), 1606–1611.
102. Kaiser, E.T. and Lawrence, D.S. Chemical mutation of enzyme active sites. *Science*, 1984, 226(4674), 505–511.
103. DeSantis, G. and Jones, J.B. Chemical modification of enzymes for enhanced functionality. *Curr. Opin. Biotechnol.*, 1999, 10(4), 324–330.
104. Bender, I.B. and Bender, A.L. Dental observations in Gaucher's disease: Review of the literature and two case reports with 13- and 60-year follow-ups. *Oral Surg. Oral Med. Oral Pathol. Oral Radiol. Endod.*, 1996, 82(6), 650–659.
105. Friedman, B., Vaddi, K., Preston, C., Mahon, E., Cataldo, J.R., and McPherson, J.M. A comparison of the pharmacological properties of carbohydrate remodeled recombinant and placental-derived beta-glucocerebrosidase: Implications for clinical efficacy in treatment of Gaucher disease. *Blood*, 1999, 93(9), 2807–2816.
106. Grabowski, G.A., Barton, N.W., Pastores, G., Dambrosia, J.M., Banerjee, T.K., McKee, M.A., Parker, C., Schiffmann, R., Hill, S.C., and Brady, R.O. Enzyme therapy in type 1 Gaucher disease: Comparative efficacy of mannose -terminated glucocerebrosidase from natural and recombinant sources. *Ann. Intern. Med.*, 1995, 122(1), 33–39.
107. Devine, E.A., Smith, M., rredondo-Vega, F.X., Shafit-Zagardo, B., and Desnick, R.J. Chromosomal localization of the gene for Gaucher disease. *Prog. Clin. Biol. Res.*, 1982, 95, 511–534.
108. Ginns, E.I., Choudary, P.V., Tsuji, S., Martin, B., Stubblefield, B., Sawyer, J., Hozier, J., and Barranger, J.A. Gene mapping and leader polypeptide sequence of human glucocerebrosidase: Implications for Gaucher disease. *Proc. Natl. Acad. Sci. U S A*, 1985, 82(20), 7101–7105.
109. Barranger, J.A. and O'Rourke, E. Lessons learned from the development of enzyme therapy for Gaucher disease. *J. Inherit. Metab. Dis.*, 2001, 24(Suppl. 2), 89–96.
110. Grabowski, G.A. and Hopkin, R.J. Enzyme therapy for lysosomal storage disease: Principles, practice, and prospects. *Annu. Rev. Genomics Hum. Genet.*, 2003, 4, 403–436.
111. Connock, M., Burls, A., Frew, E., Fry-Smith, A., Juarez-Garcia, A., McCabe, C., Wailoo, A., Abrams, K., Cooper, N., Sutton, A., O'Hagan, A., and Moore, D. The clinical effectiveness and cost-effectiveness of enzyme replacement therapy for Gaucher's disease: A systematic review. *Health Technol. Assess.*, 2006, 10(24), iii-136.
112. Barranger, J.M. and Novelli, E.A. Gene therapy for lysosomal storage disorders. *Expert Opin. Biol. Ther.*, 2001, 1(5), 857–867.
113. Schneider, E.L., Ellis, W.G., Brady, R.O., McCulloch, J.R., and Epstein, C.J. Infantile (type II) Gaucher's disease: In utero diagnosis and fetal pathology. *J. Pediatr.*, 1972, 81(6), 1134–1139.

114. Latham, T.E., Theophilus, B.D., Grabowski, G.A., and Smith, F.I. Heterogeneity of mutations in the acid beta-glucosidase gene of Gaucher disease patients. *DNA Cell Biol.*, 1991, 10(1), 15–21.
115. Sibille, A., Eng, C.M., Kim, S.J., Pastores, G., and Grabowski, G.A. Phenotype/genotype correlations in Gaucher disease type I: Clinical and therapeutic implications. *Am. J. Hum. Genet.*, 1993, 52(6), 1094–1101.
116. Barton, N.W., Brady, R.O., Dambrosia, J.M., Di Bisceglie, A.M., Doppelt, S.H., Hill, S.C., Mankin, H.J., Murray, G.J., Parker, R.I., and Argoff, C.E.,. Replacement therapy for inherited enzyme deficiency—macrophage-targeted glucocerebrosidase for Gaucher's disease. *N. Engl. J. Med.*, 1991, 324(21), 1464–1470.
117. Germain, D.P., Shabbeer, J., Cotigny, S., and Desnick, R.J. Fabry disease: Twenty novel alpha-galactosidase A mutations and genotype-phenotype correlations in classical and variant phenotypes. *Mol. Med.* 2002, 8(6), 306–312.
118. Germain, D.P. Fabry disease: Recent advances in enzyme replacement therapy. *Expert Opin. Investig. Drugs*, 2002, 11(10), 1467–1476.
119. Blom, D., Speijer, D., Linthorst, G.E., Donker-Koopman, W.G., Strijland, A., and Aerts, J.M. Recombinant enzyme therapy for Fabry disease: Absence of editing of human alpha-galactosidase A mRNA. *Am. J. Hum. Genet.*, 2003, 72(1), 23–31.
120. Linthorst, G.E., Hollak, C.E., Donker-Koopman, W.E., Strijland, A., and Aerts, J.M. Enzyme therapy for Fabry disease: Neutralizing antibodies toward agalsidase alpha and beta. *Kidney Int.*, 2004, 66(4), 1589–1595.
121. Hopkin, R.J., Bissler, J., and Grabowski, G.A. Comparative evaluation of alpha-galactosidase A infusions for treatment of Fabry disease. *Genet. Med.*, 2003, 5(3), 144–153.
122. Wilcox, W.R., Banikazemi, M., Guffon, N., Waldek, S., Lee, P., Linthorst, G.E., Desnick, R.J., and Germain, D.P. Long-term safety and efficacy of enzyme replacement therapy for Fabry disease. *Am. J. Hum. Genet.*, 2004, 75(1), 65–74.
123. Ding, E., Hu, H., Hodges, B.L., Migone, F., Serra, D., Xu, F., Chen, Y.T., and Amalfitano, A. Efficacy of gene therapy for a prototypical lysosomal storage disease (GSD-II) is critically dependent on vector dose, transgene promoter, and the tissues targeted for vector transduction. *Mol. Ther.*, 2002, 5(4), 436–446.
124. Amalfitano, A., Bengur, A.R., Morse, R.P., Majure, J.M., Case, L.E., Veerling, D.L., Mackey, J., Kishnani, P., Smith, W., Vie-Wylie, A., Sullivan, J.A., Hoganson, G.E., Phillips, J.A., III; Schaefer, G. B., Charrow, J., Ware, R.E., Bossen, E.H., and Chen, Y.T. Recombinant human acid alpha-glucosidase enzyme therapy for infantile glycogen storage disease type II: Results of a phase I/II clinical trial. *Genet. Med.*, 2001, 3(2), 132–138.
125. Van den Hout, J.M., Reuser, A.J., de Klerk, J.B., Arts, W.F., Smeitink, J.A., and Van der Ploeg, A.T. Enzyme therapy for pompe disease with recombinant human alpha-glucosidase from rabbit milk. *J. Inherit. Metab Dis.*, 2001, 24(2), 266–274.
126. Reuser, A.J., Van Den, H.H., Bijvoet, A.G., Kroos, M.A., Verbeet, M.P., and Van der Ploeg, A.T. Enzyme therapy for Pompe disease: From science to industrial enterprise. *Eur. J. Pediatr.*, 2002, 161 (Suppl. 1), S106–S111.
127. Navarro, J. and Munck, A. Perspective of use of rhDNase and new therapeutics for cystic fibrosis. *Arch. Pediatr.*, 1995, 2(7), 682–684.
128. Zahm, J.M., Girod, d.B., Deneuville, E., Perrot-Minnot, C., Dabadie, A., Pennaforte, F., Roussey, M., Shak, S., and Puchelle, E. Dose-dependent in vitro effect of recombinant human DNase on rheological and transport properties of cystic fibrosis respiratory mucus. *Eur. Respir. J.*, 1995, 8(3), 381–386.
129. Shah, P.L., Scott, S.F., Knight, R.A., Marriott, C., Ranasinha, C., and Hodson, M.E. In vivo effects of recombinant human DNase I on sputum in patients with cystic fibrosis. *Thorax*, 1996, 51(2), 119–125.
130. Pan, C.Q., Dodge, T.H., Baker, D.L., Prince, W.S., Sinicropi, D.V., and Lazarus, R.A. Improved potency of hyperactive and actin-resistant human DNase I variants for treatment of cystic fibrosis and systemic lupus erythematosus. *J. Biol. Chem.*, 1998, 273(29), 18374–18381.
131. Zahm, J.M., Debordeaux, C., Maurer, C., Hubert, D., Dusser, D., Bonnet, N., Lazarus, R.A., and Puchelle, E. Improved activity of an actin-resistant DNase I variant on the cystic fibrosis airway secretions. *Am. J. Respir. Crit. Care Med.*, 2001, 163(5), 1153–1157.
132. Dwyer, M.A., Huang, A.J., Pan, C.Q., and Lazarus, R.A. Expression and characterization of a DNase I-Fc fusion enzyme. *J. Biol. Chem.*, 1999, 274(14), 9738–9743.
133. Merkus, P.J., de Hoog, M., van Gent, R., and de Jongste, J.C. DNase treatment for atelectasis in infants with severe respiratory syncytial virus bronchiolitis. *Eur. Respir. J.*, 2001, 18(4), 734–737.

134. Nasr, S.Z., Strouse, P.J., Soskolne, E., Maxvold, N.J., Garver, K.A., Rubin, B.K., and Moler, F.W. Efficacy of recombinant human deoxyribonuclease I in the hospital management of respiratory syncytial virus bronchiolitis. *Chest*, 2001, 120(1), 203–208.
135. Konstan, M.W., Stern, R.C., Trout, J.R., Sherman, J.M., Eigen, H., Wagener, J.S., Duggan, C., Wohl, M. E., and Colin, P. Ultrase MT12 and Ultrase MT20 in the treatment of exocrine pancreatic insufficiency in cystic fibrosis: Safety and efficacy. *Aliment. Pharmacol. Ther.*, 2004, 20(11–12), 1365–1371.
136. Konstan, M.W., Davis, P.B., Wagener, J.S., Hilliard, K.A., Stern, R.C., Milgram, L.J., Kowalczyk, T.H., Hyatt, S.L., Fink, T.L., Gedeon, C.R., Oette, S.M., Payne, J.M., Muhammad, O., Ziady, A.G., Moen, R. C., and Cooper, M.J. Compacted DNA nanoparticles administered to the nasal mucosa of cystic fibrosis subjects are safe and demonstrate partial to complete cystic fibrosis transmembrane regulator reconstitution. *Hum. Gene Ther.*, 2004, 15(12), 1255–1269.
137. Walkowiak, J. and Lisowska, A. Pancreatic enzyme therapy and gastrointestinal symptoms in patients with cystic fibrosis. *J. Pediatr.*, 2005, 147(6), 870–871.
138. Kim, W., Erlandsen, H., Surendran, S., Stevens, R.C., Gamez, A., Michols-Matalon, K., Tyring, S.K., and Matalon, R. Trends in enzyme therapy for phenylketonuria. *Mol. Ther.*, 2004, 10(2), 220–224.
139. Ledley, F.D., Grenett, H.E., and Woo, S.L. Biochemical characterization of recombinant human phenylalanine hydroxylase produced in *Escherichia coli. J. Biol. Chem.*, 1987, 262(5), 2228–2233.
140. Wilde, M.I. and McTavish, D. Insulin lispro: A review of its pharmacological properties and therapeutic use in the management of diabetes mellitus. *Drugs*, 1997, 54(4), 597–614.
141. Volund, A., Brange, J., Drejer, K., Jensen, I., Markussen, J., Ribel, U., and Sorensen, A.R., Schlichtkrull, J. In vitro and in vivo potency of insulin analogues designed for clinical use. *Diabet. Med.*, 1991, 8(9), 839–847.
142. Akeboshi, H., Chiba, Y., Kasahara, Y., Takashiba, M., Takaoka, Y., Ohsawa, M., Tajima, Y., Kawashima, I., Tsuji, D., Itoh, K., Sakuraba, H., and Jigami, Y. Production of recombinant beta-hexosaminidase A, a potential enzyme for replacement therapy for Tay-Sachs and Sandhoff diseases, in the methylotrophic yeast Ogataea minuta. *Appl. Environ. Microbiol.*, 2007, 73(15), 4805–4812.
143. Brudnak, M.A., Rimland, B., Kerry, R.E., Dailey, M., Taylor, R., Stayton, B., Waickman, F., Waickman, M., Pangborn, J., and Buchholz, I. Enzyme-based therapy for autism spectrum disorders—is it worth another look? *Med. Hypotheses*, 2002, 58(5), 422–428.
144. Veitch, N.C. Horseradish peroxidase: A modern view of a classic enzyme. *Phytochemistry*, 2004, 65(3), 249–259.
145. Gazaryan, I.G., Chubar, T.A., Ignatenko, O.V., Mareeva, E.A., Orlova, M.A., Kapeliuch, Y.L., Savitsky, P.A., Rojkova, A.M., and Tishkov, V.I. Tryptophanless recombinant horseradish peroxidase: Stability and catalytic properties. *Biochem. Biophys. Res. Commun.*, 1999, 262(1), 297–301.
146. Levin, G., Mendive, F., Targovnik, H.M., Cascone, O., and Miranda, M.V. Genetically engineered horseradish peroxidase for facilitated purification from baculovirus cultures by cation-exchange chromatography. *J. Biotechnol.*, 2005, 118(4), 363–369.
147. Ryan, B.J. and O'Fagain, C. Effects of single mutations on the stability of horseradish peroxidase to hydrogen peroxide. *Biochimie*, 2007, 89(8), 1029–1032.
148. Du, H., Schiavi, S., Levine, M., Mishra, J., Heur, M., and Grabowski, G.A. Enzyme therapy for lysosomal acid lipase deficiency in the mouse. *Hum. Mol. Genet.*, 2001, 10(16), 1639–1648.
149. Du, H., Heur, M., Duanmu, M., Grabowski, G.A., Hui, D.Y., Witte, D.P., and Mishra, J. Lysosomal acid lipase-deficient mice: Depletion of white and brown fat, severe hepatosplenomegaly, and shortened life span. *J. Lipid Res.*, 2001, 42(4), 489–500.
150. Blackburn, M.R., Volmer, J.B., Thrasher, J.L., Zhong, H., Crosby, J.R., Lee, J.J., and Kellems, R.E. Metabolic consequences of adenosine deaminase deficiency in mice are associated with defects in alveogenesis, pulmonary inflammation, and airway obstruction. *J. Exp. Med.*, 2000, 192(2), 159–170.
151. Blackburn, M.R., Aldrich, M., Volmer, J.B., Chen, W., Zhong, H., Kelly, S., Hershfield, M.S., Datta, S. K., and Kellems, R.E. The use of enzyme therapy to regulate the metabolic and phenotypic consequences of adenosine deaminase deficiency in mice. Differential impact on pulmonary and immunologic abnormalities. *J. Biol. Chem.*, 2000, 275(41), 32114–32121.
152. Jonsson, J.J., Foresman, M.D., Wilson, N., and McIvor, R.S. Intron requirement for expression of the human purine nucleoside phosphorylase gene. *Nucleic Acids Res.*, 1992, 20(12), 3191–3198.
153. Jonsson, J.J., Habel, D.E., and McIvor, R.S. Retrovirus-mediated transduction of an engineered intron-containing purine nucleoside phosphorylase gene. *Hum. Gene Ther.*, 1995, 6(5), 611–623.

154. Maynes, J.T., Yam, W., Jenuth, J.P., Gang, Y.R., Litster, S.A., Phipps, B.M., and Snyder, F.F. Design of an adenosine phosphorylase by active-site modification of murine purine nucleoside phosphorylase. Enzyme kinetics and molecular dynamics simulation of Asn-243 and Lys-244 substitutions of purine nucleoside phosphorylase. *Biochem. J.*, 1999, 344(Pt 2), 585–592.
155. Witt, S., Singh, M., and Kalisz, H.M. Structural and kinetic properties of nonglycosylated recombinant Penicillium amagasakiense glucose oxidase expressed in *Escherichia coli*. *Appl. Environ. Microbiol.*, 1998, 64(4), 1405–1411.
156. Pollegioni, L., Campaner, S., Raibekas, A.A., and Pilone, M.S. Identification of a reactive cysteine in the flavin-binding domain of Rhodotorula gracilis D-amino acid oxidase. *Arch. Biochem. Biophys.*, 1997, 343(1), 1–5.
157. Raibekas, A.A. and Massey, V. Glycerol-assisted restorative adjustment of flavoenzyme conformation perturbed by site-directed mutagenesis. *J. Biol. Chem.*, 1997, 272(35), 22248–22252.
158. Raibekas, A.A., Fukui, K., and Massey, V. Design and properties of human D-amino acid oxidase with covalently attached flavin. *Proc. Natl. Acad. Sci. U S A*, 2000, 97(7), 3089–3093.
159. Hickman, D., Palamanda, J.R., Unadkat, J.D., and Sim, E. Enzyme kinetic properties of human recombinant arylamine N-acetyltransferase 2 allotypic variants expressed in *Escherichia coli*. *Biochem. Pharmacol.*, 1995, 50(5), 697–703.
160. Cutler, D., Bode, C., and Runge, M.S. The promise of new genetically engineered plasminogen activators. *J. Vasc. Interv. Radiol.*, 1995, 6(6 Pt 2 Su), 3S–7S.
161. Lijnen, H.R. and Collen, D. Remaining perspectives of mutant and chimeric plasminogen activators. *Ann. N Y Acad Sci.*, 1992, 667, 357–364.
162. Robinson, J.H., Browne, M.J., Carey, J.E., Chamberlain, P.D., Chapman, C.G., Cronk, D.W., Dodd, I., Entwisle, C., Esmail, A.F., Kalindjian, S.B. A recombinant, chimeric enzyme with a novel mechanism of action leading to greater potency and selectivity than tissue-type plasminogen activator. *Circulation*, 1992, 86(2), 548–552.
163. Wirsching, F., Luge, C., and Schwienhorst, A. Modular design of a novel chimeric protein with combined thrombin inhibitory activity and plasminogen-activating potential. *Mol. Genet. Metab.*, 2002, 75(3), 250–259.
164. Verstraete, M. Thrombolytic agents in development. *Drugs*, 1995, 50(1), 29–42.
165. Verstraete, M. Newer thrombolytic agents. *Ann. Acad. Med. Singapore*, 1999, 28(3), 424–433.
166. Gillis, J.C., Wagstaff, A.J., and Goa, K.L. Alteplase. A reappraisal of its pharmacological properties and therapeutic use in acute myocardial infarction. *Drugs*, 1995, 50(1), 102–136.
167. Gulba, D.C., Bode, C., Runge, M.S., and Huber, K. Thrombolytic agents—an overview. *Ann. Hematol.*, 1996, 73(Suppl. 1), S9–27.
168. Wnendt, S., Janocha, E., Steffens, G.J., and Strassburger, W. A strong thrombin-inhibitory prourokinase derivative with sequence elements from hirudin and the human thrombin receptor. *Protein Eng.*, 1997, 10(2), 169–173.
169. Huber, K. Increased efficacy of thrombolytic therapy in acute myocardial infarction by improved properties of new thrombolytic agents. *Wien. Klin. Wochenschr.*, 2000, 112(17), 742–748.
170. Wooster, M.B. and Luzier, A.B. Reteplase: A new thrombolytic for the treatment of acute myocardial infarction. *Ann. Pharmacother.*, 1999, 33(3), 318–324.
171. Nagai, N., Demarsin, E., Van, H.B., Wouters, S., Cingolani, D., Laroche, Y., and Collen, D. Recombinant human microplasmin: Production and potential therapeutic properties. *J. Thromb. Haemost.*, 2003, 1(2), 307–313.
172. Du, H., Schiavi, S., Wan, N., Levine, M., Witte, D.P., and Grabowski, G.A. Reduction of atherosclerotic plaques by lysosomal acid lipase supplementation. *Arterioscler. Thromb. Vasc. Biol.*, 2004, 24(1), 147–154.
173. Du, H. and Grabowski, G.A. Lysosomal acid lipase and atherosclerosis. *Curr. Opin. Lipidol.*, 2004, 15(5), 539–544.
174. Draganov, D.I. and La Du, B.N. Pharmacogenetics of paraoxonases: A brief review. *Naunyn Schmiedebergs Arch. Pharmacol.*, 2004, 369(1), 78–88.
175. Hoshi, M., Takashima, A., Murayama, M., Yasutake, K., Yoshida, N., Ishiguro, K., Hoshino, T., and Imahori, K. Nontoxic amyloid beta peptide 1–42 suppresses acetylcholine synthesis. Possible role in cholinergic dysfunction in Alzheimer's disease. *J. Biol. Chem.*, 1997, 272(4), 2038–2041.
176. Butterfield, D.A., Abdul, H.M., Opii, W., Newman, S.F., Joshi, G., Ansari, M.A., and Sultana, R. Pin1 in Alzheimer's disease. *J. Neurochem.*, 2006, 98(6), 1697–1706.

177. Uyama, O., Matsuyama, T., Michishita, H., Nakamura, H., and Sugita, M. Protective effects of human recombinant superoxide dismutase on transient ischemic injury of CA1 neurons in gerbils. *Stroke*, 1992, 23(1), 75–81.
178. Maksimenko, A.V. Experimental antioxidant biotherapy for protection of the vascular wall by modified forms of superoxide dismutase and catalase. *Curr. Pharm. Des*, 2005, 11(16), 2007–2016.
179. Mor, T.S., Sternfeld, M., Soreq, H., Arntzen, C.J., and Mason, H.S. Expression of recombinant human acetylcholinesterase in transgenic tomato plants. *Biotechnol. Bioeng.*, 2001, 75(3), 259–266.
180. Goldstein, J., Siviglia, G., Hurst, R., Lenny, L., and Reich, L. Group B erythrocytes enzymatically converted to group O survive normally in A, B, and O individuals. *Science*, 1982, 215(4529), 168–170.
181. Goldstein, J. Preparation of transfusable red cells by enzymatic conversion. *Prog. Clin. Biol. Res.*, 1984, 165, 139–157.
182. Goldstein, J. Conversion of ABO blood groups. *Transfus. Med. Rev.*, 1989, 3(3), 206–212.
183. Lenny, L.L., Hurst, R., Zhu, A., Goldstein, J., and Galbraith, R.A. Multiple-unit and second transfusions of red cells enzymatically converted from group B to group O: Report on the end of phase 1 trials. *Transfusion*, 1995, 35(11), 899–902.
184. Kruskall, M.S., AuBuchon, J.P., Anthony, K.Y., Herschel, L., Pickard, C., Biehl, R., Horowitz, M., Brambilla, D.J., and Popovsky, M.A. Transfusion to blood group A and O patients of group B RBCs that have been enzymatically converted to group O. *Transfusion*, 2000, 40(11), 1290–1298.
185. Olsson, M.L., Hill, C.A., de, l., V; Liu, Q.P., Stroud, M.R., Valdinocci, J., Moon, S., Clausen, H., and Kruskall, M.S. Universal red blood cells—enzymatic conversion of blood group A and B antigens. *Transfus. Clin Biol.*, 2004, 11(1), 33–39.
186. Shiffman, M.L., Sutton, A., Bacon, E.R., Nelson, D., Harley, H., Solá, R., Shafran, S.D., Barange, K., Lin, A., and Soman, A., Accelerate Investigators Peginterferon alfa-2a and ribavirin for 16 or 24 weeks in HCV genotype 2 or 3. *N. Engl. J. Med.*, 2007, 357(2), 124–134.
187. Carithers, R.L. Effect of interferon on hepatitis B. *Lancet*, 1998, 351(9097), 157.
188. Edwards, L., Ferenczy, A., Eron, L., Baker, D., Owens, M.L., Fox, T.L., Hougham, A.J., and Schmitt, K. A. Self-administered topical 5% imiquimod cream for external anogenital warts. HPV Study Group. Human PapillomaVirus. *Arch. Dermatol.*, 1998, 134(1), 25–30.
189. Loquai, C., Nashan, D., Hensen, P., Luger, T.A., Sunderkötter, C., and Schiller, M. Safety of pegylated interferon-alpha-2a in adjuvant therapy of intermediate and high-risk melanomas. *Eur. J. Dermatol.*, 2007, 18(1), 29–35.
190. Clavien, P.A. Interferon: The magic bullet to prevent hepatocellular carcinoma recurrence after resection? *Ann. Surg.*, 2007, 245(6), 843–845.
191. Pittock, S.J. Interferon beta in multiple sclerosis: How much benefit? *Lancet*, 2007, 370 (9585), 363–364.
192. Druker, B.J., Guilhot, F., O'Brien, S.G., Gathmann, I., Kantarjian, H., Gattermann, N., Deininger, M.W., Silver, R.T., Goldman, J.M., Stone, R.M., Cervantes, F., Hochhaus, A., Powell, B.L., Gabrilove, J.L., Rousselot, P., Reiffers, J., Cornelissen, J.J., Hughes, T., Agis, H., Fischer, T., Verhoef, G., Shepherd, J., Saglio, G., Gratwohl, A., Nielsen, J.L., Radich, J.P., Simonsson, B., Taylor, K., Baccarani, M., So, C., Letvak, L., and Larson, R.A., IRIS Investigators Five-year follow-up of patients receiving imatinib for chronic myeloid leukemia. *N. Eng. J. Med.*, 2006, 355(23), 2408–2417.
193. Buerger, C., Nagel-Wolfrum, K., Kunz, C., Wittig, I., Butz, K., Hoppe-Seyler, F., and Groner, B. Sequence-specific peptide aptamers, interacting with the intracellular domain of the epidermal growth factor receptor, interfere with Stat3 activation and inhibit the growth of tumor cells. *J. Biol. Chem.*, 2003, 278(39), 37610–37621.
194. Karamouzis, M.V., Konstantinopoulos, P.A., and Papavassiliou, A.G. The role of STATs in lung carcinogenesis: An emerging target for novel therapeutics. *J. Mol. Med.*, 2007, 85(5), 427–436.
195. Kotzia, G.A., Lappa, K., and Labrou, N.E. Tailoring structure-function properties of L-asparaginase: Engineering resistance to trypsin cleavage. *Biochem. J.*, 2007, 404(2), 337–343.
196. Benny, P.J., Muraleedhara, K.G., Sreejith, K., and Jayashree, G. Metabolism of proteins and glycoproteins in tumour bearing mice treated with Aeromonas L-asparaginase. *Indian J. Biochem. Biophys.*, 1996, 33(6), 527–530.

Part IV

Special Expression Systems

18 Production of Pharmaceutical Compounds in Plants

Miguel Angel Gómez Lim

CONTENTS

18.1 INTRODUCTION

Plants are gaining widespread acceptance as a suitable system for the large-scale production of recombinant proteins. As molecular farming has come of age, there have been technological developments on many levels, including transfection methods, control of gene expression, protein targeting, the use of different crops as production platforms [1], and modifications to alter the structural and functional properties of the recombinant product. Over the last few years, there has been a continuing commercial development of novel plant-based expression platforms accompanied by significant success in tackling some of the limitations of plants as bioreactors, such as low yields and inconsistent product quality that have limited the approval of plant-derived pharmaceuticals. Indeed, one of the most important driving factors has been yield improvement, as product yield has a significant impact on economic feasibility. Strategies to improve the recombinant protein yield in plants include the development of novel promoters, the improvement of protein stability and accumulation, and the improvement of downstream processing technologies [2]. Attention is now shifting from basic research towards commercial exploitation, and molecular farming is reaching the stage at which it may challenge established production technologies based on bacteria, yeast, and cultured mammalian cells. There are already several plant-produced proteins on the market [3] including one at a large scale [4]. Several plant-derived recombinant pharmaceutical proteins are reaching the final stages of clinical evaluation, and more are in the development pipeline. In this chapter, I highlight recent progress in molecular farming, its potential for commercial drug development and production, and the potential impact in developing countries.

18.2 MUCOSAL IMMUNIZATION

There is now substantial evidence supporting the existence of at least two immune systems, a "peripheral" immune system and a "mucosal" immune system [5]. These systems operate separately and simultaneously in most species including humans. Unless a vaccine stimulates the appropriate system or a combination of systems, immunity might not be complete. Mucosal surfaces comprise a large surface area (over 400 m^2) that is vulnerable to infection by pathogens and consists of sites where antigens are encountered and processed via mucosa-associated lymphoid tissue. It is not surprising, then, that the vast majority of infections occur at a mucosal surface [6].

Research on mucosal immunization, especially by the oral route, is currently accelerating, stimulated by recent information on the mucosal immune system. This system is involved in the local defense against pathogens at mucosal sites [7]. Exploiting the commonality of the mucosal immune system by oral immunization is an exciting possibility to develop mucosal vaccines. The seeming linked nature of mucosal immunity may allow for delivery of a vaccine at one mucosal surface to prime other mucosal surfaces.

The adaptive immune system has evolved to differentiate antigens and pathogens that gain access to the body via mucosal surfaces from those that are introduced directly into tissues or the bloodstream by injury or by injection, in the case of vaccines. It is becoming increasingly clear that local mucosal immune responses are important for protection against disease. Immune responses at the mucosal level are most efficiently induced when vaccines are directly administered on to mucosal surfaces. Injected vaccines, on the other hand, are poor inducers of mucosal immunity and are consequently less effective to prevent infections at mucosal sites. However, vaccine research has focused for the most part on parenteral delivery of antigens, and most vaccines in use today are, therefore, administered by injection. One of the reasons is that because by injection, a known quantity of antigen can be delivered and the immune response can be readily measured in blood samples. By contrast, the amount of antigen that actually enters the body via a mucosal surface cannot be accurately determined and measurements of mucosal immune responses are more complicated. Other reasons include that it has often proved difficult in practice to stimulate strong mucosal IgA immune and protection by oral administration of antigens, and the results to date of mucosal vaccination efforts using soluble protein antigens have been, with a few notable exceptions, rather disappointing [6–8].

The use of subunit vaccines for oral delivery has also been resisted because of the obvious likelihood of protein degradation in the gut. Furthermore, even if the protein were to survive within an oral delivery system, there is no certainty that trafficking the protein to the gut would be sufficient to mount an immune response. Consequently, only a handful of mucosal vaccines approved for human use are administered mucosally. These include oral vaccines against poliovirus, *Salmonella typhi*, *Vibrio cholerae*, and an oral BCG live vaccine used in Brazil against tuberculosis. Two additional mucosal vaccines against rotavirus and against influenza virus were withdrawn after a short time on the market because of unforeseen side effects.

The main feature of the mucosal immune response is the local production of secretory immunoglobulin A (IgA) and an associated mucosal immunologic memory. Furthermore, although not so well understood, there is some evidence for an important role of the cellular arm of the immune response including mucosal $CD8^+$ cytotoxic T lymphocytes, $CD4^+$ T helper cells as well as natural killer cells [9].

On the other hand, there are also important practical and logistic reasons to favor oral vaccines over injected vaccines. Oral vaccines would be easier to administer and they would be generally inexpensive. No specialized equipment is required, such as the needles and syringes required for injections and trained medical personnel would be unnecessary to administer the vaccine. Thus, equipment and personnel savings can greatly reduce the costs for oral vaccines. Furthermore, they carry less risk of transmitting the infections still associated with the use of injectable vaccines, such as hepatitis B virus and HIV.

Nevertheless, oral vaccination may also present some problems. The most obvious is the potential degradation of the antigen in the gastrointestinal tract. The vast majority of antigen molecules may be degraded before reaching effector sites on the surface of the gastrointestinal tract. To prevent this, either a very large amount of antigen needs to be delivered, or the antigen should be protected from proteolysis. Both approaches can add significantly to vaccine production costs. Oral delivery of antigens in attenuated bacterial strains such as *Salmonella* can afford protection, but raises safety concerns over the delivery vehicles.

Oral tolerance, the diminished capability to develop an immune response when re-exposed to the same antigen introduced by a parenteral injection, has also been considered as a potential obstacle for oral immunization. Although the mucosal immune system is the site of priming for two opposite responses, i.e., tolerance and mucosal immunity, the usual response of the gastrointestinal tract to antigens is tolerance rather than immunity [10]. This allows us to eat large amounts of foreign food proteins without inducing harmful systemic immune responses. Nevertheless, oral tolerance and induction of mucosal immunity in response to mucosal immunization are not mutually exclusive. Although they usually involve different types of antigens, sometimes the same mucosal immunization may give rise both to a local significant IgA antibody formation and to tolerance or suppressed immune responsiveness peripherally [11].

The mechanisms of oral tolerance remain unclear. It is thought that tolerance can occur through active suppression (cells or cytokines that induce nonresponsiveness), anergy (live, nonresponsive cells), or apoptosis (signal programs cell death) of antigen-reactive cells [6]. Induction of tolerance can be beneficial as immunotherapy for various autoimmune diseases. The preferred approach has been the genetic fusion of B subunit of cholera toxin to various auto-antigens, which has been shown to enhance the tolerogenic effect over that of mucosal administration of auto-antigen alone, and to suppress the development of various autoimmune diseases in animal models [6,12,13]. Current efforts are mainly directed towards finding more efficient means of delivering appropriate antigens to the mucosal immune system, and towards developing effective, safe mucosal adjuvants, or immunoregulatory agents. In this sense, plants represent invaluable tools for delivery of antigens to the mucosal immune system including mucosal adjuvants. Originally, a concern was expressed that humans ingesting transgenic plants expressing a foreign antigen might not respond to immunization because the same plant species was probably a component part of their regular diet. However, the results have suggested that an immune response against an antigen can be elicited even if contained in plants, which are regularly consumed, for instance, lettuce.

18.3 PLANTS AS SYSTEMS FOR EXPRESSION OF RECOMBINANT PROTEINS

The fact that proteins can be used as diagnostic reagents, vaccines and therapeutic agents have created a strong demand for the production of recombinant proteins on an industrial scale. The development of recombinant DNA technology has allowed the expression of heterologous recombinant proteins in different systems and the technology has been so successful from the economic and safety standpoints that it has replaced many native sources of proteins [14].

Most of the early work was directed to expression of therapeutic proteins in prokaryotic hosts, because of the low overall cost and short production timescale. However, limitations of prokaryotic systems in the classes of proteins that can be economically produced and in the lack of posttranslational modifications, has turned the focus to eukaryotic hosts: yeast, insect, and mammalian cell cultures, and transgenic animals. Even though mammalian cell cultures and yeast are being widely used, they have some downsides in terms of cost, scalability, risk of pathogenicity, and authenticity [15]. Therefore, a simple and inexpensive system allowing large-scale production of recombinant

TABLE 18.1
Advantages of Plant-Based Oral Vaccines

Lower cost of raw material for production of recombinant proteins
Rapid scale-up or down
Polyvalent vaccines are quite feasible
Plant cells provide protection for the antigen in the gut
Raw material easy to store and transport without the need for a cold chain
Reduced concerns over contamination with human pathogens in vaccine preparations
Eliminate syringes and needles and consequently medical assistance for administration
Eliminate concern over blood-borne diseases through needle reuse

proteins would be highly advantageous. In this sense, plants provide an attractive expression vehicle for numerous proteins. Plant bioreactors have been estimated to yield over 10 kg of therapeutic protein per acre in tobacco, maize, soybean, and alfalfa [16,17]. In comparison with conventional bioreactors and mammalian cells or microorganisms, the cost of producing a protein under good manufacturing practice conditions is reduced to perhaps one-tenth [18,19].

Genetically engineered plants have many advantages as sources of recombinant proteins (Table 18.1). First, the cost of producing raw material on an agricultural scale is low. Second, the use of plants offers reduced capitalization costs relative to fermentation methods. Third, production can be rapidly upscaled. Fourth, unlike bacteria, plants can produce multimeric proteins, such as antibodies, in the correct. Fifth, plant proteins are considered to be safer, as plants do not serve as hosts for human pathogens such as HIV, prions, hepatitis viruses, and so on. Sixth, the plant material could potentially be administered directly. Seven, the antigen can be protected from proteolysis in the gut when is contained inside the plant tissue [20].

Depending upon the promoters used, the recombinant proteins can accumulate throughout the plant, in specific organs (e.g., in seeds) or in specific organelles (e.g., chloroplasts) within a plant cell. Considering the low content of protein in plant leaves and fruits, which ranges from 0.2 to 7.5 g per 100 g fresh weight [21], seeds have been identified as a target for recombinant protein accumulation. Furthermore, proteins produced in seed exhibit high stability; for example, enzymes and antibodies expressed in seed and stored for more than 3 years in the refrigerator retain full enzymatic or binding activity [18].

Some potential issues of concern for plant-based protein production can be identified. Among them are possible allergic reactions to plant protein glycans and other plant compounds. There may be some contamination of the plant and possibly the product by pesticides, mycotoxins, and herbicides. In addition, there is some regulatory uncertainty, particularly for proteins requiring approval for human use [22]. It is important that these regulatory issues be resolved before pursuing product licensure. So far, only one plant-derived vaccine targeting a viral disease of poultry has reached licensure (Dow Agrosciences, United States). Although vaccines intended for use in humans will probably require a more laborious path to licensure, this milestone establishes a credible foothold for plant-derived vaccines [23].

18.3.1 Methods for Gene Transfer into Plants

There are three main approaches for expression of recombinant proteins in plants. One is by nuclear transformation and regeneration of transgenic plant lines, the second involves the transfer and expression of transgenes into the chloroplast genome and the third is by the use of plant viral vectors (transient expression). Each will be examined in turn.

Successful expression of transgenes in plants is possible; thanks to the unique capability of single plant cells to regenerate into whole plants while keeping all the genetic features of the parent plant. In the transgenic plant, foreign genes are stably incorporated into the plant genome, transcribed, and inherited in a Mendelian fashion [24].

In addition, it was found that gene transfer into a plant could be mediated by a plant-infecting bacterium, *Agrobacterium tumefaciens*, which is able to transfer DNA into the plant genome [25]. Subsequently, additional procedures such as microinjection, electroporation, and microparticle bombardment were developed to deliver foreign genes into the plant genome [24]. These procedures are based on the use of purified plasmid DNA.

Microprojectile bombardment or biolistics (direct DNA transfer) has been a method for gene transfer into plants extensively employed [24,26]. The method is so versatile that multiple genes (>10) coding for complex recombinant macromolecules can be transferred simultaneously into the plant genome. Interestingly, even though the genes may be delivered on different plasmids, such multiple transgenes are frequently inherited in a linked fashion [27]. Nicholson et al. [28] successfully introduced four genes coding for components of a secretory antibody into rice and approximately 20% of the resulting plants contained all four genes. This represents an advantage over alternative gene transfer methods that involve the stepwise introduction of individual components, followed by successive rounds of crossing to generate plants containing the fully assembled molecule. Direct DNA transfer also allows the introduction into plants of minimal expression cassettes containing only the promoter, open reading frame, and terminator sequences. As no vector backbone sequences are transferred, this approach increases transgene stability and expression levels considerably by preventing the integration of potentially recombinogenic sequences [29,30]. The described methods are applicable to a wide range of species and this explains why the majority of plant-derived recombinant pharmaceutical proteins have been produced by nuclear transformation and regeneration of transgenic plant lines.

Although nuclear gene transfer is now routine in many species, the main disadvantage is the production timescale, which has prompted the development of alternative plant-based production technologies. One consists in the vacuum infiltration of leaves with recombinant *Agrobacterium tumefaciens*, resulting in the transient transformation of many cells [31]. High levels of protein expression can be obtained for a short time but then declines sharply as a result of post-transcriptional gene silencing [32]. Co-expression inhibitor proteins of gene silencing can increase the expression levels of recombinant proteins at least 50-fold [33]. Researchers at Medicago Inc. have described how agroinfiltration of alfalfa leaves can be scaled up to 7500 leaves per week, producing micrograms of recombinant protein each week [34].

An alternative approach to express recombinant proteins in a stable manner is by gene transfer into the plant chloroplasts. Chloroplasts are plant cellular organelles with their own genome and transcription–translation machinery. The chloroplast genome is a highly polyploid, circular double-stranded DNA 120–180 kb in size, encoding approximately 120 genes [35]. Each chloroplast carries a number of identical genome copies, which are attached to membranes in clusters called nucleoids. A tobacco leaf cell may contain 100 chloroplasts, with 10–14 nucleoids each, and about 10,000 copies of the genome per cell [36]. A gene may be introduced into a spacer region between the functional genes of the chloroplast by homologous recombination, targeting the foreign gene to a precise location. Gene silencing has not been observed with chloroplast transformation, whereas it is a common phenomenon with nuclear transformation. Additionally, the presence of chaperones and enzymes within the chloroplast may help assemble complex multi-subunit recombinant proteins and correctly fold proteins containing disulfide bonds, which should eliminate expensive in vitro processing of recombinant proteins [35], thereby drastically reducing the costs of in vitro processing. Despite such significant progress in chloroplast transformation, this technology has not been extended to major crops. This obstacle emphasizes the need for chloroplast genome sequencing to increase the efficiency of transformation and conduct basic research in plastid biogenesis and function [37,38].

Chloroplast transformation has been achieved in several plant species such as carrot [35], tomato [39], *Brassica oleracea* [40], petunia [41], soybean [42], lettuce [43], the liverwort

Marchantia polymorpha [44], and the green algae *Chlamydomonas reinhardtii* [45], but transformation is routine only in tobacco. Plastid transformation in *Arabidopsis thaliana, Brassica napus*, and *Lesquerella fendleri* has been achieved but at low efficiency [35]. The ability to transform the chromoplasts of fruit and vegetable crops represents an interesting possibility for the expression of subunit vaccines [46].

There are several advantages of chloroplast transformation: high-level expression of the recombinant proteins, the recombinant proteins will accumulate within the chloroplast thus limiting toxicity to the host plant, multiple genes can be expressed in operons [47], and the absence of functional chloroplast DNA in pollen of most crops may provide natural transgene containment [48]. Transgene expression in tobacco plastids reproducibly yields protein levels in the 5%–20% range [35] (Table 18.2); however, levels up to 47% of the total soluble protein can be achieved [49].

The diversity of the genes expressed successfully in tobacco chloroplasts suggests the compatibility of the plastid's protein synthetic machinery with mRNAs from diverse sources, making unnecessary the optimization of genes for chloroplast expression. Since all codons are frequently used in chloroplasts, codon optimization only results in a moderate 2- to 3-fold increase in expression levels [50,51]. On the other hand there is no protein glycosylation in chloroplasts [52]. The functionality of chloroplast-derived antigens and therapeutic proteins has been demonstrated by several in vitro assays and animal protection studies [53,54].

Another tobacco transient-expression technology is based on the use of plant viruses as expression vectors. Viruses can gain entry into a plant cell where they can accumulate and

TABLE 18.2
Proteins Produced in the Chloroplasts of Transgenic Plants

Antigen	Species System	Yield % TSP	References
Cholera toxin B	Tobacco	4–5	[19]
Thermostable xylanase	Tobacco	6	[176]
Human somatotrophin	Tobacco	7	[91]
Human serum albumin	Tobacco	11.1	[177]
Human serum albumin	Tobacco	8	[178]
Tetanus toxin fragment	Tobacco	25	[179]
Heat-labile toxin from *Escherichia coli*	Tobacco	2.5	[180]
Rotavirus VP6 protein	Tobacco	3	[181]
Nontoxic mutant heat-labile toxin from *Escherichia coli*	Tobacco	3.7	[182]
Polyhydroxybutyrate	Tobacco	<7	[183]
2L21 epitope of the VP2 protein from the canine parvovirus fused to the cholera toxin B subunit	Tobacco	31	[184]
Spike protein (S1) of the severe acute respiratory syndrome coronavirus	Tobacco	NR	[185]
Foot-and-mouth disease virus VP1 gene	Tobacco	2–3	[186]
Foot-and-mouth disease virus VP1 protein fused with cholera toxin B subunit	*Chlamydomonas reinhardtii*	3	[45]
Sweet protein monellin	Tobacco	2	[187]
Human epidermal growth factor	Tobacco	NR	[188]
Gal/GalNAc lectin of *Entamoeba histolytica*	Tobacco	6.3	[189]
Protective antigen of *Bacillus anthracis*	Tobacco	14.2	[53]
HPV-16 L1 capsid protein	Tobacco	11	[190]

Note: Expression levels are as reported in the literature and indicate percentage of total soluble protein (%TSP) unless indicated otherwise. NR, not reported.

then spread throughout the entire plant. Plant viruses remain in the cell cytoplasm throughout infection and do not incorporate into the genome of the susceptible host and, thus, are not inherited by the next generation. Several features make plant viruses well suited as transient expression vectors: high-level expression of the introduced genes (up to 2 g/kg of plant tissue) within a short period (1–2 weeks after inoculation), rapid accumulation of the appropriate products, and the fact that more than one vector can be used in the same plant, allowing multimeric proteins to be assembled [55] with the added benefit of resulting in biological containment on the viral sequences. Post-transcriptional gene silencing can be avoided by expressing a replicase from some inducible promoters [56]. Viral cassettes can be integrated into the plant genome resulting also in high levels of expression but there would not be containment in this case.

Plant viruses have been used to produce a wide range of pharmaceutical proteins, including vaccine candidates and antibodies. Some plant viruses have a wide host range and are easily transmissible by mechanical inoculation, spreading from plant to plant, making large-scale infections feasible (Chapter 19).

Plant-virus genomes can be composed of DNA or RNA but the main virus systems from which efficient expression systems are being developed mainly consist of positive-sense RNA genome, single-stranded DNA geminiviruses, and double-stranded nonintegrating DNA pararetroviruses [57], but RNA viruses can multiply to very high titers in infected plants, which makes them better suited vectors for protein expression vectors. For genetic manipulation, viral RNA genomes are reverse transcribed in vitro and cloned as full-length cDNAs. There are at least three approaches for insertion of foreign genes into plant viral genomes: (1) gene replacement, when nonessential viral genes, like those coding for coat proteins, are replaced by the gene of interest. Since some viruses have limitations as to the size of the molecules than can be incorporated into their genomes this would be the best strategy; (2) gene insertion, where the gene of interest is placed under the control of an additional promoter; this approach would be advisable where large coding sequences have to be expressed; and (3) gene fusion, when the gene of interest is translationally fused with a viral gene; the use of coat-protein genes has allowed an efficient method for presentation of foreign peptide sequences on the surfaces of viral particles [58].

For inoculation of plants, an in vitro transcribed RNA is usually inoculated mechanically by gently rubbing the leaves with a mild abrasive. Extracts from these infected plants can also be used for the subsequent inoculation of additional plants.

Even though the coat protein of tobacco mosaic virus (TMV) has been the most utilized system for expression of antigenic epitopes, other viruses such as alfalfa mosaic virus, plum pox virus, potato virus X, and tomato bushy stunt virus have also been extensively employed [58].

One of the reasons for the use of various plant viruses was to overcome the apparent size limitation (<1 kb) that prevented inclusion of large peptides and inhibited virus assembly (Avesani et al. 2007). For example, only peptides of less than 25 amino acids had been successfully introduced into the tobacco mosaic virus coat protein whereas the alfalfa mosaic virus could accept an additional 37-amino acid peptide at its N-terminus forms and still be able to form particles [58]. Werner et al. [59] have recently shown that large inserts (>1 kb) can be fused to the coat protein and expressed, provided that suitable linkers are included.

The demonstration of the utility of chimeric plant viruses as carriers for antigens was first reported in 1994 when purified chimeric CPMV particles expressing epitopes derived from human rhinovirus 14 (HRV-14) and HIV-1 elicited a strong immune response in rabbits [60]. Numerous reports have since confirmed that plant viruses can be effective vectors for expression of antigens (Table 18.3) and can provide complete protection in challenge trials [61].

On the basis of the approach described above for gene replacement, Icon Genetics (Halle, Germany) has developed viral replicons that can be delivered through infection with *Agrobacterium*, a process termed "magnifection" [62]. Using this system, foreign protein can be transiently expressed at up to 80% of total soluble protein, including oligomeric proteins [63].

TABLE 18.3
Proteins Expressed in Plants Using Plant Viruses as Vectors

Plant Virus	Expressed Sequence	Expression Levels	Plant Used for Infection	References
Tobacco mosaic virus	Hypervariable region 1 of envelope protein 2 of hepatitis C virus fused to cholera toxin B subunit	0.04%	Tobacco	[138]
	Single-chain Fv fragment of immunoglobulin	0.003%	Tobacco	[191]
	Structural protein VP1	0.02%	NR	[192]
	Peptide of outer membrane (protein F) from *P. aeruginosa*	NR	Tobacco	[193]
	Peptides of circumsporozoite protein (*Plasmodium falciparum*)	0.003%	Tobacco	[194]
	Peptide of glycoprotein S (Murine hepatitis virus)	NR	Tobacco	[195]
	Betv1, a major birch pollen antigen	200 μg/g	*Nicotiana benthamiana*	[196]
	Core neutralizing epitope of porcine epidemic diarrhea virus	5.0%	Tobacco	[197]
	gDc from bovine herpes virus type 1 (BHV-1)	20 μg/g	*Nicotiana benthamiana*	[198]
	Two T-cell protective cancer epitopes		Tobacco	[199]
	15 amino acids of the poliovirus peptide (PVP)	0.05%	Tobacco	[200]
Potato virus X	HIV-gp41	NR	Tobacco	[201]
	D2 peptide of fibronectin-binding protein FnBP from *Staphylococcus aureus*	0.0003%	Tobacco	[202]
	Human papilloma virus E7 protein	0.0004%	Tobacco	[203]
Potato virus X/ Cowpea mosaic virus	Hepatitis B virus core antigen	10 μg/g	Potato and cowpea	[204]
Cowpea mosaic virus	VP1 antigen from foot-and-mouth disease virus	NR	Cowpea	[205]
	VP2 protein from Min enteritis virus	0.0002%	Cowpea	[61]
	HIV-gp41	NR	Cowpea	[60]
	HIV-gp41	0.002%	Cowpea	[206]
	VP1 protein from human rhinovirus	NR	Cowpea	[60]
	D2 peptide of fibronectin-binding protein FnBP from *Staphylococcus aureus*	0.005%	Cowpea	[202]
	Peptide of outer membrane (protein F) from *P. aeruginosa*	0.005%	Cowpea	[207]
	VP2 protein from canine parvovirus fused to β-glucuronidase	0.002%	Cowpea	[208]
Alfalfa mosaic virus	Peptides of glycoprotein and nucleoprotein from rabies virus	0.0005%	Spinach	[209]
	V3 loop of HIV-gp120	NR	Tobacco	[210]
	Peptides of G protein from respiratory syncytial virus	0.006%	Tobacco	[211]
Tomato bushy stunt virus	V3 loop of HIV-gp120	0.03%	Tobacco	[212]
	Nucleocapsid protein HIV-p24	0.4%	Tobacco	[213,214]
Plum pox virus	Major structural protein VP60 of rabbit hemorrhagic disease virus	NR	Tobacco	[215]
	VP2 protein from canine parvovirus fused to β-glucuronidase	NR	Tobacco	[215]

TABLE 18.3 (continued)
Proteins Expressed in Plants Using Plant Viruses as Vectors

Plant Virus	Expressed Sequence	Expression Levels	Plant Used for Infection	References
Cucumber mosaic virus	Synthetic peptide derived from sequences of the HCV envelope protein E2 (R9 mimotope)	10 mg/100 g	Tobacco	[216–218]
	Neutralizing epitopes of the fusion protein the hemagglutinin-neuraminidase (HN) from the Newcastle disease virus	NR	*Nicotiana benthamiana*	[219]
Bean yellow dwarf virus	Norwalk virus capsid protein	up to 1.2% TSP	Tobacco NT1 cell	[220]

Note: Expression levels are as reported in the literature and indicate percentage of total soluble protein (%TSP) unless indicated otherwise. NR, not reported.

18.4 PRODUCTION OF ANTIBODIES IN PLANTS (PLANTIBODIES)

Antibodies were the first bioactive molecules expressed in transgenic plants [64]. Since then, a number of different antibodies have been produced in a variety of plants, and expressed in different tissues and different subcellular compartments for different purposes (Table 18.4). This is not surprising, as there is a wide range of uses for antibodies in the medical, industrial, research, and diagnostic fields. Why use plant to produce antibodies? The main reason is cost. Currently, transgenic plants represent the most productive and economical system for making recombinant antibodies [65]. The cost of producing monoclonal antibodies in plants is significantly less than traditional fermentation methods. Smith and Glick [66] have made an estimate that assuming a best-case yield of 2 g of recoverable protein per liter of bacteria, the minimum cost of production

TABLE 18.4
Antibodies Produced in Transgenic Plants

Streptococcal antigen I or II	Tobacco	500 μg/g	[73]
Streptococcus mutans adhesion protein antibody	Tobacco	80 μg/g	[76]
Herpes simplex-2 glycoprotein B antibody	Tobacco	NR	[74]
Herpes simplex-2 glycoprotein B antibody	Rice	NR	[295]
Recombinant T84.66/GS8 diabody	Tobacco	~ 0.5–1.5 μg/g	[75]
Single-chain Fv antibodies	Tobacco	4%–6.8%	[112]
Non-Hodgkins lymphoma idiotypes	Tobacco	30.2 μg/g	[191]
Human IgG	Alfalfa	1.0%	[17]
Human carcinoembryonic antigen	*Nicotiana benthamiana*	<1%	[55]
Human carcinoembryonic antigen	Rice	29.0 μg/g leaves; 32.0 μg/g seed	[78]
Human carcinoembryonic antigen	Wheat	0.1 μg/g leaves; 1.5 μg/g seed	[78]
Several scFv antibodies	Tobacco	3%–6.8%	[112]
ScFv antibodies against a phytoplasma	Tobacco	0.04%	[296]
Anti-zearalenone single-chain Fv (scFv)	*Arabidopsis*	11.2 ng/mL	[297]

Note: Expression levels are as reported in the literature and indicate percentage of total soluble protein (%TSP) unless indicated otherwise. NR, not reported.

(assuming $2 per litre for the growth medium and purification costs) would be approximately $1000 per kilogram, not including salaries. Since multimeric proteins, such as antibodies, are assembled with very low efficiency in bacteria (Chapter 15), the yield could be much lower which would have a direct impact on costs. It is not surprising, then, that the cost of antibodies produced using traditional microbial fermentation can be as high as $1000 per gram [67]. Production in plants is expected to reduce dramatically the cost of antibodies. The land, infrastructure, and expertise necessary for harvesting and processing large volumes of plant material already exist, which would result in drastically reduced capital costs.

The costs of producing an IgG from alfalfa grown in a 250 m^2 greenhouse are estimated to be within $500–$600 per gram, compared with $5000 per gram for the hybridoma produced antibody [17]. In another study, the cost of producing 1 g of purified IgA in plants was estimated to be well below $50 per gram, which compares favorably with the costs of cell culture ($1000 per gram) or transgenic animal production systems ($100 per gram) [19].

Corn has been the system of choice for production of many recombinant proteins including antibodies. The main reason is the high protein content of the corn kernel, which accounts for approximately 10% of the dry weight. Even if the antibodies accounted for only 0.5% of the total grain protein, it would cost less than $200 to produce 1 kg of antibody in corn kernels; however, it has already been shown that corn can produce recombinant proteins up to 5.7% of the total protein [68]. An additional advantage is that plantibody production could be scaled up or down depending on the demand by increasing or decreasing the acreage of the antibody-producing plants. Plant-produced antibodies are predicted to cost as little as $10–$100 per gram, approximately 10–100 times less expensive than antibodies produced in bacteria [67].

Almost certainly, the biggest component of cost of antibodies will be purification. In theory, purification of plantibodies should be straightforward, using standard procedures. In most cases, the initial processing steps of plant material will benefit from technology employed for food processing, whereas the final steps will consist of standard chromatography procedures. Protein degradation that may occur during extraction can be minimized by the addition of protein stabilizers and proteinase inhibitors, although proteolysis *in planta* represents a challenge [69]. Stevens et al. [70] had suggested that proteolytic degradation in leaves is, in part, linked to the natural process of senescence. This indicates that the physiological state of the plant might have an impact on antibody integrity.

Expression in seeds is ideal because it assures adequate storage properties and flexibility in processing management and batch production. Indeed, efficient purification schemes have already been described [31]. Alternative methods based on the use of oleosin- or polymer fusions to facilitate purification of recombinant proteins are also available [71].

On the other hand, it is conceivable that for some uses the plantibody would not even need to be purified. Delivery of the antibody could be achieved by direct consumption of the plant tissue containing the plantibody [72,73]. This is very important from the standpoint of safety as plants do not serve as hosts for human viruses or prions, unlike hybridomas.

There are no plantibodies yet in commercial production, however, there are several candidates that are potentially useful as human therapeutics. The most advanced is a chimeric secretory IgG–IgA antibody called CaroRx*TM*, against a surface antigen of *Streptococcus mutans*, the bacteria that causes tooth decay, which prevents it from binding to teeth, thereby reducing cavities [73]. Unexpectedly, application protects from recolonization and this may last for up to 2 years, although the antibody was applied for only 3 weeks and functional antibody was detected on the teeth for only 3 days following the final application of the antibody. This antibody has reached a pilot phase II trial. Phase I/II confirmatory clinical trials are underway.

Another antibody, a humanized anti-herpes-simplex virus (HSV) antibody prepared in soybean, was effective in the prevention of vaginal HSV-2 transmission in mouse [74]. A third antibody was developed in rice and wheat against a carcinoembryonic antigen [75]. This antigen, a cell-surface glycoprotein, is one of the best-characterized tumor-associated antigens and antibodies against it are usually employed for diagnostic and therapy. Levels of the antibody in seeds did not show a significant decline after storage at room temperature for 6 months [75].

Finally, a tumor-specific vaccine was prepared in tobacco for the treatment of lymphoma using a modified plant virus [76]. The antibody genes for expression of an scFv were derived from a mouse B-cell lymphoma. Mice were immunized with the plant-produced scFv and anti-idiotypic antibodies (antibodies against the binding portion of the antibody) were generated. The mice were protected against infection by the lymphoma that produced the original antibody [76].

Plant viral vectors have also been employed to produce therapeutically useful antibodies in plants, including an antibody against the colorectal cancer-associated antigen GA733 [55], which upon immunization in mice elicited a comparable humoral response to that using antigen produced in insect cell culture [77].

Although plant antibodies are normally found to be properly folded and functional [73], differential glycosylation by the plant still remains a major constraint for applications in human healthcare. There is still concern over the potential immunogenicity of plant-specific complex N-glycans, which are present on the heavy chain of plant-derived antibodies (Section 18.8). There have been several approaches to prevent addition of complex N-glycans to recombinant antibodies when glycosylation-dependent effector functions are not needed. One approach is the removal of peptide recognition sequences for N-glycosylation; another is the addition of an endoplasmic reticulum (ER) retention C-terminal (KDEL in amino acid code) sequence which avoids Golgi-mediated modifications [78]. A final approach is the humanization of plant glycans, and to this end, human β-1,4-galactosyltransferase was stably expressed in tobacco plants which were crossed with plants expressing a murine antibody which resulted in a plantibody with partially galactosylated N-glycans [79]. Interestingly, the glycosylation profile of endogenous proteins and of a recombinant immunoglobulin in tobacco leaves also seems to be affected by senescence [70].

Recently a novel strategy was reported involving the plant-based production of a fusion molecule of an antigen and the corresponding antibody [80]. The HIV-p24 antigen was expressed as a genetic fusion with two constant region sequences from human Ig α-chain and targeted to the endomembrane system. This allowed to increase the expression approximately 13-fold higher than with HIV-p24 expressed alone and to enhance the immunological properties. The fusion elicited T-cell and antibody responses in immunized mice [80].

18.5 PRODUCTION OF ANTIGENS IN PLANTS

Vaccines have represented the most efficient medical intervention to prevent disease for many years. It is estimated that in the next 5–15 years, new vaccines and new vaccine delivery technology will fundamentally change how clinicians prevent and treat disease, which will have a substantial impact on public health [81]. Vaccines have been based on live, attenuated organisms, purified antigens or DNA coding for specific antigens. Purified antigens (subunit vaccines) are usually delivered at a set dose and have constituted a relatively simple and uniform material for administration, generally prepared from recombinant sources. There are several systems for expression of antigens as discussed in Section 3.1. Considerable developments have occurred since Charles Arntzen first envisaged the idea of transgenic plant vaccines in the early 1990s. There are now a number of examples demonstrating the successful expression of subunit candidate vaccines both for humans and animals in transgenic plants (Table 18.5). These include antigens from bacterial and viral sources that infect humans, domestic or wild animals and representing several classes of protein, including secreted toxins and cell or viral surface antigens. Levels of expression vary greatly depending on the protein expressed and the species of plant used to achieve expression (Table 18.5). As discussed in Section 18.7 there are several factors that affect the final yield. Different approaches have resulted in high levels of expression of several antigens but it is difficult to make comparisons since specific antigens have rarely been tested in multiple systems. Exceptions include the subunit B of the cholera toxin and the closely related heat-labile enterotoxin from *E. coli*, which

TABLE 18.5
Antigens Expressed in Transgenic Plants

Antigen	Plant System	Expression Levels	Reference
Yersinia pestis F1-V antigen	Tomato	0.9%–4.6%	[221]
Yersinia pestis F1, V, and F1-V antigens	*Nicotiana benthamiana*	1 mg/g	[222]
Cholera toxin B subunit protein	Potato	0.3%	[223,224]
	Tomato	0.02%–0.04%	[225,226]
	Tomato	0.081%	[227]
Cholera toxin B subunit-human glutamate decarboxylase fusion protein	Potato	0.4%	[137]
Human secreted alkaline phosphatase	Tobacco NT1 cell	27 mg/L	[228]
Polyepitope combining tandem repeats of a protective loop-forming B-cell epitope (H386-400) of the measles virus hemagglutinin protein with a human promiscuous, measles-unrelated T-cell epitope (tt830-844)	Carrot	NR	[229]
Measles virus hemagglutinin glycoprotein	Carrot	2%	[230]
Major structural protein VP60 of rabbit hemorrhagic disease virus	Potato	0.30%	[231,232]
VP1 protein, an epitope and the coat protein of enterovirus 71	Tomato	NR	[10]
Heat labile enterotoxin from *E. coli*	Maize	0.8%	[233]
	Potato	−4 to 10 μg/g	[234]
	Somatic embryos of Siberian ginseng	0.36%	[197]
	Lettuce	1.0%–2.0%	[235]
	Soybean	2.4%	[236]
Rotavirus VP7 fused to the cholera toxin B subunit gene	Potato	0.01%	[237]
Rotavirus VP7 protein	Potato	3.6–4.0 μg/mg	[238]
	Potato	3.84 g/mg	[239]
Rotavirus VP6 protein	Alfalfa	0.28%	[240]
eBRV4 antigen of the VP4 protein from bovine rotavirus fused to the *uidA* gene	Alfalfa	0.4 and 0.9 mg/g	[241]
Human lactoferrin	Potato	0.1%	[92]
Human milk protein β-casein	Potato	0.01	[93]
Fusion of a gut adhesion molecule and linear B- and T-cell epitopes from a fish virus	Potato	NR	[242]
M. tuberculosis ESAT6 antigen	*Nicotiana benthamiana*	800 μg/g	[243]
	Tobacco	0.5%–1%	[244]
Epitope from foot-and-mouth disease virus fused to the glucuronidase (*uidA*) reporter gene	Alfalfa	NR	[245]
Epitope of the foot-and-mouth disease virus fused with a hepatitis B core protein	Tobacco	NR	[246]
VP1 antigen from foot-and-mouth disease virus	Tobacco, alfalfa	2%–3%	[186,192]
Middle (M) and major/small (S) surface proteins from hepatitis B virus	Potato	NR	[247]
Hepatitis B virus surface antigen	Cherry tomatillo	300 ng/g in leaves and 10 ng/g in fruits, fresh weight	[248]

TABLE 18.5 (continued)
Antigens Expressed in Transgenic Plants

Antigen	Plant System	Expression Levels	Reference
	Tobacco	0.01%	[249]
	Lupine and lettuce	0.15 and 0.0055 μg/g, respectively	[89]
	Potato	0.002%	[250]
	Potato	~8 μg/g tuber protein	[251]
	Soybean and tobacco suspension cultures	20–22 mg/L and 2 mg/L, respectively	[252]
	Nicotiana benthamiana	7.14%	[253]
21-mer epitope from parvovirus	*Arabidopsis*	3%	[254]
Human growth hormone	*Nicotiana benthamiana*	10%	[255]
UreB antigen gene from a new *Helicobacter pylori*	Tobacco, rice	NR	[256]
ApxIIA, a bacterial exotoxin *Actinobacillus pleuropneumoniae*	Tobacco	0.1%	[38]
A1 leukotoxin from *Mannheimia haemolytica*	White clover	1%	[257]
Immunogenic proteins of Newcastle disease virus	Potato	0.3–0.6 μg/mg of total leaf protein	[258]
Fusion protein of Newcastle disease virus	Maize	0.95%–3%	[164]
Major F4ac fimbrial subunit protein from *E. coli*	Tobacco	1%	[259]
Fimbrial adhesin FaeG from *E. coli*	Potato	1%	[260]
Major K99 fimbrial subunit, FanC from *E. coli*	Soybean	0.5%	[261]
Middle protein of HBV containing the surface S and preS2 antigen	Potato	<0.1%	[262]
Carboxy-terminal host cell-binding domain of *E. coli* O157:H7 intimin	Tobacco	3–13 μg/g of total plant material	[263]
Synthetic neutralizing epitope gene of Porcine epidemic diarrhea virus	Tobacco	2.1%	[264]
HIV-Tat	Spinach	300–500 μg/g of leaf tissue	[265]
	Tomato	1 μg/mg dry weight	[122]
HIV-ENV and GAG fused to the surface protein antigen of hepatitis B virus	Tomato	~0.3 μg/g	[266]
HIV-1 p24 capsid protein	Tobacco	3.5 mg/g	[214]
Fusion HIV-1 gp120 V3/cholera toxin B subunit	Potato	0.002%–0.004%	[140]
SIV Gag	Potato	0.006%–0.014%	[141]
Spike protein (S1) of the severe acute respiratory syndrome coronavirus	Tobacco	NR	[185]
Hemagglutinin protein of rinderpest virus	Peanut	10 μg/g of peanut leaves	[267]
HPV16 E7	*Nicotiana benthamiana*	0.4 μg/g	[150]
Rabies glycoprotein	Tomato	<1%	[268]
Rabies nucleoprotein	Tomato	1%–5%	[173]
Fusion of epitopes from rabies virus glycoprotein and nucleoprotein	Alfalfa	0.4 ± 0.07 mg/g of fresh leaf tissue	[209]
Cholera toxin B subunit-ubiquitin fusion	Tobacco	0.9%	[269]

(*continued*)

TABLE 18.5 (continued)
Antigens Expressed in Transgenic Plants

Antigen	Plant System	Expression Levels	Reference
	Tobacco	0.2%	[270]
Rotavirus VP2 and VP6 proteins	Tomato	1%	[271]
Fusion (F) protein from respiratory syncytial virus	Tomato	0.003%	[272]
Porphyromonas gingivalis fimbrial antigen fused to the cholera toxin B subunit	Potato	0.33%	[273]
Sunflower seed albumin	Lupine	5%	[274]
Colorectal cancer antigen GA733-2	Tobacco	0.4%	[77]
Structural subunit A of the bundle-forming pilus from *E. coli*	Tobacco	7.7%	[275]
Immunocontraceptive epitope fused to the heat labile enterotoxin from *E. coli*	Tomato	37.8 μg/g	[142]
AB5 toxin from *E. coli*	Tobacco NT-1 cells	6.5–8.2 μg/g	[276]
Fusion protein gene of Newcastle disease virus	Potato	0.25–0.55 g/100 g of fresh tuber tissue	[277]
Murine rotavirus VP6 protein	Potato	0.01%	[278]
Fusion of the cholera toxin B and A2 subunits with a rotavirus NSP4 enterotoxin and *E. coli* fimbrial antigen genes	Potato	1 μg/3 g fresh weight	[279]
Heat-shock protein A from *H. pylori*	Tobacco	<1%	[280]

Note: Expression levels are as reported in the literature and indicate percentage of total soluble protein (%TSP) unless indicated otherwise. NR, not reported.

have been expressed in multiple plant systems with a range of expression from 0.2% of total soluble protein in potato tubers [82], 4%–5% of total soluble protein in tobacco leaves using a chloroplast expression system [19] up to 12% of total soluble protein in corn seed [83]. Expression in tobacco chloroplasts usually results in high yields (Section 3.1).

High level of expression of antigens in plants is one of the main goals as this will result in downstream processing and purification at low cost. On the other hand, in cases where the plant tissue is edible, a defined antigen dose capable of inducing protection can be administered in a well-defined amount of plant material. In the case of the heat-labile enterotoxin from *E. coli* expressed in corn, the plant tissue contained a sufficiently high concentration of the antigen such that a 1 mg dose, estimated to be the ideal oral dosing, corresponded to approximately 2 g of edible tissue [84]. A reduction in the amount of material to be consumed is anticipated by increasing the yield, because usually, oral immunization requires higher doses of antigen than parenteral immunizations [7].

18.5.1 Human Clinical Trials with Oral Plant-Based Vaccines

The functionality of plant-based antigens has been tested in a number of experiments in mice and other animals. In the vast majority of the cases, plant-based antigens have been able to elicit a strong immune response and to confer protection against challenge with the pathogen. These results have paved the way for several clinical trials aimed to assessing human immune responses to plant-produced recombinant proteins. So far, three plant-produced antigens have been tested in phase I human clinical trials, the heat-labile enterotoxin from *E. coli*, the capsid protein from Norwalk virus, and the surface antigen from the hepatitis B virus.

The heat-labile enterotoxin from *E. coli* has been tested twice in human clinical trials, in the first trial, 14 adult volunteers ingested three doses of transgenic potatoes (containing 3.7–15.7 μg/g of antigen) or control wild-type potato. Serum antibody responses were detected in 10 out of the 11 volunteers and 8 out of the 11 developed neutralization titers of more than 1:100 [85]. This trial was the proof of principle that humans could develop serum and mucosal immune response to antigen delivered in transgenic plants. Recently, the same antigen was delivered to volunteers in transgenic corn with seven out of the nine volunteers developing increases in serum IgG and four of the volunteers also developing stool IgA [86].

In another trial, 20 human volunteers ate two or three doses of transgenic potatoes expressing 215–751 μg of the Norwalk virus capsid protein [87]. Out of the 20 volunteers, 19 developed significant increases in the numbers of specific IgA antibody-secreting cells, 4 developed serum IgG and 6 developed specific stool IgA, although the levels of serum antibody were not high. This variation in immune response was probably due to the inconsistent assembly of the antigen into virus-like particles and to the possible presence of preexisting antibodies to the antigen, which might have had an effect [23].

The surface antigen from the hepatitis B virus was utilized in a randomized, placebo-controlled, double-blind trial conducted in volunteers previously immunized, 1–15 years earlier, with the licensed hepatitis B vaccine [88]. A total of 42 subjects were enrolled in the study and received either three doses of placebo potatoes or transgenic potatoes (two or three doses). None of the volunteers who ingested control potatoes had any change in their antibody titers during the study. On the other hand, 63% of volunteers (10 out of 16) who consumed three doses of HBsAg-containing transgenic potatoes showed marked increases in antibody titers compared with titers at day 0 [88]. Thus, an antigen from a nonenteric pathogen was capable to elicit a immune response with no buffering of stomach pH and without the presence of a mucosal adjuvant. The same antigen was expressed in lettuce and fed to three seronegative volunteers [89]. The volunteers received about 0.2–1 μg of antigen in a 200 g dose of lettuce. In comparison, the commercial hepatitis B vaccine contains 10 μg of antigen per adult dose [88]. Two of the three volunteers produced short-lived anti-HBs antibody titers, which were detectable 2 weeks after the second immunization and were no longer detectable after the additional 2 weeks.

Mucosal immunologists recognize that the heat-labile enterotoxin from *E. coli* and its relative the B subunit of cholera toxin are particular antigens with highly immunogenic properties. The fact that two different antigens were also immunogenic in humans after oral administration was rather encouraging.

As discussed in Section 18.2, concerns have been raised about the potential for induction of oral tolerance to the protein contained in a transgenic plant, since oral tolerance is the usual result when the mucosal system encounters food proteins. Although the mechanisms of oral tolerance remain unclear, it is likely that this concern is more relevant for pathogens transmitted by the parenteral route (for instance hepatitis B and malaria) than pathogens whose natural route of transmission is via the gastrointestinal tract [88]. Immune tolerance to parenterally administered proteins can occur after multiple small oral doses of the protein [88]; therefore, it is unlikely that oral vaccination with plant-based antigens would result in tolerance to parenterally administered protein since the number of doses of the oral vaccine would be very small. As the phase I studies have demonstrated, the plant-based antigen is recognized and processed as an antigen and elicits an immune response. The possibility of incorporating mucosal adjuvants may considerably improve the immune response.

18.6 PRODUCTION OF BIOPHARMACEUTICAL PROTEINS IN PLANTS

The utility of plants as biofactories is not restricted to antigens and antibodies. A number of proteins of pharmaceutical and industrial importance have been produced in transgenic plants, including glucocerebrosidase and granulocyte–macrophage colony stimulating factor (Table 18.6) two of the world's most expensive drugs [90].

The main concern has always been that plant-based proteins be properly folded and processed. In this respect Staub et al. [91], when successfully expressing human somatotrophin, showed that tobacco chloroplasts correctly process proteins requiring disulfide bonds for biological activity. Even though somatotrophin can be produced in recombinant *E. coli,* plants are able to produce somatotrophin that needs no postextraction chemical processing.

TABLE 18.6
Pharmaceutical and Industrial Proteins Produced in Transgenic Plants

Protein	Host	Expression Levels	References
Human enkephalins	*Arabidopsis*	0.10% seed protein	[100]
Human lactoferrin	Potato	0.10% TSP	[92]
Human aprotinin	Maize	NR	[90]
Human epidermal growth factor	Tobacco	0.001%	[281]
Human acid β-glucosidase	Tobacco	3%	[282]
Human hemoglobin α, β	Tobacco	0.05%	[101]
Human serum albumin	Tobacco	0.02%	[100]
Human granulocyte–macrophage colony-stimulating factor	Tobacco	NR	[99]
Human hirudin	Canola (*Brassica napus*)	0.30%	[101]
Human interferon-α	Rice	Turnip (*Brassica rapa*)	[101]
Human interferon-β	Tobacco	<0.01%	[100]
Human erythropoietin	Tobacco	<0.01	[100]
Human α-1-antitrypsin	Rice	NR	[90]
Human protein C	Tobacco	<0.01%	[101]
Human placental alkaline phosphatase	Tobacco	~3%	[283]
Angiotensin-converting enzyme	Tobacco, tomato	NR	[90]
α-Trichosanthin from TMV-U1 subgenomic coat protein	*Nicotiana benthamiana*	2.00%	[90]
Glucocerebrosidase	Tobacco	1.00%–10.00%	[101]
Transglutaminase	Rice	NR	[284]
Human interleukin-12	Tobacco	17.5–40 ng/g of fresh weight	[102]
Mouse interleukin-12	Tomato	3.4–7.3 μg/g fresh weight	[103]
E. coli β-glucuronidase and chicken egg-white avidin	Maize	0.7% and 5.7% respectively	[68]
Human insulin	Potato	0.05%	[105]
Human cytotoxic T-lymphocyte antigen 4-immunoglobulin	Rice cell suspension culture	31.4 mg/L	[285]
Human growth hormone	Tobacco	0.16%	[104]
	Nicotiana benthamiana	10%	[255]
Acetylcholinesterase	Tomato	100 mU/g leaf (specific activity)	[286]
Human homotrimeric collagen I	Tobacco	10 mg/100 g of powered plants	[94]
Murine adenosine deaminase	Maize	0.8%	[287]
Human muscarinic cholinergic receptors	Tobacco	2.0–2.5 pmol [^{3}H]QNB bound per milligram membrane protein	[288]

Note: Expression levels are as reported in the literature and indicate percentage of total soluble protein (%TSP) unless indicated otherwise. NR, not reported.

Various proteins of industrial interest such as the human milk proteins lactoferrin [92] and β-casein [93] have been produced in transgenic plants to be employed as a supplement for infant formulas to enhance nutrition, digestibility, and antimicrobial properties.

Recently, tobacco was modified with the human collagen I gene *pro1*α(*I*) [94]. Collagens are very important molecules employed in the cosmetics, medical, and food industries. They are generally extracted from animal tissues and may represent a contamination risk if the tissue is infected. Several expression systems have been developed that produce procollagens, but these have to be chemically modified ex vivo to produce mature collagens [94]. The procollagen produced in tobacco cells was spontaneously processed into mature collagen during extraction, which clearly represents a significant advantage for large-scale, low-cost production of collagen.

Plants are not only considered as a source of pharmaceuticals and nutrients but they are being increasingly seen as a source for biofuels of which methanol is currently the most popular [95]. There have been several biodegradable polymers such as polyhydroxyalkanoates as well as a protein-based polymers produced in tobacco [96,97]. One of them is a polymer similar to elastin, one of the strongest natural fibers [98]. These polymers might be used as transducers, super-absorbents, and biodegradable plastics, or in various medical applications such as tissue reconstruction surgery [90].

There are a number of proteins with important roles in the stimulation or modulation of the immune responses such as cytokines, chemokines, etc. Many of these compounds have been synthesized successfully in plants such as the human granulocyte–macrophage colony-stimulating factor [99], human interferon-α and -ß [100,101], interleukin (IL)-12 both human [102] and murine [103], etc. Similarly, some human hormones such as the growth hormone [104], erythropoietin [100], and insulin [105] have been produced in plants.

One area where plant-based recombinant proteins are having a dramatic impact is in the area of diagnostics reagents and avidin is a case study. Avidin is widely used as a diagnostic reagent and is normally found in egg white, from which it is purified. The cDNA was expressed in maize and could be reproducibly produced at 230 mg/kg of maize seed, which was, in the authors' estimation, 10-fold less expensive than avidin extracted from eggs [106,107]. The maize avidin is functional and now commercially available (Sigma-Aldrich product # A8706). Similarly, plant-based β-glucuronidase and aprotinin are now also commercially available [108,109].

There is a large demand for many pharmaceutical proteins and this poses a burden on any transgenic production system to meet the demand. Transgenic plants could be rapidly scaled up to field scale cultivation. For example, the worldwide demand for human serum albumin (about 550 metric tons per year) could be met by 30,000 hectares of land (assuming an expression level of 1% TSP in tobacco), which is less than one thousandth of the total cultivated soil in the United States [110].

18.7 STRATEGIES TO IMPROVE PROTEIN YIELDS

Production costs are the key issues for recombinant protein production and they include the generation, growing and harvesting of the recombinant material, and downstream processing and purification, which tend to increase with tissue complexity. If the recombinant product is obtained in high levels, the amounts of biomass required and the processing and purification of the product can be greatly reduced. Thus, the achievement of high expression levels is a major goal in all systems. In the case of plants, several strategies have been developed to increase the levels of recombinant proteins and they have focused on transcription, transcript stability, and translation.

18.7.1 Transcription

Increasing the level of transcription of integrated sequences has been the main approach to boost expression. To this end, the choice of an adequate promoter is critical. Several promoters have been tested from pathogens and plants as well as some synthetic promoters. The promoters tested included those of plant viruses and from genes highly expressed in dicot leaves. The cauliflower mosaic virus 35S promoter [111] has been the main promoter employed to drive the expression of recombinant proteins in a number of plants. Using this promoter, fairly high levels of accumulation of recombinant proteins have been achieved [103,112]. Other plant virus promoters employed include those of cassava vein mosaic virus [113], the C1 promoter of cotton leaf curl Multan virus [114], and the promoter of component 8 of milk vetch dwarf virus [115]. Promoters from *Agrobacterium tumefaciens* have also been used to express recombinant proteins in plants, such as the nopaline synthase promoter [116].

Some of the plant promoters employed include that of the small subunit of ribulose-bisphosphate carboxylase [117] and the maize polyubiquitin-1 promoter [118].

Since many pharmaceutical proteins may have possible toxic side effects in plant cells, expression of recombinant proteins should be restricted to the target tissue to prevent toxicity. In this sense, tissue-specific promoters have been employed such as the embryo-specific maize globulin-1 promoter [119], the endosperm-specific maize 27 kDa zein and barley D hordein promoters [120,121], and the tomato fruit-specific E8 promoter [122]. Inducible promoters also allow for expression in a target tissue following a specific treatment, such as chemical spraying [123,124]. In this case, protein accumulation can be restricted to define a timeframe to limit detrimental effects on plant health.

Expression in the chloroplast requires the use of promoters of highly expressed plastid genome sequences, including the constitutive promoter of the rRNA operon [19,52]. The high expression level reached is probably a consequence of having up to 10,000 plastid genomes per leaf cell.

Another approach to boost transcript levels has been to stack transcription units. They combine the most active sequences of well-characterized promoters such as the cauliflower mosaic virus 35S promoter or the *Agrobacterium* Ti plasmid mannopine synthetase promoter. The synthetic promoter may be several-fold more active than the natural promoter [125,126].

Another way to increase transcriptional activity is to include regulatory sequences on expression cassettes such as scaffold attachment or matrix attachment regions. These sequences have been reported to interact with plant nuclear scaffolds in vitro, and are thought to be involved in the recruitment of transcription factors to promoters [127]. For that reason there have been several attempts at using them to increase the level of transcription or recombinant proteins. Although these sequences sometimes can boost the expression of plant genes by an order of magnitude [128], their presence does not necessarily result in an increase in gene expression [129] but rather in a reduction in position effect or in gene silencing [130,131].

18.7.2 Stabilizing the Message and Ensuring Correct Message Processing

As a recombinant message is transcribed, the stability of that message is important to ensure high-level expression. The nontranslated region located downstream of the translation stop codon is critical for processing and should include signals targeting the message for polyadenylation. Message destabilizing sequences in the downstream region can greatly affect stability [132,133], and these sequences must be avoided when preparing gene constructs to express high levels of recombinant proteins. Several plant viral and plant 3′ untranslated regions have been utilized to process messages for recombinant proteins without destabilizing these messages. They include the cauliflower mosaic virus 35S terminator and the potato proteinase inhibitor II terminator [106]. For plant-based expression systems, eukaryotic genes are generally resynthesized or engineered to remove introns. Sequences such as consensus intron splice sites, message destabilizing sequences, and transcript termination sequences should be avoided in constructing synthetic genes. However, some plant intron sequences contribute positively to the expression level observed for their native genes, and these sequences can boost expression if inserted as synthetic introns into genes [134].

18.7.3 Translation

To increase translational efficiency of recombinant sequences several approaches have been tested. The use of both plant and viral 5′ nontranslated regions [135], optimization of sequences located around the translation start to fit the consensus initiation sequence of plants [136], codon-optimization to suit the expression host, removal of predicted mRNA secondary structures that might hinder translation, of internal ribosome entry sites that might compete for translation of the full message, and of rare codons, together contribute to increasing the translatability of the message.

18.7.4 Protein Fusions

Expression of a foreign protein in plants as a fusion to a plant protein present at high levels or to a recombinant protein that has been shown to be stably expressed in plants, usually results in the stable accumulation of the foreign protein. In addition, it presents a number of advantages in terms of stability, subcellular targeting, purification, and activity of product. In the case of plant-based vaccines, the fusion may enhance the immunogenicity of the incorporated antigens. Since the cholera toxin and its close relative the heat-labile toxin from *E. coli* have been shown to be effective in enhancing mucosal immune responses to vaccines [13], they have been employed extensively for fusing recombinant proteins. Fusion is usually done at the C-terminus of the B subunit.

Cholera toxin is one of the most potent mucosal adjuvants known. It is composed of an A and B subunit. The A subunit is a 28 kDa protein that ADP-ribosilates the stimulatory Gs protein of adenylate cyclase and is the toxic subunit. The B subunit is a homopentamer whose five units form a ring-like structure. Cholera toxin attaches itself to the intestinal epithelial cell via the B subunits. Feeding cholera toxin in combination with proteins may abrogate oral tolerance to the fed protein, elicit strong intestinal IgA and plasma IgG responses, and moreover induce dendritic cells (DCs) migration within Payer's patches [13]. In order to be effective as an adjuvant, the toxin has to be administered by a mucosal route simultaneously with the antigen. A closely related adjuvant to CT is the heat-labile toxin with similar structure and similar activities. Both of these toxins have been used successfully as mucosal adjuvants plant-based recombinant proteins. Among the antigens fused to either of these toxins are human insulin [105], human glutamate decarboxylase [137], a potentially neutralizing epitope of the hepatitis C virus [138], rotavirus NSP4 protein [139], HIV gp120 protein [140], the SIV Gag p27 protein [141], an immunocontraceptive epitope [142], the conserved galactosylceramide-binding domain (including the ELDKWA-neutralizing epitope) of the HIV-1 gp41 envelope protein [143], anthrax lethal factor [144], colony-forming fimbrial antigen CFA/I from enterotoxigenic *E. coli* [145], tuberculosis antigen ESAT-6 [146], and foot-and-mouth disease virus VP1 protein [147].

Fusions of antigens to other proteins have been also produced in plants and include the HIV-Tat gene fused to a rotavirus enterotoxin [148], ricin-B subunit from *Ricinus communis* fused to rotavirus NSP4 [149], and the E7 coding sequence from HPV16 fused to β-1,3-1,4-glucanase of *Clostridium thermocellum* [150].

As a potential therapy against diabetes, nonobese-diabetic mice were fed with plant-based cholera toxin B subunit gene genetically fused to human insulin [105] or to a nucleotide sequence encoding three copies of tandemly repeated diabetes-associated autoantigen, the B chain of human insulin InsB3 [151], which led to a substantial reduction in pancreatic islet inflammation (insulitis), and a delay in the progression of clinical diabetes. An alternative approach was also successful to inhibit the development of diabetes in nonobese diabetic mouse, in which murine glutamic acid decarboxylase [152] or human GAD65 and murine IL-4 were expressed in tobacco plants [153].

18.8 GLYCOSYLATION

Most of the therapeutic proteins are subjected to several posttranslational modifications, of which glycosylation is the most common. Glycosylation is equally performed by plant and animal cells. The oligosaccharide chain can be either *N*- or *O*-linked. *N*-glycosylation occurs in the endoplasmic reticulum (ER) and the primary oligosaccharide chain is further processed during its exit from the ER and passage through the Golgi apparatus, where *O*-glycosylation occurs. Glycosylation of a polypeptide chain may have a great impact on immunogenicity, specific activity, and the ligand–receptor interaction. Although plants represent a promising system for production of recombinant proteins, the late events of *N*-glycan modification, resulting in the presence of the plant-specific residues α-1,3-fucose and β-1,2-xylose, might pose problems for the production of fully functional therapeutic proteins. However, *N*-glycan modification may not occur in all cases as it has been shown that a recombinant antibody against the β-subunit of human chorionic gonadotropin retained in the endoplasmic reticulum of transformed plants lacked β-1,2-xylose and α-1,3-fucose residues [154].

Since the general eukaryotic protein synthesis pathway is conserved between plants and animals, folding and assembly, as well as transfer of an oligosaccharide precursor to *N*-glycosylation sites can be correctly accomplished in transgenic plant systems [155]. Analysis of plant-derived mouse IgG monoclonal antibody fused to KDEL ER-retention signal revealed that it was *N*-glycosylated homogeneously throughout the plant with mostly high-mannose-type *N*-glycans [156]. Both full-sized antibodies and various functional antibody derivatives have also been produced successfully in plants, including Fab fragments, scFvs, bispecific scFvs, single domain antibodies, and antibody fusion proteins (Section 18.4). Several studies have shown that while there were some differences in the glycan groups present on the recombinant antibody, neither the antibody nor the glycans were immunogenic [15]. Plant antibodies are normally found to be properly folded and functional [73].

Human lactoferrin is an 80 kDa glycoprotein which contains three potential *N*-glycosylation sites, although only the first two sites have complex *N*-glycans. Comparative analysis of *N*-glycosylation of authentic human lactoferrin and lactoferrin produced in maize and in tobacco has revealed differences in glycosylation, indicating that the first steps of *N*-glycosylation are similar in plants and humans and that the observed differences only arise from the specificity of the Golgi plant glycosyltransferases and from degradations of the matured plant *N*-glycans [157].

On the other hand, it is still not clear if recombinant proteins bearing plant *N*-glycans are indeed immunogenic. In an elegant study, Chargelegue et al. [158] tested the immunogenic effect in mice of plant glycans of transgenic murine monoclonal IgG antibodies and horseradish peroxidase. Because the same mouse strain was used as for generating the original mAb, the study specifically compared the immunogenicity of a self-protein and a plant protein displaying foreign plant glycans, with the self-protein displaying mammalian glycans. Encouragingly, the plant glycans of both the self and foreign (horseradish peroxidase) proteins were poorly immunogenic even when parenterally administered with alum adjuvant.

18.9 PRODUCTION OF VIRUS-LIKE PARTICLES IN PLANTS

Antigen presentation to the mucosa-associated lymphoid tissue is a key element in the response of the mucosal immune system [5]. A variety of mucosal antigen delivery systems has been tested, including inert systems as well as attenuated bacterial or viral vector systems [6]. Among the inert vaccine delivery systems tested extensively in the last 10-year period are various lipid-based structures such as liposomes, ISCOMs and cochlates, different types of biodegradable particles such as starch or co-polymers of poly(lactideco-glycolide) and different mucosa-binding proteins, including plant lectins and bacterial toxins such as cholera toxin and its close relative the heat-labile toxin from *E. coli*, to which antigens have been linked as gene fusion proteins (Section 7.4). Two types of live bacterial vectors have been tested: those based on attenuated pathogens such as *Salmonella*, BCG, or *Bordetella* and those based on commensal bacteria, such as lactobacilli, streptococci, and staphylococci [6]. The viral vectors developed include vaccinia, other poxviruses such as canary poxvirus, and adenoviruses [6]. Even though many of these systems have shown promise there is still no vector approved for human use.

One major obstacle in the development of mucosal vaccines is to be able to induce systemic as well as mucosal responses. Apart from immunizing with live viruses, this has proven to be a challenge but one way to overcome it has been by using adjuvants, which can promote the generation of antibodies to an antigen following immunization. However, many of these adjuvants do not enhance priming of cytotoxic T lymphocytes. The reason for this lies in the existence of two alternative antigen processing pathways, leading to stimulation of $CD4^+$ T cells, and in turn to the generation of antibodies, or stimulation of $CD8^+$ CTL. Thus, to generate a CTL response following immunization it is necessary to feed peptides into the correct processing pathway.

Recombinant virus-like particles (VLPs) represent a safe and highly immunogenic alternative for antigen presentation (Chapters 6, 9, 10, and 11). These self-assembling, nonreplicating viral core structures consisting of one or more viral coat proteins, can act as an adjuvant by carrying peptide sequences inside the antigen presenting cells and feeding into the endogenous processing pathway [159], a phenomenon known as "cross-priming." They are more immunogenic than recombinant proteins alone and are able to stimulate both the humoral and cellular arms of the immune system. VLPs can target dendritic cells and this represents a very important advantage, as targeting of this cell type is essential for activation of the innate and adaptive immune responses [160]. Exogenous VLPs can also be taken up and processed via the MHC class 1 pathway (cross-presentation) for activation of $CD8^+$ T cells, which are essential for clearance of intracellular pathogens such as viruses [160]. The efficacy of immunization with VLPs is best illustrated by the success of the HBV-like particles produced in *Saccharomyces cerevisiae*, which was the first recombinant vaccine developed [161].

VLPs are especially interesting from a mucosal vaccine point of view as they offer the opportunity to deliver the virus employing the natural route of transmission [162]. Oral delivery of VLP has been shown to induce both systemic and mucosal IgA responses both to the virus particles and to foreign epitopes expressed as chimeric proteins on the VLP surface, and did not require any external adjuvants [163]. Intranasal delivery of VLP seems more efficient and requires lower doses of antigen than oral delivery [162,164].

Chimeric particles can be prepared by incorporating homologous or heterologous antigens or epitopes into VLPs, but there may be some limitations as to the size of the molecules incorporated. Some VLPs, such as those based on the hepatitis B core antigen and the cowpea mosaic virus can only carry a small number of amino acids, but other VLPs can carry long N- or C-terminal extensions without disrupting the structure of the VLPs or altering their antigenic properties [165]. Antigens may be linked by genetic fusion or by chemical cross-linking of peptide epitopes to reactive sites on the VLPs [166].

VLPs can be tailored depending on whether the vaccine is to act as a prophylactic or therapeutic vaccine. The effectiveness of VLPs as therapeutic vaccines may be aided by the addition of adjuvants. The use of synthetic oligodeoxynucleotides containing immunostimulatory CpGs motifs as adjuvant, which stimulate dendritic cells via the Toll-like receptors, can be included by packaging CpGs into the VLPs [167]. However, when VLPs are administered with adjuvants such as alum, alhydrogel, SAF-MF, or algamulin, CTL priming does not take place, although the antibody response to the VLPs is enhanced [159]. On the other hand, the effectiveness of VLPs as T cell-based vaccines can be increased when administered as part of a DNA prime-VLP boost protocol [168].

VLPs have been produced from the capsid or envelope components of a wide variety of viruses to study virus assembly and for development of vaccines [160]. Vaccines from HBV and HPV VLPs have been successful, but VLPs from pathogens affecting immune cells and those that successfully evade the immune system, such as HIV-1 and hepatitis C virus have proven to be more challenging.

Several expression systems for production of VLPs have been tested [169] and include various mammalian cell lines, either transiently or stably transfected with viral expression vectors, the baculovirus/insect cell system, various species of yeast including *Saccharomyces cerevisiae* and *Pichia pastoris*, *E. coli* and other bacteria and green plants. *E. coli* does not allow for glycosylation, while yeast and baculovirus are limited to high mannose glycoprotein modification, and this is sometimes inconsistent [160]. Mammalian cell culture systems are favored for appropriate modifications and authentic assembly, but are a less controllable system and more costly for production [170]. Plants remain as the best production system for the reasons described in Section 18.3.

The production of properly folded VLPs in plants has been extensively reported. Antigens having the ability to assemble VLPs have been generally employed such as the surface and core antigens of Hepatitis B virus, the capsid protein of Norwalk virus, the L1 protein of human papillomavirus, the hemagglutinin/neuraminidase of paramyxoviruses, the Gag protein of HIV, several capsid proteins from rotavirus, etc. (Table 18.7). Many of these antigens as well as several capsid or core proteins from various plant viruses have also been employed as carriers to express a wide variety of antigenic peptides. However, not all antigens may form VLPs in plants. The hepatitis E virus capsid protein, which assemblies readily into VLPs in a baculovirus system, did not assembly adequately into VLPs in potato and this lead to a failure to elicit detectable antibodies in mice serum [171].

Chimeric VLPs offer enormous potential in specific, multi-epitope presentation but their success will be dependent on a judicious selection of the most relevant epitopes for vaccine efficacy. Although this knowledge is still lacking for many diseases, the rapid development and application of the genomic technologies is allowing the discovery of a number of potential vaccine candidates from many pathogenic organisms.

A wide variety of plant-based VLPs have been characterized by sucrose density gradients and by electron microscopy. In most cases, they have been shown to mimic the immunological properties of native VLPs and stimulate antibody and T-cell responses in mice (Santi et al. 2006).

TABLE 18.7
Virus-Like Particles Produced in Transgenic Plants

Antigen	Plant System	References
Hepatitis B virus surface antigen	Tobacco	[249]
	Potato	[250]
	Tobacco cells	[289]
	Lettuce	[89]
	Lupine callus	[89]
	Cherry tomatillo	[248]
	Soybean	[252]
Modified hepatitis B virus surface antigen	Tobacco cells	[290]
	Nicotiana benthamiana	[291]
Middle protein hepatitis B	*Nicotiana benthamiana*	[247]
	Potato	[291]
	Potato	[262]
Hep. B virus core antigen	Tobacco	[292]
	Nicotiana benthamiana	[253]
Norwalk virus capsid protein	Potato	[293]
	Tobacco	[293]
	Tomato	[291]
Human papilloma virus 11-L1 protein	Potato	[294]
Human papilloma virus 16-L1 protein	Tobacco	[294]
	Potato	[165]
	Tobacco	[165]
	Tobacco	[190]
Rotavirus VP2 and VP6 proteins	Tomato	[271]
Chimeric HIV (ENV and GAG) and hepatitis B virus (surface protein antigen)	Tomato	[266]

However, even though it has been anticipated that manufacturing considerations may limit the practical utility of many VLP approaches [160], recent developments in plant viral vectors for transient expression have greatly increased expression levels and indicate that plants can compete with other systems for production of VLPs.

Although research has mainly focused on oral delivery of minimally processed plant material, purification of VLPs for parenteral delivery is also a highly realistic approach.

18.10 SOME ISSUES REGARDING PROTEIN PRODUCTION IN PLANTS

There are several issues that need to be addressed before plant-based pharmaceuticals can be considered practical alternatives and additions to established programs of vaccination. These issues differ depending on whether the vaccine candidate is to be purified from plant tissues prior to formulation and delivery or whether it is to be administered orally as recombinant plant material. If purification is involved there are several approaches that could be implemented; one is to engineer the protein to be secreted into the culture medium. Secretion systems are convenient because no disruption of plant cells is necessary during protein recovery; hence, release of phenolic compounds is avoided. Nevertheless, the recombinant proteins might be unstable in the culture medium. Another approach is the use of affinity tags to facilitate the recovery of proteins. This strategy can be useful as long as the tag can be removed after purification to restore the native structure of the protein. In either case, good manufacturing practices will be needed and possible lot-to-lot variability will need to be closely monitored. By contrast, if the recombinant antigen is to be delivered in a processed plant product as an oral vaccine, production would be based on food processing technology rather than protein purification schemes, but good manufacturing practices will still apply.

Consistency of product (homogeneity) is very important for plant-based vaccines as well as for purified antigens. Therefore, rather than administering whole plant organs (fruits, or grains) directly, as it was originally envisaged, it may be better to process the plant material into a uniform state and to be stable to the food processing technology. This has been assessed using recombinant corn expressing the B subunit of the heat labile toxin, and the antigen has been shown to be stable to milling and modified extrusion conditions and to be evenly distributed in the products [83]. This will be very important as the product is to be blended with nontransgenic material or with transgenic material expressing different antigens, or even protein adjuvants, for even dosing.

Stability of antigens over time in processed food products stored at different temperatures will also need to be assessed. The B subunit of the heat labile toxin and the S glycoprotein of transmissible gastroenteritis virus have been shown to be stable for at least a year, even when stored at ambient temperatures [84]. This emphasizes the redundancy of a cold chain during storage and distribution of plant-based products. This feature is particularly important in developing countries with limited resources to provide a cold chain and the equipment and personnel needed for injections. The low cost of plant-based vaccines make them ideal for large-scale programs in developing countries.

Oral vaccines are potentially applicable to any vaccine formulation based on or including a subunit component. The oral delivery of a subunit vaccine is particularly suited to protect against pathogens that infect via the mucosal surfaces but oral vaccines may also be a viable option to combat pathogens that typically invade via the circulatory system, such as rabies [172,173]. Alternatively, plant vaccines may be used as a booster vaccine, as the results of the clinical trials with as a hepatitis B have shown.

Expression levels for several proteins are already sufficiently high for economic production [106,108,109] but further improvements in expression are necessary before other vaccine candidates can be considered practical. The various methods available for increasing the expression of recombinant proteins in plants (Section 18.7) should make this task easier. However, when implementing new strategies and combining currently available approaches, care must be taken to minimize potential negative influences, in particular gene silencing [174].

18.11 FUTURE PROSPECTS

Plants as bioreactors for the production of foreign proteins in plants with a goal of commercial production have attracted considerable attention over the past decade. Only a few small-scale products have so far reached the market although there are more approaching commercialization [175], after meeting the technological challenges and clearing the regulatory hurdles. In the near term, enzymes for large-scale industrial processes and antigens for oral animal vaccines are the most likely plant-expressed products to be commercially viable as the first ever licensure of a plant-derived vaccine targeting a viral disease of poultry described before has confirmed. Attention will be required to ensure correct post-translational processing and protein stability in plant tissues. Advances are required to boost expression further and stacking of many of the available tools discussed above will probably be required to produce commercial products, although this can be limited by access to the necessary intellectual property (IP). This is an issue that sooner or later will have to be resolved.

As with all biotechnological developments, the technology for plant-derived vaccines has been patented in industrialized countries, where there exist high levels of protection. Poor countries, which usually have a high disease burden, often have poor or inexistent IP protection rules and lack of adequate knowledge and infrastructure to protect and commercialize a biotechnological product. It has been postulated that plant-derived vaccines may be approved in an industrialized country and then be more broadly used in poor countries [23]. It remains to be seen how this approach would be implemented.

In many reviews on plant made-pharmaceuticals a point is made about the need (some authors have even called it a moral imperative [175]) to provide low-cost medicines and vaccines to poor countries. This need is used as an important justification for the development of plant-based vaccines. Plant-based pharmaceuticals could offer a new model for vaccine development, which may allow a wider participation, beyond the well-established multinational pharmaceutical companies [175]. Poor countries would potentially be involved, although it is still not well defined how, and the focus would be on specific regional diseases that do not feature in current drug development programs [175]. It is hoped that this technology will eventually help those who needed it the most and that the issue of IP does not represent an insurmountable obstacle. Putting the collective benefit ahead of the personal gains, will be the key for the full realization of this technology.

ACKNOWLEDGMENT

Research support from CONACYT, The Welcome Foundation and The Rockefeller Foundation in the author's laboratory is gratefully acknowledged.

REFERENCES

1. Twyman, R.M. et al., Molecular farming in plants: Host systems and expression technology, *Trends Biotechnol.*, 21, 570, 2003.
2. Menkhaus, T.J. et al., Considerations for the recovery of recombinant proteins from plants, *Biotechnol. Prog.*, 20, 1001, 2004.
3. Howard, J.A., Commercialization of plant-based vaccines from research and development to manufacturing, *Anim. Health Res. Rev*, 5, 243, 2004.
4. Woodard, S.L. et al., Maize (Zea mays)-derived bovine trypsin: Characterization of the first large-scale, commercial protein product from transgenic plants, *Biotechnol. Appl. Biochem.*, 38, 123, 2003.
5. Ogra, P.L., Faden, H., and Welliver, R.C., Vaccination strategies for mucosal immune responses, *Clin. Microbiol. Rev.*, 14, 641, 2001.

6. Holmgren, J. et al., Mucosal immunisation and adjuvants: A brief overview of recent advances and challenges, *Vaccine*, 21, S89–S95, 2003.
7. Ogra, P.L., Mucosal immunity: Some historical perspective on host-pathogen interactions and implications for mucosal vaccines, *Immunol. Cell Biol.,* 81, 23, 2003.
8. Dietrich, G. et al., Experience with registered mucosal vaccines, *Vaccine*, 21, 678, 2003.
9. Cheroutre, H., Starting at the beginning: New perspectives on the biology of mucosal T cells, *Ann. Rev. Immunol.*, 22, 217, 2004.
10. Chen, H.F. et al., Oral immunization of mice using transgenic tomato fruit expressing VP1 protein from enterovirus 71, *Vaccine*, 24, 2944, 2006.
11. Neutra, M.R. and Kozlowski, P.A., Mucosal vaccines: The promise and the challenge, *Nat. Immunol.*, 6, 148, 2006.
12. Choy, E.H. et al., Control of rheumatoid arthritis by oral tolerance, *Arthritis Rheum.*, 44, 1993–1997, 2001.
13. Stevceva, L. and Ferrari, M.G., Mucosal adjuvants, *Curr. Pharm. Des.*, 11, 801, 2005.
14. Frank, B.H., The development of recombinant human insulin, *Dev. Biol. Stand.*, 96, 91, 1998.
15. Balen, B. and Krsnik-Rasol, M., N-glycosylation of recombinant therapeutic glycoproteins in plant systems plant specific N-glycosylation, *Food Technol. Biotechnol.*, 45, 1, 2007.
16. Austin, S. et al., An overview of a feasibility study for the production of industrial enzymes in transgenic alfalfa, *Ann. N Y Acad. Sci.,* 721, 235, 1994.
17. Khoudi, H. et al., Production of a diagnostic monoclonal antibody in perennial alfalfa plants, *Biotechnol. Bioeng.*, 64, 135–143, 1999.
18. Larrick, J.W. et al., Production of secretory IgA antibodies in plants, *Biomol. Eng.*, 18, 87, 2001.
19. Daniell, H. et al., Expression of the native cholera toxin B subunit gene and assembly as functional oligomers in transgenic tobacco chloroplasts, *J. Mol. Biol.*, 311, 1001, 2001.
20. Streatfield, S.J. and Howard, J.A., Plant production systems for vaccines, *Expert Rev. Vaccines*, 2, 763, 2003.
21. Yeoh, H.H. and Wee, W.C., Leaf protein contents and nitrogen-to-protein conversion factors for 90 plant species, *Food Chem.*, 9, 245, 1994.
22. Kirk, D.D. and Webb, S.R., The next 15 years: Taking plant-made vaccines beyond proof of concept, *Immunol. Cell Biol.*, 83, 248, 2005.
23. Thanavala, Y., Huang, Z., and Mason, H.S., Plant-derived vaccines: A look back at the highlights and a view to the challenges on the road ahead, *Expert Rev. Vacc.*, 5, 249, 2006.
24. Vain, P., Thirty years of plant transformation technology development, *Plant Biotechnol. J.*, 5, 221, 2007.
25. Gelvin, S.B., Agrobacterium-mediated plant transformation: The biology behind the "gene-jockeying" tool, *Microbiol. Mol. Biol. Rev.*, 67, 16, 2003.
26. Altpeter, F. et al., Particle bombardment and the genetic enhancement of crops: myths and realities, *Mol. Breed.*, 15, 305, 2005.
27. Chen, L.L. et al., Expression and inheritance of multiple transgenes in rice plants, *Nat. Biotechnol.*, 16, 1060, 1998.
28. Nicholson, L. et al., A recombinant multimeric immunoglobulin expressed in rice shows assembly-dependent subcellular localization in endosperm cells, *Plant Biotechnol. J.*, 3, 115, 2005.
29. Loc, N.T. et al., Linear transgene constructs lacking vector backbone sequences generate transgenic rice plants which accumulate higher levels of proteins conferring insect resistance, *Mol. Breed.*, 9, 231, 2002.
30. Agrawal, P.K. et al., Transformation of plants with multiple cassettes generates simple transgene integration patterns and high expression levels, *Mol. Breed.*, 16, 247, 2005.
31. Fischer, R. et al., Towards molecular farming in the future: Transient protein expression in plants, *Biotechnol. Appl. Biochem.*, 30, 113, 1999.
32. Voinnet, O., RNA silencing as a plant immune system against viruses, *Trends Genet.*, 17, 449, 2001.
33. Moissiard, G. and Voinnet, O., Viral suppression of RNA silencing in plants, *Mol. Plant Pathol.*, 5, 71, 2004.
34. Fischer, R. et al., Plant-based production of biopharmaceuticals, *Curr. Opin. Plant Biol.*, 7, 152, 2004.
35. Maliga, P., Plastid transformation in higher plants, *Ann. Rev. Plant. Biol.*, 55, 289, 2004.
36. Bock, R., Transgenic plastids in basic research and plant biotechnology, *J. Mol. Biol.*, 312, 425, 2001.
37. Saski, C. et al., Complete chloroplast genome sequence of Glycine max and comparative analyses with other legume genomes, *Plant Mol. Biol.*, 59, 309, 2005.

38. Lee, K.Y. et al., Induction of protective immune responses against the challenge of *Actinobacillus pleuropneumoniae* by the oral administration of transgenic tobacco plant expressing ApxIIA toxin from the bacteria, *FEMS Immunol. Med. Microbiol.*, 48, 381, 2006.
39. Nugent, G.D. et al., Plastid transformants of tomato selected using mutations affecting ribosome structure, *Plant Cell Rep.*, 24, 341, 2005.
40. Nugent, G.D. et al., Nuclear and plastid transformation of *Brassica oleracea* var. botrytis (cauliflower) using PEG-mediated uptake of DNA into protoplasts, *Plant Sci.*, 170, 135, 2006.
41. Zubko, M.K. et al., Stable transformation of petunia plastids, *Transgenic Res.*, 13, 523, 2004.
42. Dufourmantel, N. et al., Generation and analysis of soybean plastid transformants expressing Bacillus thuringiensis Cry1Ab protoxin, *Plant Mol. Biol.*, 58, 659, 2005.
43. Kanamoto, H. et al., Efficient and stable transformation of *Lactuca sativa* L. cv. Cisco (lettuce) plastids, *Transgenic Res.*, 15, 205, 2006.
44. Chiyoda, S. et al., Simple and efficient plastid transformation system for the liverwort *Marchantia polymorpha* L. suspension-culture cells, *Transgenic Res.*, 16, 41, 2007.
45. Sun, M. et al., Foot-and-mouth disease virus VP1 protein fused with cholera toxin B subunit expressed in *Chlamydomonas reinhardtii* chloroplast, *Biotechnol. Lett.*, 25, 1087, 2003.
46. Ruf, S. et al., Stable genetic transformation of tomato plastids and expression of a foreign protein in fruit, *Nat. Biotechnol.*, 19, 870, 2001.
47. Quesada-Vargas, T., Ruiz, O.N., and Daniell, H., Characterization of heterologous multigene operons in transgenic chloroplasts. Transcription, processing, and translation, *Plant Physiol.*, 138, 1746, 2005.
48. Daniell, H., Transgene containment by maternal inheritance: Effective or elusive? *Proc. Natl. Acad. Sci. U S A*, 104, 6879, 2007.
49. Daniell, H. et al., Chloroplast-derived vaccine antigens and other therapeutic proteins, *Vaccine*, 23, 1779, 2005.
50. Lutz, K.A., Knapp, J.E., and Maliga, P., Expression of bar in the plastid genome confers herbicide resistance, *Plant Physiol.*, 125, 1585, 2001.
51. Ye, G.N. et al., Plastid-expressed 5-enolpyruvylshikimate-3-phosphate synthase genes provide high level glyphosate tolerance in tobacco, *Plant J.*, 25, 261, 2001.
52. Tregoning, J.S. et al., Expression of tetanus toxin fragment C in tobacco chloroplasts, *Nucleic Acids Res.*, 31, 1174, 2003.
53. Koya, V. et al., Plant-based vaccine: Mice immunized with chloroplast-derived anthrax protective antigen survive anthrax lethal toxin challenge, *Infect. Immun.*, 73, 8266, 2005.
54. Molina, A., Veramendi, J., and Hervas-Stubbs, S., Induction of neutralizing antibodies by a tobacco chloroplast-derived vaccine based on a B cell epitope from canine parvovirus, *Virology*, 342, 266, 2005.
55. Verch, T., Yusibov, V., and Koprowski, H., Expression and assembly of a full-length monoclonal antibody in plants using a plant virus vector, *J. Immunol. Methods*, 220, 69, 1998.
56. Mori, M. et al., Inducible high-level mRNA amplification system by viral replicase in transgenic plants, *Plant J.*, 27, 79, 2001.
57. Porta, C. and Lomonossoff, G.P., Use of viral replicons for the expression of genes in plants, *Mol. Biotechnol*, 5, 209, 1996.
58. Johnson, J., Lin, T., and Lomonossoff, G., Presentation of heterologous peptides on plant viruses: Genetics, structure, and function, *Ann. Rev. Phytopathol.*, 35, 67, 1997.
59. Werner, S. et al., Immunoabsorbent nanoparticles based on a tobamovirus displaying protein A, *Proc. Natl. Acad. Sci. U S A*, 103, 17678, 2006.
60. Porta, C. et al., Development of cowpea mosaic-virus as a high-yielding system for the presentation of foreign peptides, *Virology*, 202, 949, 1994.
61. Dalsgaard, K. et al., Plant-derived vaccine protects target animals against a viral disease, *Nat. Biotechnol.*, 15, 248, 1997.
62. Gleba, Y., Klimyuk, V., and Marillonnet, S., Magnifection—a new platform for expressing recombinant vaccines in plants, *Vaccine*, 23, 2042, 2005.
63. Giritch, A. et al., Rapid high-yield expression of full-size IgG antibodies in plants coinfected with noncompeting viral vectors, *Proc. Natl. Acad. Sci. U S A*, 103, 14702, 2006.
64. Hiatt, A., Cafferkey, R., and Bowdish, K., Production of antibodies in transgenic plants, *Nature*, 342, 76, 1989.
65. Wycoff, K.L., Secretory IgA antibodies from plants, *Curr. Pharm. Des.*, 11, 2429, 2005.

66. Smith, M.D. and Glick, B.R., The production of antibodies in plants: An idea whose time has come? *Biotechnol. Adv.*, 18, 85, 2000.
67. Potera, C., EPIcyte produces antibodies in plants—Plantibodies step in to fulfill the promise of mabs, *Genet. Engineer. News,* 19, 22, 1999.
68. Kusnadi, A.R. et al., Production and purification of two recombinant proteins from transgenic corn, *Biotechnol. Prog.*, 14, 149, 1998.
69. Sharp, J.M. and Doran, P.M., Characterization of monoclonal antibody fragments produced by plant cells, *Biotechnol. Bioeng.*, 73, 338, 2001.
70. Stevens, L.H. et al., Effect of climate conditions and plant developmental stage on the stability of antibodies expressed in transgenic tobacco, *Plant Physiol.*, 124, 173, 2000.
71. van Rooijen, G.J. and Moloney, M.M., Plant seed oil-bodies as carriers for foreign proteins, *Bio/Technology*, 13, 72, 1995.
72. Artsaenko, O. et al., Potato tubers as a biofactory for recombinant antibodies, *Mol. Breed.*, 4, 313, 1998.
73. Ma, J.K. and Hein, M.B., Immunotherapeutic potential of antibodies produced in plants, *Trends Biotechnol.*, 13, 522, 1995.
74. Zeitlin, L. et al., A humanized monoclonal antibody produced in transgenic plants for immunoprotection of the vagina against genital herpes, *Nat. Biotechnol.*, 16, 1361, 1998.
75. Vaquero, C. et al., A carcinoembryonic antigen-specific diabody produced in tobacco, *FASEB J.*, 16, 408, 2002.
76. Ma, J.K.C. et al., Characterization of a recombinant plant monoclonal secretory antibody and preventive immunotherapy in humans, *Nat. Med.*, 4, 601, 1998.
77. Verch, T. et al., Immunization with a plant-produced colorectal cancer antigen, *Cancer Immunol. Immunother.*, 53, 92, 2004.
78. Stoger, E. et al., Plantibodies: Applications, advantages and bottlenecks, *Curr. Opin. Biotechnol.*, 13, 161, 2002.
79. Elbers, I.J.W. et al., Influence of growth conditions and developmental stage on N-glycan heterogeneity of transgenic immunoglobulin G and endogenous proteins in tobacco leaves, *Plant Physiol.*, 126, 1314, 2001.
80. Obregon, P. et al., HIV-1 p24-immunoglobulin fusion molecule: A new strategy for plant-based protein production, *Plant Biotechnol. J.*, 4, 195, 2006.
81. Poland, G.A., Vaccines in the 21st century, *Immunol. Allergy Clin. North America*, 23, xi, 2003.
82. Lauterslager, T.G.M. et al., Oral immunisation of naive and primed animals with transgenic potato tubers expressing LT-B, *Vaccine*, 19, 2749, 2001.
83. Streatfield, S.J., Mucosal immunization using recombinant plant-based oral vaccines, *Methods*, 38, 150, 2006.
84. Lamphear, B.J. et al., Delivery of subunit vaccines in maize seed, *J. Control Release*, 85, 169, 2002.
85. Tacket, C.O. et al., Immunogenicity in humans of a recombinant bacterial antigen delivered in a transgenic potato, *Nat. Med.*, 4, 607, 1998.
86. Tacket, C.O. et al., Immunogenicity of recombinant LT-B delivered orally to humans in transgenic corn, *Vaccine*, 22, 4385, 2004.
87. Tacket, C.O. et al., Human immune responses to a novel Norwalk virus vaccine delivered in transgenic potatoes, *J. Infect. Dis.*, 182, 302, 2000.
88. Tacket, C.O., Garden-variety vaccines: Antigens derived from transgenic plants, *Expert Rev. Vacc.*, 3, 529, 2004.
89. Kapusta, J. et al., A plant-derived edible vaccine against hepatitis B virus, *FASEB J.*, 13, 1796, 1999.
90. Giddings, G. et al., Transgenic plants as factories for biopharmaceuticals, *Nat. Biotechnol.*, 18, 1151, 2000.
91. Staub, J.M. et al., High-yield production of a human therapeutic protein in tobacco chloroplasts, *Nat. Biotechnol.*, 18, 333, 2000.
92. Chong, D.K.X. and Langridge, W.H.R., Expression of full-length bioactive antimicrobial human lactoferrin in potato plants, *Transgenic Res.*, 9, 71, 2000.
93. Philip, R. et al., Processing and localization of bovine beta-casein expressed in transgenic soybean seeds under control of a soybean lectin expression cassette, *Plant Sci.*, 161, 323, 2001.
94. Ruggiero, F. et al., Triple helix assembly and processing of human collagen produced in transgenic tobacco plants, *FEBS Lett.*, 469, 132, 2000.

95. Wyman, C.E., What is (and is not) vital to advancing cellulosic ethanol, *Trends Biotechnol.*, 25, 153, 2007.
96. Nawrath, C., Poirier, Y., and Somerville, C., Plant polymers for biodegradable plastics—Cellulose, starch and polyhydroxyalkanoates, *Mol. Breed.*, 1, 105, 1995.
97. Poirier, Y., Production of new polymeric compounds in plants, *Curr. Opin. Biotechnol.*, 10, 181, 1999.
98. Guda, C., Lee, S.B., and Daniell, H., Stable expression of a biodegradable protein-based polymer in stable tobacco chloroplasts, *Plant Cell Rep.*, 19, 257, 2000.
99. Lee, J.S. et al., Establishment of a transgenic tobacco cell suspension culture system for producing murine granulocyte–macrophage colony stimulating factor, *Mol. Cell*, 7, 783, 1997.
100. Kusnadi, A.R., Nikolov, Z.L., and Howard, J.A., Production of recombinant proteins in transgenic plants: Practical considerations, *Biotechnol. Bioeng.*, 56, 473, 1997.
101. Cramer, C.L. et al., Bioproduction of human enzymes in transgenic tobacco, *Engineer. Plants Comm. Prod. Appl.*, 792, 62, 1996.
102. Gutierrez-Ortega, A. et al., Expression of a single-chain human interleukin 12 gene in transgenic tobacco plants and functional studies, *Biotechnol. Bioeng.*, 85, 734, 2004.
103. Gutierrez-Ortega, A. et al., Expression of functional interleukin-12 from mouse in transgenic tomato plants, *Transgenic Res.*, 14, 877, 2005.
104. Leite, A. et al., Expression of correctly processed human growth hormone in seeds of transgenic tobacco plants, *Mol. Breed.*, 6, 47, 2000.
105. Arakawa, T. et al., A plant-based cholera toxin B subunit-insulin fusion protein protects against the development of autoimmune diabetes, *Nat. Biotechnol.*, 16, 934, 1998.
106. Hood, E.E. et al., Commercial production of avidin from transgenic maize: Characterization of transformant, production, processing, extraction and purification, *Mol. Breed.*, 3, 291, 1997.
107. Hood, E.E. et al., Molecular farming of industrial proteins from transgenic maize, *Chem. Via Higher Plant Bioengineer.*, 464, 127, 1999.
108. Witcher, D.R. et al., Commercial production of beta-glucuronidase (GUS): A model system for the production of proteins in plants, *Mol. Breed.*, 4, 301, 1998.
109. Zhong, G.Y. et al., Commercial production of aprotinin in transgenic maize seeds, *Mol. Breed.*, 5, 345, 1999.
110. Fischer, R. and Emans, N., Molecular farming of pharmaceutical proteins, *Transgenic Res.*, 9, 279, 2000.
111. Odell, J.T., Nagy, F., and Chua, N.H., Identification of DNA-sequences required for activity of the cauliflower mosaic virus-35S promoter, *Nature*, 313, 810, 1985.
112. Fiedler, U. et al., Optimization of scFv antibody production in transgenic plants, *Immunotechnology*, 3, 205, 1997.
113. Verdaguer, B. et al., Isolation and expression in transgenic tobacco and rice plants, of the cassava vein mosaic virus (CVMV) promoter, *Plant Mol. Biol.*, 31, 1129, 1996.
114. Xie, Y. et al., Isolation and identification of a super strong plant promoter from cotton leaf curl Multan virus, *Plant Mol. Biol.*, 53, 1, 2003.
115. Shirasawa-Seo, N. et al., The promoter of Milk vetch dwarf virus component 8 confers effective gene expression in both dicot and monocot plants, *Plant Cell Rep.*, 24, 155, 2005.
116. Shaw, C.H. et al., A functional map of the nopaline synthase promoter, *Nucleic Acids Res.*, 12, 7831, 1984.
117. Dai, Z. et al., Expression of Acidothermus cellulolyticus endoglucanase E1 in transgenic tobacco: Biochemical characteristics and physiological effects, *Transgenic Res.*, 9, 43, 2000.
118. Christensen, A.H., Sharrock, R.A., and Quail, P.H., Maize polyubiquitin genes—structure, thermal perturbation of expression and transcript splicing, and promoter activity following transfer to protoplasts by electroporation, *Plant Mol. Biol.*, 18, 675, 1992.
119. Belanger, F.C. and Kriz, A.L., Molecular basis for allelic polymorphism of the maize globulin-1 gene, *Genetics*, 129, 863, 1991.
120. Russell, D.A. and Fromm, M.E., Tissue-specific expression in transgenic maize of four endosperm promoters from maize and rice, *Transgenic Res.*, 6, 157, 1997.
121. Horvath, H. et al., The production of recombinant proteins in transgenic barley grains, *Proc. Natl. Acad. Sci. U S A*, 97, 1914, 2000.
122. Peña Ramirez, Y.J. et al., Fruit-specific expression of human immunodeficiency virus 1-Tat gene in tomato plants and its immunogenic potential in mice, *Clin. Vacc. Immunol.*, 14, 685–692, 2007.

123. Aoyama, T. and Chua, N.H., A glucocorticoid-mediated transcriptional induction system in transgenic plants, *Plant J.*, 11, 605, 1997.
124. Zuo, J., Niu, Q.W., and Chua, N.H., Technical advance: An estrogen receptor-based transactivator XVE mediates highly inducible gene expression in transgenic plants, *Plant J.*, 24, 265, 2000.
125. Comai, L., Moran, P., and Maslyar, D., Novel and useful properties of a chimeric plant promoter combining Camv-35S and Mas elements, *Plant Mol. Biol.*, 15, 373, 1990.
126. Rance, I. et al., Combination of viral promoter sequences to generate highly active promoters for heterologous therapeutic protein over-expression in plants, *Plant Sci.*, 162, 833, 2002.
127. Spiker, S. and Thompson, W.F., Nuclear matrix attachment regions and transgene expression in plants, *Plant Physiol.*, 110, 15, 1996.
128. Allen, G.C. et al., Scaffold attachment regions increase reporter gene-expression in stably transformed plant-cells, *Plant Cell*, 5, 603, 1993.
129. De Bolle, M.F. et al., The influence of matrix attachment regions on transgene expression in Arabidopsis thaliana wild type and gene silencing mutants, *Plant Mol. Biol.*, 63, 533, 2007.
130. Mlynarova, L. et al., The mar-mediated reduction in position effect can be uncoupled from copy number-dependent expression in transgenic plants, *Plant Cell*, 7, 599, 1995.
131. Allen, G.C., Spiker, S., and Thompson, W.F., Use of matrix attachment regions (MARs) to minimize transgene silencing, *Plant Mol. Biol.*, 43, 361, 2000.
132. Green, P.J., Control of mRNA stability in higher plants, *Plant Physiol.*, 102, 1065, 1993.
133. Newman, T.C. et al., DST sequences, highly conserved among plant SAUR genes, target reporter transcripts for rapid degradation in tobacco, *Plant Cell*, 5, 701, 1993.
134. Fiume, E. et al., Introns are key regulatory elements of rice tubulin expression, *Planta*, 218, 693, 2004.
135. Pooggin, M.M. and Skryabin, K.G., The 5′-untranslated leader sequence of potato virus-X RNA enhances the expression of a heterologous gene in vivo, *Mol. Gen. Genet.*, 234, 329, 1992.
136. Joshi, C.P. et al., Context sequences of translation initiation codon in plants, *Plant Mol. Biol.*, 35, 1997.
137. Arakawa, T. et al., Suppression of autoimmune diabetes by a plant-delivered cholera toxin B subunit-human glutamate decarboxylase fusion protein, *Transgenics*, 3, 51, 1999.
138. Nemchinov, L.G. et al., Development of a plant-derived subunit vaccine candidate against hepatitis C virus, *Arch. Virol*, 145, 2557, 2000.
139. Kim, T.G. and Langridge, W.H., Assembly of cholera toxin B subunit full-length rotavirus NSP4 fusion protein oligomers in transgenic potato, *Plant Cell Rep.*, 21, 884, 2003.
140. Kim, T.G., Gruber, A., and Langridge, W.H., HIV-1 gp120 V3 cholera toxin B subunit fusion gene expression in transgenic potato, *Protein Expr. Purif.*, 37, 196, 2004.
141. Kim, T.G. et al., Synthesis and assembly of SIVmac Gag p27 capsid protein cholera toxin B subunit fusion protein in transgenic potato, *Mol. Biotechnol.*, 28, 33, 2004.
142. Walmsley, A.M. et al., Expression of the B subunit of *Escherichia coli* heat-labile enterotoxin as a fusion protein in transgenic tomato, *Plant Cell Rep.*, 21, 1020, 2003.
143. Matoba, N. et al., A mucosally targeted subunit vaccine candidate eliciting HIV-1 transcytosis-blocking Abs, *Proc. Natl. Acad. Sci. U S A*, 101, 13584, 2004.
144. Kim, T.G., Galloway, D.R., and Langridge, W.H., Synthesis and assembly of anthrax lethal factor-cholera toxin B-subunit fusion protein in transgenic potato, *Mol. Biotechnol.*, 28, 175, 2004.
145. Lee, J.Y. et al., Plant-synthesized *E. coli* CFA/I fimbrial protein protects Caco-2 cells from bacterial attachment, *Vaccine*, 23, 222, 2004.
146. Rigano, M.M. et al., Production of a fusion protein consisting of the enterotoxigenic *Escherichia coli* heat-labile toxin B subunit and a tuberculosis antigen in *Arabidopsis thaliana*, *Plant Cell Rep.*, 22, 502, 2004.
147. He, D.M. et al., Stable expression of foot-and-mouth disease virus protein VP1 fused with cholera toxin B subunit in the potato (*Solanum tuberosum*), *Colloids Surf B Biointerf.*, 55, 159, 2007.
148. Kim, T.G. and Langridge, W.H., Synthesis of an HIV-1 Tat transduction domain-rotavirus enterotoxin fusion protein in transgenic potato, *Plant Cell Rep.*, 22, 382, 2004.
149. Choi, N.W., Estes, M.K., and Langridge, W.H.R., Ricin toxin B subunit enhancement of rotavirus NSP4 immunogenicity in mice, *Viral Immunol.*, 19, 54, 2006.
150. Massa, S. et al., Anti-cancer activity of plant-produced HPV16 E7 vaccine, *Vaccine*, 25, 3018, 2007.
151. Li, D. et al., Expression of cholera toxin B subunit and the B chain of human insulin as a fusion protein in transgenic tobacco plants, *Plant Cell Rep.*, 25, 417, 2006.

152. Ma, S.W. et al., Transgenic plants expressing autoantigens fed to mice to induce oral immune tolerance, *Nat. Med.*, 3, 793, 1997.
153. Ma, S. et al., Induction of oral tolerance to prevent diabetes with transgenic plants requires glutamic acid decarboxylase (GAD) and IL-4, *Proc. Natl. Acad. Sci. U S A*, 101, 5680, 2004.
154. Sriraman, R. et al., Recombinant anti-hCG antibodies retained in the endoplasmic reticulum of transformed plants lack core-xylose and core-alpha(1,3)-fucose residues, *Plant Biotechnol. J.*, 2, 279, 2004.
155. Rayon, C., Lerouge, P., and Faye, L., The protein N-glycosylation in plants, *J. Exp. Bot.*, 49, 1463, 1998.
156. Triguero, A. et al., Plant-derived mouse IgG monoclonal antibody fused to KDEL endoplasmic reticulum-retention signal is N-glycosylated homogeneously throughout the plant with mostly high-mannose-type N-glycans, *Plant Biotechnol. J.*, 3, 449, 2005.
157. Samyn-Petit, B. et al., Comparative analysis of the site-specific N-glycosylation of human lactoferrin produced in maize and tobacco plants, *Eur. J. Biochem.*, 270, 3235, 2003.
158. Chargelegue, D. et al., Highly immunogenic and protective recombinant vaccine candidate expressed in transgenic plants, *Infect. Immun.*, 73, 5915, 2005.
159. Schirmbeck, R. et al., Priming of class I-restricted cytotoxic T lymphocytes by vaccination with recombinant protein antigens, *Vaccine*, 13, 857, 1995.
160. Grgacic, E.V.L. and Anderson, D.A., Virus-like particles: Passport to immune recognition, *Methods*, 40, 60, 2006.
161. Mcaleer, W.J. et al., Human hepatitis-B vaccine from recombinant yeast, *Nature*, 307, 178, 1984.
162. Gilbert, S.C., Virus-like particles as vaccine adjuvants, *Mol. Biotechnol.*, 19, 169, 2001.
163. Niikura, M. et al., Chimeric recombinant hepatitis E virus-like particles as an oral vaccine vehicle presenting foreign epitopes, *Virology*, 293, 273, 2002.
164. Guerrero-Andrade, O. et al., Expression of the Newcastle disease virus fusion protein in transgenic maize and immunological studies, *Transgenic Res.*, 15, 455, 2006.
165. Biemelt, S. et al., Production of human papillomavirus type 16 virus-like particles in transgenic plants, *J. Virol.*, 77, 9211, 2003.
166. Jegerlehner, A. et al., A molecular assembly system that renders antigens of choice highly repetitive for induction of protective B cell responses, *Vaccine*, 20, 3104, 2002.
167. Storni, T. et al., Nonmethylated CG motifs packaged into virus-like particles induce protective cytotoxic T cell responses in the absence of systemic side effects, *J. Immunol.*, 172, 1777, 2004.
168. Schwarz, K. et al., Efficient homologous prime-boost strategies for T cell vaccination based on virus-like particles, *Eur. J. Immunol.*, 35, 816, 2005.
169. Johnson, J.E. and Chiu, W., Structures of virus and virus-like particles, *Curr. Opin. Struct. Biol.*, 10, 29, 2000.
170. Daniell, H., Streatfield, S.J., and Wycoff, K., Medical molecular farming: Production of antibodies, biopharmaceuticals and edible vaccines in plants, *Trends Plant Sci.*, 6, 219, 2001.
171. Maloney, B.J. et al., Challenges in creating a vaccine to prevent hepatitis E, *Vaccine*, 23, 1870, 2005.
172. Modelska, A. et al., Immunization against rabies with plant-derived antigen, *Proc. Natl. Acad. Sci. U S A*, 95, 2481, 1998.
173. Perea-Arango, I. et al., Expression of the rabies virus nucleoprotein in plants at high-levels and evaluation of immune responses in mice, *Plant Cell Rep.*, (in press), 2007.
174. Yu, H. and Kumar, P.P., Post-transcriptional gene silencing in plants by RNA, *Plant Cell Rep.*, 22, 167, 2003.
175. The European Union Framework 6 Pharma–Planta Consortium, *EMBO Rep.*, 6, 593, 2005.
176. Leelavathi, S. et al., Overproduction of an alkali- and thermo-stable xylanase in tobacco chloroplasts and efficient recovery of the enzyme, *Mol. Breed.*, 11, 59, 2003.
177. Fernandez-San Millan, A. et al., A chloroplast transgenic approach to hyper-express and purify human serum albumin, a protein highly susceptible to proteolytic degradation, *Plant Biotechnol. J.*, 1, 71, 2003.
178. Fernandez-San Millan, A. et al., Expression of recombinant proteins lacking methionine as N-terminal amino acid in plastids: Human serum albumin as a case study, *J. Biotechnol.*, 127, 593, 2007.
179. Tregoning, J. et al., New advances in the production of edible plant vaccines: Chloroplast expression of a tetanus vaccine antigen, TetC, *Phytochemistry*, 65, 989, 2004.
180. Kang, T.J. et al., Expression of the B subunit of *E. coli* heat-labile enterotoxin in the chloroplasts of plants and its characterization, *Transgenic Res.*, 12, 683, 2003.

181. Birch-Machin, I., Newell, C.A., Hibberd, J.M., and Gray, J.C., Accumulation of rotavirus VP6 protein in chloroplasts of transplastomic tobacco is limited by protein stability, *Plant Biotechnol. J.*, 2, 261, 2004.
182. Kang, T.J. et al., Expression of non-toxic mutant of *Escherichia coli* heat-labile enterotoxin in tobacco chloroplasts, *Protein Expr. Purif.*, 38, 123, 2004.
183. Arai, Y. et al., Production of polyhydroxybutyrate by polycistronic expression of bacterial genes in tobacco plastid, *Plant Cell Physiol.*, 45, 1176, 2004.
184. Molina, A., Hervas-Stubbs, S., Daniell, H., Mingo-Castel, A.M., and Veramendi, J., High-yield expression of a viral peptide animal vaccine in transgenic tobacco chloroplasts, *Plant Biotechnol. J.*, 2, 141, 2004.
185. Li, H.Y., Ramalingam, S., and Chye, M.L., Accumulation of recombinant SARS-CoV spike protein in plant cytosol and chloroplasts indicate potential for development of plant-derived oral vaccines, *Exp. Biol. Med.*, 231, 1346, 2006.
186. Li, Y. et al., High expression of foot-and-mouth disease virus structural protein VP1 in tobacco chloroplasts, *Plant Cell Rep.*, 25, 329, 2006.
187. Roh, K.H. et al., Accumulation of sweet protein monellin is regulated by the psbA 5′ UTR in tobacco chloroplasts, *J. Plant Biol.*, 49, 34, 2006.
188. Wirth, S. et al., Accumulation of hEGF and hEGF-fusion proteins in chloroplast-transformed tobacco plants is higher in the dark than in the light, *J. Biotechnol.*, 125, 159, 2006.
189. Chebolu, S. and Daniell, H., Stable expression of Gal/GalNAc lectin of Entamoeba histolytica in transgenic chloroplasts and immunogenicity in mice towards vaccine development for amoebiasis, *Plant Biotechnol. J.*, 5, 230, 2007.
190. Maclean, J. et al., Optimization of human papillomavirus type 16 (HPV-16) L1 expression in plants: Comparison of the suitability of different HPV-16 L1 gene variants and different cell-compartment localization, *J. Gen. Virol.*, 88, 1460, 2007.
191. McCormick, A.A. et al., Rapid production of specific vaccines for lymphoma by expression of the tumor-derived single-chain Fv epitopes in tobacco plants, *Proc. Natl. Acad. Sci. U S A*, 96, 703, 1999.
192. Wigdorovitz, A. et al., Induction of a protective antibody response to foot and mouth disease virus in mice following oral or parenteral immunization with alfalfa transgenic plants expressing the viral structural protein VP1, *Virology*, 255, 347, 1999.
193. Staczek, J. et al., Immunization with a chimeric tobacco mosaic virus containing an epitope of outer membrane protein F of *Pseudomonas aeruginosa* provides protection against challenge with *P-aeruginosa*, *Vaccine*, 18, 2266, 2000.
194. Turpen, T.H. et al., Malarial epitopes expressed on the surface of recombinant tobacco mosaic virus, *Bio/Technology*, 13, 53, 1995.
195. Koo, M. et al., Protective immunity against murine hepatitis virus (MHV) induced by intranasal or subcutaneous administration of hybrids of tobacco mosaic virus that carries an MHV epitope, *Proc. Natl. Acad. Sci. U S A*, 96, 7774, 1999.
196. Krebitz et al., Rapid production of the major birch pollen allergen Bet v 1 in *Nicotiana benthamiana* plants and its immunological in vitro and in vivo characterization, *FASEB J.*, 14, 1279, 2000.
197. Kang, T.J. et al., Mass production of somatic embryos expressing *Escherichia coli* heat-labile enterotoxin B subunit in Siberian ginseng, *J. Biotechnol.*, 121, 124, 2006.
198. Perez-Filgueira et al., Bovine herpes virus gD protein produced in plants using a recombinant tobacco mosaic virus (TMV) vector possesses authentic antigenicity, *Vaccine*, 21, 4201, 2003.
199. McCormick, A.A. et al., TMV-peptide fusion vaccines induce cell-mediated immune responses and tumor protection in two murine models, *Vaccine*, 24, 6414, 2006.
200. Fujiyama, K. et al., In Planta production of immunogenic poliovirus peptide using tobacco mosaic virus-based vector system, *J. Biosci. Bioeng.*, 101, 398, 2006.
201. Marusic, C. et al., Chimeric plant virus particles as immunogens for inducing murine and human immune responses against human immunodeficiency virus type 1, *J. Virol.*, 75, 8434, 2001.
202. Brennan, F.R. et al., A chimaeric plant virus vaccine protects mice against a bacterial infection, *Microbiology*, 145 (Pt 8), 2061, 1999.
203. Franconi, R. et al., Plant-derived human papillomavirus 16 E7 oncoprotein induces immune response and specific tumor protection, *Cancer Res.*, 62, 3654, 2002.
204. Mechtcheriakova, I.A. et al., The use of viral vectors to produce hepatitis B virus core particles in plants, *J. Virol. Methods*, 131, 10, 2006.

205. Usha, R. et al., Expression of an animal virus antigenic site on the surface of a plant-virus particle, *Virology*, 197, 366, 1993.
206. Durrani, Z. et al., Intranasal immunization with a plant virus expressing a peptide from HIV-1 gp41 stimulates better mucosal and systemic HIV-1-specific IgA and IgG than oral immunization, *J. Immunol. Methods*, 220, 93, 1998.
207. Gilleland, H.E. et al., Chimeric animal and plant viruses expressing epitopes of outer membrane protein F as a combined vaccine against *Pseudomonas aeruginosa* lung infection, *FEMS Immunol. Med. Microbiol.*, 27, 291, 2000.
208. Langeveld, J.P.M. et al., Inactivated recombinant plant virus protects dogs from a lethal challenge with canine parvovirus, *Vaccine*, 19, 3661, 2001.
209. Yusibov, V. et al., Expression in plants and immunogenicity of plant virus-based experimental rabies vaccine, *Vaccine*, 20, 3155, 2002.
210. Yusibov, V. et al., Antigens produced in plants by infection with chimeric plant viruses immunize against rabies virus and HIV-1, *Proc. Natl. Acad. Sci. U S A*, 94, 5784, 1997.
211. Belanger, H. et al., Human respiratory syncytial virus vaccine antigen produced in plants, *FASEB J.*, 14, 2323, 2000.
212. Joelson, T. et al., Presentation of a foreign peptide on the surface of tomato bushy stunt virus, *J. Gen. Virol.*, 78, 1213, 1997.
213. Zhang, G. et al., In planta expression of HIV-1 p24 protein using an RNA plant virus-based expression vector, *Mol. Biotechnol.*, 14, 99, 2000.
214. Zhang, G.G. et al., Production of HIV-1 p24 protein in transgenic tobacco plants, *Mol. Biotechnol.*, 20, 131, 2002.
215. Fernandez-Fernandez, M.R. et al., Development of an antigen presentation system based on plum pox potyvirus, *FEBS Lett.*, 427, 229, 1998.
216. Natilla, A. et al., Cucumber mosaic virus as carrier of a hepatitis C virus-derived epitope, *Arch. Virol.*, 149, 137, 2004.
217. Piazzolla, G. et al., Immunogenic properties of a chimeric plant virus expressing a hepatitis C virus (HCV)-derived epitope: New prospects for an HCV vaccine, *J. Clin. Immunol.*, 25, 142, 2005.
218. Nuzzaci, M. et al., Cucumber mosaic virus as a presentation system for a double hepatitis C virus-derived epitope, *Arch. Virol.*, 152, 915, 2007.
219. Zhao, Y. and Hammond, R.W., Development of a candidate vaccine for Newcastle disease virus by epitope display in the Cucumber mosaic virus capsid protein, *Biotechnol. Lett.*, 27, 375, 2005.
220. Zhang, X. and Mason, H., Bean Yellow Dwarf Virus replicons for high-level transgene expression in transgenic plants and cell cultures, *Biotechnol. Bioeng.*, 93, 271, 2006.
221. Alvarez, M.L. et al., Plant-made subunit vaccine against pneumonic and bubonic plague is orally immunogenic in mice, *Vaccine*, 24, 2477, 2006.
222. Santi, L. et al., Protection conferred by recombinant Yersinia pestis antigens produced by a rapid and highly scalable plant expression system, *Proc. Natl. Acad. Sci. U S A*, 103, 861, 2006.
223. Arakawa, T. et al., Expression of cholera toxin B subunit oligomers in transgenic potato plants, *Transgenic Res.*, 6, 403, 1997.
224. Arakawa, T., Chong, D.K.X., and Langridge, W.H.R., Efficacy of a food plant-based oral cholera toxin B subunit vaccine, *Nat. Biotechnol.*, 16, 292, 1998.
225. Jani, D. et al., Expression of cholera toxin B subunit in transgenic tomato plants, *Transgenic Res.*, 11, 447, 2002.
226. Jani, D. et al., Studies on the immunogenic potential of plant-expressed cholera toxin B subunit, *Plant Cell Rep.*, 22, 471, 2004.
227. Jiang, X.L. et al., Cholera toxin B protein in transgenic tomato fruit induces systemic immune response in mice, *Transgenic Res.*, 16, 169, 2007.
228. Becerra-Arteaga, A., Mason, H.S., and Shuler, M.L., Production, secretion, and stability of human secreted alkaline phosphatase in tobacco NT1 cell suspension cultures, *Biotechnol. Prog.*, 22, 1643, 2006.
229. Bouche, F.B. et al., Neutralising immunogenicity of a polyepitope antigen expressed in a transgenic food plant: A novel antigen to protect against measles, *Vaccine*, 21, 2065, 2003.
230. Marquet-Blouin, E. et al., Neutralizing immunogenicity of transgenic carrot (*Daucus carota* L.)-derived measles virus hemagglutinin, *Plant Mol. Biol.*, 51, 459, 2003.

231. Gil, F. et al., Successful oral prime-immunization with VP60 from rabbit haemorrhagic disease virus produced in transgenic plants using different fusion strategies, *Plant Biotechnol. J.*, 4, 135, 2006.
232. Martin-Alonso, J.M. et al., Oral immunization using tuber extracts from transgenic potato plants expressing rabbit hemorrhagic disease virus capsid protein, *Transgenic Res.*, 12, 127, 2003.
233. Chikwamba, R. et al., A functional antigen in a practical crop: LT-B producing maize protects mice against *Escherichia coli* heat labile enterotoxin (LT) and cholera toxin (CT), *Transgenic Res.*, 11, 479, 2002.
234. Mason, H.S. et al., Edible vaccine protects mice against *Escherichia coli* heat-labile enterotoxin (LT): Potatoes expressing a synthetic LT-B gene, *Vaccine*, 16, 1336, 1998.
235. Kim, T.G. et al., Synthesis and assembly of *Escherichia coli* heat-labile enterotoxin B subunit in transgenic lettuce (*Lactuca sativa*), *Protein Expr. Purif.*, 51, 22, 2007.
236. Moravec, T. et al., Production of *Escherichia coli* heat labile toxin (LT) B subunit in soybean seed and analysis of its immunogenicity as an oral vaccine, *Vaccine*, 25, 1647, 2007.
237. Choi, N.W., Estes, M.K., and Langridge, W.H., Synthesis and assembly of a cholera toxin B subunit-rotavirus VP7 fusion protein in transgenic potato, *Mol. Biotechnol.*, 31, 193, 2005.
238. Li, J.T. et al., Immunogenicity of a plant-derived edible rotavirus subunit vaccine transformed over fifty generations, *Virology.*, 356, 171, 2006.
239. Wu, Y.Z. et al., Oral immunization with rotavirus VP7 expressed in transgenic potatoes induced high titers of mucosal neutralizing IgA, *Virology*, 313, 337, 2003.
240. Dong, J.L. et al., Oral immunization with pBsVP6-transgenic alfalfa protects mice against rotavirus infection, *Virology*, 339, 153, 2005.
241. Wigdorovitz, A. et al., Protective lactogenic immunity conferred by an edible peptide vaccine to bovine rotavirus produced in transgenic plants, *J. Gen. Virol.*, 85, 1825, 2004.
242. Companjen, A.R. et al., Development of a cost-effective oral vaccination method against viral disease in fish, *Dev. Biol. (Basel)*, 121, 143, 2005.
243. Dorokhov, Y.L. et al., Superexpression of tuberculosis antigens in plant leaves, *Tuberculosis (Edinb)*, 87, 218, 2007.
244. Zelada, A.M. et al., Expression of tuberculosis antigen ESAT-6 in Nicotiana tabacum using a potato virus X-based vector, *Tuberculosis (Edinb)*, 86, 263, 2006.
245. Dus Santos, M.J. et al., A novel methodology to develop a foot and mouth disease virus (FMDV) peptide-based vaccine in transgenic plants, *Vaccine*, 20, 1141, 2002.
246. Huang, Y. et al., Immunogenicity of the epitope of the foot-and-mouth disease virus fused with a hepatitis B core protein as expressed in transgenic tobacco, *Viral. Immunol.*, 18, 668, 2005.
247. Ehsani, P., Khabiri, A., and Domansky, N.N., Polypeptides of hepatitis B surface antigen produced in transgenic potato, *Gene*, 190, 107, 1997.
248. Gao, Y. et al., Oral immunization of animals with transgenic cherry tomatillo expressing HBsAg, *World J. Gastroent.*, 9, 996, 2003.
249. Mason, H.S., Lam, D.M., and Arntzen, C.J., Expression of hepatitis B surface antigen in transgenic plants, *Proc. Natl. Acad. Sci. U S A*, 89, 11745, 1992.
250. Richter, L.J. et al., Production of hepatitis B surface antigen in transgenic plants for oral immunization, *Nat. Biotechnol.*, 18, 1167, 2000.
251. Kong, Q. et al., Oral immunization with hepatitis B surface antigen expressed in transgenic plants, *Proc. Natl. Acad. Sci. U S A*, 98, 11539, 2001.
252. Smith, M.L., Mason, H.S., and Shuler, M.L., Hepatitis B surface antigen (HBsAg) expression in plant cell culture: Kinetics of antigen accumulation in batch culture and its intracellular form, *Biotechnol. Bioeng.*, 80, 812, 2002.
253. Huang, Z. et al., Rapid, high-level production of hepatitis B core antigen in plant leaf and its immunogenicity in mice, *Vaccine*, 24, 2506, 2006.
254. Gil, F. et al., High-yield expression of a viral peptide vaccine in transgenic plants, *FEBS Lett.*, 488, 13, 2001.
255. Gils, M. et al., High-yield production of authentic human growth hormone using a plant virus-based expression system, *Plant Biotechnol. J.*, 3, 613, 2005.
256. Gu, Q. et al., Expression of *Helicobacter pylori* urease subunit B gene in transgenic rice, *Biotechnol. Lett.*, 28, 1661, 2006.
257. Lee, R.W. et al., Towards development of an edible vaccine against bovine pneumonic pasteurellosis using transgenic white clover expressing a Mannheimia haemolytica A1 leukotoxin 50 fusion protein, *Infect. Immun.*, 69, 5786, 2001.

258. Berinstein, A. et al., Mucosal and systemic immunization elicited by Newcastle disease virus (NDV) transgenic plants as antigens, *Vaccine*, 23, 5583, 2005.
259. Joensuu, J.J. et al., Fimbrial subunit protein FaeG expressed in transgenic tobacco inhibits the binding of F4ac enterotoxigenic *Escherichia coli* to porcine enterocytes, *Transgenic Res.*, 13, 295, 2004.
260. Liang, W. et al., Oral immunization of mice with plant-derived fimbrial adhesin FaeG induces systemic and mucosal K88ad enterotoxigenic *Escherichia coli*-specific immune responses, *FEMS Immunol. Med. Microbiol.*, 46, 393, 2006.
261. Piller, K.J. et al., Expression and immunogenicity of an *Escherichia coli* K99 fimbriae subunit antigen in soybean, *Planta*, 222, 6, 2005.
262. Joung, Y.H. et al., Expression of the hepatitis B surface S and preS2 antigens in tubers of Solanum tuberosum, *Plant Cell Rep.*, 22, 925, 2004.
263. Judge, N.A., Mason, H.S., and O'Brien, A.D., Plant cell-based intimin vaccine given orally to mice primed with intimin reduces time of *Escherichia coli* O157:H7 shedding in feces, *Infect. Immun.*, 72, 168, 2004.
264. Kang, T.J. et al., Expression of the synthetic neutralizing epitope gene of porcine epidemic diarrhea virus in tobacco plants without nicotine, *Vaccine*, 23, 2294, 2005.
265. Karasev, A.V. et al., Plant based HIV-1 vaccine candidate: Tat protein produced in spinach, *Vaccine*, 23, 1875, 2005.
266. Shchelkunov, S.N. et al., Immunogenicity of a novel, bivalent, plant-based oral vaccine against hepatitis B and human immunodeficiency viruses, *Biotechnol. Lett*, 28, 959, 2006.
267. Khandelwal, A. et al., Systemic and oral immunogenicity of hemagglutinin protein of rinderpest virus expressed by transgenic peanut plants in a mouse model, *Virology*, 323, 284, 2004.
268. McGarvey, P.B. et al., Expression of the rabies virus glycoprotein in transgenic tomatoes, *Bio/Technology*, 13, 1484, 1995.
269. Mishra, S., Yadav, D.K., and Tuli, R., Ubiquitin fusion enhances cholera toxin B subunit expression in transgenic plants and the plant-expressed protein binds GM1 receptors more efficiently, *J. Biotechnol.*, 127, 95, 2006.
270. Tuboly, T. et al., Immunogenicity of porcine transmissible gastroenteritis virus spike protein expressed in plants, *Vaccine*, 18, 2023, 2000.
271. Saldana, S. et al., Production of rotavirus-like particles in tomato (*Lycopersicon esculentum* L.) fruit by expression of capsid proteins VP2 and VP6 and immunological studies, *Viral Immunol.*, 19, 42, 2006.
272. Sandhu, J.S. et al., Oral immunization of mice with transgenic tomato fruit expressing respiratory syncytial virus-F protein induces a systemic immune response, *Transgenic Res.*, 9, 127, 2000.
273. Shin, E.A. et al., Synthesis and assembly of an adjuvanted Porphyromonas gingivalis fimbrial antigen fusion protein in plants, *Protein Expr. Purif.*, 47, 99, 2006.
274. Smart, V. et al., A plant-based allergy vaccine suppresses experimental asthma via an IFN-gamma and CD4+CD45RBlow T cell-dependent mechanism, *J. Immunol*, 171, 2116, 2003.
275. Vieira da Silva, J. et al., Phytosecretion of enteropathogenic *Escherichia coli* pilin subunit A in transgenic tobacco and its suitability for early life vaccinology, *Vaccine*, 20, 2091, 2002.
276. Wen, S.X. et al., A plant-based oral vaccine to protect against systemic intoxication by Shiga toxin type 2, *Proc. Natl. Acad. Sci. U S A*, 103, 7082, 2006.
277. Yang, Z.Q. et al., Expression of the fusion glycoprotein of Newcastle disease virus in transgenic rice and its immunogenicity in mice, *Vaccine*, 25, 591, 2007.
278. Yu, J. and Langridge, W., Expression of rotavirus capsid protein VP6 in transgenic potato and its oral immunogenicity in mice, *Transgenic. Res.*, 12, 163, 2003.
279. Yu, J. and Langridge, W.H., A plant-based multicomponent vaccine protects mice from enteric diseases, *Nat. Med.*, 19, 548, 2001.
280. Zhang, H. et al., Expression and characterization of *Helicobacter pylori* heat-shock protein A (HspA) protein in transgenic tobacco (*Nicotiana tabacum*) plants, *Biotechnol. Appl. Biochem.*, 43, 33, 2006.
281. Higo, K., Saito, Y., and Higo, H., Expression of a chemically synthesized gene for human epidermal growth-factor under the control of cauliflower mosaic virus-35S promoter in transgenic tobacco, *Biosci. Biotechnol. Biochem.*, 57, 1477, 1993.
282. Reggi, S. et al., Recombinant human acid beta-glucosidase stored in tobacco seed is stable, active and taken up by human fibroblasts, *Plant Mol. Biol.*, 57, 101, 2005.

283. Borisjuk, N.V. et al., Production of recombinant proteins in plant root exudates, *Nat. Biotechnol.*, 17, 466, 1999.
284. Claparols, M.I. et al., Transgenic rice as a vehicle for the production of the industrial enzyme transglutaminase, *Transgenic Res.*, 13, 195, 2004.
285. Lee, S.J. et al., Production and characterization of human CTLA4Ig expressed in transgenic rice cell suspension cultures, *Protein Expr. Purif.*, 51, 293, 2007.
286. Mor, T.S. et al., Expression of recombinant human acetylcholinesterase in transgenic tomato plants, *Biotechnol. Bioeng.*, 75, 259, 2001.
287. Petolino, J.F. et al., Expression of murine adenosine deaminase (ADA) in transgenic maize, *Transgenic Res.*, 9, 1, 2000.
288. Mu, J.H., Chua, N.H., and Ross, E.M., Expression of human muscarinic cholinergic receptors in tobacco, *Plant Mol. Biol.*, 34, 357, 1997.
289. Sunil Kumar, G.B. et al., Expression of hepatitis B surface antigen in tobacco cell suspension cultures, *Protein Expr. Purif.*, 32, 10, 2003.
290. Sojikul, P., Buehner, N., and Mason, H.S., A plant signal peptide-hepatitis B surface antigen fusion protein with enhanced stability and immunogenicity expressed in plant cells, *Proc. Natl. Acad. Sci. U S A*, 100, 2209, 2003.
291. Huang, Z. et al., Virus-like particle expression and assembly in plants: Hepatitis B and Norwalk viruses, *Vaccine*, 23, 1851, 2005.
292. Tsuda, S. et al., Application of the human hepatitis B virus core antigen from transgenic tobacco plants for serological diagnosis, *Vox Sanguin.*, 74, 148, 1998.
293. Mason, H.S. et al., Expression of Norwalk virus capsid protein in transgenic tobacco and potato and its oral immunogenicity in mice, *Proc. Natl. Acad. Sci. U S A*, 93, 5335, 1996.
294. Warzecha, H. et al., Oral immunogenicity of human papillomavirus-like particles expressed in potato, *J. Virol.*, 77, 8702, 2003.
295. Briggs, K. et al., Production and characterization of anti-herpes plantibodies from transgenic rice., *Abst. Papers Amer. Chem. Soc.*, 221, U110, 2001.
296. Malembic-Maher, S. et al., Transformation of tobacco plants for single-chain antibody expression via apoplastic and symplasmic routes, and analysis of their susceptibility to stolbur phytoplasma infection, *Plant Sci.*, 168, 349, 2005.
297. Yuan, Q.P. et al., Expression of a functional antizearalenone single-chain Fv antibody in transgenic Arabidopsis plants, *Appl. Environ. Microbiol.*, 66, 3499, 2000.

19 Plant Virus Biotechnology Platforms for Expression of Medicinal Proteins

Andris Zeltins

CONTENTS

19.1 INTRODUCTION

Since ancient times people have used plant resources for different needs including food, housing construction, clothing, cosmetics, cultural, and medicinal purposes. Medicinal plants play an important if not one of the central roles among these uses. The number of plant species have been used or are still used for health improvement purposes can be approximately estimated as more than 70,000 (17%) out of the 422,000 flowering plant species. The number of commercialized species was assessed as approximately 3000 [1]. The evaluation of commercially traded amounts of medicinal plants reveals the dominance of developing countries in export; China alone has exported near to 140,000 tonnes of plant material worth of $300 million between 1991 and 1998 [2].

Numerous drugs have been developed based on chemical analysis of active compounds found in medicinal plants. Historically, the first synthetic drug aspirin was introduced by Friedrich Bayer in 1897. Aspirin is the synthetic analogue of acetyl salicylic acid found in willow bark; [3] other well-known examples include digoxin originating from foxglove and morphine from opium poppy [4]. In the last century, the development of pharmacology was mostly concentrated on replacement of active ingredients found in natural extracts with their synthetic analogues. The efforts of chemistry-based pharmaceutical industry resulted in excellent achievements in treatment of different diseases. A review of the most important plant derived pharmaceuticals can be found in Ref. [5] (see also Chapter 18). Despite the success in the twentieth century, in the last decade most pharmaceutical companies scaled down their screening programs for new natural products in favor of combinatorial chemistry approaches. However, the number of new active compounds derived by this approach did not match the early expectations and some experts started promoting a return to research of natural products [6].

According to the WHO report, over one-third of the population in developing countries lack access to essential medicines due to high therapy costs and absence of modern medical infrastructure and continue to rely on their ethnobotanical remedies. Despite the highly significant developments in chemistry-based pharmacology, herbal drugs represent a significant part of the world medical market. In industrial countries, over 50% of the population have used complementary medicine at least once. According to the data from the U.S. Commission for Alternative and Complementary Medicines, $17 billion were spent in United States on alternative remedies in 2000 [7]. Another important part of marketed plant products are herbal supplements, which can be defined as simple or multicomponent herb mixture used to supplement the medical treatment. Regardless of a weak or sometimes absent scientific evidence of the efficiency and reported adverse effects, sales of herbal products in the United States doubled to $16 billion between 1994 and 2002 [8]. All these facts confirm the enormous importance and general social acceptance of different plant products in health improvement.

New biotechnology based on the fundamental molecular knowledge has led to the interest of scientists in obtaining new pharmaceuticals from plant sources. Research conducted for more than two decades by plant biotechnologists resulted in the outstanding progress in genetic technologies for obtaining plant products of medicinal relevance. This chapter summarizes the existing publicly available data and discusses perspectives of one of the most promising molecular technologies; namely, plant virus-based biotechnological systems aiming at producing recombinant proteins (rProts) for pharmaceutical industry, including vaccines, antibodies, and different therapeutic proteins.

19.2 PLANT VIRUSES: SHORT OVERVIEW

Plant viruses can be characterized as small, nanometer-sized plant infecting agents, which are not able to multiply outside of the host cells. Structurally, the viruses are built from nucleic acid (viral genome) containing genes required for the virus lifecycle and a protein coat; some species are also enveloped in lipid-containing membrane. Viral genes code at least three types of proteins: (1) polymerase for replication of the viral genome, (2) movement protein (MP) necessary for virus spread from one cell to the next, and (3) hundreds or thousands of coat protein (CP) molecules protecting the viral genome from environmental damage. Additionally, plant virus genomes can code such proteins as proteases, helicases, methyltransferases as well as proteins for long distance movement and suppressors of posttranscriptional gene silencing. Plant RNA viruses employ several different strategies for expression of encoded proteins. The most common strategies are expression of a polyprotein, which subsequently processed into mature viral proteins by virus encoded proteases, and translation from subgenomic RNA's (sgRNA). Some viruses exploit internal ribosome entry sites (IRES), "leaky-stop" codons and ribosomal frameshift signals to maximize the coding capacity of the genome.

Plant virus proteins can be multifunctional; for example, CPs can be additionally involved in movement of the virus through the plant. In nature, plant viruses are transferred from infected to uninfected plants through mechanical damage of plant tissues caused by vectors. Virus vectors are in most cases insects feeding on plant leaves or stems. However, in the agricultural environment, working people, instruments, and machines can contribute to the virus dissemination among plants. A more detailed review of plant viruses can be found in Refs. [9–11].

According to the VIIIth Report of the International Committee for the Taxonomy of Viruses, currently, more than 6000 viruses are known and plant viruses represent a significant part of them—ca. 2000 [12]. Plant viruses belong to different virus groups classified according to the structure of the incapsidated genome and can contain double-stranded (ds) and single-stranded (ss)-DNA, positive(+)- and negative(−)-sense RNA and dsRNA. The majority of plant viruses are non-enveloped (+)-sense RNA viruses. The most of identified plant viruses (77%) are isolated from cultivated plants; therefore, the number of currently known viruses represents a minor part of really existing viruses, especially viruses infecting wild plants [13]. The number of viral nucleotide sequences deposited in GenBank database is growing exponentially and a search using term "virus" at Entrez site (The Life Science Search Engine, NCBI) results in approximately 600,000 sequence records from different viruses. Taking into account recent developments in sequencing technologies it is imaginable that millions of sequences will be available in the next few years [14]. Such a large amount of sequence information together with the knowledge of gene and encoded protein functions represents an inexhaustible source for plant biotechnologists for finding new ways to defeat virus diseases of cultivars and developing innovative plant virus-based technologies for production of numerous medically relevant products.

Plant viruses have been intensively studied for more than 100 years, when Adolf Mayer from Wageningen started to investigate the tobacco mosaic disease. Since that time tobacco mosaic virus (TMV) is an object in numerous studies, several of which significantly contributed to the development of different fields of biology, biophysics and biochemistry, including molecular biology. TMV remains one of the best studied viruses. Modern virology concepts of virus architecture and infection cycle are largely based on studies carried out on this virus [15–17].

To better characterize the development of plant viruses toward the technological exploitation it is important to mention main milestones in plant virus research leading to the progress of molecular technologies. First of all, in 1956 it was found, that purified TMV viral RNA without the CP retains infectivity [18]. The achievements of subsequent decades in techniques of DNA manipulation, especially cloning, DNA sequencing, polymerase chain reaction, and RNA copying into cDNA using reverse transcriptases resulted in complete nucleotide sequencing of the TMV genomic RNA [19] and led to the development of plant viruses into gene vectors. Nearly 20 years later (1984), the finding of the infectivity of plant virus RNA was exploited by Ahlquist and his colleagues [20]. Plant virus RNA was synthesized in vitro using corresponding brome mosaic virus (BMV) cDNA clones as templates to show the infectivity of artificially obtained RNA in BMV-susceptible plants. Soon after this report, Dawson et al. [21] demonstrated the infectivity of the similarly synthesized TMV RNA in pinto bean (*Phaseolus vulgaris*) and tobacco (*Nicotiana tabacum*) leaves. These achievements served as starting points for numerous studies aiming at constructing plant virus-derived gene vectors using cDNA of different, mostly ss(+)RNA plant viruses.

The next milestone important for plant virus-based technology developments is related to the three-dimensional (3D) structure studies of plant viruses. The very first virus, for which the atomic 3D-structure at 2.9 angstrom (Å) resolution was resolved in 1978 by Harrison and collaborators, was an icosahedral plant virus—tomato bushy stunt virus (TBSV) [22]. The Southern bean mosaic virus 3D atomic structure with a similar icosahedral subunit arrangement was elucidated next in Rossman's laboratory 2 years later [23]. For the best studied plant virus, TMV, the 3D structure was determined at 3.6 Å resolution by Namba and Stubbs [24]. During the next two decades, nearly 30 different plant viruses were purified, crystallized and their 3D structures at atomic resolution were resolved. The achievements in virus macromolecule 3D structural studies contributed very significantly to the development of peptide (epitope) presentation systems on plant virus surface (the principles and examples are discussed in the following sections).

Other milestones in the understanding of virus lifecycles related to the development of plant virus-based protein expression systems can be found in the review articles cited in different sections of this chapter.

19.3 PLANT VIRUSES AND MAMMALIAN ORGANISMS

Plant viruses are not infectious for mammalian organisms including humans. In the past decades after Arntzen's formulated the idea of edible vaccines from plants [25] many scientific groups and biotech companies started a very successful exploration of protein expression and development of vaccine candidates using plant virus expression systems. However, very little is known about outcomes of introduction of plant viruses and virus-like particles (VLP's) in human and animal organisms. Recently, a metagenomic analysis of RNA viral communities in gastrointestinal tract of healthy human individuals was carried out. Surprisingly, according to the cDNA sequence analysis, researchers found a significant abundance of RNA molecules derived from different plant viruses instead of expected bacterial viruses [26]. In total, 34 plant virus species were identified, including viruses from different plants used in human diet: Pepper mild mottle virus (PMMV), Oat blue dwarf virus, Grapevine asteroid mosaic-associated virus, Grapevine red globe virus. Also, the sequences of viruses infecting plant hosts of other than human food origin were found to be present in human gastrointestinal tract, for example, TMV, Cocksfoot mottle virus, Poinsettia mosaic virus. PMMV isolated from human feces was even infectious to susceptible plant hosts. These data suggest that humans can be dissemination vehicles for plant viruses and demonstrate the stability of a number of plant viruses in gastrointestinal tract. Stability of plant viruses in mammalian organisms represents one of important selection prerequisites in creation of plant virus-based edible vaccines.

One of the most promising candidates for edible vaccines, CPMV, was studied using a mouse model to understand the fate of virus particles in feeding and intravenous injection experiments [27]. First, CPMV was shown to remain stable and infectious after incubation in simulated low-pH gastric and neutral intestinal fluids containing digestive enzymes. Second, the analysis of virus-specific RNA and fluorescently labeled virus particles in different mice organs revealed a systemic distribution and relative stability of CPMV in spleen, liver, lungs, stomach, bone marrow, and other organs independently of the route of administration. The results of experiments demonstrated the ability of CPMV to bind endothelial cells and to enter the tissue parenchyma; however, the exact mechanism how orally fed virus particles crossed the intestinal epithelium is not known yet. An additional toxicology study revealed that even 100 mg injected CPMV nanoparticles per kilogram of mouse body weight did not cause significant hematological changes, tissue degeneration, and necroses. As a conclusion, CPMV was recommended as a safe and nontoxic multimeric protein assembly for biomedical applications in vivo [28].

Next study addressing the question of the fate of plant viruses in mammalian organisms was carried out on the cell culture level using the same virus CPMV [29]. Mouse fibroblast and human epidermal carcinoma cells were shown to bind and internalize fluorescently labeled virus in significant amounts, whereas the efficiency for both processes using particles treated with virus-specific antibodies was considerably reduced. A specific protein CPMV-BP (54 kDa) was found bound on plasma membranes of both human and animal cell lines. In an attempt to explain the existence of such specific membrane binding protein in a plant virus, the authors noted structural similarities between CPMV and mammalian viruses from genus *Enterovirus* and suggested that plant viruses have an access to mammalian hosts at a higher extent than generally accepted.

19.4 BIOTECHNOLOGY PLATFORMS: INTRODUCTION

A biotechnology platform can be defined as a complex of technological measures to create, produce, purify, and evaluate properties, and the quality of marketable products derived from different biological organisms. In other words, it is a whole technological line starting from input materials and ending with a product, which meets the criteria of efficient production, authenticity, and safety.

To date, several biotechnology platforms are known that are classified accordingly to the organism type used for the production of rProts. Side-by-side comparisons of rProt production platforms can be found in some review articles [30,31].

19.4.1 Bacterial Platforms

The platforms based on use of *Escherichia coli* or other prokaryotic organisms [30,32,33] are probably best established for the production of medicinally relevant proteins. In 1973, Cohen and Boyer demonstrated for the first time cloning of a foreign gene in bacterial plasmid vector [34]. The successful development of this platform resulted in licensing of the first human rProt, insulin, in 1982. During the next two decades the application of *E. coli* in production of rProts was introduced worldwide as a generally accepted technology for research and large-scale manufacturing. Despite such advantages of the platform as long history of application, well-known manipulation techniques, low time, and financial investments during the research phase, and medium to high product yield in a short period, other parameters like product authenticity and safety are often hardly achievable. Mammalian proteins expressed in *E. coli* are not glycosylated and frequently not properly folded or even insoluble; besides, the preparations can contain bacterial endotoxins. To obtain physiologically active, correctly folded proteins without endotoxin contamination, additional refolding and purification steps have to be included in the production procedure. Additionally, large-scale production is expensive due to special requirements imposed on manufacturing facilities using genetically modified microorganisms.

19.4.2 Yeast-Based Platforms

To overcome disadvantages of rProt production from *E. coli*, different yeast strains are frequently used [35–37]. The first cloning of a gene coding for a eukaryotic protein in baker's yeast *Saccharomyces cerevisiae* was carried out in 1978 [38]. Later, the very first recombinant human vaccine against hepatitis B was obtained from *S. cerevisiae* expression platform [39] and commercialized. Yeast systems are widely used in pharmacological industry; approximately 15% of approved therapeutic proteins are produced in different yeast strains. Usage of the yeast platform ensures several technological advantages in comparison to the *E. coli* platform; rProts are more adequately folded, often higher yields are achievable in inexpensive mineral media, preparations do not contain toxins. The main problem of the platform is a completely different glycosylation pattern of proteins in yeast cells, making yeast-produced glycosylated proteins unstable, or even immunogenic in mammalian organisms and, therefore, often unsuitable for therapeutic usage for humans and animals. However, in last years substantial improvements of yeast strains have been introduced enabling to "humanize" rProts to produce therapeutic glycoproteins [40].

19.4.3 Mammalian Cells

Cell cultures as a platform for the production of rProts are known for more than two decades since human tissue plasminogen activator, the first therapeutic protein from recombinant mammalian cell line, was commercialized [41,42]. Recent evaluations show that approximately 55% of all rProts in pharmaceutical industry are produced in different mammalian cell lines [43]; immortalized Chinese hamster ovary (CHO) cells are most frequently used. The first and the main advantage of the platform is related to quality and adequacy of rProts from mammalian cells; product proteins are properly folded, assembled, and correctly glycosylated. In addition, comparably high yields can be achieved. Despite the dominance among the most used platforms for the production of rProts, mammalian cell systems have disadvantages, mainly related to very high-production costs. Such factors as limited production scale, requirement for special equipment, high-priced raw materials, and control of human or animal pathogen contaminations contribute significantly to the price of pharmaceutical products, making them expensive and hardly accessible for the most part of the world.

19.4.4 Transgenic Animals

Production of rProts from transgenic animals are under discussion already for more than 30 years, since the first transgenic mouse carrying a retrovirus was created [44–46]. Now, different transgenic mouse lines are widely used in development of different therapeutics including monoclonal antibodies. A very attractive and promising technological approach for the fabrication of pharmaceuticals is production of products in biological fluids of transgenic livestock, especially in milk due to a potentially large-scale production. The product characteristics are comparably to mammalian cell platform; namely, authentic protein folding, activity, and glycosylation. In addition, a production scale larger than using cell culture can be theoretically achieved. However, these platforms require extreme time and finance resources in the development phase; additional problems are related to the purification of the product and control of livestock pathogens. Another factor hindering the development in this area is an ethical concern and a low public acceptance of the creation and usage of transgenic animals.

19.4.5 Transgenic Plants

The research conducted over the last two decades demonstrated that plant biotechnology platforms are conceptually perspective bioreactors for pharmaceutical production. After pioneering work of Arntzen and his colleagues postulating the idea of edible vaccines [25,47] plants became highly popular not only for new vaccine production but also for other biomedical applications. An extensive list of literature sources describing the potential for transgenic plant platforms can be found in numerous review articles [48–53]. First rProts of a therapeutic interest obtained from transgenic plant biomass were human growth hormone [54], human serum albumin produced in transgenic tobacco and potato plants [55] as well as maize-derived bovine trypsin, as a first commercialized protein [56]. Since the first reports, many medicinal proteins such as pharmaceutical intermediates, enzymes of medical interest, monoclonal antibodies and vaccines, altogether more than 100 foreign proteins have been produced in different plant species [57,58].

If compared to other production platforms, the most promising feature of transgenic plants is their potential for scalability. Transgenic plants can be easily propagated, scaled up, and cultivated extensively using typical agricultural techniques; tones of biomass containing pharmaceuticals of interest can be harvested. Another advantage of plant systems is a possibility to clone large DNA fragments, allowing for inserting several genes in the plant genome, which is necessary for production of multi-subunit proteins. Plants can be cultivated in the form of cell culture. Frequently, the plant produced medical proteins are correctly folded and assembled; the glycosylation appears to be similar to that of mammalian proteins; preparations are safe and do not contain toxins. First clinical studies demonstrated good tolerance to treatments with plant-derived pharmaceuticals. The yields are rather low and vary in a range of 0.003%–1% of total soluble protein (TSP); however, recent developments demonstrate the capability of transgenic plants to produce the protein of interest of up to 25% TSP [59].

Plant-derived products are stable and can be stored at ambient temperatures, especially, if the transgenic plant platform ensures the accumulation of the protein in separate cell compartments like chloroplasts, mitochondria, vacuoles, or seeds (for review see Ref. [60]).

The first difficulties can, however, arise already at the construction stage of a transgenic plant line because the process is time consuming and expensive. The main disadvantage of large-scale production is related to environmental risks, first, in the form of the possible gene transfer from transgenic plants to wild ones and microorganisms, and second, in the form of free-growing plants with a high active pharmaceutical content influencing insects, wild animals, and people. These aspects raise concerns and negative attitude toward genetically modified organisms in society [61]. If at the early stages, the low social acceptance involved only food products from transgenic plants, now, the attitude become sceptic against otherwise generally accepted products for human health, including the use of modified plants in pharmaceutical industry. A doubtful attitude is expressed against exploitation of transgenic food-plants for the production of technical and pharmaceutical products, which can cause a risk of contaminating the food products with rProts intended for medicinal or industrial usage. The concern can be reduced only by introduction of nonfood plants in production and detailed environmental and clinical studies in each case, requiring additional time and finance resources at the development stage. Several experts stress the cultivation of transgenic plants for pharmaceutical production under containment conditions, in greenhouses, but such an approach can reduce the accessibility of the technology for the resource limited parts of the world due to increased production costs [62].

19.5 PLANT VIRUS-BASED BIOTECHNOLOGY PLATFORMS

Plant virus-based platforms can be regarded as a variation of transgenic plant platform, however with a significant difference: foreign protein genes are expressed transiently with a help of virus genomes and are not incorporated into plant genome. Virus platforms have several advantages if compared with other platforms discussed above. Established platforms can be easily and timely adapted for production of different proteins expressed at a very high level. Plant-derived products are safe and free of toxins and mammalian pathogens. rProts obtained from plant virus-derived transient expression systems are correctly folded, glycosylated, and biologically active; the usage of transgenic plant hosts is not an obligatory prerequisite. In the next sections, plant virus biotechnology platforms are discussed in more detail. The principal schemes of plant virus vectors are shown in Figures 19.1 and 19.2.

19.5.1 Construction Principles of Plant Virus Gene Vectors

Nucleic acids isolated from virus preparations, can be infective for plants as already mentioned in Section 19.2. To create a virus gene vector, virus RNA have to be converted into corresponding cDNA by standard molecular biology techniques and further manipulated, including rearrangements of the genome and insertion of foreign genes. The next step is to add an appropriate promoter to the virus cDNA to initiate transcription of the virus genome. For the initiation of viral infections from constructed cDNA, several approaches can be used: (1) infectious RNA, which can be synthesized in vitro from cDNA using bacterial virus T7 and SP6 promoters; (2) direct infection with cDNA, subcloned under strong plant promoters, as cauliflower mosaic virus (CaMV) 35S promoter; both types of nucleic acids can be delivered into plant cells through simple mechanical wounding of leaves using abrasive materials or biolistic bombardment technique [63]; and (3) delivery with a help of phytopathogenic soil bacterium *Agrobacterium tumefaciens* (agroinfection). In the last case, viral cDNA is subcloned under a plant promoter in *E. coli/Agrobacterium* shuttle plasmid, which contains T-DNA [64]. During the bacterial infection process the subcloned viral cDNA together with T-DNA is transported to the plant cell nucleus, where the virus RNA synthesis takes place and plant cells become virus-infected.

First constructed plant virus gene vectors were so-called replacement vectors where CP genes were substituted for model protein genes [65–67]. The removal of CP from BMV and TMV caused changes in virus viability; regardless of their capability to produce rProts these gene vectors were limited in movement from one cell to the next. The systemic infection and protein expression in all parts of the plant was not observed. The principal solution of the problem was demonstrated on a TMV model. The CP gene together with a subgenomic promoter from related tobamovirus ORSV was inserted into the TMV genome at 3′end. The original TMV CP was replaced by a model protein

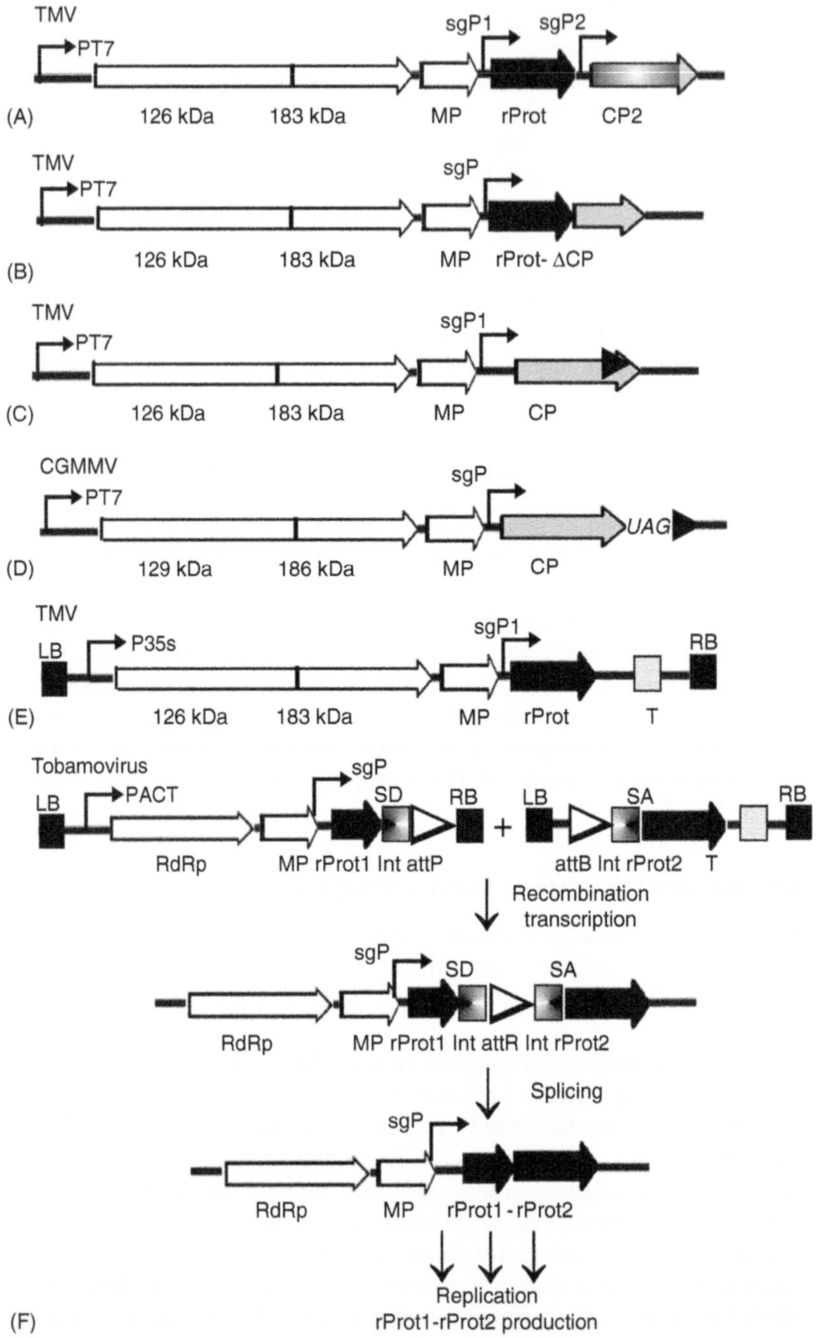

FIGURE 19.1 Tobamovirus gene vectors (A–F: See text for details). Abbreviations: TMV, tobacco mosaic virus; CGMMV, cucumber green mottle mosaic virus; MP, movement protein; rProt, recombinant protein; CP, coat protein; PT7, T7 bacterial promoter; sgP, plant virus subgenomic promoter; P35S, CaMV 35S promoter; PACT, *Arabidopsis* actin promoter; T, transcription terminator; LB and RB, left and right borders of T-DNA region in agrobacterial shuttle vector; RdRP, RNA-dependent RNA polymerase; Int, intron sequences; aatP/aatB, integrase recombination sites from *Streptomyces* phage; SD/SA, splicing donor/acceptor sites. Black triangles denote the relative position of epitope cDNA in CP genes.

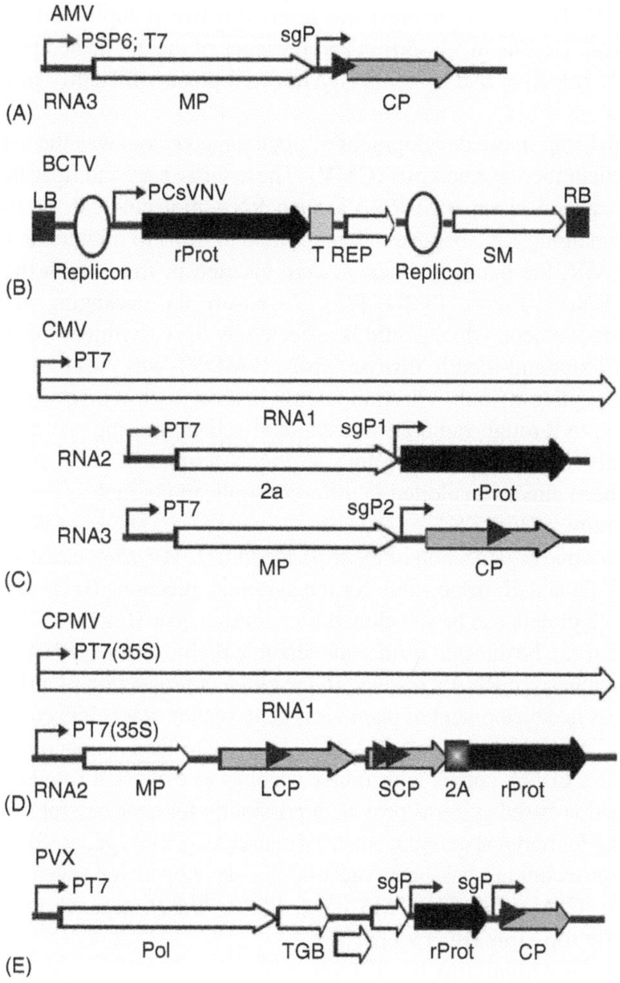

FIGURE 19.2 Examples of different plant virus gene vectors (A–E, See text for details). Abbreviations: AMV, alfalfa mosaic virus; CMV, cucumber mosaic virus; BCTV, beet curly top virus; CPMV, cowpea mosaic virus; PVX, potato virus X; MP, movement protein; rProt, recombinant protein; CP, coat protein; PSP6, PT7, SP6 and T7 bacterial promoters; sgP, plant virus subgenomic promoter; P35S, CaMV 35S promoter; LB and RB, agrobacterial shuttle vector T-DNA region left and right borders; REP, replication initiation protein of BCTV; SM, selection marker; T, transcription terminator; LCP/SCP, CPMV large and small coat proteins; 2A, self processing AA sequence; TGB, triple gene block of PVX. Black triangles denote the relative position of epitope cDNA's in CP genes.

and expressed from TMV subgenomic promoter (Figure 19.1A) [68]. The reconstructed TMV vector ensured systemic infection and the model protein synthesis in plant leaves, demonstrating advantages of this gene insertion strategy.

Later, slightly modified strategy for gene insertion in TMV genome was developed. The recombinant gene was subcloned at the 5′ end of the truncated CP gene (Figure 19.1B). The subgenomic promoter was used to drive the synthesis of the fusion protein [69,70]. Despite the loss of the two-thirds of the CP, plants inoculated with the virus vector demonstrated systemic infections, and production of rProts.

A similar strategy was used to develop another rod-shaped virus with a monopartite genome, potato virus X (PVX) as a plant virus vector platform for the foreign protein expression

(Figure 19.2E) [71,72]. The gene of interest was inserted between duplicated subgenomic promoter in the front of CP gene. Despite this insertion and presence of the duplicated promoter, recombinant PVX RNA was fully infective and systemically infected plants efficiently producing biologically active proteins.

The next principal step in the development of plant gene vectors was the use of cowpea mosaic virus (CPMV) and cucumber mosaic virus (CMV). These viruses are multipartite, e.g., viral protein-coding genes are localized in separate RNA's. One RNA molecule can be used for insertion of foreign genes; the remaining RNA's have to be provided *in trans* to initiate the infection process. In case of bipartite CPMV, the genes of interest were inserted in frame with the small coat protein (SCP) encoded by RNA2 (Figure 19.2D) [73]. To ensure the packaging of the virus and SCP functioning in virus movement, which could be affected by the C-terminal fusion of a rProt to SCP, a 2A peptide from foot-and-mouth disease virus (FMDV) was inserted into the fusion site (Figure 19.2D). 2A peptide of only 18 amino acids is able to cleave two parts of fusion protein from each other in vivo through rapid co-translational self-processing. As a result, the protein of interest contains only one additional proline at the N-terminus. Such an artificial processing of polyproteins has been already exploited in different applications such as co-expression of reporter proteins in gene therapy vectors [74].

Studies on the genomic organization of the tripartite virus CMV demonstrated that RNA2 encodes for long-distance MP 2a and 2b responsible for the systemic infection. To create the gene vector, the coding sequence for 2b protein can be substituted for a foreign gene (Figure 19.2C) and the expression of a therapeutic protein can be directed from a subgenomic promoter for sgRNA4A. Co-inoculation of all three artificially synthesized RNA's resulted in the efficient production of a rProt [75].

The latest progress in development of plant virus gene vectors was achieved using *Agrobacterium*-mediated delivery of recombinant virus cDNA in host cells. First vectors were largely based on exploitation of full-size cDNA copies with minor changes to express a foreign gene or short DNA sequences. Agroinfection-based systems provide a possibility for a successful usage of deconstructed virus vectors, allowing for removal or replacement of unnecessary virus genes (for review see Ref. [76]).

The agroinfection technique has been successfully developed for several RNA virus vectors, including TMV [64], CPMV [73,77], CMV [78], PVX [79,80], as well as for DNA containing geminivirus beet curly top virus (BCTV) [81].

Several examples of geminivirus-based foreign protein expression can be found in the extant literature [82–84]. To create the geminivirus BCTV vector, a foreign gene accompanied by a transcription terminator and a gene encoding for replication initiation protein (REP) were inserted between duplicated replication origin DNA sites encoding for a strong plant promoter (Figure 19.2B). The resulting cassette was subcloned within the *Agrobacterium* shuttle vector between left and right flanks of the T-DNA segment. The constructed BCTV vector can be used in a wide variety of plant hosts to express proteins of a medicinal interest [81].

Further developments of CPMV as a gene vector resulted in creation of an inducible system allowing for producing foreign proteins with potentially negative effects on plant growth [85]. Transgenic tobacco plants were constructed carrying a modified virus RNA2 cDNA copy together with rProt gene (construction strategy similar to the gene arrangement shown in Figure 19.2D). The virus infection and synthesis of foreign protein was induced by agroinfection with the RNA1 cDNA copy carrying CPMV replication protein genes. This system allows for synthesizing a protein of interest from RNA2 without causing the infectious process. In this case, the protein yield can be significantly increased by adding a gene silencing suppressor (HcPro from potato virus Y). Such a platform design prevents a systemic infection and can be used for the production of pharmaceutical proteins without risking viral contamination of environment.

A chemically inducible protein production platform has been constructed using cDNA copies of CMV virus RNA's [78]. This study demonstrated a different strategy to construct vectors based on multipartite viruses. All necessary vector components, including cDNA's of RNA1, RNA2, and RNA3 as well as promoters and terminators were step-by-step subcloned in a single *Agrobacterium* shuttle vector, avoiding the need to co-inoculate several vectors coding for each virus RNA. The RNA1 synthesis was directed by estradiol-inducible promoter. The application of estradiol to the plant leaves resulted in transcription of the gene responsible for synthesis of 1a protein, a component of active replicase complex, the virus infection from the vector was shown to be inducer dependent. If the CP gene was substituted by a medically relevant gene, induction of the RNA1 synthesis stimulated the efficient production of this rProt in agroinoculated leaves. Like in the case of CPMV, virus vector could not cause infection without CP gene. Therefore, this CMV gene vector represents another example of environmentally safe platform.

The strategy of the "deconstructed vector" [76] was developed using the PVX-based platform. Replacement of triple-gene block (TGB) and CP genes for a model protein gene (see Figure 19.2E) abolished the infectivity completely, but *Agrobacterium*-mediated delivery of the dramatically deconstructed virus cDNA in plant cells compensated the production of a rProt at a significant level [80]. The most efficient plant virus biotechnology platforms are developed from Tobacco mosaic virus, based on agroinfection and the strategy of "deconstructed virus."

A relatively simple TMV-derived vector was recently shown to efficiently facilitate expression of rProts in conjunction with the agroinfection technique [86]. This vector called "launch vector" was constructed by insertion of TMV genome without CP gene in the *Agrobacterium* shuttle plasmid between the left and right flanks of the T-DNA segment (Figure 19.1E). The gene coding for the protein of interest can be subcloned in place of the deleted CP. TMV constructs without CP are suggested to be more stable and subjected to recombination to a lesser extent than full-size virus vectors. As this virus vector is not able to replicate autonomously and spread outside of the agrobacterial-plant system, the enhanced efficiency, simplicity of construction, and safety for the environment makes it a promising biotechnology platform.

An efficient and flexible rProt expression system represents another new plant virus offers another new plant virus/*Agrobacterium* platform designed from cDNA modules derived from tobamovirus genomes and other components (magnifection system) [87,88]. A simplified schema of the system is shown in Figure 19.1F. A general concept for the improvement of the expression level is based on the fact that up to five independent DNA modules subcloned into separate *Agrobacterium* plasmids can be delivered in the same plant cell by a simple treatment of plant leaves with a mix of corresponding bacterial cultures. The vector system was constructed from structural blocks of two tobamoviruses, crucifer-infecting TMV, and closely related turnip vein clearing virus (TVCV). Additionally, to substantially improve the formation of RNA replicons from cDNA modules in plant cell nucleus, several intron sequences were introduced in the codon optimized virus cDNA [89].

The module system provides the possibility to assemble a complete vector from two provector modules in plant cells by site-specific recombination process with help of the module containing integrases, for example, derived from a *Streptomyces* phage. If recombination sites were flanked by a suitable intron sequences, exact mRNA coding for the fusion protein was obtained after the splicing process. As a result of subsequent replication of the joined and spliced virus RNA module, efficient production of the fusion protein (ubiquitin-green fluorescent protein as an example) was demonstrated. The module system was validated by expression of different genes coding for such medicinally relevant proteins as human cytokines, hormones, single-chain antibodies, and enzymes. This technology combines advantages of the transfection efficiency of *Agrobacterium*, efficient foreign protein production from viral vectors and posttranslational capabilities of plant cells. The magnifection platform has a high potential for scaling up to the industrial level. This concept applied to the tobamovirus-derived system can be extended to systems based on other plant viruses.

19.5.2 Plant Hosts

Host organisms are highly important components of the biotechnology platform. For plant vectors, natural or experimental hosts of corresponding viruses are used for production of proteins of interest. Several examples of hosts for plant virus gene vectors are listed in Table 19.1.

TABLE 19.1
Production of Model or Medically Relevant Proteins by Different Plant Virus Vector Platforms

Virus Vector	Host	Expressed Protein	Output	Properties	References
BCTV	*N. benthamiana*	Green fluorescent protein (GFP, 26.3 kDa)	1.5% TSP	Biochemically functional	[81]
CMV	Transgenic *N. benthamiana*	Human α-antitrypsin (AAT; 52 kDa)	240 mg kg^{-1} plant leaves	70% of AAT biologically active	[78]
CMV	*N. benthamiana*	Acidic fibroblast growth factor (aFGF, 17.5 kDa)	5%–8% TSP	Biologically active	[75]
CPMV	*N. benthamiana*	Green fluorescent protein (GFP, 26.3 kDa)	1% TSP	Biochemically functional	[196]
CPMV	*Vigna unguiculata*	Hepatitis B virus core (HBcAg, 21 kDa)	10 mg kg^{-1} plant leaves	Characteristic HBcAg structures recognized by polyclonal antibodies	[197]
CPMV	Cowpea plants	Small immune protein (SIP, 42 kDa)	2% TSP	Crude plant extracts containing SIP neutralized porcine coronavirus in tissue culture	[198]
PPV	*N. clevelandii*	Green fluorescent protein (GFP, 26.3 kDa)	250 mg kg^{-1} plant leaves	Biochemically functional	[199]
PPV	*N. clevelandii*	RHDV structural protein (VP60, 60 kDa)	—	Plant extract injections protected rabbits against RHDV challenge	[199]
PVX	*N. benthamiana*	Human lactoferrin (hLfN, 40 kDa)	0.6% TSP	Plant produced hLfN demonstrated antibacterial activity	[79]
PVX	*N. benthamiana*	Recombinant CMV coat protein (24.2 kDa) carrying Newcastle disease virus epitopes	700 mg kg^{-1} plant biomass	NDV containing VLP recognized by specific antibodies	[149]
PVX	*N. benthamiana*	Hepatitis B virus core protein (HBcAg, 21 kDa)	>100 mg kg^{-1} plant leaves	Characteristic HBcAg structures recognized by polyclonal antibodies	[197]
PVX	Transgenic *N. tabacum*	Canine oral parvovirus L1 protein (COPV L1; 57 kDa)	0.3% TSP	COPV L1 recognized by polyclonal antibodies	[113]
TMV	*N. benthamiana*	Green fluorescent protein (GFP, 26.3 kDa)	10% TSP	Biochemically functional	[69]
TMV	*N. benthamiana*	HCV epitope fused at C-termini of B subunit of cholera toxin (HVR1/CTB; 50 kDa)	0.9% TSP or 80 mg kg^{-1} plant leaves	HVR1/CTB crude extract in mice intranasal immunization elicited antibodies against HCV	[200]
TMV	*N. benthamiana*	Bovine herpesvirus glycoprotein (BHV-1 gDc; 55 kDa)	20 mg kg^{-1} plant leaves	gDc vaccinated animals protected against BHV-1	[201]

TMV	*N. benthamiana*	HIV core protein (p24; 25 kDa)	100 mg kg^{-1} plant leaves	His-tagged p24 developed strong humoral response; antibodies recognized the native p24	[202]
TMV	*Spinacea oleracea*	HIV Tat-protein (ca. 10 kDa)	300–500 mg kg^{-1}	No adverse effects in mice fed with HIV-Tat containing spinach	[70]
TMV	*N. benthamiana*	Human papilloma virus L1 protein (HPV-16 L1; ~50 kDa)	20–37 μg kg^{-1} plant leaves	HPV VLP's found in plants; weak immunogen	[203]
TMV	*N. benthamiana*	Dengue type 2 virus 2E protein domain III (D2EIII, 13.8 kDa)	0.28% TSP	D2EIII immunized mice produced neutralizing antibodies	[204]
TMV	*N. benthamiana*	Thermostable *Clostridium thermocellum* glucanase (LicKM; 27 kDa) fused with Influenza H5N1 and other epitopes	500 mg kg^{-1} plant leaves	LicKM fusions retained activity at 65°C	[86]
TMV (launch vector)	*N. benthamiana*	*Yersinia pestis* antigens F1 and LcrV fusion to *Clostridium thermocellum* glucanase (~45 and ~55 kDa)	380 and 120 mg kg^{-1} plant leaves	F1- and LcrV-fusions induced serum IgG and IgA responses and protected macaques against lethal challenge with *Y. pestis*	[121]
TMV (launch vector)	*N. benthamiana*	Human papilloma virus E7 protein fusion to *Clostridium thermocellum* glucanase (~40 kDa)	400 mg kg^{-1} plant leaves	E7-fusions induced specific IgG and T-cell responses and protected mice against challenge with tumor cells expressing E7 oncoprotein	[122]
TMV (magnifection)	*N. benthamiana*	Green fluorescent protein (GFP, 26.3 kDa)	80% TSP 5000 mg kg^{-1} plant leaves	Biochemically functional	[87]
TMV (magnifection)	*N. benthamiana*	*Yersinia pestis* antigens F1 (15.7 kDa) and V (37.2 kDa)	1000–2000 mg kg^{-1} leaves	F1/V vaccinated guinea pigs protected against *Yersinia pestis*	[205]
TMV (magnifection)	*N. benthamiana*	Hepatitis B virus core antigen (HBcAg; 21 kDa)	7.14% TSP 238 mg kg^{-1} plant leaves	Mucosally immunized mice developed HBcAg-specific IgG and IgA	[206]
TMV (magnifection)	*N. benthamiana*	Vaccinia virus B5 antigenic domain (pB5, 42 kDa)	100 mg kg^{-1} leaves	pB5 vaccinated mice protected against lethal dose of smallpox	[177]
TMV/PVX (magnifection)	*N. benthamiana*	Coexpression of antibody heavy and light chains (HC; 50 kDa, LC; 25 kDa)	500 mg kg^{-1} leaves	Full-size, functional mAb isolated from plant leaves	[127]

Note: BCTV, beet curly top virus; CMV, cucumber mosaic virus; CPMV, cowpea mosaic virus; PPV, plum pox virus; PVX, potato virus X; TMV, tobacco mosaic virus; TSP, total soluble proteins; —, data not available.

To create an efficient plant virus vector platform, it is important to understand the virus life cycle in host cells, as well as plant cell responses to the virus infection. Virus–plant interactions are reviewed extensively by several authors [90–92].

At least two factors influence considerably the expression of rProts in plant hosts: movement of virus gene vector from one cell to the next and RNA silencing as a plant defense response to the virus infection.

Plant viruses spread in the host through intercellular channels or plasmodesmata, typical organelles of plant cells; the systemic infection is facilitated by virus movement through vascular system. Virus-encoded MP bind viral nucleic acids; then host proteins are involved in transporting of MP/nucleic acid complexes to plasmodesmata. The detailed description of the virus movement mechanisms can be found in a recent review article [93].

For full-sized vectors with an added gene coding for a protein of interest, the efficiency of the movement of recombinant virus nucleic acids through plant cells can be limited due to the increased size and changes in RNA secondary structure, leading to a low level production of foreign protein. A principal possibility to improve the virus movement through the plant cells was demonstrated for TMV using the in vitro evolution or DNA shuffling techniques [94]. After three rounds of shuffling five point mutations were introduced into MP gene; the resulting TMV vector was able to move more efficiently and produce a model protein (GFP) systemically in a shorter period. Additionally, the shuffling process resulted in the TMV vector with a broader host specificity in comparison to the initial vector.

The second important factor, RNA silencing is a sequence-specific degradation of double-stranded RNA (dsRNA). The process is highly conserved among eukaryotes [95]. Segments of single-stranded virus RNA can form double-stranded-secondary structures; alternatively, dsRNA is produced during the virus replication process. In plant cells, these dsRNA structures are cleaved by Dicer-like enzymes with RNase activity releasing 21–25 nucleotide long dsRNA fragments, called small interfering RNA's (siRNA). One nucleotide strand of siRNA binds to the host RNA-induced silencing complex (RISC) and serves for the sequence specific complementary binding and subsequent degradation of the viral RNA by RNase activity of RISC [96–98].

Plant viruses have evolved a mechanism against RNA silencing, allowing them to replicate efficiently in plant cells. At least 20 different virus encoded factors including coat and MP, replication enhancers, polyprotein processing factors, and pathogenicity determinants are shown to possess RNA-silencing suppressor activity (for review see Ref. [99]). Here only some aspects are discussed, which have been used in virus vector platforms for enhancing the expression level of foreign proteins. The host-specific virus spread factor p19 from TBSV [100] was shown to be the most effective suppressor able to increase the *Agrobacterium*-mediated transient expression of foreign protein up to 50 times in *N. benamiana* plants, when the gene encoding p19 was cloned between the left and right flanks of T-DNA and introduced in plant cells by agroinfiltration procedure [101].

If a "full-sized" TMV vector was used to transiently express a model protein (GFP) together with the TBSV p19 protein, the agroinfectivity of TMV vector was increased at least 100 times, demonstrating the high-stimulating effect of p19 on expression of foreign proteins [102]. The CMV-based expression platform is another example, where the effect of p19 on the rProt expression was also observed [78].

Another suppressor of silencing, a helper component protease (HcPro) isolated from Potato virus Y (PVY) [103], was found to be effective in the inducible CPMV-based expression platform [85] (see Section 19.5.1). However, in the deconstructed tobamovirus magnifection system, a coexpression of p19 or HcPro in plant cells did not result in the enhanced production of a model protein [87].

Plant viruses are able to infect numerous hosts. For example, TMV, which is frequently used for vector construction, infects approximately 200 species from 30 plant families [104]; whereas CMV is infectious for more than 1200 species and transmitted by more than 80 aphid species [105]. In view of the technology, the virus-based vector with a broad host range is an advantage, allowing to flexibly switch from one plant host to another depending on production needs. However, the broad host specificity instigates the environmental concerns regarding the uncontrolled spread of modified plant viruses. Therefore, it was suggested to develop plant virus platforms under containment conditions, especially in case of full-sized virus vectors.

As already mentioned, deconstructed virus gene vectors represent an efficient and environmentally safe alternative. This technology is based on *Agrobacterium*-mediated transfer of plant virus gene modules into plant cells [87], restricting the host range to the *Agrobacterium* susceptible plants. Agrobacteria are soil microorganisms that are able to form galls (tumors) on plants. The gall formation is associated with integration of bacterial T-DNA into the plant genome. T-DNA is located in Ti (tumor-inducing) plasmid, which besides T-DNA contains other genes needed for gall formation. Different *Agrobacterium* strains are able to infect a rather broad spectrum of plant hosts, including plant species from over than 90 plant families [106]. Although monocotyledonous plants are not natural hosts of agrobacteria and, therefore, are considered as less sensitive to agrobacterial infections, it is shown that maize and other cereals can be infected in laboratory [107,108].

A plant host range for a certain agrobacterial strain is dependent on a complex of factors, originating both from bacterial and plant cells. For example, the host range for the strain can be altered, if corresponding Ti plasmid is replaced with a different one. The host range determinant analysis and interactions of phytopathogenic *Agrobacterium* with plant cells at the molecular level are described in a few review articles [109,110]. Selection of the *Agrobacterium* strain for a virus-based plant expression platform is usually based on ability of bacteria to efficiently transfer T-DNA into the plant host. For the tobamovirus magnifection system, only 7 species were found to support a high-level expression of rProts from 50 dicotyledonous plants tested [89].

A few plant virus platform designs require the use of transgenic plants to achieve the efficient production of rProts (see Table 19.1). Some examples were already shortly discussed in a previous section. The platform based on bipartite virus CPMV [85] was shown to efficiently produce rProts in transgenic tobacco hosts, which contain stable integrated cDNA copies of viral RNA1 or RNA2. A similar approach was used also for alfalfa mosaic virus (AMV). In this case, cDNA for the AMV RNA1 and RNA2 were integrated in *N. bentamiana* plant genome and the synthesis of recombinant VLP's was directed by RNA3 [111]. A transgenic host was developed for PVX vector. To achieve a high-level expression using PVX the silencing suppressor HcPro was produced from cDNA copy integrated in the tobacco genome [112,113].

Plant cell cultures are well-known hosts for plant viruses and efficient tools for elucidating the viral gene functions and lifecycle mechanisms. Cell cultures have been used for high-throughput experiments aiming to construct and improve the virus-based vectors [89]. Recently, several technological strategies for the virus-mediated rProt production were successfully developed in plant cell suspensions. One possibility to achieve efficient expression is to construct an appropriate transgenic plant cell line with a stably integrated viral cDNA copy together with a foreign gene. This strategy was successfully applied to tomato mosaic virus (ToMV) [114]. Another technologically interesting application of plant cell cultures was recently developed using bean yellow dwarf virus cDNA modules and agrobacterial delivery technique. The significant synthesis of a model protein was achieved when the plant cell culture was co-cultivated with two *Agrobacterium* strains, carrying both necessary cDNA modules [115].

Traditionally in agriculture, the selection of cultivars is directed toward virus-resistant plants. Contrary to it, the plant virus host selection for virus-based protein production has to be focused on environmentally stable plants that are highly susceptible to virus vector infections [116]. It is conceivable that the work on construction of different transgenic plant hosts will continue to generate hosts for efficient expression of targeted proteins, similarly to the developments in the area of microbial expression platforms.

19.5.3 Medicinal Proteins from Plant Virus-Derived Platforms

A wide variety of model proteins and proteins with a medicinal relevance have been produced from different plant virus gene vectors. In this section, the production of full-sized proteins (Table 19.1) and construction of plant VLP's carrying different epitopes is reviewed (Table 19.2). The main attention is paid to such parameters as the production output, protein folding, and glycosylation properties in connection with the biological activity, stability as well as capacity to a large-scale production.

19.5.3.1 Full-Sized Proteins

An essential economical prerequisite for the biotechnology platform is the technologically significant output of the product. The output of the process is a sum of different factors, often influencing each other. In case of plant virus platform, these factors include the efficient formation of messenger RNA (mRNA), movement of RNA from one cell to another, translation of mRNA, protein folding, posttranslational modifications, resistance to host cellular proteases, as well as downstream processing and purification procedures. As shown in Table 19.1, the yield of rProts can vary significantly from microgram to gram quantities from a kilogram of plant biomass. The expression of the model protein GFP has been tested in several virus-based platforms; the usage of full-sized virus platforms typically results in the output up to 10% of total soluble plant cell proteins (TSP). However, if the expression was performed in the deconstructed agroinfiltration based magnifection system, the output of 80% TSP can be achieved [87], demonstrating a high potential for plant virus-based platforms.

To ensure the efficient foreign mRNA synthesis, the gene of interest is often placed under the homologous or heterologous sgRNA promoter (Figure 19.1), allowing to separate the mRNA synthesis from the whole genome replication. This approach has been evaluated for both full-sized and deconstructed vector platforms. For *Agrobacterium*-mediated virus expression system, the primary formation of mRNA takes place in the cell nucleus (see also Section 19.5.2). RNA virus genomes usually are replicating in cell cytoplasm. Therefore, large viral transcripts if synthesized in cell nucleus can be degraded by RNA processing systems. As a result, low levels of mRNA are available for transporting to the cytoplasm and translation. To improve mRNA stability and foreign protein synthesis in magnifection system, the potential splicing sites were removed and intron sequences were added to the vector modules to warrant a cell nucleus-typical RNA processing. This strategy was used for successful expression of such medicinally relevant proteins as interferons, growth hormones, monoclonal, and single chain antibodies [89]. Expression of proteins can be affected also by stability of the inserted gene. A recent study on the PVX-based viral vector for the expression of several foreign genes varying in size between 0.26 to 1.76 kb (human proinsulin, interleukin, HIV-1 *nef*, expansin, and human *gad65*) demonstrated that inserted genes can be deleted during the first infection cycle. Sequence analysis revealed complete or partial deletions of foreign genes in proximity to the duplicated PVX subgenomic promoters (see Figure 19.2E). Additionally, a negative correlation between the insert size and vector stability was suggested. The vector platform was suitable only for direct production of foreign proteins from the first-generation-infected plants; recombinant PVX propagation for subsequent inoculations resulted nearly in complete loss of the rProt production in the third-generation plants [117].

Such factors as protein folding and glycosylation influence considerably the stability and biological activity. Protein glycosylation patterns in mammalian and plant cells are different; plant N-glycan molecules are deficient in galactose and sialic acids and contain fucose and xylose residues (for review see Ref. [58]). The differences in glycosylation can greatly influence the stability of rProts, produced in plants. IgG1 antibody produced in transgenic tobacco plant are more sensitive to proteolytic enzymes present in plant leaves than IgG1 obtained from mammalian cells, possibly, due to differences in folding and glycosylation pattern [118]. A proteolytic degradation of foreign proteins, especially those produced in cytosol, has been observed also in plant virus

TABLE 19.2
Plant VLP-Based Epitope Presentation Systems as Vaccine Candidates

Plant VLP	Expression Platform	Epitope	Output	Immunological Properties	References
AMV	*Spinacia oleracea*, *Nicotiana tabacum*, AMV and TMV vectors	Rabies virus glycoprotein G and N antigen chimeric peptide; 24 AA-CP fusion	60 mg VLP/kg spinach leaves	Immunized mice protected against Rabies virus; oral vaccination by ingestion of spinach leaves resulted in significant antibody responses in human volunteers	[151]
AMV	Transgenic *N. tabacum* AMV RNA3 vector	Hepatitis C virus mimotope HVR1 (R9); 27 AA CP N-terminal fusion	—	HRV1-CP fusion reacted with HRV-specific mABs and HCV infected human sera	[111]
AMV	Transgenic *N. tabacum* AMV RNA3 vector	Respiratory syncytial virus G protein epitope; 21 AA CP N-terminal fusion	500 mg VLP/kg biomass	Immunized primates generated strong T- and B-cell responses; human dendritic cells responded by significant $CD4^+$ and $CD8^+$ T cell responses *in vitro*	[152]
CGMMV	Muskmelon CGMMV vector	Hepatitis B surface antigen (HBsAg); 31 AA fused to CP C-termini via readthrough stop codon	—	CP-HBsAg fusion stimulated antibody production in cultivated peripheral blood mononuclear cells (*in vitro)*	[162]
CMV	*Nicotiana bentamiana* CMV RNA3 vector co-inoculated with RNA1&2	Newcastle disease virus (NDV) HN epitope; 8 AA; inserted in βH-βI loop	~400 mg VLP/kg biomass	HN NDV epitope was recognized by anti-NDV sera produced by NDV-challenged chicken	[146]
CMV	*N. tabacum* CMV RNA3 vector co-inoculated with RNA1&2	Hepatitis C virus (HCV) R9 mimotope; 27 AA internal insertion	100 mg VLP/kg biomass	R9-CMV fusion elicited specific humoral response in rabbits and induced significant release of interferon (IFN-γ) and interleukins in HCV patient lymphomonocyte cultures	[147,148]
CMV	*N. benthamiana* PVX vector	Newcastle disease virus (NDV) epitopes; 8, 17, 25 AA inserted in βH-βI loop	700 mg VLP/kg biomass	NDV containing VLP recognized by specific antibodies	[149]

(continued)

TABLE 19.2 (continued)
Plant VLP-Based Epitope Presentation Systems as Vaccine Candidates

Plant VLP	Expression Platform	Epitope	Output	Immunological Properties	References
CPMV	*Vigna unguiculata* CPMV RNA2 cDNA vector co-inoculated with cDNA of RNA1	Mink enteritis virus (MEV) coat protein epitope; 20 AA inserted in βB-βC loop	1000 mg VLP/kg biomass	Single-dose vaccinated minks protected against MEV challenge	[141]
CPMV	*Vigna unguiculata* CPMV RNA2 cDNA vector co-inoculated with cDNA of RNA1	*S. aureus* fibronectin-binding protein (FnBP) D2 peptide; 30 AA inserted in βB-βC loop	1200 mg VLP/kg biomass	Intranasally immunized mice generated high titers of FnBP-specific immunoglobulin G and mucosal antibodies; parenterally immunized rats were protected against endocarditis	[142,143]
JGMV	*E. coli* T7 promotor driven expression	Japanese encephalitis virus (JEV) E protein peptide, 27 AA C-terminal CP fusion	—	Recombinant VLP immunized mice were protected against lethal JEV challenge	[167]
PapMV	*E. coli*	Hepatitis C virus (HCV) E2 protein epitope; 20 AA C-terminal CP fusion	—	E2 epitope-containing VLP raised long-lasting humoral response against HCV E2 epitope in mice, whereas CP monomer with E2 epitope was not immunogen	[168]
PVA	Mammalian cell lines CP-E7 fusions in plasmid pBSC	Human papilloma virus (HPV) HPV16 E7 oncoprotein peptide; 17 AA C-terminal CP fusion	—	VLP aggregates in mammalian cells; plasmid DNA vaccinated mice were protected against tumors induced by TC-1 cells producing the E7 antigen	[190]
PVX	*N. bentamiana* PVX cDNA vector	Human immunodeficiency virus (HIV) gp41 epitope; 6 AA N-terminal CP fusion	600 mg VLP/kg biomass	Intranasally and intraperitoneally immunized mice elicited HIV specific IgG and IgA antibodies; human primary neutralizing antibodies found in immunodeficient mice reconstituted with human peripheral lymphocytes	[156]
PVX	*N. benthamiana*	Classical swine fever virus (CSFV) E2 protein epitopes; 40 or 70 AA N-terminal CP fusion	120–150 mg VLP/kg leaves	Partially purified PVX-E2 VLP's induced the immune response in rabbits	[155]

TBSV	*N. bentamiana* TBSV vector	Human immunodeficiency virus (HIV-1) protein gp120 epitope; 13 AA C-terminal CP fusion	900 mg VLP/kg biomass	HIV epitope on VLP surface reacted with specific monoclonal antibody and with sera from HIV-positive patients	[145]
TMV	*Nicotiana tabacum* TMV vector	Murine hepatitis virus (MHV) epitopes; 15 AA insertion at CP C-termini	—	Subcutaneously or nasally vaccinated mice survived challenge with a lethal dose of MHV	[163]
TMV	*N. tabacum* TMV vector	*Pseudomonas aeruginosa* outer membrane protein (OM) peptide; 14 AA insertion at CP C-termini	—	Mice immunized with peptide-containing VLP's were protected against chronic pulmonary infection	[207]
TMV	*N. tabacum* TMV vector	Poliovirus capsid VP3 and VP1 protein-derived peptide; 15 AA fused to CP C-termini via readthrough stop codon	200 mg VLP/kg leaves	Poliovirus-epitope containing VLP's induced antibodies in intraperitoneally immunized mice against VP3 of native poliovirus	[160]

Note: AMV, alfalfa mosaic virus; CGMMV, cucumber green mosaic mottle virus; CMV, cucumber mosaic virus; CPMV, cowpea mosaic virus; JGMV, Johnsongrass mosaic virus; PapMV, papaya mosaic virus; PVA, potato virus A; PVX, potato virus X; TSBV, tomato bushy stunt virus; TMV, tobacco mosaic virus; AA, amino acids; CP, virus coat protein; —, data not available.

based systems. To overcome this foreign protein sensitivity to host proteases, a strategy was adapted from transgenic plant research for targeting rProts to different cell compartments [119]. The expression of canine oral papillomavirus (COPV) primary CP L1 was undetectable in PVX platform, when L1 was synthesized in cytosol. However, when L1 was fused to peptides targeting proteins to chloroplasts, the modified L1 protein accumulated at the level of up to 0.3% TSP (Table 19.1). It is interesting that targeting L1 to apoplast and endoplasmatic reticulum (ER) resulted in protein undetectable in immuno-blot analysis. The L1 protein was suggested for the use as a model protein for testing protein stability in plants [113].

The cell compartment targeting strategy was shown to be useful also for the magnifection system. Biologically active human growth hormone (hGH) was efficiently processed and produced yielding 1 g hGH per 1 kg plant leaves when targeted to the apoplast. hGH with a corresponding targeting signal was also detected in chloroplasts, where 75% of the protein was found to be correctly processed [120].

Medically relevant proteins have to be stable to ensure their biological activity during the downstream processing and storage. The lichenase, a thermostable enzyme from *Clostridium thermocellum* was successfully used as a fusion partner for influenza H5N1 epitope, *Yersinia pestis* F1 and LcrV antigen and human papilloma virus (HPV) E7 protein production in TMV-based platform [86,121,122]. In case of H5N1 epitope fusion, the thermostability of up to 65°C was demonstrated, considerably facilitating the downstream purification process of the potential vaccine.

rProt yield can be dependent on such cultivation factors as temperature, light, and developmental stage of the plant tissue, as shown in several experiments. The best mouse antibody IgG1 production in transgenic plants was observed in young leaves growing at the top of the plant cultivated at reduced temperature and high-light conditions. Proteolytic degradation of IgG heavy chain was found to be more profound in grown-up leaf protein extracts than in young leaves [118].

Similarly to transgenic plants, plant leaves infected with full-sized plant vectors appear to accumulate rProts depending on developmental stage. PVX-based expression of human granulocyte macrophage colony-stimulating factor in *N. bentamiana* was found 2.6 times higher in young leaves in comparison to old leaves [123]. However, in the deconstructed tobamovirus magnifection system, the output of foreign protein seems to be less dependent on age of the leaf material, especially, if the protein is targeted to specific cellular compartments. As an example, in one experiment the accumulation of human growth hormone (hGH) targeted to the apoplast reached maximum after 12 days of synthesis and intact protein was detectable even in necrotic and dry leaves [120].

One of the most challenging tasks in rProt production is obtaining oligo- or multi-subunit proteins with an adequate biological activity. A full-size antibody is a multimeric protein containing two heavy and two light chains, which are connected by disulfide bonds; the heavy chain is glycosylated. Antibody production in transgenic plants has been studied nearly for two decades, since the first functionally active multimeric IgG was co-expressed from corresponding recombinant mouse hybridoma genes [124,125]. Different types of antibodies expressed in transgenic plants are discussed in Chapter 15 as well as in Ref. [31]. It was found that transgenic plants supports a low level of antibody output (0.003%–1% TSP) [58]. Additionally, the construction of corresponding transgenic plants and large-scale production is a very time-consuming process [125]. Early attempts to produce full-size antibodies using plant virus systems were only marginally successful. The genes for heavy (HC) and light chains (LC) were subcloned separately in the TMV-derived vector and co-infections with both vectors resulted in systemic production of full-sized and assembled monoclonal antibodies. However, the production level was low [126].

Recently, a highly efficient antibody production was finally achieved using a plant virus derived platform. A magnifection system developed based on tobamovirus platform was applied to virus PVX. The system was constructed by removing of CP gene and separating the virus cDNA in two agroinfectable modules. Coinfections with vector modules derived from tobamovirus and PVX demonstrated simultaneous expressions of model proteins in the same plant cell. The best antibody output (500 mg kg^{-1} leaves) was achieved when HC gene was cloned in the tobamovirus-derived vector and LC in the PVX-based module, respectively [127]. This antibody expression platform represents a principally new technological solution for obtaining gram amounts of functionally active monoclonal antibodies within a few weeks [125].

19.5.3.2 Virus-Like Particles as Epitope Carriers

VLP's are multi-subunit protein assemblies with identical or highly similar overall structure to the corresponding infectious viruses, which occasionally are found naturally in infected cells or can be produced using different expression platforms via cloning of corresponding structural genes. VLP's can contain infectious DNA or RNA depending on the expression platform. VLP may also incapsidate non-infectious host nucleic acids or may stay empty. A more detailed description of VLP of different origin can be found in chapters from Part II and in Refs. [128,129].

Due to their structural identity to human and animal viral pathogens VLP's are well accepted as vaccine candidates. Hepatitis B VLP's produced by various companies using a yeast expression platform are on the vaccine market already for more than two decades. Recently, the next VLP-based vaccine against HPV was introduced and marketed. This vaccine is a composition of four virus type HPV-like particles expressed in the yeast system. Highly ordered VLP's are antigenically indistinguishable from native virions and represent a safe and very effective vaccine [130]. The most recent developments in HPV vaccinology can be found in Chapter 11.

The efficiency of VLP in antibody induction is well known from immunological studies of different pathogenic viruses. Highly ordered and repetitive amino acid motives of structural proteins on virus surface are critical signals for B-cell activation and represent a "marker for foreignness" for the immune system [131]. Additionally, the size of VLP's (less than 50 nm) meets the range of optimum for uptake by dendritic cells [132]. Pathogenic viruses and VLP's after invading the vertebrate organisms target multiple cells in the immune system eliciting effective humoral and cellular immune responses even in the absence of adjuvant [133]. Mechanisms of immune response against viruses and VLP's are extensively reviewed in Ref. [134–136].

Plant viruses are not infectious for mammalian organisms; therefore, they cannot serve as vaccine candidates directly. However, plant viruses and VLP can be exploited as carriers of different epitopes from a variety of pathogens. These epitopes can be displayed on the particle surface to achieve a highly repetitive presentation for the efficient antibody induction. The term "epitope" can be defined as a shortest amino acid (AA) sequence from a protein able to specifically bind to the corresponding antibody molecule; epitopes can be linear or conformational. Linear epitopes represent antigenic determinants localized on the short stretch (4–10 AA) of the polypeptide independently from its three-dimensional (3D) structure. Conformational epitopes are formed from distant polypeptide regions and are dependent on protein spatial structure [128].

More than 20 years ago, epitope mapping of pathogenic virus structural proteins led to the concept that synthetic peptides containing the epitope sequences can serve as vaccines against corresponding pathogens. Immunization with peptides chemically coupled to carrier proteins demonstrated the formation of neutralizing antibodies in laboratory animals [137]. To maximize the presentation of peptides on carrier molecules, TMV CP was chosen as a carrier molecule and poliovirus antigenic peptide was inserted at C-terminus of CP (Figure 19.3C). The modified TMV CP was produced in recombinant *E. coli* cells and VLP self-assembly was accomplished in vitro. The purified TMV-like particles carrying the polio epitope on the surface induced the poliovirus neutralizing antibodies when injected to rats [138], demonstrating for the first time the potential of plant viruses as vaccine carrier candidates.

Soon after these experiments with TMV-like particle assembled in vitro, structural protein from arabis mosaic virus, Johnsongrass mosaic virus, and TMV were expressed in bacterial and insect cells, demonstrating the principal possibility to achieve self-assembly of plant virus structural proteins in different heterologous hosts. Table 19.3 summarizes the experimental results of plant VLP production in heterologous hosts.

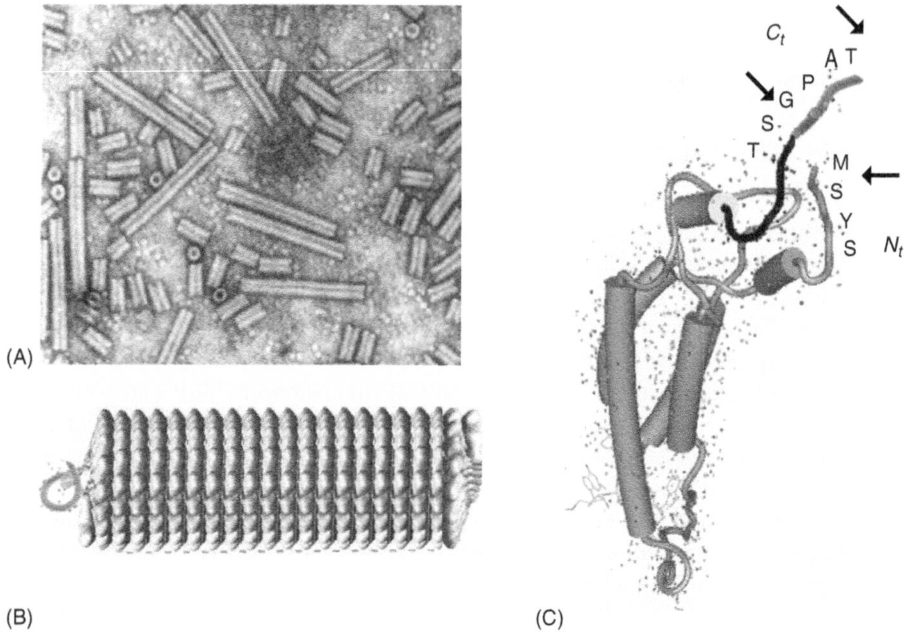

FIGURE 19.3 TMV structure. (A) Electron microscopy picture. (Courtesy of Rothamsted research, 1994. With permission.) (B) Spatial model of a fragment of TMV particle. Picture redrawn from RCSB Protein Data Bank (DOI 10.2210/pdb2tmv/pdb); (C) 3D-structure model of TMV coat protein. N- and C-terminal amino acids are shown. Arrows denote the positions for epitope presentation. The model is built using 3D Molecule Viewer program.

Further developments were made based on several plant virus platforms. As was already mentioned, the available 3D structures of several plant viruses played a significant role in creating plant virus-based vaccine carriers. CPMV presents one of the most studied plant virus epitope carriers. According to the 3D-structural model, CPMV coat is built of 60 subunits of two CPs (L and S) arranged into virion with an icosahedral symmetry (Figure 19.4B) [139]. It was found that three AA loops localized on virus surface (Figures 19.2D and 19.4C) are suitable for insertions of different foreign epitopes without a significant influence on the virus lifecycle. The studies on virus viability in respect to properties of inserted epitopes revealed the upper limit for the epitope size as less than 40 AA. Additionally, strongly positively charged epitopes with an isoelectric point (pI) below 9.0 were found to be not acceptable [73,140]. To validate CPMV-based epitope presentation system, several medically and veterinary relevant epitopes were introduced in the virus CPs (Figure 19.2D). Immunization with recombinant CPMV particles carrying surface exposed epitopes of Mink enteritis virus [141], the outer membrane protein F of *Pseudomonas aeruginosa* [142] and D2 domain of the fibronectin-binding protein (FnBP) from *Staphylococcus aureus* [143] were shown to protect laboratory animals against challenges with corresponding pathogenic agent after immunization. This plant virus expression platform produced about 1200 mg kg^{-1} of recombinant particles (Table 19.2.). As CPMV-based epitope carriers were shown to retain the infectivity for plants, a simple and efficient procedure of inactivation was recently elaborated to obtain environmentally safe preparations. Prolonged incubations at room temperature in basic ammonium sulphate solutions completely abolished the appearance of symptoms on test plants and caused an extreme degradation of recombinant CPMV RNA [144].

The analysis of crystal structure of TBSV revealed that the C-terminus of CP forms a protruding domain displayed in 180 identical copies on the virus surface. The surface localization

TABLE 19.3
Plant VLP Obtained from Heterologous Expression Platforms

Plant Virus	Expression Platform	VLP Properties	References
Bromoviruses			
Cowpea chlorotic mottle virus (CCMV)	Yeast; *Pichia pastoris*	Icosahedral particles, 26 nm in diameter	[182]
	Bacteria; *Pseudomonas fluorescens*	Icosahedral particles, 26 nm in diameter	[183]
Comoviruses			
Cowpea mosaic virus (CPMV)	Insect; baculovirus vector in *Spodoptera frugiperda* cells; coexpression of L and S proteins	Icosahedral particles	[184]
Cucumoviruses			
Cucumber mosaic virus (CMV)	Plant; PVX vector based transient expression in *N. bentamiana*	Isometric particles 13–60 nm in diameter	[149]
Furoviruses			
Indian peanut clump virus (IPCV)	Bacteria; E. coli plant; transgenic *N. Bentamiana*	Rod-shaped particles 10–140 nm (bacteria) 10–330 nm (plants)	[185]
Nepoviruses			
Arabis mosaic virus (ArMV)	Insect; baculovirus vector in *Spodoptera frugiperda* cells plant; transgenic *N. tabacum*	Isometric particles	[186]
Phytoreoviruses			
Rice dwarf virus (RDV)	Plant; transgenic rice, coexpression of two coat proteins	Isometric double-shelled particles	[187]
Potexviruses			
Papaya mosaic virus (PapMV)	Bacteria; *E. coli*	Rod-shaped particles; 50 nm in length	[188]
Potyviruses			
Johnsongrass mosaic virus (JGMV)	Bacteria; *E. coli* yeast; *S. cerevisiae*	Flexous particles of different length	[189]
Potato virus A (PVA)	Bacteria; *E. coli*	Flexous particles	[190]
Plum pox virus (PPV)	Bacteria; *E. coli*	Flexous particles	[191]
Tobacco etch virus (TEV)	Bacteria; *E. coli*	Flexous particles 12 × 90–750 nm	[192]
Sobemoviruses			
Sesbania mosaic virus (SeMV)	Bacteria; *E.coli*	Icosahedral particles, 30 nm, T = 3 symmetry	[193]
Tobamoviruses			
Tobacco mosaic virus (TMV)	Bacteria; *E. coli* coexpression of OAS[a]	Helical particles of different length	[194]
Tymoviruses			
Physalis mottle virus (PhMV)	Bacteria; *E. coli*	Icosahedral particles, 29 nm, T = 3 symmetry	[195]

[a] OAS (origin of assembly), RNA sequence found in TMV responsible for initiation of virus encapsidation.

of this domain suggested its use for foreign epitope presentation. Insertion of the epitope derived from the HIV V3 loop at the CP C-terminus resulted in chimeric TBSV particle efficiently presenting this epitope on the surface [145].

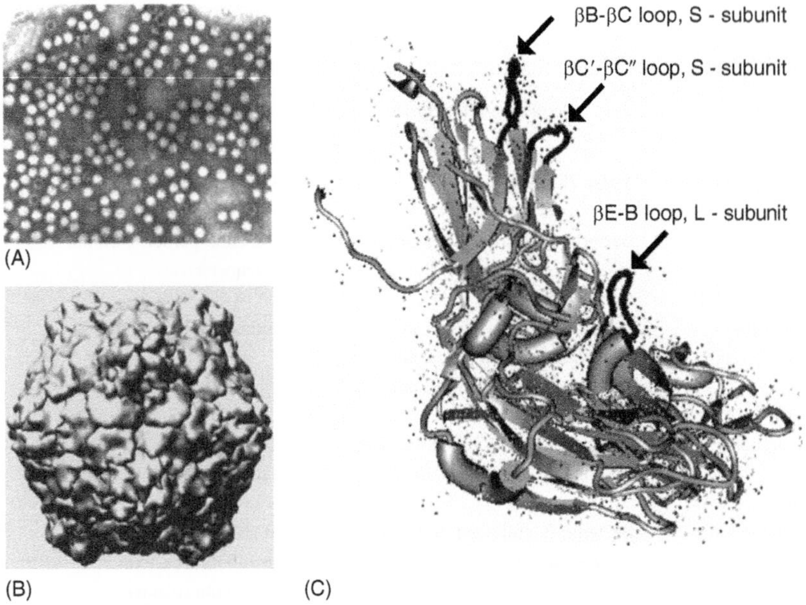

FIGURE 19.4 CPMV structure. (A) Electron microscopy picture. (Courtesy of Rothamsted research, 1994. With permission.) (B) Spatial model of CPMV particle. Picture redrawn from RCSB Protein Data Bank (DOI 10.2210/pdb1ny7/pdb); (C) 3D-structure model of CPMV large (L) and small (S) coat proteins. Arrows denote the positions of exposed loops for epitope presentation. The model is built using 3D Molecule Viewer program.

A CMV vector was shown to be efficient for the production of full-size rProts. It was found also suitable for the expression of foreign peptides on the virus surface (Figure 19.2B). Analysis of the 3D-structure of icosahedral CMV particles identified a surface exposed AA loop of the virus coat suitable for epitope insertion. Epitopes from Newcastle disease virus (NDV) [146] and hepatitis C virus [147,148] were inserted into this loop to demonstrate their immunological activity (Table 19.2). To overcome the observed genetic instability of the epitope DNA in CMV vector, a technologically interesting solution was found. The recombinant CMV CP gene carrying the epitope DNA was cloned between duplicated subgenomic promoters in a different plant virus vector based on PVX cDNA (Figure 19.2E). The PVX-derived plant platform efficiently produced CMV-like particles. Subsequent immunological analysis confirmed the display of NDV epitopes on VLP surface [149]. Similarly, 23 AA-long epitope of avian influenza A virus was engineered into the CMV surface using PVX-based expression vector [150].

Another plant virus-based epitope presentation system was constructed using AMV vector (Figure 19.2A). The AMV CP N-terminal part was shown to be suitable for insertion of epitopes derived from rabies virus G and N antigens [151] as well as the hepatitis C virus HRV1 mimotype [111] and respiratory syncytial virus G protein epitope [152].

For rod-shaped viruses with an unknown 3D structure, potential insertion sites on the virus surface can be found experimentally using immunological and radiolabeling methods. Tritium-labeling analysis of PVX CP peptides obtained by protease digestion revealed a preferential tritium incorporation in the N-terminally located peptides, whereas the C-terminus was inaccessible. Immunological analysis with antibodies against PVX confirmed that first 56 AA represent a highly immunogenic virus CP region [153], suggesting the exposition of the PVX CP N-terminal part on the virus surface.

The capacity of PVX for accommodating long AA stretches at the N-terminus was demonstrated using the model protein GFP (27 kDa). As GFP is of a similar size to the PVX CP and can affect the virus assembly, a self-processing peptide 2A (see Section 19.5.1) was inserted between the GFP and CP. This peptide facilitated a partial cleavage of GFP-CP fusion proteins into a mixture of GFP-CP and free CP. The expressed proteins were efficiently assembled into particles and encapsidated the viral RNA without any significant effect on the virus viability [154].

The 2A-strategy has been exploited to construct vaccine candidates against classical swine fewer virus (CSFV) by cloning E2 protein epitopes in PVX vector at the 5′-part of CP gene (Table 19.2; Figure 19.2E). Despite a large insert size (between 40 and 70 AA), the recombinant PVX particles were produced at the level of 120–150 mg kg^{-1}. Partially purified VLP's induced immune response in laboratory animals [155].

It is also possible to present foreign epitopes on the PVX surface without using the 2A peptide. However, the size of these epitopes should be limited to the small number of AA. Such short insertions as the 6 AA HIV epitope [156] or 14 AA combined epitope from beet necrotic yellow vein virus [157] resulted, however, in the production of PVX particles, in which each CP molecule contained the epitope.

As seen from the TMV 3D-structure studies, both the N- and C-termini of CP are located on the virus surface (Figure 19.3) [158]. Tobamovirus-based epitope presentation systems are well known for more than a decade, since angiotensin-converting enzyme inhibitor (ACEI) peptide was expressed using a TMV vector. The insertion of a cDNA segment containing a read-through stop codon and ACEI coding sequence next to the CP 3′-end resulted in the efficient TMV infection. As a result of the 5% read-through translation, the relative amount of CP-ACEI fusion molecules to free CP was about 1 to 20 in TMV particles [159]. The same vector design was used to express *Poliovirus* epitopes [160] and the larvicide peptide [161]. The read-through strategy to obtain the CP C-terminal fusions with a peptide of interest was exploited also for cucumber green mottle mosaic virus (CGMMV). A peptide derived from HBsAg was subcloned after the leaky "stop"-codon in the CGMMV vector (Figure 19.1D). The virus particles contained the corresponding CP-fusions at an approximate rate 1:1 to wild-type CP [162].

To achieve epitope presentation on each CP molecule at the TMV surface, a 15 AA epitope from murine hepatitis virus (MHV) was inserted between Ser and Gly in proximity to the CP C-terminus (Figure 19.3C) [163]. The resulted recombinant virus was mostly found in insoluble fraction. However, despite this fact, a sufficient amount of TMV particles carrying the MHV epitope was obtained. Immunization of experimental animals with this preparation showed the MHV neutralization activity. Expression of longer peptides on the TMV surface was recently achieved by removing the six C-terminal AA from CP and introducing a 25 AA long epitope of FMDV [164]. Another important factor for the successful presentation of epitopes on TMV surface is the isoelectric point (pI) of inserted peptides. Similar to the CPMV display system, it was shown that the TMV assembles into particles only with inserted acidic epitopes. To maintain competence for systemic infection TMV particles composed of hybrid CPs should have the overall negative charge [165].

A new strategy for presenting long peptides on the TMV surface was developed using the already discussed magnifection platform. The introduction of 15 AA flexible or helical linkers at the CP C-terminus allowed for expressing a fragment of the protein A (133 AA) on the TMV surface at the highest density reaching 2100 copies per viral particle [166]. These hybrid particles can bind up to 2 g monoclonal antibodies per gram of the particles and, therefore, are a very efficient tool for purification of monoclonal antibodies. A simple purification using this particles as the immunoadsorbent results in mAb preparations of 90% purity.

There are several examples of producing plant virus-based vaccine carriers from heterologous expression platforms; e.g., the Johsongrass mosaic virus [167] and papaya mosaic virus (PapMV) [168] rod-shaped VLP's carrying medically relevant epitopes were obtained after cloning the corresponding CP genes in *E. coli* (Table 19.2). It is important to note that immunogenicity of PapMV-like particles containing the HCV E2-epitope depends on the aggregation state of E2-CP. A long-lasting humoral response was only observed in animals immunized with VLP, whereas the E2-CP monomer was not immunogenic.

Another way to display peptides or proteins on the plant VLP surface is a chemistry-based approach, which requires the exposure of AA easily amenable to chemical modifications on the VLP surface. Some VLP's expose such AA as Lys or Cys on their surface. In some cases, though, such residues need to be introduced by site-directed or random mutagenesis. The TMV CP gene was subjected to random mutagenesis to introduce Lys at the exposed N-terminus. The TMV VLP's containing Lys on the surface were biotinylated and subsequently incubated with streptavidin–GFP fusion protein. The resulted aggregates contained approximately 2200 GFP molecules per TMV particle [169].

A similar chemical coupling approach was used also for the CPMV capsids. Different proteins and protein fragments were attached to the CPMV VLP surface with a high occupancy to target these modified particles to specific cell types [170]. This alternative technology can be used for the display of whole proteins on the VLP surface, which is useful for development of vaccine candidates.

VLP can be used not only as a carrier for medically relevant proteins and peptides. Over the last years, different nanotechnological applications have been also proposed. VLP can be used in preparation of different inorganic nanomaterials with defined magnetic or electric properties. The cowpea chlorotic mottle virus particles were shown to encapsidate mineralization of paratungstate ion ($H_2W_{12}O_{10}{}^{42-}$) and decavanadate ions ($V_{10}O_6{}^{28-}$) [171], while the BMV-like particles can encapsidate nanometer-scale gold particles [172]. TMV was used as a core for producing platinum biomaterial with unique electronic memory properties [173]. Some VLP's can find applications in non-invasive imaging for early detection and treatment of different diseases, including cancers [174].

Additionally, plant virus VLP's potentially represent delivery vehicles of anionic drug molecules incapsidated in nanosized protein cages [175]. If VLP surface is modified to contain addresses to specific cells, the targeted drug delivery can be accomplished, avoiding the necessity to treat patients systemically.

19.5.4 Large-Scale Production Aspects

As already discussed above, several plant virus-based platforms have a great potential for usage in a large-scale biotechnological production. One of the most advanced plant virus platform has been developed by Large Scale Biology Corporation (LSBC) exploiting full-size TMV vectors (GENEWARE). This platform has been validated to be stable and efficient in open field trials. As recombinant TMV vectors are stable for several generations, it was possible to prepare the inoculum for field experiments in greenhouses. A special tractor-mountable high-pressure spray device was constructed and used to inoculate field plants with a mixture of TMV and abrasive material. The spraying technique permitted to treat an acre in only 30 min with an efficiency of 95% infected plants. One microgram of recombinant TMV was sufficient for establishing the systemic infection. As demonstrated using TMV carrying an anti-malarial epitope, the harvest of infected tobacco biomass was approximately 8 tonnes from an acre, which corresponds approximately to 1 kg $acre^{-1}$ for this malaria vaccine candidate [116]. This platform was used also for the production of full-size medicinal proteins like aprotinin, α-galactosidase, lysosomal acid lipase, and personalized non-Hodgkin's-lymphoma monoclonal antibodies [176]. Despite these really successful developments, the LSBC reduced their activities due to financial difficulties (LSBC News release, 2006) and currently no data are publicly available on further developments toward industrial production of medicinally relevant proteins from the plant virus-based platform.

Another company, Icon Genetics, is very close to producing medicinal proteins in plants on industrial scale at greenhouse conditions. As already mentioned in Section 19.5.1, this company's proprietary magnifection system has the highest potential of producing rProts in plants at the rate of up to 5 g kg^{-1} or 80% TSP. This system is reported to be indefinitely scalable, if the special apparatus for high throughput vacuum-infiltration of whole plants is employed for rapid plant inoculations with corresponding recombinant *Agrobacterium* strains. Protein production in tobacco plants can be completed in only few days; even gram quantities of a protein of interest can be produced within 3–4 weeks. Additionally, the stability studies demonstrated that, for example, plant material containing smallpox antigens can be stored at ambient temperatures without significant loss of rProt [177]. Validations of the platform demonstrated a successful and high yield production of such proteins as somatotropin, α-interferon, bacterial antigens, and various single-chain antibodies. Calculations of expected upfront and production costs demonstrated potential investments in a GMP-compliant facility for plant platform is less risky than for microbial or mammalian cell platforms [88]. Recently, a large pharmaceutical corporation, Bayer, acquired the Icon Genetics with an aim to continue the development of the magnifection system as an industrial technology for different medicinal proteins including monoclonal antibodies against cancers (Bayer Innovation and Icon Genetics, News releases, 2006, 2007).

19.5.5 Clinical Studies

For proteins intended for medical use, the safety and therapeutic or prophylactic effects on human organisms need to be evaluated. Clinical studies or trials are divided in several phases. In a simplified way, trial phases can be defined as follows: (1) in phase-I studies, the safety and tolerability of the drug is mostly tested on limited number of healthy persons; (2) phase-II studies evaluates the efficiency of the drug in regard of dosage in larger groups of patients; (3) phase-III trial involves large groups of patients at different medical institutions and aims at entering market in the case of success. More detailed information can be found at clinical trial home page of National Institute of Health (http://clinicaltrials.gov/).

Despite the numerous examples of vaccine candidates, monoclonal antibodies and therapeutic proteins developed using both transgenic plants and plant virus-based platforms, the limited number of products for human consumption are tested for safety and efficiency in clinics. One of the main problems seems to be the different glycosylation pattern of plant derived proteins compared to the mammalian proteins [58]. Injectable therapeutic proteins from plants with non-authentic glycosylation pattern can be rapidly cleared from blood due to action of glycan-specific antibodies. In addition, the immune response against plant-derived glycoproteins can result in allergic reactions. On the other hand, glycosylated proteins from plants can serve as efficient vaccine candidates without added adjuvants due to high immunogenicity of foreign glycan residues [178].

Several clinical phase-I studies revealed a safety and good tolerability of plant produced proteins, including monoclonal antibodies and vaccines. Such medicinally relevant proteins from transgenic plants as gastric lipase for cystic fibrosis treatment, *E. coli* labile toxin (vaccine against diarrhea), hepatitis B surface antigen, Norwalk virus vaccine, and lactoferrin (gastrointestinal infection treatment) are close to commercialization [53]. A monoclonal antibody against tooth decay causing bacterium *Streptococcus mutans* (CaroRx) is a first plant-derived monoclonal antibody that is clinically tested in phase-II. It is intended for topical use to prevent bacterial colonization and is probably the most developed candidate towards commercial use [179].

Also, few rProts from plant virus platforms have been tested for safety, tolerability and efficiency in human trials. Some years ago Large Scale Biology Corporation tested individualized single-chain antibodies against non-Hodgkin's lymphoma obtained using the TMV vector-based platform. Immunization of 15 patients with glycosylated single-chain antibodies did not result in serious adverse reactions (data from Ref. [179]), demonstrating acceptable tolerance against rProts with plant-specific glycans in humans. Another example is an edible vaccine against rabies produced in spinach leaves by expressing rabies epitopes on surface of the AMV-like particles (Table 19.1) [151]. This experimental vaccine demonstrated a lack of any adverse reaction. A significant immune response and formation of neutralizing antibodies against rabies virus were observed, especially, if vaccine containing spinach leaves were used as oral booster after conventional vaccination.

19.6 CONCLUDING REMARKS

Different plant expression platforms including plant virus-based platforms have been shown to efficiently produce different proteins with medicinal relevance. The existing production capacity of microbial and cell culture platforms currently dominating the biopharmaceutical industry cannot satisfy the demand for vaccines, monoclonal antibodies, and recombinant therapeutic proteins. The price of recombinant products is a special concern for the resource limited parts of the world. Plant platforms represent a highly valuable alternative capable to replace or, at least, to complement the existing technologies. Even though the transgenic plant production typically results in low or medium level outputs, the required amount of medicinal proteins can be obtained by simple scaling up, which is nearly unlimited. However, such an approach requires cultivation of transgenic plants at open field conditions, causing concerns regarding uncontrolled spread of transgenes in the environment and negative attitude in society.

Plant virus-based platforms can be regarded as a variant of transgenic plants, however, with a significant difference from transgenic plants; namely, in a virus-based platform the transgene is not incorporated in plant genome and the expression of rProt is transient. Recent developments in this field demonstrate several important advantages of the plant virus-based platform compared to other protein expression platforms:

1. Output of the transient expression has been shown to reach levels characteristic for microbial platforms in the comparable period of cultivation, making the technology commercially attractive.
2. High yields enable the production even at contained greenhouse conditions, which is also important for reducing environmental risks and for promotion of the technology as a safe way to produce medications in plants.
3. Established virus-based platforms can be easily switched from production of one rProt to another one.
4. Option to produce multimeric proteins like full-size monoclonal antibodies, as recently demonstrated for the platform consisting of two deconstructed plant virus vectors; first clinical studies demonstrate the safety and tolerability of plant virus derived antibodies for humans.
5. Edible vaccine production in non-transgenic food plants, which are safe and immunogenic for humans. The problem of dosage and storage for edible vaccines can be solved by production of plant material in a form of capsules, containing standardized, lyophilized material.
6. Calculated prices for input materials can be as low as $1/g for medicinal proteins [88].

Developments for more than two decades clearly demonstrate the suitability of plant virus-based technology for a large-scale production of medicinal proteins; however, the technology has not yet been widely accepted by pharmaceutical industry. Taking into account costs for the introduction of new pharmaceutical products, large companies are cautious regarding new technologies. One example shows how big the costs can be; the recently marketed vaccine against papilloma virus was reported to require one billion dollar investments from each GlaxoSmithKline and Merck [180].

In addition to the cost, several other factors delaying the introduction of the technology can be identified. First of all, it is a new technology with principally different manufacturing techniques, and most companies seem to be doubtful about predictability and sustainability of the technology. Second factor is related to authenticity of plant produced pharmaceuticals. The glycosylation and correct folding of the proteins are still under discussion, despite the first successful clinical trials. Additional case-by-case studies are needed to obtain a more general information about behavior of plant-derived glycosylated proteins in mammalian organisms.

Production of pharmaceuticals from plants is new technology not only for companies, but also for regulatory authorities. To date, the regulations for plant-made pharmaceuticals are more complicated than for conventional product, and, for example, the production in United States has to be legalized and monitored not only by Federal Drug Administration but also by the U.S. Department of Agriculture. Several experts suggest the introduction of special regulations for this area of biotechnology [47].

It is understandable that the successful growth in this field requires more clinical studies and trials to clearly demonstrate all advantages of the plant biotechnology in manufacturing, safety, and specific activity of medicinal products. Recently, Dow Agrosciences achieved the world's first registration for plant-made vaccine against Newcastle disease in chickens (Dow Agrosciences; Press release, 2006). This success and further developments can serve as an important demonstration that plant-derived rProts can become real marketable products. Should this vaccine not be commercialized; it still remains to be a precedent for companies working on human and animal vaccines from plants [43].

Plant virus-based platforms have already gained an attention from large pharmaceutical companies. The magnifection platform, currently one of the leading technologies for plant-made pharmaceuticals, is under further development at Bayer Innovation [181]. It is expected that in next years, after completing wide-ranging clinical tests and marketing authorization procedures the plant virus-based biotechnology platforms will be used as an industrial technology for the production of different pharmaceuticals for human consumption.

ACKNOWLEDGMENTS

I wish to thank Prof. P. Pumpens, Prof. E. Grens, Dr. A. Sharipo, Dr. A. Kazaks, Dr. J. Jansons, Dr. K. Tars, Prof. Y. Gleba for helpful discussions and support during the preparation of this chapter. Rothamsted Research is acknowledged for the virus electron microscopy pictures. The writing of the review was supported by Latvian State Science Program VZP12 and EU Project No.VPD1/ERAF/CFLA/05/APK/2.5.1./000019/P.

REFERENCES

1. Schipmann, U. et al., A comparison of cultivation and wild collection of medicinal and aromatic plants under sustainability aspects, In: *Medicinal and Aromatic Plants*, R.J. Bogers, L.E. Craker, and D. Lange, (Eds.), Springer Verlag, 2006, pp. 75–95.
2. Lange, D., The role of east and southeast Europe in the medicinal and aromatic plants' trade, *Med. Plant Conserv.*, 8, 14–18, 2002.
3. Pierpoint, W.S., Salicylic acid and its derivatives in plants: Medicines, metabolites and messenger molecules, *Adv. Bot. Res.*, 20, 163–235, 1994.
4. Vickers, A. and Zollman, C., ABC of complementary medicine: Herbal medicine, *BMJ*, 319, 1050–1053, 1999.
5. Raskin, I. et al., Plants and human health in the twenty-first century, *Trends Biotechnol.*, 20, 522–531, 2002.
6. Newman, D.J. and Cragg, G.M., Natural products as sources of new drugs over the last 25 years, *J. Nat. Prod.*, 70, 461–477, 2007.
7. World Health Organization, Traditional medicine, *Fact sheet*, No.134. 2003.
8. Kumar, S., Kumar, D., and Prakash, O., Herbal supplements: Regulation and safety aspects. *Phcog. Mag.*, 3, 65–72, 2007.
9. Lazarowitz, S.G., Plant viruses. In: *Fundamental Virology*. 4th edn., D.M. Knipe and P.M. Howley, (Eds.), Lippincot, Williams & Williams, Philadelphia., 2001.
10. van der Want, J.P.H. A historical outline of plant virology. In: *Handbook of Plant Virology*. J.A. Khan and J. Dijkstra, (Eds.), Food Products Press, New York, London, Oxford, 2006, pp. 1–10.
11. Miller, W.A. and White, K.A. Long-distance RNA-RNA interactions in plant virus gene expression and replication, *Annu. Rev. Phytopathol.*, 44, 447–467, 2006.
12. Stanley, J. et al. Family *Geminiviridae*. In: *Virus Taxonomy: VIIIth Report of the International Committee on Taxonomy of Viruses*, Fauquet, C.M., Mayo, M.A., Maniloff, J., Desselberger, U., and Ball, L.A., (Eds.), San Diego, Elsevier Academic Press, 2005, p. 310–326.

13. Wren, J.D. et al., Plant virus biodiversity and ecology, *PLoS Biol.*, 4, e80, 2006.
14. Fauquet, C.M. and Fargette, D., International Committee on Taxonomy of Viruses and the 3,142 unassigned species, *Virol. J.*, 2, 64, 2005.
15. Harrison, B.D. and Wilson, T.M.A., Milestones in the research on tobacco mosaic virus, *Phil. Trans. R. Soc. Lond. B*, 354, 521–529, 1999.
16. Okada, Y., Historical overview of research on the tobacco mosaic virus genome: Genome organization, infectivity and gene manipulation, *Phil. Trans. R. Soc. Lond. B*, 354, 569–582, 1999.
17. Scholthof, K.B., Tobacco mosaic virus: A model system for plant biology, *Annu. Rev. Phytopathol.*, 42, 13–34, 2004.
18. Gierer, A. and Schramm, G., Infectivity of ribonucleic acid from tobacco mosaic virus, *Nature*, 177, 702–703, 1956.
19. Goelet, P. et al., Nucleotide sequence of tobacco mosaic virus RNA, *Proc. Natl. Acad. Sci. U S A*, 79, 5818–5822, 1982.
20. Ahlquist, P. et al., Multicomponent RNA plant virus infection derived from cloned viral cDNA, *Proc. Natl. Acad. Sci. U S A*, 81, 7066–7070, 1984.
21. Dawson, W.O. et al., cDNA cloning of the complete genome of tobacco mosaic virus and production of infectious transcripts, *Proc. Natl Acad. Sci. U S A*, 83, 1832–1836, 1986.
22. Harrison, S.C. et al., Tomato bushy stunt virus at 2.9 Å resolution, *Nature*, 276, 368–373, 1978.
23. Abad-Zapatero, C. et al., Structure of southern bean mosaic virus at 2.8 Å resolution, *Nature*, 286, 33–39, 1980.
24. Namba, K. and Stubbs, G., Structure of tobacco mosaic virus at 3.6 Å resolution: Implications for assembly, *Science*, 231, 1401–1406, 1986.
25. Mason, H.S., Lam, D.M.K., and Arntzen, C.J., Expression of hepatitis B surface antigen in transgenic plants. *Proc. Natl. Acad. Sci. U S A*, 89, 11745–11749, 1992.
26. Zhang, T. et al., RNA viral community in human feces: Prevalence of plant pathogenic viruses, *PLoS Biol.*, 4, e3, 2006.
27. Rae, C.S. et al., Systemic trafficking of plant virus nanoparticles in mice via the oral route, *Virology*, 343, 224–235, 2005.
28. Singh, P. et al., Bio-distribution, toxicity and pathology of cowpea mosaic virus nanoparticles in vivo, *J. Control. Release*, 120, 41–50, 2007.
29. Koudelka, K.J., et al., Interaction between a 54 kilodalton mammalian cell surface protein and Cowpea mosaic virus, *J. Virol.*, 81, 1632–1640, 2007.
30. Fischer, R. and Emans N., Molecular farming of pharmaceutical proteins, *Transg. Res.*, 9, 279–299, 2000.
31. Ma, J.K., Drake, P.M., and Christou P., The production of recombinant pharmaceutical proteins in plants, *Nat. Rev. Genet.*, 4, 794–805, 2003.
32. Baneyx, F. and Mujacic, M., Recombinant protein folding and misfolding in Escherichia coli, *Nat. Biotechnol.*, 22, 1399–1408, 2004.
33. Panda, A.K., Bioprocessing of therapeutic proteins from the inclusion bodies of Escherichia coli, *Adv. Biochem. Eng. Biotechnol.*, 85, 43–93, 2003.
34. Cohen, S.N. et al., Construction of biologically functional bacterial plasmids in vitro, *Proc. Natl. Acad. Sci. U S A*, 70, 3240–3244, 1973.
35. Gerngross, T.U., Advances in the production of human therapeutic proteins in yeasts and filamentous fungi, *Nat. Biotechnol.*, 22, 1409–1414, 2004.
36. Wildt, S. and Gerngross T.U., The humanization of N-glycosylation pathways in yeast, *Nat. Rev. Microbiol.*, 3, 119–128, 2005.
37. Gellissen G. et al., New yeast expression platforms based on methylotrophic *Hansenula polymorpha* and *Pichia pastoris* and on dimorphic *Arxula adeninivorans* and *Yarrowia lipolytica*—A comparison, *FEMS Yeast Res.*, 5, 1079–1096, 2005.
38. Nasmyth, K., Eukaryotic gene cloning and expression in yeast, *Nature*, 274, 741–743, 1978.
39. McAleer, W.J. et al., Human hepatitis B vaccine from recombinant yeast, *Nature*, 307, 178–180, 1984.
40. Hamilton, S.R. et al., Production of complex human glycoproteins in yeast, *Science*, 301, 1244–1246, 2003.
41. Hesse, F. and Wagner, R., Developments and improvements in the manufacturing of human therapeutics with mammalian cell cultures, *Trends Biotechnol.*, 18, 173–180, 2000.
42. Wurm, F.M., Production of recombinant protein therapeutics in cultivated mammalian cells. *Nat. Biotechnol.*, 22, 1393–1398, 2004.

43. Walsh, G., Biopharmaceutical benchmarks 2006, *Nat. Biotechnol.*, 24, 769–776, 2006.
44. Jaenisch, R., Fan, H., and Croker, B., Infection of preimplantation mouse embryos and of newborn mice with leukemia virus: Tissue distribution of viral DNA and RNA and leukemogenesis in the adult animal. *Proc. Natl. Acad. Sci. U S A*, 72, 4008–4012, 1975.
45. Wolf, E. et al., Transgenic technology in farm animals—progress and perspectives, *Exp. Physiol.*, 85, 615–625, 2000.
46. Melo, E.O. et al., Animal transgenesis: State of the art and applications, *J. Appl. Genet.*,48, 47–61, 2007.
47. Arntzen, C., Plotkin, S., and Dodet, B., Plant-derived vaccines and antibodies: Potential and limitations, *Vaccine*, 23, 1753–1756, 2005.
48. Gelvin, S.B., Agrobacterium and plant genes involved in T-DNA transfer and integration, *Annu. Rev. Plant Physiol. Plant Mol. Biol.*, 51, 223–256, 2000.
49. Giddings, G. et al., Transgenic plants as factories for biopharmaceuticals, *Nat. Biotechnol.*, 18, 1151–1155, 2000.
50. Mason, H.S., et al., Edible plant vaccines: Applications for prophylactic and therapeutic molecular medicine. *Trends Mol. Med.*, 8, 324–329, 2002.
51. Gelvin, S.B., Agrobacterium-mediated plant transformation: The biology behind the "gene-jockeying" tool, *Microbiol. Mol. Biol. Rev.*, 67, 16–37, 2003.
52. Kermode, A.R., Plants as factories for production of biopharmaceutical and bioindustrial proteins: Lessons from cell biology, *Can. J. Bot.*, 84, 679–694, 2006.
53. Ma, J.K.C. et al., Molecular farming for new drugs and vaccines, *EMBO Rep.*, 6, 593–599, 2005.
54. Barta, A. et al., The expression of a nopaline synthase human growth hormone chimeric gene in transformed tobacco and sunflower callus tissue, *Plant Mol. Biol.*, 6, 347–357, 1986.
55. Sijmons, P.C. et al., Production of correctly processed human serum albumin in transgenic plants, *Biotechnology*, 8, 217–221, 1990.
56. Woodard, S.L. et al., Maize derived bovine trypsin: Characterization of the first large-scale commercial protein product from transgenic plants, *Biotechnol. Appl. Biochem.*, 38, 123–130, 2003.
57. Horn, M.E., Woodard, S.L., and Howard J.A., Plant molecular farming: Systems and products, *Plant Cell Rep.*, 22, 711–720, 2004.
58. Chen M. et al., Modification of plant N-glycans processing: The future of producing therapeutic protein by transgenic plants, *Med. Res. Rev.*, 25, 343–360, 2005.
59. Tregoning, J.S. et al., Expression of tetanus toxin fragment C in tobacco chloroplasts, *Nucleic Acids Res.*, 31, 1174–1179, 2003.
60. Streatfield, S.J., Approaches to achieve high-level heterologous protein production in plants. *Plant Biotech. J.*, 5, 2–15, 2007.
61. Howard, J.A., Commercialization of biopharmaceutical and bioindustrial proteins from plants. *Crop. Sci.*, 45, 468–472, 2005.
62. Kirk, D.D. et al., Risk analysis for plant-made vaccines, *Transgen. Res.*, 14, 449–462, 2005.
63. Helenius, E. et al., Gene delivery into intact plants using the Helios™ Gene Gun, *Plant Mol. Biol. Rep.*, 18, 287a–287l, 2000.
64. Turpen, T.H. et al., Transfection of whole plants from wounds inoculated with Agrobacterium tumefaciens containing cDNA of tobacco mosaic virus, *J. Virol. Methods*, 42, 227–239, 1993.
65. French, R., Janda. M., and Ahlquist, P., Bacterial gene inserted in an engineered RNA virus: Efficient expression in monocotyledonous plant cells, *Science*, 231, 1294–1297, 1986.
66. Dawson, W.O., Bubrick P., and Grahtham G.L., Modifications of the tobacco mosaic virus coat protein gene affecting replication, movement and symptomatology, *Phytopathology*, 78, 783–789, 1988.
67. Dawson, W.O. et al., A tobacco mosaic virus-hybrid expresses and loses an added gene, *Virology*, 172, 285–292, 1989.
68. Donson, J. et al., Systemic expression of a bacterial gene by a tobacco mosaic virus-based vector, *Proc. Natl. Acad. Sci. U S A*, 88, 7204–7208, 1991.
69. Shivprasad, S. et al., Heterologous sequences greatly affect foreign gene expression in tobacco mosaic virus-based vectors, *Virology*, 255, 312–323, 1999.
70. Karasev, A.V. et al., Plant based HIV-1 vaccine candidate: Tat protein produced in spinach, *Vaccine*, 23, 1875–1880, 2005.
71. Baulcombe, D.C., Chapman, S., and Santa Cruz, S., Jellyfish green fluorescent protein as a reporter for virus infections, *Plant J.*, 7, 1045–1053, 1995.

72. Saito, H., Production of antimicrobial defensin in *Nicotiana benthamiana* with a potato virus X vector, *Mol. Plant-Microbe Interact.*, 14, 111–115, 2001.
73. Liu, L. et al., Cowpea mosaic virus-based systems for the production of antigens and antibodies in plants, *Vaccine*, 23, 1788–1792, 2005.
74. De Felipe, P., Skipping the co-expression problem: The new 2A "CHYSEL" technology. *Genet. Vacc. Therapy*, 2, 13, 2004.
75. Matsuo, K. et al., Development of Cucumber mosaic virus as a vector modifiable for different host species to produce therapeutic proteins, *Planta*, 225, 277–286, 2007.
76. Gleba Y., Marillonnet S., and Klimyuk V., Engineering viral expression vectors for plants: The "full virus"and the "deconstructed virus" strategies, *Curr. Opinion Plant Biol.*, 7, 182–188, 2004.
77. Liu, L. and Lomonossoff, G.P., Agroinfection as a rapid method for propagating Cowpea mosaic virus-based constructs, *J. Virol. Methods*, 105, 343–348, 2002.
78. Sudarshana, M.R. et al., A chemically inducible cucumber mosaic virus amplicon system for expression of heterologous proteins in plant tissues, *Plant Biotechnol. J.*, 4, 551–559, 2006.
79. Li, Y. et al., Expression of a human lactoferrin N-lobe in *Nicotiana benthmiana* with potato virus X-based agroinfection, *Biotechnology Lett.*, 26, 953–957, 2004.
80. Komarova, T.V. et al., New viral vector for efficient production of target proteins in plants, *Biochemistry (Moscow)*, 71, 846–850, 2006.
81. Kim, K.I. et al., Improved expression of recombinant GFP using a replicating vector based on Beet curly top virus in leaf-disks and infiltrated *Nicotiana benthamiana* leaves, *Plant Mol. Biol.*, 64, 103–112, 2007.
82. Palmer, K.E., Thomson, J.A., and Rybicki E.P., Generation of maize cell lines containing autonomously replicating maize streak virus-based gene vectors, *Arch. Virol.*, 144, 1345–1360, 1999.
83. Mor, T.S. et al., Geminivirus vectors for high-level expression of foreign proteins in plant cells, *Biotechnol. Bioeng.*, 81, 430–437, 2002.
84. Hefferon, K.L. and Fan, Y., Expression of a vaccine protein in a plant cell line using a geminivirus-based replicon system, *Vaccine*, 23, 404–410, 2004.
85. Cañizares, M.C. et al., A bipartite system for the constitutive and inducible expression of high levels of foreign proteins in plants, *Plant Biotech. J.*, 4, 183–193, 2006.
86. Musiychuk, K. et al. A launch vector for the production of vaccine antigens in plants. *Influenza*, 1, 19–25, 2007.
87. Marillonnet, S. et al., In planta engineering of viral RNA replicons: Efficient assembly by recombination of DNA modules delivered by Agrobacterium, *Proc. Natl. Acad. Sci. U S A*, 101, 15545–15546, 2004.
88. Gleba, Y., Klimyuk, V., and Marillonnet, S., Magnifection - a new platform for expressing recombinant vaccines in plants, *Vaccine*, 23, 2042–2048, 2005.
89. Marillonnet S. et al., Systemic Agrobacterium tumefaciens-mediated transfection of viral replicons for efficient transient expression in plants, *Nat. Biotechnol.*, 23, 718–723, 2005.
90. Ruiz, M.T., Voinnet, O., and Baulcombe, D.C., Initiation and maintenance of virus-induced gene silencing, *Plant Cell*, 10, 937–946, 1998.
91. Whitham, S.A. and Wang Y., Roles for host factors in plant viral pathogenicity. *Curr. Opinion Plant Biol.*, 7, 365–371, 2004.
92. Boevink, P. and Oparka, K.J., Virus-host interactions during movement processes. *Plant Physiol.*, 138, 1815–1821, 2005.
93. Lucas W.J., Plant viral movement proteins: Agents for cell-to-cell trafficking of viral genomes, *Virology*, 344, 169–184, 2006.
94. Toth, R.L., Pogue, G.P., and Chapman S., Improvement of the movement and host range properties of a plant virus vector through DNA shuffling, *Plant J.*, 30, 593–600, 2002.
95. Hammond, S.M., Caudy, A.A., and Hannon, G.J., Post-transcriptional gene silencing by double-stranded RNA, *Nat. Rev. Genet.*, 2, 110–119, 2001.
96. Fire, A. et al., Potent and specific genetic interference by double-stranded RNA in *Caenorhabditis elegans*, *Nature*, 391, 806–811, 1998.
97. Baumberger, N. and Baulcombe, D.C., Arabidopsis ARGONAUTE1 is an RNA slicer that selectively recruits microRNAs and short interfering RNAs, *Proc. Natl. Acad. Sci. U S A*, 102, 11928–11933, 2005.
98. Canto, T. et al., Translocation of tomato bushy stunt virus P19 protein into the nucleus by ALY proteins compromises its silencing suppressor activity, *J. Virol.*, 80, 9064–9072, 2006.

99. Voinnet, O., Induction and suppression of RNA silencing: Insights from viral infections, *Nat. Rev. Genet.*, 3, 206–220, 2005.
100. Voinnet, O., Pinto, Y.M., and Baulcombe, D.C., Suppression of gene silencing: A general strategy used by diverse DNA and RNA viruses, *Proc. Natl. Acad. Sci. U S A*, 96, 14147–14152 1999.
101. Voinnet, O. et al., An enhanced transient expression system in plants based on suppression of gene silencing by the p19 protein of tomato bushy stunt virus. *Plant J.*, 33, 949–956, 2003.
102. Lindbo, J.A., High-efficiency protein expression in plants from agroinfection-compatible Tobacco mosaic virus expression vectors, *BMC Biotechnol.*, 7, 52, 2007.
103. Hamilton, A. et al., Two classes of short interfering RNA in RNA silencing, *EMBO J.*, 21, 4671–4679, 2002.
104. Shew, H.D. and Lucas, G.B., Eds., *Compendium of Tobacco Diseases*. The American Phytopathological Society, Minnesota (EUA), 1991, p. 68.
105. Sacristán, S. et al., An analysis of host adaptation and its relationship with virulence in Cucumber mosaic virus, *Phytopathology*, 95, 827–833, 2005.
106. DeCleene, M. and DeLey, J., The host range of crown gall, *Bot. Rev.*, 42, 389–466, 1976.
107. Ishida, Y. et al., High efficiency transformation of maize (Zea mays L.) mediated by *Agrobacterium tumefaciens*, *Nat. Biotechnol.*, 14, 745–750, 1996.
108. Chung, S.M., Vaidya, M., and Tzfira, T., Agrobacterium is not alone: Gene transfer to plants by viruses and other bacteria, *Trends Plant Sci.*, 11, 1–4, 2006.
109. Stanton, B.G., Agrobacterium-mediated plant transformation: The biology behind the "Gene-jockeying" tool, *Microbiol. Mol. Biol. Rev.*, 67, 16–37, 2003.
110. Citovsky, et al., Biological systems of the host cell involved in Agrobacterium infection, *Cell Microbiol.*, 9, 9–20, 2007.
111. El Attar, A.K. et al., Expression of chimeric HCV peptide in transgenic tobacco plants infected with recombinant alfalfa mosaic virus for development of a plant-derived vaccine against HCV, *Afr. J. Biotechnol.*, 3, 588–594, 2004.
112. Mallory, A.C. et al., The amplicon-plus system for high-level expression of transgenes in plants, *Nat. Biotechnol.*, 20, 622–625, 2002.
113. Azhakanandam, K. et al., Amplicon-plus targeting technology (APTT) for rapid production of a highly unstable vaccine protein in tobacco plants, *Plant Mol. Biol.*, 63, 393–404, 2007.
114. Dohi, K. et al., Inducible virus-mediated expression of a foreign protein in suspension-cultured plant cells, *Arch. Virol.*, 151, 1075–1084, 2006.
115. Collens, J.I., Mason, H.S., and Curtis, W.R., Agrobacterium-mediated viral vector-amplified transient gene expression in *Nicotiana glutinosa* plant tissue culture, *Biotechnol. Prog.*, 23, 570–576, 2007.
116. Pogue, G.P. et al., Making an ally from an enemy: Plant virology and the new agriculture, *Annu. Rev. Phytopathol.*, 40, 45–74, 2002.
117. Avesani, L. et al., Stability of Potato virus X expression vectors is related to insert size: Implications for replication models and risk assessment, *Transgen. Res.*, 16, 587–597, 2007.
118. Stevens, L.H. et al., Effect of climate conditions and plant developmental stage on the stability of antibodies expressed in transgenic tobacco, *Plant Physiol.*, 124, 173–182, 2000.
119. Doran, P.M., Foreign protein degradation and instability in plants and plant tissue cultures, *Trends Biotechnol.*, 24, 426–432, 2006.
120. Gils, M. et al., High-yield production of authentic human growth hormone using a plant virus-based expression system, *Plant Biotechnol. J.*, 3, 613–620, 2005.
121. Mett, V. et al., A plant-produced plague vaccine candidate confers protection to monkeys, *Vaccine*, 25, 3014–3017, 2007.
122. Massa, S. et al., Anti-cancer activity of plant-produced HPV16 E7 vaccine, *Vaccine*, 25, 3018–3021, 2007.
123. Zhou, F. et al., Efficient transient expression of human GM-CSF protein in *Nicotiana benthamiana* using potato virus X vector, *Appl. Microbiol. Biotechnol.*, 72, 756–762, 2006.
124. Hiatt, A., Cafferkey, R., and Bowdish, K., Production of antibodies in transgenic plants. *Nature*, 342, 76–78, 1989.
125. Hiatt, A. and Pauly, M., Monoclonal antibodies from plants: A new speed record, *Proc. Natl. Acad. Sci. U S A*, 103, 14645–14646, 2006.
126. Verch, T., Yusibov, V., and Koprowski, H., Expression and assembly of a full-length monoclonal antibody in plants using a plant virus vector, *J. Immunol. Methods*, 220, 69–75, 1998.

127. Giritch, A. et al., Rapid high-yield expression of full-size IgG antibodies in plants coinfected with noncompeting viral vectors, *Proc. Natl. Acad. Sci. U S A*, 103, 14701–14706, 2006.
128. Pumpens, P. and Grens, E. Artificial genes for chimeric virus-like particles. In: *Artificial DNA: Methods and Applications*, Y.E. Khudyakov and H.A. Fields, Eds., CRC Press, Boca Raton, 2002, pp. 249–327.
129. Noad, R. and Roy, P., Virus-like particles as immunogens, *Trends Microbiol.*, 11, 438–444, 2003.
130. Bryan, J.T., Developing an HPV vaccine to prevent cervical cancer and genital warts, *Vaccine*, 25, 3001–3006, 2007.
131. Bachmann, M.F., Hengartner, H., and Zinkernagel, R.M., T helper cell-independent neutralizing B cell response against vesicular stomatitis virus: Role of antigen patterns in B cell induction? *Eur. J. Immunol.*, 25, 3445–3451, 1995.
132. Fifis, T. et al., Size-dependent immunogenicity: Therapeutic and protective properties of nano-vaccines against tumors, *J. Immunol.*, 173, 3148–3154, 2004.
133. Lenz, P. et al., Interaction of papillomavirus virus-like particles with human myeloid antigen-presenting cells, *Clin. Immunol.*, 106, 231–237, 2003.
134. Bachmann, M.F., Zinkernagel, R.M., and Oxenius, A., Immune responses in the absence of costimulation: Viruses know the trick, *J. Immunol.*, 161, 5791–5794, 1998.
135. Bachmann, M.F. and Dyer, M.R., Therapeutic vaccination for chronic diseases: A new class of drugs in sight, *Nat. Rev. Drug Discov.*, 3, 81–88, 2004.
136. Bertoletti, A. and Gehring, A.J., The immune response during hepatitis B virus infection, *J. Gen. Virol.*, 87, 1439–1449, 2006.
137. Chow, M. et al., Synthetic peptides from four separate regions of the poliovirus type 1 capsid protein VP1 induce neutralizing antibodies, *Proc. Natl. Acad. Sci. U S A*, 82, 910–914, 1985.
138. Haynes, J.R. et al., Development of a genetically-engineered, candidate polio vaccine employing the self-assembling properties of the tobacco mosaic virus coat protein, *Biotechnology*, 4, 637–641, 1986.
139. Lin, T. et al., The refined crystal structure of cowpea mosaic virus at 2.8 Å resolution, *Virology*, 265, 20–34, 1999.
140. Porta, C. et al., Cowpea mosaic virus-based chimaeras: Effects of inserted peptides on the phenotype, host range, and transmissibility of the modified viruses, *Virology*, 310, 50–63, 2003.
141. Dalsgaard, K. et al., Plant-derived vaccine protects target animals against a viral disease, *Nat. Biotechnol.*, 3, 248–252, 1997.
142. Brennan, F.R. et al., A chimaeric plant virus vaccine protects mice against a bacterial infection, *Microbiology*, 145, 2061–2067, 1999.
143. Rennermalm, A. et al., Antibodies against a truncated Staphylococcus aureus fibronectin-binding protein protect against dissemination of infection in the rat, *Vaccine*, 19, 3376–3383, 2001.
144. Phelps, J.P., Dang, N., and Rasochova, L., Inactivation and purification of cowpea mosaic virus-like particles displaying peptide antigens from *Bacillus anthracis*. *J. Virol. Methods*, 141, 146–153, 2007.
145. Joelson, T. et al., Presentation of a foreign peptide on the surface of tomato bushy stunt virus, *J. Gen. Virol.*, 78, 1213–1217, 1997.
146. Zhao, Y. and Hammond, R.W., Development of a candidate vaccine for Newcastle disease virus by epitope display in the Cucumber mosaic virus capsid protein, *Biotechnol. Lett.*, 27, 375–382, 2005.
147. Natilla, A. et al., Cucumber mosaic virus as carrier of a hepatitis C virus-derived epitope, *Arch. Virol.*, 149, 137–154, 2004.
148. Piazzolla, G. et al., Immunogenic properties of a chimeric plant virus expressing a hepatitis C virus (HCV)-derived epitope: New prospects for an HCV vaccine, *J. Clin. Immunol.*, 25, 142–151, 2005.
149. Natilla, A., Hammond, R.W., and Nemchinov, L.G., Epitope presentation system based on cucumber mosaic virus coat protein expressed from a potato virus X-based vector. *Arch. Virol.*, 151, 1373–1386, 2006.
150. Nemchinov, L.G. and Natilla, A., Transient expression of the ectodomain of matrix protein 2 (M2e) of avian influenza A virus in plants, *Protein Expr. Purif.*, doi:10.1016/j.pep.2007.05.015, 2007.
151. Yusibov, V. et al., Expression in plants and immunogenicity of plant virus-based experimental rabies vaccine, *Vaccine*, 20, 3155–3164, 2002.
152. Yusibov, V. et al., Peptide-based candidate vaccine against respiratory syncytial virus, *Vaccine*, 23, 2261–2265, 2005.

153. Baratova, L.A. et al., The topography of the surface of potato virus X: Tritium planigraphy and immunological analysis, *J. Gen. Virol.*, 73, 229–235, 1992.
154. Santa-Cruz, S. et al., Assembly and movement of a plant virus carrying a green fluorescent protein overcoat, *Proc. Natl. Acad. Sci. U S A*, 93, 6286–6290, 1996.
155. Marconi, G. et al., In planta production of two peptides of the Classical Swine Fever Virus (CSFV) E2 glycoprotein fused to the coat protein of potato virus X, *BMC Biotechnol.*, 6, 29, 2006.
156. Marusic, C. et al., Chimeric plant virus particles as immunogens for inducing murine and human immune responses against human immunodeficiency virus type 1, *J. Virol.*, 75, 8434–8439, 2001.
157. Uhde, K., Fischer, R., and Commandeur, U., Expression of multiple foreign epitopes presented as synthetic antigens on the surface of Potato virus X particles, *Arch. Virol.*, 150, 327–340, 2005.
158. Namba, K., Pattanayek, R., and Stubbs, G., Visualization of protein–nucleic acid interactions in a virus: Refined structure of intact tobacco mosaic virus at 2.9 A resolution by X-ray fiber diffraction, *J. Mol. Biol.*, 208, 307–325, 1989.
159. Hamamoto, H. et al., A new tobacco mosaic virus vector and its use for the systemic production of angiotensin-I-converting enzyme inhibitor in transgenic tobacco and tomato, *Biotechnology*, 11, 930–932, 1993.
160. Fujiyama, K. et al., In Planta production of immunogenic poliovirus peptide using tobacco mosaic virus-based vector system, *J. Biosci. Bioeng.*, 101, 398–402, 2006.
161. Borovsky, D. et al., Expression of Aedes trypsin-modulating oostatic factor on the virion of TMV: A potential larvicide, *Proc. Natl. Acad. Sci. U S A*, 103, 18963–18968, 2006.
162. Ooi, A.S. et al., The full-length clone of cucumber green mottle mosaic virus and its application as an expression system for hepatitis B surface antigen, *J. Biotechnol.*, 121, 471–481, 2006.
163. Koo, M. et al., Protective immunity against murine hepatitis virus (MHV) induced by intranasal or subcutaneous administration of hybrids of tobacco mosaic virus that carries an MHV epitope, *Proc. Natl. Acad. Sci. U S A*, 96, 7774–7779, 1999.
164. Jiang, L. et al., A modified TMV-based vector facilitates the expression of longer foreign epitopes in tobacco, *Vaccine*, 24, 109–115, 2006.
165. Bendahmane, M. et al., Display of epitopes on the surface of tobacco mosaic virus: Impact of charge and isoelectric point of the epitope on virus-host interactions, *J. Mol. Biol.*, 290, 9–20. 1999.
166. Werner, S. et al., Immunoabsorbent nanoparticles based on a tobamovirus displaying protein A, *Proc. Natl. Acad. Sci. U S A*, 103, 17678–17683, 2006.
167. Saini, M. and Vrati, S., A Japanese encephalitis virus peptide present on Johnson grass mosaic virus-like particles induces virus-neutralizing antibodies and protects mice against lethal challenge, *J. Virol.*, 77, 3487–3494, 2003.
168. Denis J. et al., Immunogenicity of papaya mosaic virus-like particles fused to a hepatitis C virus epitope: Evidence for the critical function of multimerization, *Virology*, 363, 59–68, 2007.
169. Smith, M.L. et al., Modified tobacco mosaic virus particles as scaffolds for display of protein antigens for vaccine applications, *Virology*, 348, 475–488, 2006.
170. Chatterji, A. et al., Chemical conjugation of heterologous proteins on the surface of Cowpea mosaic virus. *Bioconjug. Chem.*, 15, 807–813, 2004.
171. Douglas, T. and Young, M., Host–guest encapsulation of materials by assembled virus protein cages. *Nature*, 393, 152–155, 1998.
172. Sun, J. et al., Core-controlled polymorphism in virus-like particles. *Proc. Natl. Acad. Sci. U S A*, 104, 41354–1359, 2007.
173. Tseng, R.J. et al., Digital memory device based on tobacco mosaic virus conjugated with nanoparticles, *Nat. Nanotechnol.*, 1, 72–77, 2006.
174. Manchester, M. and Singh, P., Virus-based nanoparticles (VNPs): Platform technologies for diagnostic imaging, *Adv. Drug Deliv. Rev.*, 58, 1505–1522, 2006.
175. Ren, Y. et al., In vitro-reassembled plant virus-like particles for loading of polyacids, *J. Gen. Virol.*, 87, 2749–2754, 2006.
176. Groenewegen, A.S., Plant molecular farming: Policy perspectives from Canada and the United States, www.fw.ucalgary.ca/pharmingthefuture/resources/Groenewegen.doc, 2006.
177. Golovkin, M. et al., Smallpox subunit vaccine produced in Planta confers protection in mice, *Proc. Natl. Acad. Sci. U S A*, 104, 6864–6869, 2007.

178. Faye, L. et al., Protein modifications in the plant secretory pathway: Current status and practical implications in molecular pharming, *Vaccine*, 23, 1770–1778, 2005.
179. Ma, J.K.C. et al., Plant-derived pharmaceuticals—the road forward, *Trends Plant Sci.*, 10, 580–585, 2005.
180. Tonks, A., A spoonful of antigen, *BMJ*, 335, 180–182, 2007.
181. *Bayer Innovation*, Health-bringing tobacco, Bayer Scientific Magazine, 18, 37–41, 2006.
182. Brumfield, S. et al., Heterologous expression of the modified coat protein of Cowpea chlorotic mottle bromovirus results in the assembly of protein cages with altered architectures and function, *J. Gen Virol.*, 85, 1049–1053, 2004.
183. Phelps, J.P. et al., Expression and self-assembly of cowpea chlorotic mottle virus-like particles in Pseudomonas fluorescens, *J. Biotechnol.*, 128, 290–296, 2007.
184. Shanks, M. and Lomonossoff, G.P., Co-expression of the capsid proteins of Cowpea mosaic virus in insect cells leads to the formation of virus-like particles, *J. Gen. Virol.*, 81, 3093–3097, 2000.
185. Bragard, C. et al., Virus-like particles assemble in plants and bacteria expressing the coat protein gene of Indian peanut clump virus, *J. Gen. Virol.*, 81, 267–272, 2000.
186. Bertioli, D.J. et al., Transgenic plants and insect cells expressing the coat protein of arabis mosaic virus produce empty virus-like particles, *J. Gen. Virol.*, 72, 1801–1809, 1991.
187. Zheng, et al., Assembly of double-shelled, virus-like particles in transgenic rice plants expressing two major structural proteins of Rice dwarf virus, *J. Virol.*, 74, 9808–9810, 2000.
188. Tremblay, M.H. et al., Effect of mutations K97A and E128A on RNA binding and self assembly of papaya mosaic potexvirus coat protein, *FEBS J.*, 273, 14–25, 2006.
189. Jagadish, M.N. et al., Expression of potyvirus coat protein in Escherichia coli and yeast and its assembly into virus-like particles, *J. Gen. Virol.*, 72, 1543–1550, 1991.
190. Pokorna, D. et al., DNA vaccines based on chimeric potyvirus-like particles carrying HPV16 E7 peptide (aa 44–60), *Oncology Rep.*, 14, 1045–1053, 2005.
191. Jacquet, C. et al., Use of modified plum pox virus coat protein genes developed to limit heteroencapsidation-associated risks in transgenic plants, *J. Gen Virol.*, 79, 1509–1517, 1998.
192. Voloudakis, A.E. et al., Structural characterization of Tobacco etch virus coat protein mutants, *Arch. Virol.*, 149, 699–712, 2004.
193. Lokesh, G.L. et al., A molecular switch in the capsid protein controls the particle polymorphism in an icosahedral virus, *Virology*, 292, 211–223, 2002.
194. Hwang, D.J., Expression of tobacco mosaic virus coat protein and assembly of pseudovirus particles in Escherichia coli, *Proc. Natl. Acad. Sci. U S A*, 91, 9067–9071, 1994.
195. Sastri, M. et al., Assembly of physalis mottle virus capsid protein in *Escherichia coli* and the role of amino and carboxy termini in the formation of the icosahedral particles, *J. Mol. Biol.*, 272, 541–552, 1997.
196. Gopinath, K. et al., Engineering cowpea mosaic virus RNA-2 into a vector to express heterologous proteins in plants, *Virology*, 267, 159–173, 2000.
197. Mechtcheriakova, I.A. et al., The use of viral vectors to produce hepatitis B virus core particles in plants, *J. Virol. Methods*, 131, 10–15, 2006.
198. Monger, W. et al., An antibody derivative expressed from viral vectors passively immunizes pigs against transmissible gastroenteritis virus infection when supplied orally in crude plant extracts, *Plant Biotechnol. J.*, 4, 623–631, 2006.
199. Fernandez-Fernandez, M.R. et al., Protection of rabbits against Rabbit hemorrhagic disease virus by immunization with the VP60 protein expressed in plants with a Potyvirus-based vector, *Virology*, 280, 283–291, 2001.
200. Nemchinov, L.G. et al., Development of a plant-derived subunit vaccine candidate against hepatitis C virus, *Arch. Virol.*, 145, 2557–2573, 2000.
201. Pérez-Filgueira, D.M. et al., Bovine herpes virus gD protein produced in plants using a recombinant tobacco mosaic virus (TMV) vector possesses authentic antigenicity, *Vaccine*, 21, 4201–4209, 2003.
202. Pérez-Filgueira, D.M. et al., Preserved antigenicity of HIV-1 p24 produced and purified in high yields from plants inoculated with a tobacco mosaic virus (TMV)-derived vector, *J. Virol. Methods*, 121, 201–208, 2004.
203. Varsani, A. et al., Transient expression of Human papillomavirus type 16 L1 protein in *Nicotiana benthamiana* using an infectious tobamovirus vector, *Virus Res.*, 120, 91–96, 2006.

204. Saejung, W. et al., Production of dengue 2 envelope domain III in plant using TMV-based vector system, *Vaccine*, doi:10.1016/j.vaccine.2007.06.029, 2007.
205. Santi, L. et al., Protection conferred by recombinant Yersinia pestis antigens produced by a rapid and highly scalable plant expression system, *Proc. Natl. Acad. Sci. U S A*, 103, 861–866, 2006.
206. Huang, Z. et al., Rapid, high-level production of hepatitis B core antigen in plant leaf and its immunogenicity in mice, *Vaccine*, 24, 2506–2513, 2006.
207. Staczek, J. et al., Immunization with a chimeric tobacco mosaic virus containing an epitope of outer membrane protein F of *Pseudomonas aeruginosa* provides protection against challenge with *P. aeruginosa*, *Vaccine*, 18, 2266–2274, 2000.

20 Alphaviruses: Multiplicity of Vectors and Their Promising Application as Vaccines and Cancer Therapy Agents

Anna Zajakina, Baiba Niedre-Otomere*, Ekaterina Alekseeva, and Tatyana Kozlovska*

CONTENTS

20.1 INTRODUCTION

The latest virus-based technologies contribute significantly to the progress of modern medicine, which, today, is directed to the gene therapy, targeted cancer therapy, vaccine, and drug design. Among other heterologous gene expression systems based on recombinant viral replicons the alphavirus-driven vectors are gaining wide use because of their high efficiency in delivering and

* Equal contribution to the manuscript.

expressing the cloned gene of interest. During the last decade the alphaviral vectors were significantly modified to increase their potential for application in vitro and in vivo. The most remarkable application of alphavirus-based vectors is for vaccine development and gene therapy including promising results in cancer treatment. The immunogenicity and protective efficacy of recombinant hybrid alphavirus particles against different infectious disease agents, such as viruses, bacteria, and parasites have been demonstrated in vivo. A high-level antigen expression, lack of DNA stage precluding genome integration, cytoplasmic amplification through double-stranded RNA intermediates that stimulate innate immune responses, direct targeting of dendritic cells or affecting by cross-priming mechanism, and the absence of anti-vector immunity are some important advantages of the alphavirus-based vaccines.

Along with the recombinant particles a promising new approach in vaccine development is the use of plasmid DNA for immunization. In difference to the conventional plasmid DNA, protein expression directed by alphaviral DNA vectors has a suicidal effect, since the alphavirus replication causes a strong cytopathic effect. Therefore, the use of these vectors eliminates undesirable consequences of DNA integration into the host genome. Moreover, because of a strong apoptotic effect alphaviral vectors are considered for the application in cancer therapy. A high-affinity 67-kd laminin receptor, whose expression is increased in many types of human cancers, is known to mediate alphaviral infection of mammalian cells. This important feature of alphavirus entry made this vector system an attractive research target for specific cancer treatment.

In this chapter we describe a variety of alphavirus vectors and their application for the development of new vaccines and specific cancer treatments.

20.2 BIOLOGY OF ALPHAVIRUSES: OVERVIEW

The Genus Alphavirus belonging to the Togaviridae family includes a group of human and animal pathogens that cause transient febrile illness or more severe disease such as encephalitis. The natural life cycle of alphaviruses is associated with transmission to vertebrate hosts via mosquitoes. Therefore alphaviruses are representatives of a large class of arthropod borne viruses (*Arboviruses*). At present, there are described at least 26 members of alphaviruses [1,2]. Several of them, including Semliki Forest virus (SFV) and Sindbis (SIN) virus, are widely used in laboratories to study the life cycle of alphaviruses, and to generate heterologous gene expression system.

20.2.1 Virion Structure

The structure of the mature alphavirus virion has been resolved by cryo-electron microscopy and image processing techniques [3–5]. In brief, the virions resemble a small sphere of 70 nm in diameter (Figure 20.1). The particle has a $T=4$ icosahedral symmetry and is composed of 240 copies of the capsid monomer encapsidating a single molecule of the positive-strand RNA genome. The nucleocapsid is surrounded by an envelope containing 240 copies of the transmembrane glycoproteins (E2 and E1). Complexes of three E1/E2 heterodimers form the viral spikes.

20.2.2 Virus Entry

Since alphaviruses have a very wide host range, infecting many species of mosquitoes and other hematophagous insects as well as many species of higher vertebrates, these viruses use multiple receptors on host cells for infection. Several membrane molecules such as glycosaminoglycan heparan sulfate [6] and laminin [7] were found to serve as cell receptors enhancing the infection process.

At present, two mechanisms of alphavirus penetration are proposed. According to the first mechanism the virus penetrates through the receptor-mediated virion endocytosis followed by membrane fusion triggered by endosome acidification [8,9]. This pathway is widely used by

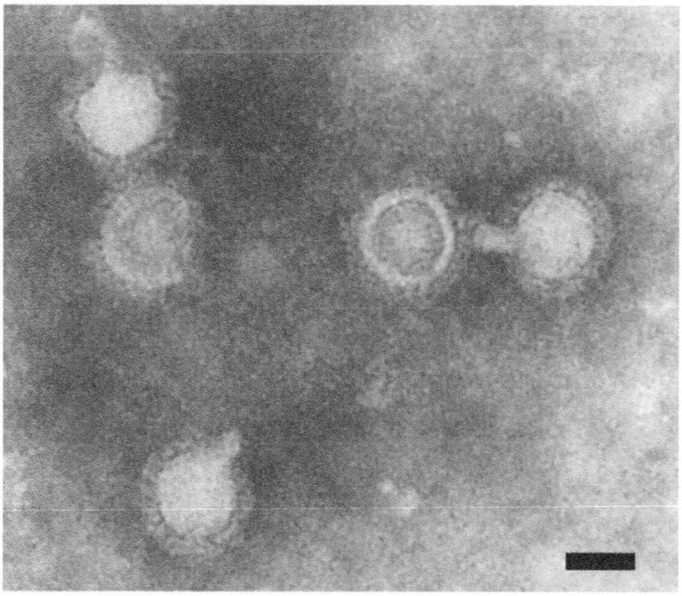

FIGURE 20.1 Electron microscopy of recombinant SFV particles. Negative staining, bar 50 nm.

many membrane-containing viruses [10]. It was shown that the key role in the membrane fusion plays the alphavirus envelope E1 protein, a class II virus membrane fusion protein [3,11,12]. Recently, an alternate model for alphavirus penetration via the generation of protein pores in the cell plasma membrane was proposed. This model is strongly supported by the observation that alphaviruses can penetrate cells using a mechanism that does not require endocytosis, exposure to acidic pH or high concentrations of cholesterol in the target membrane [13,14]. Moreover, the fact that the alphavirus infection can proceed at low temperatures, conditions which prevent endocytosis and membrane fusion [15], additionally confirms the possible pore-mediated virus penetration. A similar model of pore formation as a result of membrane modification by virus proteins has been also proposed for poliovirus [16] and human rhinovirus [17]. The mechanism of virus uncoating and liberation of the RNA genome into the cytosol of the host cell is still unknown.

20.2.3 RNA Genome Replication and Virus Assembly

The RNA genome of alphaviruses is a 5′-capped and 3′-polyadenylated RNA molecule of about 12 kb long. Since the RNA has a positive polarity, it is infectious, capable to start replication, and translation when introduced into the cell cytoplasm. Functionally the genome is divided in two parts coding the nonstructural and structural proteins correspondingly (Figure 20.2A).

Two-thirds of the 5′ end of the RNA genome encodes a polyprotein, which is processed into four viral nonstructural proteins, nsP1–4. Protein nsP1 has a methyltransferase activity for cytoplasmic capping of the viral RNAs, as well as initiation of minus-strand synthesis [18]. Protein nsP2 is the autoprotease that cleaves the virus nonstructural polyprotein [19]. Protein nsP3 is phosphoprotein involved in the replication complex formation [20,21], but its precise function is still debated. Protein nsP4 carries the RNA polymerase activity [22].

Once the nsP proteins have been synthesized, they are responsible for the replication of the plus-strand (42S) genome into full-length minus strands. These molecules then serve as templates for the production of new 42S genomic RNAs and subgenomic 26S RNAs. This about 4000 nucleotide-long subgenomic RNA is collinear with the last 3′-one third of the genome, and its synthesis is internally initiated at the 26S promoter on the 42S minus RNA strands (Figure 20.2B).

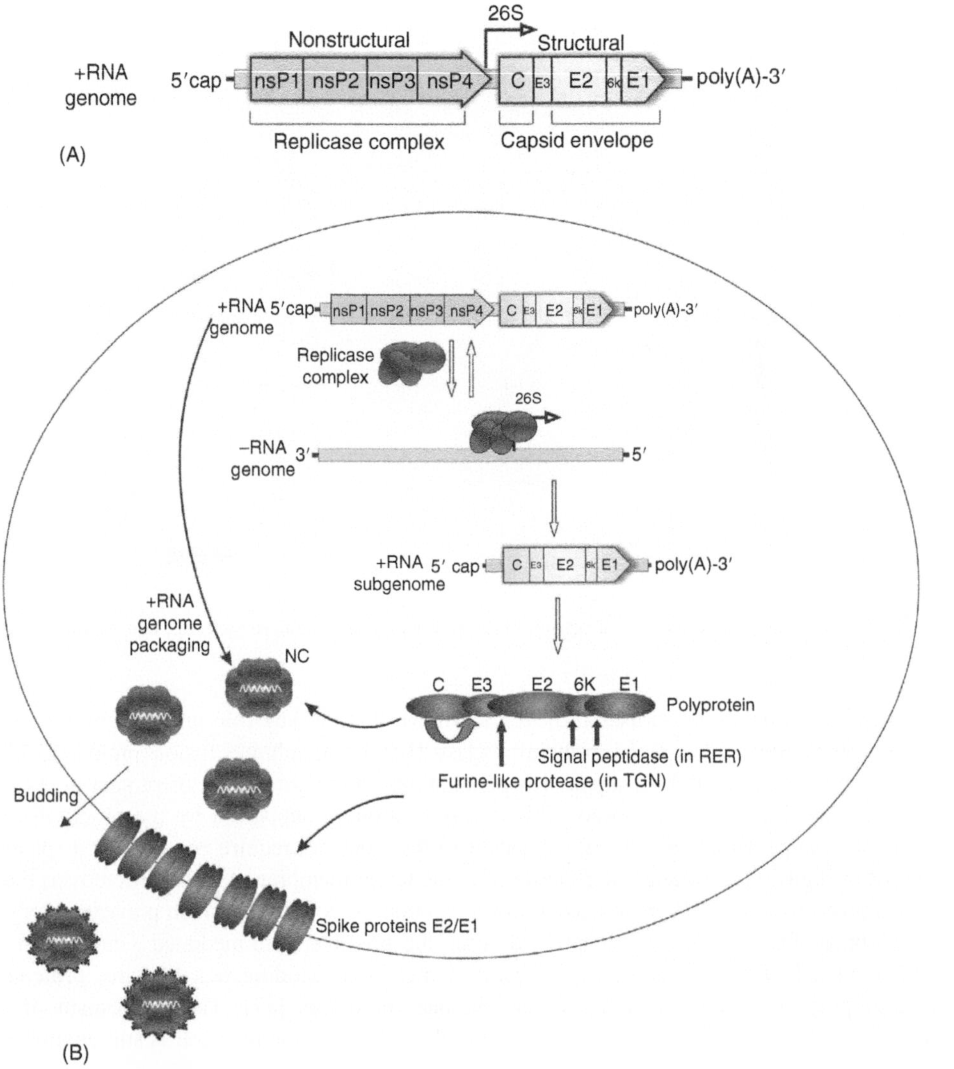

FIGURE 20.2 Genome structure (A) and replication cycle of alphaviruses (B) (for review see Schlesinger, S. and Schlesinger, M.J., *Fields Virology*, 4th edition, pp. 895–916, edited by D.M. Knipe and P.M. Howley, Philadelphia: Lippincott Williams & Wilkins, 2001). Rough endoplasmic reticulum is indicated as RER; trans Golgi network, TGN; nucleocapsid, NC.

The subgenomic RNA codes the structural proteins of the virus, which also are synthesized as a polyprotein precursor in the order C-E3-E2-6K-E1. Once the capsid (C) protein has been synthesized, it acts as an autoprotease, cleaving itself off the nascent chain [23,24]. At the endoplasmic reticulum (ER) membrane the nascent chain is co-translationally translocated and cleaved further by signal peptidase to the three structural membrane proteins, p62 (precursor of E3/E2), 6K, and E1.

After synthesis, the C proteins complex with genomic RNA into nucleocapsid structures in the cell cytoplasm. Usually, only the genomic RNA is packaged. The encapsidation signal for SIN was mapped within the nsP1 gene, while for SFV the encapsidation signal has been proposed to be located in the nsP2 gene [25].

The membrane proteins undergo extensive posttranslational modifications within the biosynthetic transport pathway of the cell. The precursor protein p62 undergoes during transport to the cell surface a proteolytic cleavage to form the mature envelope glycoprotein E2 [26]. The p62 forms a heterodimer with E1 in the ER [27]. This dimer is transported to the plasma membrane, where virus budding occurs via spike nucleocapsid interactions, for review of the budding process, see Ref. [28]. At a very late (post-Golgi) stage of transport, the p62 protein is cleaved to E3 and E2 by host furin-like proteases. This cleavage activates the host cell-binding function of the virion as well as the membrane fusion potential of E1. In the absence of p62 cleavage, virus particles are noninfectious. This feature was used for the construction of conditionally infectious particles (see Section 20.4.1). In SFV, E3 remains part of the mature virion [29], whereas it is shed from the spike in SIN [30].

20.3 PRINCIPLE OF THE ALPHAVIRUS-BASED EXPRESSION SYSTEM

The biological features of alphaviruses made them ideally fitting for the foreign gene transfer and expression both in vitro and in vivo. The most commonly used vectors were generated on the basis of three types of alphaviruses: Semliki Forest virus (SFV) [31], Sindbis virus (SIN) [32], and Venezuelan equine encephalitis virus (VEE) [33].

Since all alphaviruses have very similar biology, the principle of the vector development is identical. The main structural elements of the expression plasmids are shown in Figure 20.3. The expression system is based on the full-length cDNA clone of the corresponding alphavirus. The classical vectors were generated in such a way that the heterologous insert replaces structural genes under 26S subgenomic promoter. Therefore, the vectors contain only the nonstructural coding region, which is required for the production of the nsP1–4 replicase complex, 26S subgenomic promoter, and a multiple cloning site with several unique restriction sites for the foreign gene insertion. Because the RNA replication is dependent on short sequence elements located at the 5′- and 3′-ends of the genomic RNA [34,35], these parts are also included in the vector construct.

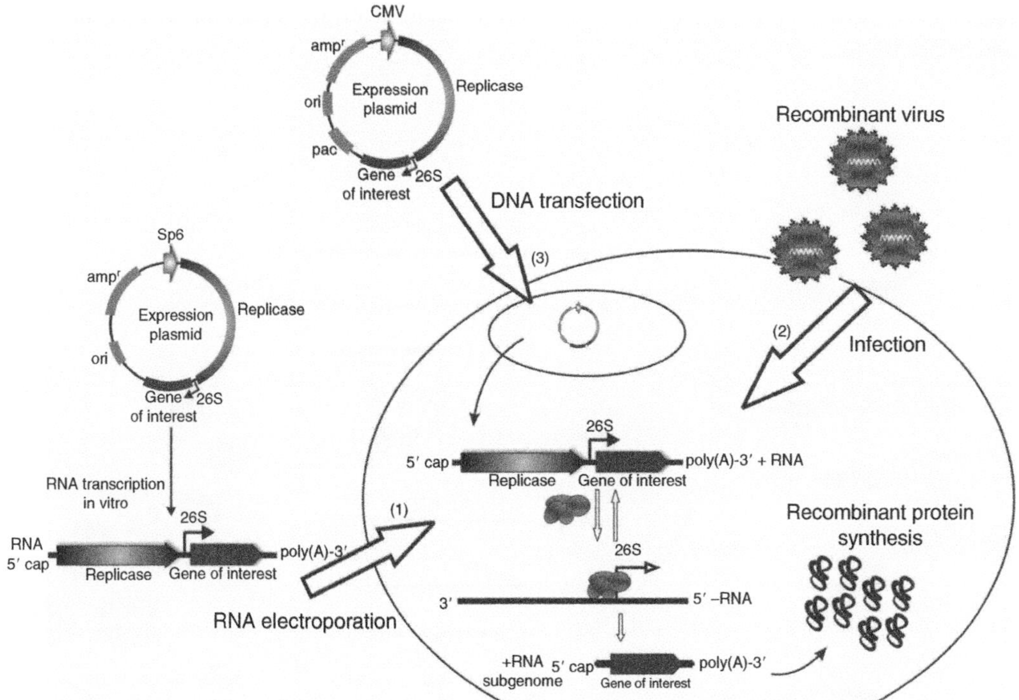

FIGURE 20.3 Three ways of the expression of gene of interest by alphaviruses: (1) cell electroporation with synthesized in vitro recombinant RNA; (2) cell infection with recombinant virus particles; (3) cell transfection with DNA construct. Details see in the text.

There are three ways for the recombinant protein expression by the alphavirus replicon (Figure 20.3).

(1) Cell transfection with recombinant RNA: The recombinant RNA is transcribed in vitro from SP6 RNA polymerase promoter. For this purpose the DNA construct is linearized with unique restriction site located at the end of the poly(A) sequence of the vector. The use of the cap analogue in the transcription reaction provides the synthesis of the completely active plus-strand RNA in vitro. The cell transfection with this RNA (usually by electroporation) turns on the replication of the recombinant RNA in the cell cytoplasm and expression of the foreign gene.

This way is the safest in the context of replication-competent alphavirus contaminations, since no virus formation could occur in the absence of viral structural genes. However, their use is limited by the efficiency of RNA transfection, which varies greatly among different cell types. Moreover, the RNA transcription in vitro is expensive and time-consuming process.

(2) Cell infection with recombinant particles: To construct infectious particles, the genes encoding structural proteins can be provided *in trans*. This is a central part of the alphavirus expression technology representing the packaging of recombinant RNAs into infectious particles using a helper construct encoding the viral structural genes. In this procedure, in vitro-made recombinant and helper RNAs are co-transfected into animal cells (Figure 20.4). The recombinant RNA codes for the RNA replicase needed for the amplification of both incoming RNA species and gene of interest, whereas the helper RNA encodes the structural proteins for the assembly of new virus particles. The helper vector is constructed by deleting a large portion of the nonstructural genes, retaining the 5′- and 3′-signals needed for RNA replication. Since almost the complete nsP region of the helper is deleted, RNA produced from this construct will not replicate in the cell, due to the lack of a functional replicase complex. When helper RNA is cotransfected with recombinant RNA, the helper construct provides the structural proteins to assemble new virus particles, while the recombinant construct provides the nonstructural proteins for RNA replication of both recombinant and helper RNAs.

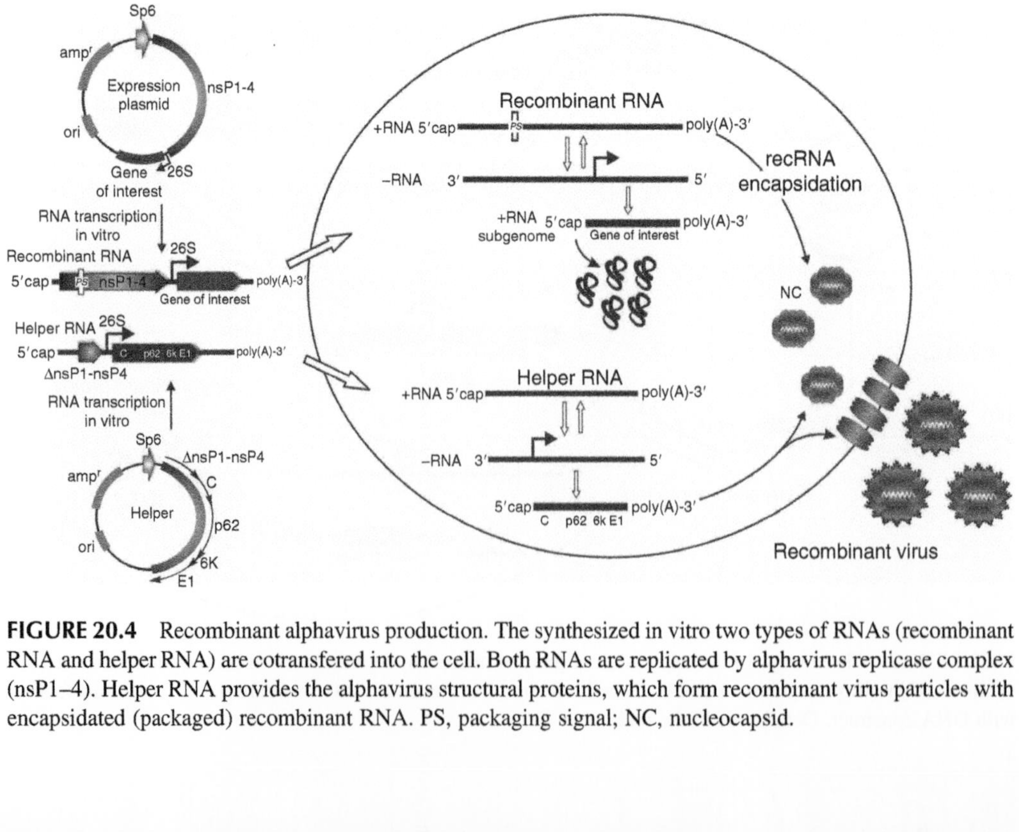

FIGURE 20.4 Recombinant alphavirus production. The synthesized in vitro two types of RNAs (recombinant RNA and helper RNA) are cotransfered into the cell. Both RNAs are replicated by alphavirus replicase complex (nsP1–4). Helper RNA provides the alphavirus structural proteins, which form recombinant virus particles with encapsidated (packaged) recombinant RNA. PS, packaging signal; NC, nucleocapsid.

The goal for this trans-complementation process is selective packaging of only recombinant RNAs into virus particles, because the helper vector lacks RNA sequence signals recognized by the capsid protein. This pack-signal is located on the replicase coding sequence. The produced recombinant virus stock, therefore, contains only recombinant genomes, and when such virus particles are used to infect animal cells, no helper proteins are expressed, providing a one-step virus infection.

In contrast to the one-step, replication-deficient particles production, the synthesis of recombinant replication-competent virus can be achieved by cell transfection with a single construct [36,37]. These vectors contain two 26S promoters leading to synthesis of two subgenomic mRNAs: one controls expression of the heterologous product, and the other controls synthesis of virus structural proteins (Figure 20.5, see the structure of the replication-competent vector). This vector

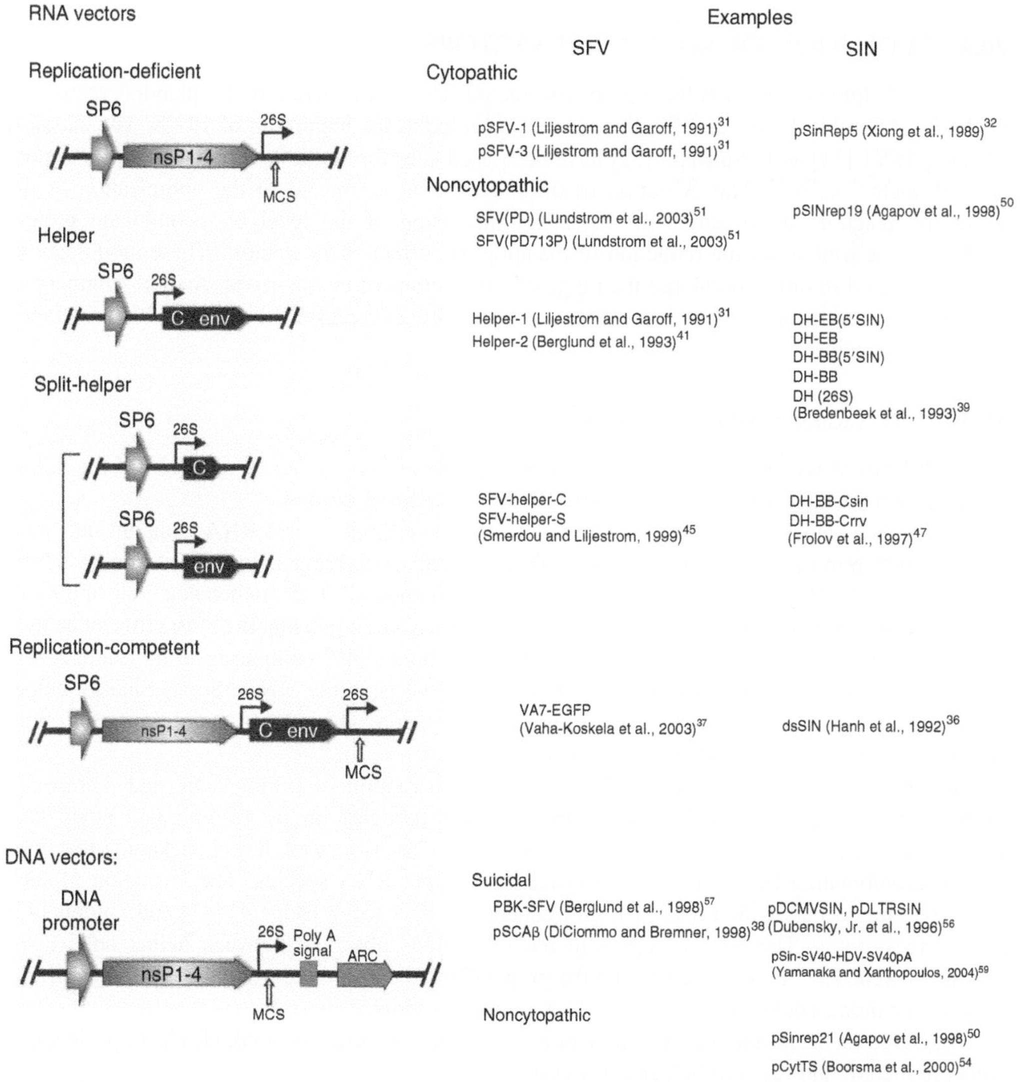

FIGURE 20.5 Schematic diagram of recombinant constructs developed on the basis of alphaviruses. The main classes of vectors are shown: RNA vectors, including replication-deficient (with helper systems) and replication-competent vectors, and DNA/RNA layered vectors. Alphavirus recombinant region is shown. The examples of each type of vector are indicated for Semliki Forest (SFV) and Sindbis (SIN) viruses. MCS, multiple cloning site; ARC, antibiotic resistance cassette.

is self-replicating, produces infectious virus particles, and can spread from cell to cell in a manner similar to that of the parental virus. Obvious advantage of this replication-competent vector is the increased efficacy of in vivo gene delivery, which should allow for spreading in infected tissue and, therefore, enhancing the therapeutic effect.

(3) Cell transfection with DNA-based vectors: To allow the direct application of plasmid DNA, the SP6 RNA polymerase promoter has been replaced by a DNA promoter (e.g., CMV IE, RSV LTR). In this case, transient transfection of plasmid DNA will result in expression of gene of interest. DNA vectors could contain also a selection marker for stable transfection. It is also possible to cotransfect this DNA with a DNA-based helper vector, and to obtain recombinant particles. However, the virus titer will be significantly lower in this system than in the case of RNA-based particle production [38].

20.4 EVOLUTION OF ALPHAVIRAL VECTORS

A variety of alphaviral vectors have been constructed. The first generation of alphaviral vectors on the basis of Semliki Forest and Sindbis viruses appeared at the beginning of 1990s. The classical vectors pSFV1 [31] and pSinRep5 [32] have been used later for the development of new and more optimal vector systems. The evolution of these vectors was directed to the optimization of the packaging reaction, improvement of biosafety, increasing of the level of recombinant protein production, as well as for the reduction of the apoptotic effect of the vectors. These modifications were introduced in order to enlarge the range of applications of alphaviruses. A short summary of recombinant constructs developed on the basis of alphaviruses is presented in Figure 20.5; a detailed description is below.

20.4.1 Packaging System Development

The high level of recombinant virus production depends on the efficient expression of the helper construct and packaging of the recombinant RNA into secreted virions.

To design a helper system for packaging of recombinant Sindbis virus RNA replicons into virus particles, different helper constructs derived from naturally occurring defective interfering RNAs [35] were tested [39]. The constructed Sindbis defective helpers (DH) contained deletions of the nsP region affecting the viral titer. Bredenbeek and colleagues have shown that the most efficient helpers are the DH RNAs with $tRNA^{Asp}$ sequence at the 5′ terminus [40] comparing to the same RNAs having the 5′ terminus identical to that of Sindbis virus genomic RNA. Some Sindbis helper RNAs are incorporated into particles along with the replicon. These RNAs may be useful under conditions in which the spread of an infection is of interest.

However, for other applications, for example, construction of prophylactic and therapeutic vaccines, the presence of replication-competent virus particles in the recombinant stock is a disadvantage. This biosafety problem arises if wild-type virus is formed during packaging reactions due to recombination between the recombinant and helper RNA species. The formation of such wild-type virus occurs with a relative high frequency in SFV (SFV Helper1) [41] and Sindbis [42] packaging systems. The rates of apparent recombinations observed between helper and vector constructs were 10^{-3} to 10^{-6}. Therefore the preparations of recombinant virus in the range of 10^9 particles, or more, could contain a replication-competent virus.

To overcome the problem of recombination two strategies were proposed: (1) cleavage-deficient virus production and (2) split-helper RNA system.

(1) It was shown that conditionally infectious particles (dependent on protease activation for infectivity) can be produced using a cleavage-deficient variant of SFV [26,43]. This variant has a mutation (substitution of R to L, so-called, mL mutation), which blocks the spike protein, p62, cleavage into E2 and E3 proteins in a post-Golgi compartment. The mL mutation does not significantly affect virus maturation and budding, but the virus produced is noninfectious, due to

reduced binding and uptake into endosomes, as well as due to reduced fusion ability between the viral and endosomal membranes. To infect cells with packaged virions, the virus stock can be activated using chymotrypsin.

Moreover, to reduce the potential for genetic reversion to wild-type infectivity, another mutant, SFV-SQL, was constructed in which all the arginine residues at the cleavage site were changed. Another helper plasmid, SFV Helper2, was generated by introducing these mutations into p62 cleavage site, providing enhanced biosafety [41]. However, even these mutations can reverse to functional p62 cleavage site, or additional mutations could regenerate the infectivity [44]. The measured frequency of genetic reversion of the SQL mutations was about 10^{-4}. Theoretically, the occurrence in the same molecule of both genetic reversion and recombination between helper and vector RNAs could happen, but would be very rare.

(2) In order to reduce the chance of regeneration of replication-competent virus by recombination, a next generation packaging system has been developed in which capsid and envelope genes are produced from separate vectors [45]. Since the capsid gene contains a translational enhancer [46], this sequence was inserted in front of the spike sequence p62-6K-E1. On the other hand, to provide cotranslational removal of the enhancer sequence and normal biosynthesis of the spike complex, a sequence coding for the foot-and-mouth disease virus 2A autoprotease was inserted in frame between the capsid translational enhancer and the spike genes. The autoprotease activity of the capsid protein was abolished by a mutation, further increasing the biosafety of the system. Co-transfection of cells with both helper RNAs (SFV-helper-C and SFV-helper-S) and the SFV vector replicon carrying foreign gene led to production of recombinant particles with high titers (up to 8×10^8 particles per 10^6 cells). An empirical frequency of replication-competent virus appearance in this system would be 10^{-13}, emphasizing the high biosafety of the system based on two-helper RNAs.

A similar strategy of the two-helper RNAs system was developed on the basis of Sindbis virus replicon [47]. In this system the Sindbis spike genes were fused to the capsid gene of Ross River virus containing deletions in the RNA-binding domain which maintained both the translation-enhancing and the self-cleaving activities. The same bipartite helper packaging system has also been described for VEE [48], but in this case the spike proteins were expressed without the capsid translation enhancer, which apparently is not needed in the VEE context.

The separation of the capsid and envelope glycoprotein genes into distinct cassettes was used also for production of a stably transfected packaging cell line with undetectable level of the replication-competent virus [49]. Interestingly, Sindbis virus-derived packaging cell lines were shown to package both Sindbis virus and SFV vectors, and the packaging level was about 10^7 infectious units/mL.

20.4.2 Expression Vector Development

The original SFV expression system provides several vectors (pSFV1, pSFV3), which contain the multiple cloning site at different positions [31]. Although the recombinant genomes derived of these vectors produce significant amounts of heterologous proteins, the expression levels obtained have been only about one-tenth of that of SFV structural proteins in a wild-type infection. An improved version of the SFV expression vector, in which the C gene is retained in front of the heterologous gene, was shown to increase heterologous protein synthesis to that of the wild-type SFV structural polyprotein [46]. In this vector, pSFVC, the target product is obtained after processing of SFV core-heterologous protein fusion. Figure 20.6 illustrates the efficiency of recombinant protein synthesis in pSFVC vector comparing to the pSFV1 and noncytopathic DNA vector pCytTS (see below). At present, this expression mode remains the most efficient among other cytopathic expression vectors.

In the last few years, much attention has been given to the engineering of novel mutant alphavirus vectors with reduced cytotoxicity, prolonged duration of transgene expression, and improved survival of host cells. Such novel noncytopathogenic vectors have been developed for

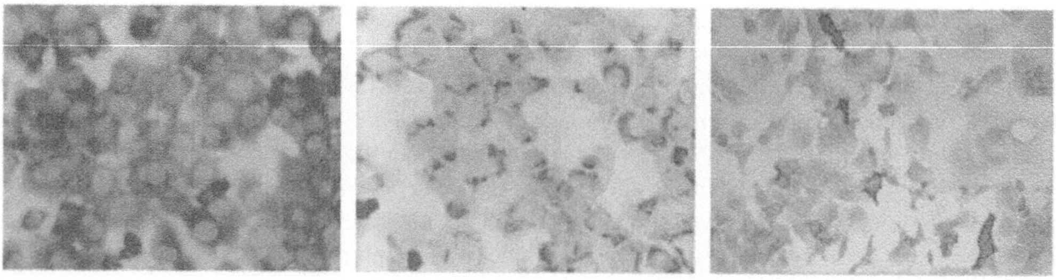

FIGURE 20.6 Comparison of recombinant hepatitis B virus core (HBc) gene expression by three vectors (pSFVC, pSFV1, pCytTS). BHK cells were infected with recombinant SFV1/HBc, SFVC/HBc particles, or transfected with pCytTS/HBc construct (6 days of thermoinduction). HBc protein was detected by immunostaining.

both Sindbis [50] and SFV [51]. The selection of noncytopathic mutants was done by analysis of cells transfected with alphavirus replicon expressing the puromycin-resistance (*pac*) gene as a dominant selectable marker. It was demonstrated that the populations of cells, expressing *pac*, and, at the same time, surviving the process of alphavirus replication, contained replicons bearing adaptive mutations, which reduced the toxicity of the replicon in these cells. The causal mutations were mapped to nonstructural protein nsP2 [52,53]. On the basis of these noncytopathic replicons new vectors were constructed providing a long-term expression of heterologous genes. However, noncytopathic Sindbis vectors demonstrated a limitation in some cell types used for expression. Cell types that have failed to support the noncytopathic replicons include primary chicken embryo fibroblasts, MDBK, MDCK, 293, HeLa, and PC12 cells. In contrast, the SFV(PD) vector showed a substantially reduced cytotoxicity and expression ability in all host cells tested [51]. Moreover, the SFV noncytopathic vectors are more promising for the large-scale production of recombinant proteins than classical SFV1 vector system.

For further development of alphavirus system additional mutations leading to the temperature-sensitive replication were introduced into the vectors. On the basis of SFV(PD), carrying two mutations in nsP2 (S259P and R650D), the triple mutant vector SFV(PD713P) was constructed by adding a point mutation L713P into nsP2 [51]. Transgene expression by this vector occurs at 31°C but is limited or absent at 37°C. Although the in vivo application of such temperature-regulated vectors is limited, the high inducibility allows for expressing highly toxic proteins upon temperature shift.

Temperature-regulated gene expression was designed also on the basis of Sindbis replicon [54]. Boorsma et al. have constructed an inducible DNA-based expression system, designated pCytTS, by site-directed mutagenesis of nsP2 (noncytopathic mutation P726S) and nsp4 (mutation G153E). The amino acid exchange in nsP4 renders the polymerase activity to temperature sensitive [55]. In this vector the viral replicase genes (nsP1–4) are placed under RSV LTR promoter, whereas the gene of interest is, as usually, under subgenomic 26S RNA promoter. In the nucleus of transfected cells the recombinant RNA copy is transcribed from the RSV LTR promoter and transported to the cytoplasm where the replicase genes (nsP1–4) are translated. Only at temperatures below 35°C (usually 29°C) does the replicase complex catalyze the synthesis of negative-strand RNA, from which new recombinant genomic RNA and subgenomic RNA coding the gene of interest are transcribed. New CytTS vectors with autonomous *pac* cassette allow easy selection of stably transfected clones with following accumulation of protein of interest in inducible conditions (Figure 20.7). Although noncytopathic replicons express less protein per unit of time than their cytopathic counterparts, their ability for prolonged expression can lead to accumulation of very high levels of protein.

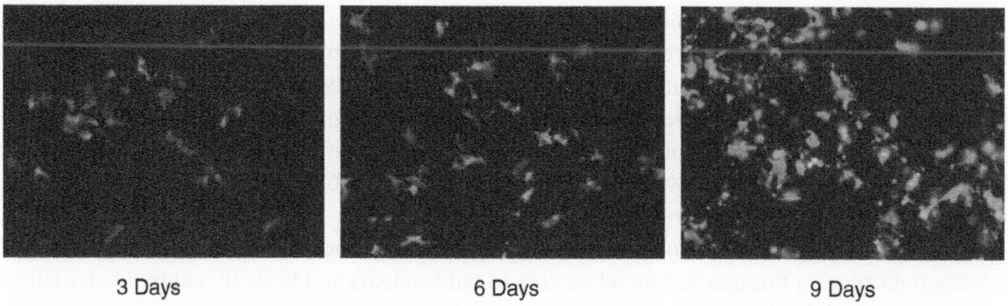

FIGURE 20.7 Kinetic of GFP protein expression in BHK cells transfected with pCytTS/EGFP. Induction time is indicated (3, 6, and 9 days).

Besides the advanced noncytopathic DNA vectors, the classical cytopathic replicons were also placed under DNA promoters [56–59]. Such, so-called suicidal, expression vectors are most adequate for DNA vaccine development. Compared to the traditional DNA vaccine strategies in which vectors are persistent and the expression constitutive, the expression mediated by the alphaviral vector was transient and caused apoptosis. As a result, biosafety risks such as chromosomal integration, and the induction of immunological tolerance, could be avoided. Furthermore, the strategy of the delay of suicidal DNA-induced cell death using coexpression of antiapoptotic proteins, suggested by Kim and coworkers [60], may greatly enhance the potency of suicidal DNA vaccines.

To achieve cell-specific expression the production of a chimeric virus with address molecules on the surface, and development of vectors with tissue-specific promoters could be applied in the future. Perspective hybrid vector for treatment of liver cancer was generated recently by Guan and colleagues [61]. This vector combines the high infectivity feature of adenoviruses to infect hepatic cells and proapoptotic feature of SFV. Moreover, cell-specific promoter (α-fetoprotein) increases the specificity of the system.

A promising strategy of the combination of genetic elements from different alphaviruses, VEE and SIN, in one vector was suggested recently [62]. Novel VEE/SIN chimeric vectors caring VEE replicase with SIN packaging signal (and opposite), which was encapsidated into SIN particles, demonstrated an optimal gene-based vaccine potency and safety in context of replication-competent virus formation.

Therefore, combining desirable elements into an expression vector may be a valuable approach toward the goal of developing vaccine/anticancer vectors with optimal in vivo potency, ease of production, and safety.

20.5 MAIN FEATURES OF ALPHAVIRUS EXPRESSION SYSTEM

Efficient gene expression: The recombinant RNA molecule is of positive polarity, and codes for its own RNA replicase, which in turn drives efficient self-replication. Within a few hours of infection, up to 200,000 copies of the plus-RNAs are made in a single cell. Because of the abundance of these molecules, practically all ribosomes of the infected cell are enrolled in the synthesis of the virus encoded proteins, thus overtaking host protein synthesis. Consequently, using the alphaviral gene expression system, 25% or more of the total cell protein can be produced.

Cytoplasmic replication: The replication of alphaviruses occurs in the cell cytoplasm, where the virus replicase transcribes and caps the subgenomes for production of the structural proteins. Therefore, many problems that are encountered in the conventional "nuclear" cDNA expression systems, such as mRNA splicing, limitations in transcription factors, problems with capping efficiency, and mRNA transport are eliminated.

Cytopathogenic effects: Replication of alphaviruses leads to cell death by apoptosis (will be considered in detail below). However, the cytopathogenic effect in infected cells appears rather late during infection. Thus, there is an extensive time window from about 4–24 h after infection during which a very high expression level of the recombinant protein can be achieved. On the other hand, dramatic shut down of endogenous gene expression, which contributes to the cell death, does not allow studies where a prolonged survival of the host cell is favorable. In this context, noncytopathic vectors are called to solve this problem.

Broad host range: This phenomenon is probably a consequence of the normal life cycle which includes transmission through arthropod vectors to wild rodents and birds in nature. Under laboratory conditions, alphaviruses infect a broad range of different cultured cells: mammalian, avian, reptilian, amphibian, and insect cells [32] (updated by Lundstrom [63]). At present, alphaviral vector systems have been used to express cytoplasmic, membrane, nuclear, as well as secreted proteins. We could add to the long list of cell lines infectable with alphaviruses also human hepatoma cell lines (Huh-7, HepG2) and primary tupaia hepatocytes, which were successfully infected with recombinant SFV particles (unpublished). Interestingly, the uninfectable cells can be infected with alphaviruses by polyethylene glycol [64], hygroscopic polymer promoting the infectivity.

Mistranslation of inserted genes: Recently, it was shown that cells expressing genes inserted into alphaviral vectors generate a large fraction of defective ribosomal products due to frequent initiation on downstream Met residues, so-called, mistranslation of inserted genes [65,66]. Such expression of a protein outside of its normal context can potentially affect their yield, therapeutic efficacy, immunogenicity, and other features, that have to be taken into account during expression of the gene of interest.

Convenience to use: A relatively small size of alphavirus vectors (10–12 kb) allows its easy handling and propagation. A simple construction of recombinant plasmids, which does not need any recombination step, represents a significant advantage of the alphavirus model. Once the required recombinant plasmids have been constructed, recombinant virus stocks with high titers (10^8–10^9 particles/mL) can be prepared within one day.

Low pathogenicity: The number of fatalities due to alphavirus infections is low and often alphavirus infections of humans remain subclinical. Because of their low pathogenicity in humans, these viruses are considered less of a hazard to laboratory personnel and were classified as a biosafety level 2 practices in the European Union and biosafety level 3 in the United States. Expression vectors, which lack the structural genes, are classified as biosafety level 2 in both EU and United States.

20.6 ALPHAVIRUS VECTORS AS VACCINE VEHICLES AND CANCER THERAPY TOOLS

In gene therapy two areas of alphavirus applications are most prominent. The first is immunotherapy, where alphavirus vectors are used as vaccine platforms and carry genes for viral, bacterial, or protozoal proteins with the aim to elicit a specific immune response and protection against challenge. Another major field is cancer therapy, where they serve for delivery of tumor antigens or immunostimulatory cytokine genes. All three alphaviruses from which efficient vector systems have been developed—Semliki Forest virus, Sindbis virus, and Venezuelan equine encephalitis virus vectors—have been tested for these purposes.

20.6.1 ALPHAVIRAL CYTOTOXICITY: ADVANTAGE OR A DRAWBACK FOR IN VIVO APPLICATIONS?

Before the applications of alphaviral vectors can be reviewed, it is essential to touch upon an inherent property of alphaviruses and hence the first generation vectors derived from them in the context of exploitation of these vectors for studies in vivo—capability to induce apoptosis of the infected cells. It, depending on the desired application or purpose, can be advantageous or can present a limitation for in vivo applications.

Cell death which follows alphavirus infection is a consequence of apoptosis or programmed cell death and is caused by alphaviruses—Semliki Forest virus [67,68], Sindbis virus [69,70], and Venezuelan equine encephalitis virus [71]—from which efficient vector systems have been developed. Although cytotoxicity of alphaviruses for in vivo applications can be viewed as a benefit or disadvantage, a link between alphavirus-induced apoptotic death of transfected cells and improved efficacy of the alphavirus replicon-based vaccine and enhanced response to an aggressive tumor challenge has been demonstrated [72,73]. This resides in the immunostimulatory effect of apoptotic death caused by viral infection and this contributes to an adjuvant effect by activating innate and adaptive immunity [74].

In the perspective of alphavirus-derived vectors for genetic vaccine applications and tumor therapy several findings are of great importance, for example, generation of T-cell epitopes has been demonstrated by processing of apoptotic bodies [75] and it has been shown that dendritic cells acquire antigen from apoptotic bodies, thus, stimulating class I-restricted $CD8^+$ cytotoxic lymphocytes, which is the basis for phenomenon of cross-priming or cross presentation [76].

Indirect evidence of cross-priming in case of immunization with recombinant Semliki Forest virus particles was obtained by Huckriede and colleagues [77]. The authors demonstrated that transfection of dendritic cells by recSFV was extremely inefficient in vitro, thus, concluding that other cell populations are the preferential target of recSFV particles, which, after undergoing apoptosis, may present a source of apoptotic bodies and play a role in CTL induction after immunization with SFV. In contrast, some Sindbis variants transfect dendritic cells efficiently [78]. In the case of VEE, it was shown that Langerhans cells, the resident dendritic cells in the skin, are the first cells to be infected after a subcutaneous injection in mice of a suicidal VEE vector, which travel to the draining lymph node after activation [79].

In case of a virus, which does not infect antigen presenting cells, cross-presentation is crucial for the development of effective adaptive immunity [80,81]. Plesa with coworkers [82] have concluded that a greater cytopathic capacity leads to a more efficient cross-priming to $CD8^+$ T cells and enhanced short-term humoral and cellular responses, being in line with observations from several other authors [83–85], who have provided evidence that apoptosis induction mediates greater cross-priming.

Alphavirus infection induces cellular caspases as demonstrated by several studies [86,87], including one by Ying et al.[73], who showed that caspase-dependent apoptotic death of transfected cells facilitated the uptake of these apoptotic cells by dendritic cells, thus stimulating these professional antigen-presenting cells, and resulting in improved efficacy and immunogenicity of the vaccine in vivo.

It has also been suggested that the effectiveness of suicidal DNA vaccines might be diminished because of the cytotoxicity and apoptotic cell death [60]. To delay cytotoxic effect and improve immune response to human papilloma virus type 16, the E7 antigen was fused with an antiapoptotic member of the BCL-2 family, BCL-xL. The authors have provided an evidence that for the enhancement of antigen-specific $CD8^+$ T-cell responses the antiapoptotic property of BCL-xL was crucial, and showed that E7/BCL-xL fusion protein generated significantly higher specific $CD8^+$ T-cell-mediated immune responses and better antitumor effects than the vector wild-type E7 gene.

Leitner et al. [72], however, have demonstrated that induction of apoptosis by replicase-based nucleic acid vaccines is essential for activation of antigen presenting cells and necessary for increased effectiveness of replicase-based vaccines. The authors co-delivered antiapoptotic gene (Bcl-xL) with the melanocyte/melanoma differentiation antigen TRP-1 and observed that protection against aggressive tumor challenge was diminished in case of co-delivery of antigen with the pro-survival Bcl-xL, despite of increased antigen production and improved antibody response in vivo.

Generally improved immunogenicity and increased efficacy of alphaviral vaccines have been attributed not to increased antigen production, but to induction of apoptosis and stimulation of innate antiviral pathways.

Furthermore, the double-stranded RNA intermediates which are formed during the replication process of alphavirus replicon play a certain role in induction of apoptosis and stimulation of innate immunity mechanisms as evidenced by autophosphorylation of dsRNA-dependent PKR [88]. It initiates antiviral response by stimulating innate immune system via 2′,5′-linked oligoadenylate synthetase antiviral pathway. Authors showed that in mice deficient for the RNase L enzyme, key component of the 2′,5′-linked oligoadenylate synthetase pathway, which degrades both viral and cellular RNA, predisposing the cell to apoptosis, the immunogenicity and the antitumor activity of the vaccine are blocked.

As we see, the alphavirus replication in target cells turns on specific cellular mechanisms leading to the apoptosis. And, finally, antiviral response enhances the immunogenic and lytic properties of the vectors that can be considered as beneficial for vaccine and cancer therapy applications.

20.6.2 Transgene Delivery and Distribution In Vivo

Two major types of alphavirus vectors used as delivery vehicles of desired genes can be outlined: propagation competent viruses, where new generation infectious viruses are produced in the host, and propagation deficient particles, which undergo only a single infection/replication cycle in the host and are therefore termed suicidal. The use of replication proficient viruses is limited by safety restrictions, and most studies in recent years have focused on the use of suicidal replicons in the form of recombinant particles and DNA-based vectors.

The issue of transgene distribution and persistence in vivo is of special importance when applications in gene and cancer therapy are considered; therefore, in order to evaluate persistence and distribution of recombinant Semliki Forest virus particles (recSFV), SFV-replicon-based DNA plasmid and conventional DNA plasmid after intramuscular injection mouse and chicken model was employed [89]. Presence of transgene was detected by RT-PCR. Recombinant SFV particles persisted for 7 days at the injection site, while SFV replicon-based plasmid 93 days and conventional DNA plasmid could be detected up to 246 days at the injection site. In chickens transgene could be detected up to 1 day at the injection site in case of recombinant Semliki Forest virus particles and up to 17 days for the SFV-based plasmid and 25 days for the conventional DNA plasmid. In mice lymph node the recombinant SFV was detectable for 1 day, and both plasmids for 3 months. Both plasmids could be detected up to 3 months in the tissues distal from the site of injection, indicating dissemination.

Localization and persistence of replicon RNA is also dependent on the route of injection as shown by Colmenero et al. [90]. Intravenous administration resulted in a systemic distribution, and the reporter gene was detectable in spleen and lymph nodes as well as in nonlymphoid tissues. Subcutaneous injection leads to a local distribution in the draining lymph node and skin surrounding the injection site, while intramuscular to local lymph nodes and injection site. This study confirmed the transient nature of Semliki Forest virus particles in vivo, since the reporter gene was almost undetectable by day 6 after injection by all examined injection routes. Intratumoral injection of recSFV leads to localization of recSFV-RNA in the tumor cells and draining lymph node only [91]. The short persistence renders the recombinant replicon particles as a safe vaccine tool, but not relevant for applications where prolonged gene expression in vivo is desired.

20.6.3 Targeting Infectious Diseases

In this section we review the most interesting and promising applications of alphaviruses for the development of genetic vaccine candidates, which are summarized in Table 20.1.

TABLE 20.1
Potential Alphavirus-Based Vaccines for Infectious Agents

Vector			Gene/Antigen	Mode of Delivery	References
			Viral Targets		
	Family	**Virus**			
SFV	Birnaviridae	IBDV	*VP2*	Particles/DNA	[144]
	Flaviviridae	BVDV	*NS3(p80)*	DNA	[143]
		CSFV	*E2*	DNA	[142]
		Hepatitis C	*NS3*	Particles	[117]
			cAg, E2	Particles/DNA	[116]
		LIV	*prME, NS1*	Particles, Particles/DNA	[89,130,131]
	Orthomyxoviridae	Influenza A	*HA, NP*	Particles, DNA, RNA	[57,96,110,114]
	Papillomaviridae	HPV 16	*E7-Hsp70*	DNA	[185]
			E6-E7	Particles	[184,186]
	Paramyxoviridae	RSV	*F, G*	RNA, Particles	[128,129]
	Picornaviridae	FMDV	*P1*	DNA	[145]
	Retroviridae	HIV-1	*Env*	Particles/RNA, RNA, Particles	[100,101,108]
			HIVA	DNA	[109]
			Clade C antigens	Particles	[102]
		SIV	*Gp160*	Particles	[106]
	Togaviridae	Rubella virus	*E1, E2*	Particles	[134]
SIN	Bunyaviridae	Seoul virus	*M, S*	Particles/DNA	[136]
		RVFV	4D4 epitope	Rep.-comp. virus (VLP)	[93]
	Flaviviridae	JEV	*prM-E, NS1–2A*	Rep.-comp. virus	[95]
	Papillomaviridae	HPV 16	*E7-VP22*	RNA, Particles	[187,188]
			E7-Hsp70	RNA	[187]
	Herpesviridae	HSV-1	*gpB*	DNA	[127]
	Hepadnaviridae	HBV	*S, C*	DNA	[140]
VEE	Arenaviridae	Lassa virus	*N*	Particles	[48]
			NP, GP	Particles	[118]
	Arteriviridae	EAV	*G(L), M*	Particles	[141]
	Caliciviridae	Norwalk-like virus	*VLP*	Particles	[138]
	Coronaviridae	SARS coronavirus	*S, N*	Particles	[126]
	Filoviridae	Ebola virus	*NP, GP*	Particles	[118–121]
			VP24, 30, 35, 40	Particles	[122]
			NP, GP, VP24, 30, 35, 40	Particles	[123]

(*continued*)

TABLE 20.1 (continued)
Potential Alphavirus-Based Vaccines for Infectious Agents

Vector	Family	Virus	Gene/Antigen	Mode of Delivery	References
		Viral Targets			
	Family	**Virus**			
		Marburg virus	*GP, NP, VP35*	Particles	[124,125,150]
	Herpesviridae	CMV	*pp65, IE1, gB*	Particles	[133]
	Orthomyxoviridae	Influenza A	*HA*	Particles	[111]
	Retroviridae	HIV-1	*MA/CA*, env	Particles	[103,104,107]
			Gag	Particles	[105]
		SIV	*MA/CA*	Particles	[103,139]
			MA/CA, env	Particles	[103,137]
VEE/SIN[a]	Retroviridae	HIV-1	*env, gag*	Particles	[62,135]
SIN/VEE[b]	Paramyxoviridae	PIV3	*HN*	Particles	[132]
	Retroviridae	HIV-1	*Gag*	Particles	[62]
		Bacterial Targets			
SFV		*Brucella abortus*	*sodC*	Particles	[148]
		Chlamydia pneumoniae	*MOMP*	Particles	[147]
SIN		*Mycobacterium tuberculosis*	*Ag85A*	DNA	[151]
VEE		*Bacillus anthracis*	*MAT-PA*	Particles	[150]
		Botulinum toxin	*H (C)*	Particles	[150]
		Staphylococcus	*enterotoxin B*	Particles	[149]
		Protozoal Targets			
SFV		*Plasmodium falciparum*	*Ag Pf332*	Particles/DNA/RNA	[154]
SIN		*Plasmodium yoelii*	*CS*	Rep.-comp. virus	[155]

Note: BVDV, Bovine viral diarrhea virus; CMV: Cytomegalovirus; CSFV, Classical swine fever virus; EAV: Equine arteritis virus; FMDV: Foot and mouth disease virus; HBV, Hepatitis B virus; HIV, Human immunodeficiency virus; HPV, Human papilloma virus; HSV, Herpes simplex virus; IBDV, Infectious bursal disease virus; JEV: Japanese encephalitis virus; LIV: Louping ill virus; MVEV: Murray Valley encephalitis virus; PIV3 – Parainfluenza virus type 3; Rep.-comp. virus, replication-competent virus; RSV: Respiratory syncytial virus; RVFV: Rift valley fever virus; SARS: severe acute respiratory syndrome; SIV: Simian immunodeficiency virus; VLP, virus like particle.

[a] VEE replicase, SIN envelope.

[b] SIN envelope, VEE replicase.

20.6.3.1 Viral Targets

A wide selection of viral proteins have been successfully expressed by alphaviral replicons delivered packed in recombinant alphavirus particles, as RNA or layered DNA plasmid, and their potential to induce specific immune response has been evaluated in numerous studies, reviewed by Lundstrom [92]. Indeed, alphaviruses are used to improve the existing vaccines, as well as for vaccine development against viruses for which there are not yet vaccines available. Different alphavirus-based vectors summarized in Figure 20.5 were tested for vaccine production. Moreover, the alphaviruses themselves can be considered as an antigen presenting units.

In 1992 a report by London et al. [93] was published describing generation of an infectious chimeric Sindbis virus, bearing 4D4 epitope derived from the external G2 glycoprotein of a Rift-Valley fever virus within structural precursor protein p62, which is a precursor of E3 and E2 membrane protein, thus supporting the approach of heterologous peptide incorporation in the Sindbis virus envelope and opening up the way for the future generation of alphaviruses targeting specific cell and tissue types [94]. Partial protection of mice from a lethal challenge of Rift-Valley fever virus was achieved in mice immunized with RVFV-Sindbis chimeras.

One of the first reports that demonstrated feasibility of Sindbis virus as a vaccine vehicle is by Pugachev et al. [95] who have exploited a double subgenomic vector coding either for prME or prM-E-NS1 of Japanese encephalitis virus; the respective genes were placed under the control of the second subgenomic promoter, either upstream or downstream of the structural protein genes, thus, making use of a replication-competent virus. Protective immunity against a lethal virus challenge was achieved.

One of the first studies, where suicidal SFV replicon was used, was by Zhou [96]. In this study recSFV particles, as well as naked recombinant RNA encoding influenza A virus nucleoprotein induced specific humoral responses with high antibody titers in mice. Besides, authors demonstrated that only 100 infectious units were sufficient in order to induce a potent class I-restricted cytotoxic T-cell response. It is also worth noting that a second, booster, immunization was permitted, which often presents a problem using other viral vectors, like adenoviruses [97] and vaccinia viruses [98], where host immune response may prevent repeated administration of the vector. The same group [99] demonstrated that CTL memory was also established, and the response was still significant after 40 days, and a strong IgG humoral response was induced which lasted at least 136 days. The authors used 10^6 IU of recSFV particles for immunization.

Attempts have been undertaken to target HIV infection with a genetic vaccine by making use of alphaviral replicons [100–105]. Infection of macaques with Simian immunodeficiency virus (SIV) has served as a model for HIV, and the first study exploiting alphavirus replicons encoding SIV *env* gene product was published in 1996 [106]. For vaccine generation recSFV particles were tested. Pigtail macaques were immunized at weeks 0, 5, 16, and 26 with 2.7×10^8 IU per animal. It was shown that a recombinant Env vaccine based on suicidal recSFV particles could protect animals from lethal SIV challenge, but not from virus infection. Reduction of virus load in plasma was achieved. T-cell proliferative responses were evaluated, however, SIV-specific T-cell proliferation failed to be detected.

Caley et al. [107] aimed at examining a potential of alphaviral vector to generate a complete immune response against an HIV immunogen. The authors used an attenuated VEE virus strain engineered as a replication-competent vector for the expression of matrix/capsid (MA/CA) coding region of the HIV-1 *gag* gene, known to contain murine cytotoxic T-lymphocyte (CTL) epitopes. It was cloned under the control of an additional VEE subgenomic promoter, thus, a propagation-competent double subgenomic construct was generated. Upon subcutaneous injection, the vector replicated in the draining lymph nodes, and HIV-1 MA/CA was expressed. Anti-MA/CA IgG and IgA antibodies were present in serum of all immunized mice and specific cytotoxic T-lymphocyte responses were detected. Besides, IgA antibodies specific for MA/CA were detected in mucosal secretion from vaginal washes.

Brand and coworkers [108] could demonstrate that the humoral response generated to HIV-1 envelope was most pronounced in mice immunized with recSFV particles when compared to DNA vaccination with the same Env glycoprotein.

Recently, Nordstrom et al. [109] compared the immunogenicity of two GMP quality-controlled vaccine batches provided for the study by the International AIDS Vaccine Initiative—a conventional DNA plasmid and SFV-based plasmid DNA vaccine encoding HIVA—the gene codes for a scrambled HIV-1 clade A Gag protein fused to a sequence of HIV-1 class I epitopes recognized by human, murine, and rhesus macaque cytotoxic T lymphocytes. A single immunization of 10 μg of both plasmids resulted in HIVA specific T-cell responses in mice immunized with the SFV replicon vaccine and no response in conventional plasmid DNA vaccinated mice, emphasizing the superiority of SFV replicon-based vaccine over conventional DNA vaccine.

Influenza A virus has also been included in the array of viral targets of alphavirus vector candidate vaccines. Hemagglutinin (HA) and nucleoprotein (NP) genes have been expressed from alphavirus vectors, delivered either as suicidal recombinant particles [110,111], naked RNA [96,112], DNA [57], or replication-competent viruses [113]. RecSFV particles coding for NP and HA of influenza virus induced potent humoral and cellular immune responses, with the serum IgG maintaining high titers up to 16 months when the study was terminated. Alongside humoral and cellular immunity intranasal (i.n.) route of immunization generated secretory IgA at the site of pulmonary mucosa [110]. Naked RNA coding for NP was administered and was capable of inducing a CTL response specific for the immunodominant epitope of the NP, and immunized mice were protected against influenza virus challenge to the same extent as the group immunized with plasmid DNA [114]. When replication-competent VEE vector encoding influenza virus HA gene was employed expression of HA in the draining lymph node, as well as induction of serum IgG and IgA after subcutaneous immunization of mice was shown. Besides, significant protection at the level of the nasal mucosa upon i.n. homologous influenza virus challenge was provided [113]. Further studies employing VEE vectors encoding influenza virus antigens have been accomplished with suicidal VEE particles [48,111] and are briefly covered in Chapter 8. One property which renders VEE especially suitable for vaccine applications is the preferential targeting of the lymph nodes draining the site of inoculation and subsequent replication there [115], more specifically, dendritic cells in the lymph nodes were identified as primary targets.

Perri et al. [62] generated chimeric VEE/SIN replicon particles encoding HIV p55*gag*. Induction of Gag-specific CD8 T-cell responses was demonstrated, although VEE and chimeric particles with VEE replicon RNA and SIN envelope stimulated a significantly greater number of Gag-specific $CD8^+$ T cells than SIN and chimeric particles bearing SIN replicon and VEE envelope.

Hepatitis C virus is another pathogen of enormous importance, against which a vaccine is not yet obtainable. Among targets for vaccine purposes virus core, envelope glycoprotein E2, and nonstructural protein 3 (NS3) have been selected [116,117]. A study where the potential of recSFV particles, SFV-based DNA plasmid, as well as conventional DNA plasmid alone and in a prime-boost regimen with recSFV particles was assessed revealed a surprising incapability of SFV-based DNA plasmids to induce detectable levels of anti-E2 antibodies. Anticore antibodies were below the detection limit in all immunization regimens employed. Immunization with recSFV particles did not result in detectable core-specific CTL. The most efficient anticore CTL response was exhibited by a group vaccinated with conventional DNA plasmid only. Similar pattern of immune response was observed for E2 antibody titers—the highest titers were detected in the group vaccinated with the conventional DNA plasmid only—with mice who received two rSFV injections after DNA priming or rSFV only three times displaying extremely low or undetectable antibody titers. Another comparative study where SFV-based vaccine platform encoding NS3 protein was compared to a conventional DNA vaccine in the capacity of inducing NS-specific cellular responses described a similar $CD8^+$ CTL response in mice vaccinated with recSFV particles and DNA plasmid.

Attempts have been under way to develop vaccine vehicles based on replication-deficient VEE viruses for Lassa and Ebola viruses [118–123], which cause acute hemorrhagic fever diseases with extremely high mortality rate. Alongside using VEE-like replicon particles delivering Lassa virus glycoprotein and nucleoprotein, Pushko et al. [118] also exploited a bivalent vaccine approach, either by administering a mixture of VEE-like replicon particles consisting of particles carrying gene for Lassa virus glycoprotein and particles encoding glycoprotein of Ebola virus, or constructing a dual expression replicon with two 26S subgenomic promoters, each of which drive expression of Lassa and Ebola glycoprotein genes. Guinea pigs were protected against a lethal challenge of Lassa virus, and against Lassa and Ebola viruses when the bivalent vaccine approach was adapted, suggesting the potential of alphavirus replicons expressing bivalent or multivalent antigens from the same RNA.

Vaccination of guinea pigs with VEE-like particles coding for glycoprotein and nuclear protein of Marburg virus, another hemorrhagic fever causing virus, provided complete protection of animals to virus challenge; but only with glycoprotein or a combination of individual VEE particles coding for glycoprotein and nucleoprotein encoding VEE-like particles vaccinated animals were completely protected [124]. In a study where various vaccine approaches delivering Marburg virus glycoprotein were compared, namely killed virus, live attenuated virus, soluble glycoprotein expressed by baculovirus recombinants, conventional DNA vaccine or VEE replicon particles [124,125]; VEE replicon particles generated significantly higher titer antibodies completely protected from virus challenge than other candidates.

Another global public health threat—SARS (severe acute respiratory syndrome) agent—coronavirus has been targeted with an alphavirus-based vaccine [126]. VEE replicon particles encoded glycoprotein (S) or the nucleocapsid protein (N) from the same strain. Mice were immunized with 10^5 IU of individual vaccine preparations or with a mixture of both (S and N) VEE replicon particles, and challenged intranasally 54 weeks later after boost to evaluate a long-term protection against the homologous virus challenge. Both VEE-S and combination of VEE-S+N, but not VEE-N, provided a complete long-term protection against challenge with the homologous vaccine strain of SARS-CoV. Short-term protection to a heterologous virus challenge 7 weeks post boost was also assessed using a genetically modified virus with S glycoprotein from highly divergent strain, GDO3, again with VEE-S and VEE-S+N conferring protection, not VEE-N.

Several other viral targets not mentioned above have been evaluated using alphavirus vectors. Broad spectrum of immune responses were induced against glycoprotein B of Herpes simplex virus [127], envelope protein F and attachment protein G of respiratory Syncytial virus [128,129], prME envelope protein and nonstructural protein NS1 of Louping ill virus, a model flavivirus [128,130,131], hemagglutinin-neuraminidase glycoprotein of Parainfluenza virus type 3 [132], pp65, IE1 and gB proteins of cytomegalovirus [133], E1 and E2 envelope proteins of Rubella virus [134], and other virus targets [135–141] indicated in Table 20.1. Moreover, alphaviruses were tested as vaccine candidates for virus-mediated diseases of domestic animals: E2 glycoprotein of classical swine fever virus (pig vaccine) [142]; NS3(p80) of Bovine viral diarrhea virus (bovine vaccine) [143]; VP2/VP4/VP3 polyprotein of infectious bursal disease virus (avian vaccine) [144]; the capsid precursor polypeptide (P1) of foot-and-mouth disease virus (pig, bovids, and others vaccine) [145].

Phase I clinical trials for the following vaccines have been initiated and sponsored by U.S.-based company Alphavax, whose core technology is based on suicidal VEE replicon particles produced in certified Vero cells (http://www.alphavax.com/technology/system.aspx): influenza A vaccine, where the VEE particles direct the expression of Influenza A hemagglutinin; a cytomegalovirus vaccine, where a VEE replicon expresses three CMV proteins—gB, pp65, and IE; HIV-1 Subtype C gag vaccine (http://clinicaltrials.gov).

20.6.3.2 Bacterial Targets

A few studies have been conducted using alphavirus replicons in attempts to assess efficacy of these vaccines to control bacterial diseases [146–151]. One of them is tuberculosis, caused by

Mycobacterium tuberculosis, which still presents a major global health burden. Currently, a live attenuated vaccine *Mycobacterium bovis* BCG exists for the prevention and control of the infection, but it is only moderately effective and degree of protection varies [152,153]. Therefore, there are continuous attempts to develop a highly effective and safe vaccine with enhanced protection. Among different approaches, a gene-based strategy has been also exploited. A layered DNA plasmid based on Sindbis virus RNA replicase encoding antigen 85A (Ag85A) was tested in a study by Kirman et al. [151]. An enhanced long-term protection against *M. tuberculosis* was achieved compared to the conventional DNA vector. Sindbis replicon-based plasmid showed to be highly immunogenic at much lower doses than the conventional DNA plasmid—at doses from 2 to 5 μg the IFN-γ production equaled to that of the conventional DNA plasmid at 100 μg.

RecSFV particles coding the major outer membrane protein or outer membrane protein 2 of *Chlamydia pneumoniae* were used alone or in a combination with a conventional DNA plasmid; namely, mice were primed with DNA and boosted with SFV particles [147]. Partial protection of challenged mice was observed in both modes of immunization with antigen encoding constructs, although the most prominent decrease in the *C. pneumoniae* culture after challenge was observed in DNA prime/recSFV boosted mice. The IFN-γ production profile of draining lymph nodes 10 days after challenge in this immunization group also was increased when compared to recSFV alone.

Use of replication-deficient VEE to elicit a protective immune response in mice against mutagenized staphylococcal enterotoxin B was demonstrated in a study by Lee et al. [149]. T-cell response was Th1 based, and the vaccine provided protection of mice to a lethal challenge with a wild-type staphylococcal enterotoxin B.

Bacillus anthracis and *Clostridium botulinum* have been targeted in an attempt to develop multi-agent vaccine by immunizing mice with a mixture of VEE replicon particles [150]—individual VEE replicon particles expressed mature 83 kDa protective antigen (MAT-PA) from *Bacillus anthracis*, or the H(C) fragment from botulinum neurotoxin. Mice were immunized with a mixture of VEE replicon particles expressing MAT-PA and H(C) fragment from botulinum neurotoxin and showed 80% protection from a *B. anthracis* challenge and 100% protection from a sequential botulinum toxin challenge. Importantly, antibody responses were not reduced in mice immunized with two VEE vaccines relative to mice that received only one individual VEE vaccine, stressing the capability to generate protective immunity to multiple agents.

20.6.3.3 Protozoal Targets

Efforts have been undertaken towards development of the alphavirus vector-based malaria vaccine. A comparative study was carried out using conventional DNA plasmid and SFV replicons, either delivered as naked RNA or packaged in recSFV particles, encoding a vaccine candidate—part of *Plasmodium falciparum* antigen Pf332 [154]. Pf332 is synthesized during the early trophozoite stage of *P. falciparum*; transported through the erythrocyte cytoplasm it associates with the host erythrocyte membrane. Immunological memory in mice was induced using all delivery platforms. However, the conventional DNA plasmid elicited a higher antibody response than SFV replicon-based constructs.

A study by Tsuji et al. [155] reported the use of a replication-competent recombinant Sindbis virus expressing a $CD8^+$ T-cell epitope of the circumsporozoite (CS) protein of *Plasmodium yoelii*. The $CD8^+$ response observed in this study was of the greatest magnitude, when compared to previous studies using recombinant vaccinia, influenza, and adenovirus vectors [156,157]. Challenged mice exhibited robust inhibition of parasite development in the liver, as determined by 80% decrease of parasite RNA.

20.6.4 Cancer Therapy

For cancer therapy employing alphaviral replicons as anticancer agents several approaches are used. One of them is to use alphavirus replicons as a genetic vaccine carrier to stimulate immunity to tumor-associated antigens. Apoptosis induction is another strategy; this being achieved either by the inherent ability of alphaviruses to cause apoptosis as well as by delivery of proapoptotic proteins expressed from the vector. Alphavirus replicons coding for cytokines and other immunoregulatory proteins that enhance antitumor immune responses and inhibit tumor cell growth is an additional promising strategy. The possible cancer vaccine candidates are summarized in Table 20.2.

Sindbis virus vectors were tested in terms of specificity and tumor targeting [158–161]. Specific marking of primary and metastatic tumor cells was achieved, leading to tumor suppression and eradication throughout the body of mice without adverse effects. Such specific alphavirus tumor targeting is explained with differences in abundance of high affinity laminin receptor, LAMR, which has been described as Sindbis virus receptor [7] on normal and tumor cell surface. LAMR is reported to be upregulated in several cancers and remains relatively unoccupied by laminin [162].

The gene delivery of multiple tumor suppressors can provide an efficient tumor therapy. Recombinant SFV (rSFV) containing three antitumor genes (rSFV-Agt/p53/PTEN) were found to efficiently transduce and express each antitumor gene in glioblastoma cells [163]. Combined or synergistic effects of angiogenesis inhibition by angiostatin and apoptosis induction by p53, PTEN, and the rSFV particle itself were shown in nude mice.

On the other hand, the ability of SFV to cause p53 independent apoptosis of target cells [164] has attracted a special attention for cancer therapy application. Intratumoral injection of 5×10^8 IU of suicide recSFV particles encoding EGFP resulted in 5-fold reduction in tumor growth in comparison to controls [165]. This study used p53-deleted human lung carcinoma cell line tumors (H358a cells) that were subcutaneously implanted as xenografts in nu/nu mice. The results indicate that SFV recombinant particles can inhibit the growth of human lung carcinoma cells and induce tumor regression by apoptosis induction, even in the absence of specific antitumor gene expression.

A similar antitumor effect of a replication-competent SFV vector expressing EGFP (VA7-EGFP) was achieved in other tumor models—mouse subcutaneous A549 human lung adenocarcinoma and rat intracranial BT4C glioma [166]. When subcutaneous mouse tumors were injected three times with VA7-EGFP, intratumorally treated animals showed almost complete inhibition of tumor growth. Since this vector is based on avirulent SFV strain (A7), it did not display any signs of abnormal behavior or encephalitis in animals. The same vector demonstrated its oncolytic capacity in human melanoma xenografts [167]. However, in this study the histological analysis revealed the presence of virus not only in all treated tumors but also in the brain of treated mice, causing progressing neuropathology beginning at day 16 after infection. Thus, such an application raises biosafety concerns.

In another study, death of 40%–60% prostate cancer cells 24–72 h after infection with recSFV-LacZ virus as well as expression of β-gal in duct epithelial cells of human prostate tissue and subsequent apoptosis were observed [168]. All these studies show the antitumor activity of the alphavirus replication upon intratumoral injection of the virus even without therapeutic gene expression.

For tumor immunotherapy and tumor vaccine production the application of a number of tumor-associated antigens (TAA) have been suggested; for the latest review see Refs. [169,170]. SFV was used to express a tumor antigen P815A, a weak transplantation antigen from mouse P815 mastocytoma, which served as a model of human MAGE-type tumor antigens. It was demonstrated that this construct induces the P815A specific and strong CTL responses after intravenous injection of 10^6 IU, detectable 100 days after administration of the vector. Also, two i.v. injections of 10^6 IU conferred protection from a lethal tumor challenge [171]. The authors conclude that recSFV is a promising vector system which might be further optimized in order to activate immune responses against cancer. In a later study the authors extended their observation and demonstrated that recSFV encoding P815A and recSFV coding for cytokine IL-12 significantly reduce tumor growth [91], outlining the potential of a cytokine-mediated approach that was confirmed also by other authors [172–176].

TABLE 20.2
Potential Alphavirus-Based Cancer Vaccines

Vector	Target	Gene	Mode of Delivery	References
SFV	Tumor antigen	*P1A*	Particles	[171]
	Tumor (CT26.CL25 cells)	*LacZ*	RNA	[73]
	Brain tumor (B16)	*Endostatin*	Particles	[178]
	Tumor (P815)	*IL-12, P1A*	Particles	[91]
	Cervical cancer (tumor antigen)	HPV *E7-Hsp70*	DNA	[185]
		HPV *E6-E7*	Particles	[184,186]
	Subcutaneous A549 human lung adenocarcinoma, intracranial BT4C glioma	*EGFP*	Propagation-competent vector	[166]
	Human melanoma xenografts	*EGFP*	Propagation-competent vector	[167]
	Human glioblastoma cells U251MG	*Agt/p53/PTEN*	Particles	[150]
	MC38 colon adenocarcinoma	*IL-12*	Particles	[172]
	Tumor (K-BALB tumor, CT26 tumor)	*IL-12*	Particles	[173]
	Tumor (K-BALB tumor, CT26 tumor)	*IL-18*	Particles	[174]
SIN	Cervical cancer (tumor antigen)	HPV *E7-VP22*	RNA	[187]
		HPV *E7-VP22*	Particles	[188]
		HPV *E7-Hsp70*	RNA	[187]
	Tumor antigen	*Tyr-related prot-1*	DNA	[72,88]
	s.c. tumor (BHK), mouse Pan02 pancreatic cancer cells, lung tumor (BHK), ovarian cancer (ES-2 human ovarian cancer cells), intrapancreatic tumor (BHKSinLuc2), spontaneous fibrosarcomas (*RGR/p15*$^{+/-}$transgenic mice)	*Firefly luciferase, LacZ, IL-12*	Particles	[159]
	Ovarian cancer (human ES-2 ovarian cancer cells, mouse MOSEC ovarian cancer cells)	*Renilla luciferase, LacZ, IL-12, IL-15,*	Particles	[159]
	Brain tumor (B16)	*Gp100, IL-18*	DNA	[176]
	Cervical tumor (HeLaS3, C33A cells), ovarian tumor (OMC-3 cells)	*GFP*	Rep.-comp. virus	[160]
	Cervical cancer (tumor antigen)	*Calreticulin*-HPV *E7*	Particles	[161]
VEE	Cervical tumor (tumor antigen)	HPV *E7, GFP*	Particles	[182]
		HPV *E6-E7*	Particles	[183]

More than 80% regression of murine colon adenocarcinomas was reported after administration of recSFV particles driving expression of IL-12 [172]. In order to elevate levels of IL-12 expression, each subunit of IL-12 was expressed from an independent subgenomic promoter fused to capsid translation enhancer of SFV. Remarkably, the SFV-based vectors were superior at eliminating tumor than a first-generation adenovirus vector expressing IL-12 [172].

In a study published in 1999 [177], 70%–90% growth inhibition and tumor necrosis of a 10 day established subcutaneous B16 tumor in mice was attributed to inhibition of tumor vascularization as monitored by Doppler ultrasonography, and it was experimentally proven that antitumor effect of SFV-IL-12 was independent of CTL and NK cell activity. In this study mice were injected intratumorally with 10^7 IU of SFV-IL12. Reduction of intratumoral vascularization was observed also in mouse with an established B16 brain tumor after intratumoral injection of SFV delivering endostatin [178].

Moreover, prophylactic and therapeutic immunization with recSFV particles encoding murine vascular endothelial growth factor receptor-2 (VEGFR-2) resulted in considerable inhibition of tumor growth and angiogenesis in two murine tumor models—CT26 colon carcinoma and 4T1 metastasizing mammary carcinoma. Antibodies against VEGFR-2 were detected, thus generating immunity against tumor endothelial cells [179]. Gene therapy with IL-12, endostatin, and VEGFR-2 delivered via SFV may be candidates for the development of new cancer therapy.

Many TAAs represent self antigens, and are subject to the constraints of immunologic tolerance. Intrinsic tolerance to *neu*, rat self TAA, could be overcome in a rat mammary tumor model after administration of VEE replicon, which is known to have in vivo tropism for immune system dendritic cells, and establishment of effective Th1 type antitumor immunity with immunological memory [180].

Wang et al. [181] reported VEE replicon particles coding for gene HER2/*neu* that inhibits breast cancer growth and tumorigenesis. Antigen-specific IgG and specific CTL responses were generated. Mice were protected from tumor challenge when HER2/*neu*-expressing tumor cells were injected into a mammary fat pad. Authors also exploited HER2/*neu* transgenic mice, which spontaneously develop tumor in breasts, and demonstrated the lack of tumor development in the VEE vaccinated animals.

Antitumor potential of VEE replicon particles coding for human papilloma virus 16 E7 oncogene, which has been one of the targets of therapeutic vaccination for cervical carcinoma, was demonstrated [182,183]. Velders with coworkers [182] vaccinated mice twice with 3×10^5 IU at 2 week interval and challenged with HPV 16 transformed cells 2 weeks after the final immunization. Class I restricted $CD8^+$ T-cell response was induced with a lasting T-cell memory as evidenced by protection from challenge for 3 months after the final administration of VEE particles. In a therapeutic setting, elimination of a 7 day established tumor was reported in almost 70% of vaccinated mice, which were administered with the same dose of VEE replicon particles subcutaneously [182].

Furthermore, in HPV E6/E7 transgenic mice, which are immunosuppressed and immunotolerant, where CTL cannot be induced by protein or conventional DNA vaccination, SFV expressing E6 and E7 was able to induce E6/E7-specific cytotoxic T cells [184–187]. One strategy to enhance the spread of propagation defective alphavirus particles was to generate a fusion protein of the target antigen HPV E7 with the VP22, a herpes simplex virus type 1 tegument protein [188]. SIN particles encoding this fusion protein E7/VP22 generated increased number of E7-specific $CD8^+$ T-cell precursors and a potent antitumor effect compared to wild-type SIN-E7 replicon particles.

Encapsulation of recombinant SFV particles in liposomes targeting tumors was used as an approach to enable a systemic delivery of the alphavirus particles and to "hide" from the immune system of the host. Liposome-mediated delivery presents several advantages, among which are the prolonged time in circulation and cell membrane permeability leading to enhanced viral delivery in tumor tissue. This strategy has been proven feasible. On the basis of preclinical studies using animal models, encapsulation of SFV replicon particles coding for IL-12 in cationic liposomes has been taken to a phase I/II clinical trials [189] in adult patients with glioblastoma multiform, an incurable brain tumor. A phase I clinical trial has been conducted employing encapsulated SFV particles expressing IL-12 on melanoma and kidney carcinoma patients [92].

The use of alphavirus-transduced dendritic cells as an approach to vaccination has also been explored. In a therapeutic model a single vaccination of mice with dendritic cells, which have been transduced with VEE replicon particles ex vivo, induced the regression of large established tumors overexpressing the *neu* oncoprotein. Immunization with dendritic cells expressing a truncated *neu* oncoprotein generated strong *neu*-specific $CD8^+$ T-cell and IgG responses [190].

Currently, Alphavax is recruiting participants for phase I and II clinical trials of a VEE replicon vaccine based on expression of human carcinoembryonic antigen (CEA) in patients with advanced or metastatic, CEA expressing malignancies. CEA is overexpressed in nearly all colorectal and pancreatic cancers, as well as in some lung and breast cancers, and uncommon medullary thyroid cancer (http://clinicaltrials.gov).

20.7 ADVANTAGES OF ALPHAVIRUSES FOR IN VIVO APPLICATION

Alphaviral delivery platforms have numerous advantages, which render them attractive tools for immunotherapeutical genetic vaccine and cancer therapy.

Safety: When the suicidal replication-deficient alphavirus particles are used, viral structural genes are not present and the infectious virus capable of infecting new target cells cannot be produced in the immunized host. Moreover, RNA replicon-based vaccines are not prone to random integration into the host genome, thus, avoiding the risk of cell transformation and development of tolerance or anti-DNA antibodies due to persistence, which present a limitation for the conventional DNA vaccines [191]. Additionally, apoptosis of the transfected cells is another safety feature of the vaccine based on suicidal alphavirus DNA vectors.

No preexisting immunity: There is no widespread immunity in the human and animal populations to alphaviruses, which is a limitation for other viral expression systems [192,193]. Therefore, the use of alphaviruses for in vivo expression of heterologous genes is not limited by the immunity to the vector.

Repeated administration: The viral structural genes are not intracellularly expressed. Therefore, alphavirus replicons can be repeatedly administered since the host immune response to the vector itself is not to such extent as to cause rejection of the booster immunization. However, this factor is a limitation for other vector systems.

Stimulation of immune response: Due to induction of apoptosis, gene expression is transient and lytic. The induced apoptosis assists in the uptake of transfected cells by dendritic cells and, subsequently, facilitates activation and stimulation of these cells. Additionally, dsRNA has immunostimulatory effect on dendritic cells [194] and on innate immunity [88].

20.8 CONCLUDING REMARKS

Application of alphavirus replicon systems induces broad and robust humoral and cellular immune responses to a wide array of viral, bacterial, parasitic, as well as tumor antigens and confers protection against a multitude of infectious agents or tumor challenges as was demonstrated in numerous studies employing several animal models such as mice, guinea pigs, chickens, sheep, rabbits, hamsters, and macaques. Alphavirus replicons are easily amenable to generating polyvalent vaccines. The demonstrated ability of alphavirus-vectored vaccines to break immunological tolerance to self antigens is crucial for cancer therapy. Indeed, these vaccines were found effective in cancer therapy models both in prophylactic and therapeutic setting. Plethora of successful preclinical studies already performed and future vector developments and improvements in vector delivery and targeting will contribute to widening the range of alphavirus vector applications and paving the way for extremely versatile tools for future immunotherapy and gene therapy.

ACKNOWLEDGMENTS

We thank Velta Ose for the electron microscopy assistance and Ruta Bruvere for the immunocytochemical analysis. Authors would like to thank also Henrik Garoff (Stockholm) for giving us the opportunity to work with alphaviruses and kind scientific support, Paul Pumpens (Riga) for critical reading of the manuscript and helpful comments, as well, we thank S. Gruen-Bernhard for the informational assistance. Our work was supported by the European Social Fund (ESF) and European Regional Development Fund (ERDF), project Nr. VPD1/ERAF/ CFLA/05/APK/2.5.1./000018/P.

REFERENCES

1. Strauss, J.H. and Strauss, E.G., The alphaviruses: Gene expression, replication, and evolution. *Microbiol. Rev.*, 58, 491, 1994.
2. Powers, A.M. et al., Evolutionary relationships and systematics of the alphaviruses. *J. Virol.*, 75, 10118, 2001.
3. Mukhopadhyay, S. et al., Mapping the structure and function of the E1 and E2 glycoproteins in alphaviruses. *Structure*, 14, 63, 2006.
4. Mancini, E.J. et al., Cryo-electron microscopy reveals the functional organization of an enveloped virus, Semliki Forest virus. *Mol. Cell*, 5, 255, 2000.
5. Forsell, K. et al., Membrane proteins organize a symmetrical virus. *EMBO J.*, 19, 5081, 2000.
6. Byrnes, A.P. and Griffin, D.E., Binding of Sindbis virus to cell surface heparan sulfate. *J. Virol.*, 72, 7349, 1998.
7. Wang, K.S. et al., High-affinity laminin receptor is a receptor for Sindbis virus in mammalian cells. *J. Virol.*, 66, 4992, 1992.
8. DeTulleo, L. and Kirchhausen, T., The clathrin endocytic pathway in viral infection. *EMBO J.*, 17, 4585, 1998.
9. Glomb-Reinmund, S. and Kielian, M., The role of low pH and disulfide shuffling in the entry and fusion of Semliki Forest virus and Sindbis virus. *Virology*, 248, 372, 1998.
10. Marsh, M. and Helenius, A., Virus entry: Open sesame. *Cell*, 124, 729, 2006.
11. Markosyan, R.M., Kielian, M., and Cohen, F.S., Fusion induced by a class II viral fusion protein, semliki forest virus E1, is dependent on the voltage of the target cell. *J. Virol.*, 81, 11218, 2007.
12. Liao, M. and Kielian, M., Functions of the stem region of the Semliki Forest virus fusion protein during virus fusion and assembly. *J. Virol.*, 80, 11362, 2006.
13. Koschinski, A. et al., Rare earth ions block the ion pores generated by the class II fusion proteins of alphaviruses and allow analysis of the biological functions of these pores. *J. Gen. Virol.*, 86, 3311, 2005.
14. Paredes, A.M. et al., Conformational changes in Sindbis virions resulting from exposure to low pH and interactions with cells suggest that cell penetration may occur at the cell surface in the absence of membrane fusion. *Virology*, 324, 373, 2004.
15. Wang, G. et al., Infection of cells by Sindbis virus at low temperature. *Virology*, 362, 461, 2007.
16. Belnap, D.M. et al., Molecular tectonic model of virus structural transitions: The putative cell entry states of poliovirus. *J. Virol.*, 74, 1342, 2000.
17. Hewat, E.A., Neumann, E., and Blaas, D., The concerted conformational changes during human rhinovirus 2 uncoating. *Mol. Cell.*, 10, 317, 2002.
18. Li, M.L. and Stollar, V., Distinct sites on the Sindbis virus RNA-dependent RNA polymerase for binding to the promoters for the synthesis of genomic and subgenomic RNA. *J. Virol.*, 81, 4371, 2007.
19. Hardy, W.R. et al., Synthesis and processing of the nonstructural polyproteins of several temperature-sensitive mutants of Sindbis virus. *Virology*, 177, 199, 1990.
20. LaStarza, M.W., Lemm, J.A., and Rice, C.M., Genetic analysis of the nsP3 region of Sindbis virus: Evidence for roles in minus-strand and subgenomic RNA synthesis. *J. Virol.*, 68, 5781, 1994.
21. Frolova, E. et al., Formation of nsP3-specific protein complexes during Sindbis virus replication. *J. Virol.*, 80, 4122, 2006.
22. Sawicki, D. et al., Temperature sensitive shut-off of alphavirus minus strand RNA synthesis maps to a nonstructural protein, nsP4. *Virology*, 174, 43, 1990.
23. Schlesinger, S. and Schlesinger, M.J., Togaviridae: the viruses and their replication. In *Fields Virology*, 4th edn, pp. 895–916. Edited by D.M. Knipe and P.M. Howley. Philadelphia: Lippincott Williams & Wilkins, 2001.

24. Hahn, C.S. and Strauss, J.H., Site-directed mutagenesis of the proposed catalytic amino acids of the Sindbis virus capsid protein autoprotease. *J. Virol.*, 64, 3069, 1990.
25. Frolova, E., Frolov, I., and Schlesinger, S., Packaging signals in alphaviruses. *J. Virol.*, 71, 248, 1997.
26. Lobigs, M. and Garoff, H., Fusion function of the Semliki Forest virus spike is activated by proteolytic cleavage of the envelope glycoprotein precursor p62. *J. Virol.*, 64, 1233, 1990.
27. Barth, B.U., Wahlberg, J.M., and Garoff, H., The oligomerization reaction of the Semliki Forest virus membrane protein subunits. *J. Cell Biol.*, 128, 283, 1995.
28. Garoff, H., Sjoberg, M., and Cheng, R.H., Budding of alphaviruses. *Virus Res.*, 106, 103, 2004.
29. Garoff, H. et al., The signal sequence of the p62 protein of Semliki Forest virus is involved in initiation but not in completing chain translocation. *J. Cell Biol.*, 111, 867, 1990.
30. Welch, W.J. and Sefton, B.M., Two small virus-specific polypeptides are produced during infection with Sindbis virus. *J. Virol.*, 29, 1186, 1979.
31. Liljestrom, P. and Garoff, H., A new generation of animal cell expression vectors based on the Semliki Forest virus replicon. *Biotechnology* (*N.Y.*), 9, 1356, 1991.
32. Xiong, C. et al., Sindbis virus: An efficient, broad host range vector for gene expression in animal cells. *Science*, 243, 1188, 1989.
33. Davis, N.L. et al., In vitro synthesis of infectious Venezuelan equine encephalitis virus RNA from a cDNA clone: Analysis of a viable deletion mutant. *Virology*, 171, 189, 1989.
34. Kuhn, R.J., Hong, Z., and Strauss, J.H., Mutagenesis of the 3′ nontranslated region of Sindbis virus RNA. *J. Virol.*, 64, 1465, 1990.
35. Levis, R. et al., Deletion mapping of Sindbis virus DI RNAs derived from cDNAs defines the sequences essential for replication and packaging. *Cell*, 44, 137, 1986.
36. Hahn, C.S. et al., Infectious Sindbis virus transient expression vectors for studying antigen processing and presentation. *Proc. Natl. Acad. Sci. U S A*, 89, 2679, 1992.
37. Vaha-Koskela, M.J. et al., A novel neurotropic expression vector based on the avirulent A7(74) strain of Semliki Forest virus. *J. Neurovirol.*, 9, 1, 2003.
38. Diciommo, D.P. and Bremner, R., Rapid, high level protein production using DNA-based Semliki Forest virus vectors. *J. Biol. Chem.*, 273, 18060, 1998.
39. Bredenbeek, P.J. et al., Sindbis virus expression vectors: Packaging of RNA replicons by using defective helper RNAs. *J. Virol.*, 67, 6439, 1993.
40. Monroe, S.S. and Schlesinger, S., RNAs from two independently isolated defective interfering particles of Sindbis virus contain a cellular tRNA sequence at their 5′ ends. *Proc. Natl. Acad. Sci. U S A.*, 80, 3279, 1983.
41. Berglund, P. et al., Semliki Forest virus expression system: Production of conditionally infectious recombinant particles. *Biotechnology* (*N.Y.*), 11, 916, 1993.
42. Raju, R., Subramaniam, S.V., and Hajjou, M., Genesis of Sindbis virus by in vivo recombination of nonreplicative RNA precursors. *J. Virol.*, 69, 7391, 1995.
43. Salminen, A. et al., Membrane fusion process of Semliki Forest virus. II: Cleavage-dependent reorganization of the spike protein complex controls virus entry. *J. Cell Biol.*, 116, 349, 1992.
44. Tubulekas, I. and Liljestrom, P., Suppressors of cleavage-site mutations in the p62 envelope protein of Semliki Forest virus reveal dynamics in spike structure and function. *J. Virol.*, 72, 2825, 1998.
45. Smerdou, C. and Liljestrom, P., Two-helper RNA system for production of recombinant Semliki forest virus particles. *J. Virol.*, 73, 1092, 1999.
46. Sjoberg, E.M., Suomalainen, M., and Garoff, H., A significantly improved Semliki Forest virus expression system based on translation enhancer segments from the viral capsid gene. *Biotechnology* (*N.Y.*), 12, 1127, 1994.
47. Frolov, I., Frolova, E., and Schlesinger, S., Sindbis virus replicons and Sindbis virus: Assembly of chimeras and of particles deficient in virus RNA. *J. Virol.*, 71, 2819, 1997.
48. Pushko, P. et al., Replicon-helper systems from attenuated Venezuelan equine encephalitis virus: Expression of heterologous genes in vitro and immunization against heterologous pathogens in vivo. *Virology*, 239, 389, 1997.
49. Polo, J.M. et al., Stable alphavirus packaging cell lines for Sindbis virus and Semliki Forest virus-derived vectors. *Proc. Natl. Acad. Sci. U S A*, 96, 4598, 1999.
50. Agapov, E.V. et al., Noncytopathic Sindbis virus RNA vectors for heterologous gene expression. *Proc. Natl. Acad. Sci. U S A*, 95, 12989, 1998.

51. Lundstrom, K. et al., Novel Semliki Forest virus vectors with reduced cytotoxicity and temperature sensitivity for long-term enhancement of transgene expression. *Mol. Ther.*, 7, 202, 2003.
52. Dryga, S.A., Dryga, O.A., and Schlesinger, S., Identification of mutations in a Sindbis virus variant able to establish persistent infection in BHK cells: The importance of a mutation in the nsP2 gene. *Virology*, 228, 74, 1997.
53. Fazakerley, J.K. et al., A single amino acid change in the nuclear localization sequence of the nsP2 protein affects the neurovirulence of Semliki Forest virus. *J. Virol.*, 76, 392, 2002.
54. Boorsma, M. et al., A temperature-regulated replicon-based DNA expression system. *Nat. Biotechnol.*, 18, 429, 2000.
55. Hahn, Y.S., Strauss, E.G., and Strauss, J.H., Mapping of RNA-temperature-sensitive mutants of Sindbis virus: Assignment of complementation groups A, B, and G to nonstructural proteins. *J. Virol.*, 63, 3142, 1989.
56. Dubensky, T.W., Jr. et al., Sindbis virus DNA-based expression vectors: Utility for in vitro and in vivo gene transfer. *J. Virol.*, 70, 508, 1996.
57. Berglund, P. et al., Enhancing immune responses using suicidal DNA vaccines. *Nat. Biotechnol.*, 16, 562, 1998.
58. Kohno, A. et al., Semliki Forest virus-based DNA expression vector: Transient protein production followed by cell death. *Gene Ther.*, 5, 415, 1998.
59. Yamanaka, R. and Xanthopoulos, K.G., Development of improved Sindbis virus-based DNA expression vector. *DNA Cell Biol.*, 23, 75, 2004.
60. Kim, T.W. et al., Enhancement of suicidal DNA vaccine potency by delaying suicidal DNA-induced cell death. *Gene Ther.*, 11, 336, 2004.
61. Guan, M. et al., Increased efficacy and safety in the treatment of experimental liver cancer with a novel adenovirus-alphavirus hybrid vector. *Cancer Res.*, 66, 1620, 2006.
62. Perri, S. et al., An alphavirus replicon particle chimera derived from Venezuelan equine encephalitis and Sindbis viruses is a potent gene-based vaccine delivery vector. *J. Virol.*, 77, 10394, 2003.
63. Lundstrom, K., Biology and application of alphaviruses in gene therapy. *Gene Ther.*, 12, Suppl 1, S92, 2005.
64. Arudchandran, R. et al., Polyethylene glycol-mediated infection of non-permissive mammalian cells with Semliki Forest virus: Application to signal transduction studies. *J. Immunol. Methods*, 222, 197, 1999.
65. Berglund, P. et al., Viral alteration of cellular translational machinery increases defective ribosomal products. *J. Virol.*, 81, 7220, 2007.
66. Zajakina, A. et al., Translation of hepatitis B virus (HBV) surface proteins from the HBV pregenome and precore RNAs in Semliki Forest virus-driven expression. *J. Gen. Virol.*, 85, 3343, 2004.
67. Glasgow, G.M. et al., Death mechanisms in cultured cells infected by Semliki Forest virus. *J. Gen. Virol.*, 78(Pt 7), 1559, 1997.
68. Scallan, M.F., Allsopp, T.E., and Fazakerley, J.K., bcl-2 acts early to restrict Semliki Forest virus replication and delays virus-induced programmed cell death. *J. Virol.*, 71, 1583, 1997.
69. Ubol, S. et al., Neurovirulent strains of alphavirus induce apoptosis in bcl-2-expressing cells: Role of a single amino acid change in the E2 glycoprotein. *Proc. Natl. Acad. Sci. U S A*, 91, 5202, 1994.
70. Lewis, J. et al., Alphavirus-induced apoptosis in mouse brains correlates with neurovirulence. *J. Virol.*, 70, 1828, 1996.
71. Jackson, A.C. and Rossiter, J.P., Apoptotic cell death is an important cause of neuronal injury in experimental Venezuelan equine encephalitis virus infection of mice. *Acta Neuropathol.*, 93, 349, 1997.
72. Leitner, W.W. et al., Apoptosis is essential for the increased efficacy of alphaviral replicase-based DNA vaccines. *Vaccine*, 22, 1537, 2004.
73. Ying, H. et al., Cancer therapy using a self-replicating RNA vaccine. *Nat. Med.*, 5, 823, 1999.
74. Restifo, N.P., Building better vaccines: How apoptotic cell death can induce inflammation and activate innate and adaptive immunity. *Curr. Opin. Immunol.*, 12, 597, 2000.
75. Bellone, M. et al., Processing of engulfed apoptotic bodies yields T cell epitopes. *J. Immunol.*, 159, 5391, 1997.
76. Albert, M.L. et al., Immature dendritic cells phagocytose apoptotic cells via alphavbeta5 and CD36, and cross-present antigens to cytotoxic T lymphocytes. *J. Exp. Med.*, 188, 1359, 1998.
77. Huckriede, A. et al., Induction of cytotoxic T lymphocyte activity by immunization with recombinant Semliki Forest virus: Indications for cross-priming. *Vaccine*, 22, 1104, 2004.

78. Gardner, J.P. et al., Infection of human dendritic cells by a Sindbis virus replicon vector is determined by a single amino acid substitution in the E2 glycoprotein. *J. Virol.*, 74, 11849, 2000.
79. MacDonald, G.H. and Johnston, R.E., Role of dendritic cell targeting in Venezuelan equine encephalitis virus pathogenesis. *J. Virol.*, 74, 914, 2000.
80. Chen, W. et al., Cross-priming of CD8+ T cells by viral and tumor antigens is a robust phenomenon. *Eur. J. Immunol.*, 34, 194, 2004.
81. Rovere-Querini, P. and Dumitriu, I.E., Corpse disposal after apoptosis. *Apoptosis*, 8, 469, 2003.
82. Plesa, G. et al., Immunogenicity of cytopathic and noncytopathic viral vectors. *J. Virol.*, 80, 6259, 2006.
83. Nowak, A.K. et al., Induction of tumor cell apoptosis in vivo increases tumor antigen cross-presentation, cross-priming rather than cross-tolerizing host tumor-specific CD8 T cells. *J. Immunol.*, 170, 4905, 2003.
84. Racanelli, V. et al., Dendritic cells transfected with cytopathic self-replicating RNA induce crosspriming of CD8+ T cells and antiviral immunity. *Immunity*, 20, 47, 2004.
85. Schaible, U.E. et al., Apoptosis facilitates antigen presentation to T lymphocytes through MHC-I and CD1 in tuberculosis. *Nat. Med.*, 9, 1039, 2003.
86. Rheme, C., Ehrengruber, M.U., and Grandgirard, D., Alphaviral cytotoxicity and its implication in vector development. *Exp. Physiol.*, 90, 45, 2005.
87. Nava, V.E. et al., Sindbis virus induces apoptosis through a caspase-dependent, CrmA-sensitive pathway. *J. Virol.*, 72, 452, 1998.
88. Leitner, W.W. et al., Alphavirus-based DNA vaccine breaks immunological tolerance by activating innate antiviral pathways. *Nat. Med.*, 9, 33, 2003.
89. Morris-Downes, M.M. et al., Semliki Forest virus-based vaccines: Persistence, distribution and pathological analysis in two animal systems. *Vaccine*, 19, 1978, 2001.
90. Colmenero, P. et al., Recombinant Semliki Forest virus vaccine vectors: The route of injection determines the localization of vector RNA and subsequent T cell response. *Gene Ther.*, 8, 1307, 2001.
91. Colmenero, P. et al., Immunotherapy with recombinant SFV-replicons expressing the P815A tumor antigen or IL-12 induces tumor regression. *Int. J. Cancer*, 98, 554, 2002.
92. Lundstrom, K., Alphavirus vectors for vaccine production and gene therapy. *Expert.Rev.Vaccines*, 2, 447, 2003.
93. London, S.D. et al., Infectious enveloped RNA virus antigenic chimeras. *Proc. Natl. Acad. Sci. U S A*, 89, 207, 1992.
94. Ohno, K. et al., Cell-specific targeting of Sindbis virus vectors displaying IgG-binding domains of protein A. *Nat. Biotechnol.*, 15, 763, 1997.
95. Pugachev, K.V. et al., Double-subgenomic Sindbis virus recombinants expressing immunogenic proteins of Japanese encephalitis virus induce significant protection in mice against lethal JEV infection. *Virology*, 212, 587, 1995.
96. Zhou, X. et al., Self-replicating Semliki Forest virus RNA as recombinant vaccine. *Vaccine*, 12, 1510, 1994.
97. Liu, Q. and Muruve, D.A., Molecular basis of the inflammatory response to adenovirus vectors. *Gene Ther.*, 10, 935, 2003.
98. Kundig, T.M. et al., Vaccination with two different vaccinia recombinant viruses: Long-term inhibition of secondary vaccination. *Vaccine*, 11, 1154, 1993.
99. Zhou, X. et al., Generation of cytotoxic and humoral immune responses by nonreplicative recombinant Semliki Forest virus. *Proc. Natl. Acad. Sci. U S A*, 92, 3009, 1995.
100. Giraud, A. et al., Generation of monoclonal antibodies to native human immunodeficiency virus type 1 envelope glycoprotein by immunization of mice with naked RNA. *J. Virol. Methods*, 79, 75, 1999.
101. Forsell, M.N. et al., Increased human immunodeficiency virus type 1 Env expression and antibody induction using an enhanced alphavirus vector. *J. Gen. Virol.*, 88, 2774, 2007.
102. Sundback, M. et al., Efficient expansion of HIV-1-specific T cell responses by homologous immunization with recombinant Semliki Forest virus particles. *Virology*, 341, 190, 2005.
103. Davis, N.L. et al., Alphavirus replicon particles as candidate HIV vaccines. *IUBMB Life*, 53, 209, 2002.
104. Dong, M. et al., Induction of primary virus-cross-reactive human immunodeficiency virus type 1-neutralizing antibodies in small animals by using an alphavirus-derived in vivo expression system. *J. Virol.*, 77, 3119, 2003.

105. Gupta, S. et al., Characterization of human immunodeficiency virus Gag-specific gamma interferon-expressing cells following protective mucosal immunization with alphavirus replicon particles. *J. Virol.*, 79, 7135, 2005.
106. Mossman, S.P. et al., Protection against lethal simian immunodeficiency virus SIVsmmPBj14 disease by a recombinant Semliki Forest virus gp160 vaccine and by a gp120 subunit vaccine. *J. Virol.*, 70, 1953, 1996.
107. Caley, I.J. et al., Humoral, mucosal, and cellular immunity in response to a human immunodeficiency virus type 1 immunogen expressed by a Venezuelan equine encephalitis virus vaccine vector. *J. Virol.*, 71, 3031, 1997.
108. Brand, D. et al., Comparative analysis of humoral immune responses to HIV type 1 envelope glycoproteins in mice immunized with a DNA vaccine, recombinant Semliki Forest virus RNA, or recombinant Semliki Forest virus particles. *AIDS Res. Hum. Retroviruses*, 14, 1369, 1998.
109. Nordstrom, E.K. et al., Enhanced immunogenicity using an alphavirus replicon DNA vaccine against human immunodeficiency virus type 1. *J. Gen. Virol.*, 86, 349, 2005.
110. Berglund, P. et al., Immunization with recombinant Semliki Forest virus induces protection against influenza challenge in mice. *Vaccine*, 17, 497, 1999.
111. Schultz-Cherry, S. et al., Influenza virus (A/HK/156/97) hemagglutinin expressed by an alphavirus replicon system protects chickens against lethal infection with Hong Kong-origin H5N1 viruses. *Virology*, 278, 55, 2000.
112. Dalemans, W. et al., Protection against homologous influenza challenge by genetic immunization with SFV-RNA encoding Flu-HA. *Ann. N Y Acad. Sci.*, 772, 255, 1995.
113. Davis, N.L., Brown, K.W., and Johnston, R.E., A viral vaccine vector that expresses foreign genes in lymph nodes and protects against mucosal challenge. *J. Virol.*, 70, 3781, 1996.
114. Vignuzzi, M. et al., Naked RNA immunization with replicons derived from poliovirus and Semliki Forest virus genomes for the generation of a cytotoxic T cell response against the influenza A virus nucleoprotein. *J. Gen. Virol.*, 82, 1737, 2001.
115. Grieder, F.B. et al., Specific restrictions in the progression of Venezuelan equine encephalitis virus-induced disease resulting from single amino acid changes in the glycoproteins. *Virology*, 206, 994, 1995.
116. Vidalin, O. et al., Use of conventional or replicating nucleic acid-based vaccines and recombinant Semliki forest virus-derived particles for the induction of immune responses against hepatitis C virus core and E2 antigens. *Virology*, 276, 259, 2000.
117. Brinster, C. et al., Hepatitis C virus non-structural protein 3-specific cellular immune responses following single or combined immunization with DNA or recombinant Semliki Forest virus particles. *J. Gen. Virol.*, 83, 369, 2002.
118. Pushko, P. et al., Individual and bivalent vaccines based on alphavirus replicons protect guinea pigs against infection with Lassa and Ebola viruses. *J. Virol.*, 75, 11677, 2001.
119. Wilson, J.A. and Hart, M.K., Protection from Ebola virus mediated by cytotoxic T lymphocytes specific for the viral nucleoprotein. *J. Virol.*, 75, 2660, 2001.
120. Pushko, P. et al., Recombinant RNA replicons derived from attenuated Venezuelan equine encephalitis virus protect guinea pigs and mice from Ebola hemorrhagic fever virus. *Vaccine*, 19, 142, 2000.
121. Geisbert, T.W. et al., Evaluation in nonhuman primates of vaccines against Ebola virus. *Emerg. Infect. Dis.*, 8, 503, 2002.
122. Wilson, J.A. et al., Vaccine potential of Ebola virus VP24, VP30, VP35, and VP40 proteins. *Virology*, 286, 384, 2001.
123. Olinger, G.G. et al., Protective cytotoxic T-cell responses induced by Venezuelan equine encephalitis virus replicons expressing Ebola virus proteins. *J. Virol.*, 79, 14189, 2005.
124. Hevey, M. et al., Marburg virus vaccines based upon alphavirus replicons protect guinea pigs and nonhuman primates. *Virology*, 251, 28, 1998.
125. Hevey, M. et al., Marburg virus vaccines: Comparing classical and new approaches. *Vaccine*, 20, 586, 2001.
126. Deming, D. et al., Vaccine efficacy in senescent mice challenged with recombinant SARS-CoV bearing epidemic and zoonotic spike variants. *PLoS. Med.*, 3, e525, 2006.
127. Hariharan, M.J. et al., DNA immunization against herpes simplex virus: Enhanced efficacy using a Sindbis virus-based vector. *J. Virol.*, 72, 950, 1998.

128. Fleeton, M.N. et al., Self-replicative RNA vaccines elicit protection against influenza A virus, respiratory syncytial virus, and a tickborne encephalitis virus. *J. Infect. Dis.*, 183, 1395, 2001.
129. Chen, M. et al., Vaccination with recombinant alphavirus or immune-stimulating complex antigen against respiratory syncytial virus. *J. Immunol.*, 169, 3208, 2002.
130. Fleeton, M.N. et al., Recombinant Semliki Forest virus particles expressing louping ill virus antigens induce a better protective response than plasmid-based DNA vaccines or an inactivated whole particle vaccine. *J. Gen. Virol.*, 81, 749, 2000.
131. Fleeton, M.N. et al., Recombinant Semliki Forest virus particles encoding the prME or NS1 proteins of louping ill virus protect mice from lethal challenge. *J. Gen. Virol.*, 80(Pt 5), 1189, 1999.
132. Greer, C.E. et al., A chimeric alphavirus RNA replicon gene-based vaccine for human parainfluenza virus type 3 induces protective immunity against intranasal virus challenge. *Vaccine*, 25, 481, 2007.
133. Reap, E.A. et al., Cellular and humoral immune responses to alphavirus replicon vaccines expressing cytomegalovirus pp65, IE1, and gB proteins. *Clin. Vaccine Immunol.*, 14, 748, 2007.
134. Callagy, S.J. et al., Semliki Forest virus vectors expressing the H and HN genes of measles and mumps viruses reduce immunity induced by the envelope protein genes of rubella virus. *Vaccine*, 25, 7481, 2007.
135. Xu, R. et al., Characterization of immune responses elicited in macaques immunized sequentially with chimeric VEE/SIN alphavirus replicon particles expressing SIVGag and/or HIVEnv and with recombinant HIVgp140Env protein. *AIDS Res. Hum. Retroviruses*, 22, 1022, 2006.
136. Kamrud, K.I. et al., Comparison of the protective efficacy of naked DNA, DNA-based Sindbis replicon, and packaged Sindbis replicon vectors expressing Hantavirus structural genes in hamsters. *Virology*, 263, 209, 1999.
137. Johnston, R.E. et al., Vaccination of macaques with SIV immunogens delivered by Venezuelan equine encephalitis virus replicon particle vectors followed by a mucosal challenge with SIVsmE660. *Vaccine*, 23, 4969, 2005.
138. Harrington, P.R. et al., Systemic, mucosal, and heterotypic immune induction in mice inoculated with Venezuelan equine encephalitis replicons expressing Norwalk virus-like particles. *J. Virol.*, 76, 730, 2002.
139. Fluet, M.E. et al., Effects of rapid antigen degradation and VEE glycoprotein specificity on immune responses induced by a VEE replicon vaccine. *Virology*, 2007.
140. Driver, D.A. et al., Layered amplification of gene expression with a DNA gene delivery system. *Ann. N Y Acad. Sci.*, 772, 261, 1995.
141. Balasuriya, U.B. et al., Alphavirus replicon particles expressing the two major envelope proteins of equine arteritis virus induce high level protection against challenge with virulent virus in vaccinated horses. *Vaccine*, 20, 1609, 2002.
142. Li, N. et al., Protection of pigs from lethal challenge by a DNA vaccine based on an alphavirus replicon expressing the E2 glycoprotein of classical swine fever virus. *J. Virol. Methods*, 144, 73, 2007.
143. Reddy, J.R. et al., Semiliki forest virus vector carrying the bovine viral diarrhea virus NS3 (p80) cDNA induced immune responses in mice and expressed BVDV protein in mammalian cells. *Comp. Immunol. Microbiol. Infect. Dis.*, 22, 231, 1999.
144. Phenix, K.V. et al., Recombinant Semliki Forest virus vector exhibits potential for avian virus vaccine development. *Vaccine*, 19, 3116, 2001.
145. Yu, X. et al., Enhanced immunogenicity to food-and-mouth disease virus in mice vaccination with alphaviral replicon-based DNA vaccine expressing the capsid precursor polypeptide (P1). *Virus Genes*, 33, 337, 2006.
146. Thomas, C.E. et al., Vaccination of mice with gonococcal TbpB expressed in vivo from Venezuelan equine encephalitis viral replicon particles. *Infect. Immun.*, 74, 1612, 2006.
147. Penttila, T. et al., DNA immunization followed by a viral vector booster in a *Chlamydia pneumoniae* mouse model. *Vaccine*, 22, 3386, 2004.
148. Onate, A.A. et al., An RNA vaccine based on recombinant Semliki Forest virus particles expressing the Cu, Zn superoxide dismutase protein of *Brucella abortus* induces protective immunity in BALB/c mice. *Infect. Immun.*, 73, 3294, 2005.
149. Lee, J.S. et al., Immune protection against staphylococcal enterotoxin-induced toxic shock by vaccination with a Venezuelan equine encephalitis virus replicon. *J. Infect. Dis.*, 185, 1192, 2002.
150. Lee, J.S. et al., Multiagent vaccines vectored by Venezuelan equine encephalitis virus replicon elicits immune responses to Marburg virus and protection against anthrax and botulinum neurotoxin in mice. *Vaccine*, 24, 6886, 2006.

151. Kirman, J.R. et al., Enhanced immunogenicity to *Mycobacterium tuberculosis* by vaccination with an alphavirus plasmid replicon expressing antigen 85A. *Infect. Immun.*, 71, 575, 2003.
152. Orme, I.M., Beyond BCG: The potential for a more effective TB vaccine. *Mol. Med. Today*, 5, 487, 1999.
153. Martin, C., The dream of a vaccine against tuberculosis; new vaccines improving or replacing BCG? *Eur. Respir. J.*, 26, 162, 2005.
154. Andersson, C. et al., Comparative immunization study using RNA and DNA constructs encoding a part of the Plasmodium falciparum antigen Pf332. *Scand. J. Immunol.*, 54, 117, 2001.
155. Tsuji, M. et al., Recombinant Sindbis viruses expressing a cytotoxic T lymphocyte epitope of a malaria parasite or of influenza virus elicit protection against the corresponding pathogen in mice. *J. Virol.*, 72, 6907, 1998.
156. Murata, K. et al., Characterization of in vivo primary and secondary CD8+ T cell responses induced by recombinant influenza and vaccinia viruses. *Cell Immunol.*, 173, 96, 1996.
157. Rodrigues, E.G. et al., Single immunizing dose of recombinant adenovirus efficiently induces CD8+ T cell-mediated protective immunity against malaria. *J. Immunol.*, 158, 1268, 1997.
158. Tseng, J.C. et al., In vivo antitumor activity of Sindbis viral vectors. *J. Natl. Cancer Inst.*, 94, 1790, 2002.
159. Tseng, J.C. et al., Systemic tumor targeting and killing by Sindbis viral vectors. *Nat. Biotechnol.*, 22, 70, 2004.
160. Unno, Y. et al., Oncolytic viral therapy for cervical and ovarian cancer cells by Sindbis virus AR339 strain. *Clin. Cancer Res.*, 11, 4553, 2005.
161. Cheng, W.F. et al., Sindbis virus replicon particles encoding calreticulin linked to a tumor antigen generate long-term tumor-specific immunity. *Cancer Gene Ther.*, 13, 873, 2006.
162. Liotta, L.A. et al., The laminin receptor and basement membrane dissolution: Role in tumour metastasis. *Ciba Found. Symp.*, 108, 146, 1984.
163. Lee, J.S. et al., Growth inhibitory effect of triple anti-tumor gene transfer using Semliki Forest virus vector in glioblastoma cells. *Int. J. Oncol.*, 28, 649, 2006.
164. Glasgow, G.M. et al., The Semliki Forest virus vector induces p53-independent apoptosis. *J. Gen. Virol.*, 79(Pt 10), 2405, 1998.
165. Murphy, A.M. et al., Inhibition of human lung carcinoma cell growth by apoptosis induction using Semliki Forest virus recombinant particles. *Gene Ther.*, 7, 1477, 2000.
166. Maatta, A.M. et al., Evaluation of cancer virotherapy with attenuated replicative Semliki forest virus in different rodent tumor models. *Int. J. Cancer*, 121, 863, 2007.
167. Vaha-Koskela, M.J. et al., Oncolytic capacity of attenuated replicative semliki forest virus in human melanoma xenografts in severe combined immunodeficient mice. *Cancer Res.*, 66, 7185, 2006.
168. Hardy, P.A. et al., Recombinant Semliki forest virus infects and kills human prostate cancer cell lines and prostatic duct epithelial cells ex vivo. *Int. J. Mol. Med.*, 5, 241, 2000.
169. Morris, L.F. and Ribas, A., Therapeutic cancer vaccines. *Surg. Oncol. Clin. N. Am.*, 16, 819, 2007.
170. Tabi, Z. and Man, S., Challenges for cancer vaccine development. *Adv. Drug Deliv. Rev.*, 58, 902, 2006.
171. Colmenero, P., Liljestrom, P., and Jondal, M., Induction of P815 tumor immunity by recombinant Semliki Forest virus expressing the P1A gene. *Gene Ther.*, 6, 1728, 1999.
172. Rodriguez-Madoz, J.R., Prieto, J., and Smerdou, C., Semliki forest virus vectors engineered to express higher IL-12 levels induce efficient elimination of murine colon adenocarcinomas. *Mol. Ther.*, 12, 153, 2005.
173. Chikkanna-Gowda, C.P. et al., Regression of mouse tumours and inhibition of metastases following administration of a Semliki Forest virus vector with enhanced expression of IL-12. *Gene Ther.*, 12, 1253, 2005.
174. Chikkanna-Gowda, C.P. et al., Inhibition of murine K-BALB and CT26 tumour growth using a Semliki Forest virus vector with enhanced expression of IL-18. *Oncol. Rep.*, 16, 713, 2006.
175. Yamanaka, R. et al., Marked enhancement of antitumor immune responses in mouse brain tumor models by genetically modified dendritic cells producing Semliki Forest virus-mediated interleukin-12. *J. Neurosurg.*, 97, 611, 2002.
176. Yamanaka, R. and Xanthopoulos, K.G., Induction of antigen-specific immune responses against malignant brain tumors by intramuscular injection of Sindbis DNA encoding gp100 and IL-18. *DNA Cell Biol.*, 24, 317, 2005.

177. sselin-Paturel, C. et al., Transfer of the murine interleukin-12 gene in vivo by a Semliki Forest virus vector induces B16 tumor regression through inhibition of tumor blood vessel formation monitored by Doppler ultrasonography. *Gene Ther.*, 6, 606, 1999.
178. Yamanaka, R. et al., Induction of therapeutic antitumor antiangiogenesis by intratumoral injection of genetically engineered endostatin-producing Semliki Forest virus. *Cancer Gene Ther.*, 8, 796, 2001.
179. Lyons, J.A. et al., Inhibition of angiogenesis by a Semliki Forest virus vector expressing VEGFR-2 reduces tumour growth and metastasis in mice. *Gene Ther.*, 14, 503, 2007.
180. Nelson, E.L. et al., Venezuelan equine encephalitis replicon immunization overcomes intrinsic tolerance and elicits effective anti-tumor immunity to the self tumor-associated antigen, neu in a rat mammary tumor model. *Breast Cancer Res. Treat.*, 82, 169, 2003.
181. Wang, X. et al., Alphavirus replicon particles containing the gene for HER2/neu inhibit breast cancer growth and tumorigenesis. *Breast Cancer Res.*, 7, R145, 2005.
182. Velders, M.P. et al., Eradication of established tumors by vaccination with Venezuelan equine encephalitis virus replicon particles delivering human papillomavirus 16 E7 RNA. *Cancer Res.*, 61, 7861, 2001.
183. Cassetti, M.C. et al., Antitumor efficacy of Venezuelan equine encephalitis virus replicon particles encoding mutated HPV16 E6 and E7 genes. *Vaccine*, 22, 520, 2004.
184. Riezebos-Brilman, A. et al., Induction of human papilloma virus E6/E7-specific cytotoxic T-lymphocyte activity in immune-tolerant, E6/E7-transgenic mice. *Gene Ther.*, 12, 1410, 2005.
185. Hsu, K.F. et al., Enhancement of suicidal DNA vaccine potency by linking Mycobacterium tuberculosis heat shock protein 70 to an antigen. *Gene Ther.*, 8, 376, 2001.
186. Daemen, T. et al., Immunization strategy against cervical cancer involving an alphavirus vector expressing high levels of a stable fusion protein of human papillomavirus 16 E6 and E7. *Gene Ther.*, 9, 85, 2002.
187. Cheng, W.F. et al., Enhancement of Sindbis virus self-replicating RNA vaccine potency by linkage of herpes simplex virus type 1 VP22 protein to antigen. *J. Virol.*, 75, 2368, 2001.
188. Cheng, W.F. et al., Cancer immunotherapy using Sindbis virus replicon particles encoding a VP22-antigen fusion. *Hum. Gene Ther.*, 13, 553, 2002.
189. Ren, H. et al., Immunogene therapy of recurrent glioblastoma multiforme with a liposomally encapsulated replication-incompetent Semliki forest virus vector carrying the human interleukin-12 gene–a phase I/II clinical protocol. *J. Neurooncol.*, 64, 147, 2003.
190. Moran, T.P. et al., Alphaviral vector-transduced dendritic cells are successful therapeutic vaccines against neu-overexpressing tumors in wild-type mice. *Vaccine*, 25, 6604, 2007.
191. Medjitna, T.D. et al., DNA vaccines: Safety aspect assessment and regulation. *Dev. Biol.* (*Basel*), 126, 261, 2006.
192. Bangari, D.S. and Mittal, S.K., Current strategies and future directions for eluding adenoviral vector immunity. *Curr. Gene Ther.*, 6, 215, 2006.
193. McCoy, K. et al., Effect of preexisting immunity to adenovirus human serotype 5 antigens on the immune responses of nonhuman primates to vaccine regimens based on human- or chimpanzee-derived adenovirus vectors. *J. Virol.*, 81, 6594, 2007.
194. Cella, M. et al., Maturation, activation, and protection of dendritic cells induced by double-stranded RNA. *J. Exp. Med.*, 189, 821, 1999.

Index

A

B

C

D

I

J

K

Q

R

S

T

U

V

W

X

Y

Printed and bound by CPI Group (UK) Ltd, Croydon, CR0 4YY
23/10/2024
01778226-0015